OXFORD MONOGRAPHS ON GEOLOGY AND GEOPHYSICS NO. 29

Series editors

H. Charnock S. Conway Morris J. F. Dewey
A. Navrotsky E. R. Oxburgh R. A. Price
　　　　　　　B. J. Skinner

OXFORD MONOGRAPHS ON GEOLOGY AND GEOPHYSICS

1. De Verle P. Harris: *Mineral resources appraisal: mineral endowment, resources and potential supply: concepts, methods and cases*
2. J. J. Veevers (ed.): *Phanerazoic earth history of Australia*
3. Yang Zunyi, Cheng Yuqi, and Wang Hongzhen (eds.): *The geology of China*
4. Lin-gun Liu and William A. Bassett: *Elements, oxides and silicates: high pressure phases with implications for the earth's interior*
5. Antoni Hoffman and Matthew H. Nitecki (eds.): *Problematic fossil taxa*
6. S. Mahmood Naqvi and John J. W. Rogers: *Precambrian geology of India*
7. Chih-Pei Chang and T. N. Krishnamurti (eds.): *Monsoon meteorology*
8. Zvi Ben-Avraham (eds.): *The evolution of the Pacific Ocean margins*
9. Ian McDougall and T. Mark Harrison: *Geochronology and thermochronology by the $^{40}Ar/^{39}Ar$ method*
10. Walter C. Sweet: *The Conodonta: morphology, taxonomy, paleoecology, and evolutionary history of a long-extinct animal phylum*
11. H. J. Melosh: *Impact cratering: a geologic process*
12. J. W. Cowie and M. D. Brasier (eds.): *The Precambrian–Cambrian boundry*
13. C. S. Hutchinson: *Geologocal evolution of south-east Asia*
14. Anthony J. Naldrett: *Magmatic sulfide deposits*
15. D. R. Prothero and R. M. Schoch (eds.): *The evolution of perissodactyls*
16. M. Menzies (ed.): *Continental mantle*
17. R. J. Tingey (ed.): *Geology of the Antarctic*
18. Thomas J. Crowley and Gerald R. North: *Paleoclimatology*
19. Gregory J. Retallack: *Miocene paleosols and ape habitats of Pakistan and Kenya*
20. Kuo-Nan Liou: *Radiation and cloud processes in the atmosphere: theory, observation, and modeling*
21. Brian Bayly: *Chemical change in deforming materials*
22. A. K. Gibbs and C. N. Barron: *The geology of the Guiana Shield*
23. Peter J. Ortoleva: *Geochemical self-organization*
24. Robert G. Coleman: *Geologic evolution of the Red Sea*
25. Richard W. Spinrad, Kendall L. Carder, and Mary Jane Perry: *Ocean Optics*
26. Clinton M. Case: *Physical principles of flow in unsaturated porous media*
27. Eris B. Kraus and Joost A. Businger: *Atmosphere-ocean interaction, second edition*
28. M. Solomon and D. J. Groves: *The geology and origin of Australia's mineral deposits*
29. R. L. Stanton: *Ore elements in arc lavas*

Ore Elements in Arc Lavas

R. L. STANTON

*Formerly Professor of Geology,
University of New England, Armidale, New South Wales*

CLARENDON PRESS · OXFORD
1994

Oxford University Press, Walton Street, Oxford OX2 6DP
Oxford New York Toronto
Delhi Bombay Calcutta Madras Karachi
Kuala Lumpur Singapore Hong Kong Tokyo
Nairobi Dar es Salaam Cape Town
Melbourne Auckland Madrid
and associated companies in
Berlin Ibadan

Oxford is a trade mark of Oxford University Press

Published in the United States
by Oxford University Press Inc., New York

© R. L. Stanton, 1994

All rights reserved. No part of this publication may be
reproduced, stored in a retrieval system, or transmitted, in any
form or by any means, without the prior permission in writing of Oxford
University Press. Within the UK, exceptions are allowed in respect of any
fair dealing for the purpose of research or private study, or criticism or
review, as permitted under the Copyright, Designs and Patents Act, 1988, or
in the case of reprographic reproduction in accordance with the terms of
licences issued by the Copyright Licensing Agency. Enquiries concerning
reproduction outside those terms and in other countries should be sent to
the Rights Department, Oxford University Press, at the address above.

This book is sold subject to the condition that it shall not,
by way of trade or otherwise, be lent, re-sold, hired out, or otherwise
circulated without the publisher's prior consent in any form of binding
or cover other than that in which it is published and without a similar
condition including this condition being imposed
on the subsequent purchaser.

A catalogue record for this book is available from the British Library

Library of Congress Cataloging in Publication Data
Stanton, R. L.
Ore elements in arc lavas/R. L. Stanton.
(Oxford monographs on geology and geophysics; no. 29)
Includes bibliographical references and index.
1. Ore deposits. 2. Volcanism. 3. Geochemistry. 4. Island arcs.
I. Title. II. Series.
QE390.S73 1994   553'.1–dc20   93-42959
ISBN 0 19 854050 7

Typeset by EXPO Holdings, Malaysia

Printed in Great Britain by
Bookcraft (Bath) Ltd.
Midsomer Norton, Avon

Preface

The study that constitutes the central theme of this book has a long history, extending back almost to the beginning of the author's professional life. Its purpose has been twofold: to investigate the patterns of behaviour of the 'ore' and associated elements during the crystallization of a volcanic island-arc lava series and, in so doing, to contribute to a solution of the problem as to whether exhalative ore deposits of the volcanic regime are derived from the melt by magmatic processes or by much later sea-floor leaching. It is also hoped that the volume will be a contribution to knowledge of volcanic island-arc geochemistry generally, and will provide a firmer basis for an understanding of the behaviour of the ore elements in magmatic differentiation processes — a field recognized long ago by Lindgren, Fenner, Bowen, Niggli, and others, but so far little investigated by modern analytical methods.

In an investigation of the Palaeozoic conformable ore province of south-eastern New South Wales begun in the late 1940s (see Stanton 1955a,b) it became apparent that all the small sulphide deposits of that region occurred in shallow-water (limestone-bearing) sediments in one or other of two stratigraphic positions: a lower 'horizon' of copper–zinc ores associated with basaltic–andesitic volcanic materials, and an upper horizon of (copper)–zinc–lead ores associated with dacitic–rhyolitic lavas and pyroclastic rocks. This, by happy coincidence, combined with a concurrent study (1950–1) of one of the modern south-west Pacific volcanic island festoons — the Solomon Islands — to lead to the conclusion that the comformable ores were genetically tied to Palaeozoic island-arc volcanism, and to the suggestion that variation in the nature of the volcanic materials might cause variation in the nature of the associated exhalative ores.

Shortly after this the author began work in Canada (1956–8) and became aware of the conformable copper–zinc deposits associated with basaltic and metabasaltic rocks in Newfoundland, and the large (copper)–zinc–lead orebodies of some of the felsic volcanic districts of Newfoundland (e.g. Buchans) and the Bathurst–Newcastle area of New Brunswick. The apparent association of these deposits with Schuchert and Dunbar's 'New Brunswick Geanticline' (Schuchert and Dunbar 1934; Stanton 1961) seemed once again to indicate an island-arc volcanic environment, and again the copper–zinc:basalt and zinc–lead:dacite–rhyolite associations were apparent. The evidence seemed to be indicating not only that *ore formation* was a manifestation of island-arc volcanism and volcanic sedimentation, but that *ore type* might be related to differentiation of the lavas concerned.

A little later (1959), and in the company of J. D. Bell, the author began the mapping and investigation of the Pleistocene-to-Recent lavas of the New Georgia Group of the Solomon Islands. It quickly became apparent that these lavas constituted an extraordinarily wide spectrum of volcanic products ranging from highly olivine-rich picritic lavas to felsic andesites.

The earlier suspicion that different volcanic–exhalative ore types might have systematic genetic ties with particular stages of lava differentiation, combined with this observation of what appeared to be a particularly extensive sequence of differentiation in a young volcanic arc milieu, led quickly to the thought that it might be interesting to examine the behavior of the traces of the 'ore metals' — copper, zinc, lead, etc. — in the minerals and groundmasses of the New Georgia lavas as a differentiation series.

This investigation was begun in 1963 using wet chemical and X-ray fluorescence methods. It soon became clear, however, that the analytical techniques then available were inadequate for such a comprehensive study, and the project was postponed in the hope that better and more rapid methods of analysis might become available in the next few years. A minor, semi-exploratory, effort was made in the period 1974–8 (using mainly atomic absorption spectroscopy) with the assistance of a postgraduate student, W. R. H. Ramsay. While this investigation was very limited, particularly from the point of view of mineral, as distinct from whole-rock, geochemistry, it did give a very clear indication that the incidence of the trace metals in the Solomons lavas was systematic and hence probably amenable to detailed geochemical investigation.

By 1980, automated and highly accurate X-ray fluorescence analysis had become firmly established, and the electron-probe microanalyser had been developed to the point where, given appropriately long counting times, it was capable of accurate trace analysis of minerals and glass *in situ*. After a pause of some seventeen years, it appeared that the necessary techniques had now become available, and the investigation could be resumed.

Two periods aggregating about four months were devoted to meticulous collection in the relevant areas of the Solomon Islands. The author had collected extensively in 1959 and 1963–4, the Solomons Island Geological Survey staff collected on his behalf throughout 1965–77, and three postgraduates working under his supervision had collected further during the period 1974–6, but it was deemed desirable to make a new collection designed specifically for a very much more detailed and comprehensive attack on the problem. This was carried out by the author in 1980 and 1981. Of about 140 specimens obtained, 122 were finally selected for use. (Any containing even the most minute traces of sulphide were discarded, and the final 122 samples were free of microscopically detectable sulphide minerals.) All were sectioned for transmitted and reflected light microscopy, and crushed for whole-rock analysis, for mineral:groundmass separation, and for analysis of the relevant separations. A total of approximately 900 X-ray fluorescence, neutron activation, and ICP–MS (inductively coupled plasma emission spectrometry/mass spectrometry) analyses of whole rock and mineral and groundmass separations were carried out by Dr B. W. Chappell of the Australian National University, and just over 12 000 electron-microprobe analyses of mineral grains and glass *in situ* were carried out by the writer using the facilities of the Research School of Earth Sciences, Australian National University, under the supervision of N. G. Ware.

The principal results of this investigation, set in the context of the geochemistry of other island arc lavas and those of the mid-ocean ridges, are reported in this volume.

Following the Introduction, Chapters 2–4 are intended to 'set the scene' and to give the reader at least an outline understanding of some of the problems in volcanic ore petrology that led to a detailed investigation of the patterns of abundance of the trace 'ore elements' in an island-arc lava series. Chapter 2 is a brief account of the principal features of the 'conformable' or 'stratiform' exhalative orebodies of the arc regime — deposits now widely referred to as 'volcanic massive sulphides'. Chapters 3 and 4 give, respectively, a short history of the development of ideas on the origin of these ores and their environments, and an outline of current controversies concerning the precise nature and derivation of the exhalations from which the ores are precipitated. Chapter 3 is substantially historical. Chapter 4 has a historical component, stemming from its concern with the development and course of a controversy that remains current. The purposes of these two chapters should not, however, be confused. Chapter 3 is concerned with the derivation of the ores from the volcanic rocks through the medium of exhalations. Chapter 4, on the other hand, is concerned with the precise nature of these exhalations, and the nature and timing of the processes by which they abstracted the metals from the lavas and transported them to the sea floor. Chapter 5 is a somewhat more detailed 'scene-setter', and presents the broad petrological and geochemical features of the Solomons lava series as a basis for the more detailed geochemical chapters that follow.

Chapters 6–19 are concerned with the geochemistry and abundance behaviour of each of fourteen major, minor, and trace elements chosen on the basis of their incidence in exhalative ores and their place in magmatic crystallization. They are intended to provide a reasonably detailed view of the behaviour of the selected elements in the melt, in turn to provide a basis for considering their incidence in exhalative ores and, in particular, the association of specific ore types with specific lava types.

Chapter 20 is intended to provide at least an initial view of the behaviour of the hyperfusible elements such as sulphur, chlorine, and fluorine, of minor metals such as silver, gold, molybdenum, cadmium, and uranium, and of the semi-metals arsenic, antimony, and bismuth. In particular it is intended to demonstrate some of the linkages between (1) lithophile elements, such as potassium, rubidium, zirconium, and strontium, that are generally regarded as normal, intrinsic parts of the melt, and the varying abundances of which are widely recognized as indicators of melt processes; (2) elements such as barium, well recognized as ubiquitous large-ion lithophile trace constituents of volcanic lavas but also established as major elements of exhalative ores; and (3) 'ore elements' such as lead, the igneous systematics of which are only sketchily known, but which constitute important components of many ores closely associated with volcanic rocks.

The order of presentation of the fourteen elements in chapters 6–19 may at first sight appear a little quaint. Copper, zinc, lead, and barium are presented first since they are well known to be the principal components of the ores apart from iron and sulphur. Strontium appears next, since, in spite of its close chemical relationship with barium, it is conspicuous by its absence from volcanic exhalative ores. This sharp contrast in the incidence of two closely related elements looks as if it may bear some eloquent clues concerning ore-forming processes, and so the two are considered in sequence. The remaining nine elements are then treated in order of increasing atomic number.

In presenting the data on the Solomons lavas, the writer has tried to provide a context — a framework of reference for the reader — by giving brief accounts of the early acquisition of trace-element data for volcanic and related igneous rocks, and information on the incidence of the relevant elements in mid-ocean ridge basalts (MORBs) and associated lavas in other island arcs. The arcs range from fully intra-oceanic arcs such as Vanuatu and Tonga that, like the Solomons, have no connection with continental crust, to at least partly continental arcs such as the Aleutians, the Lesser Antilles, and the southern part of the Tonga–Kermadec festoon. On addition to providing a context for the Solomons data, presentation of information on MORBs and these other arcs has had as a secondary purpose the provision of a readily consulted source of information on the incidence of the ore elements in marine lavas. A collection of such information should have many uses and, in showing the rather fragmentary nature of present knowledge, indicate areas for further investigation.

In Chapter 21 the petrological and geochemical information of Chapters 5–20 is applied to examining the abundance patterns developed in a crystallizing volcanic melt of island-arc type — as represented by the Solomon Islands Younger Volcanic suite. This involves a consideration of the probable effects of fractional crystallization, and of loss in the volatile phase, on the enrichment and impoverishment of the ore elements in residual volcanic melts. Leading on from this, Chapter 22 is concerned with observation of volcanic sublimates, condensates, gases, and plumes, with current knowledge of volatile formation in volcanic melts, with melt–vapour partitioning of elements, and hence with possible patterns of volatile loss from volcanic magma chambers. Chapter 23 is concerned with the derivation and development of the Solomons lavas, with particular emphasis on the basalt–picrite and basalt–andesite–dacite lineages. Chapter 24 considers the petrogenesis of exhalative ores in the light of all the features of lava evolution and geochemistry observed in the Solomon Islands Younger Volcanic Suite. This leads to a consideration of possible relations between crystallization, lava type, and ore type, and to some reflections on the leaching hypothesis.

The account concludes with some thoughts on the possible significance of ore-element geochemistry for island-arc petrogenesis in general, and for the refining of mineral exploration methods in ancient island-arc terrains.

Finally, but far from least, a purpose of the book has been to show how intimately and systematically some ore deposits may be related to fundamental geological processes — in this case magmatic differentiation. As it was put so simply by Thomas Crook some three-quarters of a century ago, 'The origin of ore deposits is ... inseparably bound up with the origin of rocks' (Crook 1914, p. 78). It is hoped that this book will serve to emphasize Crook's perceptiveness, and add substance to the work of Lindgren, Fenner, Bowen, Niggli, and Buddington, who saw so clearly the importance of magmatic differentiation in generating the materials of many ore deposits.

ACKNOWLEDGEMENTS

It is inevitable that, in an investigation begun in its first form some thirty-four years ago, I should have received help and support from many people. The first is John Grover, formerly Director of the Geological Survey of the British Solomon Islands, who, in 1950, introduced me to the fascination of the Melanesian arcs, and gave me the opportunity to stumble over the fact that many of the world's great ore deposits are products of island-arc volcanism and sedimentation. The second is Dr B. W. Chappell, of whose fastidious work all of the present X-ray fluorescence, neutron activation, and related analyses of rocks and mineral separations are the products. The phase of the work commencing in 1982 was in fact begun as a collaborative effort, and although this later succumbed to the pressures of other commitments, it is a pleasure to acknowledge that Dr Chappell's superb analyses provide much of the basis of the account that follows. For their never-failing assistance over a period of more than thirty years I should like to give special thanks to Mrs W. P. H. Roberts (who assisted with the early X-ray fluorescence analyses in 1963, and has given much analytical and related help throughout the years since), Mrs H. M. Roan, and Mr J. S. Cook. I should like to acknowledge and thank most warmly the Australian Mineral Industries Research Association (especially its member companies Aberfoyle, BHP, Billiton, CRA, Geopeko, Pasminco, and Western Mining Corporation Ltd), and its Director Mr J. R. May, for all of the financial support of the work since 1982; my many friends in the mining geological world, particularly the late Haddon King, and Messrs R. Woodall, I. R. Johnson, N. Herriman, O. N. Warin, and D. F. Fairweather, for their support and encouragement over many years; Dr. B. D. Hackman for many discussions on Solomon Islands geology, J. H. Hill for his early assistance (in 1959) on New Georgia and for much help in the years since; Dr J. D. Bell for his company and his help in a memorable early collaboration (1959–69) on the New Georgia problem; Frank Coulson and Dr Peter Dunkley for kind help in 1981–2; my old Solomon Islander friends and helpers of the years 1950–88: William Saemanea, Silas Selo, David Haolo, Zephania Sala, John Arabola, Nelson Legua, John Tietala, Kuva, and others of whom I have the happiest and most grateful recollections; Miss Catherine Brown for much help in programme of mineral and groundmass separation involving more than 700 separations and extending over almost three years; N. G. Ware for his skilful, patient, and invaluable help with the ANU electron microprobes, and Professors A. E. Ringwood and Kurt Lambeck for their permission to use these instruments; Professor R. W. Hutchinson, Professor R. J. Arculus, Professor R. R. Large, and Dr D. J. Swaine and Dr. J. D. Bell for so kindly reading the manuscript and commenting on it; Dr J. G. Holland for providing some 1500 individual Lesser Antilles lava analyses, and Dr Stephen Eggins, for making his unpublished Vanuatu analyses available to me. I should like to thank Mrs R. M. Moloney for her enthusiasm, skill, and care in the final word-processing of text and tables, and Mrs Sandra Kelly for early help with word-processing. The draughting of the line drawings — which I think speak for themselves — is the work of Michael Roach of the Cartographic Department of the University of New England, and the microphotographs were taken by my colleague Dr Peter Flood and Miss Shirley Dawson.

Armidale, New South Wales   R. L. S.
March 1993

Contents

List of plates ... xii

1. Introduction ... 1

2. Volcanic rocks and exhalative ores: the general nature of the ores and their environments ... 7

3. Historical: development of ideas on the origin of exhalative ores ... 10

4. Derivation of the exhalations ... 18
   The leaching hypothesis. Persistence of the magmatic hypothesis. Concluding statement.

5. Petrology of the Solomon Islands Younger Volcanic Suite (SIYVS) ... 26
   Principal rock types. Major element chemistry. Principal minerals. Lava nomenclature. Concluding statement.

6. Copper ... 53
   Copper in lavas of the marine environment. Chemical properties and crystal chemistry of copper. The incidence of copper in the principal mineral species of the Solomon Islands Younger Volcanic Suite. Crystal:melt partitioning. The incidence of whole-rock copper in the Solomon Islands Younger Volcanic Suite. Concluding statement.

7. Zinc ... 75
   Zinc in lavas of the marine environment. Chemical properties and crystal chemistry of zinc. The incidence of zinc in the principal mineral species of the Solomon Islands Younger Volcanic Suite. Crystal:melt partitioning. The incidence of whole-rock zinc in the Solomon Islands Younger Volcanic Suite. Concluding statement.

8. Lead ... 96
   Lead in lavas of the marine environment. Chemical properties and crystal chemistry of lead. The incidence of lead in the principal mineral species of the Solomon Islands Younger Volcanic Suite. Crystal:melt partitioning. The incidence of whole-rock lead in the Solomon Islands Younger Volcanic Suite. Concluding statement.

9. Barium ... 108
   Barium in lavas of the marine environment. Chemical properties and crystal chemistry of barium. The incidence of barium in the principal mineral species of the Solomon Islands

x Contents

>   Younger Volcanic Suite. Crystal:melt partitioning. The incidence of whole-rock barium in the Solomon Islands Younger Volcanic Suite. Concluding statement.

10. Strontium — 120

    Strontium in lavas of the marine environment. Chemical properties and crystal chemistry of strontium. The incidence of strontium in the principal mineral species of the Solomon Islands Younger Volcanic Suite. Crystal:melt partitioning. The incidence of whole-rock strontium in the Solomon Islands Younger Volcanic Suite. Concluding statement.

11. Phosphorus — 135

    Phosphorus in lavas of the marine environment. Chemical properties and crystal chemistry of phosphorus. The incidence of phosphorus in the principal mineral species of the Solomon Islands Younger Volcanic Suite. Crystal:melt partitioning. The incidence of whole-rock phosphorus in the Solomon Islands Younger Volcanic Suite. Concluding statement.

12. Calcium — 148

    Chemical properties and crystal chemistry of calcium. The incidence of whole-rock calcium in the Solomon Islands Younger Volcanic Suite. Concluding statement.

13. Titanium — 154

    Titanium in lavas of the marine environment. Chemical properties and crystal chemistry of titanium. The incidence of titanium in the principal mineral species of the Solomon Islands Younger Volcanic Suite. Crystal:melt partitioning. The incidence of whole rock titanium in the Solomon Islands Younger Volcanic Suite. Concluding statement.

14. Vanadium — 169

    Vanadium in lavas of the marine environment. Chemical properties and crystal chemistry of vanadium. The incidence of vanadium in the principal mineral species of the Solomon Islands Younger Volcanic Suite. Crystal:melt partitioning. The incidence of whole-rock vanadium in the Solomon Islands Younger Volcanic Suite. Concluding statement.

15. Chromium — 183

    Chromium in lavas of the marine environment. Chemical properties and crystal chemistry of chromium. The incidence of chromium in the principal mineral species of the Solomon Islands Younger Volcanic Suite. Crystal:melt partitioning. The incidence of whole-rock chromium in the Solomon Islands Younger Volcanic Suite. Concluding statement.

16. Manganese — 203

    Manganese in lavas of the marine environment. Chemical properties and crystal chemistry of manganese. The incidence of manganese in the principal mineral species of the Solomon Islands Younger Volcanic Suite. Crystal:melt partitioning. Manganese–zinc relations in the Solomons minerals. The incidence of whole-rock manganese in the Solomon Islands Younger Volcanic Suite. Concluding statement.

Contents   xi

| 17. | Iron | 223 |

Chemical properties and crystal chemistry of iron. The incidence of whole-rock iron in the Solomon Islands Younger Volcanic Suite. Concluding statement.

| 18. | Cobalt | 228 |

Cobalt in lavas of the marine environment. Chemical properties and crystal chemistry of cobalt. The incidence of cobalt in the principal mineral species of the Solomon Islands Younger Volcanic Suite. Crystal:melt partitioning. The incidence of whole-rock cobalt in the Solomon Islands Younger Volcanic Suite. Concluding statement.

| 19. | Nickel | 244 |

Nickel in lavas of the marine environment. Chemical properties and crystal chemistry of nickel. The incidence of nickel in the principal mineral species of the Solomon Islands Younger Volcanic Suite. Crystal:melt partitioning. The incidence of whole-rock nickel in the Solomon Islands Younger Volcanic Suite. Concluding statement.

| 20. | Sundry elements | 264 |

Sulphur. Chlorine. Fluorine. Rubidium. Zirconium. Arsenic, antimony, bismuth. Molybdenum. Gold and silver. Cadmium. Gallium. Tungsten. Uranium. Concluding statement.

| 21. | Abundance patterns in the crystallizing melt | 280 |

Paragenesis and context of crystallization. Principal minerals and their patterns of trace-element acceptance. Principal trace-element abundance patterns in the Solomons lavas. Crystal subtraction and element abundance. Concluding statement.

| 22. | The ore elements in volcanic exhalations | 304 |

Historical: c. 1853–c. 1963. The modern era: c. 1963 to the present. Theoretical and experimental aspects. Concluding statement.

| 23. | Petrogenesis: 1: The lavas | 325 |

Relationships of the three lava groups: a single series? Provenance of the suite. Development of the principal lava groups. Concluding statement.

| 24. | Petrogenesis: 2: Exhalative ores | 341 |

Broad considerations. Discussion. Summary of discussion: active magmatism versus passive leaching. Prognostications on the possible influence of magmatic processes on patterns of exhalative ore formation. Concluding statement.

Epilogue   361
Bibliography   366
Author index   385
Subject index   389

Plates

1. Principal lava types of the Solomon Islands Younger Volcanic Suite.
2. Hornblende and its opaque rims in the Solomon Islands Younger Volcanic Suite.
3. Electron microprobe traverse across a single crystal of hornblende and its opaque rim in the Solomon Islands Younger Volcanic Suite.

1

Introduction

It may seem trite to say that an understanding of ore-element geochemistry in volcanic rocks should be necessary for the proper understanding of the formation of volcanic orebodies. Surprisingly, however, little is known of the behaviour of the principal ore metals in lavas and other volcanic rocks, and there is no systematic documentation of their incidence in the lavas of island arcs, which are the principal locale of volcanic ore deposits in ancient terrains.

The suggestion that certain metallic ore deposits might be products of volcanic activity is of comparatively ancient vintage. In his Presidential Address delivered to the Geological Society of France in July 1847 Elie de Beaumont referred to deposits derived 'd'émanations volcaniques et métallifères', and in 1851 Henry de la Beche proposed that certain 'chemical deposits' might form on the sea floor as the result of submarine volcanic activity. That both of these great contemporaries should refer to the idea at about the same time indicates that it may have had fairly general currency when they wrote. Since that time the 'volcanic–exhalative theory', as it has come to be known, has passed through several cycles of popularity and, in two principal forms, is now firmly established as an important element of ore-genesis theory.

Although the early ideas of de Beaumont and de la Beche re-emerged from time to time during the following one hundred years, they lay in substantial eclipse for most of this period. However their reappearance in the 1950s could almost — and perhaps appropriately — be said to have been explosive. Clear statements of the theory began to emerge in the English language in 1954–5, and in 1955 there came the first suggestion that volcanic–sedimentary ore formation might constitute a part of volcanic island arc formation — that some exhalative ore deposits were an intrinsic feature of the petrotectonic evolution of the volcanic arc regime. It was not long after this that volcanic associations, and particularly calc-alkaline volcanic palaeogeography, were being applied in base metal mineral exploration, and the first important discoveries stemming from these hypotheses had been made in Eastern Canada by the end of the decade.

Since that time a wide variety of volcanic ore types and associations have been recognized. The first, and still probably the most important, are the essentially stratiform deposits of base metal 'massive sulphides'. Various more-or-less stratiform precious metal deposits, previously diagnosed as stratigraphically controlled replacement orebodies, are now recognized as volcanic–exhalative sediments. The widespread 'stratiform skarn' (calc-silicate) lenses containing base metal sulphides and, particularly, cassiterite and scheelite, are now viewed as being of substantially exhalative origin, as are the smaller banded iron formations, bedded barite, tourmalinite, garnet quartzite, and chemical carbonate lenses and other 'exhalites' commonly found associated with exhalative orebodies. All these deposit types represent materials that have escaped from the volcanic milieu to be immediately captured by the sedimentary milieu. Other ore types, perhaps most importantly the precious metal tellurides and native elements, have formed within the volcanic edifice itself; and yet others, notably the porphyry copper, gold, and molybdenum deposits, have been emplaced in the sub-surface, but none the less relatively shallow, sub-volcanic environment.

With the first suggestion that it was *island-arc* volcanism that was of particular significance in volcanic ore formation came the first indication that in this environment ore type might be genetically linked to an associated volcanic rock type. By 1955 it was suspected that within a given exhalative ore province, copper-rich ores were associated with earlier, more mafic volcanism, whereas the zinc-rich and zinc–lead ores were associated with later, more felsic, acitivity. A little latter (the early 1970s) the basalt-associated copper-rich ores (of 'Cyprus type') came to be regarded as being of deep-sea-floor–mid-ocean ridge origin, whereas the calc-alkaline associated, more zinc- and lead-rich ores (of 'Kuroko type') were seen to be of arc origin. By the late 1970s it had, however, been proposed that at least many of the Cyprus-type ores, though certainly associated with deep-sea-floor basaltic rocks, had been formed in the very earliest, basaltic, stages of island arc formation, rather than at mid-

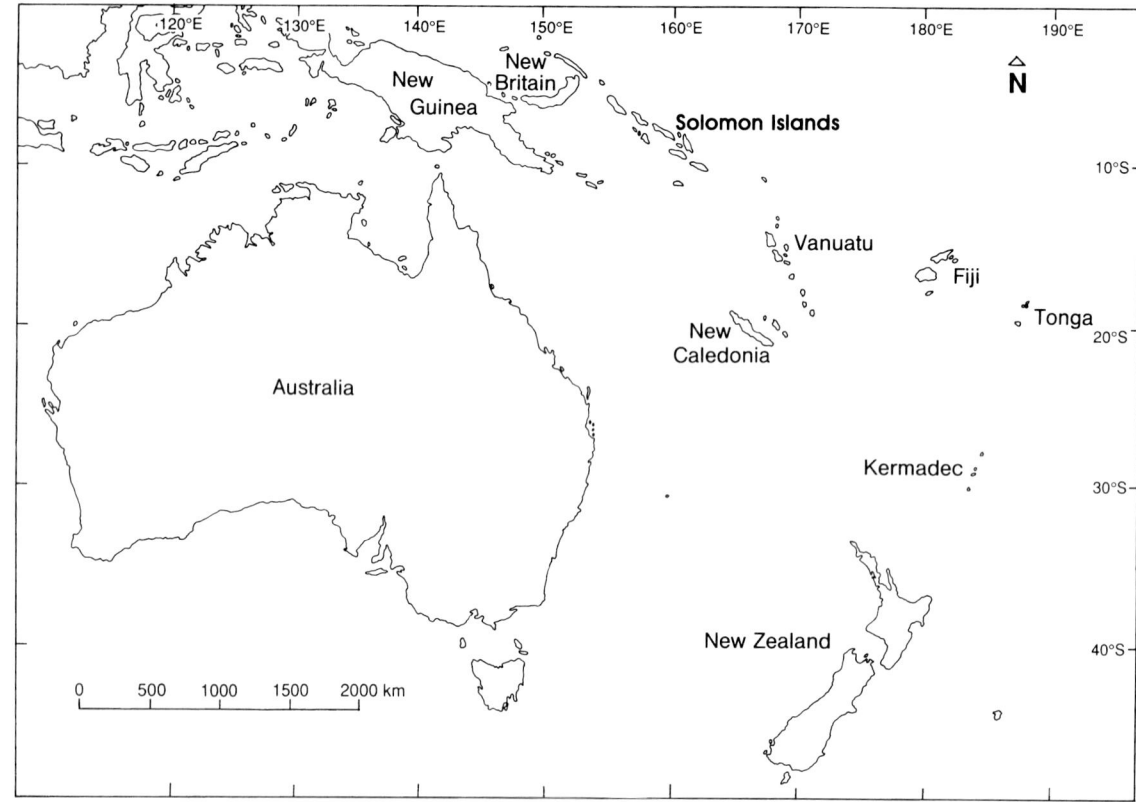

Fig. 1.1. The Solomon Islands as a component of the Outer Melanesian volcanic island-arc system of the south-west Pacific. Other important volcanic festoons are those of New Britain, Vanuatu, the Fijian archipelago, and the Tonga–Kermadec islands trending southwards to the North Island of New Zealand. From Stanton (1991).

ocean ridges. Porphyry copper and their related deposits are associated with the sub-volcanic manifestations of andesitic to dacitic volcanism and hence tend to form as part of the later stage, calc-alkaline phases of arc development — in some cases at stages following incorporation of the arc within a continental mass. The precious metal deposits seem to accompany more alkaline volcanism, and to appear at almost any stage of arc development.

Although the derivation of the aqueous component of the exhalations — juvenile magmatic versus meteoric — had been a matter of considerable controversy in the time of Elie de Beaumount, re-emergence of the exhalative theory in the 1950s seems to have involved a tacit assumption that the metals and their transporting solutions were entirely juvenile. However by the late 1960s the idea that surface waters — particularly sea water convecting deeply within, and leaching, sub-sea-floor basalts — constituted the major part of the exhalations, re-appeared and gained almost immediate dominance. As the end of the twentieth century approaches both of these hypotheses are current, with that involving sub-sea-floor leaching of lavas by heated, convecting sea-water the more generally favoured.

Extensive stable isotope measurement and limited experiment appear to support the leaching hypothesis, though this evidence focuses on the aqueous phase and the sulphide sulphur, rather than on the metals themselves. On the other hand, comparatively little has been done to investigate evidence concerning possible active magmatic derivation, particularly of the metals. Almost sixty years ago both Bowen and Fenner pointed to magmatic differentiation, leading to the concentration and loss of volatiles, as the prime mechanism of derivation of the metals from igneous rocks. This does not, however, seem to have been followed up in any specific or systematic way. While it is now generally accepted that most 'massive sulphide' and related stratiform precious metal deposits, precious metal telluride ores, and porphyry coppers have at least some kind of volcanic connection, and that there are some very pronounced ties between ore type and volcanic association, comparatively little has been done to investigate ways in which such features might relate to progressive crystallization, magmatic differentiation, and lava evolution in the petrotectonic environment of the volcanic arc.

The present volume is concerned with such an investigation. The principal base metals of the volcanic sulphide ore association are iron, copper, zinc, and lead, with minor and sporadic amounts of cobalt and nickel. The principal precious metals are gold and silver, both of which occur as traces that may or may not be associated with significant quantities of the base-metal sulphides. Apart from the anion-forming 'hyperfusible' elements such as sulphur, selenium, tellurium, and fluorine, other commonly conspicuous elements are manganese, calcium, barium, and phosphorus. Elements that on the other hand may be said to be conspicuous by their paucity in ores of volcanic affiliation are chromium, vanadium, titanium,

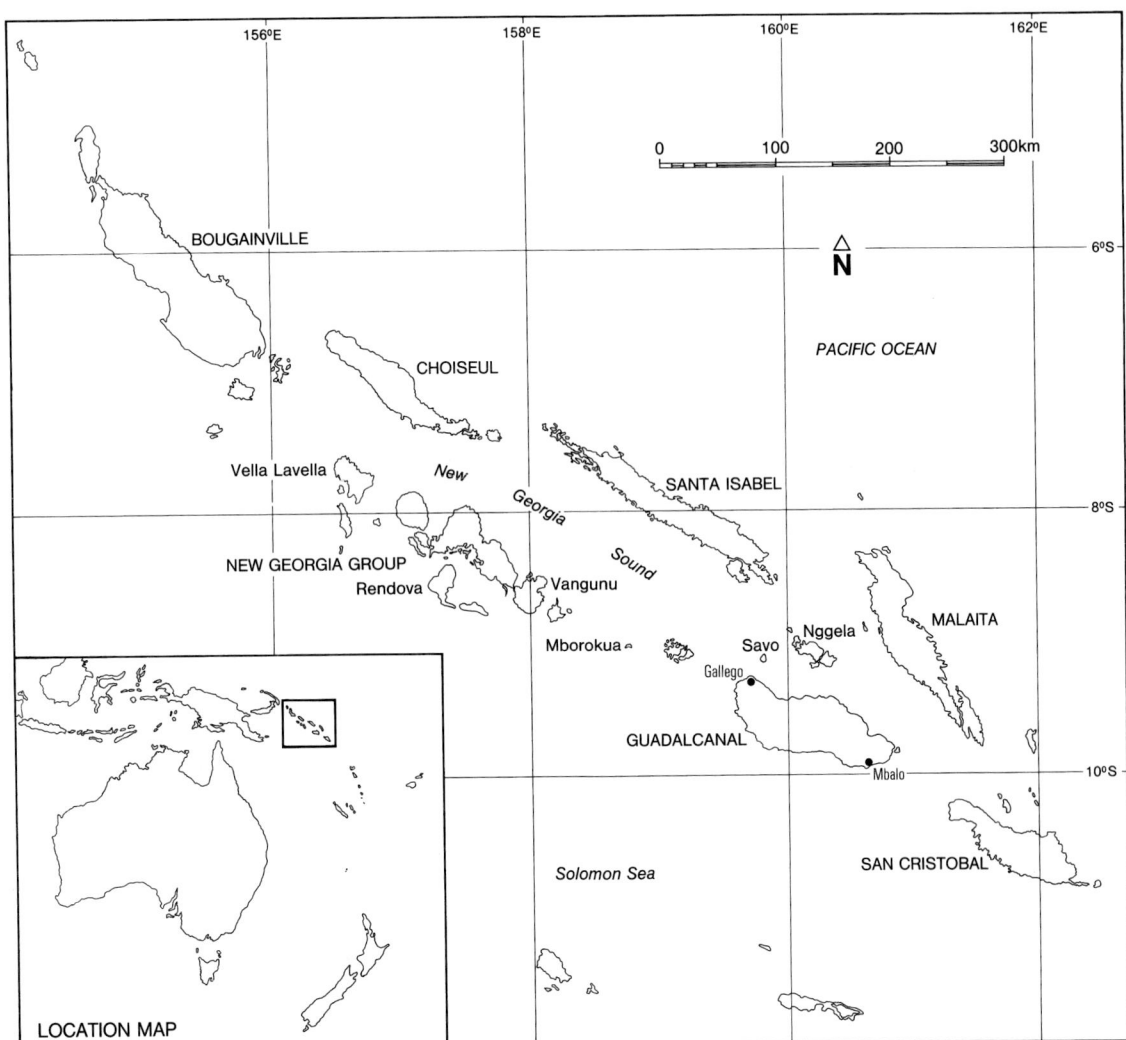

Fig. 1.2. The principal islands of the Solomons festoon. Lavas of the Solomon Islands Younger Volcanic Suite occur with varying prominence on the islands of the New Georgia Group, Bougainville, Choiseul, Guadalcanal, Mborokua, and Savo; they may occur on San Cristobal; but are unknown on Santa Isabel, Malaita, and Nggela. The present contribution is chiefly concerned with occurrences from Vella Lavella to Mbalo on Guadalcanal, but refers also to the analyses of Rogerson *et al.* (1989) of material from Bougainville.

and strontium. Variation of chemical and physical properties among these various elements is substantial and certainly such as to be capable of inducing considerable variation in their behaviour in geochemcial processes — including, in particular, igneous crystallization — and hence to lead to their separation or fractionation as such processes proceed. Patterns of concentration or impoverishment of these elements should thus provide a key to processes through which they might have passed.

With this in mind the abundance behaviour of fourteen major, minor, and trace elements has been studied in an island-arc lava suite displaying a very large evolutionary spectrum — from highly basic picritic basalts to calc-alkaline dacites — within a single young, undeformed, sulphide-poor volcanic sequence; that of the Pliocene-to-Recent 'Younger Volcanic Suite' of New Georgia–Guadalcanal, in the Solomon Islands of the south-west Pacific (Figs. 1.1–1.3; see also Fig. 5.1). This 'Younger Suite' is one of two broad groups into which the volcanic rocks of the Solomons arc may be divided:

1. The Cretaceous–Lower Oligocene ocean floor tholeiitic lavas constituting the oceanic basement, and hence the oldest exposed units, of the Solomons Islands stratigraphic sequence. These lavas are dominantly basaltic but include minor andesites, and are commonly prominently pillowed. At least many of the basalts contain very low levels of potassium (see Stanton and Ramsay 1980). We may term this group of lavas 'The Solomon Islands Older Volcanic Suite'.
2. The Pliocene-to-Recent lavas and associated pyroclastic rocks, ranging from picritic rocks of a high-magnesium basalt suite to dacitic lavas of a typical island-arc calc-alkaline suite. These rocks occur here and there from north-western Bougainville to Mbalo on Guadalcanal (and perhaps to San Cristobal), are best developed on the islands of New Georgia, and most commonly are components of still-recognizable cones and volcanic complexes. Basaltic members of the group commonly contain 1–2 per cent K_2O, and some contain >2.5 per cent K_2O. We may term this group of lavas 'The Solomon Islands Younger Volcanic Suite'. That part of the present volume concerned with the behaviour of the ore elements in the Solomon lavas refers exclusively to materials of this 'younger suite'.

The elements selected are copper, zinc, lead, barium, strontium, phosphorus, calcium, vanadium, chromium, titanium, manganese, iron, cobalt, and nickel; and, treated less comprehensively in Chapter 20, sulphur, chlorine, fluorine, rubidium, zirconium, arsenic, antimony, bismuth, molybdenum, gold, silver, cadmium, gallium, tungsten, uranium, and thorium. Some of the fourteen elements of Chapter 6–19, such as copper, zinc, lead, and barium, were chosen because they are characteristically present in exhalative ores. Others, such as chromium, nickel, and strontium, were included because they are characteristically — and conspicuously — extremely low or absent in such ores. It was thought that the consistent absence of an element, know to be abundant in a volcanic melt, from the associated exhalative ores might have as much to tell about processes as the consistent presence of others. Elements such as cobalt, manganese, and phosphorus are present in some exhalative deposits but not in others; and some occur in noteworthy associations — for example the zinc–lead–manganese–phosphorus grouping of many large Lower Proterozoic stratiform orebodies. Again, some of these elements, such as nickel and chromium, are rapidly incorporated in the first-separated olivine and spinel crystals of the volcanic regime and are hence removed very early from the melt. Others, such as lead and barium — the 'large-ion' or 'incompatible' elements — tend not to be accommodated in the more abundant volcanic crystals and remain within the melt right up to the residual stages. It seemed that the behaviour of these elements in the melt might bear some systematic relationship with their occurrence in the ores.

The principal reasons for the choice of the Younger lava sequence of the Solomon islands were:

1. The Solomons volcanic chain is an intra-oceanic festoon: it has therefore not formed in a continental crustal association, and none of its products have been subjected to compositional complications inherent in continental contamination. This minimizes the number of 'variables' involved in elucidating geochemical processes.
2. The younger lava sequence spans a very wide compositional range — from picrite basalts ($SiO_2 \approx 46$ per cent; $MgO \approx 29$ per cent) to dacites ($SiO_2 \approx 68$ per cent; $MgO \approx 1.0$ per cent) — and hence provides a long evolutionary series in which relationships between progressive crystallization and changing patterns of distribution of the minor and trace elements may be investigated.
3. Most of the younger lavas are notably porphyritic in one or more of olivine, ortho- and clinopyroxene, hornblende, feldspar, and, in some cases, magnetite. This demonstrates the progressive nature of their crystallization, and permits the separation of crops of crystals and groundmasses for trace-element analysis.
4. In contrast to the members of the Solomons Islands Older Volcanic Suite, many of which contain visible, apparently primary, sulphide, the lavas of the Younger Suite generally do not contain visible

Introduction 5

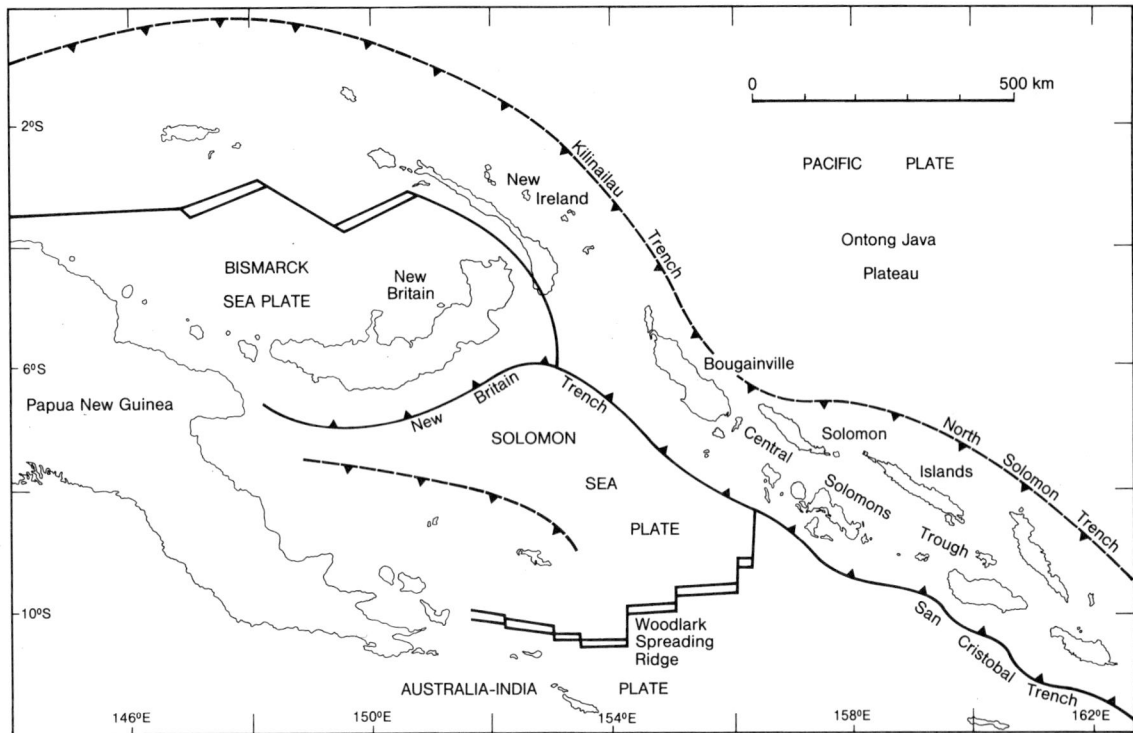

Fig. 1.3. Current view of the tectonic context of the Solomon Islands (after Vedder and Colwell 1989). There is some similarity between the abutting of the Woodlark Spreading Ridge against the New Britain–San Cristobal Trench opposite New Georgia, on which picrites are abundant, and the abutting of the d'Entrecasteaux Spreading Ridge against the Vanuatu Trench opposite the island of Aoba, on which occur very similar picrites of very similar age (see Fig. 23.2).

sulphide and upon analysis yield very low total sulphur. The younger lavas were therefore appropriate for investigating the behaviour of the metals in silicate–oxide systems containing negligible sulphur.

5. The lavas are young — Pliocene-to-Recent — and they have not been involved in any deformational or metamorphic events. All specimens were collected from still-recognizable cones, some from very young and well-preserved volcanoes in which hot spring activity persists. All materials are thus now essentially as they were at the time of extrusion and solidification.

6. The lavas were originally mapped by the author (with J. D. Bell of Oxford) and this, together with subsequent more detailed mapping by members of the British Geological Survey, provided a firm knowledge of field relationships.

The patterns of incidence of the fourteen elements in the full range of whole rocks and in their constituent olivines, spinels (chromite, ferrochromite, magnetite, titanomagnetite), orthopyroxenes, clinopyroxenes, hornblendes, feldspars, microcrystalline groundmass, and glass have been determined by X-ray fluorescence and electron-microprobe methods. Other elements have been determined as indicated throughout the text.

The result has been the demonstration of many clear ties between particular trace elements and particular mineral groups, and between the level of abundance of these elements and variation in major element chemistry within each mineral group. Some of these relationships were already known, some predicted but unsubstantiated, and some previously unknown. As a result of this quite highly systematic preferential incorporation of particular elements into particular major minerals, there has been corresponding enrichment and impoverishment of the different elements in the residual melt as this has evolved — a state of affairs anticipated long ago by Goldschmidt, Bowen, Fenner, and many others, and now substantiated and quantified at least to some extent. What is, however, of principal significance in the present context is that these abundance patterns appear to correspond well with the established relations, outlined above, between ore type and volcanic rock type in older, ore-bearing, terrains. It is therefore concluded that, contrary to much current opinion, derivation of the exhalative metals of volcanic milieux — as distinct from the solutions in which they arrive at the sea floor — is probably chiefly the result of active magmatism rather than later, passive, leaching.

2

Volcanic rocks and exhalative ores: the general nature of the ores and their environments

In view of its derivation the term 'exhalative' (from English *exhale*; French *exhaler*; Latin *(ex)halare*, 'to breathe out': and *exhalation*; Latin *exhalatio*, 'breath, vapour') might be seen to be most appropriately applied to ores deposited from a gas phase as this was emitted at the Earth's surface. Perhaps the only materials fitting such an exclusive definition are those deposited from volcanic gases at and very close to the surface in areas of fumarolic activity. In fact the term is applied much more loosely, and while it certainly embraces those ores deposited directly from gases at and close to the contemporary surface it is used principally to refer to those deposited from hot aqueous solutions emitted on to the sea floor, or on to the floors of lakes and lagoons. Exhalation of this kind takes place in a variety of geological environments, the most prominent of which (though, it must be emphasized, certainly not the only one) is volcanic. It is these exhalative ores of the volcanic environment that constitute the background to the present volume.

There are in fact three important ore types of the volcanic milieu which may be said to represent different parts of the volcanic exhalative regime. They are, as noted in the Introduction, the porphyry copper deposits, volcanic (epithermal) base and precious metal vein systems and, pre-eminently, the stratiform 'massive sulphide' and precious metal bodies of volcanic–sedimentary environments.

The first clear recognition that at least some porphyry copper deposits were of a sub-volcanic nature came in 1964 with the discovery of the large Panguna deposit on Bougainville Island, in the south-west Pacific. That many such deposits in older terrains occurred in intrusions of distinctly volcanic context had been known for at least thirty years prior to that, but the now-recognized common close genetic tie between intrusion and intruded volcanic rocks was not widely perceived until the 1970s. The discovery of large numbers of porphyry copper deposits in the western Americas, the western Pacific, and elsewhere since 1970 has, however, revealed a clear tie with young volcanic-arc structures and this, together with an increasing understanding of geochemical and petrological relations between mineralized intrusions and associated volcanic lavas and pyroclastic rocks, has led to the recognition of the comparatively shallow, intrinsically volcanic nature of most porphyry copper deposits and the closely related porphyry gold and molybdenum deposits.

The volcanic and sub-volcanic nature of many precious metal vein systems was first recognized in the late nineteenth and early twentieth centuries, with the discovery of the gold and silver deposits associated with volcanic plugs and diatremes in the Mesozoic and Tertiary terrains of the western United States. It was such deposits that constituted a large part of Lindgren's epithermal class of hydrothermal ore deposits, and shallow, epithermal vein systems are now regarded as an important locale for the occurrence of precious metal ores. The intrinsically volcanic nature of such deposits has been amply confirmed by discoveries in modern, and in some cases still active, volcanic centres in Fiji, the Lihir Islands, and elsewhere in the south-west Pacific and Indonesia. Closely related deposits of the base metals are well-established entities of the epithermal environment.

Neither of these ore types is, however, exhalative in the strict sense, in that they have not been fully emitted from their source. If, as will be indicated below, such ores are indeed derived through active magmatism from the volcanic materials with which they are closely associated, they have been caught, not after their escape, but in the *process* of their emission from their source. The epithermal precious and base-metal vein systems probably represent solution transport of material from one part of the parental volcanic centre to another.

Within our broad class the exhalative ores *par excellence* are without doubt the broadly conformable, or stratiform, masses of 'volcanic–sedimentary' or 'exhalative–sedimentary' materials that occur in association with a wide variety of lavas, pyroclastic, and volcaniclastic rocks of volcanic terrains of all ages. They occur not only within the continents but appear also as part of modern island arcs and since about 1980 have been recognized as forming on the modern sea floor. The many thousands of known deposits of this type range in size from less than 10 tonnes to over 2×10^8 tonnes. If the large quantities of iron sulphide they commonly contain are taken into account, some of these deposits attain sizes of over 10^9 tonnes. In some cases ore-bearing 'feeder zones' occur beneath the stratiform masses, indicating in principle the genetic ties between these and the epithermal and sub-volcanic ore types.

EXHALATIVE ORE TYPES AND THEIR VOLCANIC ASSOCIATIONS

One of the first to recognize distinctiveness of ore type in conformable deposits of volcanic affiliation was the American geologist W. H. Weed (1902, 1911). Working in the Appalachian province he noted a volcanic association and observed that the orebodies themselves conformed with a world-wide type — the Kieslager of the German ore geologists — in both their form and their mineralogy. From his detailed experience in the Ducktown district of the southern Appalachians he stated that the ores contained pyrite or pyrrhotite or both, variable quantities of chalcopyrite as an accessory, and, generally, sphalerite and galena in small amounts. He also noted that while pyrrhotite might or might not be present, 'pyrite and chalcopyrite appear never to be absent'.

Had Weed concentrated his attention on the Northern Appalachian ore province rather than the Southern he would have become aware of two principal ore types: the pyrite/ pyrrhotite–chalcopyrite type to which he had referred, and a second in which sphalerite and galena were much more abundant and pyrrhotite present much less commonly. Had he looked very closely he would have discerned a third, intermediate, type: a pyrite (pyrrhotite)–chalcopyrite–sphalerite assemblage of the kind referred to by Hutchinson (1973).

All of this resolves itself into a simple pattern of metal sulphide occurrence: Fe (as pyrite or pyrrhotite); Fe + Cu; Fe + Cu + Zn; Fe + Cu + Zn + Pb.

As is only too well known to those involved in mineral exploration, there are many stratiform sulphide masses of volcanic affiliation that contain iron sulphide — either pyrite or pyrrhotite or both — as their only sulphide component. These are commonly referred to as 'barren' pyritic or pyrrhotitic lenses and perhaps occur most frequently in the more mafic volcanic terrains. Others consist of pyrite–chalcopyrite or pyrite/pyrrhotite–chacopyrite, but copper sulphide does not appear without iron. Yet others contain sulphides of iron, copper, and zinc, but never zinc without iron and copper; and others contain iron, copper, zinc, and lead — lead however never appearing without the other three metals. As a very general rule, pyrrhotite tends to decrease in abundance and frequency of occurrences as the ores become more complex; i.e. it is common and often abundant in barren iron sulphide lenses and, as first pointed out by Weed (1911), in iron-sulphide-chalcopyrite bodies, but it is relatively uncommon and of low abundance in stratiform Cu–Zn and Zn–Pb ores of volcanic affiliation.

In many stratiform ore districts assemblages of only one type occur. For example most of the deposits of the Archaean 'greenstone' terrains of Canada and Australia are copper–zinc ores containing very little lead. Those of Cyprus are all pyritic copper ores with generally lesser zinc, and those of the Kuroko province are chiefly zinc–lead ores with only minor admixed copper. Other volcanic-stratiform ore provinces, notably those of 'Appalachian–Caledonide' age and type, exhibit distinct copper, and zinc–lead–minor copper, types. In such cases these two ore types may occur separately and on distinct horizons, or separately on a single volcano-stratigraphic unit. In other cases they may be contiguous and interdigitate or, as first pointed out by Kraume et al. (1955) at Rammelsberg, the different ore types may be arranged stratigraphically within the one sulphide lens: iron sulphide at the base, overlain successively by iron plus copper sulphides, iron plus copper plus zinc and, at the top, Fe + Zn + Pb (+ Ba).

Consideration of the end members occurring at a distance, e.g. the copper-rich Cyprus ores with their mafic basaltic association, and the zinc–lead-rich Kuroko ores of Japan with their dacitic–rhyolite association, might yield the conclusion that the two were products of different kinds of volcanic source. This could be seen, perhaps, as not being contradicted by the much greater geographical closeness prevailing in relatively small areas such as the Bathurst district of New South Wales (Stanton 1955b), where the copper-rich ores occur in association with more mafic volcanic rocks on a lower horizon, the zinc–lead-rich ores in association with more felsic rock types on a higher horizon.

Where, however, the different ore types occur not far from each other on a single volcano-stratigraphic horizon, distinctiveness of source is not easy to sustain. The difficulty is compounded where the different ore types are contiguous, and difference in source simply cannot be entertained where the different mineral assemblages interdigitate along bedding, or change one into the other across the stratigraphy of the orebody concerned.

The beginning of an explanation may lie in a consideration of (1) the simple pattern of incidence of the four principal sulphide metals already noted with (2) the spectrum of ore-type:lava-type associations mentioned above.

In the Cyprus-type basaltic environment only the iron–copper-(± variable, often minor, zinc) assemblages are found. Iron sulphide lenses bearing no more than trace copper (and lesser zinc) are common. The iron sulphide is usually pyrite and the mass may consist of almost 100 per cent sulphide, thus constituting a very high sulphur 'ore type'. In Kuroko-type and related dacitic–rhyolitic environments, on the other hand, the whole spectrum of iron–copper–zinc–lead sulphide assemblages may occur. The Iberian Pyrite Belt possesses several pyritic copper (–zinc) orebodies essentially similar in their sulphide mineralogy (both assemblages and textures), and form, to typical Cyprus-type copper orebodies; but the Spanish ores are associated, not with basalts, but with dacitic and rhyolitic pyroclastic rocks. This association of cupriferous pyritic bodies of 'Cyprus' mineralogical and chemical type with felsic environments is in fact quite common, and in may cases they occur here in close association with pyritic lead–zinc ores; e.g. Kidd Creek, Heath Steele in Canada, Avoca in Eire.

Thus, whereas the basaltic environment may be said to play host only to Cyprus-type ores, the more felsic — including the bimodal — environment exhibits the whole range from barren iron sulphide, cupriferous pyrite right through to high lead lead–zinc ore types. The Cyprus mineralogical ore type thus constitutes a thread that runs through the whole gamut of basaltic to rhyolitic environments. The progressively more zinc- and lead-rich ores are simply *added* to the assemblage as the volcanic regime becomes progressively more felsic.

Two further features of the exhalative ore-volcanic rock association are apparent and, as will be shown later, are likely to have critical significance.

1. As an increasingly felsic nature of volcanism is accompanied by increasing pyroclastic emission, the exhalative ores of felsic association tend to be accompanied by pyroclastic rocks rather than lavas. Cyprus-type ores are generally associated with basaltic *lavas*; copper–zinc ores of andesitic association by both lavas and tuffs; and the generally more complex ores of more felsic association, by dacitic and rhyolitic tuffs and breccias. The felsic volcanic and exhalative regime is clearly characterized by a much higher degree of explosiveness, due at least in part to the emission of much larger quantities of gas.
2. The size of the larger known exhalative deposits tends to increase with increase in the felsic index of the associated volcanic materials. The largest Cyprus-type orebodies are of the order of 10^7 tons. Those of andesite–dacite association are $2-4 \times 10^7$ tons, and those of dacite–rhyolite association may be $1-2 \times 10^8$ tons. The felsic volcanic regime thus appears capable of the localized generation of much larger quantities of exhalations than does the basaltic.

CONCLUDING STATEMENT

The foregoing presents a brief but general view of the nature of exhalative ores, their volcanic associations, and of relations between ore types and different volcanic–sedimentary environments, particularly in the marine environment. The features described indicate that the range of ores and associations are not unsystematic, semi-random phenomena but are, on the contrary, a well-defined, if broad, petrological grouping through which runs a strong connecting and ordering thread. Questions of their origin have thus invited systematic scientific investigation, and this now has a long and detailed history. It is appropriate, therefore, to look briefly at the development of ideas and at some of the still-unanswered questions that have led to the investigation of lava geochemistry reported in this volume.

3

Historical: development of ideas on the origin of exhalative ores

The early contributions of de Beaumont and de la Beche have already been referred to. They were however but two — though a particularly perspicacious pair — of a long line of geologists who have contributed to present understanding of stratiform ores of the aquatic–volcanic milieu.

Some of the earliest observations were made in the Caledonide Province of Norway, now recognized as one of the classical regions of exhalative ore occurrence in the volcanic environment. The first serious work appears to have been that of E. R. Vargas Bedemar published in 1819, which drew attention to the layered or bedded nature of the ores of the Røros district of south-central Norway. Vargas Bedemar considered these orebodies to be disrupted parts of a single layer or bed of sulphides. He did not speculate on the origin of this layer, but it is clear that he recognized the stratiform, or conformable, nature of the deposits.

The first recorded speculation on the genesis of the Norwegian Caledonide ores appears to be that of Ström, who in 1825 considered that the sulphide layers, like the rocks that enclosed them, were sedimentary.

Elie de Beaumont's prognostications did not spring from the observation and description of a particular deposit or locality, but were of a more general kind based on wide observation and theoretical deduction. He considered that volcanic activity gave rise to two distinct classes of materials: those that were erupted in the molten state and those that were exhaled as gases:

The volcanic eruptions bring to the surface of the globe, on the one hand, molten rocks, lavas, and all their adjuncts; on the other, matter volatilized or carried away in the molecular state; of vapour of water, of gas, such as hydrochloric acid, hydrosulphuric acid, carbonic acid; of the salts, such as hydrochlorates of soda, of ammonia, of iron, of copper, etc. These volatilized substances are released, now from active craters, now from cooling lavas, now from fissures about volcanoes, like the Etuves of Nero, the geysers, and one naturally finds connections with other jets of hot vapours which are being given off at greater or lesser distances from active volcanoes, like the soffioni and the lagonis of Tuscany, as well as hot springs and most mineral springs. These emanations from the interior of the globe generally give rise to more or less solid masses such as the sulphur and salts of solfataras, the deposits from mineral waters, etc.

One can thus distinguish two classes of volcanic products, those which are *volcanic in the manner of lavas*, and those which are *volcanic in the manner of sulphur, of sal ammoniac etc.*

In all of the epochs of the history of the globe, these eruptive phenomena have yielded products falling in these two classes, but the nature of one and the other have varied with time ... one sees the materials *volcanic in the manner of lavas* becoming richer and richer in silica ... At the same time one sees the materials *volcanic in the manner of sulphur* becoming more and more varied. I designate these products under the heading of 'émanations volcaniques et métallifères' (de Beaumont 1847 pp. 1249–50).

A little further on in his Presidential Address de Beaumont (see Chap. 1) went on to say:

... one considers ... that the substances spread on the surface have come from the interior of the earth; that they have been carried along, either by mineral waters, or sometimes by aqueous vapours; that they have been deposited in part in the fissures through which these emanations passed, and that the remainder only of what has passed through the fissures, and is in part held there, has been spread into the surface waters and has been finally deposited by these (de Beaumont 1847, p. 1284).

While de Beaumont could not couch his views in modern terms, it is clear that he considered metalliferous emanations and solutions to be important and distinctive accompaniments of deep-seated volcanism, that some of the metals they contained were deposited on the walls of fissures up which the emanations ascended, and that those materials remaining in solution as the emanations reached the surface of the Earth were in some cases contributed to the bottom waters of oceans and lakes, there to be precipitated as metalliferous beds.

This appears to be the first clear enunciation of the volcanic–sedimentary, or 'exhalative–sedimentary' theory of ore formation.

de Beaumont's great English contemporary, Sir Henry de la Beche, also clearly saw the potential importance of seafloor volcanism in the formation of bedded chemical deposits. In his book *The geological observer* (1851) he noted under the heading 'Chemical deposits in seas':

> ... we have to reflect that the volcanic action which we know has been set up upon the seafloor, sometimes throwing up matter above the surface of the sea, forming islands, must as a whole have caused no small amount of soluble matter to be vomited forth. Looking at the gases evolved and substances sublimed from sub-aerial volcanoes, we should expect many combinations formed and decompositions to arise ... without entering further upon this subject, we would merely desire to point out that, in volcanic regions, the sea may not only receive saline solutions marked by the presence of certain substances not so commonly thrown into it by rivers elsewhere, but that also submarine volcanic action may be effective in producing chemical deposits, either directly or indirectly, which, under ordinary conditions, would either not be formed, or not so abundantly (de la Beche 1851, pp. 126-7),

adding as a footnote that:

> It would be very desirable to ascertain points of this kind, so far as examining the seawaters around volcanic regions may enable the observer to do so; and more especially when, by any fortunate chance, opportunities may be afforded after any submarine volcanic action may be evident or supposed (de la Beche 1851, p. 127).

Not only did de la Beche clearly recognize, over 140 years ago, the likely significance of sea-floor volcanic exhalation in the formation of bedded chemical deposits, but he saw the desirability of investigating the hypothesis by examination of volcanic activity on the modern sea floor. He had anticipated by well over 120 years the current investigations of sea-floor hydrothermal activity now known to be associated with mid-ocean ridges and related structures.

The latter half of the nineteenth century saw several further allusions to the formation of stratiform ores of various kinds by exhalative sedimentary processes. Some observers noted only that the ores were bedded like the sediments enclosing them, and deduced a sedimentary influence in their deposition; others recognized a volcanic association as well, and pointed to the possibility of a volcanic exhalative contribution to an environment of reduced detrital sedimentation.

J. D. Whitney (1854), in speculating on the origin of some of the great bedded chert:iron mineral deposits (banded iron formations) of the Lake Superior region suggested that 'The iron ore may have been introduced ... by sublimation of metalliferous vapours from below during the deposition of the siliceous particles ...' In the preceding year he had remarked on the similarities among, and general concordancy of, a number of the base metal sulphide deposits of Ducktown, Tennessee (Whitney 1854). Ansted (1857) and Hitchcock (1878) also emphasized this feature of many of the Southern Appalachian deposits, though neither seems to have speculated on its possible genetic connotations. At about the same time the Norwegian, Helland (1873), was confirming and extending the earlier work of his compatriot, Ström. Investigating base metal deposits between the Hardanger and Trondheim districts of Norway, Helland noted the concordant nature of the sulphides and the substantially volcanic nature of the metasediments in which the bedded ores occurred, and he concluded that the latter were chemical sediments that had been precipitated from volcanic exhalations contributed to the Palaeozoic sea floor. Similar views were propounded by the Norwegian J. H. L. Vogt in papers published in 1889 and 1890.

At about the same time exhalative–sedimentary ideas were being put forward to explain the 'manto' copper deposits of the Boleo district of Baja California, in Mexico. These orebodies occur as stratiform lenses within tuffaceous sediments of the Pliocene Boleo Formation, and were identified as exhalative sediments by Tinoco (1885), by Bouglise and Cumenge (1885) and by Fuchs (1886). The latter considered the deposits to have formed by 'muddy eruptions' and that they had 'the double character of eruptivity, with respect to their origin, and of sedimentation, with respect to their form, their mode of deposition, and their extent' (Fuchs 1886, pp. 90–1, as reported by Wilson 1955, p. 83.).

Although put forward in clearer form by Ohashi some fifteen years later, the importance of submarine exhalative activity in the formation of the Kuroko ores was well understood by the Japanese geologists Fukuchi and Tsujimoto (1902). Their paper 'Ore beds in the Misaka Series' was published in Japanese, and in what was undoubtedly a rather obscure journal so far as Western economic geologists were concerned. In consequence, their ideas had virtually no impact on the mainstream of exhalative ore genesis theory.

The volcanic exhalative idea was, however, still alive in the Western world: Thomas and MacAlister (1909) recognized a class of deposits which they referred to as 'gold in sinters' and they observed that 'Gold in the native state occurs as precipitations in certain highly siliceous accumulations formed at the surface under hydrothermal conditions' (Thomas and MacAlister 1909, p. 328). They suggested that the gold was probably transported in an iron sulphate solution which decomposed on reaching the surface, depositing the gold together with gelatinous silica and the iron oxides. Thomas and MacAlister gave as an example an occurrence at Mount Morgan, Queensland, where a siliceous sinter associated with rhyolite was 'impregnated with haematite and yielded as much as 170 ounces of gold to the ton' (1909, p. 328). They were clearly aware of modern analogues, remarking that similar deposits, though of less economic value, occur in modern volcanic hot spring localities in Nevada and Yellowstone Park. They pointed out that such sinters were being deposited from highly alkaline, silica-rich solutions connected with rhyolitic rocks.

Not long after this van Hise, Leith, and Mead (1911), in a detailed study of the Lake Superior banded iron formations, concluded that 'the iron and silica, especially the silica, were precipitated from concentrated solutions coming directly from the magma' (1911, p. 514). They noted the local presence of acidic rocks between the iron formations and abundant associated basalts, and that these acidic lavas were slightly later than the basalts. This suggested that the materials of the iron formations might be related to the felsic lavas rather than to the basalts and that 'The iron salts and the acidic phases then might represent the extreme differentiation products of a primary magma of which the basalt was the first extrusion' (van Hise *et al.* 1911, p. 514). We shall see below just how prescient that observation may have been.

The 1910–20 decade concluded with the important paper by Ohashi, 'On the origin of the *Kuroko* of the Kosaka Copper Mine, Northern Japan', published in Japanese in 1919 and in English in 1920. Ohashi stated 'the proposed submarine sinter theory' in simple terms:

Regarding the origin of the deposits of Kosaka the author holds a view that they are none other than sinter formed on the seafloor by hot springs.

Rhyolite must have been the rock composing the very part of the seafloor. Hot springs flowed out through this rhyolite, and precipitation of $(Zn, Fe)S$, PbS, $BaSO_4$ and SiO_2 took place on the surface. The mud mixed with those minerals during this process.

The finely laminated bed of *kuroko*, with more or less admixture of mud, must have been formed first upon the floor.

An accumulation of soft mud became thicker and thicker, and on this account the free outflowing of springs was impounded. The spring water, therefore, made its ways in all directions through the mud and deposited the sinter as concretionary masses (Osashi 1920, pp. 15–16).

Ohashi thus might almost be said to have ushered in — just 100 years after the early observations of Vargas Bedemar — the modern understanding of exhalative ores in the volcanic milieu: he saw the lead, zinc, and barium as derived from the rhyolite melt, and then localized by both sedimentary and hydrothermal–diagenetic processes.

It is difficult to discern at this distance whether Ohashi's work was noted in Europe or America. In Norway Carstens (1920, 1923) espoused exhalative–sedimentary principles in connection with the 'vasskis' — thin iron sulphide-rich layers associated with magnetite and chlorite-rich layers of minor iron formation — which he attributed to sea-floor biochemical precipitation of iron supplied by volcanic exhalation. Indeed it seems that, as far as this type of mineralization in volcanic environments was concerned, Norwegian opinion was that an exhalative–sedimentary origin was more or less self-evident. In continental Europe geologists such as P. Niggli, Kurek, and Schneiderhohn were quite conversant with the exhalative idea and the theory persisted there, in somewhat muted form, as the central tenet of the 'European hot spring school' during the period from about 1920 to the beginning of the Second World War.

By the end of the War the sub-surface hydrothermal replacement theory as applied to 'massive sulphide' and related ores was completely in the ascendant. In 1948, however, the German geologist Hegeman published a short paper, 'Uber sedimentäre Lagerstätten mit submariner vulkanischer Stoffzufuhr' ('On sedimentary ore deposits with submarine volcanic derivation'), which in hindsight may be seen to have been a sign of things to come. Hegeman noted that many sedimentary and metasedimentary iron and manganese deposits had long been recognized as exhalative by Schneiderhohn and others, and proposed that the principle might be extended to a much wider range of bedded deposits of volcanic-sedimentary environments. Among these he included conformable zinc and lead–zinc ores (mentioning, particularly, Franklin, New Jersey and Broken Hill, New South Wales), and copper, antimony, arsenic, and perhaps mercury sulphide deposits. He also suggested that some stratiform corundum occurrences might have had an origin of this kind and observed that 'The accompanying minerals of rich exhalative–sedimentary ore deposits e.g. rutile, apatite, fluorite, barite, quartz, ferruginous quartz, morion, jasper and others have acquired their chemical components mainly through submarine emanations. In part this also holds for the surrounding beds of siliceous and clayey character, perhaps also for dolomite similar explanations are applicable' (Hegeman 1948, p. 55).

Hegeman's paper appears to have passed unnoticed in the English-speaking world and exerted virtually no influence outside Europe and Scandinavia. However it led on to two remarkable contributions, one by Ehrenberg, Pilger, and Schroder (1954) on the great stratiform orebodies of Meggen, the other by Kraume, Dahlgrun, Ramdohr, and Wilke (1955) on those of Rammelsberg.

Of Meggen, Ehrenberg, and his colleagues deduced that magmatic 'aqueous solutions of sulphides were responsible for the formation of the ... deposit, ascending as highly concentrated colloidal solutions and debouching onto the anoxic muds of the volcanic seafloor. Due to the sudden cooling, and the seawater's electrolyte content, the highly concentrated metal sols flocculated, to settle and accumulate as bodies of amorphous gels.

'The entire ore-body represents a very flat mound-cone ...

'As a result of quite complete geological evidence this deposit can be referred to as a submarine, magmatogenic, exhalative, hydrothermal formation' (*Ehrenberg et al.* 1954, pp. 335–56).

The Ehrenberg group's conclusions were thus quite clear and unequivocal: the Meggen deposit was derived from magmatic emanations and laid down as mixed gels on the contemporary volcanic sea floor.

Kraume *et al.* (1955), whose investigations must be regarded as having been contemporary with those of the Ehrenberg group, came to essentially similar conclusions concerning the Rammelsberg deposits. However as well as identifying a volcanic derivation of the ore solutions, they noted a distinct chemical stratigraphy within the orebody and deduced from this that 'The chemical composition of the solutions changed so that at first iron, then copper, zinc, lead and finally barium were present in larger concentration' (Kraume *et al.* 1955, p. 351).

In the same year the present author (Stanton 1955*a, b*) noted several striking relations between conformable ore occurrence and features of stratigraphy, sedimentary facies, and volcanic activity in the Lower to Middle Palaeozoic eugeosynclinal terrain (now known as the 'Lachlan Fold Belt') of south-eastern Australia. He concluded that the sulphides had formed, by a combination of volcanic and sedimentary/diagenetic factors, in near-shore sediments about volcanic highs that had earlier constituted components of volcanic island arcs. He proposed from this that many of the world's great 'metallogenetic provinces' might in fact be old island-arc structures.

The author noted that the south-eastern Australian deposits were restricted to one or other of two stratigraphic horizons: a lower one, associated with basaltic–andesitic volcanics, which was copper to copper-zinc rich, and an upper horizon, associated with dacitic to rhyolitic materials, which was more zinc-lead rich. This led on to the recognition of some general world-wide metal sulphide and lead isotope abundance relationships, indicating that the metallic component of the ores came via volcanism from some quite fundamental source (Stanton 1958; Stanton an Russell 1959).

The potential usefulness of volcanic stratigraphy in the search for stratiform orebodies was soon perceived by mineral explorers, and Holyk (1956) quickly showed the applicability of the principle in delineating a striking pattern of exhalative–sedimentary ore occurrence in the New Brunswick province of the Northern Appalachians. Soon after Oftedahl (1958) drew attention to the possible particular importance of felsic — 'granitic' — melts in generating metal-rich magmatic gases. (The Norwegians have never been far from the lead in exhalative ore genesis theory.) With a special interest in ignimbrites and their implications concerning magmatic degassing, and strongly influenced by personal contacts with Hegeman, Ehrenberg, and Kraume, Oftedahl proposed that rhyolitic sub-volcanic melts might, here and there, degas directly on to the sea floor forming exhalative accumulations of metallic sulphides and other materials (Oftedahl, discussions with the author, 1979). In three later contributions (Stanton 1959, 1960, 1961) the author drew increasingly specific attention to the fact that many conformable ore provinces, such as those of the Appalachians, Caledonides, Cascades, Australian Eastern Highlands Belt and the Urals appeared to be related to old island arcs, and that island-arc volcanic exhalation must have been an important generator of stratiform ores all over the world.

Perhaps ironically, the initial discovery of large-scale exhalative metal deposition on the modern seafloor — the Red Sea brine deposits, first reported by Miller *et al.* (1966) — involved not a volcanic arc but a spreading basaltic 'ocean' floor. In the same year Bostrom and Peterson (1966) discovered sediments rich in iron and manganese hydroxides on the present-day sea floor of the East Pacific Rise. They noted a marked coincidence of these chemical sediments with areas of high heat flow, and concluded that the precipitates were derived from ascending solutions of deep-seated origin, probably related to magmatic processes occurring at depth along the Rise: 'The Rise is considered to be a zone of exhalation from the mantle of the earth, and these emanations could serve as the original enrichment in certain ore-forming processes' (Bostrom and Peterson 1966, p. 1258).

All major discoveries of exhalative sulphide deposition on the modern sea floor — and there now have been many of them, some of spectacular size — have been made on mid-ocean ridges or on sea-floor structures related to them. No significant discoveries have yet been made in volcanic arc provinces — perhaps a surprising state of affairs given that the world's major volcanic exhalative deposits of older terrains occur chiefly in felsic arc, rather than basaltic ridge, environments. This may well represent bias in selection of ocean drilling sites and submersible observation areas rather than any actual state of affairs, however, and current investigation of the Melanesian sea floor seems likely to demonstrate this.

For the remainder of the 1960s and into the 1970s these modern sea-floor observations, together with increasing investigation of the young, unmetamorphosed basalt-associated pyritic copper ores of Cyprus and the dacite: rhyolite-associated zinc–lead ores of the Kuroko of Japan, held centre stage in exhalative ore genesis theory. Clark (1971) was the first to draw attention to the likelihood that these latter two exhalative ore associations might represent two distinctive petrogenetic milieux: that deposits of Cyprus type were generated in an oceanic environment in association with ophiolite complexes, whereas those of Kuroko type were generated in a continental edge environment in association with felsic volcanism. Clark noted that:

It is an interesting question whether there is a spectrum of deposits whose physical and chemical properties and nature of occurrence lies between the above described examples [Cyprus-type and Kuroko-type — author] and for which they may form end members of a series. Alternatively, base metal sulphide deposits of volcanic affinity may fall into two discrete groups. In either case the nature of the genetically associated volcanic rocks would seem to play a very important role ... (1971, p. 214).

At about the same time Sillitoe (1972) was drawing attention to the likely importance of mid-ocean (spreading) ridges and of subduction zones as two distinctive loci for exhalative ore formation. In the following year he brought his ideas into somewhat sharper focus, dividing 'volcanogenic massive sulphide deposits' into two principal categories based — as was Clark's slightly earlier classification — on ore type, volcanic association and environment of formation: (1) cupriferous pyrite deposits formed by sea-floor hydrothermal activity associated with tholeiitic volcanism along mid-ocean ridges; deposits of 'spreading centre situation' and (2) polymetallic sulphide deposits formed by similar activity associated with felsic (calc-alkaline) volcanism along island arcs; deposits of 'island arc and continental margin situations' (Sillitoe 1973). Sillitoe suggested — as had Clark two years previously — that the deposits of the first group, i.e. those of ophiolite association, were probably best represented by the pyritic copper ores of the Troodos Complex of Cyprus and the second group by the Kuroko ores of the Green Tuff of Japan. This quickly led to a fashionable two-fold categorization of volcanic exhalative ores into those of 'Cyprus type' and those of 'Kuroko type'.

Such a classification did not take into account a very large category of copper-bearing, zinc-rich lead-poor ores that tended to be associated with andesitic–dacitic rather than basaltic or dacitic–rhyolitic volcanic activity, i.e. ore types and environments intermediate between those of Cyprus and Kuroko types.

This was quickly perceived by Hutchinson (1973), who, on the basis of extensive observation of volcanic-associated deposits in Canada, proposed a third grouping: the 'pyrite–sphalerite–chalcopyrite ores'. He regarded these as typically associated with sequences of differentiated lavas such as those of the Abitibi and similar 'greenstone' belts of the Canadian Precambrian. Of 'volcanogenic' sulphides as a whole, Hutchinson observed:

Three distinct varieties of such deposits can be distinguished by their compositions, relative and absolute ages, and rock association. Pyrite–sphalerite–chalcopyrite bodies are found in differentiated, mafic-to-felsic volcanic rocks; pyrite-galena-sphalerite–chalcopyrite bodies occur in more felsic, calc-alkaline volcanic rocks, and pyrite-chalcopyrite bodies occur in mafic, ophiolitic volcanic rocks (Hutchinson 1973, p. 1223).

At this time the present author was pursuing two threads; a continuation of his investigations of stratiform ores of volcanic association, and his old interest in the petrotectonic evolution of the Melanesian volcanic arcs. He observed (Stanton 1967, 1978) that while some of the exhalative ores of the volcanic milieu did occur with sea-floor basalts, such deposits were usually rather small; that the most abundant, and by far the larger, of the exhalative orebodies occurred with andesites, dacites, and rhyolites and their pyroclastic equivalents — which are minor or absent on the deep-sea floor. He also noted that vast volumes of sea-floor basalt — in both young and ancient terrains — often occured with very little accompanying ore, whereas comparatively minute associated sequences of andesite, dacite, and rhyolite contained very substantial bodies of metallic sulphides. In addition there was much evidence that many major exhalative orebodies were laid down in comparatively shallow water rather than in the deep oceans.

Such evidence, combined with considerations of the incidence of the ore metals in modern volcanic rocks, led to the conclusion that relations between exhalative sulphide ore type and volcanic silicate rock type reflected two parallel, closely interlocking evolutionary paths: intimately related components of the overall petrotectonic evolution of the parent volcanic arc (Stanton 1978). The pyritic copper and copper–zinc ores were products of exhalation during the earlier, basaltic stages of arc development. The more zinc-rich, and then lead–zinc ores were products of later calc-alkaline volcanism, when the arc was partially emergent and exhalation commonly occurred at shallower water depths. The progression from smaller iron-copper rich deposits to larger zinc–lead-rich deposits was thus seen as a manifestation of progressive change in lava type and water depth as the volcanic arc evolved from a deep submarine sea-floor swell to a semi-emergent array of rises. The principal influences on exhalation — as seen towards the end of the 1970s — were therefore magmatic differentiation leading to a progressive evolution of the lavas, and a lessening of water depth yielding a reduction in confining pressure:

'... while it is certainly possible to point to a conspicuous group of pyritic copper deposits of basaltic association (Cyprus-type) and a similarly conspicuous group of lead-zinc deposits of dacitic to rhyolitic association (Kuroko-type), this ignores the fact that there are many copper-zinc and zinc-copper ores often associated ... with more intermediate andesitic to dacitic rocks, and tends to conceal what appears to be an essentially continuous connection between volcanic rock type and the dominant metals of the associated ores.

... It may therefore be proposed that this whole spectrum of stratiform ore types and their associated volcanic rocks are of arc provenance; that the 'ophiolites' here simply represent an early stage of volcanic arc formation, and that the pyritic copper ores of this affiliation are no more and no less than the earliest formed ores of the arc regime. Ores of this type but richer in zinc appear a little later, and zinc becomes the important metal as soon as the lava sequence develops an andesitic, calc-alkaline aspect. Zinc, and then lead, increase as the volcanic products become more felsic still, and lead achieves prominence as volcanism achieves a dacitic to rhyodacitic stage (Stanton 1977; 1978, p. 11).

Although he used the term 'arc' here, the author was in fact referring to all instances of marine volcanism in which initial tholeiitic basaltic activity constituted the early stages, giving way later to a calc-alkaline type of progressively more felsic and explosive nature. Whether this happened to be of arcuate form, as shown so well by many Palaeozoic to Recent examples, of linear form as in the present Solomons and New Hebrides festoons, or as essentially non-linear groups of domes and volcanic islands, as in the Philippines and Fiji at the present day and in the Keewatin 'greenstone' belts of the Precambrian of Canada, was of no significance in principle. All constituted basaltic, dacitic, and rhyolitic volcanism in the marine environment — the apparent chief habitat of the volcanic sedimentary exhalative ores. Linking the ores and their associated volcanic rocks in this way the author brought the Canadian copper–zinc ores (and their counterparts in Australia, India, and elsewhere) into a median relationship between the Cyprus and Kuroko ore types. He thus suggested that Hutchinson's three groups of ores were simply different members of a single spectrum, in agreement with the astute suspicion expressed by Clark in 1971.

All this indicated that an important influence in the generation of different exhalative–sedimentary ore types was *source*. Basaltic volcanism or volcanic piles appeared to be capable of generating only iron- and copper-rich exhalations, with minor zinc in some cases. These also tended to be sulphur-rich. More felsic volcanism — or its products — appeared capable of generating all of iron-, copper-, zinc- and lead-rich exhalations. These were, in some cases, less sulphur-rich.

It was soon recognized, however, that there commonly occurred a separation of ore-type within particular volcano-stratigraphic horizons as these appeared in a single district or small area and this looked, on the other hand, to reflect variation of *conditions of deposition*. This likelihood was emphasized by those occurrences where different ore types interdigitated.

An early explanation for this (Stanton 1960) was that the different ore types were products of contiguous sedimentary facies. Many of the copper ores of felsic association contain conspicuous pyrrhotite in addition to pyrite and minor magnetite. This led to the suggestion that the copper ores were deposited under conditions characterized by a slightly higher Eh than those conducive for the formation of the lead–zinc ores: that the pyrrhotite–chalcopyrite–(minor magnetite) ores may have been deposited in sedimentary environments that varied from *just* oxidizing to *just* reducing. While conditions were oxidizing, there was sufficient oxygen for the formation of the ferroso-ferric oxide magnetite but insufficient for the formation of haematite; when conditions were reducing, sulphur activity was very low — sufficient for the precipitation of the monosulphides, but not for any large quantity of pyrite. For the pyritic sphalerite–galena–chalcopyrite ores, on the other hand, the environment was more consistently and more highly reducing. This led to the higher concentration of total sulphide, the occurrence of much disulphide, and the general paucity or absence of magnetite.

This hypothesis appeared to be supported by the nature of the associated silicates and carbonates. Those associated with the pyritic zinc–lead ores, deposited under conditions of relatively high sulphide activity, were iron-poor; iron-rich chlorite and other iron-bearing silicates were subordinate to absent, and any carbonate was virtually iron-free. The pyrrhotitic copper ores on the other hand often contained much green chlorite and other iron-bearing silicates, and any carbonate tended to be sideritic, indicating that sulphide activity was low to the point of being insufficient to satisfy much of the iron of the depositional environment. Hence it was thought that the relative positioning of the two ore types might be seen, in terms of James's (1954) 'facies of iron formation', thus: haematite facies → magnetite facies → carbonate:silicate facies → pyrrhotitic copper ores (nearer-shore sulphide facies) → pyritic zinc–lead ores (distal sulphide facies).

It has, however, also been suggested (e.g. R. R. Large 1977) that separation of the ore types might have been brought about by variations in physico-chemical conditions stemming not from simple sea-floor *sedimentary* factors, but from variations induced by the interplay of metal:sulphide ion-bearing volcanic hydrothermal solutions with the normal bottom waters of the local volcanic sea floor. According to Large (1977) consideration of mineral–solution equilibria in the Fe–S–O system indicates that many volcanic exhalative sulphide ores deposited close to or within the source orifice are precipitated from high-temperature (> 275 °C), mildly acid, and reduced chloride-rich solutions that mix with sea water at the surface of the volcanic pile. Mixing results in an increase in solution pH, oxygen activity, and total sulphur, together with a fall in temperature, all of which leads to the precipitation of (1) chalcopyrite–pyrrhotite (±pyrite ±magnetite) assemblages in the sub-surface feeder fissures and lower massive ore zones developed beneath the contemporary local sea floor and (2) pyrite–sphalerite–galena assemblages on the sea floor around or at some distance from the hydrothermal vent. In some cases the hydrothermal solutions may debouche on to the sea floor carrying essentially their full load, in which case the copper-rich assemblage deposited adjacent to the hydrothermal orifice to form *proximal* orebodies and the zinc–lead-rich assemblages deposited under cooler conditions slightly further away to form *distal* orebodies.

This constitutes a very plausible explanation not only for the variety of stratiform — and stratigraphically equivalent — copper and zinc–lead sulphide lenses that occur as separated, contiguous, or interdigitating masses, but also for the systematic metal sulphide stratigraphy now known to be developed in many exhalative orebodies.

More recently it has been suggested that differences in confining pressure, in turn stemming from differences in water depth under which exhalation takes place, may be responsible for some instances of variation in ore type (e.g. Finlow-Bates and Large 1978; Plimer and Finlow-Bates 1978; Plimer 1981). For example where exhalation occurs under relatively shallow conditions and hence low confining pressure, boiling may occur leading to the rapid precipitation of chalcopyrite within the hydrothermal feeder fissures beneath the seafloor. At greater depths and under higher confining pressure, on the other hand, boiling cannot occur, copper remains in solution until debouchment, and chalcopyrite is then precipitated as a component of stratiform ore deposited on the adjacent sea floor. On this general basis Plimer (1981) has proposed that shallow-water exhalation (low confining pressure; boiling) tends to produce stratiform (Fe–Mn, Ba, Zn–Pb) ores associated with an underlying zone of vein, disseminated or stockwork (Cu ± Au, Sn) mineralization. Where exhalation has occurred under a deep-water column, on the other hand, underlying 'footwall' precipitation does not occur and the exhalative deposits, which contain (Fe–Mn, Ba–Zn–Pb–Cu), are stratiform.

While variations in water depth and resultant confining pressure may well explain some observed variations in the metal assemblages of some stratiform ores, and particularly the variable incidence of sulphide-bearing 'feeder zones', it is unlikely to explain within-bed and small-scale stratigraphical differences in ore type. It still seems most likely that these stem from the influence of sedimentary microfacies as suggested by Stanton (1960), hydrothermal-sedimentary facies suggested by R. R. Large (1977), and progressive change in the composition of the exhalations as first suggested by Kraume et al. (1955).

Before leaving the subject of ore type it should be mentioned that recent investigation of stratiform gold occurrence in the volcanic environment (e.g. the Noranda area of Quebec) has shown that the gold deposits occur in a zone distinctly further from the principal volcanic centre than the related zone of base metal sulphide deposits; i.e. the gold deposits are even more distal with respect to the active centre than are the distal base metal occurrences, though there appears to be some overlap. Most of the gold deposits occur in fine, siliceous, pyritic laminated tuffs and in this connection it is interesting to note the observation of R. R. Large (1977) that following precipitation of the base metal sulphides, the residual ore solutions roll out further along the sea floor depositing 'silica and pyrite to form the laminated cherty tuffs which are commonly associated with the ores' (Large 1977, p. 549). It is these laminated cherty tuffs, with their commonly abundant pyrite, that frequently constitute hosts to the gold. The consequent relation between the base metal sulphide concentrations and these cherty pyritic gold ores thus appears to be a genuine example of the separation of exhalative ore type as a manifestation of hydrothermal-sedimentary facies patterns in the marine volcanic environment.

Since the late 1970s attention has largely focused on three principal aspects: (1) the nature of the ore solutions, (2) observation and investigation of current hydrothermal activity and ore formation on the seafloor, and (3) the range of environments — marine and non-marine — in which exhalative ores have been deposited.

Ore-solution studies have been concerned chiefly with determining the relative contributions of magmatic, sea-water, and meteoric sources and have been based very largely on stable isotope measurement and interpretation. Such investigations commonly indicate that sea water or meteoric waters have dominated, and this is often interpreted as indicating that the metals of the ores have been derived simply by the leaching of the rocks these waters have traversed. This is countered — as we shall see in the following chapter — by the view that the bulk of the metals is contributed by what may be a comparatively minor proportion of magmatic water, with which large quantities of near-surface waters have, virtually inevitably, mixed.

Investigation of modern sea-floor hydrothermal activity and metal sulphide–barite deposition substantially began with the dredging and coring of the Red Sea deposits in the middle 1960s, and has now evolved to sophisticated observation and collection using personnel-carrying submersibles and a wide range of recording techniques. Perhaps perversely, while almost all the major discoveries of exhalative orebodies of marine affiliation continue to be made among felsic terrains of ancient arc environments, most theoretical and modern ocean-floor research has until recently continued to emphasize basaltic environments, and the mid-ocean ridge milieu. In the late 1980s to early 1990s, however, attention has moved to the western and south-western Pacific arcs and their associated basins — much to the present author's satisfaction. The recent observations of hydrothermal emission and sedimentation in the Lau, Manus, and Woodlark basins and in the Marianas–Bonin region are an enlightened move towards what look to be the 'real' modern analogues of most ancient marine exhalative ore environments. (The change of emphasis is also an ironic one: it was the area of Simbo and Vella Lavella in the Solomons that, in 1955, the author, with I. R. Kaplan and L. G. M. Baas-Becking, sought to have dredged for evidence of modern sea-floor hot spring sulphide deposition. The proposal was greeted with amused derision.)

The more recent — and proliferating — investigation of ancient depositional environments has shown these to range from the sea floor and island arcs, to those of epicontinental rifts and the saline lakes of continental graben. It is of course with those of marine, and quite clearly volcanic, association that the present volume is concerned.

CONCLUDING STATEMENT

Observation and investigation of conformable and stratiform ores of volcanic association — the 'exhalative' or 'exhalative–sedimentary' deposits of the aquatic volcanic milieu — thus shows them to constitute a broad spectrum of types and settings. Their common features are that they are derived from metal-bearing solutions and gases that are contributed to the floors of bodies of water, principally the sea but also lakes and lagoons; that they develop conformable, often stratiform, configurations within the sediments, pyroclastic rocks, and lavas that enclose them; and that they are invariably associated with volcanic products of one kind and another. Their differences lie in their constituting a spectrum of ore types involving substantial variations in chemical constitution, mineralogy, and style of volcanic association. This spectrum of features and types may reflect differences in the tectonic setting of the different styles of associated volcanism (mid-ocean ridge versus island arc) or it may reflect the tectonic and related petrological evolution of a single tectonic entity, e.g. a single intra-oceanic volcanic arc.

All of this pertains to the nature, setting and *broad* origin of the metals; the ores are closely and systematically associated with volcanic rocks, so presumably they are derived from them. However, by what means, and at what stage of volcanic history, did this derivation occur? What, precisely, are the exhalations and how are they related to the volcanic rocks from which they appear to have arisen?

This constitutes one of the principal problems of ore genesis theory in the latter part of the twentieth century. The resulting controversy is briefly reviewed in the following chapter.

4

Derivation of the exhalations

There seems no reason to doubt[*] that de Beaumont and de la Beche assumed the ore-bearing emanations to result from degassing of the associated molten volcanic rocks, and hence to have a common, deep, origin with the latter. Indeed, in differing from the views of Werner, de Beaumont asserted that 'One admits with him that mineral substances have been deposited by the action of waters ... but one does not admit that it may have been by supergene solutions; one considers, on the contrary, that the substances spread on the surface have come from the interior of the earth ... ' (de Beaumont 1847, p. 1284). One possible deviation from this assumption was the statement by N. H. and H. V. Winchell, in connection with the origin of the Lake Superior banded iron ores, that sea water, on encountering recently erupted basalts, would '... become surcharged with soluble silica and iron, obtaining the latter from the augitic minerals of the basic lavas, and possibly from masses of erupted metallic iron' (Winchell and Winchell 1891, p. 396).

This was, however, no more than an incidental and very minor speculation, and concerned an ore type very much peripheral to the exhalative ores considered now. Although unstated, it seems clear that the long line of exponents of the volcanic exhalative–sedimentary idea — Fukuchi and Tsujimoto, Ohashi, Niggli, Schneiderhohn, Hegemann, the Ehrenberg and Kraume groups, Stanton, Oftedahl, Clark, and Hutchinson — all assumed that exhalative ore formation was an intrinsic part of the volcanic process; that the materials of the ores were products of degassing that constituted part of the active magmatism responsible for the formation of the associated volcanic rocks (Fig. 4.1) — and hence that they were, like the latter, of some indeterminate, deep, igneous origin.

Such a view followed naturally from the work of the igneous petrologists of the late nineteenth and early twentieth centuries, and especially from the investigations of Fenner (1926, 1933) and Bowen (1928, 1933). Fenner drew attention to the large amount of mineral matter,

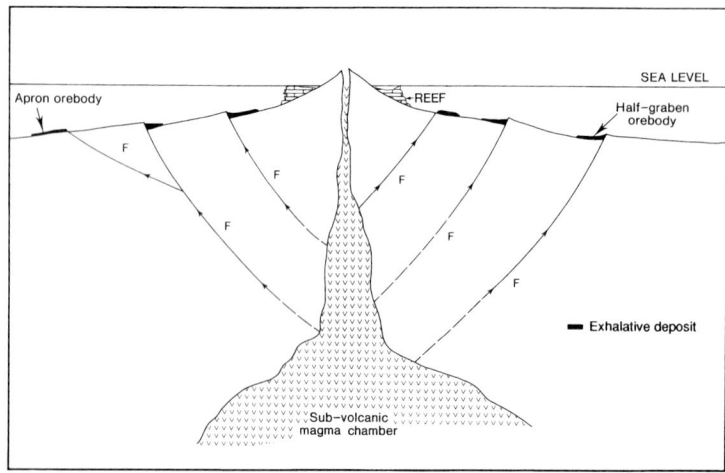

Fig. 4.1. The principal mode of derivation of volcanic exhalative sulphide ores ('volcanic massive sulphide' or simply 'VMS' deposits) as seen to about 1970. The ore materials were assumed to have derived from active subvolcanic magma chambers, transported in volcanic gases and/or hydrothermal solutions leading from magma chamber to sea floor, and precipitated as aprons and ponded accumulations on contact with the cool bottom waters (of sea, lake, or lagoon) surrounding the flanks of the relevant volcanic centre.

---

[*]This chapter, like the preceding one, has a substantial historical component, stemming inevitably from the controversial nature of the subject. Chapter 4 is, however, concerned with the *derivation* of the ores, as distinct from their nature, setting, associations, etc., which were the concerns of Chapters 2 and 3.

including the metallic oxides and sulphides, that occurred in contact deposits and attributed this to metal-bearing gases given off by igneous masses as these cooled. He cited as proof of this proposition that many compounds of iron, copper, zinc, lead, arsenic, antimony, etc. could be observed subliming from volcanic gases in fissures and at the surface around Vesuvius, Stromboli, Etna, and other volcanoes. Bowen formalized this in his elegant statement on progressive crystallization and differentiation of igneous melts, and the concentration of the 'hyperfusible' (potentially volatile) components in the final phases of the latter. All of the evidence — and theoretical considerations as well — appeared to indicate that the metallic sulphides and their immediate associates were products of active magmatism and hence had a deep origin. This was supported by early interpretations of lead isotope abundances (Stanton and Russell 1959) which indicated derivation from a very uniform source, probably the mantle.

Thus increasing acceptance of the volcanic–exhalative–sedimentary theory in the late 1950s and early 1960s was accompanied by the tacit assumption that the exhalations were a manifestation of active magmatism and hence were of deep — perhaps mantle — origin.

THE LEACHING HYPOTHESIS

All this changed with the discovery and geochemical investigation of the Red Sea brine deposits. Indeed, although unrecognized at the time, early evidence indicating that the derivation of the exhalations might be more complex than simple degassing of mantle-derived volcanic melts began to appear about 1962. At this time more precise measurements of lead isotope ratios in galenas from volcanic exhalative ores began to show small but significant deviations from the original $^{207}Pb/^{204}Pb$:$^{206}Pb/^{204}Pb$ growth curve of Stanton and Russell (1959). These generally took the form of a slight apparent deficiency in ^{207}Pb (or excess ^{206}Pb), which was most readily explained by the mixing of two ordinary leads, of different ages.

The real change came however with the publication in 1969 of several isotopic and related geochemical studies of the Red Sea heavy metal deposits and their associated warm brines. In view of the prominent volcanism of the Red Sea rift and the known basaltic nature of the sea floor, it was presumed initially that the warm bottom waters, and the sulphide and other minerals precipitated from them, were of volcanic exhalative origin. Craig (1969), on the basis of $D/^{18}O$, salinity and trace-element data, showed, however, that the brine waters were apparently not magmatic, but were likely to have originated as near-surface sea water from the southern areas of the Red Sea. According to Craig these waters may have passed downward through the sea floor and travelled northwards along aquifers provided by shales and evaporitic sediments of the Red Sea floor. In doing so they dissolved sulphate and chloride from the evaporites and heavy metals from the shales. When the resultant solutions finally debouched, as hot spring waters, into block-faulted deeps of the median rift of the Red Sea, sulphates, sulphides, and a wide variety of associated authigenic minerals were precipitated and accumulated. Simultaneous investigations of sulphur isotope ratios of sulphides and sulphates of the Red Sea deeps by Kaplan *et al.* (1969) generally supported the proposals of Craig; and investigations of lead isotopes (Cooper and Richards 1969) were compatible with them.

At about the same time Sangster (1968) published evidence indicating that the sulphide sulphur of many older volcanic-associated stratiform sulphide deposits might have been derived from sea water rather than from juvenile volcanic fluids.

Following the work of Thode and Monster (1965), which showed that the isotopic constitution of the sulphur of petroleum of different geological ages closely paralleled, on the lower $\delta^{34}S$ side, that of the contemporary sea-water sulphate, Sangster applied the same principle to deduce the origin of the sulphide sulphur of marine stratiform ores.

Thode and Monster had shown that the isotope ratios of petroleum sulphur of a wide range of ages appeared to result from the bacterial reduction of sea-water sulphate, leading to substantial time variations in the isotope ratios of the fractionated 'light' sulphur of the petroleum which paralleled the time variations in the isotopic constitution of sea-water sulphate, as the latter had been deduced from measurements on coeval evaporite deposits.

Through a survey of the literature up to that time Sangster found what appeared to be similar behaviour on the part of the sulphide sulphur of volcanic-associated stratiform ores. The sulphur isotope ratios of the latter were, as expected, always on the light side of the contemporary evaporite sulphate sulphur, but showing a remarkably parallel variation from one geological age to the next. From this Sangster concluded that a significant portion of the sulphide sulphur in 'volcanic type stratabound deposits' was derived from contemporaneous sea-water sulphate and was reduced to sulphide by bacterial action. While Sangster suspected a minor contribution of volcanic sulphur to the total sulphide of these ores, the indication, again, was that the ore-forming exhalations were not of a simple, wholly juvenile, origin.

Sangster's suggestions soon received support from the work of Corliss (1971), who viewed with concern what seemed to him the then very imprecise ideas about the identity of the exhalations to which volcanogenic sedimentary ores were attributed. Corliss noted that several processes had been proposed for the transfer of elements from submarine basaltic magmas and observed that 'A ... mechanism, often cited but never explicitly defined, is the action of 'hydrothermal exhalations' or 'volcanic emanations' that accompany submarine volcanic activity, which transport elements from the magma into sea water' (Corliss 1971, p. 8128).

Corliss set out to examine the problem with greater rigour, and did so by the careful investigation and comparison of the compositions of holocrystalline cores and glassy selvages of deep-sea pillow basalts. He demonstrated, by analysis for sixteen major and trace elements in a suite of mid-Atlantic ridge basalts, that the slowly cooled interior portions of flows and pillows are depleted in Fe, Mn, Co, the rare earths, and other elements (those commonly enriched in pelagic sediments and manganese nodules) relative to the rapidly quenched, glassy margins. He noted that several of these elements tended to be excluded from the solid phases that crystallize from the melt, thus concentrating in the residual liquids. Corliss also noted that other elements are mobilized during deuteric alteration of early formed olivine and the formation of immiscible sulphide liquids. He suggested that all of these components of the melt occupy accessible sites (e.g. intergranular boundaries) in the hot solid rock mass, and might be mobilized by dissolution as chloride complexes in sea water introduced along contraction cracks that formed during cooling and solidification. These solutions, Corliss proposed, might constitute the metal-bearing 'hydrothermal exhalations' or 'volcanic emanations' that accompany submarine volcanism and 'which are often cited as a source of metals into the pelagic environment'. He considered that reasonable estimates of the amount of material involved indicated that a significant fraction of the mass of these elements that reside in pelagic sediments could have been supplied by a process such as this.

Closely following as it did Sangster's suggestion that the sulphides of exhalative ores were chiefly products of sea-water sulphate reduction, Corliss' work quickly led to a widespread view that it was deeply convected sea water, rather than degassing volcanic melts, that constituted the prime supplier of metals for the exhalative deposits. By about 1973 it had become conventional wisdom that volcanism had not played any *active* role in the generation of ore solutions; it simply provided materials — the accumulated basaltic lavas of the oceanic floors — which, long after their emission, become *passive* providers of metals for leaching.

Many contributions concerned with the substantiation of this idea quickly followed. Most of these involved measurement and interpretation of stable isotope ratios (sulphur, oxygen, carbon, hydrogen: see e.g. Garlick and Dymond 1970; Spooner and Fyfe 1973; Rye and Ohmoto 1974; Bischoff and Dickson 1975; Sheppard 1977; Heaton and Sheppard 1977; Spooner 1977) in fluid inclusions, hydrous phases such as clays, chlorites, and secondary micas resulting from hydrothermal precipitation and alteration, associated minerals such as gypsum and barite, and the sulphides themselves. Materials examined ranged from samplings of basalt, sediment, sulphide, and exhaled solutions from sites of modern sea-floor hydrothermal discharge, through analogous samplings of young undeformed ores and ore environments such as those of Cyprus and the Kuroko of Japan, to the more complex and less unambiguous environments of stratiform ore provinces in old terrains.

The evidence of D/H, $^{18}O/^{16}O$, $^{13}C/^{12}C$, $^{34}S/^{32}S$, and, in association, $^{86}Sr/^{87}Sr$, $^{207}Pb/^{206}Pb$, and $^{208}Pb/^{206}Pb$ ratios, together with fluid inclusion compositions, generally indicated a substantial sea-water component in the ore solutions, features of the volcanic rocks consistent with alteration by sea water at elevated temperatures, and deposition of the ores in the 150–300 °C range. This led to a general view of the convection–alteration–leaching–ore deposition process essentially as shown in Fig. 4.2, a diagrammatic model proposed by Heaton and Sheppard (1977) for sea-water circulation through the Troodos crust of Cyprus — a mechanism postulated to have led both to metamorphism of the sub-sea-floor basalts and to deposition of the sulphide ores. The principle is very similar to that of 'lateral secretion' proposed by Sandberger and others in the 1870s and re-activated from time to time during the twentieth century. The only difference between the older and more modern theories is that in the former deposition was seen to take place within the relevant fault conduits and hence beneath the contemporary surface, whereas in the latter the conduits conveyed the major part of their load right to the sea floor, sulphide deposition thus occurring at the surface of the lithosphere.

PERSISTENCE OF THE MAGMATIC HYPOTHESIS

At the end of the 1980s the leaching hypothesis prevailed. It was considered as established that the major component of the solutions constituting the exhalations was sea water, that this was involved in sub-sea-floor convection cells of the order of kilometres in size, that such convection cells were induced and driven by magmatic 'hot spots' commonly related to mid-ocean ridges and associated structures, and that the major part of the metals was

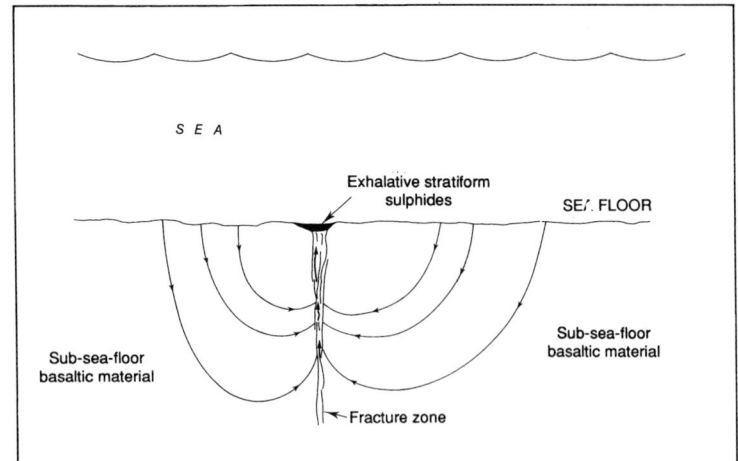

Fig. 4.2. The currently favoured hypothesis for derivation of volcanic exhalative sulphide ores, as illustrated by the model of Heaton and Sheppard (1977): penetration of sea floor by cold bottom water followed by sub-sea-floor convection, hydrothermal leaching of chalcophile metals from accumulated basaltic lavas, and the deposition of exhalative sulphides in areas of resultant hot spring discharge.

derived when the descending sea water became heated and reduced, altering and leaching olivines, pyroxenes, and magnetites and taking their trace copper, zinc, etc. into solution. In continuing along the convection path dictated by the relevant cell, such solutions eventually rose above the focus of the hot spot, ultimately to debouch on the sea floor and to deposit their load of solute on contact with the cold, alkaline sea water, as indicated in Fig. 4.2.

In fairness it should be noted that most proponents of the leaching hypothesis have stated only that the solutions were composed *principally* of sea water; not that they were *entirely* sea water. In almost all cases it had been acceded that there may have been a component of magmatic or metamorphic water, though the evidence has indicated that this was subordinate.

However, while the leaching hypothesis currently dominates ideas on exhalative ore solutions, it is by no means universally accepted. Doubts have centred on two main points: (1) the correctness of some of the interpretations of isotope ratios, and (2) the source of *the sulphide metals*, as distinct from *the solutions that transported them*.

Early questioning of sulphur isotope interpretations came from Sasaki (1970) who asked why, if the sulphur of the volcanogenic stratiform sulphide ores were a mixture of bacterially reduced sea-water sulphate and lesser volcanic (magmatic) sulphur as suggested by Sangster (1968), should the isotopic constitution of the sulphide sulphur of these deposits show such a remarkable parallelism with coeval sea-water sulphate? If the average isotopic composition of volcanic sulphur be taken as near-meteoritic, regardless of geological age, the mixing of the two kinds of sulphur must have occurred in a remarkable pattern of proportions to achieve the isotopic pattern now observed.

Sasaki proposed that a much more likely explanation of the parallelism noted by Sangster was isotopic exchange between original volcanic sulphide sulphur in the ores and the sulphate of the overlying sea water. He suggested that as the amount of sulphur in the sulphate of the sea would be vastly larger than the amount of volcanically derived sulphur in the accumulating stratiform sulphide ores, the isotopic constitution of the sulphur of the sulphides would be substantially controlled by that of the sea-water sulphate. Thus the remarkable parallelism observed between the sulphur isotopic constitutions of ancient sea-water sulphate and coeval exhalative sulphide deposits might be more satisfactorily explained. Sasaki noted that the parallelism was closest when the reported *minimum* $\delta^{34}S$ value of each ore deposit was compared with that of the contemporary sea-water sulphate, perhaps indicating that in most cases the exchange equilibrium had been approached but rarely accomplished. Such an interpretation appeared to be consistent with the common tendency for sulphide deposition temperatures calculated on the basis of SO_4^{2-}–H_2S equilibrium to be slightly higher than temperatures derived by other means.

Sasaki's work therefore indicated that the apparent close relation between exhalative sulphide sulphur and coeval sea-water sulphate was due not to an original direct derivation of sedimentary sulphide from the latter, but to a slightly later equilibration of volcanically derived sulphide with sea-water sulphate. This was one line of argument diminishing the case for sea water as the principal ore solution.

Urabe and Sato (1978) added to this in their examination of the Kuroko ores of the Kosaka Mine in northeastern Honshu. Working on the closely related Matsumine and Shakanai deposits, Ohmoto and Rye (1974) had earlier deduced, on the basis of D/H and

$^{18}O/^{16}O$ measurements, ore solutions consisting principally of sea water, possibly contaminated with up to 25 per cent magmatic water. On the basis of salinities of fluid inclusions, consideration of the sizes of potential heat sources, and lead isotope ratios of ore and basement rocks, Urabe and Sato concluded that convecting sea water was unlikely to have constituted the principal component of the Kosaka ore solutions. They proposed 'a model in which the aqueous phase released from an acidic magma plays an essential role in the formation of the Kuroko deposits' — a magma that probably constituted the final product of a large-scale magmatic differentiation now represented by the white rhyolite domes with which the Kuroko ore deposits are so closely and so characteristically associated. Sato and Urabe deduced that the ore solutions had approximately twice the salinity of sea water, transported the base metals as chloride complexes, and reached the sea floor at temperatures as high as 300 °C. They concluded that 'an aqueous phase once in contact with acid magma is suggested to have played an essential role in the formation of the Kuroko deposits. An alternative convective seawater model is not preferred because it fails to explain the high salinities of the ore solution, the sufficient heat sources to drive the system, and the intimate association of the deposits with the well-differentiated white rhyolite domes' (Urabe and Sato 1978, p. 177).

Sawkins and Kowalik (1981) examined the problem from another aspect: the mass balance of lead as between a large Kuroko-type deposit and its associated volcanic rocks. In investigating the derivation of the large Palaeozoic exhalative deposit of Buchans, Newfoundland, Sawkins and Kowalik noted that while the orebody contained at least 2.7 million tons of lead metal, the underlying rocks — chiefly basalts — contained no more than 6 ppm of this element. Their mass-balance calculations indicated that if lead were leached from the basalts at 50 per cent efficiency, it would require a volume of some 300 km^3 of such basalt to supply the lead of the orebody. If, as might be expected, leaching were restricted to the lower half of the convective system, a single convection cell occupying 600 km^3 would be required — an order of magnitude larger than typical well-developed modern geothermal systems. On the other hand, a cooling felsic stock with a volume of only 100 km^3 could have supplied the lead of the Buchans deposit.

In a more general examination of the problem Sawkins (1986) accepted that isotopic and geochemical data, together with experimental results on hydrothermal rock–water interaction, indicated a 'large scale involvement of fluids of ultimate seawater origin' in the life-history of many exhalative ore-forming systems. He considered, however, that the evidence indicated a process less simple than sea-water leaching alone, and he proposed a hybrid process involving magmatic derivation of metal-bearing hydrothermal solutions followed by mixing of these with convecting sea water at higher levels in the crust. His principal points were:

1. Kuroko-type deposits exhibit a pronounced tendency to be restricted to very limited stratigraphic intervals within the intermediate-to-felsic volcanic sequences in which they occur. Although chemical and isotopic studies indicate active involvement of sea water in the ore-generating systems, investigation of the history of alteration shows in many cases that hydrothermal activity continued on long after the ore-forming event.

2. If, in the context of the leaching hypothesis, most of the convective fluid flow is confined to joints and faults, effective water:rock ratios will be large and alteration selvages will soon armour the fracture walls. This will inhibit further leaching unless continuing fracturing events constantly provide fresh channelways. Structural studies indicate that this has not occurred in the case of the Japanese Kuroko province.

3. Salinity data from fluid inclusions in quartz and sphalerite from feeder zones indicate a range of values varying from close to, to greater than twice that of sea water (*vide* Urabe and Sato 1978 above). Recent experimental evidence indicates, however, that, under the conditions of low water/rock ratios needed for effective alteration and metal leaching, chlorine tends to be *lost* rather than gained from solution.

4. The δD values of inclusion fluids are consistently negative and fall within the range -5 to -40 ‰. The combination of this with high salinities is difficult to explain on the basis of a simple sea-water model: if hydration reactions are proposed in order to reduce the amount of H_2O relative to NaCl (i.e. to increase salinity) in the ore fluid, such reactions will tend to drive the deuterium values of the fluid to higher rather than lower (negative) values. This has been confirmed by sampling and measurement of modern mid-ocean ridge hydrothermal systems. Observed $\delta D:\delta^{18}O$ relations in ore fluids preserved as fluid inclusions appear best explained by mixing of sea water and magmatic water.

5. Theoretical calculations of heat required to initiate and drive large (sub-sea-floor) sea-water convection systems indicate that relatively large magmatic bodies would be required to occur beneath any significant ore-generating cell.

6. Rare-earth element (REE) abundance studies of ore-associated and non-ore-bearing felsic volcanic sequences in a variety of Archaean volcanic terrains

have demonstrated distinctive patterns for the two categories of rocks. Those with which significant ore deposits are associated exhibit patterns that may be interpreted as indicating advanced degrees of magmatic differentiation — a process that might be expected to lead to the generation of metal-rich residual hydrothermal solutions.

These various lines of evidence led Sawkins to propose a hybrid model: one involving relatively long-lived sea-water convection systems on which were superimposed short-term additions — pulses — of metal-rich, late-magmatic fluids (Fig. 4.3, after Sawkins 1985). Such an hypothesis appeared to explain what is certainly a 'strong sea-water geochemical imprint' on the ore solutions, but at the same time clear geochemical deviations from systems involving sea water alone.

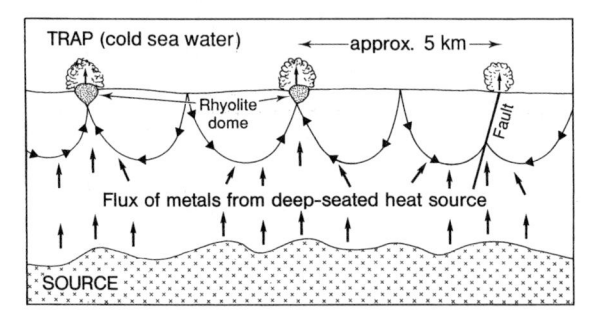

Fig. 4.3. A hybrid mechanism for derivation of volcanic exhalative sulphide ores, as proposed by Sawkins (1986): short-term pulses of metal-rich, late-magmatic fluids are superimposed on relatively long-lived sea-water convection systems.

In a related vein, a number of investigators (see, e.g., Campbell *et al.* 1981; Lesher *et al.* 1986; Lesher and Campbell 1987, 1990) were proposing that shallow subvolcanic sills and related intrusions might be critical to — and perform a dual role in — ore-forming exhalative activity; that such intrusions (1) themselves contribute, as a manifestation of primary igneous de-gassing, an increment of the exhalative discharge and (2) act as a localized heat source to drive relatively shallow sea-water convection cells that in turn induce leaching of associated lavas and pyroclastic materials. Again the suggested process is hybrid. Considerable emphasis has been placed on the importance of *shallow* emplacement of the intrusion, both to satisfy isotopic constraints and because deeper intrusions are seen as having a lesser capacity to drive what are essentially sub-volcanic sea-water convection cells.

To this end substantial effort has been directed to the application of rare-earth and other trace-element abundances to the elucidation of the petrogenesis of ancient intrusions, and their depth of consolidation, in attempts to identify potential ore-bearing volcanic piles.

At much the same time the author was also proposing a hybrid mechanism (Stanton 1985). He pointed out that there were at least three quite elementary pieces of evidence indicating that sea-water leaching was unlikely to be the principal mechanism of derivation and transport of exhalative ores of volcanic association:

1. The association of particular ore types with particular volcanic rock types. As already noted, the characteristic exhalative ore of the basaltic milieu is pyritic copper — the ores of 'Cyprus type'. Zinc is most commonly inconspicuous to minor in these ores, and lead is essentially absent. On the other hand the exhalative ores of the andesite–rhyolite association are conspicuously rich in zinc and commonly possess abundant lead, in addition to copper. Basalts, however, contain as much if not more zinc than andesites, dacites, and rhyolites and, what is more, contain it in at least partly more leachable form, i.e. in olivine. It is thus not easy to see, on the basis of the leaching hypothesis, why basalts should not yield zinc deposits just as vast (e.g. Kidd Creek, Ontario) as those associated with some calc-alkaline sequences. In addition basalts contain about one-third as much lead as dacites and on the basis of a leaching mechanism it is not at all clear why the basalt-associated orebodies should be essentially devoid of lead.

    In the same general connection basalts contain nickel and cobalt in the same order of abundance as copper and zinc. The nickel and cobalt of basalts is located principally in olivine, the most readily decomposed component of these rocks, and nickel and cobalt form insoluble sulphides. On the basis of the leaching hypothesis it is therefore not easy to see why the basalt-associated Cyprus type ores do not habitually contain nickel and cobalt sulphides as major constituents.

2. The relative sizes of orebodies of the different volcanic associations.
    (a) The total volume of basalts in the marine environment is many orders of magnitude greater than that of marine andesites, dacites, and rhyolites.
    (b) On an approximate estimate basalts contain about twice as much trace Cu + Zn + Pb (say $\approx$ 180 ppm) as dacites (say $\approx$ 90 ppm).
    (c) Most marine basalts are erupted on to the deep sea floor and remain on and beneath the ocean

floor for geologically long periods of time. The andesites, dacites, and rhyolites on the other hand generally do not appear until the volcanic edifice approaches the surface of the sea, and for the most part are erupted in shallow water just before volcanism becomes subaerial. The marine basalts thus bulk enormously larger, contain more base metal, and are exposed to sea-floor leaching to a far greater degree than materials of the marine andesite–rhyolite series. This would suggest that the exhalative orebodies associated with basalts should be far larger and more numerous than those associated with andesites, dacites, and rhyolites. Exactly the opposite is the case. Apart from a very few possible exceptions in Japan, Cyprus, and Norway, the basalt-associated deposits are small in size and relatively few in number. The world's great volcanogenic orebodies — and there are many of them — are almost invariably associated with rocks of the andesite–rhyolite series and their pyroclastic and volcaniclastic derivatives.

3. The incidence of groups of exhalative orebodies within single volcanic districts or in association with individual volcanic piles is not random. As earlier pointed out by the author (Stanton 1954, 1955a, b), and by many others since, orebodies of exhalative type in the volcanic environment are almost invariably confined to one or perhaps two horizons, or restricted stratigraphical intervals, within the relevant volcanic sequence. This is illustrated very clearly in such areas as Noranda and Bathurst in Canada (Spence and de Rosen-Spence 1975; Holyk 1956), the Rio Tinto district of Spain (the 'Iberian pyrite belt'), the Bathurst area of New South Wales (Stanton 1955a, b), the Troodos district of Cyprus, and the Kuroko province of Japan, and is the rule in virtually every volcanic–sedimentary stratiform ore province known.

Were the deposits the result of essentially continuous sub-sea-floor convection and leaching it might be expected that there would be a relatively even time distribution throughout the relevant volcanic sequence, perhaps with an increased incidence near the top.

This is clearly not the case: the deposits are always clustered in time, suggesting that exhalation results from some relatively brief, clearly defined event rather than from a long-continued process such as sea-floor leaching. Such an abrupt event would seem most likely to be related to volcanism itself: perhaps the attainment of a particular stage of crystallization, differentiation of a sub-volcanic melt, a sudden decrease in confining pressure with consequent degassing, abrupt ingress of external water to the sub-volcanic magma chamber, or perhaps a combination of two or more of these.

More recently Urabe (1987) — again investigating the Japanese Kuroko province — has cast further doubt on the adequacy of the leaching hypothesis to explain features of the occurrence and geochemistry of these ores. He re-emphasized that Kuroko deposition took place as a remarkably sharply defined event around 15 million years ago and that it coincided with a peak of bimodal volcanic activity, a major shift in the tectonic stress field and a maximum degree of subsidence, and a sharp change in style of volcanicity (from lava flows, felsic hyaloclastites, and lithic tuff-breccias prior to mineralization to extensive post-ore pumiceous pyroclastic flows). Urabe reiterated the view that oxygen and hydrogen evidence indicated that sea water could not be the sole component of the ore solutions, and that the remarkable between-deposit homogeneity of lead, strontium, and sulphur isotope abundances of the ores negated the possibility that the ore metals were largely derived from the various basement rocks by leaching.

Urabe noted several other lines of geological reasoning apparently militating against a dominantly leaching origin for the principal constituents of the Kuroko ores:

1. Deposition of the Kuroko deposits was associated with a peak of felsic (as opposed to basaltic) igneous activity during the opening of the Japan Sea approximately 15 Ma BP.
2. Comparison of the Kuroko volcanic succession with that of the modern Yellowstone caldera indicates that a molten felsic pluton existed at depth during the formation of the Kuroko deposits.
3. As well as the previously mentioned change in style of volcanism, Kuroko formation was also accompanied by a change in the normative composition of the volcanic rocks from meta-aluminous to alkaline, i.e. a notable igneous differentiation event.
4. The rhyolitic melt contemporaneous with ore formation was a product of this differentiation process and must itself have had the potential for the production — as a further phase of the differentiation sequence — of metal-rich aqueous solutions during its final crystallization.

Urabe concluded that the ultimate source of the solid materials of the ores was therefore an actively differentiating melt. The magmatic solutions carrying these materials mixed with circulating sea water during their ascent, yielding the very consistent, but in part hybrid, geochemical features the ores now display.

These considerations of Urabe and Sato, Sawkins, Stanton, Urabe, Campbell, and others therefore indicate that while the geochemical evidence certainly indicates the involvement of seawater in the ore solutions, it indicates with equal certainty that sea water was not the only major aqueous component. Other evidence, together with some simple geological reasoning, seems to indicate that the sea water was an addition to solutions of some other derivation — probably magmatic — and that the major solid components of the ores were principally derived from an actively crystallizing melt.

CONCLUDING STATEMENT

On this basis the various hybrid geochemical features may have been acquired:

1. by mixing of magmatic solutions with circulating seawater at some depth between the source intrusive and the floor of the sea;
2. by such mixing at and closely following discharge of magmatic solutions on the sea floor, isotope and element exchange occurring at elevated temperatures in either or both environments.

This hybridism was in fact presaged by Corliss (1971) himself, and is now acknowledged by all. What is not agreed however is the relative abundance of magmatic *vis-à-vis* sea (and perhaps meteoric) water, and the source of metals and associated materials that they carry.

If the metals are derived principally by leaching of pre-existing volcanic rocks it would seem reasonable to suggest:

1. the constitutions of the ores would vary according to the nature of the volcanic materials from which they were derived;
2. the frequency of occurrence and size of the orebodies would correspond approximately to the incidence of underlying source rocks — e.g. those derived from, and hence now associated with, basalts being most common and largest.
3. the relevant ores would contain in significant abundance all those chalcophile metals occurring in leachable form within the parent lavas that form insoluble sulphides, e.g. Fe, Cu, Zn, Pb, Ni, and Co;
4. In view of the general association of strontium with the other alkaline earths calcium and barium, and its greater abundance than barium in most volcanic rocks, strontium compounds would be frequently occurring constituents of exhalative ores.

If on the other hand the metals and associated elements were derived by volatile loss from an actively crystallizing and differentiating melt it would be expected that:

1. the constitution of the ores would tend to reflect the behaviour of the relevant elements during progressive differentiation of the source melt, rather than the simple abundance of those elements in volcanic materials undergoing leaching;
2. related to this, the constitution of the ores would reflect the stage of differentiation reached by the melt at the time of volatile loss;
3. the frequency and size of orebodies would be related to the amount and nature of volatile matter concentrated in the residual melt at the relevant stage of differentiation and time of escape. It would not necessarily be related to, and in many cases would not correspond with, the amount and nature of materials available for leaching.

The third factor — the amount and nature of volatile matter concentrated in the relevant residual melt — is difficult or impossible to estimate in the present state of knowledge. What is eminently possible, however, is the investigation of the first two factors: (1) the patterns of concentration of the ore elements in a parent basaltic melt as this crystallizes and progressively differentiates to more felsic residue, and (2) the ways in which these might correspond with the now well-established patterns of association of ore and lava types.

5

Petrology of the Solomon Islands Younger Volcanic Suite

Before proceeding to examine the behaviour of the ore elements — the principal ore-forming cations — in the spectrum of Solomon Islands younger lavas, it is necessary to consider briefly the more important petrological and petrochemical features and the geographical distribution of the rock types involved.

As noted in Chapter 1, the main area of study has been in the New Georgia–Guadalcanal segment of the island festoon, and it is here that the most conspicuous development of the full range of lava types occurs (Figs. 1.2 and 5.1). Minor intrusions of dacite have, however, been reported from San Cristobal (Stanton and Ramsay 1980) and major areas of andesitic rocks, including the prominent volcanic centre Mt. Matambe, appear on the island of Choiseul. In addition the major part of the surface of Bougainville is composed of andesitic and dacitic lavas of the Solomon Islands Younger Volcanic Suite (Rogerson *et al.* 1989). As far as the writer is aware, however, no members of the Suite occur on the islands of Santa Isabel, Malaita, and Nggela, the very substantial volcanic components of which are composed entirely of materials of the Older Volcanic Suite. The incidence of the younger

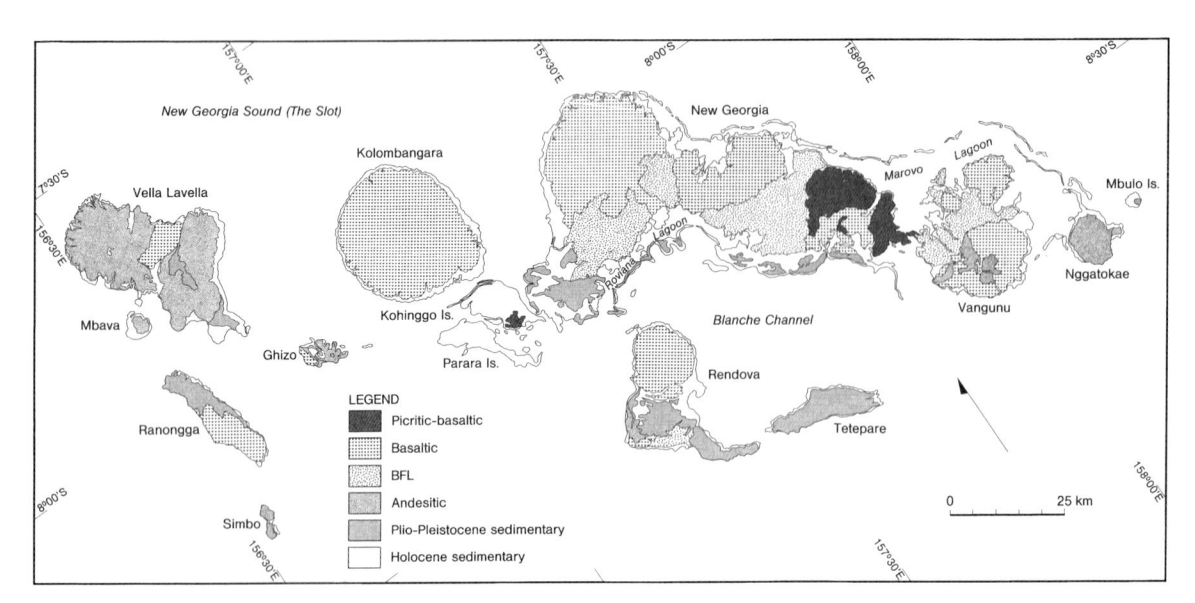

Fig. 5.1. Distribution of the principal rock types of the Solomon Islands Younger Volcanic Suite on the islands of the New Georgia Group. *BFL*, big-feldspar (feldsparphyric clinopyroxene–olivine)-basalts and basaltic andesites. All rock units comprise both lava and pyroclastic materials, the latter dominant in many instances. Minor occurrences are omitted for reasons of scale and clarity: e.g. the predominantly basaltic rocks of Kolombangara are punctured by younger satellite cones of hornblende-andesite, particularly in the southern half of the island, as is the lobe of picritic material constituting the eastern extremity of the main island of New Georgia. Basaltic members of the Suite are proportionately much more abundant in the New Georgia Group than elsewhere: members occurring on Bougainville, Choiseul, Savo, Guadalcanal, and elsewhere are dominantly, through not entirely, andesitic to dacitic.

lavas may thus be seen as taking the form of a greatly elongated ellipse, the centre of which lies in the New Georgia Group. Here the Younger Volcanic Suite covers the whole of the land surface and exhibits its broadest spectrum of lava type and its greatest complexity of composition and occurrence. Outwards, towards Guadalcanal, Choiseul, and Bougainville, the lavas change from basalt-dominated to andesite:(dacite)-dominated, exhibit lesser compositional complexity, and apart, perhaps, from Bougainville, occupy only part of the prevailing land surface.

Although the primary focus of the present contribution has been on the lavas of the New Georgia–Guadalcanal segment, it has transpired that the recent observations on the Solomon Islands Younger Volcanic Suite as this appears on Bougainville by Rogerson *et al.* (1989) has provided whole-rock chemical data that neatly complements the New Georgia–Guadalcanal analyses at the more felsic end of the Solomons lava spectrum. Where appropriate, the 119 Rogerson *et al.* whole-rock analyses have therefore been combined with the 122 analyses of the present study to improve the statistical reliability of elemental variation diagrams in the more felsic portion of the overall lava compositional range. Although extensive sampling, particularly on Guadalcanal and Savo, yielded some fifty andesites and dacites, the New Georgia–Guadalcanal suite is clearly weighted towards the basaltic end of the spectrum, with a mode in the 48–52 per cent SiO_2 interval (filled bars, Fig. 5.2(a)). The Bougainville collection of Rogerson *et al.* is, on the other hand, equally clearly weighted towards the andesite – dacite portion of the compositional spectrum, with a mode in the 56–8 per cent SiO_2 interval (open bars, Fig. 5.2(a)). The two distributions combined have a mode in the 54–6 per cent SiO_2 interval and give a considerably enlarged representation of lavas in the upper (56 to > 64 per cent) SiO_2 range (Fig. 5.2(b)). In the chapters that follow many of the diagrams depicting abundance relations generated by the 122 whole-rock analyses of the New Georgia– Guadalcanal lavas are supplemented by additional figures incorporating the full 241 analyses representing the Bougainville–New Georgia–Guadalcanal segment — and hence almost the full length of the Solomons festoon. Increased regularity of curves at higher SiO_2 values ensues, and is obvious, in almost all cases.

PRINCIPAL ROCK TYPES

The principal rock types, in order of increasingly felsic nature, are picrites to picritic basalts, olivine–clinopyroxene-to-clinopyroxene–olivine- (in many cases ankaramitic) basalts, feldsparphyric (clinopyroxene–

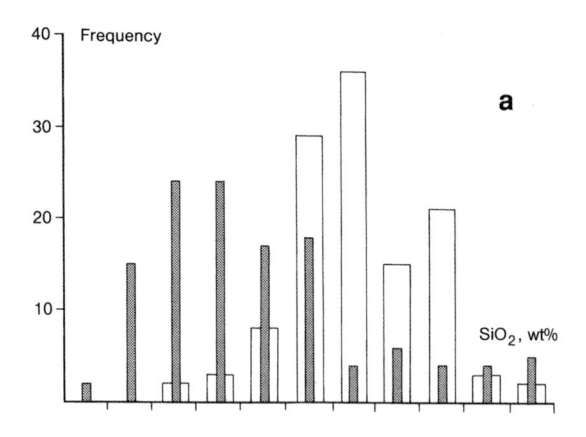

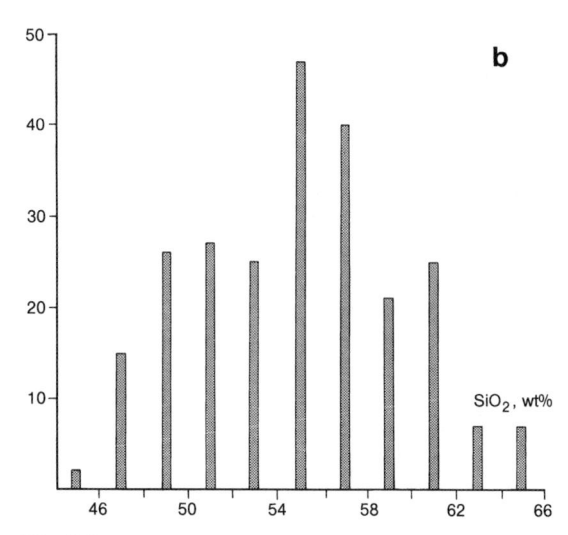

Fig. 5.2. Frequency distribution of SiO_2 abundances in (a) the 122 analyses of lavas of the Solomon Islands Younger Volcanic Suite forming the basis of the present work (filled bars, mode at 48–52 per cent SiO_2), and the 119 analysis of Bougainville lavas by Rogerson *et al.* (1989) (open bars, mode at 56–8 per cent SiO_2); (b) the two distributions combined, with mode in the 54–6 per cent interval — i.e. essentially middle andesite range.

olivine)-basalts to basaltic andesites, hornblende-andesites, and very subordinate dacitic to rhyolitic materials. There is also a lesser category of two-pyroxene basalts to andesites, most conspicuous on Simbo, but also appearing here and there throughout the Younger Volcanic province.

Stanton and Bell (1963, 1969) originally divided the spectrum of Younger Volcanics — as these appeared on New Georgia (Fig. 5.1) — into five major types: (1) picrite basalts; (2) porphyritic pyroxene–olivine-basalts; (3)

feldsparphyric basaltic andesites; (4) non-porphyritic basalts, and (5) hornblende basaltic andesites. At that stage of investigation hornblende basaltic andesites and andesites (SiO_2 generally <≈55 per cent) appeared to be the most felsic members of the Suite. Since that time, however, fully andesitic rocks have been found in some abundance in the New Georgia Group and elsewhere, and further investigation of Savo and some of the very young volcanic centres on Guadalcanal and elsewhere has revealed the presence of dacites and minor rhyolitic lavas and tuffs.

While the Younger Volcanics do constitute a very broad chemical and mineralogical spectrum, and while they exhibit two distinctive chemical groupings — 'high magnesium' and 'calc-alkaline' — several lines of evidence indicate that they constitute a single consanguineous suite:

1. The whole range of rock types is closely associated in the New Georgia group of islands and in several instances virtually the full spectrum appears within single volcanic centres or cones. While the high magnesium and feldsparphyric varieties are generally the older, and the hornblende-andesites and occasional more felsic types are almost invariably the younger products, there is not a clear, constant sequence of

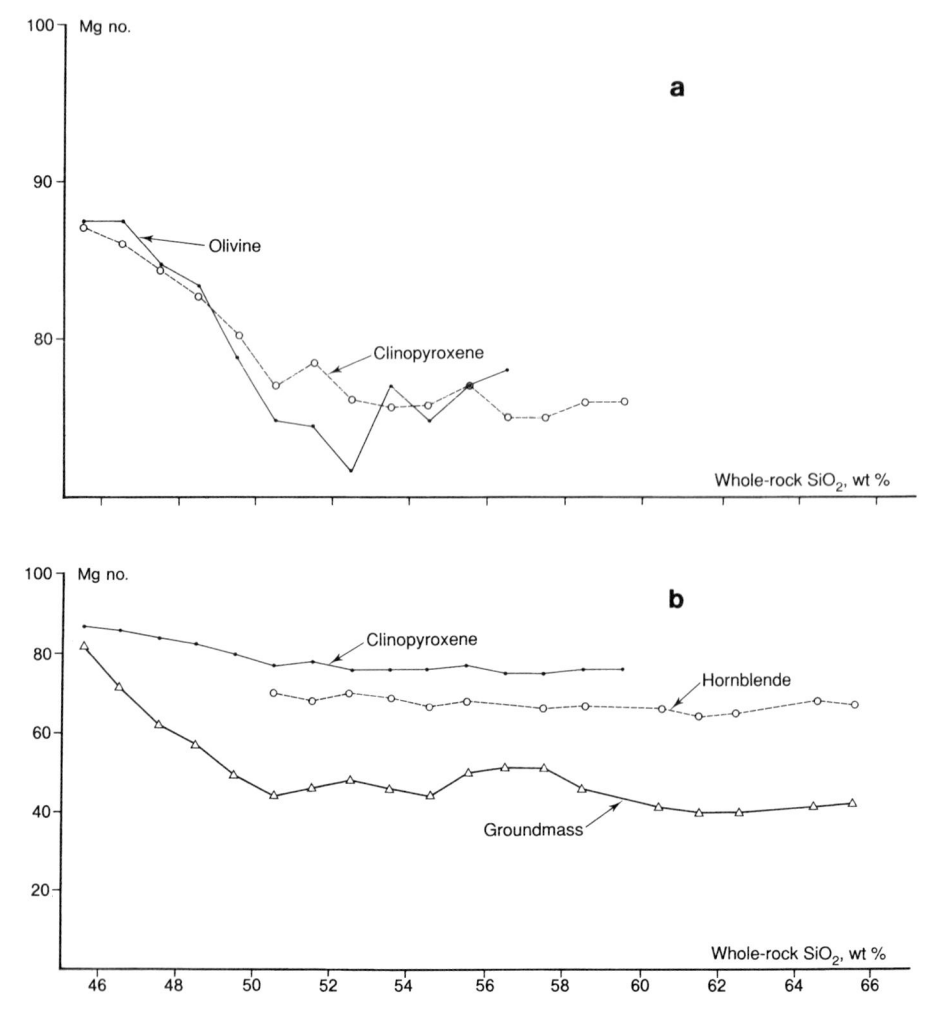

Fig. 5.3. Variation in mean Mg number of (a) olivine and clinopyroxene and (b) clinopyroxene, hornblende, and groundmass, with respect to one per cent intervals of whole-rock SiO_2 in the containing lavas of the Solomons suite.

emission of the different types. In several instances volcanic cones consist almost entirely of high-magnesium and/or feldsparphyric lavas, but possess a core or minor satellite cones of hornblende-andesite. In some cases, e.g. the north flank of the Kolo centre of the Marovo area, high-magnesium lavas contain minor inter-layers of andesite, and on the east flank of the Parasso cone of Vella Lavella minor units of high-magnesium to picritic rocks occur within a volcanic edifice dominated by hornblende-basaltic andesites and andesites.

Spatial, temporal, and volcanological ties between all lava types thus appear close and substantially systematic.

2. The full range of rock types, from picrites to dacites, though of greatly varying appearance and mineralogical constitution, constitutes in many respects a chemical continuum. Various manifestations of this are shown in the following pages. Numerous graphs demonstrating co-variation of element or elemental ratios in the lava group as a whole exhibit high degrees of correlation and relatively even densities of points. Where inflexions occur at a change from high-magnesium to calc-alkaline suites, the plots for the two groups always join neatly, and always with a number of points (transitional rock types) common to each.

3. The almost ubiquitous incidence of a characteristic green diopsidic pyroxene constitutes a mineralogical thread running through the lava group from picrite basalts to relatively felsic hornblende-andesites. A notable compositional consistency among these pyroxenes through the lava suite as a whole is illustrated by Fig. 5.3. Taking 1 per cent whole-rock SiO_2 intervals, pyroxene Mg numbers decline gradually from 87 to 77 with rise in SiO_2 from 45 to 50 per cent, and then remain essentially constant (between 75 and 78) to 60 per cent SiO_2, above which clinopyroxene occurs in very small amount or not at all. The only discontinuity of any kind in this plot (involving 122 rocks and 1075 individual pyroxene analyses) is the very gentle inflexion at 50–1 per cent SiO_2. The continuity from this point to 59–60 per cent SiO_2 (well into the hornblende-andesite range) seems far too remarkable to ascribe to chance.

As is shown below, a number of other features of mineralogy and mineral chemistry indicate that the Younger Volcanics constitute a continuum rather than an assemblage of approximately coeval, but discrete, rock groups. For present convenience, however, the rock suite is loosely divided into three principal groups (Fig. 5.4):

1. Olivine–clinopyroxene-basalts, picritic basalts, picrites, and ankaramites. These are characterized by

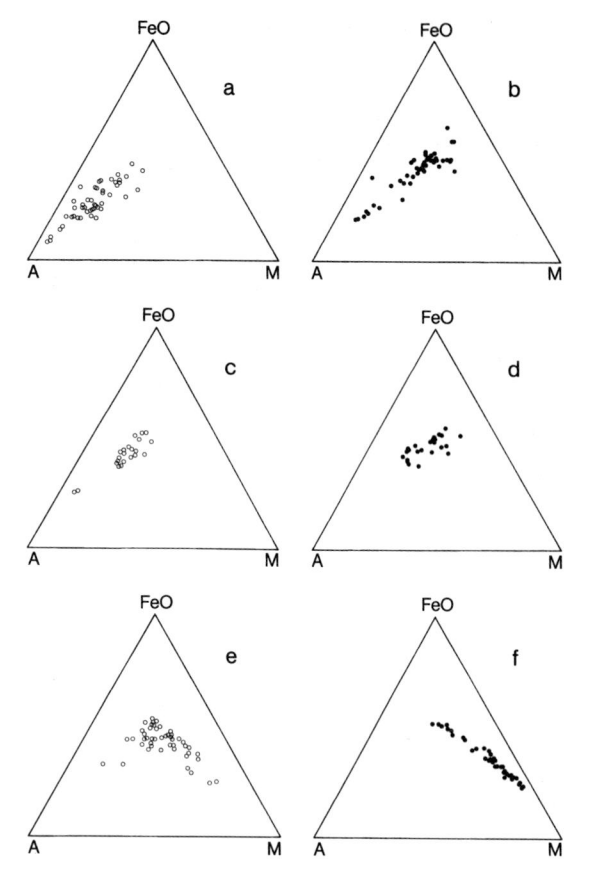

Fig. 5.4. Total Fe as FeO–MgO–($Na_2O + K_2O$) relations in (f) olivine–clinopyroxene-basalt, and (e) their groundmasses; (d) big-feldspar lavas and (c) their groundmasses; (b) hornblende-andesites and dacites (and subordinate basalts) and the hornblende-andesite group, and (a) their groundmasses. The three points at the upper right of (b) are basalts of the hornblende-andesite group (see text). Three features of the whole-rock component (right-hand triangles, filled circles) stand out: (1) the pronouncedly linear nature of the olivine–clinopyroxene-basalt *cum* picrite field; (2) the intermediate identity of the big-feldspar lavas *vis-à-vis* the olivine–clinopyroxene-basalts and hornblende-andesites; and (3) the neat fit between the most felsic olivine–clinopyroxene-basalts and the most mafic hornblende-andesites.

widely varying proportions of olivine and pyroxene phenocrysts, absent to sparse phenocrystic feldspar and iron oxide, and glassy to finely crystalline groundmass. Phenocrysts and groundmass are normally sharply distinct. High magnesium members (MgO $\geq$ 12.0 per cent) are abundant and lavas containing >28.0 per cent MgO are not uncommon. Total

30 *Ore Elements in Arc Lavas*

Table 5.1 Chemical and mineralogical compositions of representative lavas of the Soloman Islands Younger Volcanic Suite

Fld no.	1 143/2	2 64/80	3 63/80	4 39/81	5 45/81	6 32/81	7 26/81	8 57/80	9 71/80
SiO_2	44.99	46.29	45.89	46.03	47.62	47.25	49.40	47.96	49.15
TiO_2	0.33	0.31	0.31	0.35	0.43	0.43	0.50	0.61	0.58
Al_2O_3	6.54	6.60	6.49	7.05	8.88	8.85	11.76	13.16	13.62
Fe_2O_3	2.67	2.16	2.52	3.31	2.48	3.35	<0.01	3.82	3.34
FeO	7.29	6.83	6.74	7.06	7.26	7.35	9.08	6.76	5.71
MnO	0.19	0.15	0.15	0.17	0.16	0.22	0.17	0.21	0.14
MgO	28.10	28.07	28.06	25.02	22.48	18.76	14.42	12.52	10.14
CaO	6.56	6.83	6.77	7.81	8.28	9.37	10.65	10.23	12.73
Na_2O	0.92	0.83	0.78	1.19	1.45	1.52	1.95	1.92	1.87
K_2O	0.57	0.73	0.66	0.61	0.83	1.02	0.98	0.67	1.19
P_2O_5	0.08	0.14	0.15	0.15	0.16	0.18	0.21	0.20	0.21
H_2O^+	1.06	0.69	0.79	0.62	0.50	0.71	0.47	0.93	0.98
H_2O^-	0.30	0.49	0.70	0.37	0.19	0.30	0.17	0.70	0.46
CO_2	–	0.24	0.22	0.12	0.09	0.12	0.08	0.14	0.18
Total	99.60	100.36	100.23	99.86	100.81	99.43	99.84	99.83	100.30
Ba		62	110	74	100	96	125	72	190
Sr		205	206	382	348	469	560	421	447
Pb		<1	<1	2	2	2	3	2	2
V		148	146	158	177	209	198	200	263
Cr	2325	1770	1820	1380	1310	1000	670	595	380
Mn		1400	1380	1540	1460	1950	1500	1870	1280
Co		101	101	112	86	81	60	61	46
Ni	1180	1270	1410	920	820	535	376	372	126
Cu		42	44	66	65	56	97	32	85
Zn	85	62	63	71	72	76	72	69	68
Rb		8.5	6.5	5	9.5	16	11.5	10	16
Zr		8	7	4	15	7	11	21	19
Olivine	52.0	36.6	36.6	35.1	30.7	24.2	11.4	14.6	1.8
Cpx	7.1	2.4	2.5	8.4	5.4	21.0	19.6	13.4	22.8
Hb									
Biot									
Plag							6.6	1.6	8.2
Spinel	1.1	0.8	1.0	0.4	0.6	0.5	tr	tr	tr
G'mass	39.8	60.1	59.9	56.0	63.1	54.2	62.2	70.3	67.1

1–10 = representatives of the olivine–clinopyroxene-basalt group, in order of decreasing MgO content. Analysis 1 (field number 143/2), picrite from the Hire River section of the Kolo volcanic centre, Marovo (see Fig. 5.1), is given for historical reasons: collected by Stanton and Bell in September 1959, analysed by G. M. Harral (Mineral Resources Division, British Overseas Geological Surveys) in 1961, and the first island-arc picrite to be so recognized. Similar lavas were found on the Vanuatu and Lesser Antilles arcs some 5 years later. Numbers 11–14 = representatives of the big-feldspar lava group; numbers 15–19 = representatives of the hornblende-andesite lava group. Analyst: nos 2–19, B. W. Chappell. Fld no. = field number. Major element abundances given as per cent by mass; trace-element abundances in parts per million (ppm) by mass; modal abundances of principal minerals as per cent by volume as determined by point counts (2500 to 18 000 points per lava sample).

SiO_2 range is ≈45–54 per cent. Of the 47 samplings of this lava group, 17 are picrites (whole rock Mg >75, olivine mg >85, modal olivine ⩾25 per cent), 27 are basalts (SiO_2 <52 per cent), and three are basaltic andesites (SiO_2 = 52–4 per cent). In terms of the actual volumes in which the lavas occur, this is a gross over-sampling of the basaltic andesites, and, to a lesser extent, of the picrites. Analyses of ten representative lavas (numbers 1 to 10, listed in order of decreasing MgO) are given in Table 5.1, and mean chemical composition is given in Table 5.2.

2. Big-feldspar basalts and basaltic andesites. These are characterized by conspicuous and usually abundant lath-shaped feldspar phenocrysts up to ≈20 mm long,

Petrology of the Solomon Islands Younger Volcanic Suite 31

10 60/80	11 54/81	12 62/81	13 77/80	14 49/81	15 8/80	16 36/80	17 94/81	18 30/80	19 41/80
49.74	48.75	50.95	52.24	55.31	50.86	53.01	57.29	62.86	65.36
0.71	0.89	0.87	0.83	0.72	0.85	0.78	0.54	0.36	0.27
16.85	16.19	18.42	13.99	18.32	19.98	17.55	18.47	17.48	17.85
3.45	4.14	4.46	4.42	4.29	5.40	6.52	4.37	1.68	1.42
6.15	7.05	4.91	4.95	3.47	3.26	1.91	2.42	1.37	1.03
0.22	0.17	0.18	0.17	0.16	0.15	0.14	0.14	0.07	0.07
7.93	7.11	4.33	6.91	3.54	3.44	3.96	2.99	1.56	1.06
10.44	12.11	9.62	9.25	7.96	9.93	8.69	8.56	3.61	3.27
2.66	2.38	3.22	3.11	3.94	3.53	4.00	2.97	6.38	6.40
0.81	0.91	1.71	1.97	1.56	1.16	1.48	0.92	2.33	2.05
0.24	0.13	0.23	0.38	0.30	0.19	0.20	0.17	0.19	0.10
0.78	0.40	0.73	0.96	0.50	0.73	0.66	0.56	0.18	0.42
0.53	0.20	0.60	0.94	0.41	0.42	0.69	0.34	0.63	0.12
0.27	0.11	0.25	0.18	0.10	0.07	0.11	0.08	0.63	0.11
100.78	100.54	100.48	100.30	100.58	99.97	99.70	99.82	99.33	99.53
90	120	260	265	195	220	285	160	730	640
530	408	565	560	665	550	701	401	1540	1220
2	3	5	4	3	4	5	3	16	10
224	426	246	332	215	236	243	196	68	52
253	124	<1	220	5	12	19	6	9	13
1940	1520	1580	1530	1380	929	852	852	387	387
54	35	36	38	36	26	24	18	8	5
161	44	8	100	14	8	11	5	7	5
38	182	95	209	134	96	82	40	21	23
66	80	83	82	77	74	69	60	48	46
11	11	27	18.5	25	24.5	18.5	15.5	38.5	38.5
30	32	95	63	74	74	70	49	133	129
7.0	3.2	1.8	3.9	0.1					
9.3	22.1	7.5	17.3	6.6	7.3	8.4			0.2
					3.2	1.0	19.9	4.3	3.9
									2.0
18.6	16.0	30.9	12.2	17.1	28.2	27.5	30.4	33.0	47.2
tr	2.0	2.4	0.7	1.9	2.0	2.6	1.3	1.0	0.5
65.0	56.6	57.3	65.9	74.2	59.3	49.7	48.4	61.7	46.2

variable but generally much less abundant phenocrystic clinopyroxene and lesser olivine, glassy to very finely crystalline groundmass, MgO ≈2.5–7.0 per cent, and SiO_2 ≈50 to 55 per cent. Phenocrystic titaniferous magnetite is moderately prominent (up to ≈2.5 modal per cent). Of the 26 samplings of this group, 12 are basalts, 14 are andesites (SiO_2 = 52–62 per cent) — with most of the latter best termed basaltic andesite, SiO_2 ≈52–4 per cent. Analyses of four members (numbers 11 to 14) are given in Table 5.1, and mean chemical composition of the group as a whole is shown in Table 5.2.

3. Hornblende-andesites. These are characterized by conspicuous hornblende phenocrysts (green or brown;

Table 5.2 Mean chemical compositions of the three principal lava groups of the Solomon Islands Younger Volcanic Suite

	1	2	3
n	47	26	49
SiO_2	48.73	52.73	56.79
TiO_2	0.56	0.79	0.63
Al_2O_3	11.81	17.76	17.66
Fe_2O_3	2.65	4.10	4.45
FeO	7.14	4.51	2.21
MnO	0.17	0.16	0.13
MgO	15.21	4.20	3.41
CaO	9.46	8.88	7.61
Na_2O	1.99	3.47	4.11
K_2O	1.09	1.95	1.29
P_2O_5	0.22	0.35	0.17
H_2O^+	0.67	0.75	0.77
H_2O^-	0.42	0.62	0.41
Total	100.12	100.27	99.64

1 = olivine–clinopyroxene-basalt group (including picritic members); **2** = big-feldspar lava group; **3** = hornblende-andesite group; n, number of samples analysed.

occasionally both) set in a mass of feldspar crystals and finer groundmass, lesser green clinopyroxene, minor but commonly conspicuous phenocrystic titanomagnetite, MgO from <1 per cent to about 5 per cent and $SiO_2 \approx 51$–74 per cent with most in the range 53–60 per cent. Although there is a minor component of dacitic to rhyolitic rocks, by far the major part of this group is composed of andesites. Of the 49 samplings, eight are basalts, 32 are andesites (including basaltic andesites), eight are dacites, and one is a rhyolite. Analyses of five representative members (numbers 15 to 19) are given in Table 5.1, and the mean chemical composition of the group as a whole is shown in Table 5.2.

MAJOR ELEMENT CHEMISTRY

(Total Fe as FeO)–MgO–($Na_2O + K_2O$) — i.e. 'FMA' — relations for these three broad rock groups and their groundmasses are shown in Fig. 5.4 and other major oxide co-variations are depicted in the Harker diagrams of Fig. 5.5.

The most noteworthy feature of the suite is the very high magnesium content of the more mafic members. At the time of their discovery by J. H. Hill and Stanton and Bell in 1959 they were probably the most magnesian and olivine-rich lavas to have been observed by modern methods. The Hawaiian picritic basalt containing 19.6 per cent MgO referred to by Macdonald and Katsura (1964) was easily surpassed in this respect by the New Georgia (Marovo) picrite (MgO = 28.1 per cent) reported by Stanton and Bell (1969). With a whole-rock Mg number (M) of 84 and constituent olivine of mean composition Fo_{89} these appear to have been the most picritic *eruptive* rocks then known. Later more extensive and detailed investigation has revealed an abundance of these rocks as flows in the Kolo River, western Vangunu, and Kohinggo Island areas (and as smaller areas of lava in widely scattered areas of New Georgia) and also as occasional dykes cutting other lava types here and there throughout the New Georgia Group. The more recently discovered picrites of Kohinggo Island have been found to include members with MgO >28 per cent, M = 88, and containing olivine of composition Fo_{94}.

In general the more mafic members of the suite have olivine and hypersthene in the norm, the more felsic, quartz and hypersthene. From the point of view of major element chemistry the Younger Volcanics thus appears to constitute a truncated basalt–andesite–dacite–rhyolite suite that, with a few exceptions in the rhyodacite:rhyolite field, more or less terminates around a felsic andesite-to-dacite composition. While truncated at the felsic end, the suite extends, however (as noted above), deep into the field of high magnesium rocks — picrites — at the other (Fig. 5.4(f)), and the total evolutionary span is thus large.

As first pointed out by Stanton and Bell (1969) the suite as a whole exhibits relatively low TiO_2 and though part of

Petrology of the Solomon Islands Younger Volcanic Suite 33

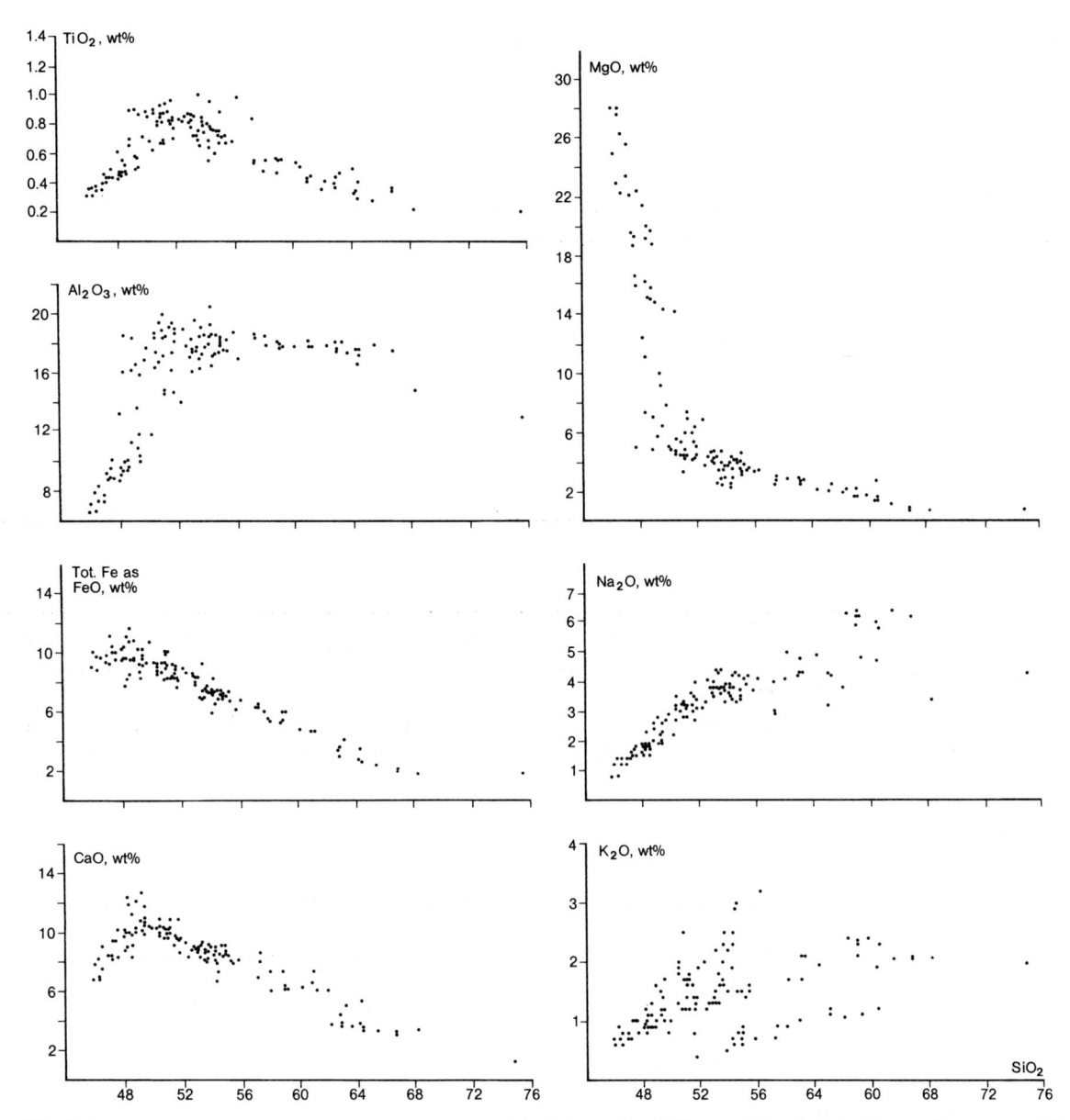

Fig. 5.5. Variation (Harker) diagrams for the major elements of the Solomon Islands Younger Volcanic Suite. The near-linear array of 19 points characterized by $K_2O \approx 0.3$–1.1 and $SiO_2 \approx 52$–61 per cent represents the relatively low-K_2O andesites of the Mt Gallego volcano.

an intra-oceanic volcanic domain its members contain notably less abundant TiO_2 than do most mid-oceanic basaltic provinces. In this respect they appear to have a circum-oceanic rather than a mid-oceanic affinity.

Lavas of the basaltic andesite to basic andesite interval tend to be notably aluminous (Fig. 5.5). Al_2O_3 contents of 17–19 per cent are virtually the rule for this group, and occasionally exceed 20 per cent. The pattern of abundance of Al_2O_3 appears to reflect initial olivine–clinopyroxene fractionation followed by abundant plagioclase crystallization.

The suite is not notably iron-rich. In spite of the sharp increase in phenocrystic magnetite at the 50–1 per cent whole-rock SiO_2 interval (Figs. 5.5, 5.6, 17.2, and 21.2),

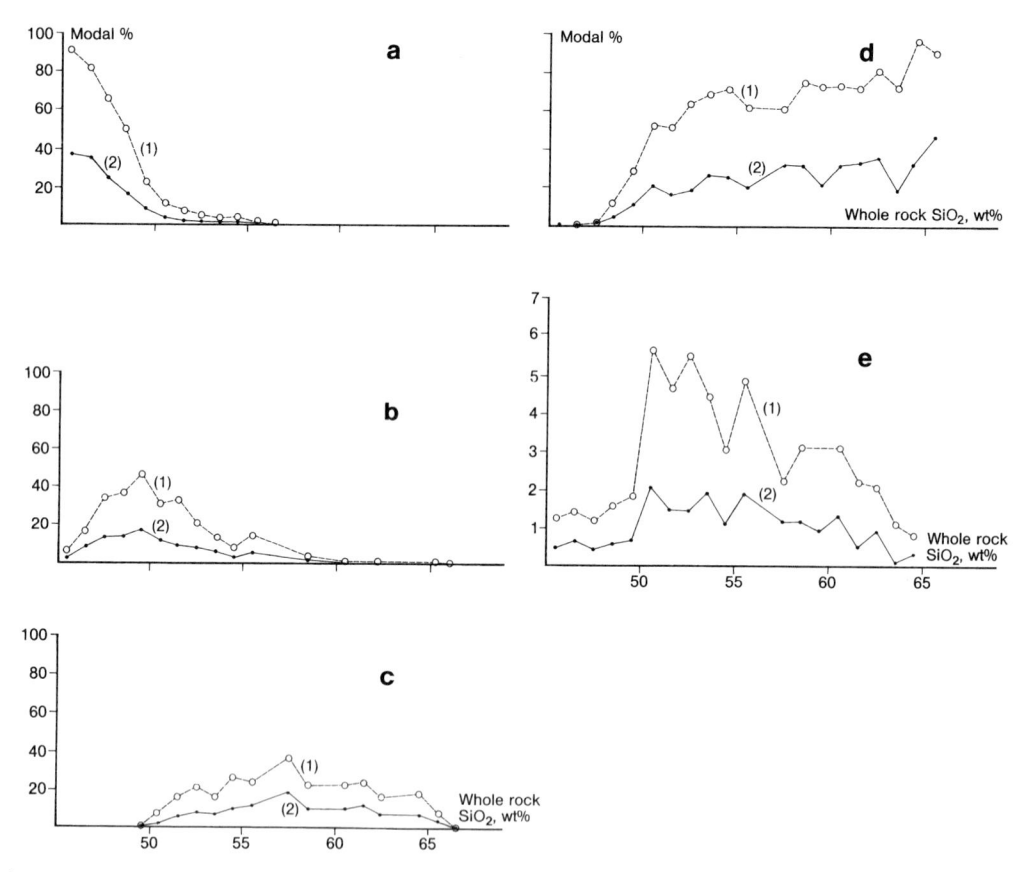

Fig. 5.6. Modal abundances of the principal phenocrystic minerals of the Solomons younger lavas as related to one per cent intervals of whole-rock SiO$_2$; (a) olivine, (b) clinopyroxene, (c) hornblende, (d) feldspar, and (e) spinel. 'Curves' labelled (1) and plotted as open circles represent the incidence of the mineral as a percentage of total phenocrysts in the relevant lavas; those labelled (2) and plotted as filled circles represent the incidence of the mineral as a percentage of the whole rock.

iron decreases with remarkable evenness from about 9.7 per cent total Fe as FeO in the most basic members to about 2.5 per cent in the felsic andesites and dacites. The groundmass at equivalent whole-rock SiO$_2$ intervals contains less iron than the whole rock throughout (Figs. 17.2 and 21.2). Following an initial rise reflecting early crystallization of high-Mg olivine and clinopyroxene, groundmass total iron exhibits a steady decrease, essentially similar to that shown by the whole-rock series, from about 8 per cent at 50–2 per cent whole rock SiO$_2$ to about 2.0 per cent at ≈64 per cent whole rock SiO$_2$. Such uniform patterns of change are not, however, shown by the oxidation indices (expressed as oxidation index = Fe$_2$O$_3$/(FeO + Fe$_2$O$_3$)) of whole rock and groundmass (Fig. 5.7). These rise from 0.27 and 0.43 respectively in the most mafic rocks, to almost coincident peaks at 0.85 and 0.88 in the 60–2 per cent whole-rock SiO$_2$ interval, and then decline in the more felsic members. While progressive rise in oxidation index up to SiO$_2$ ≈60–2 per cent accords with trends in lava series elsewhere, their apparent decline with onset of dacite formation is not, and no petrographical reason for such a reversal in behaviour is apparent. The possibility immediately arises that his feature may be spurious: a statistical artefact stemming from the low sample numbers at high SiO$_2$ levels referred to above. Such a suspicion is confirmed by inclusion of the 119 Bougainville analyses (Fig. 5.7(b)). The whole-rock oxidation index is again 0.25 at <48 per cent SiO$_2$, increases rapidly in the 48–52 per cent SiO$_2$ interval, and then exhibits a remarkably constant rate of increase well into the field of dacites. There is, as might have been expected on petrological grounds, no sign of a reversal at

Petrology of the Solomon Islands Younger Volcanic Suite 35

60–2 per cent SiO$_2$, and this feature of Fig. 5.7(a) is thus virtually certainly a reflection of sampling. The impression gained from Fig. 5.7(b) is that the trend of increasing oxidation index should continue into higher SiO$_2$ fields and approach 0.75 at the onset of rhyolite formation. The oxidation index of the suite as a whole appears to be relatively high (Fig. 17.2(b)).

The FMA diagrams for the three lava groups and their groundmasses (Fig. 5.4) indicate that, like the lava assemblages of most other volcanic arcs, those of the Solomons embrace two distinct 'series': a relatively high magnesium picrite–olivine-basalt–basalt grouping and a calcalkaline basalt–andesite–dacite grouping. The high FeO hinge at which the two relevant trends meet is occupied — somewhat diffusely — by the field of the big-feldspar lavas, the chemical and mineralogical features of which overlap those of the olivine basalts and hornblende andesites in many respects. The question of a nomenclature for these and related trends is considered later in the present chapter.

As would be expected, calcium behaves similarly to aluminium at the mafic end of the suite, and in parallel with iron and aluminium in the series as this develops at and above 48 per cent whole-rock SiO$_2$. Maxima for whole-rock and groundmass are at about 11–11.5 per cent and the two exhibit essentially similarly decreasing CaO (to 1.0 and 2.5 per cent respectively) as the series becomes more felsic. (Fig. 12.5).

The abundances of Na$_2$O and K$_2$O in the three rock groups are indicated in Figs. 5.8 and 5.9. Sodium increases fairly uniformly from mean Na$_2$O = 1.99 per cent to 3.47 per cent to 4.11 per cent in the olivine–pyroxene-basalts, big-feldspar basalts, and hornblende-andesites respectively. Potassium, however, behaves differently: K$_2$O increases from a mean of 1.09 per cent in the olivine–pyroxene-basalts to 1.95 per cent in the big-feldspar basalts, and then decreases again to 1.29 per cent in the hornblende andesites. The increased level of the alkali metals in the groundmasses behave correspondingly: mean groundmass Na$_2$O is 2.95 per cent, 3.81 per cent, and 4.99 per cent and mean K$_2$O is 1.55 per cent, 2.47 per cent and 1.81 per cent respectively. K$_2$O/Na$_2$O ratios in the whole rocks are 0.55, 0.56, and 0.28, and in the groundmasses 0.53, 0.65, and 0.36. The decrease in mean K$_2$O and in K$_2$O/Na$_2$O ratios from the big-feldspar basalts and basaltic andesites to the hornblende-andesites is thus quite marked.

As has been noted in many young volcanic provinces, adjacent, very similar volcanic centres may show subtle constitutional differences. For example the volcanoes of Mount Gallego (Western Guadalcanal) and Savo consist dominantly of very similar hornblende-andesites and lesser dacites, are of essentially of the same age and

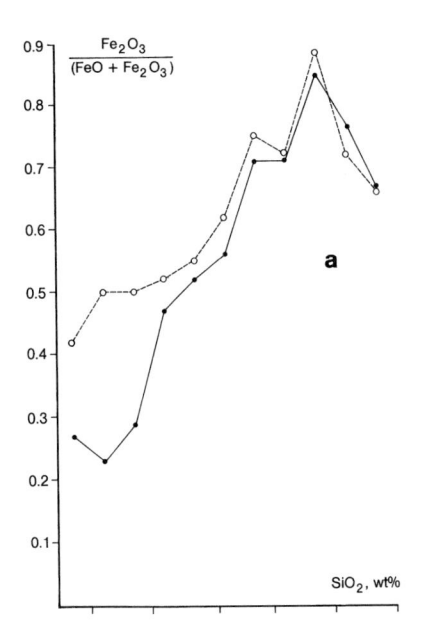

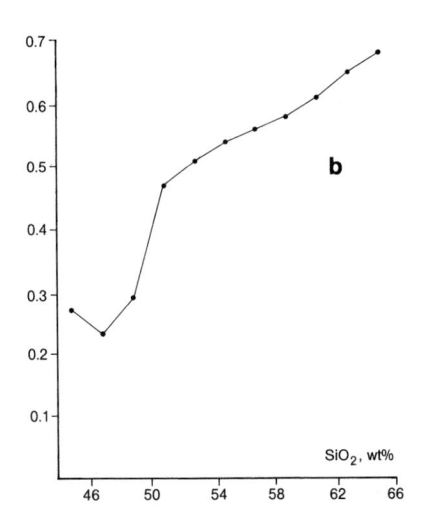

Fig. 5.7. Mean oxidation indices (Fe$_2$O$_3$ / (FeO + Fe$_2$O$_3$)) corresponding to 2 per cent whole-rock SiO$_2$ intervals <46, 46–8, 48–50 → >64 per cent in (a) the 122 lavas of the Solomons suite excluding Bougainville (filled circles) and their groundmasses (open circles), and (b) the 241 lavas of the Solomons suite including Bougainville. Reference to Fig. 5.2 indicates that the apparent decrease in oxidation index from <46 to 46–8 per cent SiO$_2$ and, in (a), from 60 per cent SiO$_2$, is probably a reflection of sampling deficiency. There is, however, no reason to doubt the genuineness of the rapid increase in oxidation index at ≈ 50 per cent SiO$_2$. This is followed by a lesser but conspicuously uniform rate of increase, which appears (b) to be continuing at SiO$_2$ = 64 per cent.

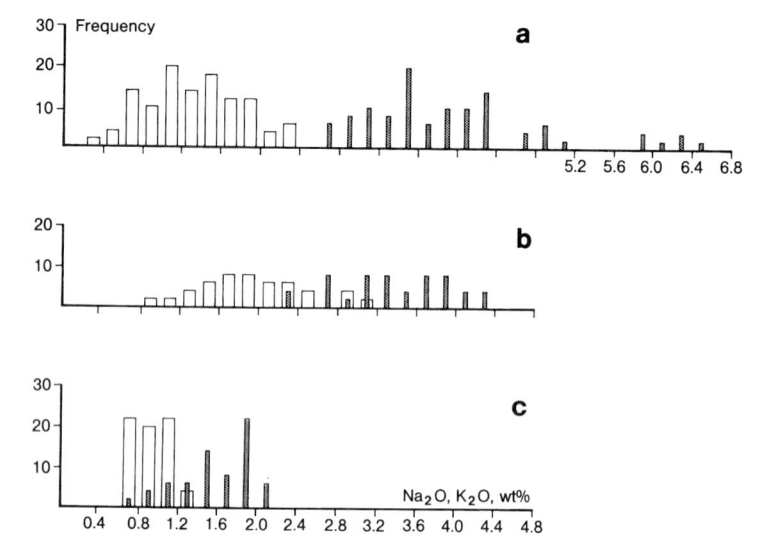

Fig. 5.8. Frequency distributions of Na$_2$O (filled bars) and K$_2$O (open bars) abundances in (a) the hornblende-andesite group, (b) big-feldspar lavas, and (c) olivine–clinopyroxene-basalts of the Solomons suite.

interdigitate at their bases beneath the intervening sea, and both are at a similar stage (hot spring) in their volcanological evolution. Of the two, however, the Savo rocks contain slightly but systematically higher K$_2$O: a feature which, as will be shown later, is reflected in the abundances of a number of the trace elements they contain.

PRINCIPAL MINERALS

Olivine. The abundance of olivine is the most conspicuous feature of the high-Mg members of the Younger Volcanic Suite. Its modal abundance in relation to whole-rock SiO$_2$ is shown in Fig. 5.6(a). The points given here are, however, the mean values for each whole-rock SiO$_2$ interval, and olivine modes for individual samplings are in some cases much higher than the average for the relevant rock group. The incidence and appearance of the olivine in picritic members of the high-Mg suite are depicted in Plate 1. In some cases the olivine occurs as large crystals (2–4 mm) adhering to each other by highly vesicular glass — spectacular rocks in which the olivine constitutes over 50 per cent of the solid matter. The grain size is commonly around 2 mm across, though large crystals may reach 4–5 mm, and groundmass olivines in some of the less olivine-rich rocks range down to a small fraction of 1 mm.

Most olivines are of generally rounded form, though some — particularly among the intermediate to large grains — display good development of faces and occur as sub-idiomorphic to near-idiomorphic crystals. Zoning is generally not obvious optically, but does appear in some crystals (not abundant but by no means uncommon) as pale brownish central areas, the coloration being due to the incidence in the earlier-formed part of the crystal of myriads of extremely fine fluid inclusions.

While overall, and particularly among the rocks containing <50 per cent SiO$_2$, there is a general relation between olivine and whole-rock compositions (Fig. 5.3), the major element chemistry of olivine is as variable as its textural features. The full range of compositions measured is Fo$_{94}$ to Fo$_{54}$. In the high-Mg rocks of picritic composition the olivines are commonly in the range Fo$_{87-89}$, with cores that are often more forsteritic than Fo$_{90}$. In the big-feldspar basalts and basaltic andesites, on the other hand, compositions are generally in the Fo$_{68}$–Fo$_{78}$ range and some (see above) are less forsteritic than Fo$_{60}$. Variation in composition core-to-rim within individual crystals is most commonly from nil to about five forsterite units, with an average of about two. Compositional differences between adjacent crystals within individual rocks are almost the rule; e.g. crystal A exhibits core to rim variation Fo$_{92}$–Fo$_{85}$, crystal B essentially no variation at Fo$_{80}$–Fo$_{80}$. These various compositional relations indicate that the olivines are essentially products of melts represented by the rocks in which they are now found, i.e. they are products of *in situ* crystallization rather than crystal accumulation, but that the conditions under which they formed were turbulent.

Small amounts of calcium and aluminium (0.10–0.60 per cent) are present in most olivines, but the incidence of these minor elements appears to show no systematic tie with the major element chemistry of either the relevant crystal or the containing whole rock. An analysis of CaO in 253 olivines with Mg numbers >Fo$_{80}$ in 27 lavas (chiefly olivine–pyroxene-basalts and picrites) and in 166 olivines

PLATES

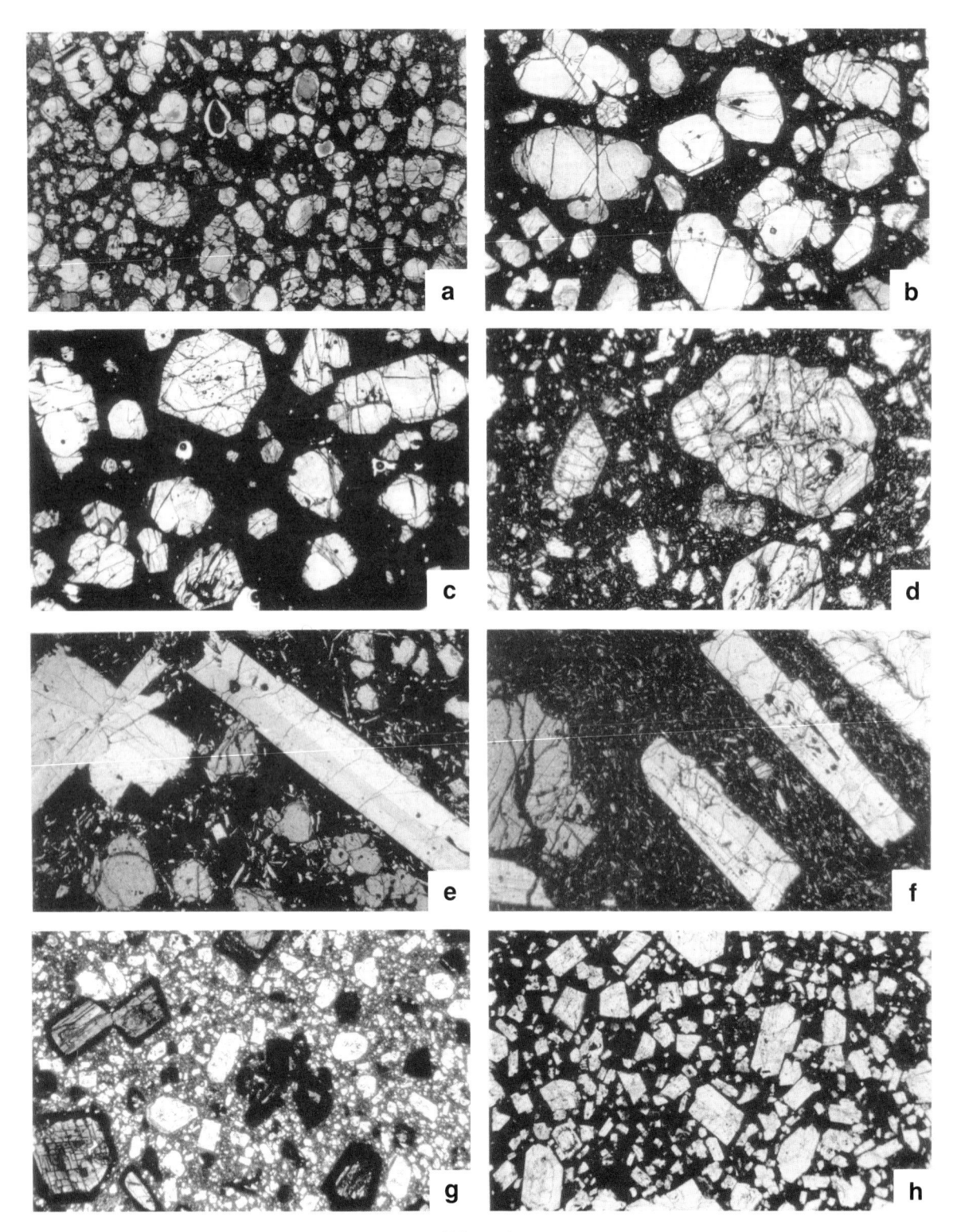

Plate 1

Plate 1. (opposite page) Microphotographs, all taken in transmitted, ordinary light, illustrating the principal lava types of the Solomon Islands Younger Volcanic Suite: (a) picrite of the olivine–clinopyroxene-basalt group, Hire River, New Georgia, × 6; (b) similar, × 14, showing variation from subidiomorphic to allotriomorphic form of the olivine phenocrysts and variation in the density of fine inclusions within them; (c) conspicuously clinopyroxene-bearing member of the olivine–clinopyroxene-basalt group, showing idiomorphic clinopyroxene (upper centre) and olivine (upper-right), Marovo, New Georgia, × 9 (see also Stanton and Bell (1969), Plates 2 and 3); (d) large, zoned, subidiomorphic clinopyroxene crystals and subordinate plagioclase in an ankaramitic member of the olivine–clinopyroxene-basalt group, Rendova Island, × 14; (e) mafic basalt of the big-feldspar lava group showing typical large and prominent plagioclase laths and associated phenocrystic clinopyroxene (lower half) and olivine (upper right), all set in a groundmass of glass and smaller feldspars, Vangunu, × 9; (f) more felsic basalt of the big-feldspar lavas group showing large plagioclase crystals containing prominent glass inclusions (see also Chapters 6 and 21, and Stanton and Bell (1969), Plate 1) and large clinopyroxenes, Roviana, × 9; (g) hornblende-andesite of the hornblende-andesite lava group showing abundant phenocrystic plagioclase and green hornblende (the latter displaying the almost ubiquitous opaque rims) set in a fine, principally crystalline, groundmass, Gallego Volcano, Guadalcanal, × 6; and (h) dacite (SiO_2 = 65.36 per cent) of the hornblende-andesite group, showing abundant phenocrystic plagioclase (but no quartz); not apparent in the photograph are small particles of brown hornblende (unrimmed) and biotite, Savo Volcano, × 6.

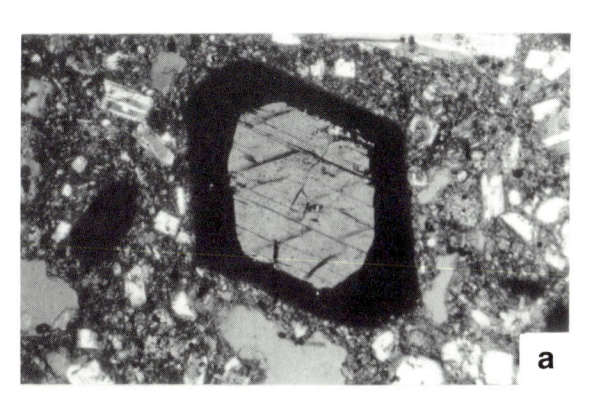

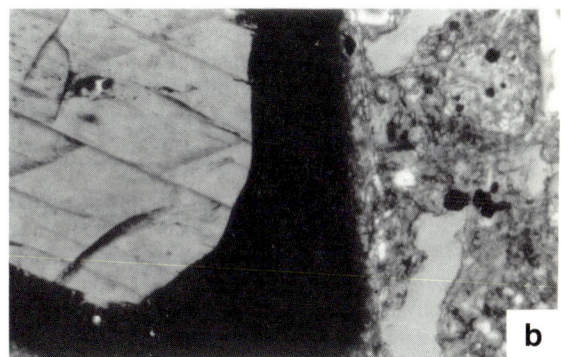

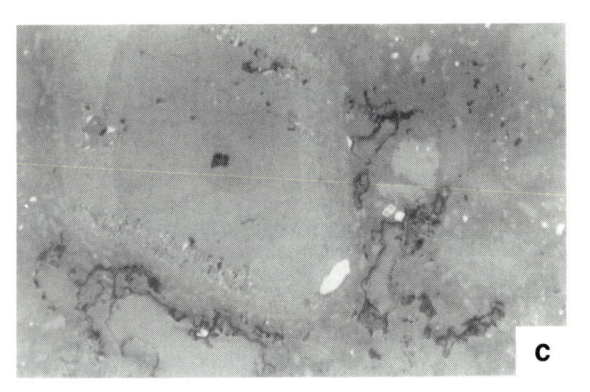

Plate 2. (a) Hornblende crystal (centre, grey) set in finely crystalline andesitic groundmass and showing typical development of opaque rim (transmitted ordinary light, × 14); (b) the same, enlargement of lower right of crystal, showing sharp inner, and diffuse outer, margins of rim (× 38), and (c), same, in reflected light (× 19), showing zoned and spongy texture of rim.

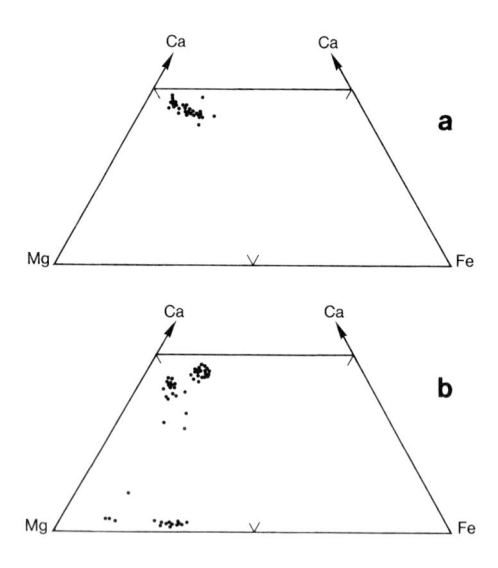

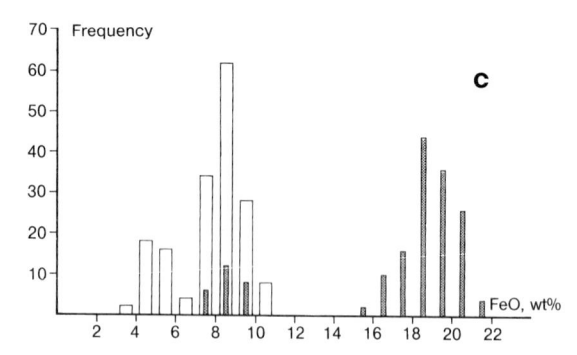

Fig. 5.10. Compositions of (a) clinopyroxene of the olivine–clinopyroxene-basalts, (b) coexisting clinopyroxenes and orthopyroxenes of 10 two-pyroxene-bearing andesites. Variation in FeO in the latter clinopyroxene–orthopyroxene pairs (clinopyroxenes, open bars; orthopyroxene, filled bars) is shown in (c).

bearing basalts (SiO_2, 50–1 per cent) to hornblende-andesites (SiO_2, 59–60 per cent). Second, for lavas between approximately 49 per cent and 53 per cent SiO_2, mean Mg numbers of the clinopyroxenes are greater than those of the accompanying olivines (Fig. 5.3(a)). At lower and higher whole-rock SiO_2 contents, the Mg numbers of the two mineral groups are essentially similar. The first feature indicates that there is a quite extended interval of lava evolution — involving a whole-rock SiO_2 increase of about 10 per cent and Fe_{tot} and MgO decreases of 4.5 per cent — over which the pyroxene has been able to maintain a constant composition. The second may imply noteworthy turbulence in the magma system at a particular stage in its evolution causing the mixing of earlier-formed pyroxene with later-formed, more fayalitic, olivine.

The two principal minor elements of the pyroxenes are Al and Ti. Where the two pyroxenes coexist Al and Ti in the clinopyroxenes (84 analyses: mean $Al_2O_3 \approx 1.60$ per cent; mean $TiO_2 \approx 0.27$) is about double that in the orthopyroxenes (81 analyses: $Al_2O_3 \approx 0.86$ per cent; $TiO_2 \approx 0.15$). In addition the two elements are distinctly higher in the clinopyroxenes of the single-pyroxene lavas than they are in the two-pyroxene varieties. Mean Al_2O_3 in clinopyroxenes of the range picrite to hornblende-andesite is generally in the range 3.0–3.5 per cent, and mean TiO_2 is 0.26 per cent. There is some indication that Al_2O_3 in these clinopyroxenes rises evenly from about 3.0 per cent in lavas with whole-rock SiO_2 <46, to about 3.8 per cent at 48–50 per cent SiO_2, then declines symmetrically to ≈ 3.0 per cent at 52–3 per cent, remaining at that level until the limit of its occurrence at 59–60 per cent whole-rock SiO_2. The behaviour of Ti in the pyroxenes is considered again in Chapter 13.

One feature of the pyroxenes that may merit closer scrutiny than the writer has been able to give it is the incidence of Fe_2O_3 in the clinopyroxenes. In many cases structural totals based on 6O consistently exceed 4, indicating the presence of significant Fe^{3+}. Using standard methods such totals may be recalculated to the ideal value

Table 5.3 The incidence of ferric iron and ferric/ferrous iron ratios in coexisting clino- and orthopyroxenes

	Clinopyroxene	Orthopyroxene
n	36	32
Mean total Fe as FeO	7.22	17.19
Mean Fe_2O_3	2.18	2.20
Mean $Fe^{3+}/Fe^{3+} + Fe^{2+}$	0.27	0.12

n = number of grains analysed by electron microprobe; all values as per cent by mass.

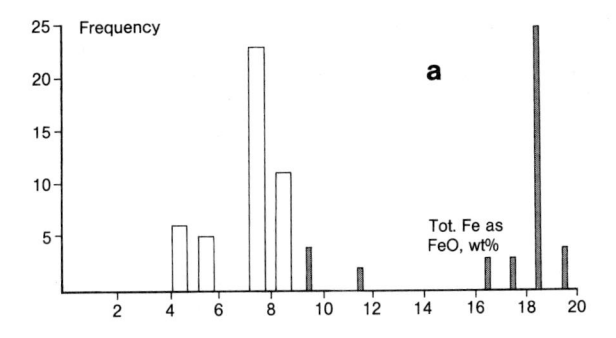

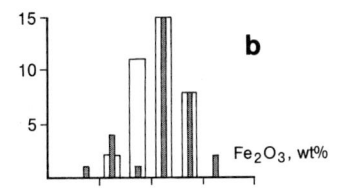

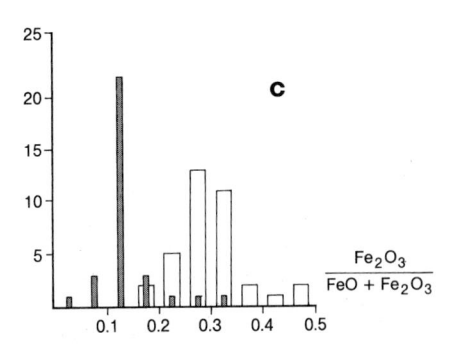

Fig. 5.11. Incidence of iron and its oxidation state in 45 pairs (90 analyses) of coexisting ortho- and clinopyroxenes (solid and open bars respectively): (a) total Fe as FeO, (b) Fe_2O_3, and (c) oxidation index, $Fe_2O_3 / (FeO + Fe_2O_3)$.

of 4, yielding the amount of Fe^{3+} present. Semi-casual observation of structural totals indicates that the clinopyroxenes of the andesitic lavas may be characterized by higher $Fe^{3+}/Fe^{3+}+Fe^{2+}$ than those of the olivine-rich basalts, but further, more systematic, analysis is required to investigate this. Investigation of Fe^{3+} in 68 coexisting clinopyroxene: orthopyroxene crystals in a two-pyroxene andesite gave the results of Table 5.3 and Fig. 5.11. Clearly the absolute amounts of Fe^{3+} in the two groups of pyroxenes are very similar.

Hornblende. This is the most characteristic component of the andesites and dacites of the Younger Volcanic Suite and is a conspicuous mineral of the more felsic centres such as Gallego, Savo, Mbalo, Vella Lavella, and Gatukai (Fig. 5.6(c), Plates 2 and 3). It is also a constituent of some of the minor andesites occurring in more mafic cones such as those of Vangunu and Kolo, and inevitably appears in many of the pyroclastic and volcaniclastic materials of andesitic–dacitic affinity.

Comparison of Figs. 5.6(a) and 5.6(c) indicates that olivine and hornblende are substantially complementary in their occurrence; indeed, in only one sample have the two minerals been found together. However there is a substantial overlap in terms of the SiO_2 contents of the lavas in which they occur: olivine appears in materials up to the 56–7 per cent SiO_2 interval, and hornblende in materials down to 49–50 per cent SiO_2. One notable feature is that the first appearance of hornblende closely coincides with sudden increases in modal plagioclase and, particularly, modal magnetite (Fig. 5.6(c),(d), and (e) respectively).

Hornblende occurs as green and brown varieties, both of which occasionally appear in a single thin section. A striking feature of its chemistry is depicted in Fig. 5.3(b): Mg numbers remain essentially constant (within the 68–70 range) through the full spectrum of hornblende-bearing rocks from basalt to dacite (whole-rock silica range 50–66 per cent). Also, in addition to its relation to the abundance of plagioclase and magnetite, the first appearance of hornblende coincides with the point of flattening of the Mg number/whole-rock SiO_2 curves for olivine and clinopyroxene — the interval 50–1 per cent SiO_2.

Many — perhaps the majority — of the hornblendes exhibit the dark rims common among andesitic and dacitic hornblendes generally. These rims are black in transmitted light, relatively highly reflecting in incident light, and in polished section may present a somewhat spongy texture, particularly towards their inner margins. Inner contacts with the parent hornblende crystal are generally sharp and well-defined, whereas the outer margins in contact with groundmass are characteristically diffuse and almost grade into the latter (Plate 2). Occasionally the opaque material occurs as cores and zones within the relevant hornblende crystal, indicating that the conditions responsible for its development did not necessarily arise only at, or following, the final stages of hornblende crystallization. In some instances, e.g. the andesitic centre of Mbalo, on the south coast of Guadalcanal, the hornblende has been completely pseudomorphed by the opaque material, the latter thus constituting 15–20 per cent of the whole rock.

As this material plays a significant role in accommodating some of the minor and trace metals of the Solomons hornblende andesites, it is appropriate to examine briefly its principal chemical features.

40 Ore Elements in Arc Lavas

Table 5.4 Chemical compositions of 'mean Solomons hornblende' and 'mean opaque rim' as determined by X-ray fluorescence analysis of mineral separations

	Mean hornblende	Mean opaque rim
n	25	7
SiO_2	43.94	47.59
TiO_2	1.63	1.05
Al_2O_3	11.66	10.06
Fe_2O_3	nd	9.01
FeO*	12.20	6.10
MnO	0.26	0.45
MgO	13.65	12.40
CaO	11.80	10.65
Na_2O	2.39	1.73
K_2O	0.45	0.31
Ba	67	40
Sr	176	64
Pb	1.5	1.8
V	434	419
Cr	120	9
Co	66	60
Ni	64	18
Cu	39	40
Zn	111	173
Total§	97.98	99.35

n = number of analyses; FeO* = total Fe as FeO for hornblende only; Total§ = total of major oxides, SiO_2 to K_2O; given as per cent by mass; all trace-element abundances, Ba to Zn, given as parts per million (ppm) by mass.

Digression on opaque rims to hornblende. Black rims of this kind are commonly referred to in the literature as 'magnetite'. While magnetite does occur as a finely distributed phase within the Solomons rim material, the latter as a whole is certainly not magnetite. It appears (in polished section in reflected light) to be composed of at least three finely intergrown phases, and extensive electron microprobe investigation (≈400 analyses) indicates that it is a mixture of materials of overall composition clearly related to, but deviating to various degrees from, that of the parent hornblende. That the latter is indeed the original, 'parent' material is indicated by the fact that the opaque material not only forms rims to the amphibole, but also complete pseudomorphs of it.

The multiphase nature of the opaque material is confirmed by X-ray diffraction: measurement of seven magnetic/heavy liquid separations of rim material showed this to consist principally of magnetite, clinopyroxene, and plagioclase, with trace quantities of haematite and quartz. Very small amounts of amphibole were detected in four of the samples, though this probably represented small inclusions of hornblende within opaque, rather than any residual amphibole structure in the latter. That the opaque material might consist of finely intergrown magnetite–pyroxene–plagioclase had in fact been hinted at earlier by microscopical observation: here and there in some of the andesites finely granular aggregates of the three minerals appear to represent coarsened residues of rim material. Most of these aggregates now appear as no more than small patches and wisps, but their derivation seems clear and seems to corroborate the X-ray diffraction evidence. The complete opacity of the rims is presumably induced by the presence of magnetite and haematite, combined with the very fine grain size of the rim intergrowth. X-ray fluorescence and wet chemical analysis of the rim separations indicated that about half of their total Fe is in ferric form (Table 5.4), a state of affairs qualitatively in keeping with the X-ray diffraction results.

Approximately 300 paired analyses of hornblende crystals and their rims, involving a range of Solomons hornblende-bearing lavas, have been carried out by electron microprobe, and representative results are given in Table 5.5 and Fig. 5.12. The transformation of amphibole to daughter product in these two individual Guadalcanal andesites involves not only a change in mean values of all major components apart from CaO, but also a pronounced flattening of frequency distributions. Both features are shown very clearly (Fig. 5.12) by the 200 paired analyses of hornblendes and their rims in Lava A of Table 5.5. In this case rim development is accompanied by clear increase in SiO_2, TiO_2, Al_2O_3, and total Fe, and concomitant

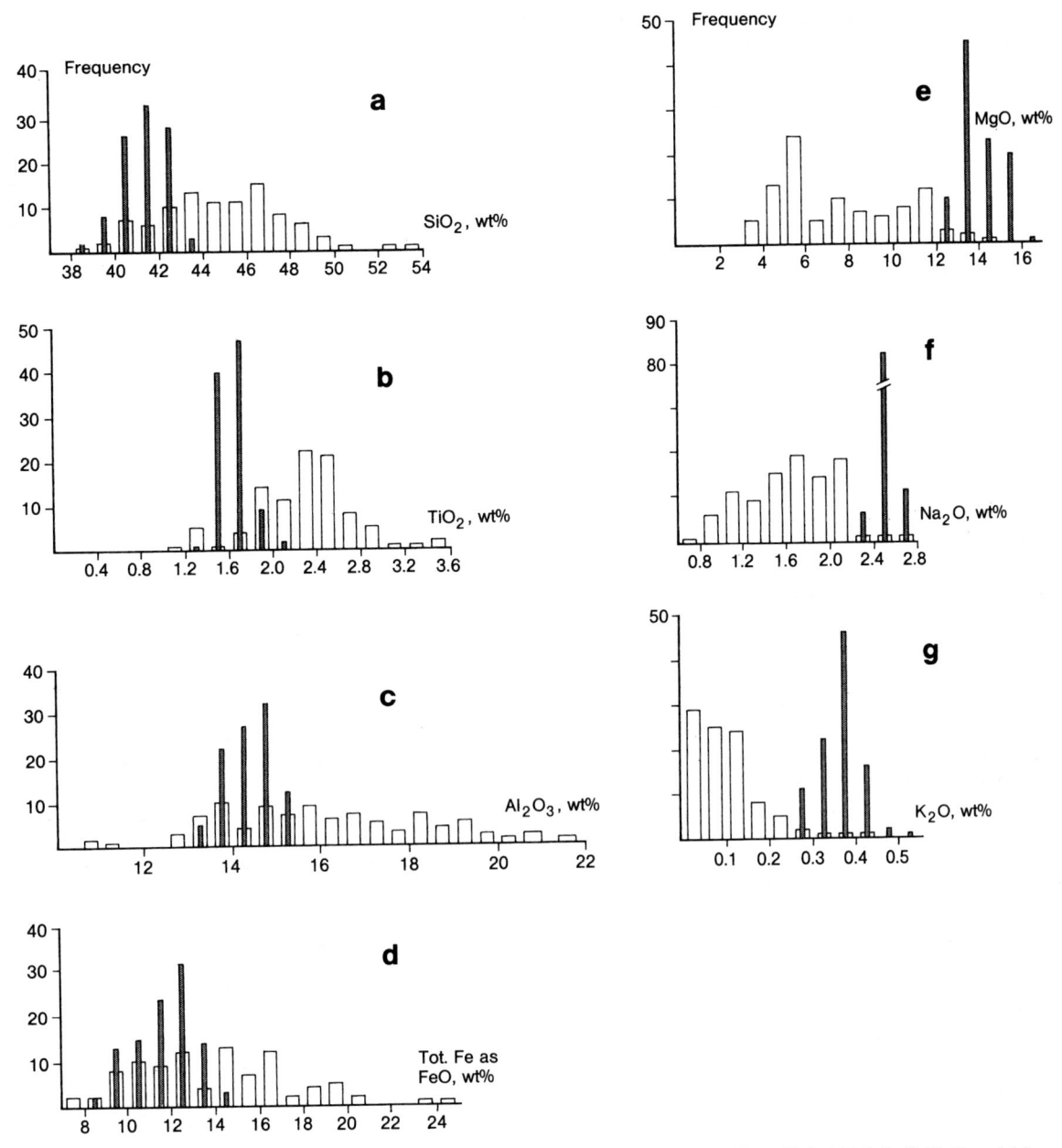

Fig. 5.12. Frequency distributions of abundances of (a) SiO_2, (b) $TiO2$, (c) Al_2O_3, (d) total Fe as FeO, (e) MgO, (f) Na_2O, and (g) K_2O in 100 pairings (200 analyses) of hornblende crystals (filled bars) and their opaque rims (open bars). All analyses by electron-microprobe using 12×12 μm raster.

decrease in MgO, Na_2O and K_2O. While mean CaO remains virtually unchanged, the frequency distribution of CaO values among the 100 rims exhibits a flattening similar to that shown by the other major components.

Such changes as they manifest themselves in an individual crystal:rim pairing are shown rather beautifully by a single large crystal and its rim sectioned from Lava A and shown in Fig. 5.13 and Plate 3. The profile consists of

42 Ore Elements in Arc Lavas

Table 5.5 Chemical compositions of 246 coexisting hornblende crystals and their opaque rims in two Solomon Islands hornblende-andesites

	A		B	
	Crystals	Opaque rims	Crystals	Opaque rims
n	100	100	21	25
SiO_2	41.36	44.80	41.80	42.04
TiO_2	1.63	2.27	1.79	2.63
Al_2O_3	14.35	16.01	13.55	14.46
FeO^*	11.67	13.86	10.56	17.41
MnO	0.15	0.26	0.12	0.21
MgO	14.02	7.62	14.60	7.92
CaO	12.41	12.47	12.60	11.91
Na_2O	2.49	1.62	2.51	1.09
K_2O	0.36	0.10	0.43	0.12
V	539	585	474	717
Cr	200	89	256	216
Co	50	40	77	58
Ni	77	90	115	81
Zn	65	382	49	326
Total§	98.44	99.01	97.96	97.79

n = number of analyses; FeO^* = total Fe expressed as FeO; Total§ = total of major oxides, SiO_2 to K_2O; all analyses by electron microprobe.

19 analyses, 13 traversing the length of the hornblende core and three traversing the rim of either end. In this individual case crystal-to-rim changes in MgO, Mg number, and the alkalis are clear (in the case of K_2O, dramatically so). Change in SiO_2 is less obvious; in total Fe as FeO it is ambiguous; and in CaO it is barely if at all apparent. Al_2O_3 shows a sharp increase in the inner side of the rim, decreasing in quite regular fashion towards the outer boundary.

The latter feature of Al_2O_3 enrichment presumably indicates an enhanced formation and concentration of plagioclase close to the front of the rim as this developed at the expense of the hornblende crystal. It has already been mentioned that the opaque rims commonly exhibit a 'spongy' appearance near their inner boundaries when viewed in reflected light. This feature, which is due to variation in reflectivity and polishing capacity, may thus be due to a greater incidence of plagioclase in this zone. With this in mind, the electron microprobe was used to determine fluorine and chlorine in adjacent hornblende, inner spongy rim, and outer smooth, more highly reflecting rim. Results are given in Table 5.6. Clearly the fluorine and chlorine build up within the inner, spongy rim and one may speculate that this is in association with the formation of the plagioclase. In parallel with Al_2O_3 the two halogens decrease again in the outer smooth portion of the rim. F/Cl ratios exhibit what appears to be a consistent increase from parent hornblende to outer rim, but as it is not known to what extent the F and Cl of the rim is inherited from the hornblende or derived by flux from the residual melt, no interpretation of abundances or ratios can yet be made.

Biotite. This occurs occasionally, and always in very small amounts (<2.5 per cent), in the most felsic andesites and in dacites, particularly on Savo. The biotite appears as sporadic small, stumpy particles and is of very uniform, typically igneous composition averaging ≈14 per cent Al_2O_3, 4 per cent TiO_2, and 0.18 per cent MnO. As noted in the relevant geochemical chapters, it is a noteworthy receptor of several of the trace elements of present concern. Because of its very low abundance it is, however, of little or no significance in influencing the patterns of distribution of these traces.

Table 5.6 Fluorine and chlorine in hornblende and associated opaque rim

	A	B	C
n	11	13	11
F	370	1640	291
Cl	217	515	68
F/Cl	1.71	3.18	4.28

A = hornblende crystal; B = inner, spongy zone of opaque rim; C = outer, smooth zone of opaque rim; n = number of analyses; all quantities in parts per million by mass.

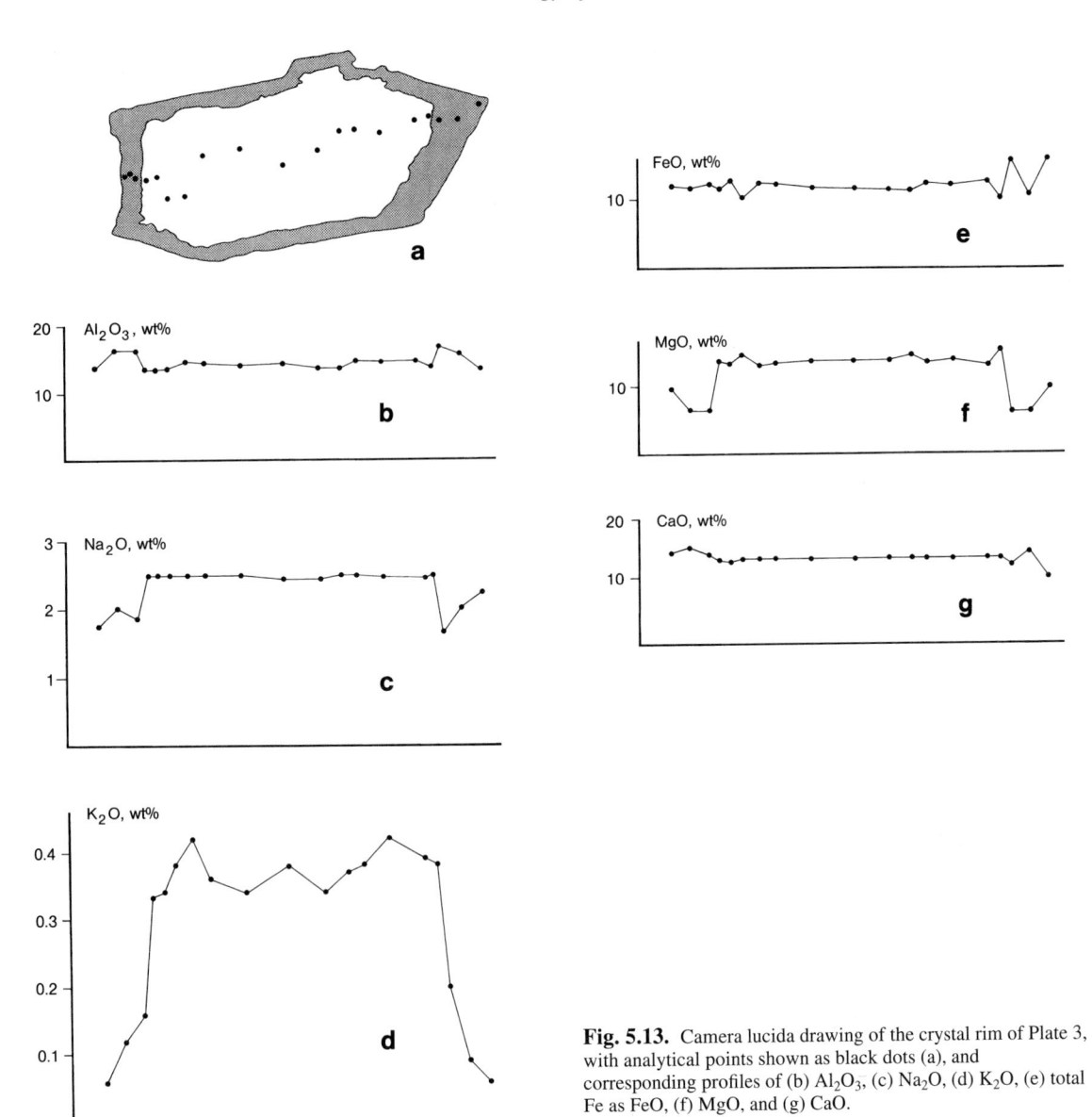

Fig. 5.13. Camera lucida drawing of the crystal rim of Plate 3, with analytical points shown as black dots (a), and corresponding profiles of (b) Al_2O_3, (c) Na_2O, (d) K_2O, (e) total Fe as FeO, (f) MgO, and (g) CaO.

Plagioclase. Apart from minor K-feldspar in the groundmasses of some of the lavas, the overwhelmingly dominant feldspars are members of the plagioclase series ranging from bytownite in the high-magnesium basalts to sodic andesine in the hornblende-andesites and more felsic types.

The patterns of abundance and compositional variation as related to whole-rock SiO_2 are shown in Figs. 5.6(d) and 5.14 respectively. Abundance increases fairly steeply between 47 per cent and 51 per cent SiO_2 (a trend accentuated in the graph of feldspar as a percentage of non-groundmass components, Fig. 5.6(d)), at which point rate of increase flattens somewhat. Mean CaO content (Fig. 5.14) decreases fairly steadily from a high point of ≈15 per cent (bytownite) in basalts of 48–50 per cent SiO_2 to ≈4 per cent (sodic andesine) in the andesitic to dacitic rocks of 65–6 per cent SiO_2. Mean *maximum* CaO (shown by the open circles and upper dashed line in Fig. 5.14) does not, however, decrease so markedly, perhaps indicating that the mean compositions of the feldspars of

44 Ore Elements in Arc Lavas

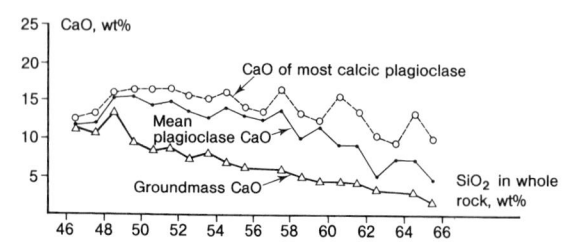

Fig. 5.14. Variation of calcium in plagioclase and groundmass with respect to one per cent intervals of whole-rock SiO_2. Filled circles, mean plagioclase CaO; open circles, CaO of the most calcic plagioclase identified by electron-probe analysis; open triangles, CaO in the coexisting groundmass. (This diagram represents X-ray fluorescence analyses of ultimate purity feldspar separations from 75 of the 122 Solomons lavas (all those from which satisfactory bulk separations could be made), together with ≈ 1100 electron-microprobe analyses of individual feldspar crystals in 109 of these lavas).

any given lava were induced by the progressive crystallization of more and more crystals and outer zones of more and more CaO-poor feldspar, rather than through the successive formation of a single crop of relatively like crystals.

A very small proportion of the feldspars is potassic, and eighteen out of 1014 feldspar analyses yielded K_2O above 1.0 per cent — the highest 4.2 per cent. All of these were found in members of the olivine–pyroxene-basalt group; perhaps surprisingly none were detected in lavas of the big-feldspar and hornblende–andesite groups. Mean K_2O for 220 analysed feldspars of the olivine–pyroxene-basalts was 0.42 per cent, and, when the above-18 high-K individuals were omitted, reduced to 0.28 per cent. Mean K_2O for 259 analyses of feldspars of the big-feldspar lavas was 0.36 per cent, and for 535 analyses of hornblende-andesite feldspars, 0.22 per cent.

All feldspars of the lavas contain small but apparently systematic quantities of iron and magnesium, as shown in Fig. 5.15. While the quantities shown agree well with those determined on the high-purity feldspar separations, all analyses used in Fig. 5.15 were obtained by electron microprobe on microscopically clear areas of feldspar; there is thus no possibility that the small quantities of iron and magnesium represent glass or other included matter.

As indicated by Fig. 5.15 the iron and magnesium decrease proceeding from the feldspars of the olivine–pyroxene-basalts, through the big-feldspar basalts and basaltic andesites to those of the hornblende-andesites. Means are 1.01 and 0.36, 0.83 and 0.18, and 0.38 and 0.16 per cent respectively. The incidence of these two elements in the feldspar is presumed to result principally from essentially random interstitial solid solution, the

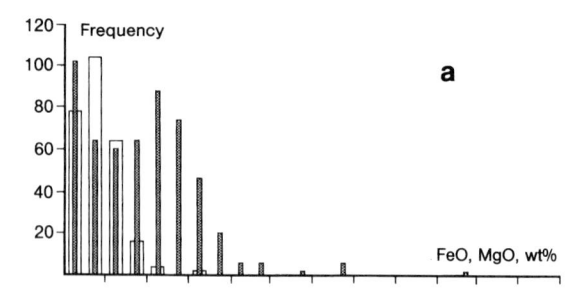

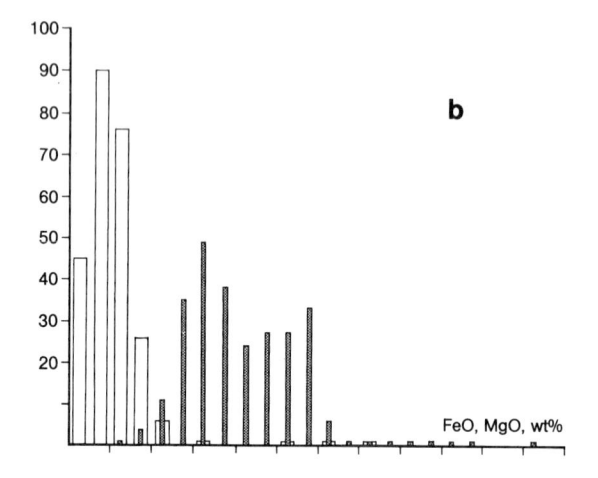

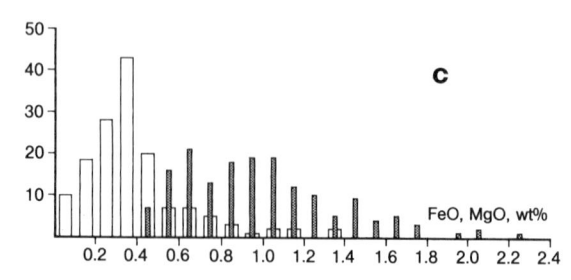

Fig. 5.15. Frequency distribution of abundances of total Fe as FeO (filled bars) and MgO (open bars) as determined by electron-probe analyses of inclusion-free plagioclase crystals in (a) hornblende-andesites, (b) big-feldspar lavas, and (c) olivine–clinopyroxene-basalts of the Solomons suite.

amounts incorporated reflecting the Fe_{tot} and Mg contents of the melt from which the relevant feldspar crystallized. Some Fe and Mg may, however, serve to balance the single charge on K^+ occupying Ca^{2+} positions in the feldspar structures. This is lent credence by the fact that, particularly in some of the olivine-pyroxene basalts those feldspars which are notably high in potassium are invariably high in iron and magnesium also. The eighteen feldspars referred to above, analysed in five of these basalts, gave a mean K_2O content of 2.31 per cent, mean FeO = 3.38 per cent, and mean MgO = 2.19 per cent. This contrasts with the above FeO and MgO contents of 202 'normal', low-K plagioclases (mean K_2O = 0.28 per cent) of the olivine–pyroxene-basalts. The molecular ratio of $K^+/(Fe^{2+} + Mg^{2+})$ in the high-K feldspars is thus 0.24 and in the 'normal' low-K feldspars is 0.13. This indicates that if there is an element of substitutional charge balancing this is neither constant nor precise, and that much of the compensation for incorporation of Fe^{2+} (and any Fe^{3+} and Mg^{2+}) must take place by vacancy formation in the feldspar.

As indicated in Figs. 5.6(d) and Plate 1(a), the most mafic lavas contain essentially no feldspar, either as phenocrysts or groundmass: the rocks consist of variable amounts and proportions of olivine and clinopyroxene phenocrysts set in a glassy to ultra-fine-grained matrix. The first appearance of feldspar may be as phenocrysts or as fine groundmass material, most commonly the former. Crystals tend to elongated lath form; they are twinned but not conspicuously zoned. As constitutions change to those of the big-feldspar basalts and basaltic andesites, olivine and clinopyroxene diminish in abundance, feldspar becomes more prominent as phenocrysts (Plate 1(d)) and to a lesser extent in the groundmass, and the rocks are increasingly characterized by conspicuous needle- to lath-shaped feldspars up to 20 mm long which, where abundant, may delineate flow-foliation. Phenocryst feldspar in the big feldspar lavas is always characterized by sharply defined uncorroded faces and a high degree of idiomorphism, though again it is not conspicuously zoned. Many of these large feldspars are quite clear and devoid of inclusions; others, however, contain conspicuous glass, which in some cases is so abundant as to give the relevant crystals — in spite of their highly idiomorphic external morphology — a distinctly spongy appearance. The compositions of such inclusions are variable: in some cases their chemistry is almost identical with that of the host crystal but most commonly they are higher in Fe and Mg than the host. The composition of the inclusion glass may on the other hand be quite different from those of the surroundings of the host crystal; some inclusions contain notable amounts of copper and comparatively high zinc (see Chapters 6 and 7).

Up to this point in whole-rock composition the feldspars are essentially as 'ordered' in their textures and compositions as the associated olivines and clinopyroxenes. With the appearance of hornblende and the development of the hornblende-andesites, however, the feldspars come to dominate both the phenocryst assemblage and the groundmass, exhibit highly variable size, shape, and zoning pattern, display variable corrosion, and, in most cases, the 'moth-eaten' appearance typical of many calc-alkaline feldspars (Plate 1(e) and (f)). Although Figs. 5.6(d) and 5.14 indicate an essentially gradual and continuous change in feldspar abundance and composition with progression from basalts to andesites and more felsic types, there appears to be a distinct textural break accompanying development of the calc-alkaline lineage.

Iron-rich spinel: 'magnetite'. 'Magnetite' may be used as a term of convenience to denote a range of spinels from chromite to titaniferous magnetites which as a group constitute well over 99 per cent of the opaque oxide fraction of the lava series as a whole. Although ilmenite appears sporadically — in two out of the 114 sections in which the oxides were subjected to detailed reflected light microscopy and probe analysis — it is very minor indeed: by far the major part of the titanium oxide occurs as a component of titanomagnetite.

The abundance of these members of the magnetite series is shown in Fig. 5.6(e). They appear at around 0.5 per cent of the whole rocks in the more mafic basalts, and then increase abruptly in abundance to ≈2 per cent at around 50–2 per cent whole-rock SiO_2, the approximate point of advent of the big-feldspar basalts and basaltic andesites. At this point, they become notably titaniferous and moderately conspicuous as members of the phenocryst assemblage. Although remaining in the 1–2 per cent range until well into the andesitic members of the lava suite, magnetite abundance decreases gradually until it is less than 0.5 per cent of the whole rock in the more felsic andesites and dacites.

The principal elements other than iron in the magnetites are Cr, Ti, Al, and Mg. Variation in the incidence of these in relation to the composition of the rock in which the magnetite occurs (as denoted by whole-rock SiO_2) is shown in Fig. 5.16. Chromium as Cr_2O_3 drops abruptly from a mean of 25.06 per cent in magnetites occurring in rocks containing <46 per cent SiO_2 to 0.3 per cent in those of lavas in the 52–4 per cent SiO_2 range. As indicated above, TiO_2 becomes prominent as whole-rock SiO_2 rises to 50–2 per cent. In contrast to Cr_2O_3 it increases initially from 3.13 per cent in the low-SiO_2 basalts to 7.59 per cent in the 50–2 per cent SiO_2 category, and then decreases fairly steadily to around 4.0 per cent in the magnetites of lavas of >64 per cent SiO_2. Like Cr_2O_3, Al_2O_3 and MgO consistently decrease from a maximum of 6.7 per cent and 6.5 per cent in the magnetites of the high-magnesium basalts to 2.1 per cent and 1.3 per cent respectively in those of the 64–6 per cent whole-rock SiO_2 interval. The marked parallelism of the Al_2O_3 and MgO 'curves' in Fig. 5.16 indicates the

46 *Ore Elements in Arc Lavas*

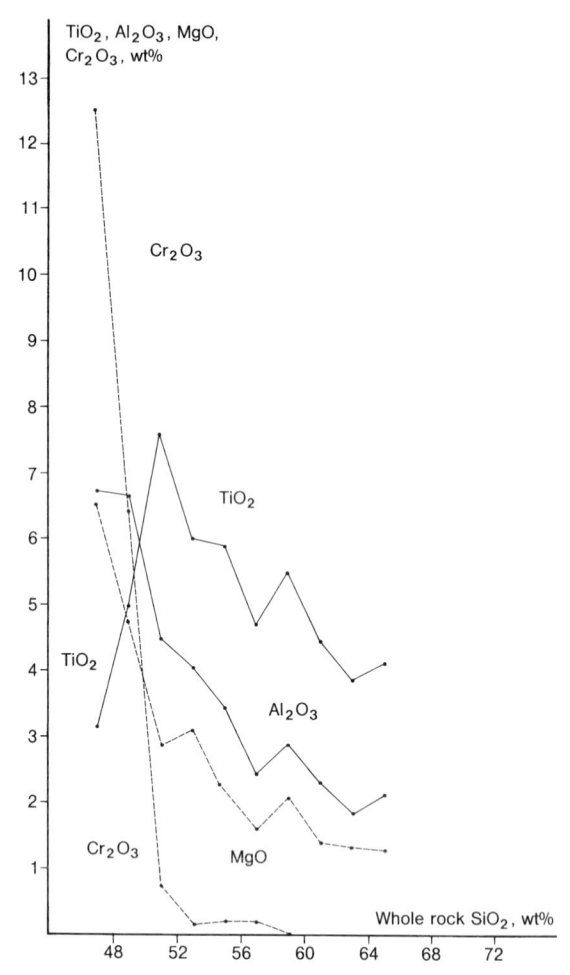

Fig. 5.16. Variation in abundances of Cr_2O_3, TiO_2, Al_2O_3, and MgO in Solomons spinels with respect to 2 per cent intervals of whole-rock SiO_2 in the containing lavas.

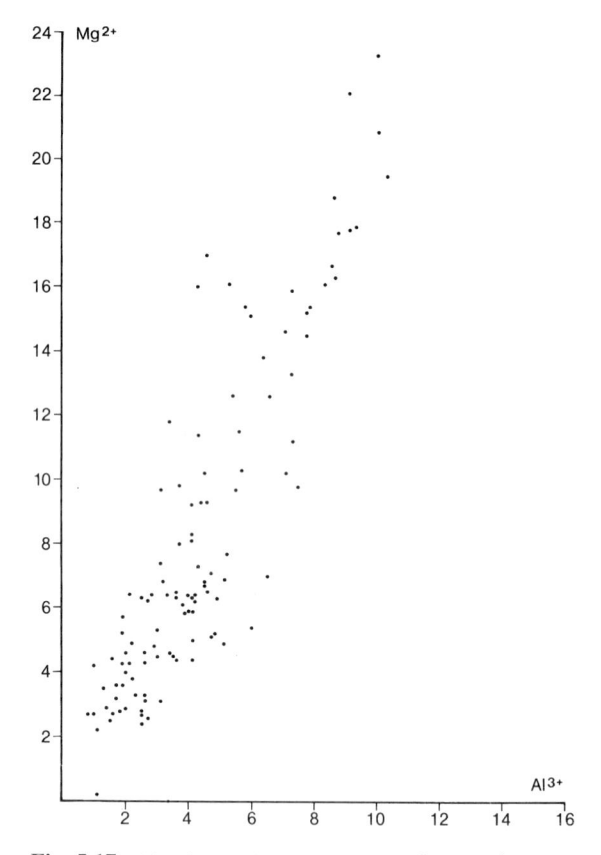

Fig. 5.17. Abundance relations between Mg^{2+} and Al^{3+} in individual spinels of the Solomons suite (atoms per cent × 100).

likelihood of a correlation between the two elements, and this is confirmed in Fig. 5.17.

Such progressive changes in spinel composition accompanying evolution of the lava series manifest themselves down to the fine scale of differences between various categories of spinel within a single lava. Fig. 5.18 depicts differences in MgO and Al_2O_3 abundances (see also table 5.7): in (a) those spinels included entirely within large clinopyroxene crystals in a single olivine–pyroxene-basalt from the northern (modern) cone of the island of Rendova; and in (b) those entirely within the groundmass of the same rock. Such differences extend to minor and trace elements of the two groups of spinel (Table 5.7), and these are referred to again in the relevant geochemical chapters.

The spinels of the high MgO-low SiO_2 lavas are thus much less 'pure magnetites' than those of the more felsic types. Also, as indicated by the progressive sharpening of the histograms of Fig. 5.19, as the containing rocks become more felsic, their compositions are much less variable — a feature that manifests itself from the broad scale of the rock group as a whole down to small areas of a single thin section. The latter is illustrated by twelve random analyses (Table 5.8) of small phenocrystic spinels of a single thin section of one of the high-magnesium olivine–pyroxene-basalts. The grains were generally too small to yield significant core-to-rim variation, and differences therefore do not simply reflect the incidence of compositional zoning. Inter-grain distances were of the order of 1.0 mm or less. Analysis 4

Table 5.7 Mean composition of magnetites included in clinopyroxene (A) and in groundmass (B) of an olivine–pyroxene-basalt from the island of Rendova

	A	B
n	45	45
TiO_2	7.81	9.36
Al_2O_3	5.97	5.20
Fe_2O_3	48.43	45.17
FeO	32.66	36.34
MnO	0.42	0.54
MgO	4.09	2.51
Total	99.38	99.12
Tot. Fe as FeO	76.24	76.98
$Fe_2O_3/(FeO + Fe_2O_3)$	0.60	0.55
V	3990	4284
Cr	731	1030
Mn	3237	4195
Ni	402	234
Cu	42	105
Zn	407	508

n = number of electron-microprobe analyses; major elements (TiO_2 to MgO) as per cent by mass; minor and trace elements (V–Zn) as parts per million by mass.

Table 5.8 Variation in spinel composition within a single lava

No.	1	2	3	4	5
TiO_2	6.18	4.13	3.82	0.60	6.87
Al_2O_3	3.39	4.67	5.34	3.77	3.06
V_2O_3	0.72	0.36	0.31	bd	0.64
Cr_2O_3	8.61	19.59	16.91	53.60	6.17
Fe_2O_3	45.64	38.57	40.02	12.55	46.87
FeO	29.95	26.51	27.32	21.22	31.60
NiO	0.23	bd	bd	bd	bd
MnO	0.45	0.21	0.28	bd	0.32
MgO	4.40	6.06	5.58	7.59	4.22
Total	99.57	100.09	99.57	99.32	99.75
Ti	0.169	0.110	0.102	0.016	0.188
Al	0.145	0.195	0.223	0.155	0.131
V	0.021	0.010	0.009	–	0.019
Cr	0.247	0.548	0.473	1.482	0.177
Fe^{3+}	1.248	1.027	1.066	0.330	1.280
Fe^{2+}	0.910	0.784	0.809	0.621	0.959
Ni	0.007	–	–	–	–
Mn	0.014	0.006	0.008	–	0.010
Mg	0.239	0.319	0.294	0.396	0.228
Total	3.000	3.000	2.984	3.000	2.992

bd = below detection limit

48 Ore Elements in Arc Lavas

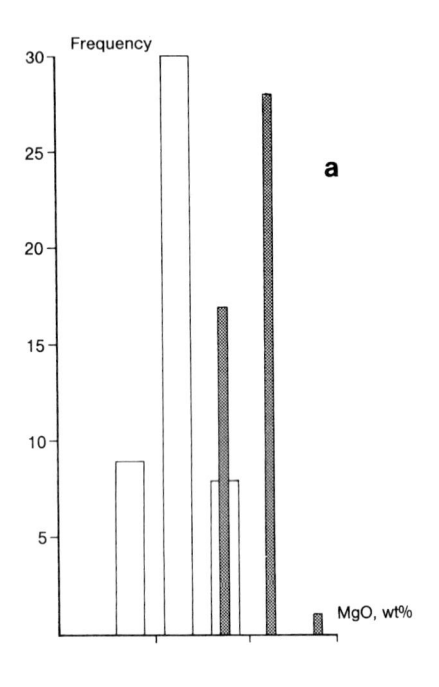

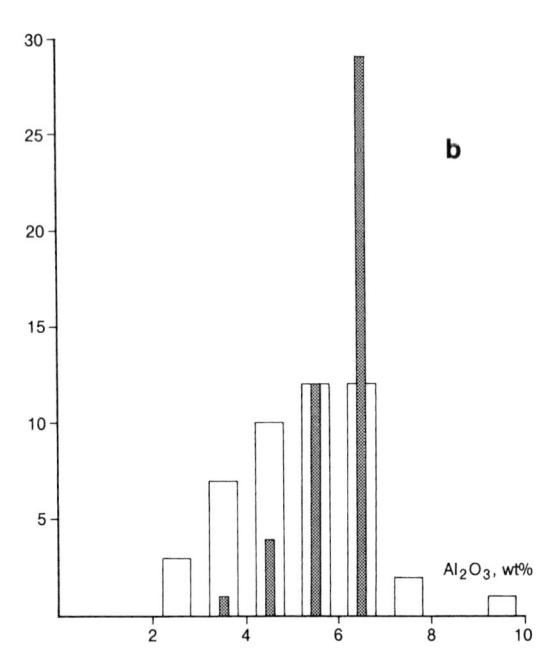

Fig. 5.18. Frequency distributions of abundances of (a) MgO and (b) Al$_2$O$_3$ in 45 spinels included in large clinopyroxene crystals (filled bars) and 45 spinels included in the coexisting groundmass (open bars) of an ankaramitic basalt from the island of Rendova.

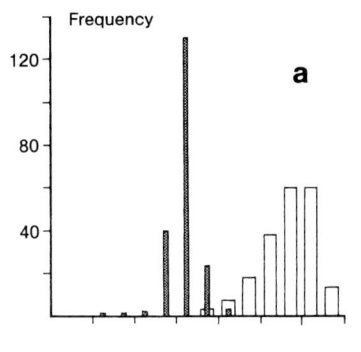

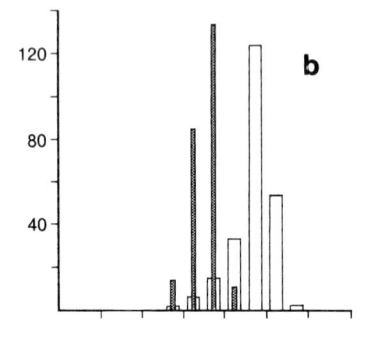

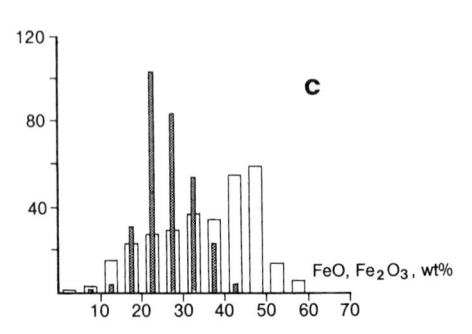

Fig. 5.19. Frequency distributions of abundances of FeO (filled bars) and Fe$_2$O$_3$ (open bars) in the spinels of (a) the hornblende-andesite group, (b) big-feldspar lavas, and (c) olivine–clinopyroxene-basalts of the Solomons suite. All analyses by electron-microprobe, with FeO and Fe$_2$O$_3$ determined by recalculation of the structural formula on the basis of 4O.

is that of a chromite, whereas that of 5 represents a chromian magnetite. The two grains in question were 0.4 mm apart. Such large compositional differences in nearly juxtaposed particles appears to indicate differences in timing of crystallization, metastable preservation, and subsequent turbulent mixing.

Groundmass. The groundmass shows a strong tendency towards a wholly to largely glassy state in the olivine–pyroxene- and big-feldspar basalts, and to a generally finely crystalline one in the hornblende-andesites. This change in groundmass crystallinity coincides quite closely with the onset of the calc-alkaline trend in the lava series as a whole, with the development of copious feldspar, and the change in feldspar textures already referred to. Small quantities of hornblende and clinopyroxene appear in the hornblende-andesite groundmasses, but these are overwhelmingly dominated by fine feldspar.

The oxidation indices and potassium contents of the groundmasses are indicated in Figs. 5.7 and 5.9 respectively. Electron-probe analysis of small areas of glass using a broadened beam indicates considerable heterogeneity on a small scale in many of the lavas. Heavy liquid separation and analysis of bulk groundmass thus yields what are simply averages, and the data in Figs. 5.7 and 5.9 reflect this. The patchy, small-scale variation of groundmass composition commonly involves the local development of very high levels of potassium, and K_2O determinations of 8–10 per cent are not uncommon.

LAVA NOMENCLATURE

The classification and genetic grouping of igneous rocks has had a long and complex history and, as part of this, efforts to identify and understand the great variety of volcanic associations and lineages continue as intensely as ever to the present day. Beginning, perhaps, with the classical work of Bowen (1928), many distinguished petrologists have contributed, and the identification and elucidation of the world's principal lava associations unquestionably constitutes one of the great fields of endeavour in the natural sciences.

Any comprehensive review of such a large area of investigation is, however, far beyond the scope of the present book. Although some reference to the evolution of ideas concerning volcanic-arc petrogenesis is made in Chapters 23 and 24, present concern is with the materials of the lavas rather than with their classification, associations, and derivation. For this the interested reader is referred, if necessary, to modern compilations such as that of Carmichael *et al.* (1974).

Much of the presentation and discussion of the following chapters does, however, require a reasonably well-defined nomenclature, and for this reason it seems worth while to attempt to 'sharpen' or even slightly redefine, two widely, but perhaps somewhat loosely, used petrological terms — and to add two new ones. The criteria applied are based essentially on features of chemical composition.

The two currently — and widely — used terms are 'calc-alkaline' (or 'calc-alkalic') and 'tholeiitic'. Something of the history of this nomenclature and of the ideas behind it is given in Chapter 23, to which such concerns are relevant. At the present stage, however, we are interested only in unambiguous definition of lava groupings for the purposes of the geochemical chapters (6–20) that follow. We shall do this substantially in terms of the FMA [i.e. (total Fe as FeO): MgO: ($Na_2O + K_2O$)] ternary diagram.

The general area and trend of lavas termed *calc-alkaline* conforms closely to the basalt–andesite–dacite–rhyolite series of Daly (1933) and subsequent authors, and approximates (as a relatively thin zone) to the line basalt–rhyolite (*BR*) in Fig. 5.20(a). In many volcanic provinces it is coupled with a trend basalt–olivine-basalt–picrite (*BPi*) which, from one province to another, extends for varying distances towards the high-magnesium field of the picrites. Clearly the spectrum of lavas of the Solomon Islands Younger Volcanic Suite provides an excellent example of such a calc-alkaline series:basalt–olivine-basalt–picrite series pairing, and the usual FMA relations between the two (Fig. 5.4). Such calc-alkaline lavas have what is regarded as a 'moderate' total iron content and the basaltic hinge area of the two series lies around the central area of the FMA diagram.

In contrast to these, the lava series termed *tholeiitic* trend into the high-iron region of the FMA diagram as indicated by the line B–Tb in Fig. 5.20(b). Total iron as FeO may reach 17 per cent or more, as compared with ≈10–12 per cent in calc-alkaline basalts. The FMA relationship of tholeiitic basalt lineages with the trend of any associated olivine and picritic basalts approximates to a straight line, in contrast to the inflected relationship developed between calc-alkaline lineages and the basalt–picrite trend. The tholeiitic basalts may also be complemented by andesites, dacites, and rhyolites, in this case trending from the region of the high-iron basalts down the zone tholeiitic basalt–rhyolite (*Tb–R*) of Fig. 5.20(b). These are often referred to as tholeiitic andesites, etc.

This brings us to the principles to which the writer has drawn attention on several previous occasions (Stanton 1967, 1978; Stanton and Ramsay 1980).

If, as depicted in Fig. 5.20(c), a parent basalt *P* begins to undergo crystallization and segregation of olivine (and, in many cases, Mg-rich clinopyroxene), there will develop a series of accumulates (basalts + olivine + clinopyroxene) trending in the direction *P–A*, and a series of derivatives (basalt minus olivine, etc.) trending in the direction *P–D*. If this proceeds in a closed system (i.e. with no gains or losses from or to its surroundings), the parent *P* and any given daughter pair *A* and *D* lie in the straight line *APD* and the mass ratio of *A* to *D* is as *DP:PA*. The first is a

50 *Ore Elements in Arc Lavas*

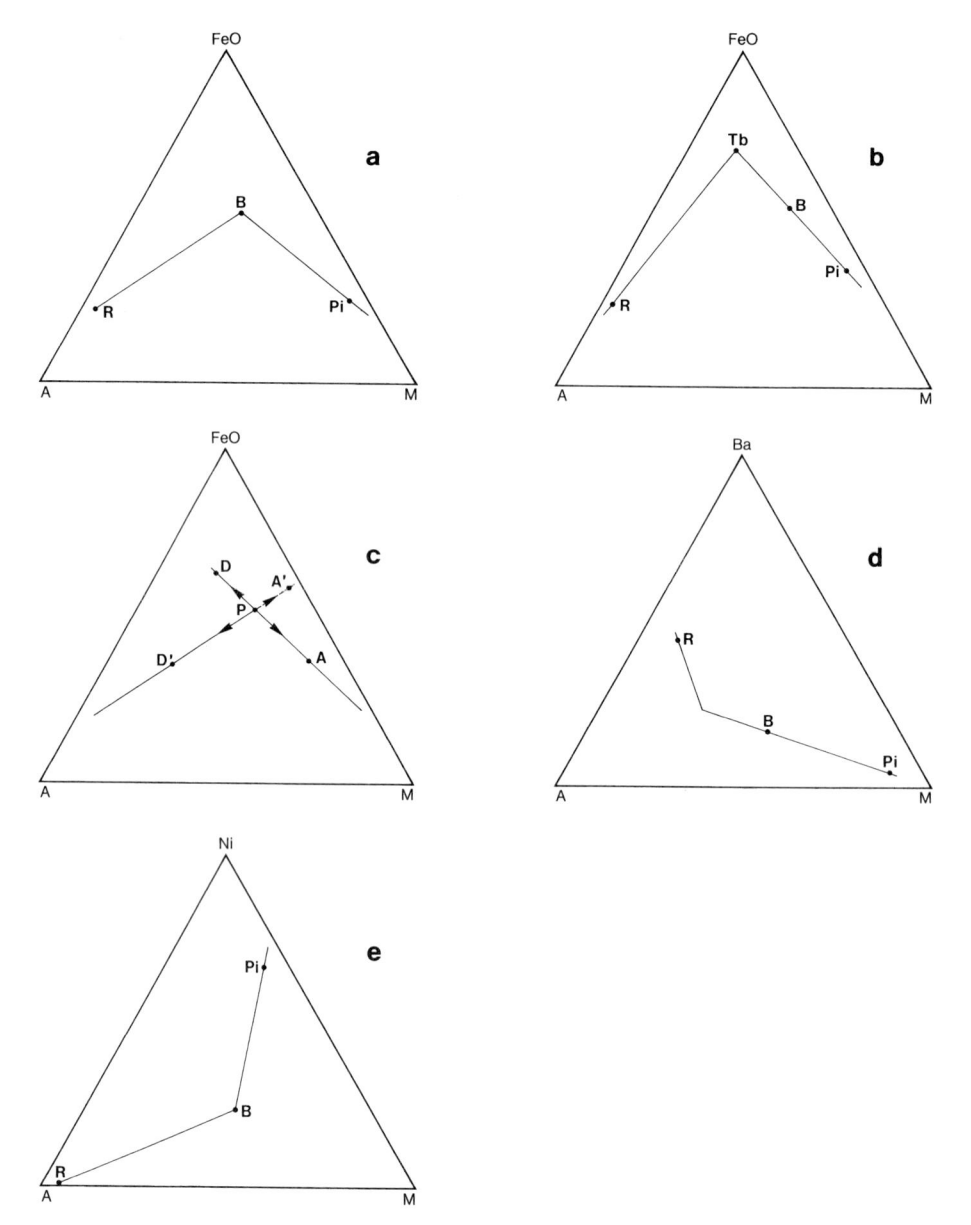

Fig. 5.20. (a) The two essentially linear basalt–picrite ('high-magnesium', *BPi*) and basalt–rhyolite ('calc-alkaline', *BR*) trends commonly developed by the ternary FeO–MgO–(Na$_2$O + K$_2$O) system in island-arc (and many other) lava suites; (b) the conspicuously FeO-enriched basalt–tholeiitic basalt (*BTb*) and tholeiitic basalt–tholeiitic andesite-rhyolite (*TbR*) trends developed in evolved tholeiitic basaltic provinces, such as that of Iceland; and (c) simplified FMA diagram showing straight-line relationship that must develop between parent *P*, accumulate *A*, and derivative *D* where *A* and *D* form by fractionation *in a closed system*, and the trend of accumulate *PA'* that would be the inevitable consequence of the formation of derivative *D'* from parent *P* *in a closed system* (note that *A*, *A'*, *D*, and *D'* may consist of several mineral components, but that the *sum of these* must fall along the trends *PD*, *PA*, etc.). The overall patterns of (a) and (b) are hypotholeiitic, and it may be seen from (c) that they may in certain cases represent the accumulative component of one distinct volcanic lineage and the derivative component of another. The upwardly inflected trend of (d) is a simplified representation of a hypertholeiitic pattern (using the Ba–MgO–(Na$_2$O + K$_2$O) system as an example), and that of (e) a simplified representation of a nichrome pattern (using the Ni–MgO–(Na$_2$O + K$_2$O) system as the example). All these patterns are developed with respect to different elements in the lava suites of volcanic island arcs.

simple property, the second a simple corollary, of this kind of diagram. The closed-system fractionation of Mg-rich olivine from a basalt such as that of *B* in Fig. 5.20(a) will tend to yield olivine-rich and ultimately picritic accumulates, and a series of increasingly tholeiitic basalts as shown in Fig. 5.20(b) and (c). Analogously the generation of the 'calc-alkaline' andesite–dacite–rhyolite series trending in the direction *P–D'* by subtraction of, say orthopyroxene, or a combination of clinopyroxene and magnetite (hornblende is arithmetically inappropriate as it contains about the same amount of $Na_2O + K_2O$ as any likely parental basalt) would yield a series of accumulates trending in the direction *P–A'*, as in Fig. 5.20(c). This is referred to again in more detail in Chapters 23 and 24.

The substantially linear and commonly extensive (high-Mg:low-Fe; low-Mg:high-Fe) arrays of points generated by many tholeiitic lava series conforms very well with this closed system model. The highly linear arrays generated by MORBs (e.g. Fig. 17.1(e); see also Fig. 24.1 depicting the extreme development of tholeiitic characteristics in MORBs of the Galapagos Ridge), the Icelandic suite (Fig. 17.1(f)), the Solomon Island Older Volcanic Suite (Stanton and Ramsay 1980), and many other analogous lava series appear to conform well with the general pattern to be expected from extensive but constant closed-system olivine–clinopyroxene fractionation of a basaltic parent.

For the purposes of the present study we shall therefore define as tholeiitic those lava series that, on ternary chemical composition (e.g. FMA) diagrams, approximate to linear arrays attributable to substantial olivine: clinopyroxene fractionation in common basaltic parents, and in which the derivative (*D* of Fig. 5.20(c)) array is extensively developed.

Thus, on the FMA diagram itself such lavas generate essentially linear fields of points extending from olivine basalts (including, in some cases, picrites), through common basaltic compositions ($\approx$ 10 per cent Fe as FeO) to the region of high iron tholeiitic basalts (15–17 per cent Fe as FeO). Such arrays *are attributable, on purely arithmetical grounds*, to olivine:clinopyroxene fractionation in a closed system, and *on this criterion alone* we now refer to them as being *tholeiitic*.

Linear fields of points diverging from and extending *below* the general trends of tholeiitic lavas on the FMA and analogous diagrams i.e. the various basalt–andesite–dacite–rhyolite series ranging from calc-alkaline to more iron-rich trends (Figs. 5.20(a), (b), and (c)) are now referred to as *hypotholeiitic* (*hypo*, from the Greek *hupo* = under, below).

Linear fields of points diverging from and extending *above* the general trends of tholeiitic lavas in the relevant ternary diagrams (Fig. 5.20(d)) are termed *hypertholeiitic* (*hyper*, from the Greek *huper* = over, above).

Doubly linear or curved (upwardly concave) fields of points extending from the right side towards the lower left apex of the ternary triangle as in Fig. 5.20(e) are henceforth referred to as exhibiting *nichrome* trends — so named because they are typically generated by many lava series when chromium or nickel are substituted for iron in the ternary diagrams (see Chapters 15 and 19).

All four definitions are thus based entirely on chemistry and arithmetic: the graphical representation of chemical compositions. They are quite independent of qualitative or subjective features of mineralogy and are free of genetic connotations. Above all, they are independent of current complex, differing, and somewhat vague conceptions of 'calc-alkaline' and 'tholeiitic'.

While such a nomenclature has the qualities of simplicity and unambiguousness, it has, however, one slightly complicating corollary that must be clearly recognized. As will be seen in the chapters that follow, when different elements are substituted for FeO in the ternary plot, a single lava series may show a variety of abundance patterns from one element to another. For example, the Solomon Islands Younger Volcanic Suite exhibits a very close approximation to a straight-line, closed system, tholeiitic character when Sr takes the place of FeO in the ternary diagram. However, FeO, MnO, and Zn generate very similar, well-developed, hypotholeiitic trends. Ba and Rb produce clear and quite unequivocal hypertholeiitic patterns. And finally Ni and Cr generate the characteristic nichrome trends referred to above. That is, a single lava suite may exhibit the whole gamut of 'trends' from one element to another.

Our nomenclature therefore requires that we specify that a particular lava suite exhibits a particular characteristic (tholeiitic, hypotholeiitic, hypertholeiitic, etc.) *with respect to a particular element*. Thus the Solomons Suite, for example, is tholeiitic with respect to Sr, hypotholeiitic with respect to Zn, hypertholeiitic with respect to Ba, and shows a very well developed nichrome trend with respect to Ni.

CONCLUDING STATEMENT

The Solomon Islands Younger Volcanic Suite thus constitutes an extensive spectrum of lava types from picrites to felsic hornblende-andesites and occasional biotite-bearing dacites, with basalts (48 <SiO_2 <52 per cent) volumetrically the most abundant. When its occurrences on Bougainville are taken into account, andesites (54 <SiO_2 <58 per cent) are seen to be volumetrically the most abundant lavas. Virtually all the lavas are highly porphyritic, the mineral chemistry of the abundant phenocrysts reflecting in many respects the long evolutionary path of the lava series in which they occur.

This outline — brief, but it is hoped not too superficial — account of the principal petrographic and mineralogical features of the Suite provides a basis for the following examination of the behaviour of the 'ore-element' traces as the series of lavas has developed.

6

Copper

With iron, copper is ubiquitous among the sulphides of volcanic-associated exhalative ores. It is also by far the dominant non-ferrous sulphide element of the sub-volcanic porphyry copper deposits, and of the feeder vein systems that underlie many exhalative deposits. The overwhelmingly dominant mineral is chalcopyrite ($CuFeS_2$), but this may be associated with small quantities of bornite (Cu_5FeS_4) in porphyry and Cyprus-type ores, and occasionally by traces of cubanite ($CuFe_2S_3$) and valeriite ($Cu_3Fe_4S_7$). Any covellite (CuS) or chalcocite (Cu_2S) in these ores is usually secondary. As noted in Chapter 2, copper — chiefly as chalcopyrite — may dominate exhalative deposits ranging from those of basalt to rhyolite and bimodal associations. It is characteristically the principal non-ferrous metal of basalt-associated Cyprus-type ores; in deposits of andesitic to rhyolitic association it is mostly commonly overshadowed by zinc and then lead.

In his book *The data of geochemistry* Clarke (1924) commented that minute traces of copper 'are often detected in igneous rocks, although they are rarely determined quantitatively'. One of the first to investigate the incidence of copper in volcanic rocks was F. F. Grout (1910), who estimated 290–320 ppm in a Keweenawan dolerite of Minnesota. Steiger (1924) determined a mean of 155 ppm in Hawaiian lavas, principally basalts, and Wells (1924) estimated what now appears to be the very high figure of 370 ppm in Columbia River basalts. Sandell and Goldich (1934), in what may be regarded as the first modern study of trace 'ore' metals in igneous rocks, found a generally inverse relationship between whole-rock silica and copper in a range of American magmatic rocks, chiefly lavas, as follows: mean SiO_2 = 48.5 per cent, mean Cu = 149 ppm; mean SiO_2 = 62 per cent, mean Cu = 38 ppm; mean SiO_2 = 72 per cent, mean Cu = 16 ppm. At the same time they estimated an average copper content of 74 ppm for 'American magmatic rocks' — rather lower than earlier estimates of 93 ppm by Clarke and Steiger (1914) and of 100 ppm by Clarke and Washington (1924).

Almost simultaneously with the work of Sandell and Goldich, Wager and Mitchell (1951) published the first preliminary results of their major work on the incidence of trace elements of the highly differentiated Skaergaard gabbroic intrusion of East Greenland. In their monumental paper (Wager and Mitchell 1951) on the distribution of trace elements during the strong fractionation of the Skaergaard melt, these investigators determined 80–150 ppm Cu (average 130 ppm) for the 'chilled marginal gabbro' — the apparent parental magma. Copper contents in the lower, more mafic layers ranged from 10 ppm (picrite gabbro) to 100 ppm, and in the higher, less mafic units, were generally in the range 150–400 ppm, with one as high as 700 ppm. These higher layers were, however, complicated by the presence of readily identifiable copper sulphides, and hence did not simply reflect the silicate:oxide abundances (see Wager *et al.* (1957) of present concern.

In a closely following contribution on the incidence of trace elements in a suite of Hawaiian lavas, Wager and Mitchell (1953) gave copper contents of a small suite ranging from picrite basalt to sodium-rich trachyte, and compared these with the principal Skaergaard layers and some of the Tertiary lavas of Skye (Table 6.1). The relatively high copper, ≈100–200 ppm, in the Hawaiian basaltic rocks, and its apparently sudden decrease to less than the lower limit of detection, i.e. <10 ppm, in the andesites and more felsic derivatives, are striking features of Wager and Mitchell's results.

At about the same time Patterson (1952) published trace analyses of a small suite of Tertiary basalts and their differentiates occurring in north-eastern Ireland. His determinations gave notably high copper contents: olivine-basalt, 400 ppm; tholeiitic basalt 180 ppm; quartz-trachyte, 50 ppm; rhyolite, 66 ppm.

The work of Wager and Mitchell, Patterson, and D. J. Swaine (the latter three all from the Macaulay Institute, Aberdeen) and their contemporaries may be said to constitute the beginning of the post-war era of trace analysis, and a large bank of data has accumulated since. Optical emission spectrography has been superseded by X-ray fluorescence and other more refined methods, and it appears that with this, copper estimations in basalts and associated lavas have reverted from the relatively high

54   Ore Elements in Arc Lavas

Table 6.1 Major and trace-element abundances in selected lavas of Hawaii and Skye, and observed and calculated abundances in Skaergaard materials

	Hawaii					Skye			Skaergaard						
	1	2	3	4	5	6	7	8	9	10	11	12	13	14	15
SiO$_2$	48.26	51.92	50.68	52.13	62.02				41.27	47.92					
TiO$_2$	2.10	3.21	2.64	2.58	0.31				1.54	1.40					
Al$_2$O$_3$	9.23	12.21	16.42	16.68	18.71				8.71	18.87					
Fe$_2$O$_3$	1.37	2.12	5.79	3.13	4.30				2.69	1.18					
FeO	10.36	9.17	6.22	7.32	0.10				10.52	8.65					
MnO	0.11	0.12	0.22	0.14	0.15										
MgO	18.85	7.86	4.25	3.16	0.40				27.09	7.82					
CaO	7.67	10.17	6.47	6.25	0.86				6.59	10.46					
Na$_2$O	1.59	2.36	4.70	5.52	6.90				0.69	2.44					
K$_2$O	0.39	0.39	2.16	2.18	4.93				0.13	0.19					
P$_2$O$_5$	0.24	0.28	0.17	0.94	0.24										
Cr	1700	470	–	–	20	–	–	–	1500	170	75	–	–	5	14
V	250	300	70	20	–	50	7	15	120	140	220	10	–	4	18
Ni	1000	80	6	12	15	3	15	15	1000	170	50	–	–	5	8
Co	70	33	14	5	2	20	3	2	90	53	35	22	13	5	5
Cu	150	170	<10	<10	<10	<10	<10	<10	100	130	220	360	440	260	20
Sr	300	800	2500	3500	100	1000	800	250	100	350	500	470	430	400	350
Ba	100	120	600	1000	800	1500	2000	3000	10	45	90	170	230	630	1400
Rb	–	–	35	50	300	20	100	150	–	–	–	10	20	50	160
Zr	70	100	350	1200	1500	200	2000	1500	30	50	80	90	180	550	850

1, Mean picrite basalt; **2**, mean basalt; **3**, andesine-andesite; **4**, mean oligoclase andesite; **5**, soda trachyte; (**1**)–(**5**), all from Hawaii; **6**, mugearite; **7**, mugearite; **8**, trachyte; (**6**)–(**8**), all from the Isle of Skye; **9**, mean gabbro picrite; **10**, chilled marginal gabbro (initial liquid), both from the Skaergaard intrusion; **11**–**15**, calculated liquids 1–5, Skaergaard intrusion. All from Wager and Mitchell (1951, 1953).

Table 6.2 Average concentrations of the 14 'ore elements' in mid-ocean ridge basalts (MORBs) and Icelandic basalts

	n	TiO$_2$	FeO*	MnO	CaO	P$_2$O$_5$	Ba	Sr	Pb	V	Cr	Co	Ni	Cu	Zn
Mid-Atlantic Ridge, mean	199	1.34	9.32	0.18	11.70	0.20	56	157	nd	244	355	44	154	66	72
Proximate Amer.–Antarctic ridge	19	1.63	9.42	0.18	10.89	0.26	33	181	nd	235	265	46	109	56	82
Southern extremity, MAR	5	1.74	9.91	0.16	10.83	nd	57	165	nd	243	304	51	135	60	105
Lat. 54.5–51°S	40	1.55	10.29	0.18	11.09	0.18	50	132	nd	269	297	49	143	59	86
Lat. 30°S	7	1.21	9.16	0.15	12.16	nd	15	125	nd	271	566	49	174	102	nd
Lat. 12.4–16.34°N	7	1.52	nd	0.17	11.18	0.19	nd	nd	nd	282	339	42	nd	nd	nd
Lat. 30°N	7	1.37	10.08	nd	10.96	nd	11	113	nd	nd	nd	nd	154	nd	nd
Lat. ~35°N ('Famous' area)	2	0.51	8.51	0.16	12.80	nd	13	80	nd	143	505	45	185	104	nd
Lat. 36°49'N	22	1.06	9.18	0.19	11.85	nd	40	101	nd	202	475	44	186	80	nd
Lat. 36°50'N	29	1.01	8.92	0.17	12.70	0.13	nd	nd	nd	223	380	42	165	78	66
Lat. 36°50'N (off-ridge)	24	0.87	8.33	0.15	12.70	0.08	60	111	nd	nd	395	nd	151	66	65
Lat. 40°N	3	0.85	10.21	0.17	12.35	0.08	19	82	nd	240	401	nd	135	nd	nd
Lat. 43°N	14	1.54	8.93	0.17	11.58	0.28	171	306	nd	261	509	45	194	66	nd
Lat. 45°50'N	7	1.39	8.55	0.15	10.40	0.19	136	280	nd	297	469	36	277	59	nd
Iceland	141	2.01	12.29	0.20	11.38	0.20	117	194	nd	323	260	53	105	102	94
Atlantic–South-west Indian Ridge	112	2.02	10.08	0.18	10.53	0.27	98	248	nd	256	163	44	79	62	98
Indian Ocean ridges	135	1.65	10.04	0.16	10.59	0.24	128	217	3.4	302	214	47	90	91	84
Eastern Pacific ridges	120	1.59	9.61	0.16	11.58	0.34	157	207	nd	289	283	37	101	61	89
All mid-ocean ridge basalts, including Iceland	707	1.68	10.27	0.18	11.16	0.23	95	200	nd	272	266	45	112	75	87
All mid-ocean ridge basalts, excluding Iceland	566	1.60	9.76	0.17	11.10	0.24	92	202	nd	263	268	44	114	71	84

FeO* = total iron as FeO; n = number of lava samples analysed; id = insufficient data for reliable estimate; nd = no data available; major oxides (TiO$_2$ to P$_2$O$_5$) given as per cent by mass; trace elements (Ba to Zn) given as parts per million (ppm) by mass. Total of 199 analyses of Mid-Atlantic ridge basalts includes 13 miscellaneous analyses not included in the list of sampling sites. NB. Number of available analyses varies from one element to another: e.g. determinations of TiO$_2$ were available for 141 Icelandic basalts, but Cu for only 54 and Zn for only 99. MAR denotes Mid-Atlantic Ridge.

values obtained by emission methods to the general range of values obtained by Sandell and Goldich (1943) and others using either techniques.

Wedepohl (1978) gives an arithmetic mean for 1674 'basalts' as 90 ppm for copper, with the mode for the relevant frequency distribution in the 40–60 ppm range. Some of the lavas included in Wedepohl's calculations have been metamorphosed; an unweighted arithmetical mean for the unmetamorphosed segment of his sample (1097 analyses) is 88 ppm copper. Wedepohl (1978) also gives arithmetic means of 55 ppm Cu for 230 andesites and 9 ppm Cu for 121 dacites, rhyolites, and obsidians.

COPPER IN LAVAS OF THE MARINE ENVIRONMENT

There is now a substantial body of data on the incidence of copper in mid-ocean ridge basalts (MORBs), and a summary of this is given in Table 6.2. The mean for 415 basalt analyses (SiO_2 <52 per cent) from the Atlantic, Pacific, and Indian Ocean ridges is 71 ppm, a figure remarkably close to that obtained by Sandell and Goldich (1943) for the broad range of mafic to felsic magmatic rocks of the North American continent. Weighted means for each of the Atlantic, Pacific, and Indian MORBs are 66, 61, and 91 ppm respectively, and for Iceland, which represents at least some sub-aerial emission, 102 ppm. Individual analyses are variable, for the most part in the 40–80 ppm Cu range, as indicated in Fig. 6.1(a). The general variation along a major ridge is indicated, using the Mid-Atlantic Ridge as an example, in Table 6.2. Copper contents of Icelandic lavas more felsic than basalts (i.e. $SiO_2 \geqslant 52$ per cent) are indicated by open bars in Fig. 6.2(b). Although the number of samples involved (9) is not large, the indicated mean copper content, 15 ppm, is very significantly lower than that of the basalts.

There is also now a considerable body of information on the copper content of basalts and more felsic lavas of volcanic island arcs. Brown *et al.* (1977) carried out major and trace-element analyses of 1518 rocks of the Lesser Antilles, including quite comprehensive samplings of the major islands Grenada, Dominica, and St Kitts. Ewart, Kay, and Gorton and Eggins have made substantial contributions on Tonga–Kermadec, the Aleutian Islands, and Vanuatu respectively, so that there is now a significant bank of data on a range of arcs from entirely intra-oceanic provinces (Vanuatu) to variably ensialic provinces (the southern extremity of the Tonga–Kermadec ridge, and the north-eastern Aleutian Islands).

The copper contents of the various categories and groups of arc lavas are given in Table 6.3 and Fig. 6.2(a) and (b). The average copper content of basalts (SiO_2 <52 per cent) of Grenada, Dominica, and St Kitts combined is 105 ppm; that of the relevant andesites (52 per cent <SiO_2 <62 per cent) is 65 ppm, and of the more felsic types (principally dacites), 42 ppm. Averages yielded by rather less comprehensive data from the Aleutian Islands are 100 ppm copper in the basalts and 43 ppm in the andesites; from Vanuatu, 138 ppm copper in the basalts, 80 ppm in the andesites, and 38 ppm in the dacites. On the basis of this data the abundance behaviour of copper in these three provinces thus corresponds in principle with the earlier finding of Sandell and Goldich (1943) — that copper in magmatic rocks tends to decrease with increase in whole-rock SiO_2. The copper content of the Tonga–Kermadec lavas does not, however, appear to behave in quite this way; averages obtained from the data of Ewart and others indicates the basalts to contain 104 ppm copper, the andesites 130 ppm, and the dacites and more felsic types 23 ppm. That is, copper here does not simply decrease with increase in the felsic nature of the lavas concerned; it appears to develop a peak in rocks within the andesite range. This is considered in more

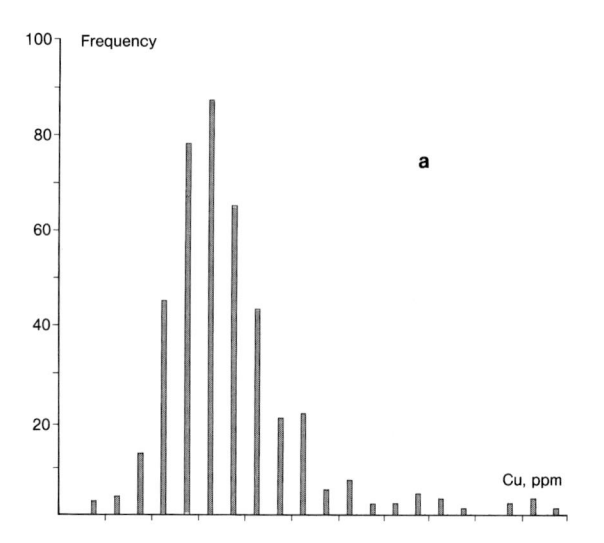

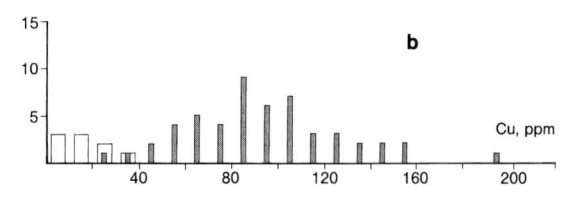

Fig. 6.1. Frequency distributions of copper abundances in (a) 415 MORBs; (b) 54 Icelandic basalts (SiO_2 < 52 per cent; filled bars); and nine Icelandic lavas more felsic than basalts ($SiO_2 \geqslant 52$ per cent; open bars).

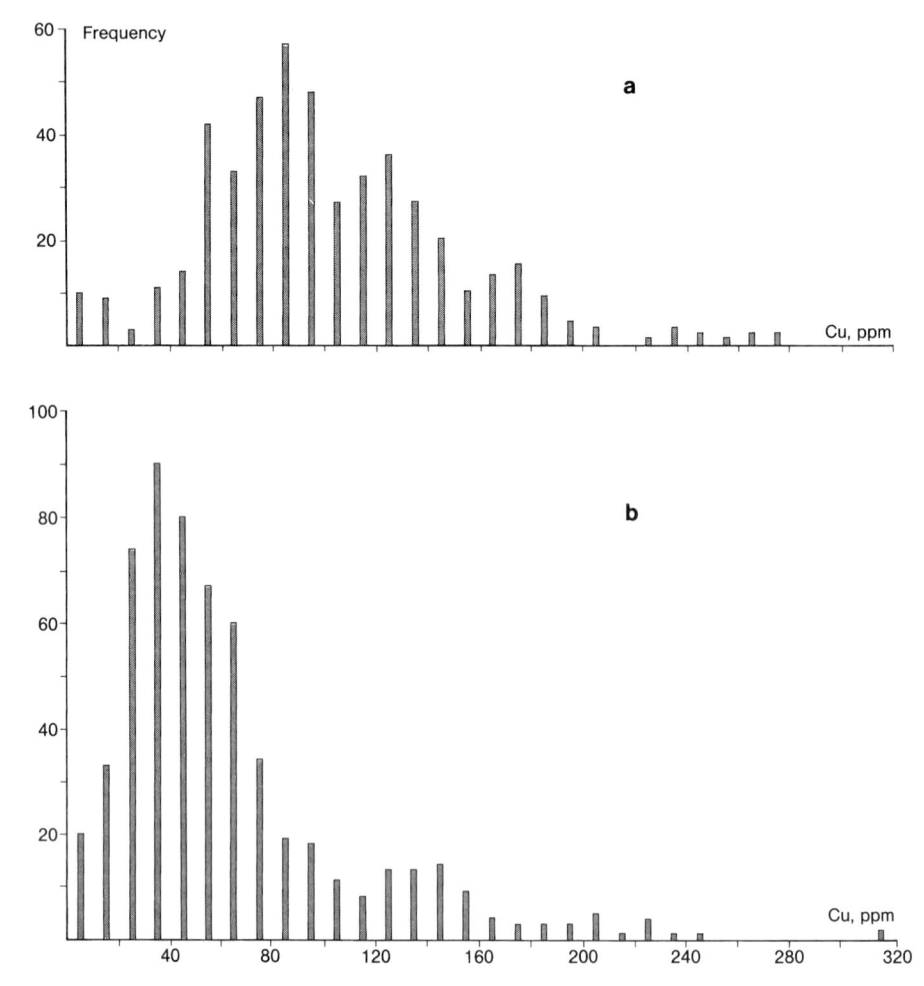

Fig. 6.2. Frequency distributions of copper abundances in (a) 490 island arc basalts ($SiO_2 < 52$ per cent) and (b) 585 island-arc lavas more felsic than basalts ($SiO_2 \geq 52$ per cent).

detail on p. 73, where it is shown that this pattern of behaviour is in fact a general feature of all five arc lava series.

It is therefore apparent that for the group of lavas that have been most comprehensively sampled — the basalts — those of the island-arc provinces have a somewhat higher copper content (≈ 108 ppm) than do MORBs (≈ 71 ppm), and a rather greater spread of abundances. In both types of province, and that of arcs particularly, the development of more felsic lava types leads to an increase in variability, and a general decrease in absolute abundance, of copper.

CHEMICAL PROPERTIES AND CRYSTAL CHEMISTRY OF COPPER

Copper is a notably chalcophile element, leading Goldschmidt to observe that the formation of sulphides so dominates the geochemistry of this element that 'it may even be doubted whether any appreciable amount of copper in a truly ionic state enters the usual silicates of magmatic rocks such as olivines, pyroxenes, amphiboles and biotite, which could accommodate ions of the size of cupric copper' (Goldschmidt 1954, p. 176). Detailed investigation of the principal mineral species of the

Table 6.3 Average concentrations of the 14 ore elements in modern island-arc lavas

	n	TiO$_2$	FeO*	MnO	CaO	P$_2$O$_5$	Ba	Sr	Pb	V	Cr	Co	Ni	Cu	Zn
Solomon Is Younger Volcanic Suite, excluding Bougainville															
Basalts (all lavas of SiO$_2$ <52%)	64	0.62	9.34	0.20	9.72	0.22	152	523	2.9	235	606	59	351	100	75
All lavas with SiO$_2$ >52%	58	0.65	6.27	0.13	7.32	0.19	344	702	5.4	187	29	22	19	87	67
Andesites (52<SiO$_2$<62%)	49	0.70	6.88	0.14	7.97	0.20	261	632	4.4	210	32	25	21	102	71
Dacites (SiO$_2$ >62%)	9	0.34	2.93	0.07	3.77	0.14	798	1082	10.9	62	10	8	8	19	46
Solomon Is Younger Volcanic Suite, cum Bougainville															
Basalts	69	0.63	9.26	0.25	9.79	0.22	156	529	3.0	235	568	id	329	102	75
All lavas with SiO$_2$ >52%	172	0.69	6.23	0.15	7.05	0.26	336	733	5.4	164	16	id	10	77	64
Andesites	158	0.72	6.49	0.16	7.31	0.27	308	715	5.0	172	17	id	11	82	66
Dacites	14	0.40	3.36	0.10	4.23	0.16	650	944	10.2	72	11	id	7	24	46
Tonga–Kermadec															
Basalts	31	0.86	10.20	0.20	11.38	0.11	91	197	1.6	297	104	35	35	104	81
All lavas with SiO$_2$ >52%	66	0.67	8.32	0.17	7.71	0.12	190	207	2.3	206	19	24	10	98	97
Andesites	40	0.74	9.71	0.19	9.58	0.09	144	201	1.9	294	28	30	15	130	93
Dacites	26	0.56	6.03	0.15	4.82	0.16	260	215	3.8	71	3	12	2.5	23	107
Lesser Antilles															
Basalts	177	0.92	9.05	0.17	11.66	0.17	272	679	nd	244	419	nd	183	105	74
All lavas with SiO$_2$ >52%	540	0.63	6.44	0.16	7.28	0.12	287	403	nd	129	112	nd	39	56	66
Andesites	329	0.74	7.19	0.17	8.10	0.13	274	429	nd	141	108	nd	41	65	72
Dacites	211	0.46	5.25	0.14	6.01	0.11	308	362	nd	83	119	nd	35	42	56
Grenada															
Basalts	129	0.91	9.05	0.17	11.79	0.20	333	822	nd	242	551	nd	242	111	76
All lavas with SiO$_2$ >52%	135	0.67	6.27	0.14	7.23	0.16	532	754	nd	138	183	nd	80	59	71
Andesites	101	0.72	6.80	0.15	7.69	0.17	487	742	nd	154	204	nd	90	67	77
Dacites	34	0.53	4.69	0.10	5.88	0.15	667	791	nd	91	120	nd	51	37	55
Aleutians															
Basalts	47	1.22	9.31	0.19	10.15	0.27	350	584	id	270	95	33	50	100	74
All lavas with SiO$_2$ >52%	71	0.67	5.88	0.14	6.38	0.22	566	522	id	id	id	id	id	id	id
Andesites	45	0.76	6.90	0.15	7.69	0.25	574	561	8.9	154	45	19	25	43	59
Dacites	26	0.52	4.12	0.12	3.71	0.17	525	377	id	id	id	id	id	id	id
Vanuatu															
Basalts	170	0.79	10.43	0.20	11.12	0.24	349	615	4.7	355	303	44	97	138	84
All lavas with SiO$_2$ >52%	29	0.76	7.38	0.16	6.38	0.31	477	478	9.8	155	17	20	8	73	89
Andesites	25	0.79	7.58	0.16	6.65	0.33	488	498	9.7	167	20	20	9	80	92
Dacites	4	0.61	6.24	0.15	4.88	0.16	330	367	10.5	94	2	15	0.1	38	79
All arcs, except Solomon Is															
Basalts	425	0.95	9.76	0.19	11.06	0.20	261	491	3.2	299	213	37	86	110	78
All lavas with SiO$_2$ >52%	706	0.69	7.04	0.16	6.98	0.19	367	408	6.1	145	47	22	19	76	77
Andesites	439	0.78	7.90	0.17	8.06	0.20	369	423	5.8	164	52	23	24	81	80
Dacites	267	0.55	5.37	0.14	4.83	0.15	364	343	7.2	80	29	14	10	32	72
All arcs, cum Solomon Is															
Basalts	494	0.89	9.66	0.20	10.80	0.20	240	498	3.1	289	284	43	134	108	77
All lavas with SiO$_2$ >52%	878	0.68	6.75	0.15	6.87	0.19	389	472	6.2	151	35	22	19	70	68
Andesites	597	0.77	7.62	0.17	7.91	0.21	357	481	5.4	167	45	24	21	81	76
Dacites	281	0.52	4.96	0.13	4.71	0.15	421	463	8.2	79	25	12	9	30	59

FeO* = total iron as FeO; n = number of lava samples analysed; id = insufficient data for reliable estimate; nd = no data available; major oxides (TiO$_2$ to P$_2$O$_5$) given as per cent by mass; trace elements (Ba to Zn) given as parts per million (ppm) by mass. All averages for Lesser Antilles (including Grenada) lavas are calculated from the averages of Brown *et al.* (1977).

Solomons Younger Volcanic Suite indicates, however, that it does, and this in a substantially systematic way.

As indicated in Table 6.4, copper has a single s electron outside the filled 3d shell. It occurs as Cu^+ and Cu^{2+}, the relative stabilities of which are indicated by the potential data:

$$Cu^+ + e = Cu \qquad E^\ominus = 0.52 \text{ V}$$

$$Cu^{2+} + e = Cu^+ \qquad E^\ominus = 0.153 \text{ V}$$

whence

$$Cu + Cu^{2+} = 2Cu^+ \quad E^\ominus = -0.37 \text{ V}; \quad K = [Cu^{2+}]/[Cu^+]^2 = \sim 10^6$$

The dipositive — cupric — state is the more important. The d^9 configuration, however, makes Cu^{2+} subject to Jahn–Teller distortion if placed in an environment of cubic (i.e. regular octahedral or tetrahedral) symmetry, and this has a substantial effect on its stereochemistry. When six-coordinate, the coordination polyhedron is severely distorted, the typical distortion being an elongation along one 4-fold axis, resulting in a planar arrangement of four short Cu–ligand bonds and two *trans* long ones (Cotton and Wilkinson 1972). On the basis of valency and ionic size, Fe^{2+} (0.082 nm) and Mn^{2+} (0.091 nm) are the abundant ions of silicate:oxide compounds most likely to be substituted for, and hence to exhibit an abundance relationship

Table 6.4 Physical and chemical properties, and estimated crustal abundances, of the 14 ore elements

Elements	Atomic no.	Atomic wt	Ground state electr. config.	Atomic/ionic radius (nm)	Electronegativity Pauling	Electronegativity Allred–Rochow	Coordination no. w.r.t. O^{2-}	Bond character (% ionic w.r.t. O)	Est. crustal abundance (ppm)
Cu	29	63.55	$[Ar]3d^{10}4s^1$	0.1278	1.90	1.75			68
Cu^+				0.096			6	51	
Cu^{2+}				0.072			6	42	
Zn	30	65.39	$[Ar]3d^{10}4s^2$	0.1332	1.65	1.66			76
Zn^{2+}				0.083			6	63	
Pb	82	207.2	$[Xe]4f^{14}5d^{10}6s^26p^2$	0.1750	2.33	1.55			13
Pb^{2+}				0.132			8–12	58	
Ba	56	137.33	$[Xe]6s^2$	0.2173	0.89	0.97			390
Ba^{2+}				0.143			12	83	
Sr	38	87.62	$[Kr]5s^2$	0.2151	0.95	0.99			384
Sr^{2+}				0.127			8–12	80	
P	15	30.97	$[Ne]3s^23p^3$	0.093 (w)*	2.19	2.06			1120
				0.115 (r)*					
P^{5+}				0.044			2	38	
Ca	20	40.08	$[Ar]4s^2$	0.1973	1.00	1.04			46600
Ca^{2+}				0.106			6–8	80	
Ti	22	47.88	$[Ar]3d^24s^2$	0.1448	1.54	1.32			6320
Ti^{4+}				0.068			6	58	
V	23	50.94	$[Ar]3d^34s^2$	0.1321	1.63	1.45			136
V^{2+}				0.072			6	74	
V^{3+}				0.065			6	66	
V^{4+}				0.061			4–6	58	
V^{5+}				0.059			4–6	51	
Cr	24	51.996	$[Ar]3d^54s^1$	0.1249	1.66	1.56			122
Cr^{3+}				0.064			6	58	
Mn	25	54.94	$[Ar]3d^54s^2$	0.0124	1.55	1.60			1060
Mn^{2+}				0.091			6	63	
Mn^{3+}				0.070			6	63	
Fe	26	55.85	$[Ar]3d^64s^2$	0.1241	1.83	1.64			62000
Fe^{2+}				0.082			6	52	
Fe^{3+}				0.067			4–6	51	
Co	27	58.93	$[Ar]3d^74s^2$	0.1253	1.88	1.70			29
Co^{2+}				0.082			6	54	
Ni	28	58.69	$[Ar]3d^84s^2$	0.1246	1.91	1.75			99
Ni^{2+}				0.078			6	54	

(w) = white phosphorus; (r) = red phosphorus. Values of atomic weights, atomic/ionic radius, electronegativity, and crustal abundance after Emsley: *The elements*, Oxford University Press 1989.

with, Cu^{2+} (0.072 nm). The latter, however, tends to form a more covalent bond with O^{2-} (~45 per cent ionic) than do Fe^{2+} (~52 per cent ionic) and Mn^{2+} (~63 per cent ionic). Thus, while preferential substitution of Fe^{2+} and Mn^{2+} by Cu^{2+} must be expected, this, and any consequent abundance relationships within relevant mineral compounds, is likely to be modified by differences in stereochemistry induced by the Jahn–Teller distortion of Cu^{2+}, and in bond character. Three other transition metal ions subject to Jahn–Teller distortion in octahedral coordination are Cr^{2+}, Mn^{3+}, and Ni^{3+} (Burns 1970). These ions are, however, uncommon in the igneous regime (Cr^{3+}, Mn^{2+}, and Ni^{2+} represent the principal valence states of these three elements in igneous melts) and copper, as Cu^{2+}, is therefore the only important 'ore element' whose mineralogical affinities are likely to be influenced by the Jahn–Teller phenomenon. This may have important implications concerning the behaviour of copper in fractionating lava systems and in the formation of volcanic ores, and is referred to again in Chapters 21 and 24.

The unipositive cuprous state probably has little importance in magmatic mineral formation (see, however, Fyfe (1964) and Candela and Holland (1984) concerning the incidence of the Cu^+ ion relative to Cu^{2+} in less-oxidized igneous melts), and the only abundant silicate element with which it might have substitutional ties is its moderately close associate (in Group I of the Periodic Table), the alkali element sodium. Cu^+ has an electron configuration of $[Ar]3d^{10}$ and an ionic radius of 0.096 nm; Na^+ an electron configuration of $[Ne]3s$, and an ionic radius of 0.098 nm. If significant amounts of Cu^+ existed in a relatively reduced melt it might thus, presumably, substitute for Na^+ in minerals such as the plagioclases and hornblende.

THE INCIDENCE OF COPPER IN THE PRINCIPAL MINERAL SPECIES OF THE SOLOMONS YOUNGER VOLCANIC SUITE

General

A moderately substantial literature on the mineralogical incidence of copper (other than in sulphides) has now accumulated.

Comparatively recent neutron activation analysis of olivines from the Lower Zone of the Skaergaard intrusion gave 21–4 ppm Cu (Vincent 1974). Fifteen volcanic olivines from Tertiary and Quaternary basalts and andesites from the Izu–Hakone district of Japan contained an average of 30 ppm Cu (Iida *et al.* 1961), and two from the tholeiitic Keweenawan basalts of Michigan contained 36 ppm Cu (Cornwall and Rose 1957).

If the relatively small number of pyroxenes analysed so far is an accurate indication, these appear to accommodate slightly more copper than do the olivines. Vincent (1974) obtained 42–59 and 47–59 ppm Cu for the Skaergaard Lower and Middle Zones respectively; Snyder (1959) a range of 44–720, and an average of 217 ppm, Cu for pyroxenes of Duluth Gabbro; Tiller (1959) a range of 49–170 (mean 106) ppm in Tasmanian dolerite pyroxenes; Cornwall and Rose (1957) a range of 38–320 (mean 130) in the pyroxenes of the Keweenawan basalts referred to above; and Ewart and his co-investigators (Ewart and Bryan 1972, 1973; Ewart 1976; Ewart *et al.* 1977) a range of 13–40 (mean 21) ppm Cu in clinopyroxenes and 11–53 (mean 27) ppm Cu in orthopyroxenes of the Tonga–Kermadec volcanic island chain.

Although there is a considerable body of data available on copper in amphiboles, all of it appears to relate to either granitic or metamorphic rocks. The overall average abundance of copper in such amphiboles is ≈30 ppm.

The general abundance of copper in feldspars reported in the literature is, in view of crystal chemical considerations, remarkably high. In his compilation of data on copper in silicates Wedepohl (1974*a*) lists three groups of volcanic feldspars which together have a mean copper content of 44 ppm. Three alkali feldspar phenocrysts from Icelandic 'pitchstones' (Carmichael and Macdonald 1961) contained, however, only 3–7 (mean, 5) ppm Cu. The meticulous work of Vincent (1974) on Skaergaard plagioclases gave 16–28 and 25–28 ppm Cu in separations from the Lower and Middle zones respectively. Ewart and his colleagues (Ewart and Bryan 1972; Ewart *et al.* 1973) found a range of 0–27 ppm and a mean of 11 ppm Cu in 22 feldspars separated from lavas of the Tonga–Kermadec chain. Although, as noted a little further on in this chapter, some of the relatively high values of copper in the feldspars have been accounted for by postulating substitution of Cu^+ for Na^+, the presence of minute, relatively high-copper, glassy inclusions in some feldspars may constitute a better explanation. The real value of intrinsic (i.e. lattice-bound) copper in most volcanic feldspars seems likely to fall in the 5–15 ppm range.

Magnetite is clearly the principal copper receptor among the minerals of present concern (biotite may accept substantial quantities of copper, but this mineral occurs in only minute — and, in the present context, quite insignificant — quantities in the Solomons suite). Five groups of volcanic magnetites, ranging in age from Precambrian to Modern, listed by Wedepohl (1978*a,b*) gave a mean of 175 ppm Cu. Three groups of Tertiary to Modern volcanic magnetites yielded a mean of 76 ppm copper. Five magnetites from the Tonga–Kermadec arc gave a range of 75–205 ppm, and a mean of 134 ppm Cu (Ewart and Bryan 1972; Ewart *et al.* 1973).

60   Ore Elements in Arc Lavas

Solomons suite

Mean values for copper in the principal silicates and in magnetites are given in Table 6.5, and spreads of individual values are depicted in the histograms of Figs. 6.3 and 6.4. Mean values for FeO and MnO (ppm Mn in the case of feldspars) are given for comparison in Table 6.5.

Fe–Mg silicates. The number of samples yielding separable quantities of orthopyroxene was small and their representation in Table 6.5 is barely adequate. This accepted, however, the incidence of copper in the four principal Fe–Mg silicates is remarkably similar. Mean values in olivine and hornblende are virtually identical. Clinopyroxene contains a little less, but the consistency of mean copper in pyroxenes representing the full spectrum of rock types from picritic basalts to felsic hornblende andesites is remarkable. The overall mean value, 31 ppm, is a little higher than the 21 ppm obtained by Ewart for the Tonga–Kermadec clinopyroxenes, but the general level of incidence is quite similar and may well be partly accounted for in the disparity of sample numbers (Solomons, 96; Tonga–Kermadec, 17). The Solomons orthopyroxenes also contain slightly higher copper (35 ppm) than those of Tonga–Kermadec (27 ppm), though again this may be partly a sampling artefact.

The consistency of copper in the clinopyroxenes at first sight appears in keeping with the remarkable consistency of the latter's Mg numbers through a large part of the range of Solomons lava types (cf Fig. 5.3). Inspection of Table 6.5 shows, however, that there is no close tie between copper and incidence of iron (and hence Fe:Mg ratios), or the amount of manganese, in the Fe–Mg silicates overall. Such constancy of copper in these silicates is all the more remarkable for the fact that whole-rock copper changes markedly through the evolution of the rock suite (cf. Figs. 6.10, 6.11, and 6.12). Among the pyroxenes, those of the big-feldspar basalts occur in lavas exhibiting the maximum abundances of whole-rock copper (about four times the level of the hornblende-andesites in the 60–2 per cent silica range), but themselves contain no more copper than the pyroxenes of the picrites and hornblende-andesites. Copper in the olivines and hornblendes similarly appears to bear no relation to the abundance of this metal in the respective coexisting melts. This is referred to again below in connection with partition coefficients.

Such a lack of any apparent tie between copper and the principal components of the Fe–Mg silicates, or with the concentration of copper in the coexisting melt, may indicate that incorporation occurs as a statistical phenomenon — either by random substitutional or interstitial solid solution — and that ≈30–50 ppm represents the

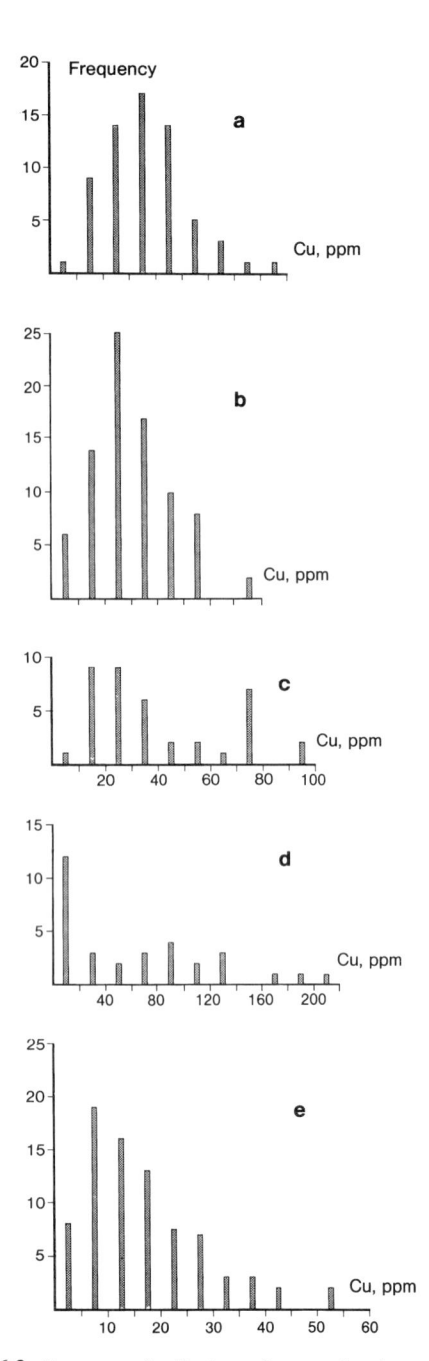

Fig. 6.3. Frequency distributions of copper abundances in magnetic/heavy liquid separations of (a) olivines, (b) clinopyroxenes, (c) hornblendes, and (e) feldspars (ultimate purity separations) of the Solomon Islands Younger Volcanic Suite. The biotite analyses of (d) were carried out using the Cameca-Camebax electron-microprobe at 25 kV, ≈ 50 nA, and a counting time of 160 seconds.

Table 6.5 Mean abundances of the 14 ore elements together with MgO and K$_2$O in the principal minerals of the Solomon Islands Younger Volcanic Suite

Mineral	n	TiO$_2$	FeO*	MnO	MgO	CaO	K$_2$O	P	Ba	Sr	Pb	V	Cr	Mn	Co	Ni	Cu	Zn
Olivine (OPB)	43	0.05	19.77	0.31	43.31	0.20	0.02	132	14	2	nd	8	346	2401	184	1511	36	156
Olivine (BFL)	7	0.03	25.64	0.79	34.51	0.18	0.04	255	18	2	nd	10	83	5838	193	545	51	285
Clinopyroxene (OPB)	43	0.38	5.13	0.13	15.79	21.78	0.05	88	10	71	nd	152	2435	1223	40	370	31	30
Clinopyroxene (BFL)	26	0.57	8.70	0.28	14.95	20.44	0.04	123	11	50	nd	244	270	2410	50	80	31	58
Clinopyroxene (HbA)	27	0.55	8.28	0.34	14.58	20.63	0.04	264	11	54	nd	215	241	2757	42	36	31	67
Orthopyroxene (HbA, including 2-pyroxene lavas)	8	0.15	18.54	0.51	26.11	1.38	<0.01	1628	nd	nd	nd	92	211	3949	98	146	35	213
Hornblende (HbA)	41	1.63	12.20	0.26	13.65	11.80	0.45	528	67	176	1.8	434	120	2207	66	64	39	111
Biotite (HbA)	2	3.67	14.85	0.33	15.26	0.06	8.35	nd	4875	nd	nd	445	52	2556	62	110	70	411
Plagioclase (OPB)	10	0.03	0.76	0.02	0.25	16.31	0.07	695	45	1160	1.2	3	0.50	147	2.1	7	7	7
Plagioclase (BFL)	23	0.05	0.85	0.01	0.23	13.26	0.38	150	141	1311	2.5	3	0.65	121	2.4	2	24	6
Plagioclase (HbA)	48	0.03	0.37	0.01	0.34	11.00	0.44	180	205	1357	3.6	4	0.07	117	0.8	2	10	3
Spinel (OPB)	20	6.32	69.11	0.40	4.89	<0.01	<0.01	176	53	51	13	3707	9.63	3689	144	943	295	490
Spinel (BFL)	23	6.46	73.75	0.50	2.80	<0.01	<0.01	308	16	8	6	4271	0.29	4250	123	236	145	487
Spinel (HbA)	48	5.09	77.70	0.69	2.27	<0.01	<0.01	792	55	22	18	3029	0.05	6003	102	103	175	643

Apart from those of orthopyroxene and biotite, analyses principally represent XRF determinations on mineral separates (analyst: B. W. Chappell), supplemented by extensive electron-microprobe analysis (analyst: R. L. Stanton) where insufficient material was present for satisfactory bulk separation. Values for orthopyroxene and biotite are averages of ≈ 100 electron-microprobe analyses of each mineral (analyst: R. L. Stanton). Pb analyses of magnetite are corrected for thallium interference and all analyses of feldspar are corrected for impurity induced by minute inclusions of glass. OPB = olivine-clinopyroxene-basalt group; BFL = big-feldspar lavas; HbA = hornblende-andesite group; FeO* = total Fe as FeO; n = number of lava samples from which mineral separations and probe sections were made; nd = not determined; major oxides TiO$_2$ to K$_2$O given as per cent by mass; trace elements P to Zn given as parts per million by mass except chromium, which, for the spinels only, is given as per cent by mass; manganese determined both as major oxide (MnO, per cent) and as a trace (Mn, ppm).

maximum that the relevant structures are capable of accepting. The marked difference between the acceptance of copper into the silicates, as compared with that of other divalent ions of comparable size, such as Zn^{2+}, Co^{2+}, and Ni^{2+}, may, as suggested above, result from the distorted nature (Jahn–Teller distortion) of the Cu^{2+} ion and its resultant poor fit in 6-fold coordination.

Feldspar. As would be expected, the incidence of copper in the feldspars is significantly lower than in the Fe–Mg silicates — by a factor of ≈2. The mean for 81 separations is 14 ppm, which compares with 11 ppm for 22 Tonga–Kermadec feldspars (Ewart 1973; Ewart and Bryan 1972).

The general incidence of small quantities of copper in feldspars has been recognized for a long time and attributed at least partly to the substitution of Cu^+ (ionic radius 0.096 nm) for Na^+ (0.098 nm). By postulating incorporation of copper in the cuprous form, both ionic size and charge seemed appropriate for substitution for sodium.

Initial analyses of the Solomons feldspars yielded unexpectedly high values for copper (weighted mean Cu for 82 separations ≈28 ppm), and although the relevant mineral separations had been carried out with uncompromising regard for purity, minute inclusions of groundmass were briefly suspected. It quickly became apparent, however, that Cu:Zn ratios exhibited by the feldspars (≈3.36) were quite different from, and distinctly greater than, those of the related groundmasses (≈1.78).

Feldspar separations were then prepared in pairs: one member of 'ultimate' purity, the other of 'penultimate purity', i.e. impure to a minute degree. It was noted that when carrying out the final centrifuging of these two in diluted tetrabromoethane, the less pure floated, i.e. was the less dense. This indicated that the impurity involved might be extremely fine glass included within some of the feldspar crystals, rather than particles of groundmass adhering to crystal surfaces. Very accurate H$_2$O determinations were then carried out on four pairings with the results given in Table 6.6. Water was in all cases higher in the less pure member, adding weight to the possibility that the impurity was finely included glass.

Approximately one hundred glass inclusions were then subjected to analysis, *in situ*, using the Cameca electron microprobe. Mean copper so determined was 1900 ppm, and the Cu:Zn ratio 5.74. If the five highest copper analyses (>1.0 per cent) were disregarded, the mean reduced to 980 ppm and the Cu:Zn ratio to 3.0.

The higher-than-expected copper in the feldspars was thus traced to fine glass inclusions of high, and in some

62 Ore Elements in Arc Lavas

Table 6.6 Apparent H_2O content of feldspars as an indicator of fine, glassy, Cu–Zn-bearing inclusions

Sample	Purity	H_2O^+	H_2O^-	CO_2	Cu (f)	Zn (f)	Cu/Zn(f)	Cu(g)	Zn(g)	Cu/Zn (g)
A	up	0.22	0.08	0.01	25	9	2.78	108	31	3.48
	pup	0.43	0.39	0.05	62	13	4.77			
B	up	0.49	0.11	0.02	50	9	5.56	75	49	1.53
	pup	0.84	0.41	0.06	86	15	5.73			
C	up	0.25	0.06	0.05	33	10	3.30	81	44	1.84
	pup	0.33	0.12	0.03	106	19	5.58			
D	up	0.27	0.03	0.10	50	9	5.56	169	70	2.41
	pup	0.33	0.04	0.07	114	13	8.77			

up = ultimate purity feldspar; pup = penultimate purity feldspar; (f) = feldspar; (g) = groundmass; H_2O^+, H_2O^-, CO_2 given as per cent by mass; Cu and Zn given as parts per million by mass.

cases quite spectacularly high, copper content. It was then found that the FeO content of these feldspar separations (as determined by XRF) was a good indicator of the quantity of glass present. In each case 10–15 analyses of completely clear feldspar in each relevant section were then carried out by Cameca microprobe to determine that feldspar's intrinsic iron content (in general, 0.10 per cent <FeO <1.20 per cent). Using the FeO determinations by XRF on each pair of separations and by microprobe analysis of the relevant clear feldspar, and the copper determinations by XRF on the pair of separations, the intrinsic copper content of the feldspar was estimated graphically. This yielded the results given in Table 6.5. (Similar corrections have been carried out for vanadium, chromium, manganese, cobalt, nickel, and zinc in the feldspars, and are referred to in the relevant chapters).

There appears to be no systematic, e.g. linear, relationship between copper and sodium in the feldspars. There is, however, some indication that the highest maximum values of copper are developed in association with the middle range of Na_2O, ≈ 1.8 per cent to 5.5 per cent. When mean values for copper corresponding to 0.5 per cent intervals of Na_2O are plotted the relationship is sharpened and it appears that low mean feldspar Cu contents (Cu ≤5 ppm) are associated with both low and high levels of Na_2O in the plagioclase (1.3 per cent < Na_2O < 9.3 per cent), whereas significantly higher mean copper (20–5 ppm) is associated with the Na_2O values between the limits 1.7 <Na_2O <6.5 per cent. The resultant 'curve' might indeed be interpreted as rising rapidly to a peak at about Na_2O = 2.0–2.5 per cent, declining steadily to Na_2O ≈ 6.0–6.5 per cent and then flattening along the high values of sodium.

This lack of any simple Na_2O–Cu relationship indicates that the copper is not present through a systematic $Cu^+:Na^+$ substitution. The higher incidence of copper in the middle range of Na/Ca leads, however, to a suspicion that enhanced incorporation may stem at least in part from lattice distortion related to Ca:Al–Na:Si coupled substitution. This might be expected to be at a minimum towards the albitic and anorthitic ends of the plagioclase spectrum, and at a maximum where the atomic ratio of Ca to Na approached unity. Such enhanced opportunity for incorporation related to degree of lattice strain might well be reinforced by the inherently distorted nature of the Cu^{2+} ion itself — Cu^{2+} being much the more likely species in the melt environment.

Inspection of Table 6.5 indicates the likelihood that there is no systematic relationship between copper and 'intrinsic' iron in the feldspars, and that there might be an inverse relationship with manganese. Plotting of individual feldspar analyses shows, in fact, that both sets of relationships are essentially random.

Spinel. As indicated in Table 6.5 and Fig. 6.4, copper is distinctly more abundant in spinel than in any of the silicates, the mean value for 91 separations being 196 ppm. This compares with a mean of 134 ppm copper determined in five Tonga–Kermadec magnetites by Ewart and Bryan (1972) and Ewart et al. (1973). The three conspicuously high-copper spinels of Fig. 6.4(a), containing 1163, 963 and 910 ppm Cu respectively, are all from the picritic lavas of Kohinggo Island (Fig. 5.1). Detailed microscopical examinations of polished sections of these spinels showed no sign of copper-rich inclusions, and similarly detailed electron microprobing of microscopically clear areas of spinel grains substantiated the high copper values obtained by X-ray fluorescence analysis of mineral separations. Electron probe analysis of 52 spinel grains in the lava yielding the separation containing 1163 ppm Cu gave the frequency distribution of copper values shown in Fig. 6.4(b) and a mean of 1234 ppm Cu. This, with analogous results from the other two Kohinggo lavas, confirms that a high copper content is a genuine and intrinsic feature of these spinels.

This immediately raises the question as to whether the feature is simply one of locality, or whether it is tied to some other feature of the particular spinels. Microscopical observation indicated no mineralogical distinctiveness, and the only notable geochemical feature is the high chromium content: 10.76 per cent Cr in the mineral separation. However, other picritic spinels of the Solomons Suite exhibit high chromium without high accompanying copper, and systematic examination indicates a lack of any general tie between the two elements in the Solomons spinels overall. As might have been expected, finer-scale investigation of the Kohinggo lava of Fig. 6.4(b) — long-counting-time analysis by electron microprobe of four high-chromium spinels (three of them zoned, one unzoned) and two low chromium magnetites — gave somewhat equivocal results (Table 6.7). All six were comparatively large grains occurring as minor phenocrysts set in the relevant groundmasses. Spinels 1 and 2, both markedly zoned with respect to chromium, are also zoned with respect to copper, the two elements occurring in antipathetic relationship. Spinel 4 also exhibits a high-chromium core and a lower

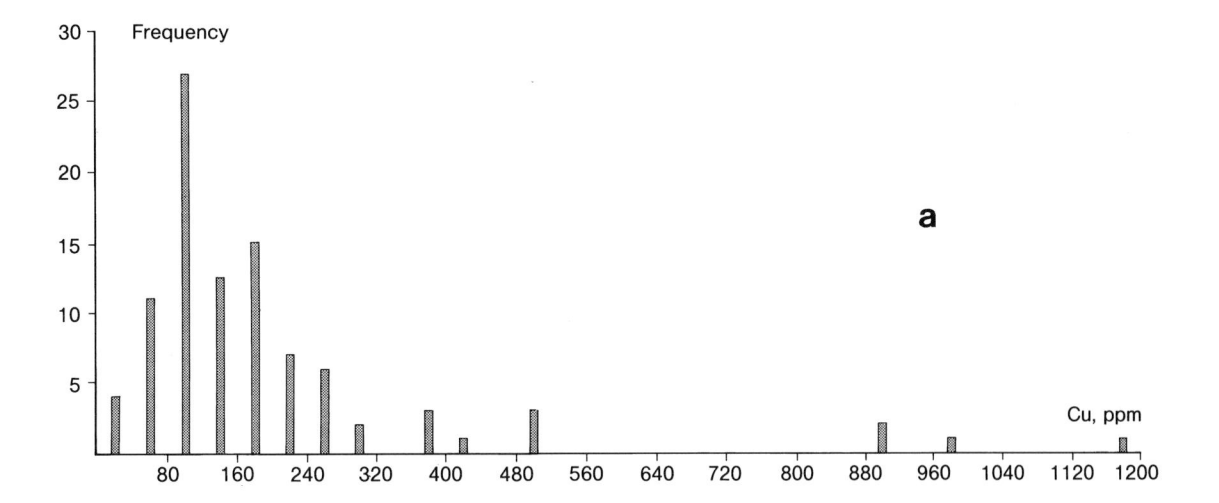

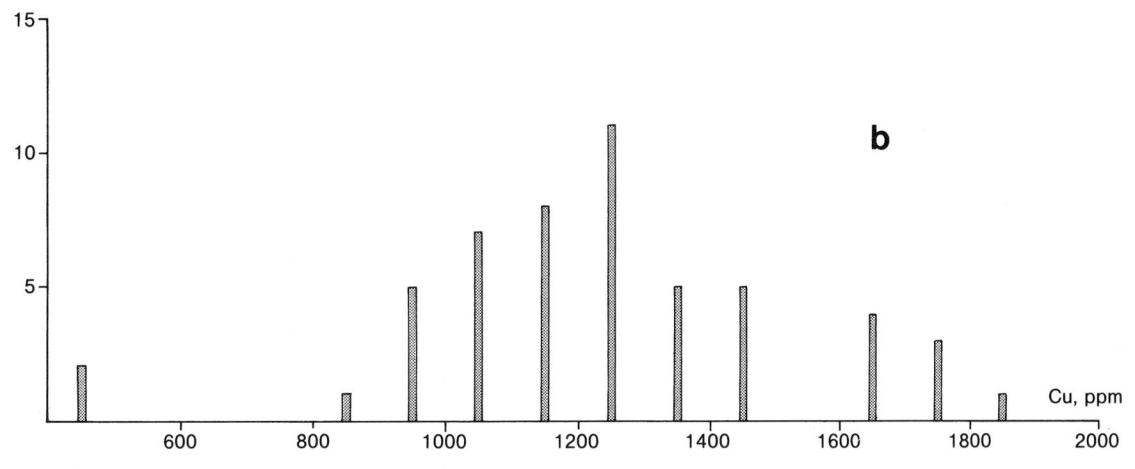

Fig. 6.4. Frequency distributions of copper abundances in spinels of (a) the Solomons suite as a whole (XRF analyses of mineral separations), and (b) Kohinggo Island picrite (electron-microprobe analyses of 52 individual grains in a single picrite thin section). Copper abundance in spinels of this picrite as determined by XRF analysis of separated material is 1163 ppm [see (a)]; copper abundance as indicated by the mean of the 52 probe analyses is 1234 ppm.

chromium rim, and although higher copper is again associated with lower chromium, the difference in copper values (1070 and 1176 ppm) is virtually negligible. Spinel 3, also possessing high-chromium but showing negligible zoning, also shows essentially no significant differences in copper — though here the slightly higher chromium core contains slightly higher copper than the lower chrome rim. Spinels 5 and 6, both of low chromium content, also both exhibit low copper — not high copper as might have been anticipated from relationships in magnetites 1 and 2. There are, however, significant differences in copper contents between core and rim, the rims showing lower values. Of the six magnetites, three show higher copper in the rim, three in the cores.

The data of Table 6.5 indicate that there is also no systematic Mn–Cu relationship, at least as indicated by bulk mineral separations. In detail, however, Table 6.7 reveals a state of affairs for Mn–Cu more or less analogous to that for Cr–Cu.

Table 6.5 does nevertheless indicate the possibility of an overall inverse relationship with FeO (and hence, from Fig. 5.19, total Fe), and this is confirmed by Fig. 6.5(a). The relationship is by no means linear: a rapid decrease in copper from ≈ 500 ppm at FeO <26 per cent to 140 ppm Cu at 30 <FeO <32 per cent then flattens to a decline to 118 Cu at 38 <FeO <40 per cent in spinel. We have already noted that spinel FeO tends to increase with increase in whole-rock SiO_2, i.e. with progressive evolution of the lava series, and this combined with the Cu:FeO relations of Fig. 6.5(a) implies that there should be a decrease in spinel copper accompanying increase in whole-rock SiO_2. This is indeed the case (Fig. 6.5(b)), and the observed spinel copper:whole rock SiO_2 'curve' is substantially analogous to those generated by magnesium, aluminium, chromium, etc. in spinel (Fig. 5.16).

The mode of accommodation of the copper within the spinel structure is not clear. MgO, Al_2O_3, Cr_2O_3, and TiO_2 range from major to 'substantial minor' constituents, and their inverse abundance relationship with iron thus results substantially from the inevitable complementariness implicit in the summing of components to one hundred. Their steady decline in abundance with evolution of lava type

Table 6.7 Core/rim compositions of six individual Kohinggo Island spinel (ferrochromite to titanomagnetite) crystals

	1 core	1 rim	2 core	2 rim	3 core	3 rim	4 core	4 rim	5 core	5 rim	6 core	6 rim
SiO_2	0.04	0.07	0.02	0.05	0.03	0.04	0.02	0.62	0.11	0.11	0.11	0.14
TiO_2	3.29	4.61	0.39	4.40	4.25	4.48	3.29	4.82	9.89	9.96	12.76	13.49
Al_2O_3	3.06	2.51	8.95	4.01	3.60	3.94	4.74	4.00	4.99	4.69	2.75	2.04
V_2O_3	0.68	0.87	0.06	0.75	0.69	0.60	0.43	0.59	0.89	0.88	1.01	1.03
Cr_2O_3	24.58	12.57	46.48	14.75	12.89	11.09	25.48	11.13	<.01	<.01	<.01	<.01
Fe_2O_3	35.62	45.42	14.88	43.09	45.65	47.15	34.57	45.25	44.74	44.85	40.82	39.93
FeO	26.79	28.63	19.31	25.76	25.97	25.11	22.95	25.60	34.54	34.99	39.32	40.48
MnO	1.08	0.86	1.24	0.85	0.86	0.83	0.90	0.77	1.21	1.21	1.27	1.31
MgO	4.39	3.98	8.39	5.92	5.59	6.36	7.20	6.77	3.52	3.24	1.85	1.45
Total	99.53	99.52	99.72	99.58	99.53	99.60	99.58	99.55	99.89	99.93	99.89	99.87
Ni	1598	1530	727	1790	1925	1830	1728	2005	121	96	73	129
Cu	1320	1821	416	1264	1284	1112	1070	1176	128	62	93	77
Zn	782	428	1061	323	425	308	637	317	580	538	746	771
Si	0.0013	0.0027	0.0008	0.0017	0.0011	0.0013	0.0006	0.0218	0.0040	0.0039	0.0040	0.0053
Ti	0.0893	0.1264	0.0100	0.1179	0.1146	0.1199	0.0867	0.1283	0.2688	0.2715	0.3543	0.3768
Al	0.1303	0.1080	0.3566	0.1684	0.1523	0.1652	0.1957	0.1668	0.2125	0.2003	0.1196	0.0892
V	0.0197	0.0255	0.0017	0.0213	0.0199	0.0172	0.0120	0.0167	0.0258	0.0255	0.0300	0.0307
Cr	0.7015	0.3623	1.2418	0.4157	0.3653	0.3121	0.7061	0.3115	0.0000	0.0000	0.0000	0.0000
Fe^{3+}	0.9674	1.2462	0.3783	1.1555	1.2312	1.2632	0.9117	1.2049	1.2162	1.2236	1.1340	1.1158
Fe^{2+}	0.8085	0.8731	0.5457	0.7677	0.7784	0.7475	0.6726	0.7575	1.0435	1.0608	1.2140	1.2573
Mn	0.0330	0.0266	0.0354	0.0256	0.0260	0.0249	0.0267	0.0231	0.0370	0.0370	0.0398	0.0413
Ni	0.0059	0.0057	0.0025	0.0065	0.0071	0.0067	0.0062	0.0073	0.0004	0.0004	0.0001	0.0005
Cu	0.0045	0.0063	0.0013	0.0043	0.0044	0.0037	0.0035	0.0039	0.0004	0.0002	0.0003	0.0003
Zn	0.0026	0.0014	0.0033	0.0011	0.0014	0.0010	0.0021	0.0010	0.0019	0.0018	0.0025	0.0026
Mg	0.2362	0.2160	0.4226	0.3144	0.2984	0.3374	0.3763	0.3573	0.1894	0.1752	0.1017	0.0802
Total	3.0001	3.0001	3.000	3.0001	3.0001	3.0001	3.0001	3.0001	3.0001	3.0001	3.0002	3.0001

Major constituents given as per cent by mass; Ni, Cu, Zn given as parts per million by mass; structural formulae based on 4O.

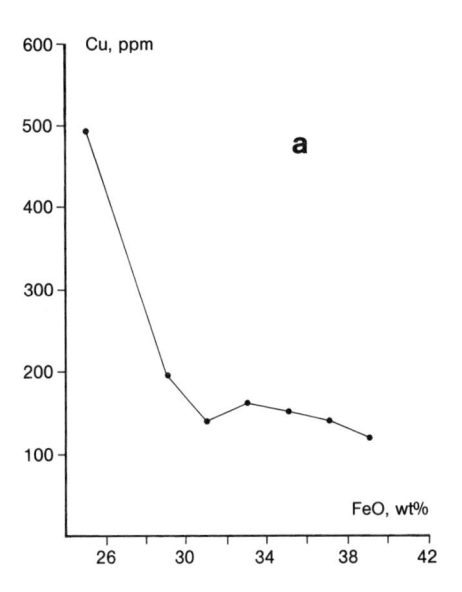

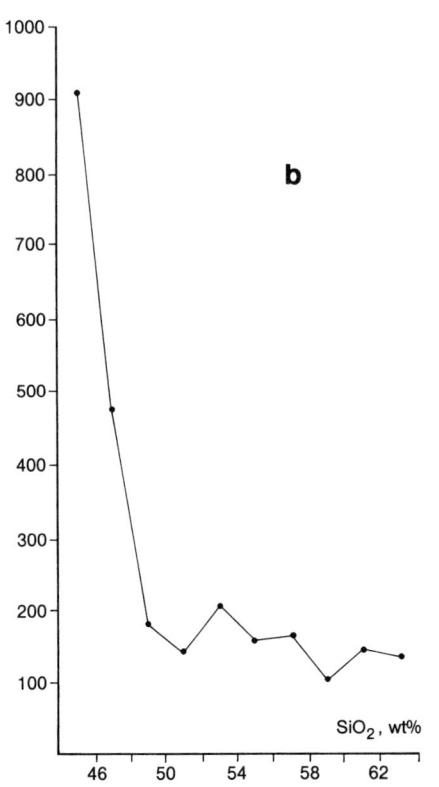

Fig. 6.5. (a) Copper abundances in the Solomons spinels as related to 2 per cent intervals of spinel FeO, and (b) copper abundances in the spinels corresponding to 2 per cent intervals of whole-rock SiO_2 in the containing lavas.

reflects at least in part magnetite:liquid equilibria and progressive subtraction of MgO, Cr_2O_3, and TiO_2 from the melt as crystallization proceeds. (Al_2O_3 does not conform with this pattern and, as is doubtless the case with the other three elements, its abundance in magnetite must be at least partly a temperature effect).

At first sight the incidence of copper might also be taken to reflect spinel:melt equilibria and the progressively reduced availability of this element in the melt from which the succession of spinels crystallized. If this were the case it might be presumed that the copper had been incorporated in the spinel structure simply by random substitution, principally in divalent sites. That the pattern of abundance of the spinels might reflect such random substitution related to concentration of copper in the melt is indicated by what at first sight might be seen as the 'anomalous' behaviour of this element in the two categories of spinel of the Rendova ankaramitic basalt referred to in Chapter 5 (see Table 5.7 and Fig. 5.18). As is shown a little further on (see Fig. 6.12) the abundance of copper in the melt does not follow a pattern of simple, progressive decline. It rises from a low point of ≈44 ppm in lavas of SiO_2 <46 per cent to a peak of ≈125 ppm in those of 50 per cent <SiO_2 <54 per cent, and then declines fairly steadily as whole-rock (and groundmass) SiO_2 increases towards the felsic andesites and dacites. The Rendova ankaramite contains 49.09 per cent SiO_2 and 83 ppm Cu and thus lies to the left of the copper maximum of Fig. 6.12. Early crystallization of the spinels now found within the clinopyroxenes would presumably have occurred at a stage when the concentration of copper in the melt was relatively low, and these magnetites contain a mean of 42 ppm Cu. Somewhat later crystallization of the groundmass spinels would have occurred when the concentration of copper in the melt was higher (ankaramite groundmass contains 51.80 per cent SiO_2 and 155 ppm Cu) and these magnetites contain a mean of 105 ppm Cu. The nature of the difference between the copper contents of the two categories of magnetites is shown in greater detail in Fig. 6.6. The very high copper content of the spinels of the most mafic basalts of Fig. 6.5 is due in large part to the very high copper concentrations in the spinels of the Kohinggo Island picrites. Had these not been included in the data from which Fig. 6.5 was constructed, the pattern of mean copper abundance in the spinels would have been somewhat similar to that of TiO_2 as shown in Fig. 5.16.

The spinel structure is anomalous in that the larger, divalent, ions Fe^{2+} and Mg^{2+} (ionic radii 0.082 and 0.078 nm respectively) occur in 4-fold coordination with O^{2-} whereas the smaller, trivalent ions Fe^{3+}, Al^{3+}, and Cr^{3+} (ionic radii 0.067, 0.057, and 0.064 nm respectively) occur in 6-fold coordination (cf. Berry and Mason 1959; Stanton 1972). With its normal propensity for 6-fold coordination

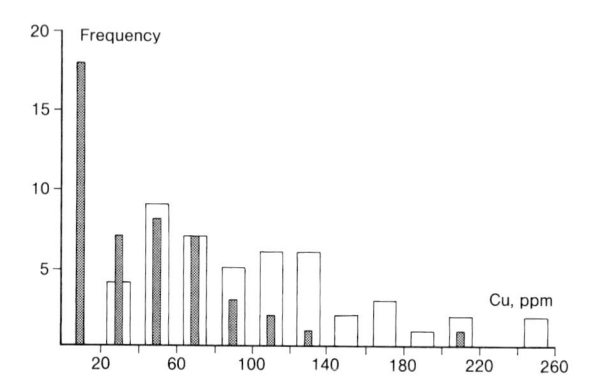

Fig. 6.6. Copper abundances in 45 spinel grains included in clinopyroxene of the Rendova ankaramitic basalt (filled bars; mean = 42 ppm) vis-à-vis those of 45 spinel grains in the groundmass of the same lava (open bars; mean = 105 ppm).

with oxygen there is the possibility that Cu^{2+} (0.072 nm) might substitute preferentially in trivalent sites, accompanied by appropriate vacancy formation. There is, however, no clear correlation with chromium (as we have seen above) or aluminium as may have been expected on the basis of systematic substitution. Ionic size and charge are, on the other hand, appropriate for substitution in divalent sites (even though radius ratio values are just as anomalous as they appear to be for Mg^{2+} and Fe^{2+}) and there is just a suggestion of muted correlation between Cu and MgO. The latter evidence is, however, very weak. Perhaps a better explanation — and one conforming with the evidence that the incidence of copper in the spinel is not consistent with its concentration in the coexisting melt — is that the distorted Cu^{2+} ion is accepted into the spinel structure most readily where the latter itself sustains maximum distortion by the incorporation of maximum amounts of Mg^{2+}, Al^{3+}, Cr^{3+}, and Ti^{4+}. Thus the incidence of copper has a loose relation with the abundances of these other elements in the spinel, rather than with the availability of copper in the melt.

CRYSTAL:MELT PARTITIONING

It follows from the foregoing mineral-chemical considerations that partitioning of copper between crystals and coexisting melt has varied widely between mineral species. We may now consider, first, the observations of previous investigators on copper partitioning between igneous crystals and melt, and then the various mineral:melt partitioning relations developed in the Solomons lava series.

General

Probably because of its habitually low abundance in most of the common igneous silicates, comparatively little work has been done on the partitioning of copper in igneous systems and almost nothing on the products of marine volcanism, the regime of present concern.

The early work of Wager and Mitchell (1951) on olivine of the highly differentiated and layered Skaergaard gabbroic intrusion gave K_D values of 0.15, 0.08, 0.7, and 0.9 in their A, B, C, and D stages respectively. Paster et al. (1974) obtained rather lower values, from 0.02 to 0.04 for the upper Zone of the Skaergaard intrusion; Bougault and Hekinian (1974) determined K_D^{Cu} for olivine of 0.11 in a Mid-Atlantic Ridge basalt; and Dostal et al. (1983) obtained a value of 0.05 for material contained in a cumulate inclusion occurring in a Lesser Antilles lava. Bird (1971: see Irving 1978) obtained experimental results for olivine:melt in the system $SiO_2–Al_2O_3–MgO–CaO$ at one atmosphere pressure as follows: 1300 °C, $K_D^{Cu}= 0.47 \pm 0.02$; 1350 °C = 0.36 ± 0.02; 1400 °C = 0.27 ± 0.04.

Comparatively little research seems to have been done on the acceptance of copper into the structures of igneous pyroxenes, and on resultant crystal:melt partitioning. In their work on the Skaergaard intrusion Wager and Mitchell determined values of K_D^{Cu}(clinopyroxene:melt) = 0.14, 0.3, 0.7 and 2.3 for their B, C, D, and E stages respectively. Paster et al. (1974) obtained a value of 0.089 for clinopyroxene of the Upper Zone of the intrusion. In their investigation of the Lesser Antilles cumulate inclusion referred to above, Dostal et al. obtained values of 0.05 and 0.08. No information appears to be available on the partitioning of copper between orthopyroxene and melt.

There is similarly little general information on the acceptance of copper into igneous amphiboles and the resulting K_D values. Dostal et al. (1983) obtained a value of 0.05 for hornblende:groundmass in the Lesser Antilles cumulate inclusion referred to above. Hendry et al. (1981, 1985) determined copper in a large number of amphibole crystals in both Solomon Islands and other intrusive rocks, but did not examine the relevant partition coefficients. Absolute values for copper were very high in some cases (775 ppm Cu in the hornblende of a biotite-granodiorite porphyry of the Christmas intrusive complex, Arizona), but the predominant range was ($\approx$) 1–10 ppm Cu. They found (Hendry et al. 1981) a maximum of 40 ppm in hornblende of the Koloula subvolcanic igneous complex of southern Guadalcanal, but again a principal range of 1–10 ppm Cu.

In contrast to the olivines, pyroxenes, and most occurrences of igneous amphiboles, biotite is capable of accepting relatively large quantities of copper into its structure — up to 5000 ppm determined by Hendry et al.

(1981) by ion probe in one Koloula biotite. However, although there have been a number of studies of copper in biotites of intrusive rocks — principally those of relevance to porphyry copper investigations — there has been little consideration of the incidence of copper in the biotites of lavas, or of partition coefficients developed by them. Had biotite constituted a significant constituent of the Skaergaard intrusion, Wager and Mitchell (1951) doubtless would have seen to it that we had good basic information on the partitioning of copper by this mineral!

In their thorough fashion, and in spite of the very low abundances of copper in feldspars, Wager and Mitchell have, however, determined K_D values of 0.1, 0.14, 0.05, 0.6, and 0.6 for the plagioclases of their A, B, C, D, and E stages respectively. Paster et al. (1974) obtained a value of 0.004 for K_D plagioclase:melt in the Upper Zone of the Skaergaard Layered Series (though their abundance data of 14 ppm Cu in plagioclase and 2143 ppm Cu in coexisting melt indicates K_D = 0.0065). Bougault and Hekinian (1974) determined a K_D plagioclase:melt of 0.17 in a mid-Atlantic Ridge basalt.

In view of its relatively high capacity — vis-à-vis the silicates — for incorporating copper in its structure, it is perhaps surprising that relatively little work has been done on the partitioning of copper by magnetite in crystallizing volcanic melts. Once again we have to turn to Wager and Mitchell (1951) for early information. Their determinations of K_D magnetite:melt for the Skaergaard layers B, C, D, and E were 0.07–0.08, 0.2, 1.2, and 0.7 respectively. Paster et al. (1974) found a K_D value of 0.54 for magnetite of the Upper Zone of the Skaergaard intrusion, and Dostal et al. (1983) obtained a value of 0.15 for magnetite in the plutonic cumulate inclusion contained within the aforementioned Lesser Antilles lava.

Solomons suite

Mean partition coefficients for each of the principal minerals as these have been developed within each of the three Solomons lava groups are given in Table 6.8. The values given are averages for what in some cases is more than 40 crystal:melt pairings; e.g. the calculation of K_D^{Cu} plagioclase:melt for the hornblende-andesite group involves 44 coexisting crystal:melt pairs. In other instances the number of pairs is much less; e.g. six in the case of K_D^{Cu} olivine:melt in the big-feldspar lava group. Such variations in sample numbers reflect variations in occurrence and in amenability of the various components to standard separation techniques. It should also be kept in mind that where large numbers of pairs are involved it is virtually inevitable that K_D values for a given pairing differ from one individual lava to another. The range of values of $K_D^{ol:melt}$ developed in the olivine–pyroxene-basalts provides an example (Fig. 6.7). Such variations display poorly developed regularity in some cases (presumably at least partly a reflection of differences in crystallization T/P) but this has not been pursued in the present study.

Table 6.8 gives two values for each mean distribution coefficient, designated $K_{D(1)}$ and $K_{D(2)}$. In each case $K_{D(1)}$ is calculated from the mean copper contents of all the mineral separations and all the groundmass (residual melt) separations available. Thus for hornblende, for

Table 6.8 Crystal:melt distribution coefficients (K_D) with respect to copper

Mineral species	Lava group	n_1	n_2	m_1	m_2	$K_{D(1)}$	N	M_1	M_2	$K_{D(2)}$
Olivine	OPB	40	44	36	101	0.36	39	33	100	0.33
	BFL	7	23	51	168	0.30	6	53	152	0.35
Clinopyroxene	OPB	43	44	31	101	0.31	42	29	99	0.29
	BFL	26	23	31	168	0.18	17	34	168	0.20
	HbA	27	47	31	64	0.48	20	34	64	0.53
Hornblende	HbA	25	47	39	64	0.61	22	36	52	0.69
Feldspar	OPB	10	44	7	101	0.07	10	7	102	0.07
	BFL	23	23	24	168	0.14	22	26	172	0.15
	HbA	46	47	10	64	0.16	44	13	63	0.21
Spinel	OPB	20	44	295	101	2.92	20	295	104	2.84
	BFL	23	23	145	168	0.86	22	147	169	0.87
	HbA	37	47	175	64	2.73	36	164	68	2.41

OPB = olivine–clinopyroxene-basalt group; BFL = big-feldspar lavas: HbA = hornblende-andesite group; n_1 = total number of mineral separations analysed; n_2 = total number of groundmass separations analysed; m_1 = mean copper content of mineral of each lava group; m_2 = mean copper content of groundmass of each lava group; $K_{D(1)}$ = mean distribution coefficient determined as m_1/m_2; N = number of coexisting mineral:groundmass pairs analysed; M_1 and M_2 = mean copper contents of coexisting mineral and groundmass respectively; $K_{D(2)}$ = distribution coefficient determined as M_1/M_2, for comparison with $K_{D(1)}$. All concentrations as parts per million by mass.

68 *Ore Elements in Arc Lavas*

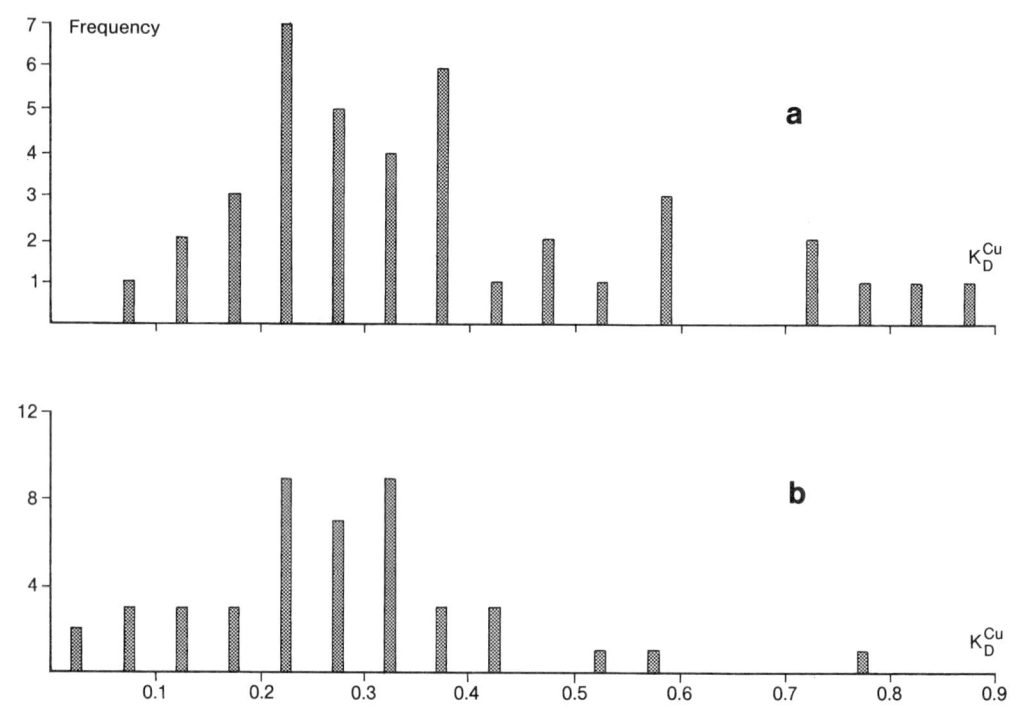

Fig. 6.7. Frequency distributions of individual K_D^{Cu} values for (a) coexisting olivine:groundmass pairs in 39 olivine–clinopyroxene-basalts (mean = 0.33), and (b) coexisting clinopyroxene:groundmass pairs in 42 such basalts (mean = 0.29).

example, there are 47 groundmass analyses, but only 25 for the mineral. All the 47 lavas represented by the separated groundmasses contained hornblende, but in 22 cases there was insufficient material for separation of an adequate sample of hornblende for XRF analysis, or there were textural problems, or — as in the majority of cases — the hornblende had been substantially converted to the opaque derivative discussed in Chapter 5. In other cases satisfactory separation of the relevant groundmass was not possible. Thus the distribution coefficients designated as $K_{D(1)}$ do not strictly present coexisting pairs. Those designated as $K_{D(2)}$, on the other hand, have been calculated entirely from analyses representing coexisting pairs — of which, in the case of hornblende, there happen to have been 22.

The Solomons olivines yield mean K_D^{Cu} values that, at 0.30–0.35, are somewhat higher than those obtained for natural materials by the earlier investigators, though they are very much in keeping with the experimental results of Bird (1971 in Irving 1978). On the other hand Fig. 6.7(a) shows that K_D^{Cu} values for the olivines of *individual* Solomons lavas (of the olivine–pyroxene-basalt group) almost precisely encompass the range of values previ-

ously obtained by other investigators. Such a range, from 0.09 to 0.89, immediately raises the possibility of a partition coefficient–temperature relationship. If, however, olivine Mg number is taken as a crude indicator of temperature of olivine crystallization, and if such Mg numbers are plotted against K_D^{Cu} values generated by the relevant olivines, no systematic relationship between the two is apparent. Nine of the sampled Solomons picrites contain olivines of mean Mg no. = 88–9, and these exhibit mean K_D^{Cu} values ranging from 0.15 to 0.89 — almost the full range of the whole group of 39 olivines of Table 6.8 and Fig. 6.7. Thus if there is a relationship between olivine K_D^{Cu} and temperature, such a simple approach as this does not reveal it.

Mean K_D^{Cu} values developed by the Solomons clinopyroxenes, 0.20 to 0.53, are of the same order as those found by Wager and Mitchell (1951) for those in the Skaergaard layers, but appear to be distinctly higher than those obtained by Paster *et al.* (1974) and Dostal *et al.* (1983). As in the case of olivine, there is a wide range of clinopyroxene K_D^{Cu} values developed from one individual lava to another (Fig. 6.7(b)), and again there is no systematic relationship between temperature of crystallization — as indicated by clinopyroxene Mg number — and the value of the partition coefficient. One aspect of clinopyroxene K_D^{Cu} values that must be kept in mind is the effect of the very consistently low acceptance of copper by members of this mineral group. With the strong tendency of the clinopyroxene to accept ≈30 ppm Cu, with an upper limit ≈50 ppm, values of partition coefficients become substantially a reflection of the copper content of the melt (Table 6.8). Thus with their characteristically high groundmass (i.e. residual melt) copper the big-feldspar lavas yield low $K_D^{cpx:melt}$ values (≈0.20) whereas the andesites and dacites of the hornblende-andesite group, with their equally characteristically lower groundmass copper abundances, tend to yield notably higher $K_D^{cpx:melt}$ values (≈0.50). This being recognized, the Solomons K_D values are essentially in the median range of those found elsewhere.

The hornblendes of the Solomons hornblende-andesite group contain slightly but distinctly more copper than the coexisting clinopyroxene, and hence develop marginally higher partition coefficients — in this case in the approximate range 0.6–0.7.

Although biotite is moderately prominent in both of the Savo biotite-bearing dacites (see Chapter 5), the amounts involved were not adequate for satisfactory mineral separation and the electron probe was therefore used for all trace analysis. The abundance and state of preservation of the biotite crystals were, however, quite sufficient for extensive and hence quite rigorous probe analysis, and the two lavas thus yielded substantial information on the acceptance of trace elements by biotite. Mean copper in the biotites (33 analyses) was 70 ppm (Table 6.5) and 24 ppm in the groundmasses, giving a mean K_D^{Cu} of 2.92. Biotite is therefore the only ferromagnesian silicate (indeed the only silicate, as feldspar accepts only very small quantities of copper) that induces $K_D^{Cu} > 1$, and that, hence, tends to partition copper out of the melt and into the solid phase. Judging by the high copper abundances found by others (e.g. Hendry et al. 1981, 1985) in biotite of felsic subvolcanic intrusions in the Solomons and elsewhere, the development of biotite $K_D^{Cu} > 1$ is probably widespread and may even be general.

Values of mean partition coefficients obtained for Solomons plagioclases, although low compared with the other silicates, are generally somewhat higher than values obtained by other investigators in other provinces. The data of Table 6.8 indicate that the Solomons plagioclase K_D^{Cu} may increase systematically proceeding from olivine–clinopyroxene-basalts to hornblende-andesites. This may, however, be an artefact of the relatively rapid loss of copper from the melt to the volatile phase in the formation of the more felsic lavas — a possibility considered in greater detail in Chapters 21, 22, and 24.

Partition coefficients (overall mean of 78 pairings, K_D^{Cu} = 2.09) induced by the Solomons spinels are somewhat higher than most of those appearing in the literature. The high copper abundances of the Solomons spinels is, however, a consistent feature, and its validity has been firmly established by detailed microscopical examination and the checking of XRF determinations on separated material by probe analysis of single, clear grains. The lower value developed by the spinel of the big-feldspar lavas is due, as indicated in Table 6.8, not so much to low copper abundance in the spinel as to the substantial increase in copper in the melt at this stage of lava evolution. This is referred to again in the following section.

Thus crystallization of all of the principal silicates tends to *concentrate* copper in the residual melt. Magnetite is the only phenocrystic mineral capable of *reducing* the copper content of the melt.

THE INCIDENCE OF WHOLE-ROCK COPPER IN THE SOLOMONS YOUNGER VOLCANIC SUITE

Mean copper in the basaltic members of the suite is 100 ppm, which, as noted earlier, compares with 71 ppm for MORBs, 105 for the Lesser Antilles arc basalts, and 100, 138 and 104 ppm for the Aleutian, Vanuatu, and Tonga–Kermadec basalts respectively. Frequency distributions of copper abundances in the Solomons basalts, and in lavas more felsic than basalts (SiO_2 >52 per cent), are given in Fig. 6.8 (cf. Fig. 6.2).

The distribution of copper in the three main lava types — olivine–pyroxene-basalts, big-feldspar basalts and basaltic andesites, and hornblende-andesites — and their groundmasses are depicted in Fig. 6.9. Fig. 6.9(c) shows values in the olivine–pyroxene-basalts and groundmasses to be substantially restricted to the range 30–160 ppm, with copper slightly more abundant in the groundmass (means 78 and 97 ppm respectively). Progression to the big-feldspar lavas is accompanied by a doubling of both the amount of copper present (means 161 and 168 ppm Cu for whole rocks and groundmasses respectively) and of the spread of values (Fig. 6.9(b)). Development of the hornblende-andesites is accompanied by a reduction in the amount of copper (mean values, 66 and 64 ppm Cu respectively) and a contraction of the spread (Fig. 6.9(a)) to one similar to that of the olivine–pyroxene-lavas.

This suggests the possibility of a relationship between copper abundance and lava type that is somewhat more specific than that diagnosed by Sandell and Goldich (1943): that copper simply decreases with increase in whole-rock SiO_2. That this might be the case has already been suggested in a preliminary way (Stanton 1978; Stanton and Ramsay 1980), and is confirmed by Figs. 6.10–6.12. In relating copper abundance to those of MgO and (Na$_2$O + K$_2$O) in an 'FMA' style of diagram, Fig. 6.10 gives an indication of the place of copper in the evolution of the three broad lava types and of the Younger Volcanic Suite as a whole. Figs. 6.10(a)–(f) depicts 'CuMA' relations for individual lavas within each group of whole rocks (filled circles) and their groundmasses (open circles). It may be seen that proceeding from the olivine–pyroxene-lava whole rocks (Fig. 6.10(f)) to their groundmasses (Fig. 6.10(e)), and then to the big-feldspar lavas and their groundmasses (Figs. 6.10(d) and (c)), Cu and (Na$_2$O + K$_2$O) proportions increase, substantially through the subtraction of olivine and clinopyroxene. With the continued crystallization of clinopyroxene, the onset of crystallization of amphibole, and the increasingly copious precipitation of feldspars —

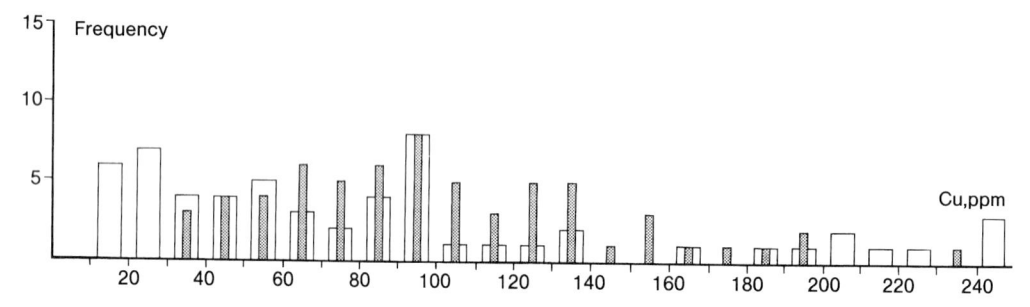

Fig. 6.8. Frequency distributions of copper abundances in Solomons basalts (SiO$_2$ < 52 per cent; filled bars), and in Solomons lavas more felsic than basalts (SiO$_2$ ⩾ 52 per cent; open bars).

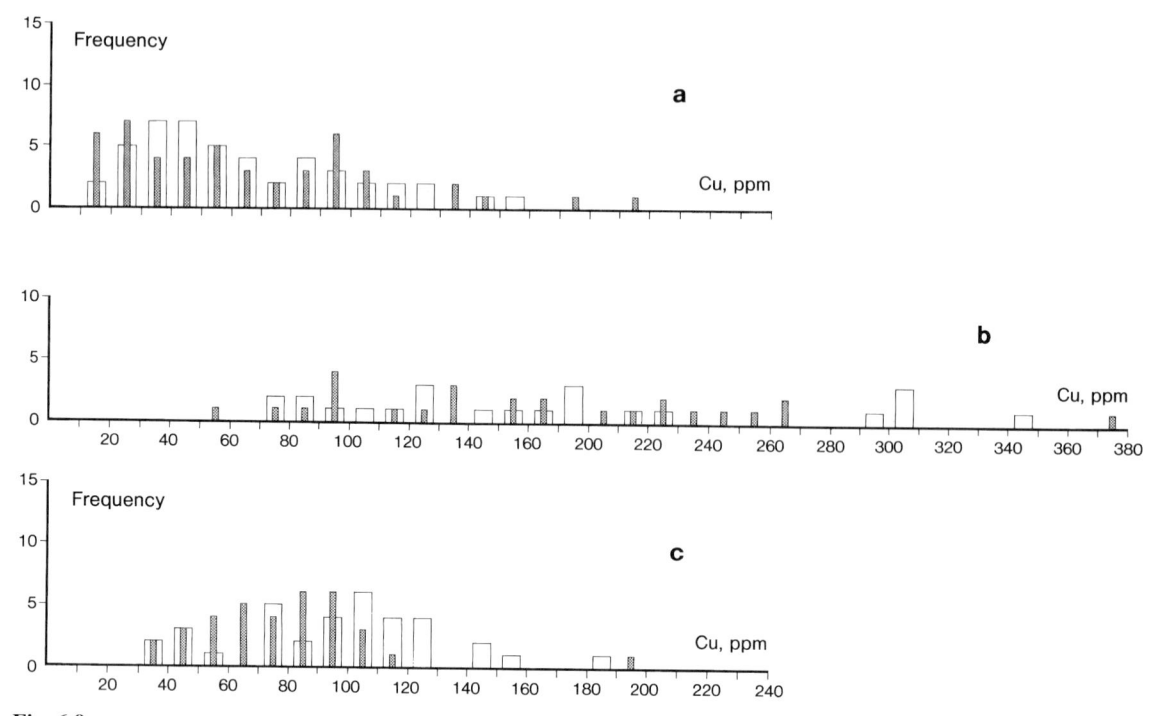

Fig. 6.9. Frequency distributions of copper abundances in whole rocks (filled bars), and groundmass (open bars) of (a) lavas of the hornblende-andesite group, (b) big-feldspar lavas, and (c) the olivine–clinopyroxene-basalts.

all silicates incorporating significantly less copper than the whole rocks of the felsic basalt–basaltic andesite stage of differentiation — the development of the hornblende-andesites and dacites might have been expected to have been accompanied by further increase in whole-rock copper. This is clearly not the case, however; (Na_2O + K_2O) continues to increase with the development of the andesites and dacites, but copper decreases markedly (Fig. 6.10(b) and (a)).

The trends of Fig. 6.10 develop with enhanced clarity when points representing mean CuMA relations calculated for 2 per cent whole-rock SiO_2 intervals (<46, 46–8, 48–50 → 64 per cent SiO_2) are plotted on the ternary diagram (Fig. 6.11). Fig. 6.11(a) shows mean values for whole rocks (filled circles) and for corresponding groundmasses (open circles) for the 122 lavas of the New Georgia–Guadalcanal domain. Fig. 6.11(b) shows the analogous sequence of points for the 241 analyses representing the full Bougainville–Guadalcanal domain. The point at the lower right of the diagram represents the high-Mg picrites (SiO_2 <46 per cent), that at the lower left the most dacitic rocks (SiO_2 >64 per cent). Fig. 6.11(c) is an analogous presentation of CuMA relations for successive 2 per cent whole-rock SiO_2 intervals for Grenada and St Kitts in the Lesser Antilles, and Fig. 6.11(d) the same for Vanuatu and Tonga–Kermadec. For comparison with the trends developed by the arc suites, Fig. 6.11(e) shows CuMA relations in individual Mid-Atlantic Ridge (MORBs) lavas and Fig. 6.11(f) those in Icelandic basalts (SiO_2 <52 per cent; filled circles) and lavas more felsic than basalts (SiO_2 >52 per cent; open circles).

The salient feature of Figs. 6.11(a)–(d) — i.e. those depicting CuMA patterns generated by the lava series of the island arcs — is a high copper inflexion analogous to the high FeO inflexions of the FMA diagrams of all volcanic island-arc (and many other) lava suites. In terms of the nomenclature set out in Chapter 5, the lavas of the Solomon Islands Younger Volcanic Suite are, as a group, clearly hypotholeiitic with respect to copper. This is also the case with the other arc assemblages depicted in Fig. 6.11. The mid-Atlantic ridge MORBs (Fig. 6.11(e)) do not appear to develop any distinctive trend, and may simply cluster about some primitive parent. The Icelandic basalts, on the other hand, display a quite pronounced tholeiitic trend with respect to copper (Fig. 6.11(f)), analogous with their well-known tholeiitic trend with respect to iron. Those Icelandic lavas more felsic than basalts (SiO_2 >52 per cent), in contrast, display a muted linearity towards the alkali apex of the triangle. The impression may be gained that had there been more analyses including copper available, the full range of Icelandic lavas might have generated a hypotholeiitic trend essentially analogous to those developed by the arc suites.

These distinctive ternary patterns formed by the arc assemblages are in fact largely a reflection of the Cu–SiO_2 relationships shown in Fig. 6.12. This depicts variation in mean copper with respect to the 2 per cent whole-rock SiO_2 intervals of Fig. 6.11; e.g. Fig. 6.12(a) shows mean values of copper for all Solomons (excluding Bougainville) lavas (and their groundmasses) within each 2 per cent SiO_2 interval, plotted against the relevant SiO_2 values. Mean copper in the Solomons picrite basalts is 44 ppm; this increases regularly to peak at ≈ 125 ppm in rocks containing 50–4 per cent SiO_2, then decreases —again quite regularly — to ≈8 ppm at SiO_2 >64 per cent. On the left-hand side of the peak, groundmass copper (mean, 103 ppm) increasingly exceeds whole-rock copper (mean, 87

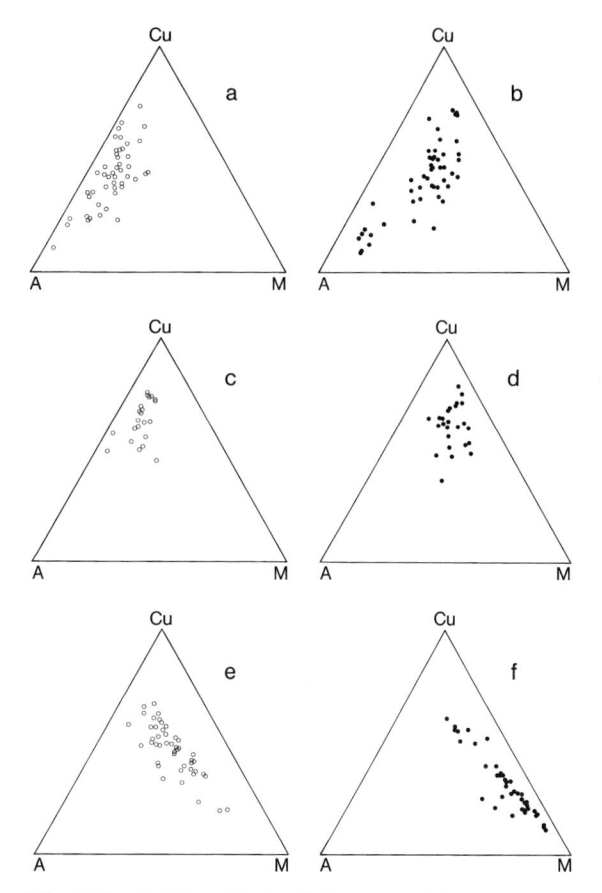

Fig. 6.10. Cu–MgO–(Na_2O + K_2O) ternary relations developed in olivine–clinopyroxene-basalt, big-feldspar lava, and hornblende-andesite whole rocks (filled circles: f, d, and b respectively), and their corresponding groundmasses (open circles: e, c, and a respectively). Cu calculated as ppm × 10^{-1}; MgO and (Na_2O + K_2O) as per cent by mass.

72 *Ore Elements in Arc Lavas*

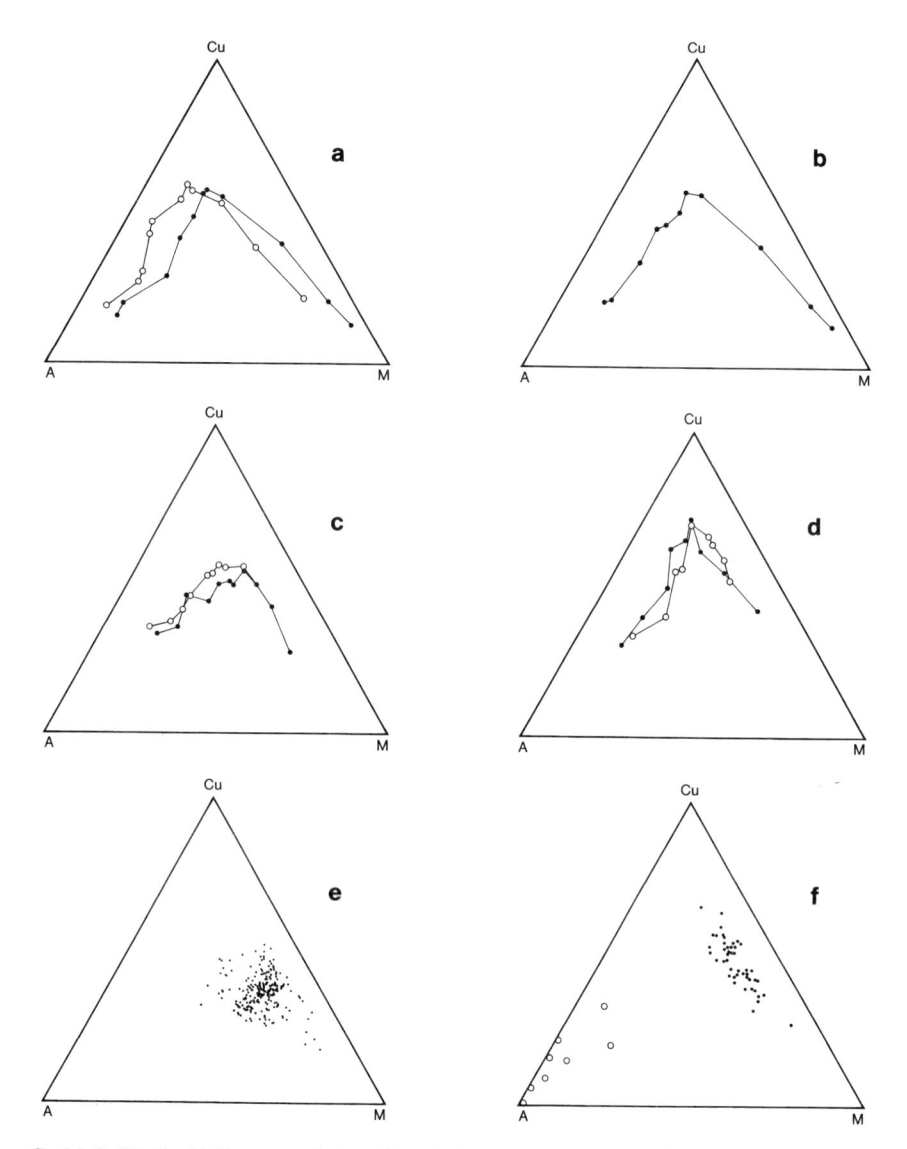

Fig. 6.11. Mean Cu–MgO–(Na$_2$O + K$_2$O) ternary relations (Cu calculated using ppm Cu × 10^{-1} corresponding to 2 per cent whole-rock SiO$_2$ intervals < 46, 46–8, 48–50 → >64 per cent in (a) whole rocks (filled circles) and corresponding groundmasses (open circles) of the Solomons suite excluding Bougainville (122 analyses); (b) whole rocks of the Solomons suite including Bougainville (241 analyses); (c) lavas of Grenada (filled circles) and the Lesser Antilles as represented by means of Grenada, Dominica, and St Kitts (open circles); and (d) lavas of Vanuatu (filled circles) and Tonga–Kermadec (open circles). Ternary relations developed by MORBs as represented by Mid-Atlantic Ridge basalts are given in (e), and in individual samplings of Icelandic basalts (SiO$_2$ < 52 per cent; filled circles) and Icelandic lavas more felsic than basalts (SiO$_2$ ⩾ 52 per cent; open circles in (f). All averages for the Lesser Autilles (including Grenada) lavas used in this and analogous diagrams in Chapters 7–20 are calculated from the averages given by Brown *et al.* (1977).

ppm) as SiO$_2$ increases. At its peak (50 per cent <SiO$_2$ <54 per cent) groundmass copper reaches 128 ppm. On the right-hand side of the peak the groundmass: whole-rock copper relationships are irregular; some groundmasses contain less copper than the whole rock, and the means for the two are essentially similar (groundmass, 73 ppm; whole rock, 69 ppm). The rise in copper corresponding to the increase in whole-rock SiO$_2$ from 44 to 50 per cent may be attributed, as noted above, largely to the subtraction of olivine and lesser clinopyroxene, the two containing an average of ≈35 as compared with the 70–100 ppm of any likely parental basalt.

This returns us to the earlier observation that copper in the Tonga–Kermadec lavas does not simply decrease with increase in whole-rock SiO_2 (as had been suggested as a possible rule by Sandell and Goldich (1943) and Goldschmidt (1954)) but develops a maximum in rocks within the andesite range. Fig. 6.12(c), plot 2 depicts mean copper vs. 2 per cent whole-rock SiO_2 intervals for all analyses available for Tonga–Kermadec. Keeping in

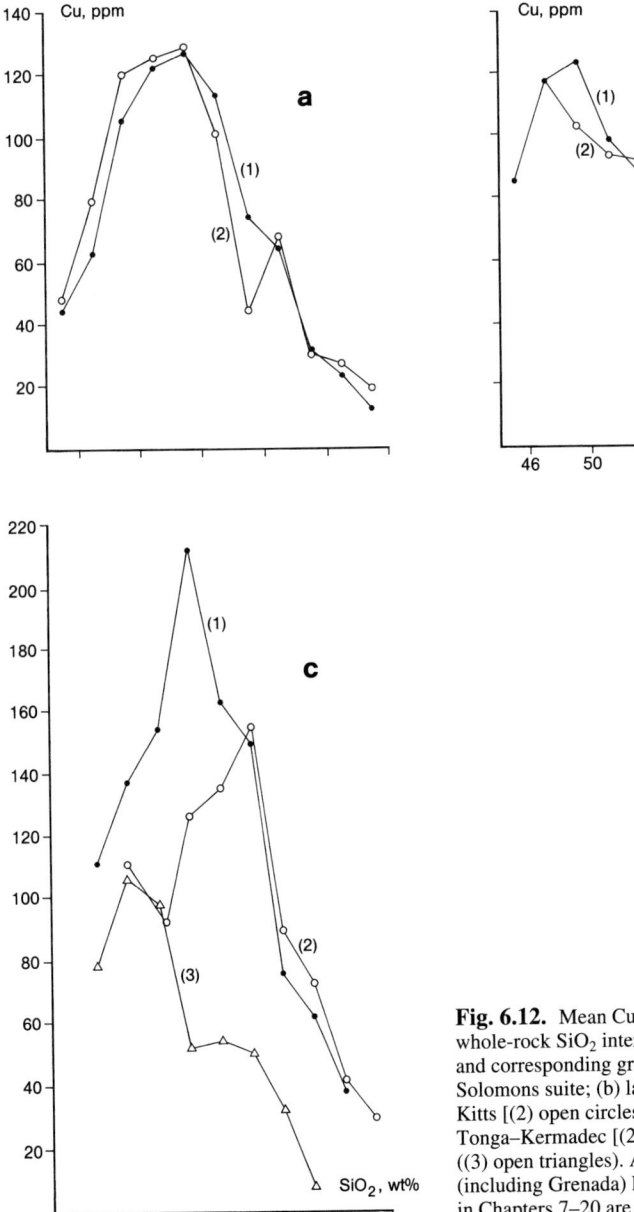

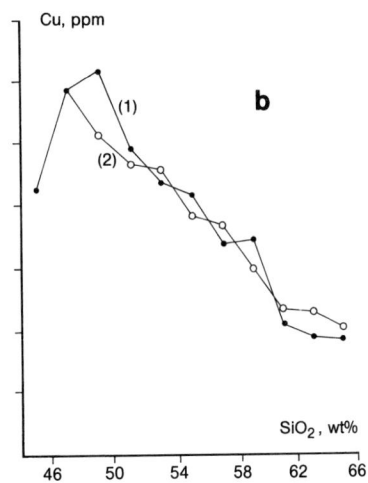

Fig. 6.12. Mean Cu:SiO_2 relations corresponding to 2 per cent whole-rock SiO_2 intervals in (a) whole rocks [(1) filled circles] and corresponding groundmasses [(2) open circles] of the Solomons suite; (b) lavas of Grenada [(1) filled circles] and St Kitts [(2) open circles]; (c) lavas of Vanuatu [(1) filled circles], Tonga–Kermadec [(2) open circles], and the Aleutians arc ((3) open triangles). All averages for the Lesser Antilles (including Grenada) lavas used in this and analogous diagrams in Chapters 7–20 are calculated from the averages given by Brown *et al.* (1977).

mind the smaller number of analyses, and hence the lesser reliability of individual points, the 'curve' here is remarkably similar to that of the Solomons in Fig. 6.12(a). The maximum for Tonga–Kermadec develops at ≈54–8 per cent SiO_2 as compared to 50–4 per cent for the Solomons but the curves appear similar in principle. Similar plots for Grenada and St Kitts, in the Lesser Antilles (Fig. 6.12(b)), Vanuatu, and the Aleutians (Fig. 6.12(c)) reveal Cu vs. SiO_2 trends clearly reminiscent of that of the Solomons and Tonga–Kermadec.

The Vanuatu suite (Fig. 6.12(c), plot 1) is clearly one of high copper: much higher in this element than all the other four arc suites. The Aleutians series (Fig. 6.12(c) plot 3) is the lowest in copper, and the ratio highest (Vanuatu):lowest (Aleutians) has a value of about two. Fig. 6.12 also shows that the lava suites of the different arcs tend to achieve their peak copper abundances at somewhat different SiO_2 levels. Four of them exhibit maxima in the 48–54 per cent SiO_2 (basalt– basaltic andesite) range, but that of Tonga–Kermadec does so in the 56–8 per cent SiO_2 interval — well into andesite compositions.

At least in the case of the Solomons Suite, the rise in copper corresponding to the increase in whole-rock SiO_2 to the left of the copper maximum (Fig. 6.12(a); see also Fig. 21.2) may be attributed largely to the subtraction of olivine and lesser clinopyroxene, the two containing an average of ≈35 ppm Cu. For a parent melt containing 70–100 ppm Cu, the subtraction of quantities of olivine and clinopyroxene of the order of those indicated by Fig. 5.6 would readily drive the copper content of the residual melt to 120–50 ppm.

The decrease in copper to the right of the maxima of Fig. 6.12 does not, however, appear capable of explanation on the basis of crystal subtraction. Continued crystallization of clinopyroxene (and the appearance of any orthopyroxene) would continue to increase the amount of copper in the residual melt. This effect would then be greatly accentuated by the crystallization of hornblende (mean Cu ≈39 ppm) and of copious quantities of feldspar (mean Cu ≈ 14 ppm) that occurs with the onset of the formation of the andesites at SiO_2 ≈52 per cent. The only major mineral capable of removing copper at concentrations greater than 120 ppm is magnetite, which incorporates Cu at ≈1.5 times this level (see discussion above of partition coefficients). Even at this concentration, however, the observed reduction of copper in the melt would require the removal of >30 per cent magnetite, for which there is no evidence in the rocks; whereas some of the highly porphyritic lavas exhibit over 50 per cent modal per cent of olivine, clinopyroxene, or feldspar phenocrysts, magnetite never exceeds 3 per cent and rarely exceeds 2.5 per cent.

This pattern of behaviour of copper is considered again in Chapters 21 and 24.

CONCLUDING STATEMENT

On the basis of current data, mean copper in MORBs is 71 ppm; in basalts of five of the world's major modern volcanic island arcs (the Lesser Antilles, Aleutians, Solomon Islands, Vanuatu, and Tonga–Kermadec) it is 108 ppm; and in those of the Solomons themselves, 100 ppm.

The feldspars of the Solomons lavas generally contain 10–25 ppm, and a mean of 14 ppm, Cu. The principal ferromagnesian minerals, olivine, ortho- and clinopyroxene, and hornblende, very consistently contain 30–40 ppm Cu. Biotite contains mean copper ≈70 ppm. All of the phenocrystic silicates therefore contain significantly less copper (in bulk, of the order of 20–5 per cent) than the apparent parental basalts. The only common (though very minor) mineral to contain more copper than the basalts is magnetite, with a mean of ≈196 ppm Cu.

When whole-rock analyses of the Solomons and other lava suites are plotted on copper–magnesium–total alkali ('CuMA') diagrams analogous to the FMA diagrams of standard petrological usage, the resultant fields of points exhibit characteristics reminiscent of those developed by total iron as FeO. This feature manifests itself with even greater sharpness when mean Cu–MgO–($Na_2O + K_2O$) points for successive 2 per cent intervals of whole-rock SiO_2 are plotted on the CuMA diagram. Plotting of analyses in this way indicates that volcanic island-arc lava assemblages are characteristically and pronouncedly hypotholeiitic with respect to copper.

When mean whole-rock copper corresponding to 2 per cent intervals of whole-rock SiO_2 is plotted against the latter, copper in the Solomons Suite rises from a relatively low level in the more mafic lavas to a peak at about 48–54 per cent SiO_2 (56–8 in the case of Tonga–Kermadec), and then decreases steadily as SiO_2 increases. This development of a well-defined copper maximum at around the basalt–andesite transition is a consistent feature of all of the five island-arc suites.

The initial rise in whole-rock copper content is readily attributed to crystallization and progressive subtraction of the silicates. However, on the basis of the copper contents and modal abundances of the principal phenocrystic minerals of the Solomons lavas, the subsequent decrease in whole-rock copper cannot be accounted for by crystal subtraction alone. The possibility that loss of copper in the volatile phase may be a contributing factor is considered in detail in Chapters 21 and 24.

7

Zinc

Apart from iron, zinc is the most abundant sulphide metal of the exhalative ores. Within the sulphides it occurs almost entirely as the cubic mineral sphalerite, ZnS, though unknown small quantities may also appear in sulphosalts. Substantial quantities of zinc exist in some exhalative orebodies as the spinel series gahnite–hercynite and as a component of olivines and other metamorphic silicates. Although the 'classical' Cyprus-type ores are often (somewhat loosely) regarded as pyrite:chalcopyrite assemblages, there are few that do not contain at least minute quantities of zinc, usually as microscopic emulsoid blebs within chalcopyrite. Some Cyprus-type ores, particularly those associated with more felsic basalts and basaltic andesites, contain quantities of zinc >1 per cent, though for commercial reasons this has commonly received little attention. Exhalative deposits associated with more felsic volcanism may contain very large quantities of zinc — of the order of 10^7 tonnes — which then dominates among the non-ferrous sulphide metals. It may be worth noting that whereas in the copper-dominated ores copper rarely exceeds 3–4 per cent, zinc in the zinc-dominated ores is commonly in the 8–12 per cent range, and not uncommonly exceeds 20 per cent. In some large exhalative orebodies, e.g. Kidd Creek, Ontario, zinc may occur as large bodies of almost massive sphalerite. The exhalation of zinc during the original deposition of such deposits must have been concentrated and copious.

One of the earlier estimations of zinc in magmatic rocks was that of Robertson (1894) who, in an investigation of lead, zinc, and copper in samples of granite, porphyry, and dolerite from the Archaean of Missouri, determined a range of 0.00139 per cent to 0.0176 per cent, and an average of 0.009 per cent, i.e. 90 ppm. Clarke and Steiger (1914) estimated 51 ppm Zn in a sampling of American magmatic rocks ranging from basalts to granites, and I. and W. Noddack (1934) proposed an overall figure of 70 ppm for 'magmatic rocks'. Sandell and Goldich (1943) found an average of 63 ppm in a large sampling of 'silicic' (SiO_2 >63 per cent) igneous rocks and 130 ppm in 'subsilicic' (SiO_2 <63 per cent) types — and an overall mean of 97 ppm. In the same study they determined a mean of ≈105 ppm Zn in the Keweenawan basalts (Kearsarge and Greenstone flows) of Michigan, though, as with copper, there was some indication of a relation with sulphur, and hence of the occurrence of at least a part of this zinc as sulphide. Their detailed analysis of a wide range of lava types from the young volcanic centre of Clear Lake, California, gave zinc contents from 75 ppm in olivine-basalt to 30 ppm in a biotitic rhyolitic pumice. An unweighted average for the group was 50 ppm, with the basalts–andesites averaging ≈67 ppm and the dacites–rhyolites ≈42 ppm.

Because of the low sensitivity of the emission spectrograph with respect to zinc (results using that technique were generally unreliable at Zn <200 ppm), the burst of geochemical activity of the 1950s was characterized by a consistent omission of zinc from published analytical tables. Thus, while the work of Wager and Mitchell on the Skaergaard intrusion and Hawaiian lavas (1951 and 1953 respectively), Patterson (1952) on the Irish lavas, and a number of their contemporary volcanic geochemists greatly expanded knowledge of the incidence of traces such as Cu, Co, Ni, Pb, V, Cr, Sr, and Ba, there was little or no extension of data on zinc during this period. This was to await the development of X-ray fluorescence, neutron-activation, atomic absorption spectroscopy, and other techniques, the application of which has led to the accumulation of a substantial data base in the years since about 1965.

Goldschmidt (1954 and earlier) emphasized the crystal-chemical similarities of zinc and ferrous iron, and hence the likelihood of sympathetic geochemical relationships. From this he drew attention to the importance of the iron and iron–titanium oxides in concentrating zinc, and observed that 'The concentration of zinc in oxidic iron ores from gabbroid rocks is shown by the fact that the ilmenite from such rocks contains up to 3000 ppm Zn, the magnetite up to 1000 ppm, and that the amount of zinc in chromite ores from olivine rocks may even surpass 10 000 ppm' (Goldschmidt 1954, p. 262). He also noted the possibility that some of the ferromagnesian silicates were likely to concentrate zinc, quoting an observation by Sandell and Goldich (1943) of an amphibole containing 160 ppm Zn within a granite containing 80 ppm. Goldschmidt (1954) estimated averages of 100–30 ppm Zn for 'gabbro, dolerite, and basalts' and 60 ppm for 'granites, granodiorites, and their volcanic equivalents'.

Wedepohl (1978a), in his compilation of data on the incidence of zinc in igneous rocks, notes that the range of scattering of the averages of zinc in basalts from different authors and areas is not large, principally between 80 and 120 ppm Zn. His arithmetical mean for the sample sets available to him (a total of 1657 analyses) at the time was close to 100 ppm Zn. The mean for 19 sets of data (660 analyses) for Tertiary to modern (i.e. completely unmetamorphosed) basalts, from both marine and continental settings, was 80 ppm Zn.

Wedepohl estimated a mean of 70 ppm Zn for 132 andesites and diorites and 80 ppm Zn for 303 dacites, rhyolites, pitchstones, and obsidians. As with the basalts, the data represent both marine and continental materials.

ZINC IN LAVAS OF THE MARINE ENVIRONMENT

As with copper, there is now a substantial body of data on the incidence of zinc in MORBs, and a summary of this is given in Table 6.2. The mean for 332 determinations of material from the Atlantic, Pacific, and Indian Ocean ridges is 84 ppm which, again as in the case of copper, is remarkably close to the estimate of 80 ppm from North American igneous rocks by Sandell and Goldich (1943). Weighted means for each of the Atlantic, Pacific, and Indian ocean MORBs are 72, 89, and 84 ppm respectively, and for the emergent and partly subaerial basalts of Iceland, 94 ppm. Over 90 per cent of all analyses lie in the 50–140 ppm range and 80 per cent in the 60–110 ppm range, as indicated in Fig. 7.1(a). Variation along and between the individual ridges of the three oceans is somewhat less than in the case of copper. Zinc contents of Icelandic lavas more felsic than basalts (SiO_2 >52 per

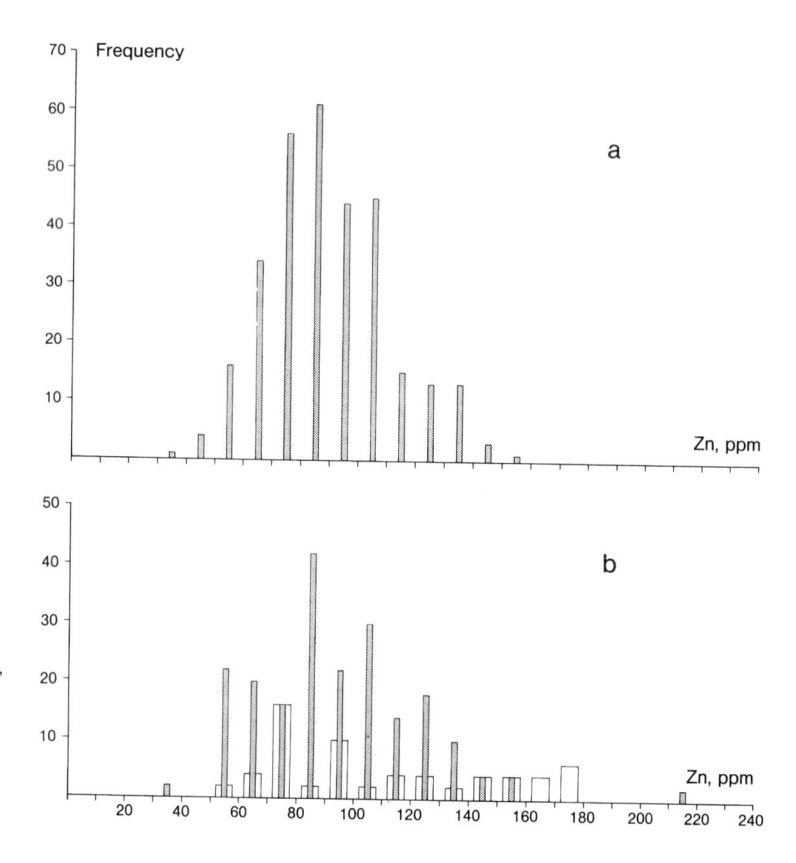

Fig. 7.1. Zinc abundances in (a) 315 analyses of mid-ocean ridge basalts (< 52 per cent SiO_2) representing Atlantic, Indian, and Pacific ocean ridges and (b) 102 analyses of Icelandic basalts. The small number of available analyses of Icelandic lavas more felsic than basalts ($SiO_2 \geqslant 52$ per cent; 31 analyses) are shown as open bars in (b).

cent), indicated by open bars in Fig. 7.1(b), do not appear to be significantly different from those of the basalts as far as the sampling goes. The samples are, however, few and of doubtful adequacy, and it is difficult not to suspect that significant differences would emerge from a more comprehensive sampling.

Zinc exhibits a clear if somewhat diffuse correlation with total iron as FeO in both MORBs and Icelandic basalts (Fig. 7.2) — a state of affairs well anticipated in general terms by Goldschmidt. The correlation and parallel behaviour of iron and zinc in MORBs, extending to some of the very high-iron–high-zinc basalts of the eastern Pacific ridges, is explored further in Chapter 24.

There are now also substantial data on the zinc content of basalts and more felsic lavas of volcanic island arcs (e.g. Brown *et al.* 1977; Ewart 1982; Gorton 1974), and this is summarized in Table 6.3 and Fig. 7.3.

The average zinc content of basalts (SiO_2 <52 per cent) of Grenada, Dominica, and St Kitts combined is 74 ppm, of the associated andesites (52 per cent <SiO <62 per cent) is 72 ppm, and of the dacitic and more felsic lavas

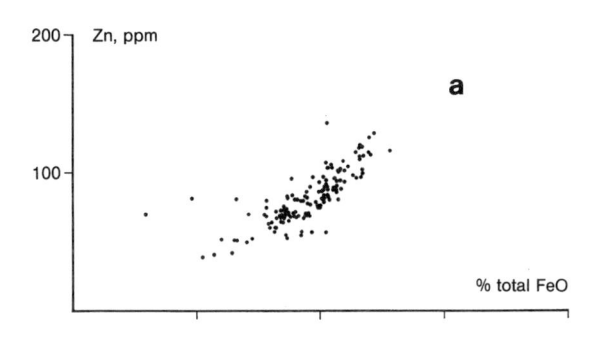

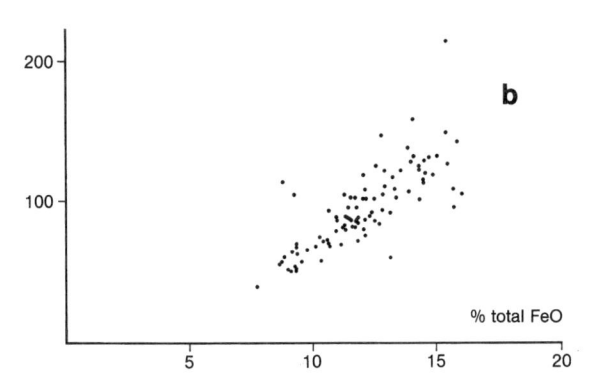

Fig. 7.2. Relations between zinc (ppm) and total iron as FeO (wt per cent) in (a) MORBs, as represented by basalts of the Mid-Atlantic Ridge and (b) Icelandic basalts.

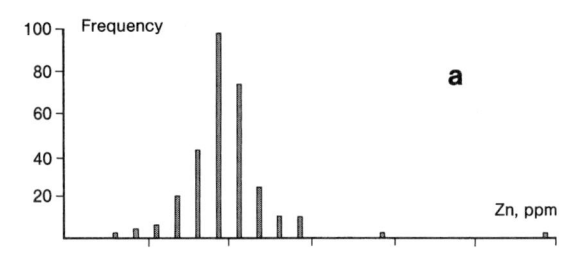

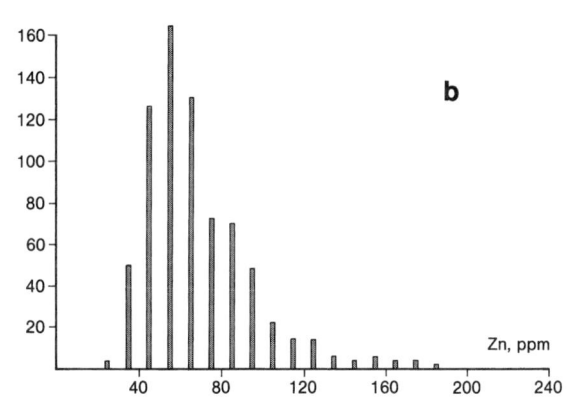

Fig. 7.3. Zinc abundances in (a) 274 analyses of island-arc basalts, and (b) 627 analyses of arc lavas more felsic than basalt ($SiO_2 \geqslant 52$ per cent).

56 ppm. The figures for basalts and andesites are clearly very close. Averages based on very much smaller samplings from the Aleutians are 74 ppm for basalts and 59 ppm for all lavas more felsic than basalts (SiO_2 >52 per cent). As with copper, zinc behaves differently in the Tonga–Kermadec lavas, the basalts containing an average of 81 ppm, andesites 93 ppm, and dacitic and more felsic types 107 ppm. Rather, however, than rising to a peak in the andesite range and then falling again in the dacitic rocks as in the case of copper, zinc in the Tonga–Kermadec lavas appears to rise consistently from basalts through andesites to dacites.

CHEMICAL PROPERTIES AND CRYSTAL CHEMISTRY OF ZINC

Zinc does not have the strong chalcophile nature that is such a notable characteristic of copper, and is found in both lithophile and chalcophile associations. For example, while the major part of the zinc of exhalative deposits occurs as the sulphide sphalerite, many such orebodies possess conspicuous, and in some cases abundant, gahnite

(Zn, Fe^{2+}) $O.Al_2O_3$, the zinc spinel. Other zinc-rich silicates such as the zincian olivine, roepperite, appear in many metamorphosed exhalative ores, and the occurrence of abundant silicate zinc in the great conformable deposits of Franklin and Sterling Hill, New Jersey, is famous. Goldschmidt (1954) observed that for the upper lithosphere zinc 'is more strongly lithophil than ferric iron, and less chalcophil than ferrous iron...'

Because of its competed d shell (Table 6.4) there is no ligand-field stabilization effect in the Zn^{2+} ion, and its stereochemistry is thus determined solely by considerations of size, electrostatic forces, and covalent bonding forces. Where Zn^{2+} coordinates O^{2-}, $R_a/R_x = 0.083/0.140 = 0.59$ (where R_a denotes the ionic radius of the cation and R_x denotes the ionic radius of the anion; Table 6.4), which is sufficiently close to the lower limit of radius ratio for 6-fold coordination to enable the zinc to assume a coordination number of 4, as in ZnO. In sulphides, $R_a/R_x = 0.083/0.185 = 0.45$, close to the upper limit (0.414) for 4-fold coordination, and we find that the sulphur is always tetrahedrally coordinated with respect to zinc.

On the basis of ionic size and valency, Fe^{2+} (0.082 nm) and Mn^{2+} (0.091 nm) are the abundant ions of silicate:-oxide compounds most likely to be substituted by, and hence to exhibit abundance relationships with, Zn^{2+} — a state of affairs similar to that already noted in connection with the stereochemistry of Cu^{2+}. However, whereas the latter tends to form a Cu–O bond considerably more covalent than Fe^{2+}–O and Mn^{2+}–O, and undergoes Jahn–Teller distortion when in 6-fold arrangement, the Zn–O bond (≈63 per cent ionic) has a degree of covalency very similar to those of Fe^{2+}–O and Mn^{2+}–O, and conforms to an undistorted coordination octahedron. As a result of these affinities, combined with its substantially lithophile nature, Zn^{2+} substitutes much more readily for Fe^{2+} and Mn^{2+} than does Cu^{2+} in the common silicates and oxides, and the three elements exhibit a number of sympathetic abundance relationships as a result.

The first to draw attention to the systematic ties between these elements in igneous minerals and rocks were Sandell and Goldich (1943), though Goldschmidt (1954), citing earlier work with N. H. Brundin and H. Horman, had observed that zinc enters the iron–magnesium silicates (such as augite, hornblende, and especially biotite) to a conspicuous degree. In their sampling of North American igneous rocks already referred to, Sandell and Goldich (1943) noted, *inter alia*, a close correlation of whole-rock zinc with iron, which, following Goldschmidt, they attributed to the similarity of ionic radii of zinc and ferrous iron. Using total iron calculated as Fe_2O_3 as their measure of whole-rock iron, Sandell and Goldich noted a general close correlation between the two elements through the full range of lava types. They also observed that MnO/Fe_2O_3 relations were linear or nearly so over a wide range of whole-rock iron. Thus they noted that a plot of zinc against MnO approached linearity with a ratio of Zn/MnO of approximately 0.07/1 for rocks with more than 0.1 per cent MnO. The ratio was 'considerably higher' in the more silicic rocks of their sample, in which MnO was generally less than 0.1 per cent. In carrying out limited analysis of mineral concentrates, Sandell and Goldich found unquantified high zinc in biotites from Texas and New Hampshire granites, 1600 ppm Zn in a hornblende from a New Hampshire granite, and 300–400 ppm in a magnetite–ilmenite concentrate from some basic rocks of Minnesota.

Since the time of Sandell and Goldich there has been a great understanding that zinc tends to follow iron and manganese in igneous rocks, that this is a consequence of systematic mineralogical affinities, and that this in turn stems from similarities of valency and ionic size. Little appears, however, to have been done in quantifying these relationships, or in examining the possible significance of bond type and stereochemistry in inducing the differences in substitutional behaviour between Cu^{2+} and Zn^{2+} in igneous, and particularly volcanic, silicates and oxides.

THE INCIDENCE OF ZINC IN THE PRINCIPAL MINERAL SPECIES OF THE SOLOMON ISLANDS YOUNGER VOLCANIC SUITE

General

As with copper, data on the incidence of zinc in the common volcanic silicates and oxides are sparse. Any metallic element recognized as a component of 'ores', and thus subject to the odium of being useful has, at least until recently, been regarded by most igneous petrologists as of little intrinsic interest to those of other than mercantile pursuit. The result has been that, apart from a small group including Sandell and Goldich, Goldschmidt, Wager, Mitchell (the soil scientists have long been aware of the importance of trace metals), and Vincent, few petrological geochemists have paid much attention to the chalcophile elements apart from cobalt and nickel. The result is a paucity of *systematic* data on these elements in the principal rock-forming silicates and oxides. One can only be philosophical and hope that enlightenment is close at hand.

In view of the well-known propensity for zinc to occur in substantial amounts in the metamorphic olivines associated with metamorphosed exhalative orebodies it is perhaps surprising that little work appears to have been done on the incidence of zinc in volcanic olivines. Wedepohl (1953) determined 51 ppm Zn in an olivine from an Etna basalt, and 50–65 (mean, 56) ppm zinc in olivines of peridotite inclusions in a Tertiary basalt of Germany. He also (see Wedepohl 1978a) determined 50–82 (mean, 74) ppm zinc in the peridotite inclusions of basaltic ejecta of Dreiser Weiher, Germany. Vincent (see Dissanayake and Vincent 1972; Vincent 1974) found 238–58 and 310 ppm Zn in the olivines of the Skaergaard Lower and Upper Zones respectively, while noting the possibility of loss in the outer zones of mineral separations due to abrasion induced in crushing and sieving. In view of the results on the Solomons olivines set out below it is likely that the variations in zinc content found by Wedepohl and Vincent are due to variation in occurrence, and concomitant Fe:Mg ratios, of the different olivines concerned.

The few available determinations of zinc in volcanic pyroxenes indicate wide variation from ≈20 to 450 ppm (see Wedepohl 1978a). Trachytic amphiboles from the Tertiary of Germany have been found to contain 257–690 (mean, 355) ppm Zn (Jasmund and Seck 1964) and plagioclases from Scottish and Icelandic pitchstones to contain 7.5–26 (mean, 11) ppm Zn (Carmichael and McDonald 1961). Vincent (1974) in his fastidious work on the mineral chemistry of the Skaergaard layers, determined 88–90, 98–118, and 135 ppm Zn in the pyroxenes, and 34–20, 28–19, and 20 ppm Zn in the plagioclases of the Lower, Middle, and Upper Zones respectively.

As in the case of copper, magnetite is by far the principal receptor of zinc among the common igneous minerals. Vincent (1974) found 817, 752–4, and 757 ppm Zn in the magnetites of the Lower, Middle, and Upper Skaergaard Zones respectively. Among the volcanic magnetites, various investigators have found a range of ≈270–2200 ppm Zn with an average for 22 samples of 1645 ppm. Compared with the present author's findings on the Solomons magnetites this is a somewhat high figure, but it is clear that the spinel structure is, as might have been expected, an important host for zinc in the volcanic milieu.

Solomons suite

Mean values of zinc in the principal silicates and in magnetites of the Solomons suite are given in Table 6.5, and spreads of values in the relevant minerals in individual lavas are indicated in the histograms of Figs. 7.4 and 7.12.

Fe–Mg silicates. In contrast to the consistency of mean copper abundance in the Solomons Fe–Mg silicates, the incidence of zinc varies widely from one mineral species to another, and within single mineral species as these occur in the different rock groups (Table 6.5). Also in

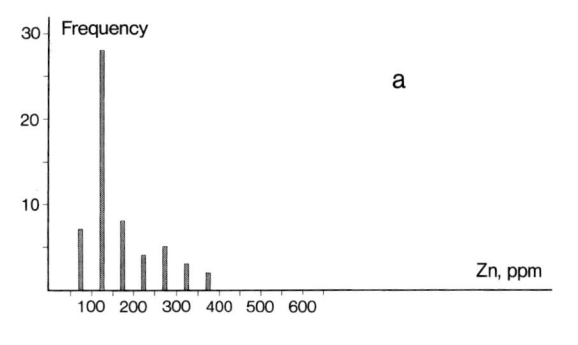

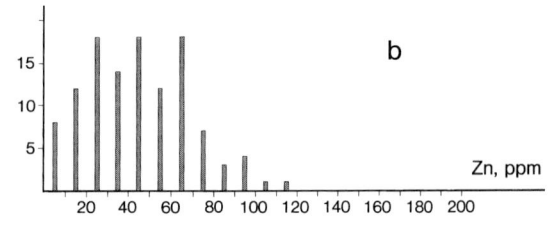

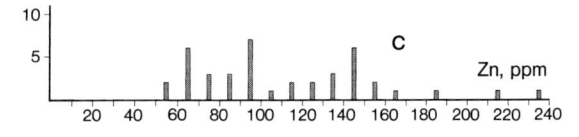

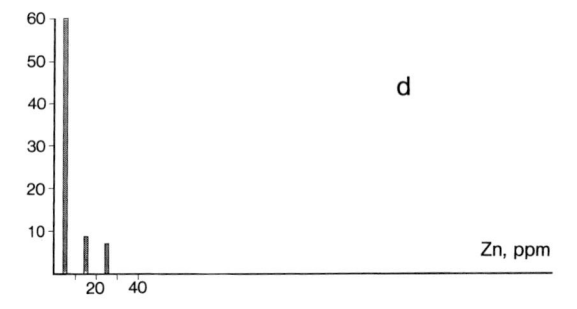

Fig. 7.4. Zinc abundances as determined by X-ray fluorescence analysis of mineral separations in (a) olivine, (b) clinopyroxene (including electron-microprobe analyses), (c) hornblende, and (d) plagioclase of the Solomon Islands Younger Volcanic Suite.

contrast to copper, and with the marked exception of biotite, zinc as expected exhibits a general correlation with total iron as FeO and, correspondingly, with MnO (Fig. 7.5). The relation between FeO and zinc in orthopyroxene *vis-à-vis* that in olivine appears slightly out of keeping with the general trend of the relationship but, as noted in the previous chapter, this may be a sampling artefact resulting from the relatively small number of orthopyroxene analyses.

80 Ore Elements in Arc Lavas

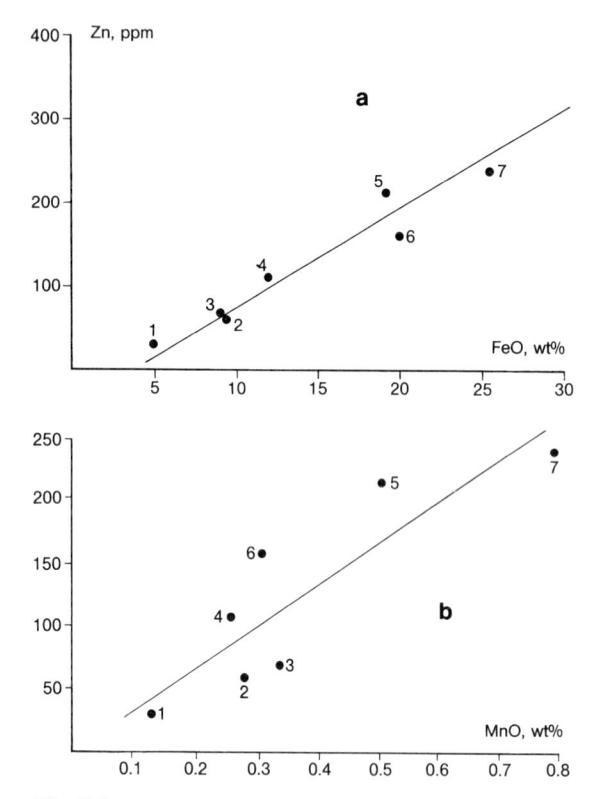

Fig. 7.5. Relations between mean zinc (ppm) and (a) mean total iron as FeO (wt per cent) and (b) mean MnO (wt per cent) in mineral separations of (1) clinopyroxene, olivine–clinopyroxene-basalts, (2) clinopyroxene, big-feldspar lavas, (3) clinopyroxene, hornblende-andesites, (4) hornblende, (5) orthopyroxene, two-pyroxene-bearing andesites, (6) olivine, olivine–clinopyroxene-basalts, and (7) olivine, big-feldspar lavas (see also Table 6.5).

Relations between Zn and FeO in individual olivines, clinopyroxenes, and hornblendes are shown in Figs. 7.6. Correlation is close in the olivines and then appears to decrease going from olivines to clinopyroxenes to hornblendes. The correlation between zinc and MnO is closer (Fig. 16.17), and appears to be at its best in the hornblendes. At least on the basis of individual electron probe analyses, correlation between zinc and iron and manganese is poor in the biotites.

Olivine. The relation between FeO and relatively high levels of zinc (>400 ppm) in olivines is a striking one which, somewhat surprisingly, does not appear to have been observed previously. The tie between nickel and the forsterite component of olivine (cf. Fig. 19.7) has been recognized for a long time, but that between zinc and the

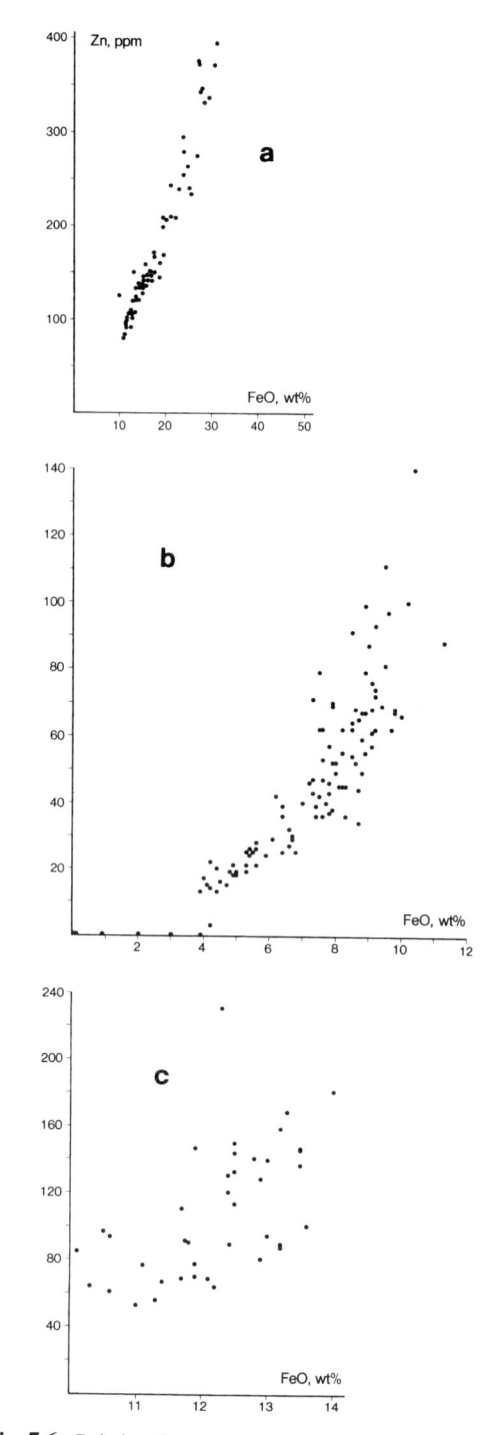

Fig. 7.6. Relations between zinc (ppm) and total iron as FeO (wt per cent) in (a) olivine, (b) clinopyroxene, and (c) hornblende in individual Solomons lavas.

fayalite component has not, to the author's acknowledge, been remarked upon. In view of the similarities of ionic size and valency of Zn^{2+} and Fe^{2+}, and the fact that $(Fe, Zn)_2SiO_4$ (roepperite) is a well-established member of the olivine family, the relationship is clearly one that might have been predicted. However, although Goldschmidt (1954) referred to the incidence of zinc in silicates 'such as pyroxenes, amphiboles, and biotites' he did not refer to its presence in olivines and hence to its strikingly close relationship with iron in the olivine minerals.

Whereas nickel does not enter the olivine structure significantly at MgO contents less than ≈30 per cent, zinc is present at about the 60–70 ppm level at 11–12 per cent olivine FeO, extrapolation (Fig. 7.6(a)) indicating first entry of zinc at around 3–4 per cent FeO. Maximum nickel in olivines is ≈2500 ppm, at ≈50 per cent MgO (i.e. ≈Fo 92), maximum zinc ≈400 ppm at ≈30 per cent FeO (i.e. ≈Fo 65). Also whereas Ni–MgO relations appear to be essentially linear between 30 per cent >MgO <50 per cent, Zn–FeO distributions are slightly curved for, if of linear nature, exhibit a slight inflection at approximately 23 per cent FeO, 230 ppm Zn (Fig. 7.6(a)).

This general close correlation between zinc *vis-à-vis* iron (and manganese) in the olivines of rocks ranging from picrites to basaltic andesites pointed to the possibility that similar relationships might hold for the various compositional zones within individual olivine crystals. Figs. 7.7(a) and (b) show, respectively, the results of across-grain electron-probe analysis of a large highly magnesian olivine of a picrite basalt, and of a more fayalitic grain occurring in a basaltic andesite. Zoning with respect to zinc is clearly present in both crystals. In the more iron-rich olivine (Fig. 7.7(b)) this is very simple and more or less directly related to that of FeO and MnO. In the more magnesium-rich crystal (Fig. 7.7(a)) zoning is present but complex, and shows an ambiguous to semi-random relationship with coexisting FeO and MnO contents. Clearly there is an overall direct tie between zinc, iron, and manganese in the Solomons volcanic olivines, but the detail of relationships on a fine scale requires further investigation.

Pyroxenes. Correlation of zinc and FeO in the clinopyroxenes is clear but not as close as in the olivines. As shown in Fig. 7.6(b), Zn falls to zero at about 2 per cent FeO, and exhibits a maximum of ≈140 ppm at ≈10.5 per cent total Fe as FeO. As in olivine, the relationship is either slightly curved, or shows a minor inflection (steepening) at ≈35 ppm Zn/8–9 per cent FeO. Data on the orthopyroxenes indicate that Zn–FeO relations in the latter are continuous with those of the clinopyroxenes, indicating (Figs. 7.5 and 7.8) that substitution of zinc in the Fe–Mg silicates is dependent more on the incidence of iron than the structure of the crystal. The degree of correlation between Mn and Zn in the clinopyroxenes is similar to that for Zn:FeO, though there may be a very slight inflection in the opposite sense at around 40 ppm Zn and 1750 ppm Mn (Fig. 16.17). Again the field for the orthopyroxenes is essentially complementary and collinear with that of the clinopyroxenes, but on the high Zn:high Mn side (Fig. 7.8(b)).

Like the olivines, the clinopyroxenes exhibit zoning with respect to iron and manganese and, to some extent concomitantly, with respect to zinc. Fig. 7.7c shows across-grain variation of FeO, MnO, and Zn in a large clinopyroxene coexisting with the olivine of Fig. 7.7(b). The clinopyroxene has a slightly higher mean Mg number (77.04) than the olivine (74.87), but lower FeO (8.14 per cent), MnO (0.22 per cent), and Zn (53 ppm) than the olivine (mean FeO = 22.70 per cent; MnO = 0.46 per cent; Zn = 133 ppm). FeO_{ol}/FeO_{cpx} is 2.79, MnO_{ol}/MnO_{cpx} is 2.09, and Zn_{ol}/Zn_{cpx} is 2.51. As with the more complex olivine of Fig. 7.7(a), however, zoning of Zn *vis-à-vis* FeO and MnO in the clinopyroxene is ambiguous: in the core and rim regions there is an indication of a sympathetic relation, but in the intermediate zones relationships appear random to inverse. As with the zoning of Zn in the more complex olivines, more work is required.

Hornblende. Apparently in keeping with its intermediate iron content, the zinc content of the hornblendes lies generally between those of the clinopyroxenes on the one hand and the orthopyroxenes and olivines on the other (Table 6.5; Fig. 7.4). The full analysed range is 53 ppm to 231 ppm, with a mean zinc content of 111 ppm.

The correlation of Zn and FeO in the hornblendes is distinctly less close than in the olivines and pyroxenes (Fig. 7.6(c)), but that between Zn and Mn in the hornblendes is closer than in the other two mineral groups (Fig. 16.7(c)). There are apparently factors more subtle than ionic size and valency influencing substitution of zinc for iron and manganese in the Fe–Mg silicates, and these are presumably related to electronegativities, consequent ratios of ionic to covalent bonding, and other elements of stereochemistry. The difference in degree of correlation, e.g. between zinc and iron in hornblende and olivines and zinc and manganese in hornblende and clinopyroxene, are so sharp as to indicate the influence of some very real and significant factor.

A striking feature of the incidence of zinc in the Solomons hornblendes is its increase, relative to the parent hornblendes, in the opaque amphibole derivative described in Chapter 5. The effects of such rim formation on the incidence of Zn and Mn in a single hornblende crystal are shown in Fig. 7.9 and, in 40 crystal-rim pair-

82 *Ore Elements in Arc Lavas*

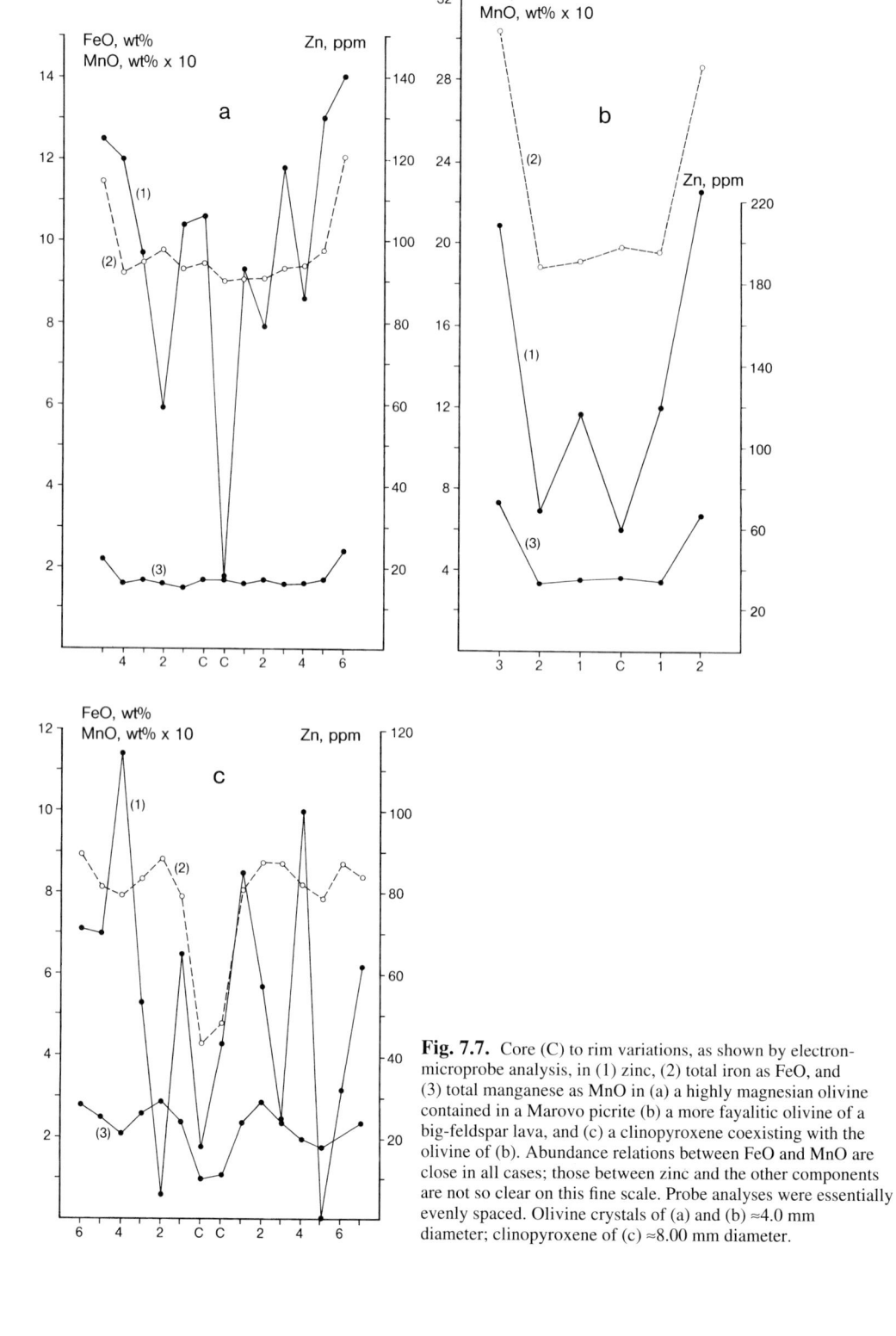

Fig. 7.7. Core (C) to rim variations, as shown by electron-microprobe analysis, in (1) zinc, (2) total iron as FeO, and (3) total manganese as MnO in (a) a highly magnesian olivine contained in a Marovo picrite (b) a more fayalitic olivine of a big-feldspar lava, and (c) a clinopyroxene coexisting with the olivine of (b). Abundance relations between FeO and MnO are close in all cases; those between zinc and the other components are not so clear on this fine scale. Probe analyses were essentially evenly spaced. Olivine crystals of (a) and (b) ≈4.0 mm diameter; clinopyroxene of (c) ≈8.00 mm diameter.

mean Zn for the parent hornblendes. The corresponding figures for MnO were 0.45 per cent and 0.26 per cent (Table 5.4).

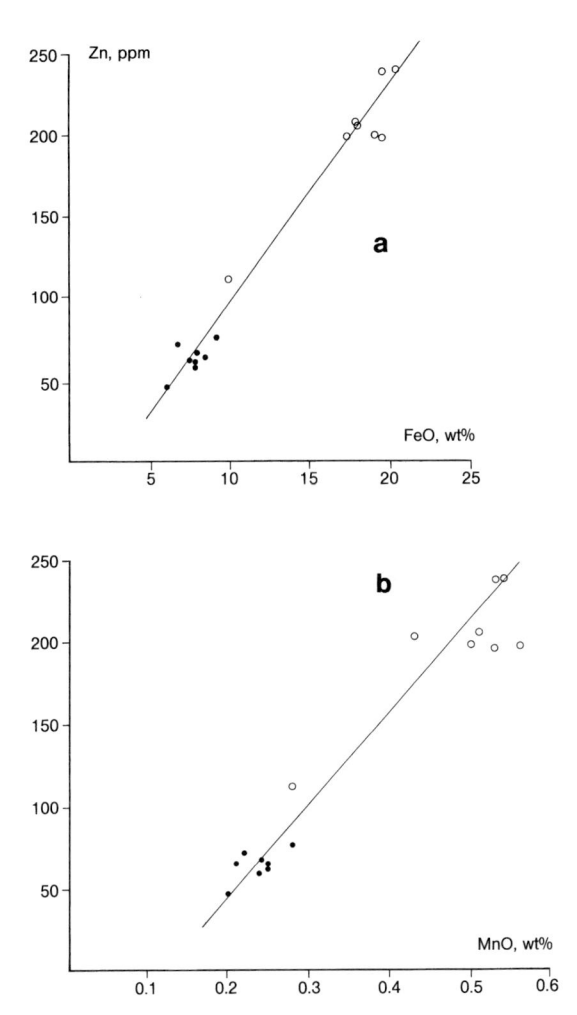

Fig. 7.8. Relations between mean zinc and (a) mean total iron as FeO and (b) mean total manganese as MnO in the coexisting pairs of ortho- and clinopyroxenes of eight Solomons two-pyroxene-bearing andesites. Clinopyroxenes denoted by filled circles, orthopyroxenes by open circles.

ings in the same lava, in Fig. 7.10. Early, but quite comprehensive, major element analysis of hornblendes and associated opaque materials in a large sampling of Solomons andesites by electronprobe had already indicated that elevation of MnO was a general feature of the opaque material. In view of this and the known close tie between zinc and manganese in the hornblendes, it was then suspected that zinc might occur at relatively high levels in the opaque material as a general feature. Electromagnetic/heavy liquid separations of such opaque material from seven of the Solomons hornblende andesites gave a mean of 173 ppm Zn as compared with the 111 ppm

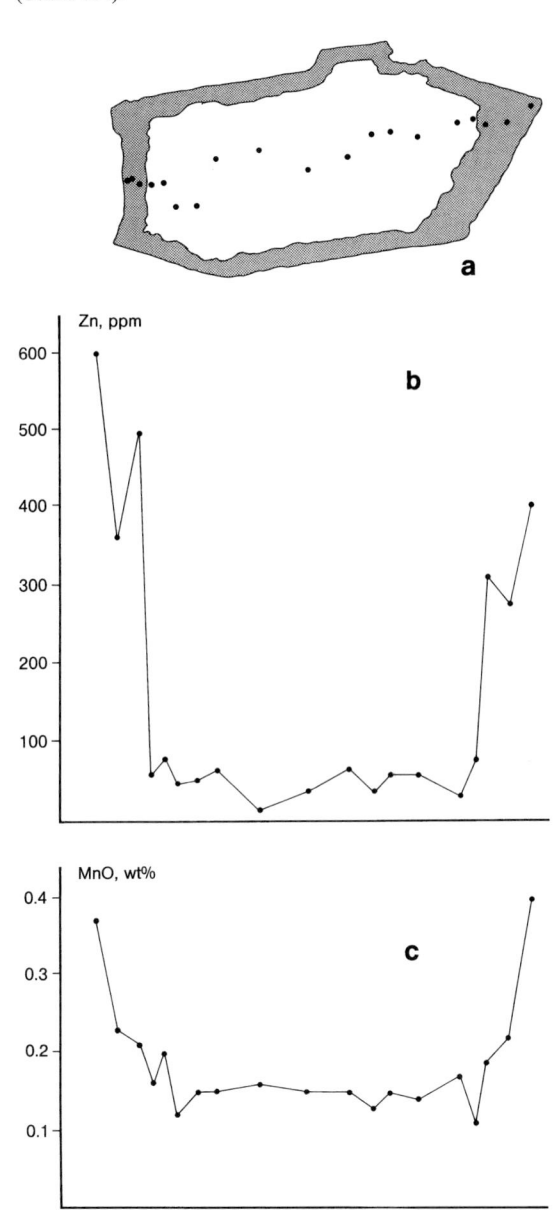

Fig. 7.9. (a) Camera lucida representation of the hornblende crystal and opaque rim of Plate 3; (b) zinc profile as revealed by the 19 probe analyses (positions indicated by black dots in (a)) along the length of the crystal/rim; and (c) total manganese as MnO profile [details as for (a)]. The diagram shows the distinctly higher abundance of zinc and manganese (particularly the former) in the rim material relative to the hornblende.

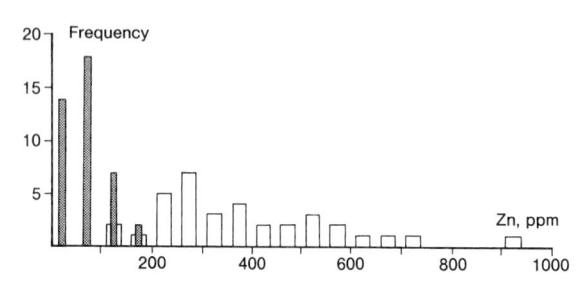

Fig. 7.10. Frequency distributions of zinc abundances in 40 paired analyses of hornblende crystals (filled bars, mean zinc = 40 ppm) and their opaque rims (open bars, mean zinc = 382 ppm) in a hornblende-andesite from Mt Gallego, Guadalcanal.

To test this more comprehensively on the scale of Fig. 7.9 'paired' analyses of 121 hornblendes and 121 adjacent opaque derivatives, in two Solomons andesites, were carried out by the electron probe (Table 5.5), confirming the results given in Table 5.4 and in Fig. 7.9 and 7.10. The results are substantially consistent, and show a sympathetic increase of FeO, MnO, and Zn in rim material relative to the parent hornblendes. FeO increases by 20–70 per cent (see also Chapter 5), MnO by factors of 2–3, and Zn by factors of 3–5 in rim relative to hornblende. As is considered in more detail in Chapter 24, this process must be important in influencing the abundance pattern of zinc in the melt as it converts a material — hornblende — accommodating zinc in only slightly higher concentration than it occurs in the parent melt to one — the opaque material — capable of subtracting zinc at over three times the concentration of this in the melt.

Biotite. This is clearly the principal receptor of zinc among the silicates. The 33 grains in each of the two Savo dacites referred to in Chapters 5 and 6 gave means of 412 and 409 ppm Zn respectively, with spreads of values as indicated in Fig. 7.11. Clearly inter-grain variation in zinc content is substantial: in the dacite of Fig. 7.11(b), for example, biotite zinc ranges from 241 to 568 ppm — a factor of more than two. The corresponding mean Mg numbers of the two groups of biotites were 0.48 and 0.53, and their mean MnO contents 0.37 and 0.29 per cent, indicating that (the almost infinitesimally) higher zinc was associated with higher total iron and manganese. As noted earlier, however, this general tie between Zn, Fe, and Mn does not reflect a high degree of correlation in individual probe analyses.

Feldspar. In view of the very low amounts of iron (and manganese) occurring as intrinsic, i.e. lattice-bound, components of the feldspars, it might have been expected that the abundance of zinc would also be low, and this is the case as indicated in Table 6.5. The overall corrected mean abundance of zinc in the 91 analysed feldspars is 4.6 ppm, with a very narrow spread of values (Fig. 7.4d).

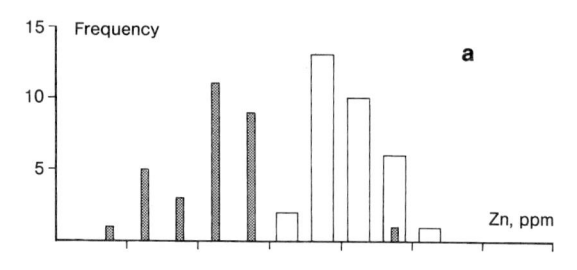

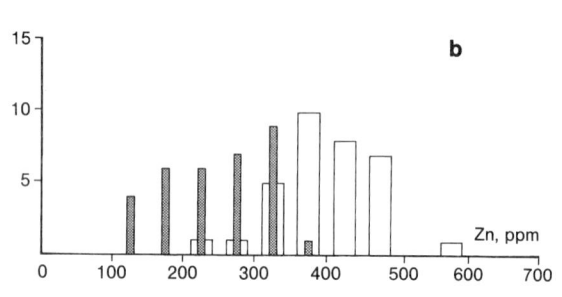

Fig. 7.11. Frequency distributions of zinc abundances in the biotites (open bars) and coexisting hornblendes (filled bars) of the two dacites from Savo. Mean zinc in biotite and hornblende of (a), 412 and 232 ppm respectively; of (b), 409 and 224 ppm respectively.

As in the case of copper, initial analyses of the Solomons feldspars yielded unexpectedly high values for zinc. The picrite basalt feldspars gave 6 ppm and those of the big-feldspar basaltic andesites and hornblende-andesites both gave 10 ppm, with a weighted mean of 9.6 ppm for the full 91 separations. Detailed electron probe analysis of the same glass inclusions as those analysed for copper (Chapter 6) revealed these also to contain conspicuous zinc (Table 6.6). The feldspar analyses were then corrected with respect to zinc by the method used for copper, yielding the results given in Table 6.5.

Spinel. As with copper, spinel — chiefly magnetite — is the principal mineral receptor for zinc in the Solomons lavas. The overall weighted mean zinc for the 91 magnetite separations is 560 ppm, and the spread of values in a total of 96 magnetites (XRF + probe analyses) is indicated in Fig. 7.12.

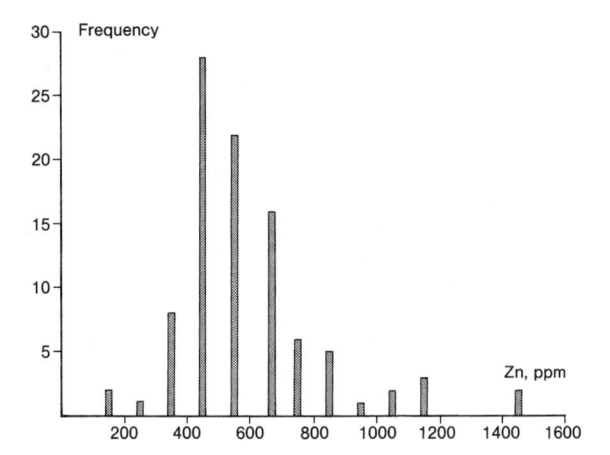

Fig. 7.12. Frequency distribution of zinc abundances in the magnetites of 96 Solomons lavas (X-ray fluorescence analyses of 91 separations; averages of probe analyses of 20 magnetite grains in each of the remaining 5 lavas).

Although the general relationship between mean zinc and mean total iron already noted in the silicates appears to continue on to magnetite (Fig. 7.13(a)), the correlation does not appear to extend to individual magnetites. The tie between zinc and manganese is conspicuously better (Fig. 7.13(b)), particularly in those magnetites containing relatively lower levels of these two elements.

The lack of a systematic relationship between zinc and iron — particularly FeO — in these magnetites is perhaps a little surprising in view of the similarities of the ions, and the close ties between the two elements in the coexisting silicates. It is also perhaps surprising in the light of charge/size considerations that correlation of zinc with manganese is closer than that with FeO. With Zn^{2+}, Fe^{2+}, and Mn^{2+} having ionic radii of 0.083, 0.082, and 0.091 nm respectively, the reverse might have been anticipated. However Zn^{2+} and Mn^{2+} have slightly closer electronegativities (1.5 and 1.4 respectively) than Fe^{2+} (1.65) and hence form M–O bonds of slightly closer ionic:covalent ratio. The resultant difference in bond type is minimal, but it may just account for the difference in Zn–Fe and Zn–Mn relations in the close-packed structure of the magnetites.

These differences are demonstrated again when the behaviour of zinc and manganese in the magnetites is examined with respect to the evolution of the lava series containing them. Fig. 7.14(a), showing change in the incidence of mean zinc, FeO, Fe_2O_3 and (FeO + Fe_2O_3) in the magnetites with respect to 2 per cent whole-rock SiO_2 intervals from <48 per cent to >64 per cent SiO_2, indicates a lack of any close relationship between zinc and iron abundances in the magnetite as the lava series evolves. The early rise of FeO, Fe_2O_3, and (FeO + Fe_2O_3), at first sight concomitant with the lower part of the zinc 'curve', is due to the early decrease in magnetite MgO, Al_2O_3, TiO_2, and, especially, Cr_2O_3 (Fig. 5.16).

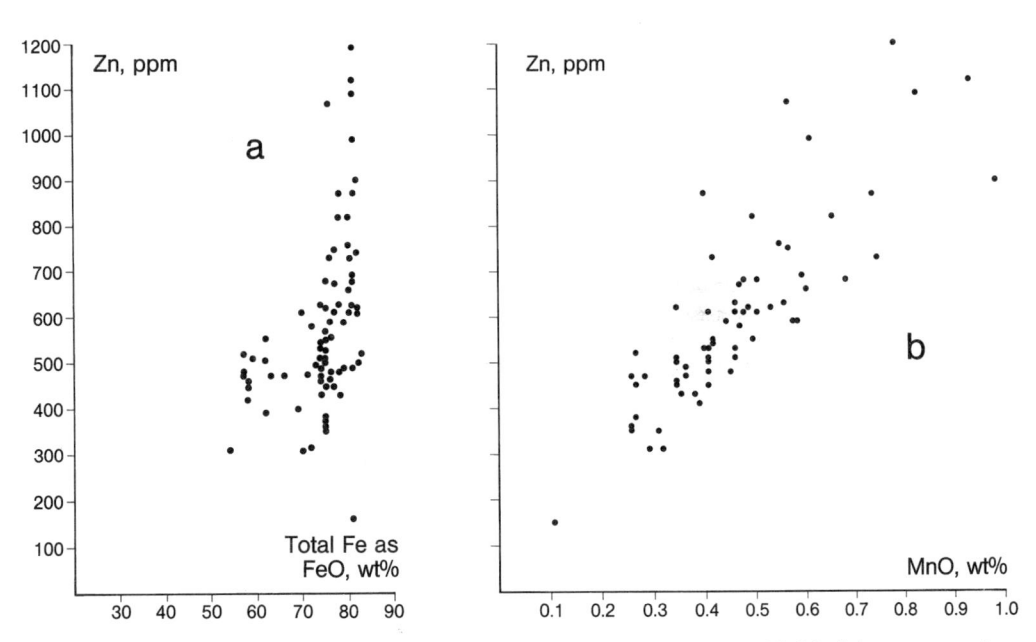

Fig. 7.13. Relations between zinc and (a) total iron as FeO and (b) total manganese as MnO in Solomons magnetites.

86 Ore Elements in Arc Lavas

While this might be suspected to have had an influence on the reception of zinc into the magnetite structure, such a suggestion is belied by relationships above $c.$ 50–2 per cent SiO_2, at which stage the zinc curve clearly diverges from those for iron.

On the other hand, Fig. 7.14(b) not only re-emphasizes the zinc/manganese relationship, but shows this to reflect, *inter alia*, the evolution of the containing lavas. As the lava series becomes progressively more felsic, zinc and manganese in the magnetites increase in parallel at least to rocks of the 64–6 per cent SiO_2 interval. While lavas with SiO_2 >66 per cent are unusual in the Solomons Younger Volcanic Suite, and obtaining them requires special search, investigation of magnetite zinc at later stages of lava evolution clearly requires extra sampling in the high SiO_2 range.

A rather elegant illustration of the tie between magnetite composition and lava evolution is provided by a comparison of the two sets of magnetites grouped according to their textural contexts and referred to in Chapters 5 and 6. It will be recalled that one set of magnetite crystals was included in the large pyroxene phenocrysts of an olivine–pyroxene-basalt, and presumably constituted the earlier-formed magnetites of this particular lava. Those in the other set were all components of the fine groundmass, and thus appeared to have crystallized later. Zinc and manganese contents of 33 grains of each category are shown in Fig. 7.15: the earlier magnetites included in the pyroxene contained mean zinc and manganese of 407 and 3237 ppm, the later, groundmass magnetites contained mean zinc and manganese of 508 and 4195 ppm respectively, apparently neatly corroborating the evidence.

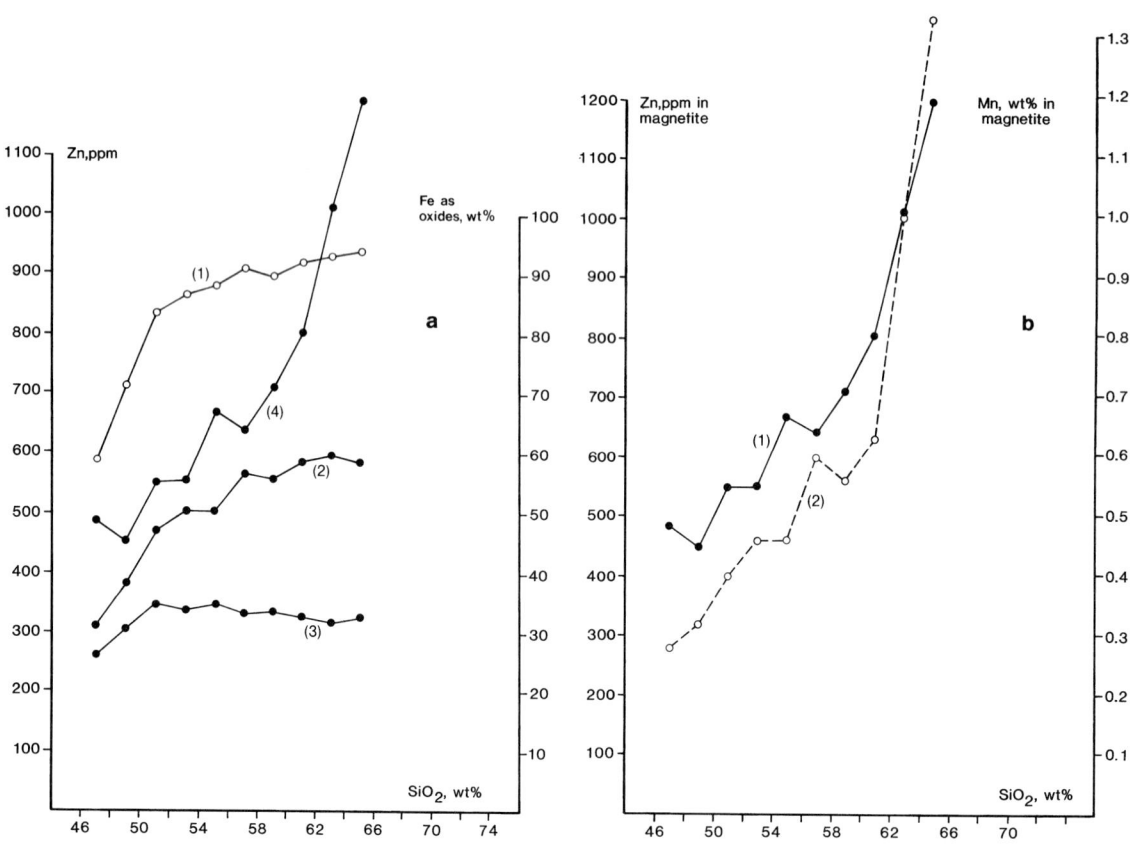

Fig. 7.14. (a) Abundances of (1) $FeO + Fe_2O_3$, (2) Fe_2O_3, (3) FeO, and (4) Zn in Solomons magnetites, with respect to 2 per cent intervals of whole-rock SiO_2 in the containing lavas; and (b) relations between (1) Zn and (2) Mn in the same magnetites, with respect to 2 per cent intervals of whole-rock SiO_2 in the containing lavas.

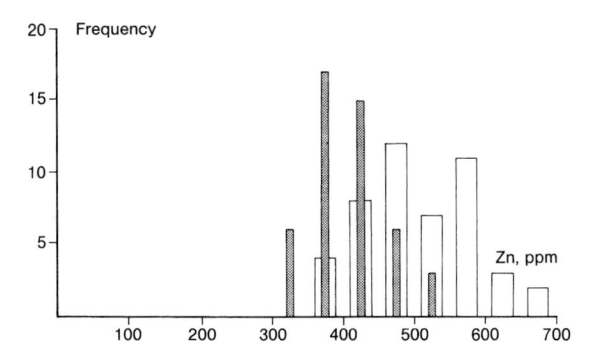

Fig. 7.15. Frequency distribution of zinc abundances in magnetites included within the clinopyroxenes (filled bars) and within the groundmass (open bars) of the Rendova ankaramitic basalt referred to in Chapters 5 and 6.

CRYSTAL:MELT PARTITIONING

The stereochemical similarities, and hence the geochemical affinities, of Zn^{2+} vis-à-vis Fe^{2+} and Mn^{2+} leads inevitably to a concentration of zinc in those minerals that tend also to concentrate iron and manganese with respect to the coexisting melt: the state of affairs predicted by Goldschmidt and first comprehensively observed and documented by Sandell and Goldich (1943). As we have seen (and as we shall see further in the following pages) this pattern of association is well developed in the principal silicates of the Solomon Island younger lavas. In view of these quite conspicuous relationships it might have been expected that much would now be known about the partitioning behaviour of zinc in igneous systems, but this is not the case — a state of affairs due in part at least to the relative insensitivity of zinc in emission spectrography, and the resultant analytical difficulties encountered by early investigators.

General

For the reasons given above, Wager and Mitchell (1951) did not determine zinc in the Skaergaard rocks and minerals even though, as it happens, zinc occurs in some of the Skaergaard olivines in abundances that might have been estimated with reasonable accuracy by emission spectrography. Although, as shown in the preceding section, the more iron-rich members of the olivine series may accept zinc in amounts at least up to 400 ppm, relatively little information on partitioning has accumulated in the literature. Gunn (1971) deduced a value of 0.95 for the olivine of an Hawaiian olivine tholeiite, Bougault and Hekinian (1974) 0.86 for that of a mid-Atlantic Ridge basalt, Paster et al. (1974) 1.8 for olivine in the Upper Zone of the Skaergaard gabbroic intrusion, and Dostal et al. (1983) 1.2 in plutonic cumulate material in a Lesser Antilles lava. At these values the olivines are more or less just at the point where they might deplete the melt in zinc.

A few results on zinc partitioning between clinopyroxene and basaltic melt have been obtained by some of the above investigators. Paster et al. (1974) determined K_D^{Zn} of 0.49 for clinopyroxene:groundmass in the Upper Zone of the Skaergaard intrusion, Bougault and Hekinian (1974) 0.50 for clinopyroxene in the Mid-Atlantic Ridge basalt, and Dostal et al. (1983) 0.31 for clinopyroxene in the Lesser Antilles cumulate inclusion referred to above.

Little appears to have been done on partitioning of zinc between hornblende and coexisting melts, and a survey of the literature has revealed only an estimate by Dostal et al. (1983), who obtained a K_D^{Zn} amphibole:melt value of 0.40 for material contained in the cumulate inclusion already referred to. In view of its clear and substantial capacity for accepting zinc into its structure — even more pronounced than that of olivine — it is remarkable that there are virtually no reliable data on the partitioning of zinc by biotite. For plagioclase, Paster et al. (1974) determined values of 0.13 and 0.14 for material of the Upper Zone of the Skaergaard layered series, and Bougault and Hekinian (1974) obtained a value of 0.11 for plagioclase in their sample of mid-Atlantic Ridge basalts.

Although there have been a number of studies of zinc in magnetite, little attention has been paid to quantitative aspects of partitioning. In the studies already referred to, Paster et al. (1974) determined K_D^{Zn} magnetite:melt = 2.6 in the Upper Zone of the Skaergaard intrusion, and Dostal et al. (1983) a value of 3.1 in the Lesser Antilles cumulate inclusion.

Solomons suite

As indicated in Figs. 7.5 and 7.6 (see also Fig. 16.17), the concentration of zinc in the principal Solomons ferromagnesian silicates appears to bear simple, essentially linear relationships with iron and manganese contents and is thus apparently independent of differences in host crystal structure. Of the four principal silicate mineral groups of the Solomons lavas, two exhibit mean $K_D^{Zn} > 1$ and two show mean $K_D^{Zn} < 1$ (Table 7.1). Thus in contrast to copper, differing patterns of silicate crystallization have the potential to induce quite different patterns of zinc abundance in the evolving melt.

K_D^{Zn} for olivine:groundmass determinations for olivine–pyroxene lavas gave a mean of 2.64 (Table 7.1), with a frequency distribution of values as shown in Fig. 7.16(a). Mean K_D^{Zn} for the olivines of the big-feldspar basalts is slightly higher at 3.86 and the overall mean is 2.80. The partition coefficient for those rocks, which might

Table 7.1 Crystal-melt distribution coefficients (K_D) with respect to zinc

Mineral species	Lava group	n_1	n_2	m_1	m_2	$K_{D(1)}$	N	M_1	M_2	$K_{D(2)}$
Olivine	OPB	40	44	156	57	2.74	39	148	56	2.64
	BFL	7	23	285	74	3.85	6	286	74	3.86
Clinopyroxene	OPB	44	44	30	57	0.53	42	24	57	0.42
	BFL	20	23	58	74	0.78	17	58	74	0.78
	HbA	25	47	67	41	1.63	20	64	45	1.42
Hornblende	HbA	25	47	111	41	2.71	22	128	43	2.98
Feldspar	OPB	10	44	7	57	0.12	10	7	57	0.12
	BFL	23	23	6	74	0.08	22	7	74	0.09
	HbA	46	47	3	41	0.07	44	5	42	0.12
Spinel	OPB	20	44	490	57	8.60	20	490	54	9.07
	BFL	23	23	487	74	6.58	22	479	74	6.47
	HbA	37	47	643	41	15.68	36	699	41	17.05

OPB = olivine–clinopyroxene-basalt group; BFL = big-feldspar lavas: HbA = hornblende-andesite group; n_1 = total number of mineral separations analysed; n_2 = total number of groundmass separations analysed; m_1 = mean zinc content of mineral of each lava group; m_2 = mean zinc content of groundmass of each lava group; $K_{D(1)}$ = mean distribution coefficient determined as m_1/m_2; N = number of coexisting mineral:groundmass pairs analysed; M_1 and M_2 = mean Zn contents of coexisting mineral and groundmass respectively; $K_{D(2)}$ = distribution coefficient determined as M_1/M_2, for comparison with $K_{D(1)}$. All concentrations as parts per million by mass.

reasonably be interpreted as later in the crystallization-differentiation sequence, and hence of lower temperature of solidification, is thus the higher — a state of affairs that would be expected if partition coefficients were temperature-dependent. Whether in this case the difference is the direct result of temperature, or an indirect result through increased iron and manganese contents at what may have been lower temperatures, is not, however, obvious.

The above values for K_D^{Zn} for olivine:melt in basaltic rocks are rather higher than those derived from other materials by previous investigators.

K_D^{Zn} values for clinopyroxene: liquid in the Solomons basaltic rocks (Table 7.1) are generally similar to those of earlier determinations. The mean K_D^{Zn} for the more mafic materials of the olivine–pyroxene-basalts is 0.42 and that for the somewhat more felsic types of the big feldspar group is 0.78. As indicated in Table 6.5, this corresponds with substantial increase in both iron and manganese in the relevant pyroxenes and it is therefore not clear whether the increase in K_D^{Zn} bears a direct or indirect tie with temperature of crystallization.

K_D^{Zn} clinopyroxene:melt is distinctly higher in the hornblende-andesites, at a mean value of 1.42. This is accompanied by a slight decrease in the mean iron of the pyroxene (and of the melt) and an increase in manganese. The data thus indicate that at the basaltic stage of lava evolution the crystallization of clinopyroxene tends to increase the concentration of zinc in the residual melt, whereas with the onset of andesite development crystallization of clinopyroxene depletes the zinc in the melt.

Probe analysis of the rim zones of coexisting olivine:-clinopyroxene in two olivine–pyroxene lavas (25 pairs of grains) gave a mean K_D^{Zn} olivine:clinopyroxene of 3.1, and in six big-feldspar lavas (60 pairs of grains), a mean of 5.9.

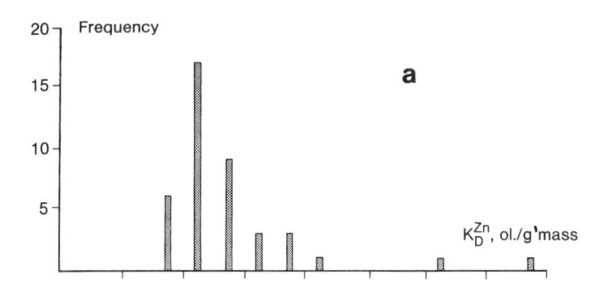

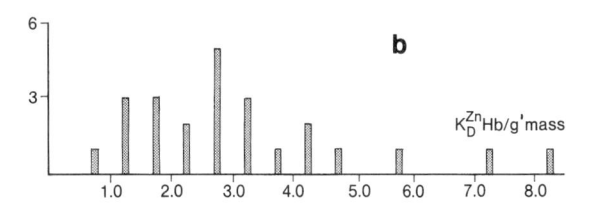

Fig. 7.16. Frequency distributions of values of (a) K_D^{Zn} olivine:groundmass in 39 olivine–pyroxene basalts, and (b) K_D^{Zn} hornblende:groundmass in 22 hornblende-andesites.

K_D^{Zn} values for orthopyroxene in basaltic andesites and andesites have been obtained by extrapolation using XRF analyses of mixed orthopyroxene/clinopyroxene separations, and are clearly higher than for the corresponding clinopyroxenes. Best estimates from data of this type are in the range 4.0–5.0. Probe analysis of the rims of 90 closely coexisting and contacting orthopyroxene:clinopyroxene pairs (180 grains in all) in eight two-pyroxene andesites gave relatively consistent results and a mean K_D^{Zn} orthopyroxene: clinopyroxene of 3.16 (Fig. 7.17). This is a little lower than might have been expected from the calculations above, though the latter involved whole-grain analyses as compared with the rims to which probe analysis was confined.

The mean value of 2.98 obtained by the present writer for 22 hornblende:groundmass pairings from the Solomons hornblende-andesites is clearly much higher than that found by Dostal *et al.* (1983). There is some scatter in the values for individual pairings (Fig. 7.16(b)) but there is substantial consistency between $K_{D(1)}$ and $K_{D(2)}$ of Table 7.1, indicating that a value of 2.7–3.0 represents a reliable mean for these materials.

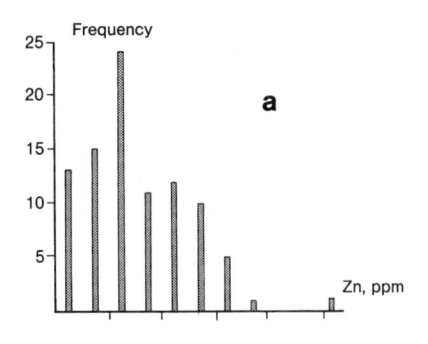

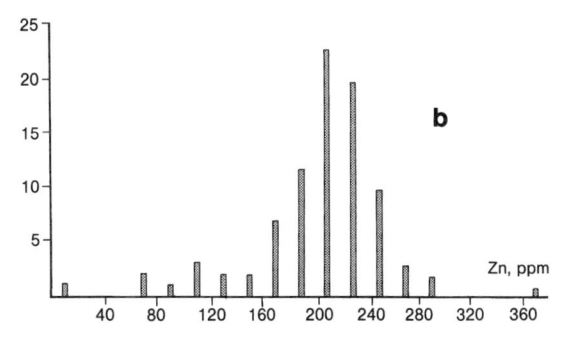

Fig. 7.17. Frequency distributions of zinc abundances in 90 pairs coexisting clino- and orthopyroxenes: (a) clinopyroxenes, mean = 63 ppm Zn, and (b) orthopyroxenes, mean = 199 ppm Zn.

On the basis of the limited data of the two Savo dacites, K_D^{Zn} for biotite:groundmass is 9, and for coexisting biotite:hornblende 1.8 (see Fig. 7.11).

The plagioclase series is an important influence on the distribution of zinc in that this element is almost completely excluded from the plagioclase structure. The minute amounts of zinc that do occur in plagioclase (Table 6.5) are presumably associated with the very minor iron and manganese of the feldspar lattice. As a result, values of K_D^{Zn} feldspar:melt obtained for the Solomons lavas are very low: for 76 feldspar:groundmass pairings the mean value is 0.11 (Table 7.1). The highest mean value, 0.12, is developed in the olivine–clinopyroxene-basalts and hornblende-andesites and the lowest, 0.09, in the big-feldspar lavas. These results correspond well with values of 0.13 obtained by Paster *et al.* (1974) for the Skaergaard Upper Zone and 0.11 obtained by Bougault and Hekinian (1974) for a Mid-Atlantic Ridge basalt.

As a major receptor of zinc in the Solomons lavas (Table 6.5) magnetite partitions this element strongly from the melt — a state of affairs reflected in the very high K_D^{Zn} magnetite:groundmass values given in Table 7.1. Partitioning is very strong in the hornblende-andesites in which the magnetite achieves its highest zinc content, the residual melt (groundmass) its lowest, and the resulting K_D^{Zn} is ≈17.

Striking evidence of the propensity of the iron-bearing minerals, especially magnetite, for sequestering zinc from the melt is provided by serial electromagnetic separations of individual groundmasses. Any separated groundmass may be simply divided into a succession of fractions by running the material through an electromagnetic separator at progressively increased current, removing the magnetic fraction at each run and then re-running the remainder at a slightly higher field. In this way the most magnetic fraction is removed first, the least last, and a series of 6–7 separations based on magnetic susceptibility is obtained.

Simple groundmass series of this kind were produced from the separated groundmasses of seven of the Solomons lavas, and the resultant fractions analysed by XRF for the standard major and trace elements. Three of these, representing groundmasses from each of the three major lava types, are shown in Fig. 7.18.

The 'primary' groundmasses of all three consist of varying proportions of glass, fine silicate (chiefly feldspar crystallites), and magnetite dust. The iron content of each fraction, and its distinctive magnetitic susceptibility, is substantially a reflection of this magnetite dust.

Fig. 7.18 shows that in all cases the relationship between zinc and iron, i.e. essentially very fine magnetite, is linear with a very high degree of correlation. That is, the rapid crystallization of any residual melt results in the

90 *Ore Elements in Arc Lavas*

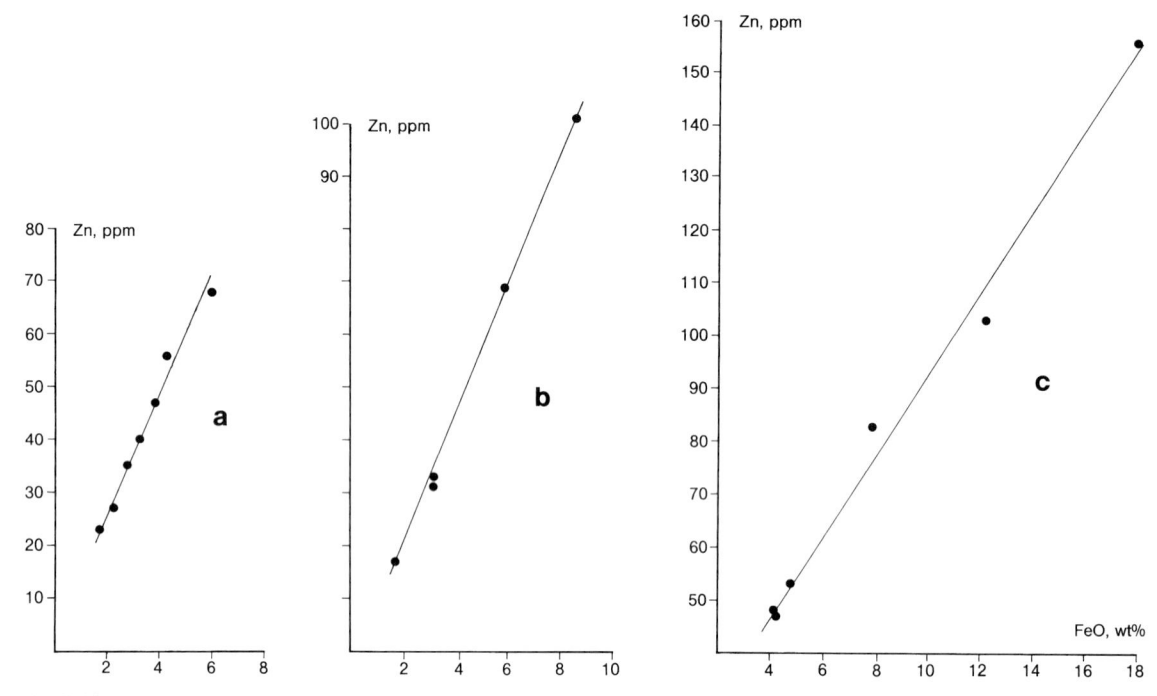

Fig. 7.18. Relations between zinc and total iron as FeO in groundmass fractions separated, according to magnetic susceptibility, from the total groundmass of a representative of each of (a) olivine–clinopyroxene-basalt (b) big-feldspar lava, and (c) hornblende-andesite.

incorporation (capturing) of virtually all of its zinc in fine crystals of magnetite — a particularly good example of highly preferential partitioning.

THE INCIDENCE OF WHOLE-ROCK ZINC IN THE SOLOMON ISLANDS YOUNGER VOLCANIC SUITE

Mean zinc for the suite as a whole is 71 ppm and for the basaltic members, 75 ppm. The latter compares with a mean of 84 ppm for all analysed MORBs (72 ppm for those of the Atlantic and 89 and 84 ppm respectively for those of the Pacific and Indian Oceans), as noted earlier in this chapter. Brown *et al*. (1977) determined a weighted mean of 72 ppm zinc in the volcanic products of the lesser Antilles as a whole and, as noted previously, the weighted mean zinc content of basalts of the three main islands is 74 ppm. Mean zinc content of the Tonga–Kermadec basalts is 81 ppm, of Vanuatu 84 ppm, and of the Aleutians 74 ppm. The overall zinc content of the Solomons lavas thus conforms quite closely with those of most other major lava bodies of the marine environment, and this is particularly the case with the basaltic members.

Frequency distributions of zinc abundances in the Solomons basalts (SiO_2 <52 per cent) and more felsic members (SiO_2 >52 per cent) are given in Fig. 7.19 (cf. also Fig. 7.3 and Table 6.3). As with the other volcanic arcs (Fig. 7.3) the two distributions are generally similar but that for the basalts is rather sharper and has a slightly higher mean (75 ppm) than the distribution for the andesites and dacites (mean, 67 ppm).

The distribution of zinc in the three principal Solomons lava types (Fig. 7.20) is a subdued analogue of that shown by copper, i.e. the lava group occupying the central position in the apparent differentiation series (the big-feldspar basalts and basaltic andesites) has slightly higher zinc (mean = 80 ppm) than do the olivine–pyroxene-basalts (mean = 72 ppm) and the hornblende-andesites (mean = 67 ppm). In addition to having lower means than those of copper, the zinc distributions have distinctly smaller spreads — a feature also exhibited by zinc in MORBs and lavas of other arcs (compare, e.g., Figs. 6.1 and 7.1).

Behaviour of zinc in the whole rocks *vis-à-vis* groundmass, i.e. phenocrysts *vis-à-vis* coexisting melt, is also somewhat different from copper. Whereas copious crystallization of clinopyroxene and olivine, both accepting little copper, led to early enrichment of this metal in the residual melts of the pyroxene–olivine-basalts

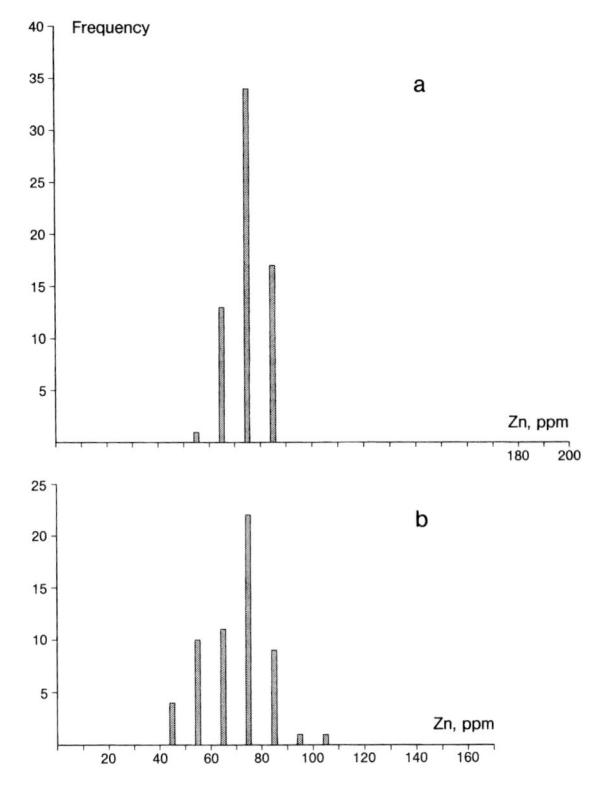

Fig. 7.19. Frequency distributions of zinc abundances in (a) basalts (SiO$_2$ < 52 per cent; 64 samples, mean = 75 ppm Zn); and (b) lavas more felsic than basalts (SiO$_2$ ⩾ 52 per cent; 58 samples, mean 67 ppm Zn) of the Solomon Islands Younger Volcanic Suite.

the reason for the development of the observed whole-rock/groundmass zinc relations here, the overall reduction of zinc against the trend of crystallization-induced differentiation is, as for copper, not readily explained on the basis of simple crystal subtraction.

(see Fig. 6.9(c); mean copper in whole-rock = 78, in groundmass = 97 ppm), crystallization of olivine of mean zinc content = 160 ppm has led to rapid impoverishment of zinc in the early residual melts (Fig. 7.20(c); mean zinc in whole rock 72 ppm, in groundmass 55 ppm). With the development of the big-feldspar lavas zinc increases slightly overall and the two distributions become substantially similar (Fig. 7.20(b); mean zinc in whole rock = 80 ppm, in groundmass = 74 ppm). Development of the hornblende-andesites is then accompanied by a slight overall reduction in zinc and a reversion to a high-zinc whole-rock/low-zinc groundmass relation (Fig. 7.20a; mean zinc in whole-rock = 67 ppm, in groundmass = 41 ppm). The reduction of zinc in the groundmass *vis-à-vis* the whole rock accompanying the development of the hornblende-andesites might be attributed to crystallization of hornblende at this point, though the copious contemporary precipitation of phenocrystic feldspar would be expected to have the reverse effect. Whatever

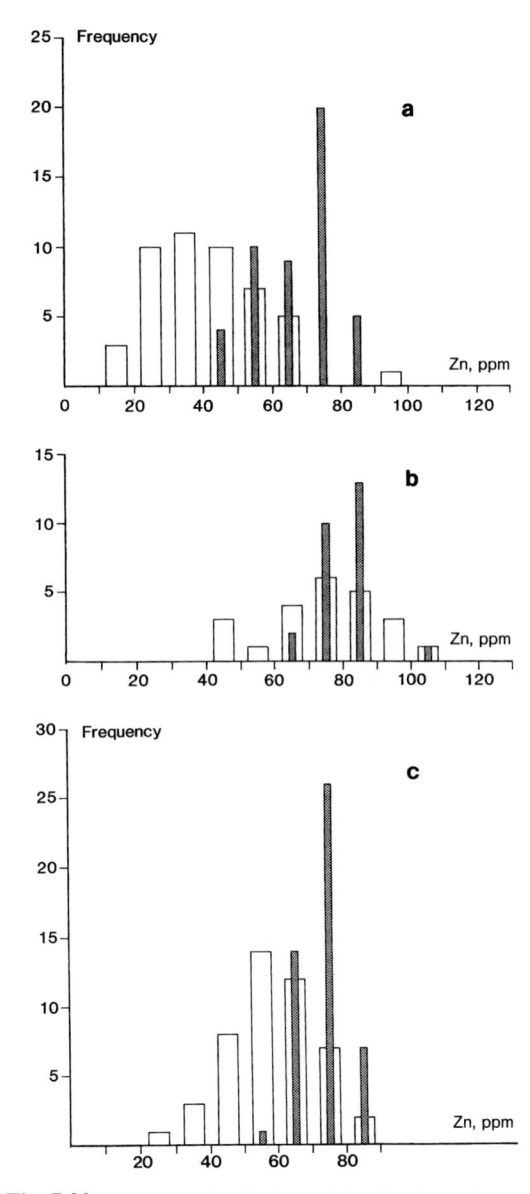

Fig. 7.20. Frequency distributions of zinc abundances in whole rocks (filled bars) and groundmasses (open bars) of (a) the hornblende-andesite group, (b) the big-feldspar lavas, and (c) the olivine–clinopyroxene-basalts of the Solomon Islands Younger Volcanic Suite.

92 Ore Elements in Arc Lavas

This brings us to a consideration of the place of zinc in the evolution of the three principal lava types and of the suite as a whole.

Ternary plots of ZnMA (Zn–MgO–(Na$_2$O +K$_2$O)) are given in Fig. 7.21. As for copper (cf. Fig. 6.10), the plots appear to show an orderly progression from the high Mg-low Zn of the forsteritic olivine-rich accumulates of the olivine–clinopyroxene-basalt series, to a peak in zinc in the big-feldspar lavas, and then to a diminution of zinc and magnesium and a progressive rise in (Na$_2$O + K$_2$O)/Zn ratios as the hornblende-andesites and dacites develop. Although once again it is clear that the field of the groundmass of one set of rocks, e.g. the olivine--pyroxene-basalts, does not constitute the field of the whole rocks for the next set, e.g. the big-feldspar lavas, the apparent progression from picrites to felsic basalts to hornblende-andesites to dacites does appear to be quite an orderly and systematic one. Indeed the order in the system Zn–MgO–(Na$_2$O + K$_2$O) is better than that of Cu–MgO–(Na$_2$O + K$_2$O): the field of Fig. 7.21(f) is a remarkably linear 'olivine subtraction line' (cf. Fig. 6.10(f)), and the fields for all six entities in Fig. 7.21 are much better defined than they are for copper. A further point of difference is that (Na$_2$O + K$_2$O)/Zn ratios in the hornblende-andesites do not attain the magnitudes of (Na$_2$O + K$_2$O)/Cu, and hence that the apparent impoverishment of Zn in the late-stage members is perhaps not quite as great as that of copper. The high degree of order in Zn–MgO–(Na$_2$O + K$_2$O) relations is emphasized when mean ternary relations developed in whole rocks and corresponding groundmasses, both corresponding to 2 per cent intervals of whole-rock SiO$_2$, one plotted as in Fig. 7.22(a). For the very substantial SiO$_2$ interval <46 to 56–8 per cent the seven relevant points describe a very flat curve (i.e. one with a large radius of curvature), almost compatible with closed-system fractionation. Following the point representing the 56–8 per cent SiO$_2$ interval there is a sharp inflexion followed by rapid increase in (Na$_2$O + K$_2$O) relative to Zn and MgO. That is, the pattern of ternary relationships in both whole rocks and groundmasses is clearly hypotholeiitic. This is confirmed very clearly with the incorporation of the 119 Bougainville analyses (and the resultant improvement in statistics at higher SiO$_2$ levels) as shown in Fig. 7.22(b). Analogous, if not quite so clearly developed, patterns are displayed by the Grenada and St Kitts suites (Fig. 7.22(c)) and by the lavas of Vanuatu (Fig. 7.22(d)). The Tonga–Kermadec suite (series (2) of Fig., 7.22(d)) is, because somewhat limited, rather ambiguous: it may be entirely tholeiitic, or may be seen as showing incipient hypotholeiitic character.

MORBs, as represented by analyses of individual mid-Atlantic ridge lavas, appear to be unequivocally tholeiitic

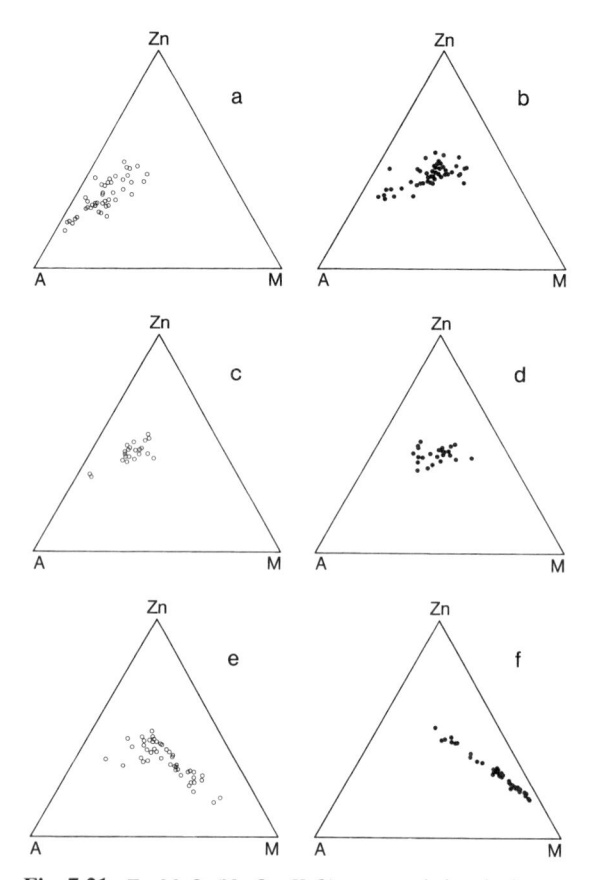

Fig. 7.21. Zn–MgO–(Na$_2$O + K$_2$O) ternary relations in the Solomons olivine–clinopyroxene-basalts, big-feldspar lavas, and hornblende-andesite (filled circles: (f), (d), and (b) respectively) and their groundmasses (open circles: (e), (c), and (a) respectively). Zn calculated as ppm × 10^{-1}; MgO and (Na$_2$O + K$_2$O) as per cent by mass.

with respect to zinc (Fig. 7.22(e)) — the slope of the ternary field being somewhat similar to that of Grenada. The Icelandic basalts (SiO$_2$ <52 per cent), on the other hand, develop a tholeiitic trend with a steep slope almost parallel to the Zn–MgO edge of the triangle (Fig. 7.22(f)). The field of points then inflects, at its high-Zn extremity, with the development of lavas of SiO$_2$ ⩾52 per cent, and the distribution of the field as a whole acquires a hypotholeiitic aspect. Indeed, in comparing the Solomons lavas (as a whole: Fig. 7.22(b)) with the Icelandic suite it might be said that the former assemblage is hypotholeiitic with respect to zinc at a low zinc abundance, the latter hypotholeiitic with respect to zinc at high zinc abundance. The generality of an ordered pattern of zinc abundance related to the evolution of these lava occurrences is clear.

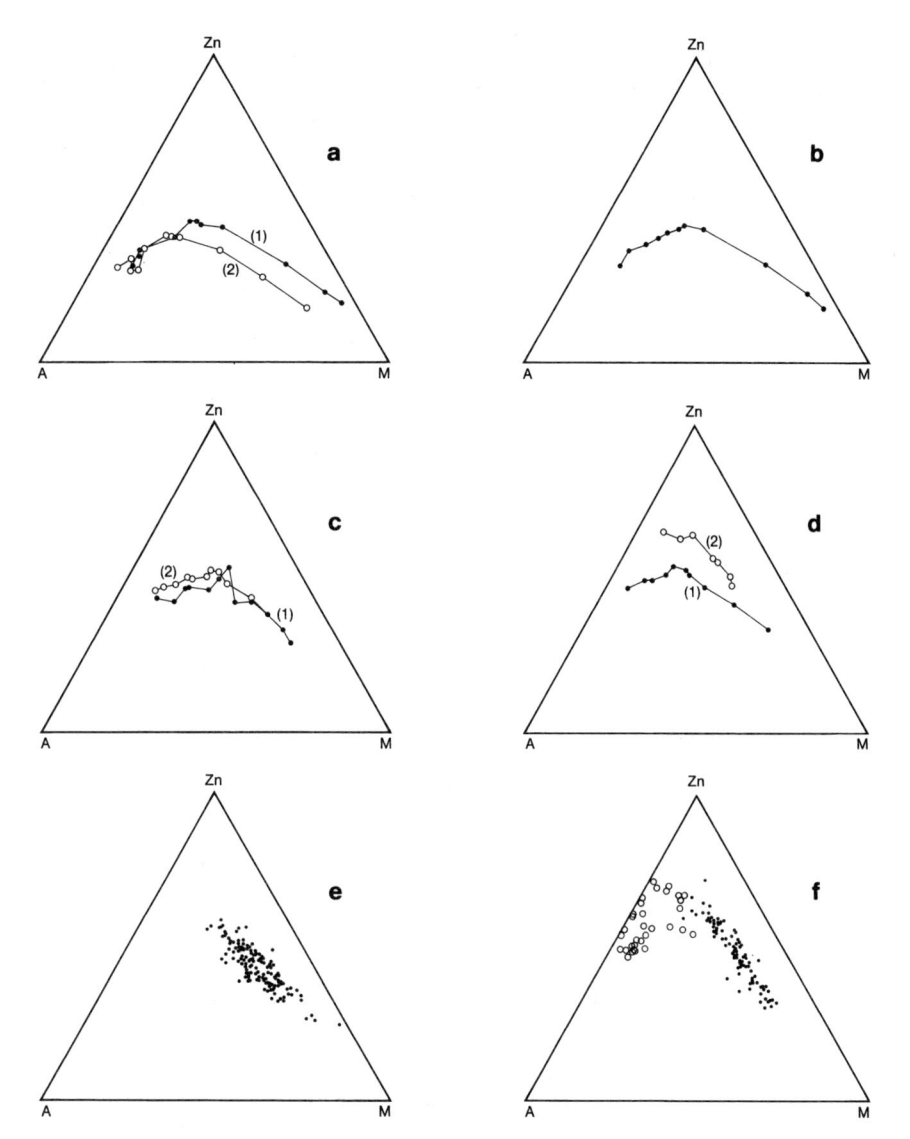

Fig. 7.22. (a) Mean Zn–MgO–(Na$_2$O + K$_2$O) ternary relations corresponding to 2 per cent whole-rock SiO$_2$ intervals < 46, 46–8, → > 64 per cent in the Solomons whole rocks (filled circles) and corresponding groundmasses (open circles); (b) the same, incorporating the full 241 analyses of Solomons *cum* Bougainville lavas; (c) the lavas of Grenada [(1) filled circles] and St Kitts [(2) open circles]; and (d) Vanuatu [(1) filled circles] and Tonga–Kermadec [(2) open circles]. Ternary relations in individual MORBs as represented by analyses of Mid-Atlantic Ridge lavas are shown in (e), and in individual Icelandic basalts (filled circles; SiO$_2$ < 52 per cent) and lavas more felsic than basalts (open circles; SiO$_2$ ⩾52 per cent) in (f). Units of mass as for Fig. 7.21.

The related two-component manifestation of such order — the simple relations between zinc abundance and lava evolution as measured by whole rock SiO$_2$ — is shown in Fig. 7.23. Zinc rises less sharply at the lower end of the SiO$_2$ spectrum than does copper (Fig. 6.12): from ≈62 ppm Zn at <46 per cent SiO$_2$ to ≈78 ppm at 50–4 per cent SiO$_2$. Then in a manner analogous to copper it decreases fairly regularly, to ≈38 ppm Zn at >64 per cent SiO$_2$. The behaviour of zinc in the groundmass is however rather different from that of copper. Whereas the latter is higher in the groundmasses than whole rocks at and to the left of the Cu maximum, and equal or less than the whole rocks to the right, zinc in the groundmass is lower than in the whole rocks throughout the evolutionary spectrum represented by the Solomons suite. As noted in the preceding section, olivine, hornblende, biotite, spinel, and, in the andesitic lavas, clinopyroxene and orthopyroxene, all develop $K_D^{Zn} > 1$. That is, all of the principal minerals other than plagioclase and the earlier-formed clinopyroxenes tend to lower zinc concentration in the melt, leading at least in part to the relative impoverishment of the groundmass depicted in Fig. 7.23(a).

A prominent feature of the groundmass zinc 'curve' of Fig. 7.23(a) is the apparent development of a second zinc maximum at ≈62–4 per cent SiO_2. Whether this is a result of poor statistics (i.e. low sample numbers at these SiO_2 levels) or a real effect is not clear, though the essential regularity of the whole-rock curve at the same SiO_2 level would suggest that the effect is real. If so, the late increase in groundmass zinc may reflect copious late-stage crystallization of feldspar, with concomitant strong partitioning of zinc into the melt. This might just be an important — and general — event in the metallogenetic evolution of island-arc lava suites, and is referred to again in Chapter 24.

In contrast to copper, $Zn:SiO_2$ relations are far from constant between arcs. The Lesser Antilles suite behaves somewhat similarly to that of the Solomons (Fig. 7.23(b)), though the maximum is not nearly so broad. The lava series of Tonga–Kermadec and Vanuatu, on the other hand, do not exhibit maxima at all — at least within the 46 <SiO_2> 64 per cent range — and appear to be still rising at the most felsic limits of lava evolution (Figs. 7.23(c) and (d)).

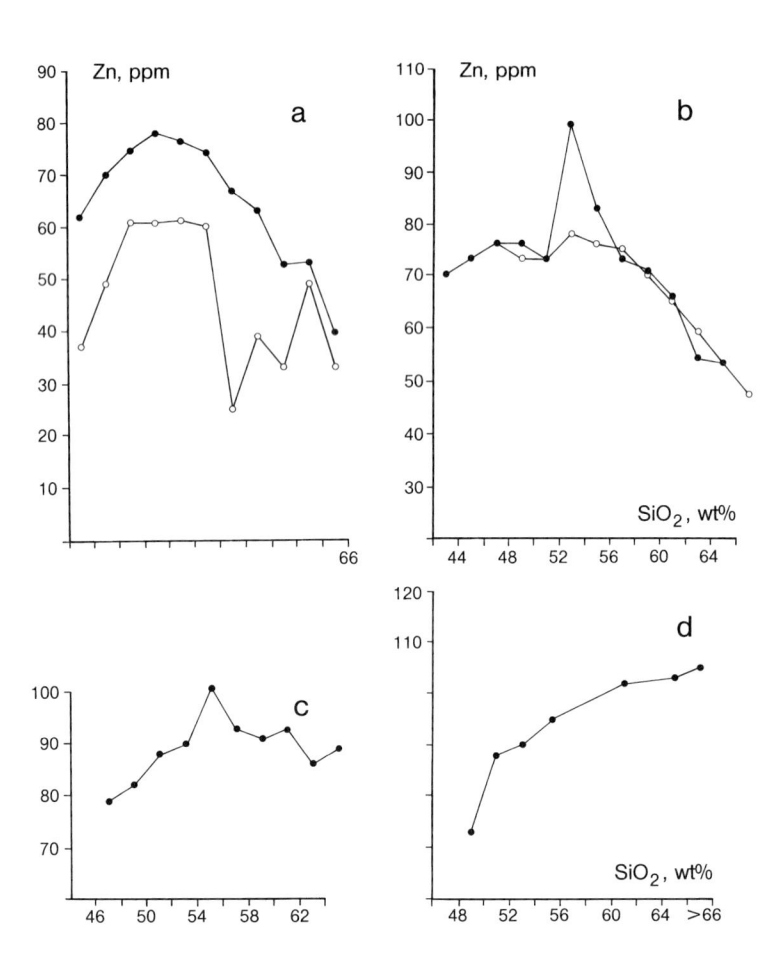

Fig. 7.23. Mean $Zn:SiO_2$ relations corresponding to 2 per cent whole-rock SiO_2 intervals in (a) the Solomons suite (filled circles, whole rocks; open circles, corresponding groundmasses); (b) Grenada (filled circles) and St Kitts (open circles); (c) Tonga–Kermadec, and (d) Vanuatu.

CONCLUDING STATEMENT

Mean zinc in MORBs is, on the basis of quite extensive data, 84 ppm — slightly higher than copper. Mean zinc in the basalts of the five major volcanic island arcs is 77 ppm, and in those of the Solomon Islands, 75 ppm. Thus in contrast to copper, zinc in arc basalts is lower than in MORBs.

Apart from the feldspars, which contain only a minute amount of zinc ($\approx$5 ppm average) in the Solomons lavas, all other silicates contain significant zinc, the abundance of which is related to iron and manganese with varying degrees of correlation. Ferromagnesian zinc ranges from 30 ppm in the clinopyroxenes of the olivine–pyroxene-basalts to 400 ppm in the biotites of the dacitic lavas. However individual iron/manganese-rich olivines and biotites have been found to contain up to 430 and 540 ppm Zn respectively. Magnetite is by far the principal acceptor of zinc among the Solomons minerals, some magnetites containing over 2000 ppm and an overall average of $\approx$550 ppm. Perhaps the two most striking geochemical relationships are those between zinc and iron (and manganese) in olivine, and between zinc (and manganese) in magnetites and the overall evolution of the lava series as expressed by increasing whole-rock SiO_2. $K_D^{Cryst:melt}$ with respect to zinc is higher than for copper for all minerals other than the feldspars, and is >1 for all minerals other than feldspars and some of the clinopyroxenes.

Plotting of individual Solomons whole-rock analyses on ternary $Zn–MgO–(Na_2O + K_2O)$ diagrams shows the development of distinct 'high-magnesium', 'tholeiitic', and 'calc-alkaline' patterns, in all cases more sharply defined than those for copper and generally bearing a striking similarity to those for total iron as FeO. Analogous plotting of mean ZnMA relations corresponding to 2 per cent whole-rock SiO_2 intervals shows the Solomons suite to be hypotholeiitic with respect to zinc in terms of our present nomenclature — a state of affairs very clearly confirmed by the inclusion of the Bougainville analyses and the resultant improved statistics in the more felsic groupings. This hypotholeiitic character also appears in the Lesser Antilles lavas, but is shown less unambiguously by those of Vanuatu and Tonga–Kermadec. MORBs as represented by the mid-Atlantic ridge analyses appear to be entirely tholeiitic, and the Icelandic lavas hypotholeiitic — but at distinctly higher zinc levels than, for example, the Solomons *cum* Bougainville assemblage.

Variation diagrams of mean zinc versus 2 per cent intervals of whole-rock SiO_2 yield varied patterns from curves involving broad maxima as in the case of the Solomons, to more or less continuously increasing Zn vs. SiO_2 in the Vanuatu and Tonga–Kermadec suites. On the basis of mineral partition coefficients the decrease in zinc accompanying increase in SiO_2 in the more felsic lavas of the Solomons and the Lesser Antilles may be ascribed to fractionation to at least a greater degree than with copper. The effect of copious plagioclase crystallization in increasing zinc concentrations in the residual melt requires more comprehensive investigation, and is considered again in Chapters 21 and 24.

8

Lead

Of the four major sulphide elements of the exhalative ores, lead is the least abundant and least widespread overall. It appears in significant quantities only when the accompanying volcanic rocks possess a dacitic or more felsic component. Where the relevant deposits possess a chemical stratigraphy it tends to occur as the uppermost, youngest component of the sulphide sequence.

As with zinc, the geochemistry of lead is greatly influenced by its behaviour both as a chalcophile and a lithophile element. Its most prominent mode of occurrence is as the common sulphide galena, but it is also well known to occur in Pb–O bonding as in K-feldspars, plagioclases, hornblendes, and other minerals.

One of the earliest estimations of lead in silicate rocks was that of Clarke and Steiger (1914), who gave what now appears to be the rather low figure of 7.5 ppm for a group of 'American magmatic rocks'. Clarke and Washington estimated a mean of ≈20 ppm for igneous rocks, and von Hevesy one of 16 ppm. Sandell and Goldich (1943), in their classic work already referred to, obtained an average of 14 ppm for 54 North American igneous rocks. These authors gave a number of determinations specifically on lavas: for the basaltic materials of the Kearsage flow of the Keweenawan of Michigan they obtained a weighted mean of 8 ppm, and for the Greenstone flow <5 ppm. For Keweenawan basalts of Minnesota they obtained 12 ppm and for two related rhyolites a mean of 15 ppm. An olivine-basalt from Palaeozoic rocks of Missouri gave 5 ppm and two rhyolites a mean of 17 ppm. The series of lavas from Clear Lake, California, already referred to, gave: olivine basalt, 8 ppm; two andesites, a mean of 9 ppm; three dacites, 14 ppm; two rhyodacites, 14 ppm; a rhyolitic pumice and an obsidian, both 20 ppm.

Since the time of Sandell and Goldich analysis of lead in silicate rocks has been greatly accelerated by improved analytical methods and by the impetus of uranium–lead and lead–lead geochronology. Wedepohl (1974b), using data obtained by numerous investigators from various parts of the world, gives the following means: 93 tholeiitic basalts, 3.7 ppm; 6 oceanic tholeiitic basalts (low K), 0.8 ppm; 95 alkali olivine-basalts, 4.3 ppm; 79 andesites, 5.8 ppm; 74 dacites, 15.2 ppm; 145 rhyolites and obsidians, 24 ppm.

LEAD IN LAVAS OF THE MARINE ENVIRONMENT

Because of its relatively low abundance and, until recently, its consequent difficulty of determination, there is much less information on the incidence of lead in marine lavas than on copper and zinc. What information is available is probably just sufficient to provide an initial insight into the abundance of lead in the principal lava types.

A survey of this very limited literature indicates mean lead in mid-ocean ridge tholeiitic basalts to be ≈2.2 ppm, alkaline olivine-basalts ≈4.0 ppm, dacites ≈4.5 ppm, and rhyolites ≈9 ppm, the last two representing Icelandic material.

Information on island-arc lavas may be a little more abundant, but is still sparse. Ewart and his co-workers have produced about fifty determinations of lead in the lavas of Tonga–Kermadec, giving means of 1.6 ppm for lavas with <52 per cent SiO_2, 1.9 ppm for 52 per cent <SiO_2 <62 per cent, and 4.1 ppm for lavas with SiO_2 >62 per cent. Very scanty data on the Aleutians chain gives a mean of 8.3 ppm Pb in andesites and 36 ppm Pb in a single rhyodacite. A few analyses (Gorton 1974) of material from the younger lavas of Vanuatu give means of 8 ppm Pb for basalts and 9 ppm for andesites. More extensive work on Gorton's material (95 samples) by B. W. Chappell (on behalf of the author for the present work) has given 4.47 ppm Pb for 72 basalts, 9.84 for 19 andesites, and 10.50 ppm for four dacites.

From this very limited data some tentative estimates may be made. Low-K MORBs probably contain about 1 ppm Pb or a little less. An average for MORBs is probably close to 2 ppm and for ocean island alkali olivine-basalts about 4 ppm. More felsic associates, such as the andesites, dacites, and rhyolites of Iceland, probably contain up to 10 ppm Pb and may average ≈8 ppm. Lead contents of analogous lava types of modern island arcs are probably close to these values, possibly one or two parts per million higher.

CHEMICAL PROPERTIES AND CRYSTAL CHEMISTRY OF LEAD

As pointed out by Wedepohl (1974b) the general oxygen fugacity of the Earth's crust leads to a prevalence of Pb^{2+} over Pb^{4+}. Estimations of the ionic radius of Pb^{2+} range from 0.118 to 0.132 nm, with a most favoured value of c. 0.132 (Emsley 1989). This compares with K^+ (0.133 nm), Sr^{2+} (0.127 nm), and Ba^{2+} (0.143 nm). A wide variety of coordinations ranging from 3 to 12 exists in Pb–O compounds. Pb^{2+}, because of its comparatively large size, tends to incorporation in silicates which contain large ion M^+ and M^{2+} structural sites. The most notable of these are K^+ structural positions: thus, as first pointed out by Goldschmidt (1954 and earlier unpublished work), trace lead usually shows good correlation with potassium, especially in igneous minerals. For reasons of size and charge, lead may also show strong correlation with trace strontium and barium. While silicates containing potassium as a major component are certainly the principal hosts for trace lead, Wedepohl (1974b) has pointed out that Na^+–Ca^{2+} structural sites may also accommodate significant lead, and that as a result the plagioclases and amphiboles may accept quantities comparable with those of some K-bearing silicates.

Goldschmidt (1954) was one of the earliest to observe that, through its substitutional relation with potassium, lead tended to follow SiO_2 in igneous rocks: as silicon and potassium are enriched in the later, more felsic differentiates, and as lead tends to 'follow' potassium, the three elements are concentrated together in the relevant residual melts. Goldschmidt noted as a striking example of this effect the behaviour of lead in the lavas of Clear Lake, California (Sandell and Goldich 1943) in which silica ranges from 56 to 76 per cent. As the zinc of this lava series was concentrated in the ferromagnesian minerals and lead in the K-bearing minerals (chiefly feldspars), zinc decreased from 75 to 35 ppm, lead increased from 8 to 20 ppm, as SiO_2 rose to its upper limit.

The electronic configuration, size, and principal properties of the lead atom and ion are given in Table 6.4. Radius ratios for Pb–O and K–O are nearly the same at 0.94 and 0.95 respectively, appropriate to 8- or 12-fold coordination. Although in K-feldspar, the principal host to trace lead, the potassium is rather irregularly coordinated by nine oxygens, the K–O distances are about 0.3 nm, so that acceptance of the lead ion is relatively easy. On the other hand the K–O bond is ≈85 per cent ionic and hence stronger than the Pb–O bond, and acceptance of Pb^{2+} in a K^+ site requires charge adjustment by vacancy formation — two factors militating against 'capture' of lead at the expense of potassium. On balance, however, substitution is favoured, leading to the K/Pb correlation observed by Goldschmidt and now well recognized as an important element of trace lead occurrence.

THE INCIDENCE OF LEAD IN THE PRINCIPAL MINERAL SPECIES OF THE SOLOMON ISLANDS YOUNGER VOLCANIC SUITE

General

Although the lead content of K-feldspars has been investigated extensively, this has been almost entirely confined to material from granites, pegmatites, and gneisses. Little has been done on volcanic feldspars and even less on the feldspars of island-arc lavas and of MORBs. One of the earliest 'modern' estimations appears to be that of Wedepohl (1956), who determined 2–46 (mean 21) ppm in emission spectrographic analyses of sanidines separated from trachytes of Eifel, in Germany. Carmichael and MacDonald (1961) found 7.5–14 (mean 9.8) ppm lead in K-feldspar phenocrysts of Icelandic pitchstones; Doe and Tilling (1967) 26.2–36.6 (mean 30.1) in K-feldspars of rhyolites and quartz-latites of the western USA; and Karamata (1969) 18–58 (mean 37.9) in sanidines of Yugoslavian quartz-latites. Cuturic et al. (1968) found 31–45 (mean 38) in sanidines of Yugoslavian latites; Leeman (1979) gave a range of 22.6–47.6 ppm lead for sanidines from a variety of quartz-latitic and rhyolitic lavas, and 12.2–24.5 ppm for the relevant coexisting plagioclases. Leeman also noted that lead contents of 0.12 and 1.40 ppm had been obtained for the plagioclases of mid-ocean ridge tholeiites by earlier authors.

There appears to be little or no information on the lead content of volcanic amphiboles and micas. The mean for amphiboles listed by Wedepohl is ≈15 ppm Pb, but the samples are all from granitoid and metamorphic rocks; no volcanic material is included. As pointed out by Wedepohl, the micas appear to accommodate less lead than the K-feldspars. Of the former, biotite tends to contain somewhat more lead than muscovite, in spite of the fact that biotite generally contains rather less potassium than muscovite. Wedepohl gives an average of 37 ppm Pb for a sample of >300 biotites, 26 ppm Pb for 32 muscovites, and notes that lead favours the biotite lattice of coexisting biotite–muscovite. As with the amphiboles, however, all Wedepohl's data is for plutonic and metamorphic material, and volcanic micas are unrepresented.

In his earlier analyses Wedepohl (1956) does, however, give lead contents for three magnetites from basalts from Lower Saxony: a range of <3–10, with a mean of ≤6.7 ppm.

Solomons suite

The only minerals to contain significant lead are hornblende, the plagioclase feldspars, and magnetite. Mean abundances are given in Table 6.5 (with K_2O, Ba, and Sr for comparison) and spreads of values are shown in Fig. 8.1.

Hornblende exhibits a range from <1 ppm to 4 ppm, the latter represented by one analysis out of 36. There is a conspicuous mode at 2 ppm and the overall arithmetic mean is 1.8 ppm. Plotting of individual analyses shows no more than a muted indication of a correlation with potassium and strontium. The precision of analysis is, however, ±1 ppm Pb: at values only slightly above the level of precision, correlations are liable to be severely blurred or concealed, and this may be the case in the present instance. Analysis at the ppb level might well reveal some degree of correlation between lead and potassium–strontium.

As most of the Solomons feldspars are low-K plagioclases they incorporate relatively little lead. Of 76 'ultimate purity' separations, most fall in the range 1–4 ppm Pb, with an overall arithmetic mean of 2.9 ppm. As indicated in Table 6.5, Pb, K_2O and Sr increase in general sympathy from the feldspars of the high-magnesium lavas to those of the hornblende-andesites. The procedure of making two feldspar separations (one

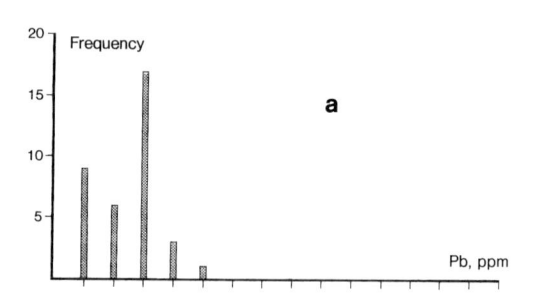

Fig. 8.1. Frequency distributions of lead abundances in (a) hornblendes, (b) feldspars, and (c) magnetites of the Solomons lavas.

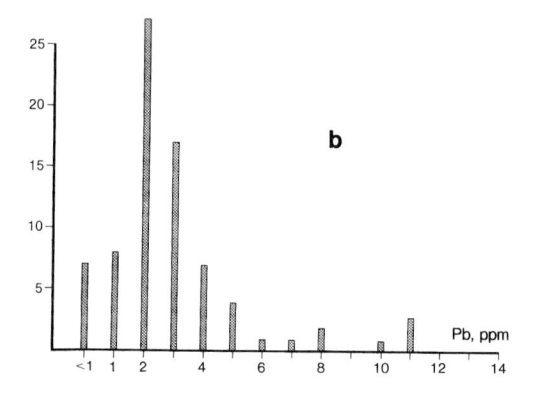

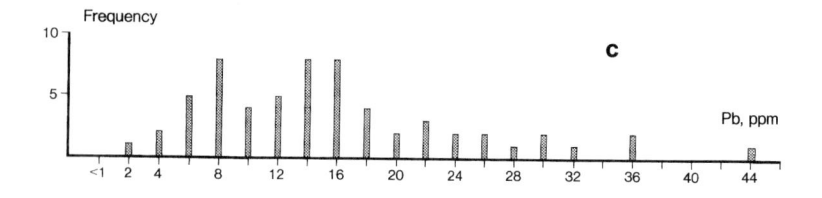

of 'ultimate purity' the other of 'penultimate purity') in each case has already been referred to in the preceding two chapters. This provides a simple technique for assessing the effect of minute glass inclusions on the apparent compositions and trace-element content of the relevant plagioclases. Fig. 8.2(a) shows Pb:K_2O relations in separations of ultimate purity; Fig. 8.2(b) the corresponding relations in separations of penultimate purity; and Fig. 8.2(c) the mean values of Pb/K_2O in the two categories of feldspars of each of the three lava types. All indicate a clear, if rather poor, correlation between Pb and K_2O: Figs. 8.2(b) and (c) show, however, that there has been a tendency towards partitioning of both elements into the contained glass with respect to the host feldspar crystals, and that such partitioning of potassium is somewhat more pronounced than that of lead.

The effect of the lead:potassium relationship, and slight but systematic differences in potassium content, on the trace lead level in the feldspars of adjacent, contemporary, and very similar volcanoes is illustrated by Fig. 8.3. The closely related volcanoes of Mount Gallego and Savo — of essentially the same age, both andesitic–

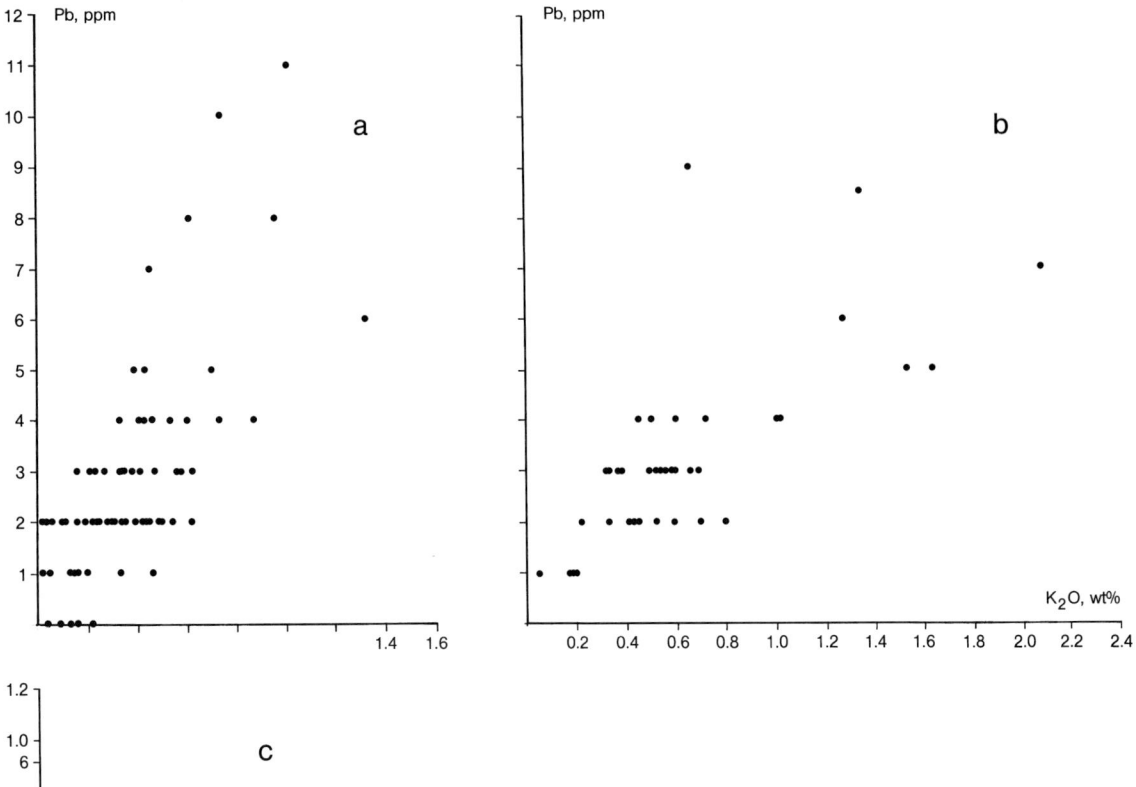

Fig. 8.2. Pb:K_2O relations in (a) 'ultimate purity' and (b) 'penultimate purity' feldspars separated from lavas of the Solomons suite. Mean values for the feldspars of each of the three principal lava groups (c) are shown as filled circles representing the ultimate purity separations and open circles representing the penultimate purity separations. In each case the least Pb/K_2O-rich feldspars are those of the olivine–clinopyroxene-basalts, and the most are those of the hornblende-andesites group (see Table 6.5). Both lead and potassium tend to partition into the glass inclusions within the feldspar crystals.

100 *Ore Elements in Arc Lavas*

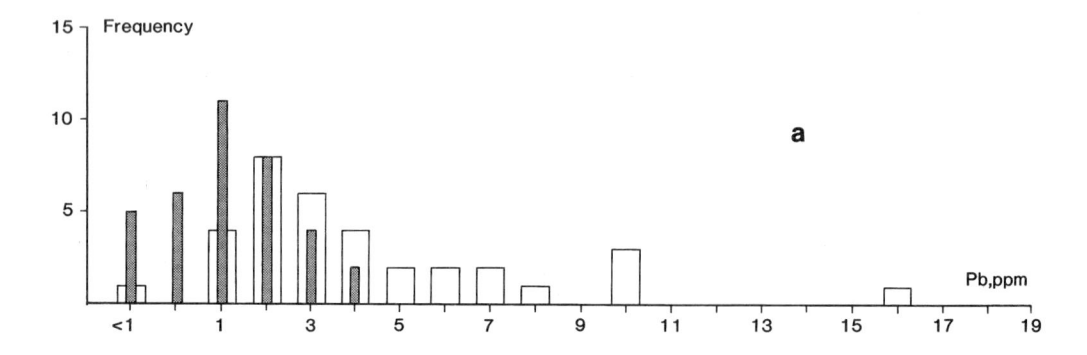

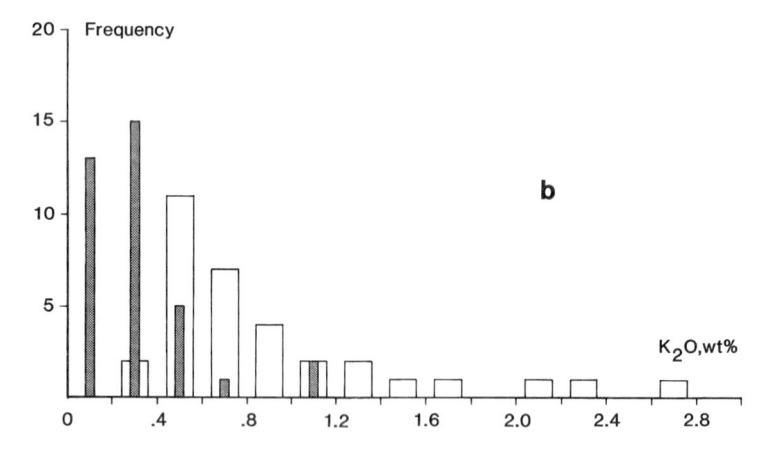

Fig. 8.3. Frequency distributions of (a) lead and (b) K$_2$O abundances in the feldspars of the andesites and dacites of the closely related Gallego (filled bars) and Savo (open bars) volcanoes.

dacitic and both currently at the hot-spring stage — interdigitate beneath the sea as noted in Chapter 5. However, the lavas of Savo — and their feldspars — exhibit a slightly but systematically higher level of potassium than do those of Mount Gallego. As shown in Fig. 8.3, this is faithfully reflected in the lead content of the two groups of feldspars. Pb/K$_2$O relations in the whole rocks and groundmasses are referred to in the following section.

Fig. 8.4 shows relations between lead and strontium in the 'ultimate purity' feldspar separations. Clearly the relation between the two traces is sympathetic but far from functional. Partitioning of strontium as between host feldspar and glass inclusions is considered in Chapter 10.

Like copper and zinc, magnetite constitutes the principal 'sink' for lead among the major mineral phases of the Solomons lava series. The very limited data available on the incidence of lead in magnetites generally indicate that the quantities of lead in most of the Solomons magnetites are quite high. Most analyses are within the 4–32 ppm range, with a mode in the 12–18 ppm range, but one sample, from the volcano of Mount Gallego on

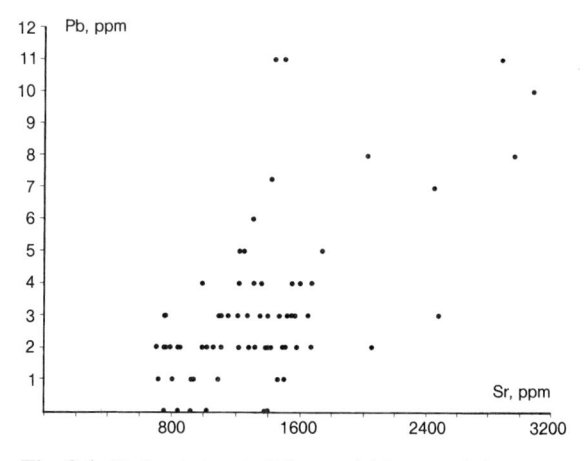

Fig. 8.4. Pb:Sr relations in Solomons 'ultimate purity' feldspar separations.

Guadalcanal, gave 44 ppm Pb (Fig. 8.1). The overall mean value for 69 magnetites is 16.7 ppm Pb. No reason is apparent for the comparatively low lead content of the big-feldspar lava magnetites (Table 6.5), an 'anomaly' analogous with those already observed for copper and zinc. On the other hand Pb may tend to follow the same path as Cr, Al, Mg, Ti, and Cu in the magnetite (Figs. 5.16, 6.5), with a reversal of the trend in the latter differentiates due to the pronounced concentration of lead in the liquid at this stage (see next section). Clearly there are no large ions for which the lead might substitute in the magnetite, and there is no obvious feature of the crystal structure that might be particularly suitable as a site for Pb^{2+}. One possibility is that slightly disordered interfaces might be provided by mosaic and lineage structures in the magnetite crystal and that small quantities of Pb^{2+} or elemental lead might be incorporated in these.

CRYSTAL:MELT PARTITIONING

General

Very little appears to have been done on the partitioning of lead in igneous systems, and, apart from limited studies involving K-feldspars and the plagioclases, the literature is almost devoid of reference to it.

Early controversy about the relative propensities of K-feldspar and plagioclase for capturing lead was resolved by Doe and Tilling (1967). Contrary to theoretical expectations it had been indicated (Howie 1955; Heier 1960; Taylor 1965) that plagioclase might, in natural systems, develop higher concentrations of lead than the coexisting K-feldspar. This was, however, based on emission spectrograph results, with their inherent uncertainties. Doe and Tilling, using isotope dilution methods on feldspars from a variety of contexts, demonstrated that trace lead in coexisting K-feldspar and plagioclase did in fact behave as predicted by theoretical considerations. Their results, obtained on three volcanic K-feldspar–plagioclase–glass groupings, are given in Table 8.1.

The most comprehensive modern work is that of Leeman (1979) who, in the work already referred to, investigated the partitioning of lead between sanidine and plagioclase feldspars, and coexisting volcanic glass. He found, in two carefully chosen series of samples, that sanidine/glass distributions were close to unity in rocks ranging from quartz-latite to rhyolite, and that plagioclase/glass distribution coefficients increased from 0.1 to 0.75 from basalt to rhyolite. While Leeman suggested that the latter relationship indicated that the distribution coefficient was dependent upon bulk-rock composition and temperature, it may be seen from the data on the Solomons feldspars (see below) that it may be due to the slight but systematic increase of minor K in the relevant feldspars. Change in whole-rock composition (including trace lead concentration in the evolving melt), decrease in temperature, and increase in plagioclase potassium would indeed proceed hand-in-hand, but the immediate cause of proportionate increase in lead in the plagioclase would be the progressive increase in plagioclase K_2O as the lava series evolved from basalt to rhyolite.

Leeman concluded that with a K_D^{Pb} value of approximately one, even the crystallization of sanidine would have little effect on the lead content of residual melts. Minerals such as plagioclase, olivine, clinopyroxene, apatite, and magnetite, with their low to very low crystal/liquid distribution coefficients for lead, would remove very little of the latter from the melt and hence contribute strongly to its concentration in the residual fractions.

Solomons suite

Results for the principal lead-bearing Solomons minerals (Table 8.2) indicate that the silicates have low partition coefficients for lead. In spite of its mean K_2O of 0.45 per cent and possession of Na^+–Ca^{2+} sites, hornblende contains a mean of only 1.5 ppm lead and develops a mean K_D^{Pb} of approximately 0.20. Results for the plagioclases may be slightly higher than those of Leeman but are generally similar. Mean K_D^{Pb} feldspar/ groundmass for the

Table 8.1 Lead abundances in coexisting K-feldspar, plagioclase, and glass in three American volcanic lavas

Rock type	Age (Ma)	Pb in K-feldspar	Pb in plagioclase	Pb in glass	K_D^{Pb}K-feldspar:plag.	K_D^{Pb}K-feldspar: glass
Rhyolite obsidian, California	< 0.06	30.2	14.4	28.8	2.1	1.05
Quartz-latite vitrophyre, Colorado	25	31.8	13.1	27.5	2.4	1.16
Rhyolite vitrophyre, Montana	50	36.8	14.4	41.6	2.6	0.88

All lead abundances in parts per million by mass. From Doe and Tilling (1967).

102 *Ore Elements in Arc Lavas*

Table 8.2 Crystal:melt distribution coefficients (K_D) with respect to lead

Mineral species	Lava group	n_1	n_2	m_1	m_2	$K_{D(1)}$	N	M_1	M_2	$K_{D(2)}$
Hornblende	HbA	25	47	1.8	6.8	0.26	21	1.5	7.5	0.20
Feldspar	OPB	10	44	1.2	3.9	0.31	10	1.2	3.6	0.33
	BFL	23	23	2.4	5.3	0.45	22	2.5	5.3	0.47
	HbA	46	47	3.6	6.8	0.53	44	3.8	7.0	0.54
Spinel	OPB	10	44	13.0	3.9	3.33	8	13.0	4.4	2.97
	BFL	23	23	6.2	5.3	1.17	15	6.2	4.9	1.27
	HbA	37	47	18.3	6.8	2.69	34	17.7	6.8	2.60

OPB = olivine–clinopyroxene-basalt group; BFL = big-feldspar lavas: HbA = hornblende-andesite group; n_1 = total number of mineral separations analysed; n_2 = total number of groundmass separations analysed; m_1 = mean lead content of mineral of each lava group; m_2 = mean lead content of groundmass of each lava group; $K_{D(1)}$ = mean distribution coefficient determined as m_1/m_2; N = number of coexisting mineral:groundmass pairs analysed; M_1 and M_2 = mean lead contents of coexisting mineral and groundmass respectively; $K_{D(2)}$ = distribution coefficient determined as M_1/M_2, for comparison with $K_{D(1)}$. All concentrations as parts per million by mass.

olivine–pyroxene-lava group is 0.33, but some individual coefficients are as low as <0.10. As may be seen from Tables 6.5 and 8.2, progressive increase in plagioclase K_D^{Pb} proceeding through olivine–pyroxene-basalts, big-feldspar basalts–andesites to hornblende-andesites (0.33, 0.47, and 0.54 respectively) conforms with the increase in mean plagioclase K_2O (0.07, 0.38, and 0.44 per cent) and mean plagioclase lead (1.2, 2.4 and 3.6 ppm) for the three lava groups.

As already remarked, the lead contents of the Solomons magnetites are relatively high, and this is reflected in the apparent partition coefficients. If the absence of large ion sites in the magnetite does mean that the lead here is accommodated in the poorly ordered zones of mosaic and lineage structure boundaries, the partitioning is not a true crystal structure/melt phenomenon. This might account for the unsystematic progression in both lead content and distribution coefficients with the progressively more felsic nature of the host rocks (Table 6.5 and 8.2). Whatever the explanation the data indicates that magnetite is the only major mineral of the Solomons lavas capable of depleting the melt in lead. As, however, modal magnetite is <2.0 per cent in most of the lavas, its absolute effect in reducing the concentration of lead in the melt must be minor to negligible.

THE INCIDENCE OF WHOLE-ROCK LEAD IN THE SOLOMON ISLANDS YOUNGER VOLCANIC SUITE

Mean lead for the suite as a whole is 4.1 ppm, which compares with ≈2.2 ppm in MOR tholeiitic basalts, ≈4.0 ppm in MOR alkali olivine-basalts, and 2.0 and 5.8 ppm in the full range of lavas of Tonga–Kermadec and Vanuatu respectively, the only other examples of arc volcanism for which significant data is available. Mean lead for the three principal Solomons lava groups is 2.5, 4.2, and 5.6 ppm for olivine–pyroxene-basalts, big-feldspar lavas, and hornblende-andesites respectively. On a two-group basis, mean lead for lavas with SiO_2 <52 per cent is 2.9 ppm; for those with SiO_2 >52 per cent it is 5.4 ppm. If the latter group is divided further, andesites

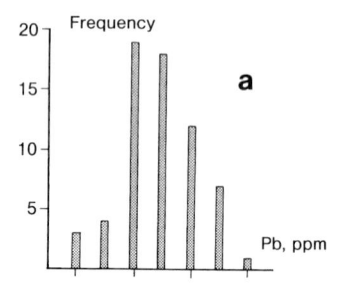

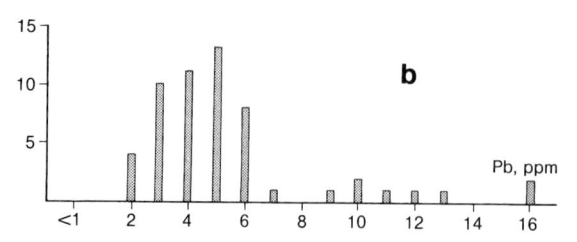

Fig. 8.5. Frequency distributions of lead abundances in (a) the basalts (SiO_2 < 52 per cent; n = 64; $\bar{x}$ = 2.9 ppm Pb) and (b) lavas more felsic than basalts ($SiO_2 \geqslant$ 52 per cent; n = 58; $\bar{x}$ = 5.4 ppm Pb) of the Solomons lava suite.

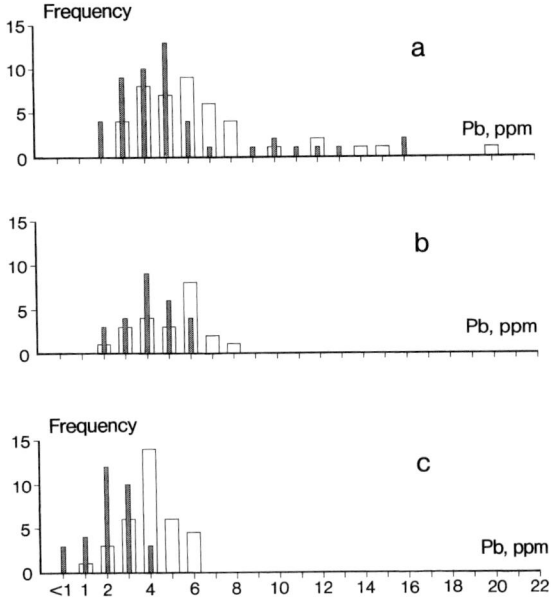

Fig. 8.6. Frequency distributions of lead abundances in whole rocks (filled bars) and groundmasses (open bars) of (a) the hornblende-andesites group, (b) the big-feldspar lavas, and (c) the olivine–clinopyroxene-basalts of the Solomons suite.

($52 < SiO_2 < 62$ per cent) contain a mean of 4.4 ppm and dacites ($SiO_2 > 62$ per cent) 10.9 ppm Pb. Frequency distributions for $SiO_2 < 52$ per cent and $SiO_2 > 52$ per cent categories are given in Fig. 8.5.

Frequency distributions of ppm Pb in the Solomons hornblende-andesites, big-feldspar basalts–andesites, and olivine–pyroxene-basalts are given in Fig. 8.6. As might be expected from Fig. 8.5, there is a substantial overlap of lead values from one group of lavas to the next, but lead increases steadily with increase in felsic nature, and there is throughout a clear tendency for fractionation of lead into groundmass.

Not surprisingly the strong tendency for lead to follow potassium observed in the feldspars also holds in the whole rocks (Fig. 8.7) and groundmasses. (The slightly but systematically higher K_2O and Pb in the Savo as compared with the Mt Gallego feldspars is, as a partial consequence, also apparent in the corresponding whole rocks and groundmasses, emphasizing the genuineness of such local variation.)

Like zinc and copper, lead exhibits a systematic pattern of behaviour in parallel with the apparent evolution of the lava suite.

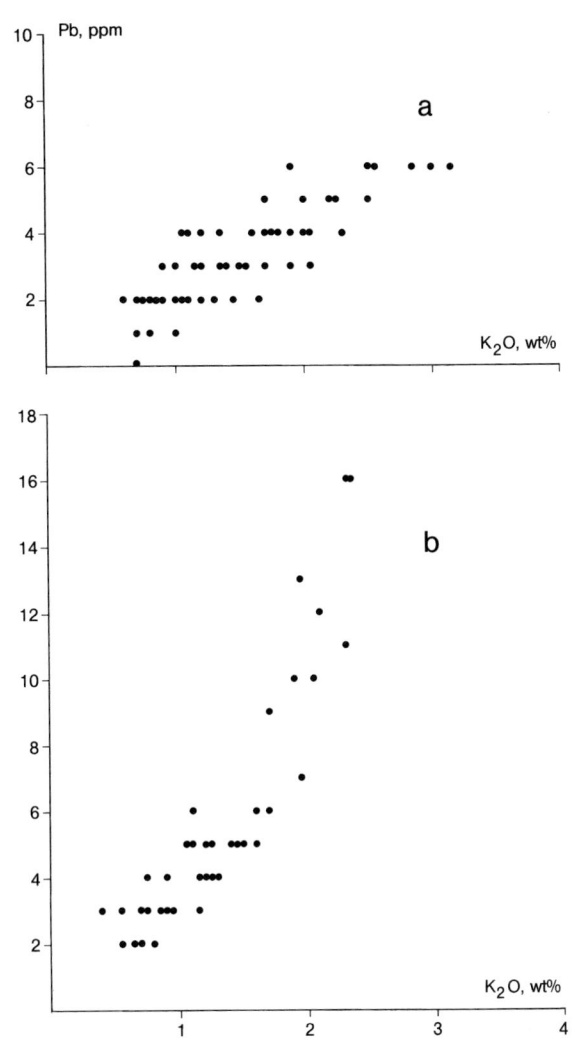

Fig. 8.7. Pb:K_2O relations in (a) the basalts ($SiO_2 < 52$ per cent) and (b) lavas more felsic than basalts ($SiO_2 \geq 52$ per cent) of the Solomons suite.

Ternary plots of PbMA (Pb–MgO–($Na_2O + K_2O$)) for the whole rocks of the three principal rock groups and for the corresponding groundmasses are given in the two sequences of diagrams comprising Fig. 8.8. Progression from the high MgO–low Pb field of the olivine–pyroxene-basalts to the low MgO–high Pb field of the hornblende-andesites follows an essentially linear pattern until, in the hornblende-andesites and their groundmasses, lead increases relative to the alkali metals at the lowest Mg numbers. The shifts in the field of the groundmasses (Fig. 8.8(a), (c), (e)) with respect to those for the whole rocks is

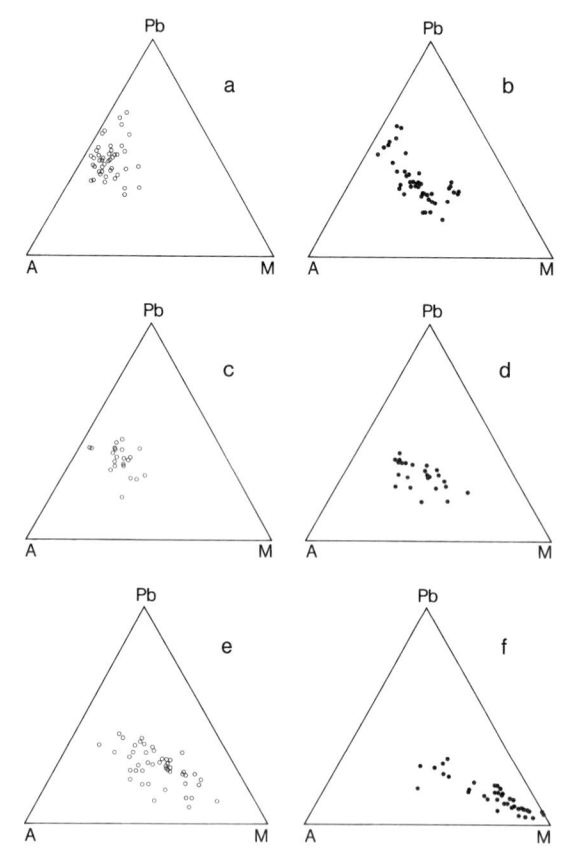

Fig. 8.8. Pb–MgO–(Na$_2$O + K$_2$O) ternary relations (lead as parts per million; MgO and the alkalis as per cent by mass) in the Solomons olivine–clinopyroxene-basalts, big-feldspar lavas, and hornblende-andesites (filled circles: (f), (d), and (b) respectively) and their groundmasses (open circles: (e), (c), and (a) respectively).

generally more in the direction of increased alkali rather than increased lead, though the latter does in fact increase in each case and in an accentuated way at the lowest Mg numbers.

The trends are made clearer where Pb–MgO–(Na$_2$O + K$_2$O) proportions corresponding to 2 per cent whole-rock SiO$_2$ intervals are plotted as in Fig. 8.9.

Proceeding from the point at the lower right of Fig. 8.9(a) (representing mean PbMA for rocks with SiO$_2$ <46 per cent), the initial nine points (to whole-rock 60 <SiO$_2$ <62 per cent) fall very close to a straight line, which is an olivine–clinopyroxene subtraction line. However, rather than showing a decrease in lead relative to (Na$_2$O + K$_2$O) at SiO$_2$ values greater than, say, 62 per cent — in a manner analogous to that exhibited by copper (Figs. 6.10, 6.11) and to a lesser extent zinc (Figs. 7.21, 7.22) — the lead line develops a sharp positive inflexion and trends upwards towards PbMA = 30–0–70. As shown in Fig. 8.9(a), the sequence of points representing the corresponding groundmasses emphasizes the feature. This pattern of behaviour contrasts with the negative inflexions that characterize the hypotholeiitic trends developed with respect to FeO, copper, and zinc, and is distinct from the straight-line patterns we have characterized as tholeiitic trends (Fig. 5.20). This positive inflexion of the lead diagram thus provides us with our first example of a hypertholeiitic trend among the ore elements.

Addition of the Bougainville analyses and plotting of points representing the full 241 Solomons analyses (Fig. 8.9(b)) confirms the pattern: a steady, essentially straight line, increase in lead-plus-(Na$_2$O + K$_2$O) to whole-rock SiO$_2$ = 60–2 per cent, i.e. the point of dacite formation, followed by an accelerated increase in lead relative to total alkalis at SiO$_2$ ≈62 per cent.

All this appears to indicate that whereas copper and zinc were concentrated to maxima by olivine/pyroxene subtraction and then, by some other process, reduced in relation to the alkali metals, lead is concentrated analogously by olivine/pyroxene subtraction along a very constant line Pb:(Na$_2$O + K$_2$O) ≈45:55 but is then further increased by some other process, trending towards Pb:(Na$_2$O + K$_2$O) ≈70:30. Figs. 8.9(c) and (d), derived from 95 analyses of a range of lavas from the Vanuatu arc for which lead determinations were available, and 59 analyses of a similar range from Tonga–Kermadec, show clear indications of similar behaviour. There thus seems to be a general indication of an increase in lead relative to total alkali metals at the more felsic end of island-arc lava suites.

A notable feature of Figs. 8.8 and 8.9(a) is that although there is a significant increase in alkalis in the groundmasses relative to the relevant whole rocks, the concomitant increase in lead is somewhat muted (see also Fig. 8.6). The most likely explanation for this is that as the evolving lavas become more felsic the corresponding feldspars (present to large extent as phenocrysts) become more potassic and hence more lead-rich. The lead content of the feldspars becomes closer and closer to that of the remaining melt, and so the difference between whole-rock and groundmass lead eventually becomes negligible.

Lead is concentrated relative to the alkali metals in the later stages of lava evolution; i.e. the lava series is hypertholeiitic with respect to lead. In contrast, there is a relative impoverishment of iron, copper, zinc, calcium, and several other elements in the lava series; i.e. the behaviour of the series is hypotholeiitic with respect to these elements. There appears to be an important point of principle here, which is considered further in Chapters 21 and 24.

Lead 105

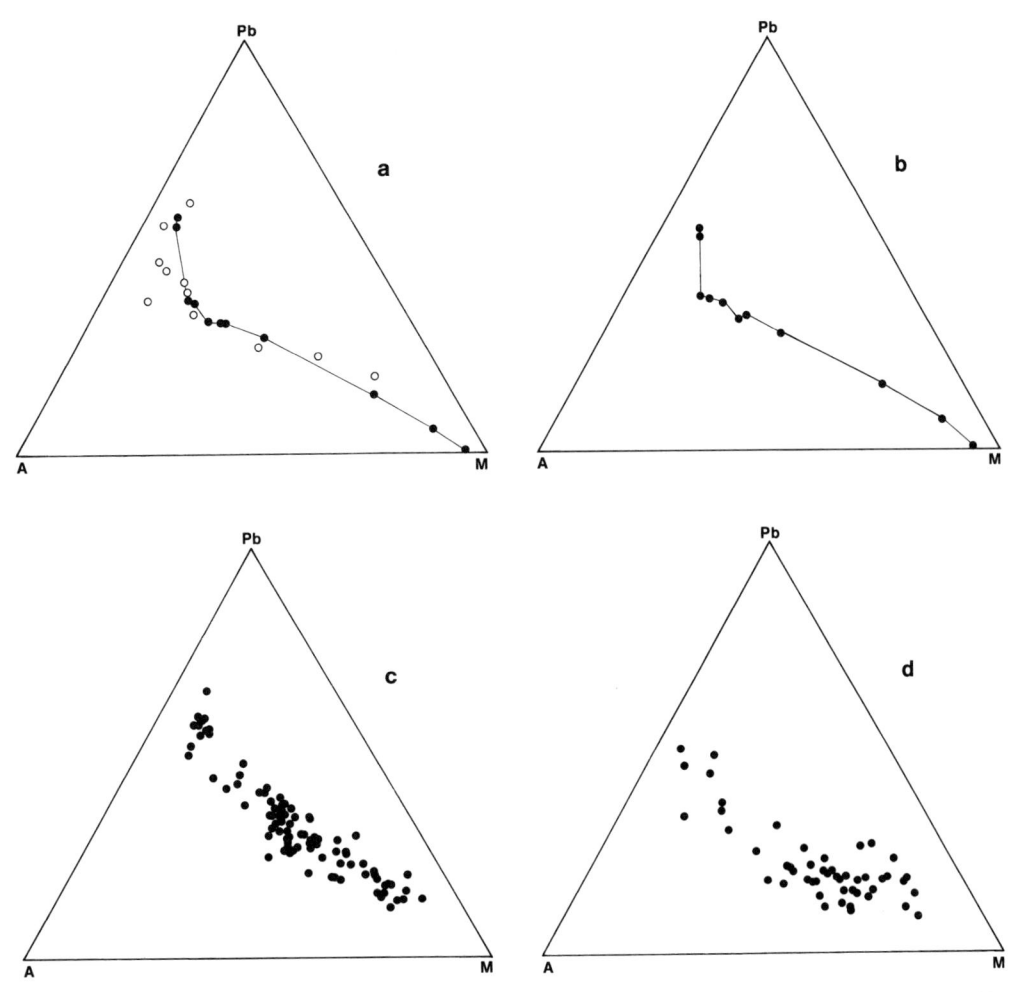

Fig. 8.9. Mean Pb–MgO–(Na$_2$O + K$_2$O) relations corresponding to 2 per cent whole-rock SiO$_2$ intervals < 46, 46–8 → > 64 per cent in (a) the Solomons whole rocks (filled circles) and corresponding groundmasses (open circles), (b) the same, incorporating the full 241 analyses of Solomons *cum* Bougainville lavas; (c) and (d), ternary relations developed in individual lavas of Vanuatu and Tonga–Kermadec respectively. Units of mass as for Fig. 8.8.

'Evolutionary' behaviour of lead with respect to whole-rock SiO$_2$ is, as might have been expected from the foregoing, quite different from that of iron, copper, and zinc at higher SiO$_2$ values (Fig. 8.10(a)). From <46 to ≈52 per cent SiO$_2$ lead content rises steadily and in a manner very similar to copper: an effect reflecting fractionation of olivine and clinopyroxene, two minerals that accept only infinitesimally small quantities of lead. At ≈52–4 per cent SiO$_2$ the curve begins to flatten, and between 54 and 60 per cent SiO$_2$ the rate of increase in lead content appears to lessen and becomes irregular. Then, rather than decreasing as in the case of copper and zinc, lead abundance suddenly increases from ≈4.5 to 12 ppm between 60 and 62 per cent SiO$_2$. In the absence of any appropriate petrographic evidence, the almost equally sudden decrease at SiO$_2$ >62 per cent was suspected to be a statistical effect resulting from the paucity of samples (in this SiO$_2$ range) already referred to. Inclusion of the 119 Bougainville analyses confirmed this. Fig. 21.2h, representing the full 241 Solomons analyses, shows that what appears to be a somewhat irregular inflexion between 54 and 60 per cent SiO$_2$ in Fig. 8.10(a) is in fact a gently and quite regularly flattening curve, and that increase in lead continues at SiO$_2$ >64 per cent. The beginning of a second flattening of the

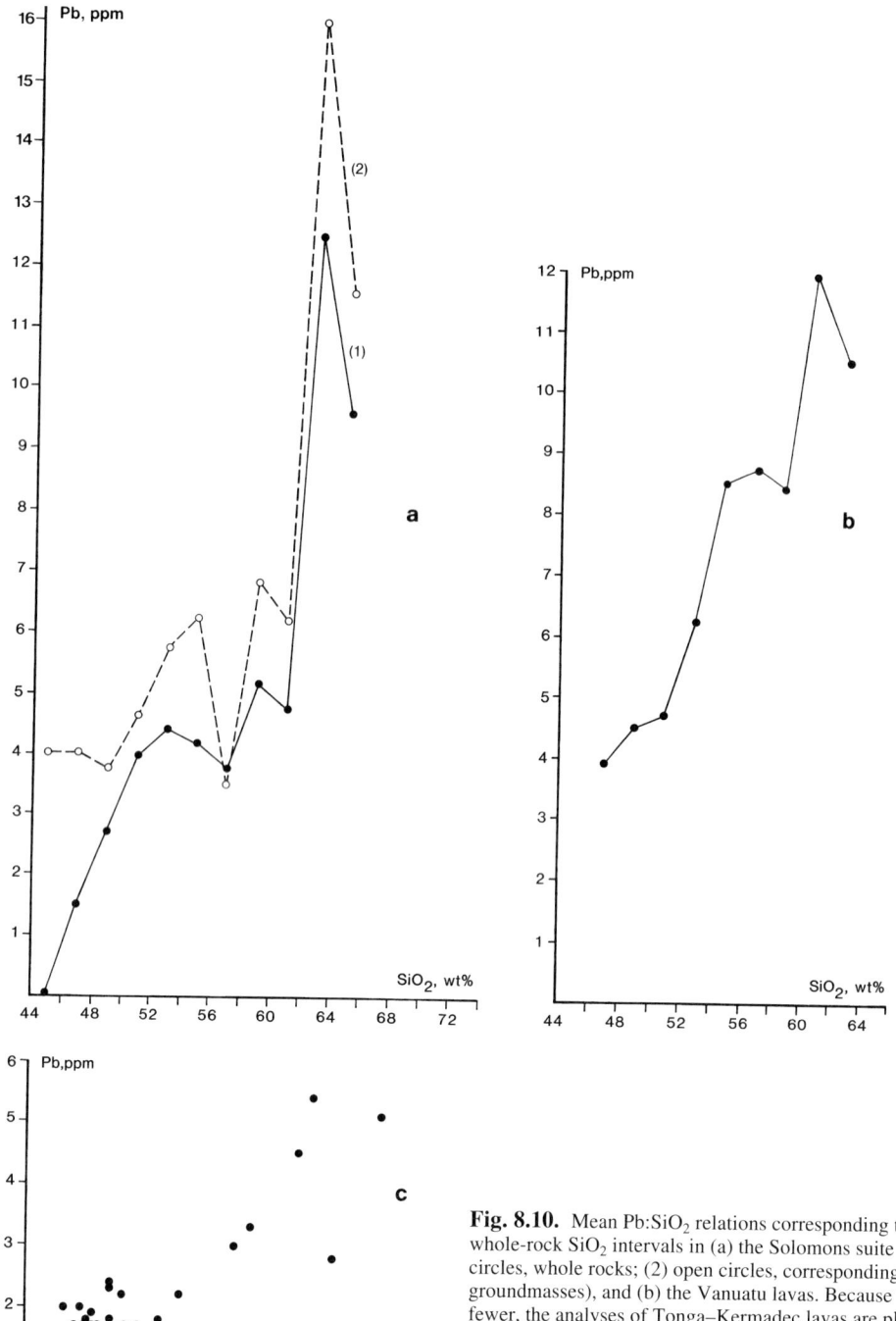

Fig. 8.10. Mean Pb:SiO$_2$ relations corresponding to 2 per cent whole-rock SiO$_2$ intervals in (a) the Solomons suite ((1) filled circles, whole rocks; (2) open circles, corresponding groundmasses), and (b) the Vanuatu lavas. Because they are fewer, the analyses of Tonga–Kermadec lavas are plotted individually in (c). The decrease in lead at SiO$_2$ >64 per cent in the Solomons lavas of (a) is not confirmed when the Bougainville analyses are incorporated in the figure [see Fig. 21.2(h)]. Plotting the relevant mean values of lead for the full array of 241 analyses indicates that lead is still increasing at this stage of evolution of the Solomons suite.

curve may be indicated by Fig. 21.2: there is clearly need for further work on lead abundances in the higher SiO$_2$ range of island-arc suites.

Lead:SiO$_2$ relations in the spectrum of Vanuatu lavas (Fig. 8.10(b); 136 analyses) are clearly reminiscent of those of the Solomons. Again there is a suggestion of an inflexion (at ≈58–60 per cent SiO$_2$) and a sharp increase in lead in the SiO$_2$ ≈60–2 per cent interval. As in the Solomons, lead appears to decrease at SiO$_2$ >65 per cent; this point on the graph of Fig. 8.10(b) represents only four analyses, however, and once again the effect may be due to sparseness of sampling.

As only a few of the available analyses of the more felsic Tonga–Kermadec lavas include lead, points representing individual lavas are plotted in Fig. 8.10(c). A general increase in lead with SiO$_2$ is indicated.

CONCLUDING STATEMENT

The mean lead content of the Solomons lavas is 4.1 ppm; that of the basaltic component is 2.9 ppm, which compares with about 2.0 ppm for MORBs and 4 ppm for ocean-island alkali olivine-basalts.

Although it is readily accepted into K$^+$ sites and shows moderate to good correlation with K$_2$O in all feldspars, whole rocks, and groundmasses, none of the silicates is enriched in lead relative to coexisting groundmass. By far the most effective acceptors among the silicates are the K-bearing plagioclases, but the mean K_D^{Pb} plagioclase:melt is low even here, at 0.45. Crystallization of all silicates thus concentrates lead in the remaining melt. The only major mineral to impoverish the melt in lead is magnetite, for which mean K_D^{Pb} = 2.30. The proportion of magnetite in the lavas is, however, too small for its crystallization to have any significant reducing effect on the lead in the remaining melt. No information is yet available on the acceptance of lead by Solomons biotites.

Lead thus concentrates in the melt to the point where, with the formation of the dacites, its rate of enrichment in the melt exceeds that of the alkalis and it develops a hypertholeiitic trend on Pb–MgO–(Na$_2$ + K$_2$O) diagrams.

This sudden increase in the rate of enrichment of lead follows a marked flattening of the Pb:SiO$_2$ curve (see Fig. 21.2 for the most reliable presentation of this feature). While the evidence so far is virtually limited to the Solomons suite, we may tentatively identify this marked change in form of the Pb:SiO$_2$ curve as 'the lead inflexion', and its point of development as the onset of dacite formation.

9

Barium

Barium occurring as barite, $BaSO_4$, is one of the principal components of many exhalative ores of volcanic affiliation. It shows a strong tendency to occur in those deposits containing lead as a significant constituent and hence in those associated with rhyodacitic to rhyolitic volcanic rocks. Where, as noted in Chapter 2, the relevant orebody exhibits chemical stratification, the barite-rich layer usually constitutes the stratigraphic top of the succession. The barite may simply constitute a component of the upper, galena-rich ore layers, or it may appear as the dominant constituent, in which case it may contain little more than disseminated, usually conspicuously argentiferous, galena.

In all cases the barite appears to be just as much a product of exhalation as the sulphides; indeed, abundant barite has long been known to occur in association with mid-ocean ridge hydrothermal activity (see, e.g. Bostrom and Petersen 1966).

As an alkaline-earth element closely related to calcium and strontium, barium occurs as a substantial trace in all lava types. Goldschmidt (1954) estimated approximately 63 ppm Ba in basalts, 233 ppm in andesites, 430 ppm in rhyolites, 520 ppm in phonolites, 1600 in trachytes, and 900–3600 in 'leucite rocks'. Because of ionic size similarities (see later in this chapter) barium tends to substitute for potassium in the common igneous silicates. However, as noted by Goldschmidt (1954) and also by Puchelt (1972), the barium is 'captured' by the earliest-formed potassium-bearing minerals and hence may in some cases become depleted in the melt prior to the formation of the final fractions. For this reason barium and potassium commonly do not show clear-cut correlation in igneous rock series. Barium may indeed be conspicuously low in some of the later, high-potassium members of igneous differentiation sequences.

Von Engelhardt (1936) analysed a variety of igneous K-feldspars and plagioclases and showed that while there was certainly a tendency for barium to be more abundant in those feldspars that were more potassium-rich, relations between the two elements were far from systematic.

Puchelt (1972) made the observation (based on a total of 399 analyses from the literature) that plots of barium concentrations for the principal members of the plagioclase solid solution series indicated that these were greatest in the intermediate (i.e. andesine) members. Mean Ba <100 ppm in the albites increased to mean Ba ≈400 in the andesines, and then decreased progressively to Ba <100 ppm again in the anorthite. This is reminiscent of the behaviour of copper in the Solomons plagioclases already referred to (p. 62), and is considered further below (p. 113).

Puchelt (1972) also emphasized the differences in barium content of basalts of different tectonic settings. He estimated means of 14.5 ppm Ba for oceanic tholeiitic basalts, 246 ppm for tholeiitic basalts of continents and oceanic islands, 253 ppm for 'average continental basaltic rock', and 613 ppm for alkali basalts. He noted that barium concentration in different igneous rock types was strongly affected by fractional crystallization and differentiation, but commented, in connection with the basaltic rocks, that the wide range of barium concentrations within similar petrographic types, and the similarity of barium values for different petrographic types, indicated that there must be substantial differences in the initial abundances of barium in the various parent melts. The development of such differences was presumably attributable to such factors as primary mantle inhomogeneity and varying degrees of partial melting.

Largely in keeping with their potassium and calcium contents, barium is much higher in intermediate and felsic lavas than in the basalts. From his survey of the literature, Puchelt (1972) estimated an average of 1177 ppm Ba for intermediate lavas and 1127 ppm for the effusive equivalents of granitic rocks.

BARIUM IN LAVAS OF THE MARINE ENVIRONMENT

A summary of available data on barium in MORBs is given in Table 6.2. The mean of 391 analyses representing the Atlantic, Atlantic–Indian, Indian, and Pacific midocean ridges is 92 ppm Ba, and if the subaerial basalts of Iceland are included (giving a total of 452 analyses) the mean value is 95 ppm. As indicated in Table 6.2, mean values for the different ridges vary considerably, from 56 ppm for the Mid-Atlantic Ridge to 157 ppm for the Eastern Pacific ridges — a factor of almost three. Variation of barium within ridges and ridge systems (e.g. the Mid-Atlantic Ridge: see Table 6.2) is also substantial, though the limitations of some samplings must be kept in mind here.

The 452 available analyses of MORBs (SiO_2 <52 per cent; mean Ba = 92 ppm) that include barium are plotted in the frequency distribution of barium abundances shown in Fig. 9.1. The analogous frequency distribution for 199 Mid-Atlantic Ridge basalts is shown in Fig. 9.1(b), for Icelandic basalts in Fig. 9.1(c), and for Icelandic lavas more felsic than basalt (SiO $\geqslant$52 per cent) in Fig. 9.1(d). The shift to higher barium abundances in the Icelandic basalts, and again in the more felsic members of the Iceland assemblage, is clear.

The body of data for island arcs is also now considerable and this is plotted in Fig. 9.2. The spreads of the distributions are broad, reflecting the wide differences in the general level of barium both between and within individual arcs, as indicated in Table 6.3. Figure 9.3, which shows the frequency distributions of values of barium in 146 lavas with SiO_2 $\geqslant$52 per cent from St Kitts, and 135 such lavas from Grenada, illustrates the latter point. Mean barium abundance in arc basalts (SiO_2 <52 per cent) is 240 ppm, which compares with 389 ppm in arc lavas more felsic than basalt (SiO_2 $\geqslant$52 per cent).

Concerning arc lavas, for which there is a large and significant amount of data for andesites and dacites as well as for basalts, it is interesting to note that mean barium abundance in arc basalts (SiO_2 <52 per cent) is 240 ppm (as above), andesites (52 <SiO_2 <62 per cent) 357 ppm, and dacites and more felsic lavas (SiO_2 >62 per cent) is 421 ppm. That is, barium increases progressively from basalts to andesites to dacites and more felsic types. This appears to be contrary to the suggestion of Goldschmidt and of Puchelt (see below) that barium, in tending to be captured in K^+ sites as soon as potassium minerals begin to crystallize, is eventually stripped from the melt leaving little for the residual stages of crystallization, even though the residual melt is likely to be rich in potassium. The present study has not, however, been concerned with plutonic rocks, fractionation patterns of which may not be entirely analogous with those of arc volcanic suites.

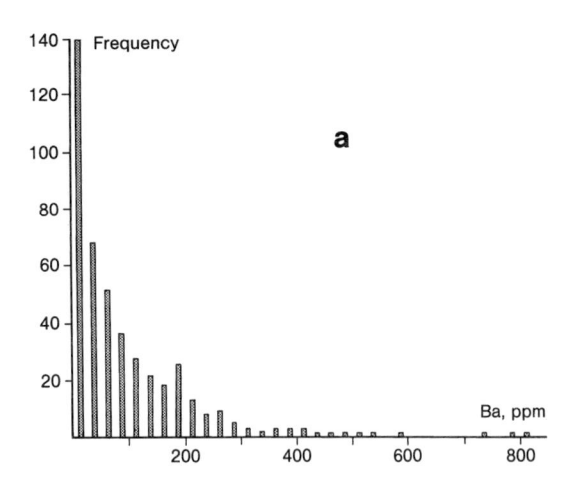

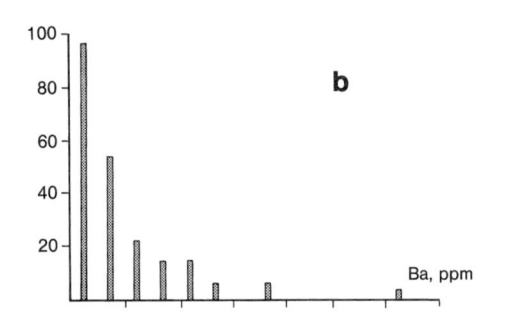

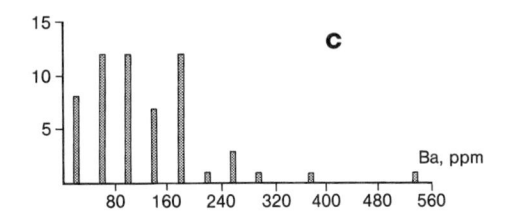

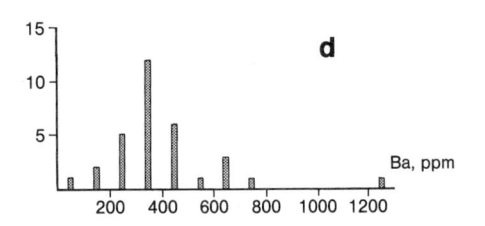

Fig. 9.1. Frequency distributions of barium abundances in (a) 447 MORBs of all major oceans; (b) 199 MORBs of the Mid-Atlantic Ridge; (c) 61 Icelandic basalts (SiO_2 < 52 per cent); and (d) 31 Icelandic lavas more felsic than basalt (SiO_2 $\geqslant$ 52 per cent).

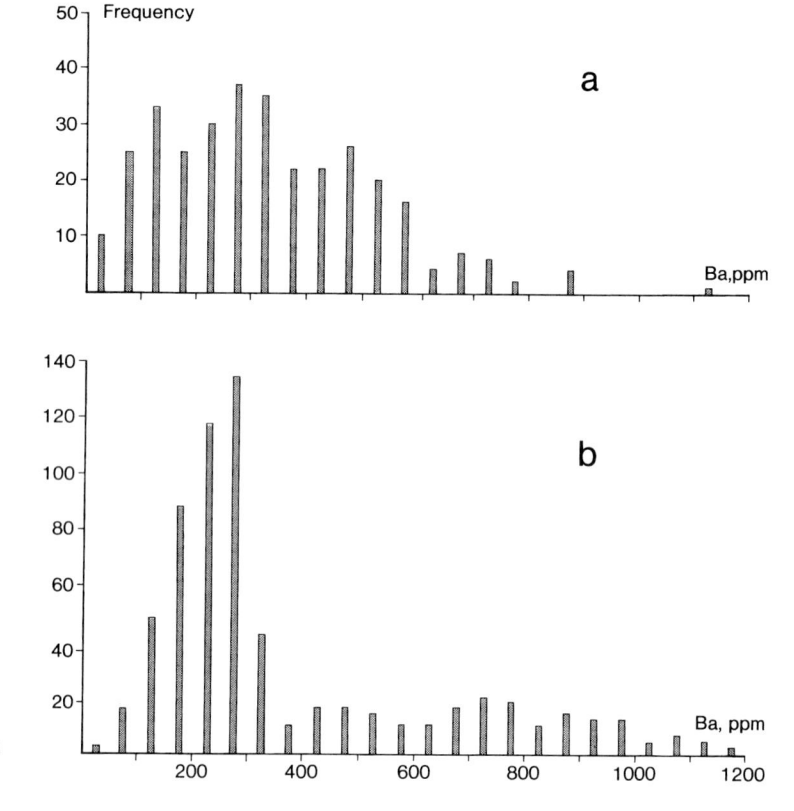

Fig. 9.2. Frequency distributions of barium abundances in (a) volcanic island-arc basalts and (b) arc lavas more felsic than basalt.

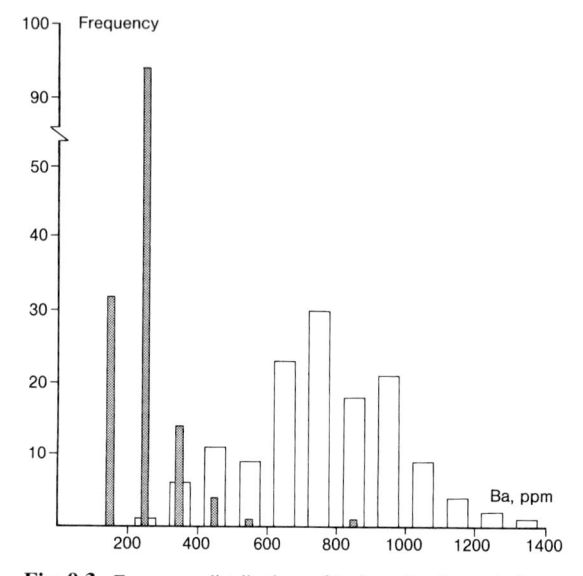

Fig. 9.3. Frequency distributions of barium abundances in lavas more felsic than basalts (i.e. $SiO_2 \geqslant 52$ per cent) of St Kitts (filled bars, 146 analyses, mean = 246 ppm Ba) and Grenada (open bars, 135 analyses mean = 760 ppm Ba) in the Lesser Antilles. (Raw data for this diagram supplied by J. G. Holland).

Figure 9.4 shows abundance relations between barium and K_2O in MORBs as represented by the basalts of the Mid-Atlantic Ridge, and in Icelandic basalts.

CHEMICAL PROPERTIES AND CRYSTAL CHEMISTRY OF BARIUM

Barium is lithophile in character and is one of the petrogenetically important 'large ion lithophile' (lil) or 'incompatible' elements. As pointed out early by Goldschmidt the ionic radii of Ba^{2+} (0.143 nm) and K^+ (0.133 nm) are close enough for barium to be accommodated extensively in potassium sites, and the relatively common occurrence of the barium-rich feldspar, celsian [$(Ba,K)AlSi_3O_8$] attests to this.

The electron configurations and related properties of barium and the Ba^{2+} ion are given in Table 6.4. With a radius ratio of approximately unity, Ba^{2+} is capable of co-ordinating twelve oxygen ions, and the bond is approximately 80 per cent ionic. However with the long interatomic distances involved with large ions such as Ba^{2+} and O^{2-} in complex silicate structures, it is often difficult to define coordination polyhedra; Fischer (1972)

chemical properties allow strontium ($Sr^{2+} = 0.127$ nm) to substitute in many crystal sites also favoured by barium. There is thus a common association of barium with strontium (and lead) in potassium-rich minerals.

THE INCIDENCE OF BARIUM IN THE PRINCIPAL MINERAL SPECIES OF THE SOLOMONS YOUNGER VOLCANIC SUITE

General

Although its significance as a large ion lithophile element has led to its extensive determination and interpretation in whole rocks (particularly in the elucidation of the origin of many lava series) comparatively little attention has been paid to the incidence of barium in volcanic minerals other than feldspars.

The K-feldspars are conspicuous among the common igneous silicates in their role as principal acceptors of Ba^{2+}. Puchelt (1972) has collated 598 analyses of K-feldspars in a wide variety of igneous rocks, obtaining a range of 3–18 000 ppm and a mean of 2626 ppm Ba. His observation that intermediate members of the plagioclase series contain notably more barium than do the pure end-members has already been referred to. Igneous andesines and labradorites have been found to range up to 1250 ppm Ba, with a mean of ≈400 ppm. In contrast, anorthites and albites most commonly contain <100 ppm Ba.

Of those ferromagnesian silicates of present concern, biotite, with its high potassium content, is the principal barium acceptor. The small number of volcanic biotite analyses available in the literature give a range of ≈1500–6500 ppm Ba, with a mean of about 5000 ppm. A similarly sparse literature on volcanic amphiboles indicates these (with K_2O an order of magnitude less than that of biotites) to contain 100–80 ppm Ba. A total of 23 analyses of volcanic pyroxenes give a mean of 18 ppm Ba. An olivine from a Vanuatu basalt (Gorton 1974) contained 0.40 ppm Ba, and the coexisting clinopyroxene 1.3 ppm. Philpotts and Schnetzler (1970) determined (by isotope dilution) 1.91 ppm Ba in an olivine from an ankaramitic basalt from Hawaii, and 5.07 ppm Ba in olivine from a similar rock from Tahiti. The coexisting pyroxenes contained 10.2 and 11.49 ppm Ba respectively.

Solomons suite

Mean abundances of barium in the ferromagnesian silicates, feldspars, and the spinels are given in Table 6.5, and spreads of abundances within each mineral group are indicated in Figs. 9.5, 9.6, and 9.8 respectively. The feldspars and biotites (Table 6.5) are clearly the principal barium receptors.

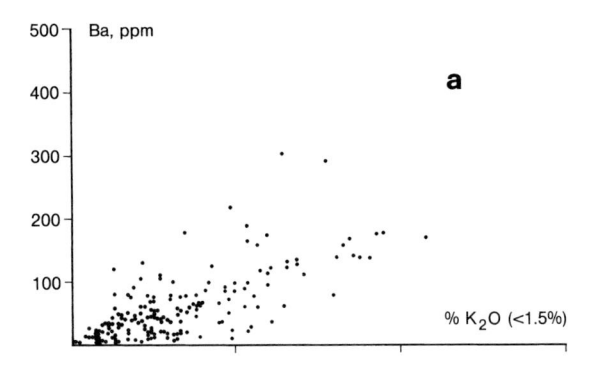

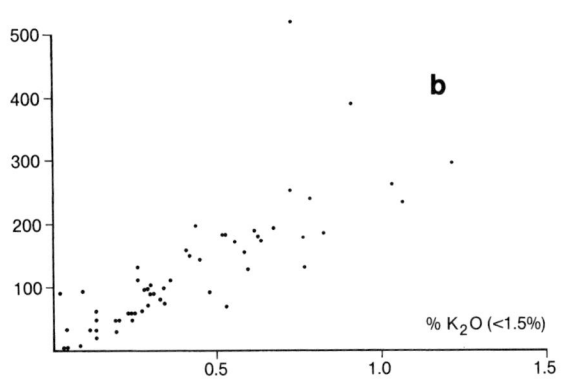

Fig. 9.4. Abundance relations between barium and K_2O in (a) MORBs of the Mid-Atlantic Ridge, and (b) Icelandic basalts.

notes that the coordination number for Ba^{2+} ranges from 6 to 12 for O, OH, and H_2O.

Owing to the large size difference involved, Ba^{2+} substitutes for calcium (ionic radius ≈0.106 nm) to only a very limited extent. Barium thus appears principally in potassium-bearing minerals such as the K-feldspars, micas, and to a lesser extent the hornblendes. In substituting for potassium, the barium is in a loose 9-fold coordination with oxygen in the feldspars, in interlayer positions in a widely spaced 6-fold coordination in the micas, and in 8-fold coordination in the hornblendes. Goldschmidt (1954) has pointed out that the relationship of barium and potassium in the magmatic crystallization of K-feldspars is analogous to the relationship between calcium and sodium in the crystallization of the plagioclases. It cannot thus be expected that there should be any simple proportionality between the percentages of barium and potassium in igneous rocks — in spite of their association in the same minerals — any more than between calcium and sodium in plagioclase-bearing rocks.

Although there is a difference in ionic radius of some 11 per cent of the larger ion, similarity of charge and

112 *Ore Elements in Arc Lavas*

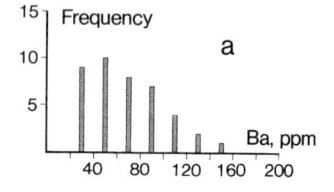

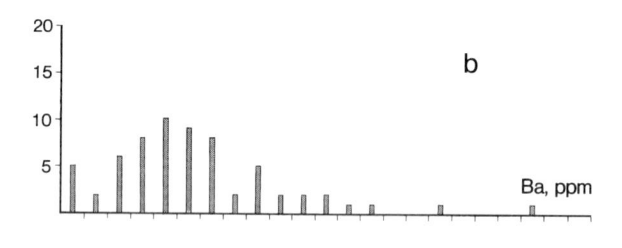

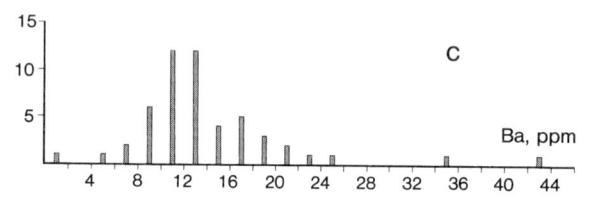

Fig. 9.5. Frequency distributions of barium abundances in (a) hornblendes, (b) clinopyroxenes, and (c) olivines of the Solomons suite.

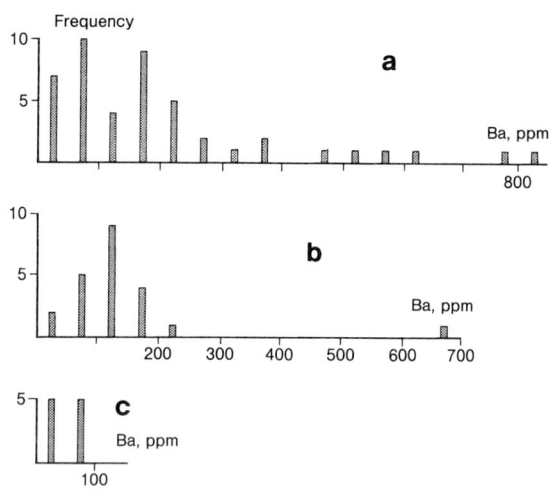

Fig. 9.6. Frequency distributions of barium abundances in ultimate purity feldspar separations from (a) hornblende-andesite group (46 feldspars; mean = 205 ppm Ba); (b) big-feldspar lavas (21 feldspars; mean = 141 ppm Ba); and (c) olivine–clinopyroxene-basalts (10 feldspars; mean = 45 ppm Ba) of the Solomon Islands Younger Volcanic Suite.

Fe–Mg silicates. Barium contents here accord quite well with earlier data from other provinces. Olivines exhibit a range from <2 to 43 ppm, with a mean of 16 ppm. The more iron-rich olivines of the big-feldspar lavas may contain slightly higher barium (mean = 18 ppm) than those of the olivine–pyroxene lavas (14 ppm), though this may be no more than a sampling artefact. The clinopyroxenes exhibit a barium range very similar to that of the olivines (Fig. 9.5(b)), though there is a much higher frequency of values in the 0–10 ppm range and hence a lower mean. The latter is very consistent for the three groups of clinopyroxenes, at about 11 ppm (Tables 6.5, 9.1). Hornblende, with a mean K_2O content of 0.45 per cent, has a distinctly higher barium range, 32–150 ppm, than the olivine and clinopyroxene, and averages 67 ppm Ba. Biotite, as represented by just the two Savo dacites, and with its K_2O content more than a magnitude higher, exhibits a range of 3615–5941 ppm, and a mean of 4875 ppm Ba.

Feldspars. The feldspars follow biotite, at values an order of magnitude lower, as acceptors of barium, though — almost certainly reflecting their generally comparatively low K_2O content — their barium contents are low compared with many occurrences of the plagioclases.

Figure 9.6, which shows the spreads of barium values in the feldspars (ultimate purity separations) of the three principal lava types, also shows a progressive shift to higher values of barium going from the feldspars of the olivine–pyroxene-basalts to those of the hornblende-andesites and dacites.

On an individual sample basis $Ba:K_2O$ correlation is far from remarkable (Fig. 9.7(a)) and apparently no greater than that between Ba and Na_2O (Fig. 9.7(b)). When the feldspars occurring in each of the three principal lava groups are averaged, however, a broad $Ba:K_2O$ correlation is clear (Table 6.5; Fig. 9.7(c)). Also, when Na_2O is averaged and plotted in the same way and using the K_2O (horizontal) scale multiplied × 10^{-1}, the correlation of $Ba:Na_2O$ is seen to be virtually precisely the same as that for $Ba:K_2O$, and the regression lines for the two pairs of points are parallel. In keeping with the impression gained from Figs. 9.7(a) and (b), the correlation between Ba and the two alkali metals is essentially identical, with Na_2O at exactly one order of magnitude greater abundance than K_2O.

This similarity of degrees of correlation appears remarkable in view of the fact that the Na^+ ion is so much smaller (0.098 nm) than the K^+ (0.133 nm) and Ba^{2+} (0.143 nm) ions. The electronegativities of the three elements (0.9, 0.8, and 0.85 respectively) are very similar, indicating essentially similar ionic:covalent bond ratios. The overall relationship might be attributed simply to a capacity of plagioclase to accept small quantities of ions much larger than Na$^+$ and Ca^{2+}, but this is belied by the fact that the incidence of Ba (and K) is directly related to the abundance of sodium. The answer may lie in the development of a mosaic structure, involving discrete domains of alkali feldspar (K,Na)AlSi$_3$O$_8$ randomly distributed in the plagioclase structure, rather than an evenly dispersed 'solid solution' of K$^+$ in an essentially continuous and uniform plagioclase crystal. The Ba^{2+} would then occur by substitution-*cum*-vacancy formation in alkali feldspar rather than plagioclase, and if the incidence of the former were related to the abundance of sodium in that plagioclase, the loose positive correlation of barium (and potassium) with sodium would follow.

In this general connection it may be noted that the tendency noted by Puchelt for highest barium to occur in the intermediate members does not hold for the Solomons plagioclases: while the degree of correlation from one individual feldspar to another is not high, barium increases with increase in sodium at least as far as the oligoclase–albite composition. Barium appears to enter the lattice first at ≈1 per cent Na$_2$O, i.e. towards the anorthite–bytownite boundary.

In this general connection Table 6.5 indicates the greater general capacity of plagioclase to incorporate barium compared with hornblende, regardless of the incidence of potassium. Mean K$_2$O of hornblende and the plagioclases of the hornblende-andesites is the same at 0.45 per cent. Mean barium in the feldspars (205 ppm) of the hornblende-andesites is, however, three times that in the hornblende (67 ppm) of the same lavas.

Spinels. Mean barium values are given in Table 6.5, and their spreads are indicated in Fig. 9.8. The range is

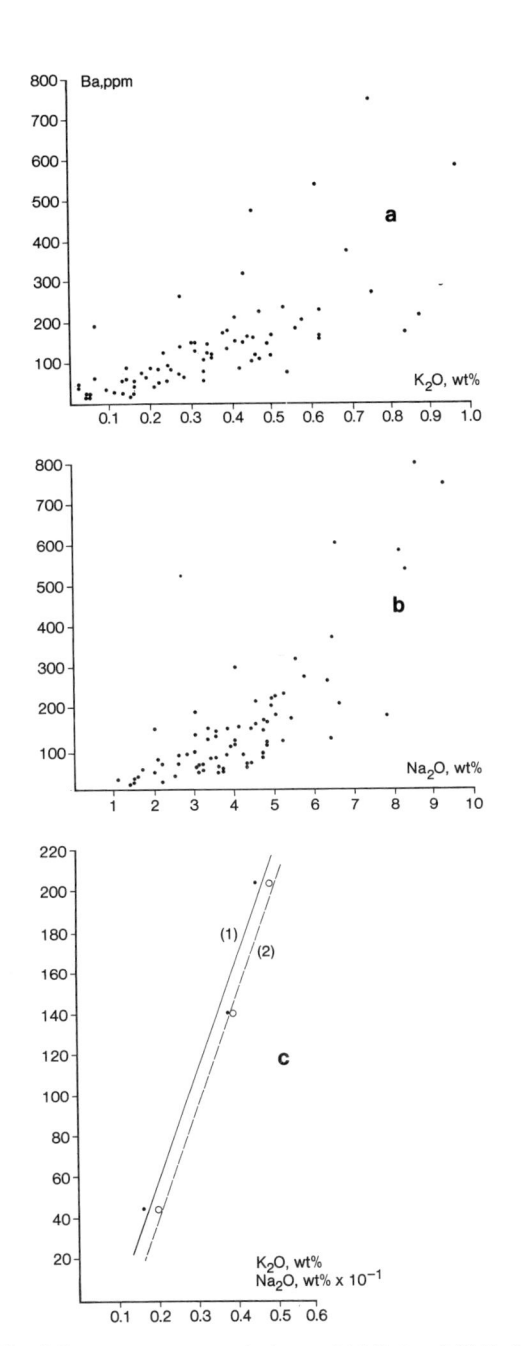

Fig. 9.7. Relations between barium and (a) K$_2$O and (b) Na$_2$O in the 77 ultimate purity feldspar separations. Mean Ba:K$_2$O and Ba:Na$_2$O relations for the feldspars of each of the three principal lavas groups are shown in (c) by lines (1) (filled circles, solid line) and (2) (open circles, broken line) respectively. In both cases correlation between individual feldspars is poor, but that between large groupings of feldspars (i.e. corresponding to the three broad lava groups) is good.

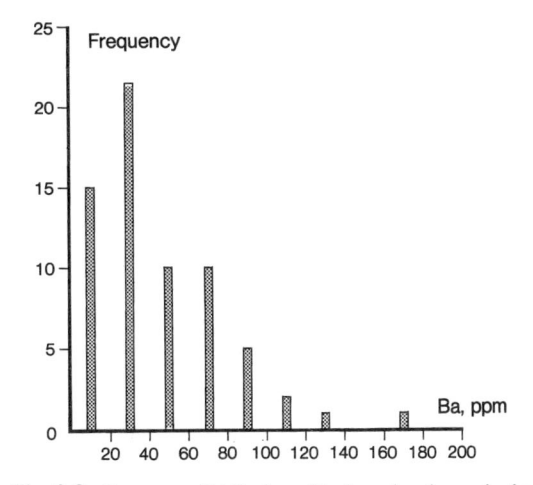

Fig. 9.8. Frequency distribution of barium abundances in the Solomons spinels.

<2–173 ppm, and the overall mean 46 ppm. No reason is apparent for the conspicuously low incidence of barium in the spinels (titanomagnetites) of the big-feldspar lavas. This is matched by a low, and equally unexplained, abundance of strontium (Table 6.5).

CRYSTAL:MELT PARTITIONING

As barium is the most conspicuous of the large-ion lithophile elements and hence one of the more important 'tracers' used in determining the derivation of igneous melts, there is now an extensive literature on its crystal:melt partitioning, particularly with the potassium feldspars and the plagioclases.

Among the more modern investigators, Carmichael (1967) found a range of K_D^{Ba} biotite:rock of 1.6–15 in a group of dacites, rhyolites, and trachytes, and Haslam (1968) determined K_D^{Ba} hornblende:melt of 0.31 in a Scottish rhyolite.

Berlin and Henderson (1969) examined the distribution of barium (and strontium) between the alkali feldspars, plagioclases, and groundmasses of a group of porphyritic trachytes and phonolites from Europe, several of the Atlantic islands, and Antarctica. Analysing multiple pairings, they found a range of K_D^{Ba} crystal:rock of 1.17–8.95 for sanidine in trachyte, 2.14–2.37 for anorthoclase in phonolite, and 0.72–3.98 for plagioclase in trachyte. Ewart and Taylor (1969) determined ranges of K_D^{Ba} for plagioclase:hornblende and clinopyroxene:melt of 0.12–0.53, 0.05–0.14, and 0.02–0.15 respectively in a series of basalts, andesites and rhyolites, from the North Island of New Zealand. Philpotts and Schnetzler (1970), in a comprehensive study of a range of young porphyritic lavas from the Pacific (ocean and margin), Atlantic, and Mediterranean provinces, obtained K_D^{Ba} crystal:melt in the following ranges: sanidine, 6.12; plagioclases, 0.054–0.589, mean = 0.236; micas, 1.09–15.3, mean = 7.583; clinopyroxenes, 0.0128–0.388, mean = 0.0734; orthopyroxenes, 0.0121–0.0141, mean = 0.0131; hornblendes, 0.1–0.731, mean = 0.4159; olivines, 0.0086–0.0112, mean = 0.0099. They also determined K_D^{Ba} for the following coexisting mineral pairs: biotite/pyroxene, 85–116; plagioclase/clinopyroxene, 4.75–18.7;and clinopyroxene/olivine 2.59–34.5.

Nagasawa and Schnetzler (1971) determined coefficients of 0.3 and 0.05 respectively for plagioclase and clinopyroxene in dacites; and in a comprehensive series of experiments in the Ab–An–Di system Drake and Weill (1975) obtained values for plagioclase ranging from 0.22–0.25 at 1340 °C, 1 atm to 0.59–0.69 at 1153 °C, 1 atm. In a series of experiments on the Ab–An–Di–H$_2$O system at high pressures Shimizu (1974) obtained very low values (c. 0.001) for clinopyroxenes, and for natural clinopyroxenes in a range of lavas from alkali basalts to rhyolites Kovalenko et al. (1986) obtained clinopyroxene:melt values of the general range 0.03–0.08.

The literature is too extensive to summarize further but, allowing for the effects of variation in mineral and melt chemistry, temperature, and pressure, ranges of K_D^{Ba} for the principal volcanic minerals are approximately: olivine 0.0001–0.03; clinopyroxene 0.001–0.01; hornblende, 0.05–0.50, with a few higher; plagioclase, 0.1–2.0; magnetite ≈0.4.

Partition coefficients for the principal Solomons minerals are given in Table 9.1. As for the previous elements, calculations are presented for 'all minerals-all groundmasses' for each category of lavas, and for coexisting pairs exclusively. In general the two correspond closely.

K_D^{Ba} olivine:melt and clinopyroxene:melt are very low (generally less than 0.1) and the clinopyroxene value is probably the lowest, at a mean of 0.05. Partition coefficients for the Solomons olivines may be a little higher than those obtained by earlier investigators, but based as they are on 39 coexisting pairs their reliability must be regarded as high. Values for hornblende are well within the range reported by others. There is little information on barium partitioning by magnetite in the literature but the mean values for the Solomons material (0.06–0.23) are well within the range reported by others.

In keeping with the higher incidence of barium in the feldspars, K_D^{Ba} values here range from a mean of 0.24 in the plagioclases of the olivine–pyroxene-basalts to 0.56 in those of the hornblende-andesites. These figures, obtained from feldspar:groundmass separations, are generally confirmed by microprobe analysis of feldspars, glass inclusions, and glassy groundmass in which the feldspars are enclosed. The glass inclusions analysed for copper and zinc (see Chapters 6 and 7) were at the same time analysed for barium, strontium, and manganese. K_D^{Ba} plagioclase:included glass for nine pairings ranged from 0.25 to 0.59. The groundmass glass, i.e. that surrounding, as contrasted with that included by, the feldspars, in representing a later fraction of the melt, characteristically contained higher barium than did the included glass. Thus, for example, feldspars giving K_D^{Ba} crystal:included glass of 0.51 and 0.59 gave K_D^{Ba} crystal:enclosing glass of 0.36 and 0.29 respectively. A limited examination of the relation between barium zoning in the feldspars and the incidence of barium in the two categories of glass was inconclusive, but more detailed examination is clearly called for.

In spite of the progressive increase in mean feldspar K_D^{Ba} proceeding from olivine–pyroxene-basalts to the hornblende-andesites and dacites (Table 6.5), there is no systematic relationship between individual feldspar:groundmass K_D^{Ba} and whole-rock SiO$_2$ (as an indicator of

Barium   115

Table 9.1 Crystal:melt distribution coefficients (K_D) with respect to barium

Mineral species	Lava group	n_1	n_2	m_1	m_2	$K_{D(1)}$	N	M_1	M_2	$K_{D(2)}$
Olivine	OPB	40	44	14	173	0.08	39	14	177	0.08
	BFL	7	23	18	302	0.06	6	18	254	0.07
Clinopyroxene	OPB	43	44	10	173	0.06	42	10	179	0.06
	BFL	26	23	11	302	0.04	17	18	296	0.06
	HbA	27	47	11	391	0.03	20	13	301	0.04
Hornblende	HbA	25	47	67	391	0.17	22	73	428	0.17
Feldspar	OPB	10	44	45	173	0.26	10	48	198	0.24
	BFL	23	23	141	302	0.47	22	141	315	0.45
	HbA	46	47	205	391	0.52	44	221	393	0.56
Spinel	OPB	10	44	53	173	0.31	8	56	240	0.23
	BFL	15	23	16	302	0.05	15	16	291	0.05
	HbA	37	47	55	391	0.14	36	53	368	0.14

OPB = olivine–clinopyroxene-basalt group; BFL = big-feldspar lavas: HbA = hornblende-andesite group; n_1 = total number of mineral separations analysed; n_2 = total number of groundmass separations analysed; m_1 = mean barium content of mineral of each lava group; m_2 = mean barium content of groundmass of each lava group; $K_{D(1)}$ = mean distribution coefficient determined as m_1/m_2; N = number of coexisting mineral:groundmass pairs analysed; M_1 and M_2 = mean barium contents of coexisting mineral and groundmass respectively; $K_{D(2)}$ = distribution coefficient determined as M_1/M_2, for comparison with $K_{D(1)}$. All concentrations as parts per million by mass.

lava evolution). Neither does there appear to be any such tie involving individual distribution coefficients for hornblende: and pyroxene:groundmass pairings.

THE INCIDENCE OF WHOLE-ROCK BARIUM IN THE SOLOMONS YOUNGER VOLCANIC SUITE

Mean barium in the basaltic members (SiO_2 <52 per cent; solid bars, Fig. 9.9) of the suite is 152 ppm, which compares with 92 ppm in MORBs, 272 ppm in the basalts of the Lesser Antilles, and 350, 91, and 349 ppm in the basalts of the Aleutians, Tonga–Kermadec, and Vanuatu arcs respectively. Mean barium in the more felsic lavas of the Solomons ($SiO_2 \geqslant 52$ per cent; open bars, Fig. 9.9) is 344 ppm, which compares with 404 ppm in 33 more felsic lavas of Iceland, and 287, 566 190, and 477 ppm Ba in lavas more felsic than basalts of the Lesser Antilles, Aleutians, Tonga–Kermadec, and Vanuatu respectively. While barium in both the basalts and more felsic types of the arc environment is clearly highly variable from one province to another (as already noted for the mid-ocean ridges), abundances in the Solomons lavas fall in about the middle of the arc range.

Frequency distributions of barium values in the three principal Solomons lava types and their groundmasses are shown in Fig. 9.10. Not only does mean barium increase proceeding from olivine–pyroxene-basalts (Fig. 9.10(c), mean = 121 ppm Ba) through the big-feldspar lavas (Fig. 9.10b, mean = 260 ppm Ba) to hornblende-andesites

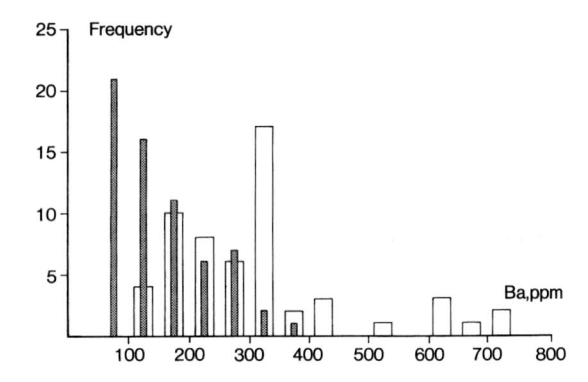

Fig. 9.9. Frequency distributions of barium abundances in Solomons basalts (SiO_2 < 52 per cent; filled bars) and in Solomons lavas more felsic than basalts ($SiO_2 \geqslant 52$ per cent; open bars).

(Fig. 9.10(a) mean 303 ppm Ba) but variability, i.e. spread of values, also increases. There is also a clear tendency (see the previous section on partition coefficients) throughout for the barium to partition into the groundmass.

This progressive increase of barium in the whole rocks with increase in the felsic nature of the rock, which develops in parallel with the increase in barium of the feldspars as these become more potassic, is at first sight out of keeping with Puchelt's observation that barium may undergo 'early capture' in the first-formed potassium-bearing minerals, and thus be impoverished in late-stage differentiates. There is, however, no contradiction

116 *Ore Elements in Arc Lavas*

essentially linear arrangement, scatter increasing progressively from the MgO apex. The series thus appears to be tholeiitic with respect to barium. When, however, mean Ba–MgO–(Na$_2$O + K$_2$O) proportions for 2 per cent intervals of whole-rock SiO$_2$ are plotted on the triangular diagram (Fig. 9.12(a)) it is seen that what at first sight appeared to be a straight line arrangement (i.e. a series tholeiitic with respect to barium) is inflected towards the Ba apex at its low-MgO extremity: that is, like lead, barium exhibits hypertholeiitic behaviour in the Solomons lava series. Analogous plots for the Solomons *cum* Bougainville, Grenada–Dominica–St Kitts, Tonga–Kermadec, Vanuatu, and the Aleutians (Figs. 9.12(b)–(d)) indicate similar hypertholeiitic trends.

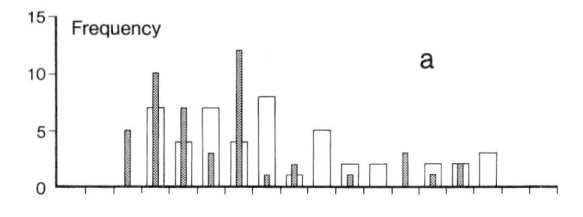

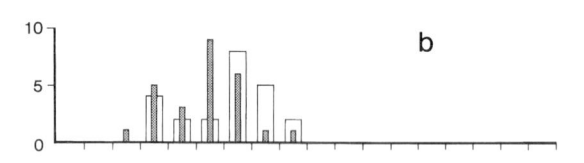

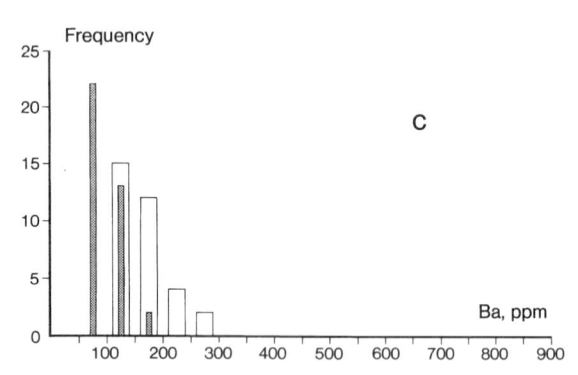

Fig. 9.10. Frequency distributions of barium abundances in whole rocks (filled bars) and groundmass (open bars) of (a) lavas of the hornblende-andesite group, (b) big-feldspar lavas, and (c) the olivine–clinopyroxene-basalts.

in principle: barium tends to accompany potassium, there is little incorporation of potassium in early-formed crystals of the Solomons lavas; as a result the major part of the potassium — and hence barium — of the Solomons lava series is left to be incorporated in the crystals and residual glass of the final differentiates.

This apparent systematic tendency for barium to accompany potassium suggests that it may, like copper, zinc, and lead, follow a fairly well-defined path during the evolution of the lavas. Plotting of the three principal lava types and their groundmasses on Ba–MgO–(Na$_2$O + K$_2$O) diagrams shows an ordered sequence of fields (Fig. 9.11) proceeding from an elongated but sharply defined group of points trending down to the MgO corner of the triangle and representing the olivine–pyroxene lavas (Fig. 9.11(f)), to a somewhat diffuse field on the Ba–(Na$_2$O + K$_2$O) side representing the hornblende-andesite groundmasses (Fig. 9.11(a)). The six fields combined present an

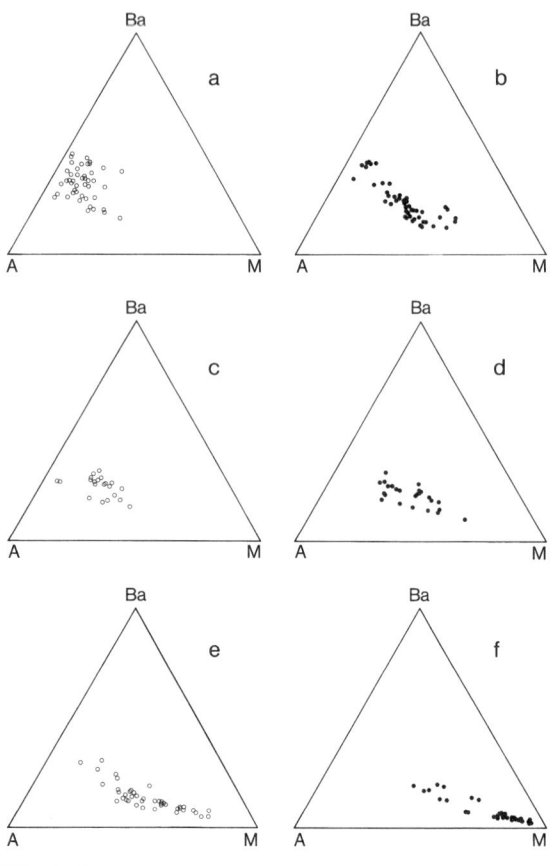

Fig. 9.11. Ba–MgO–(Na$_2$O + K$_2$O) ternary relations (Ba calculated using ppm Ba × 10^{-2}) developed in olivine–clinopyroxene-basalt, big-feldspar lava, and hornblende-andesite whole rocks (filled circles: (f), (d), and (b) respectively), and their corresponding groundmasses (open circles: (e), (c), and (a) respectively).

The plot of individual analyses of MORBs, as represented by those of the Mid-Atlantic ridge, presents an unclear and ambiguous field (Fig. 9.12(e)). It might be suggested that the beginning of a tholeiitic trend is evident, but if it is, it is diffuse. The Icelandic basalts (Fig. 9.12(f)) on the other hand show a clear tholeiitic tendency, and this continues to develop a hypertholeiitic pattern in the more felsic lavas. One $\geqslant 52$ per cent SiO_2 sample of Fig. 9.12(f) plots in the high total alkali metal region (it is tempting to suspect that this might be an artefact of analysis), but the field of points as a whole seems to indicate that the Icelandic lavas, like those of most volcanic arcs, are hypertholeiitic with respect to barium.

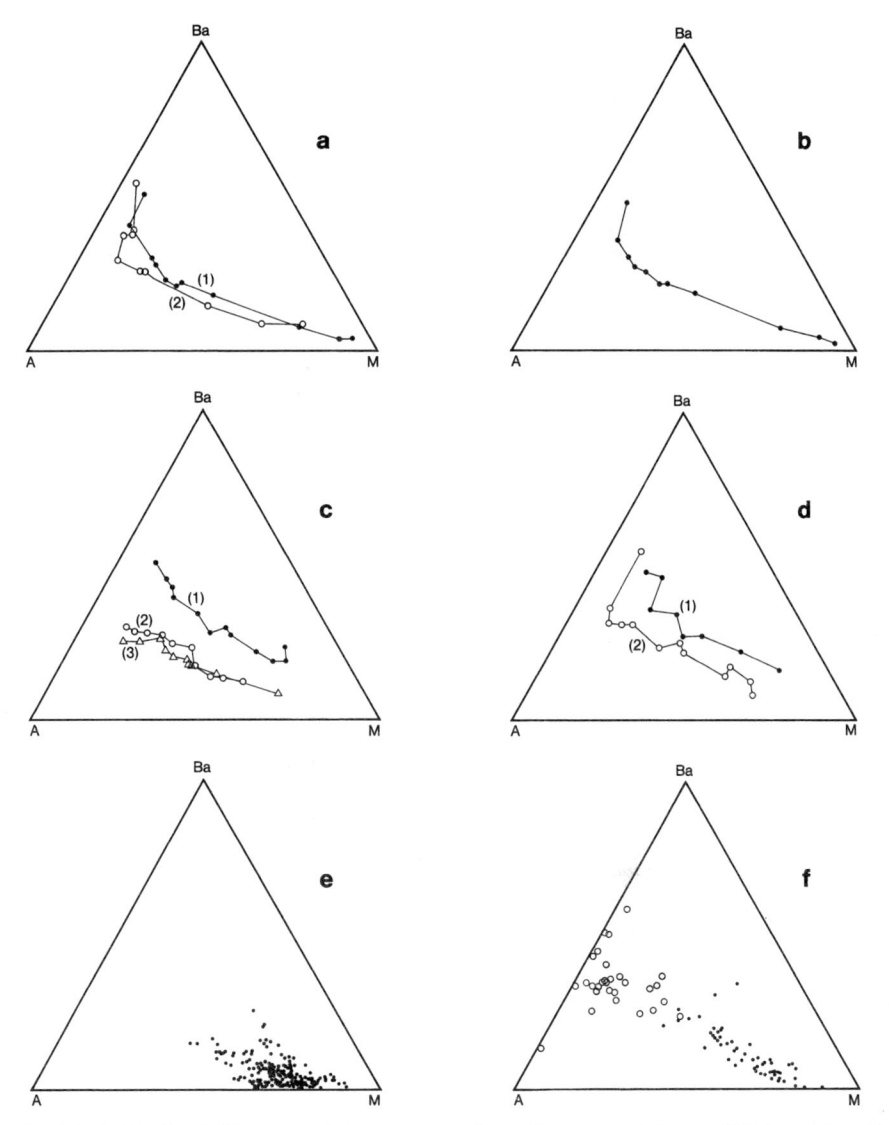

Fig. 9.12. Mean Ba–MgO–(Na$_2$O + K$_2$O) ternary relations corresponding to 2 per cent whole-rock SiO$_2$ intervals in (a) whole rocks (filled circles) and corresponding groundmasses (open circles) of the Solomons suite excluding Bougainville; (b) whole rocks of the Solomons suite including Bougainville (241 analyses); (c) lavas of (1) Grenada, (2) Dominica, and (3) St Kitts; and (d) lavas of (1) Vanuatu and (2) Tonga–Kermadec. Ternary relations developed in MORBs as represented by Mid-Atlantic Ridge basalts are given in (e), and in individual samplings of Icelandic basalts (SiO$_2$ <52 per cent; filled circles) and Icelandic lavas more felsic than basalt (SiO$_2$ $\geqslant$52 per cent: open circles) in (f).

118 Ore Elements in Arc Lavas

All this indicates that in most volcanic-arc assemblages — and in the Icelandic suite — barium continues to rise in the later products of fractionation, rather than establishing maxima in the intermediate products of lava evolution. The only apparent exception is that of the Vanuatu suite (Table 6.3, Fig. 9.13(c)), though in this case the most felsic category is represented by only four analyses, and the apparent decrease may reflect no more than inadequate sampling. The continuing increase in barium is, on the other hand, quite unequivocal in the Solomons suite — as indicated by Fig. 9.13(a), which shows the relationship between barium abundance and 2 per cent intervals of whole-rock SiO_2. This diagram is remarkably reminiscent of that for lead (Fig. 8.10) and shows that, with some minor irregularities, barium increases fairly steadily with increase in whole-rock SiO_2, i.e. with progressive evolution of the lava series.

Barium:SiO_2 relations in the respective groundmasses are also shown, as open circles, in Fig. 9.13(a). The whole-rock and groundmass enrichment trends are remarkably parallel, presumably indicating the maintenance of very constant Ba:SiO_2 ratios in the mixtures of crystals (principally olivines and pyroxenes) being subtracted.

An analogous diagram for the Lesser Antilles (Grenada, Dominica, and St Kitts; Fig. 9.13(b)) is less unambiguous, particularly in the case of Grenada. The early-formed crystals of the lava series — those which, by accumulation, generate the most mafic lavas — are olivine and clinopyroxene. As shown above, the partition coefficients of these minerals with respect to barium are almost vanishingly small, and it is difficult to see why barium should be higher in the picritic rocks of Grenada than in the associated basalts. The lavas of Dominica and St Kitts, on the other hand, show gentle progressive increases in barium with evolution of the respective series. Analogous Ba:SiO_2 plots for Vanuatu, Tonga–Kermadec, and the Aleutians (Fig. 9.13(c) and (d)) reveal whole-rock patterns rather similar to that of the Solomons; the behaviour of barium is somewhat irregular as the lavas evolve from basalts towards felsic andesites (i.e., c. 58 per cent whole-rock SiO_2), at which stage it increases rapidly and continues to do so. This is the case whatever the absolute value of barium in the particular lava series. Thus the Tonga–Kermadec series, which possesses relatively low absolute barium, exhibits much the same kind of pattern as those of the Aleutians, Vanuatu, and the Solomons, which overall contain more than twice the abundance. This is, of course, as would be expected from crystal-chemical considerations. As noted above, the apparent sudden decrease of barium in the Vanuatu lavas at SiO_2 >62 per cent is likely to be a sampling artefact; only four analyses were available in this SiO_2 interval.

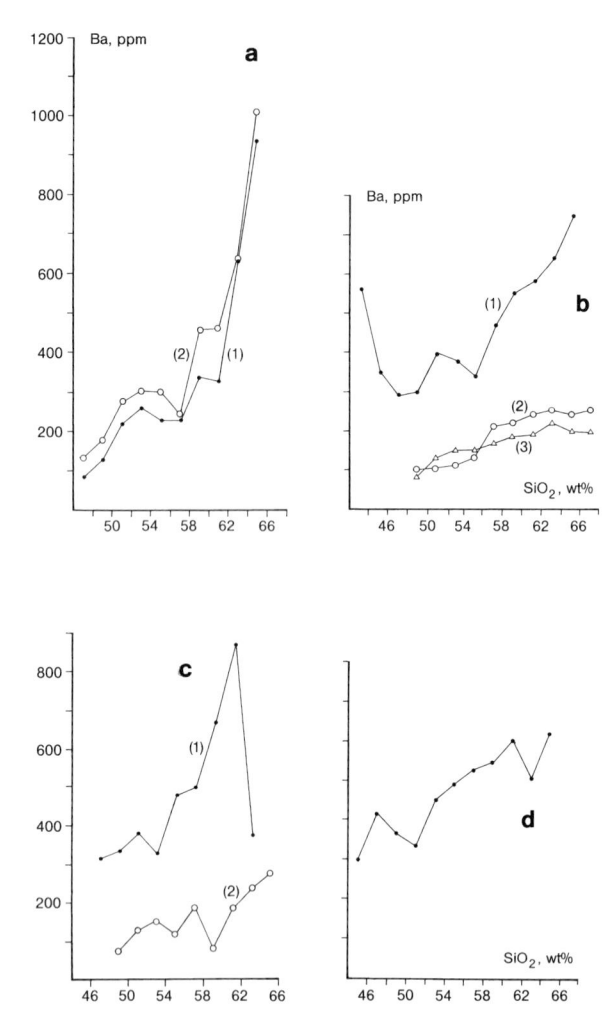

Fig. 9.13. Mean Ba:SiO_2 relations corresponding to 2 per cent whole-rock SiO_2 intervals in (a) whole rocks [(1) filled circles] and corresponding groundmasses [(2) open circles] of the Solomons suite; (b) lavas of (1) Grenada, (2) Dominica, and (3) St Kitts; (c) lavas of (1) Vanuatu and (2) Tonga–Kermadec; and (d) lavas of the Aleutian arc.

CONCLUDING STATEMENT

On the evidence of current data, mean barium for MORBs is 92 ppm and for island-arc basalts it is 240 ppm; i.e. volcanic-arc basalts contain some 2.6 times as much barium as do those of the mid-ocean ridges. Consideration of arc lavas more felsic than basalt indicates that, overall, barium rises to ≈357 ppm in andesites and ≈421 in dacites and more felsic lavas (mean = 389 ppm). Mean barium in the Solomons basalts is 152 ppm, and this increases fairly systematically to a mean of ≈340 ppm in the andesites and dacites.

Stemming from the 'incompatibility' of barium, as a large-ion lithophile element, K_D^{Ba} values for all of the principal Solomons minerals are low, ranging from a mean of 0.05 for the clinopyroxenes to one of 0.56 for the feldspars of the more felsic rocks. Biotite, which is a rare mineral in the Solomons lavas, is the exception, with a mean barium content of 4875 ppm and mean K_D^{Ba} biotite–groundmass of 3.56. Because of its rarity, however, the contribution of biotite to the subtraction of barium from the melt is infinitesimally small.

Thus, in a manner very similar to that exhibited by lead, barium increases in the Solomons lavas as these become more felsic. Expressed in terms of Ba–MgO–(Na_2 + K_2O) ternary relations, this leads to the development of an essentially linear field of points up to the stage of dacite development, where Ba:(Na_2O + K_2O) ratios suddenly increase and the trend becomes hypertholeiitic, as in the case of lead. As with the latter, this feature also appears in the lava series of other arcs.

10

Strontium

Although strontium occurs between its two closely related alkaline-earth elements calcium and barium in Group IIa of the Periodic Table, and although calcium compounds are common and sometimes abundant components of exhalative ores, strontium is conspicuous by its absence — at least in anything other than trace quantities — in deposits of this kind.

The reason for this is not clear. Although less soluble than anhydrite, celestite ($SrSO_4$), with a solubility of $\approx 10^{-3}$, is more soluble than barite ($BaSO_4$, solubility $\approx 10^{-6}$). It might therefore be suspected that in the seafloor exhalative environment the two elements were separated on a solubility basis. It has, however been known for over a hundred years that celestite may occur as an abundant constituent of carbonate sediments, and that it may co-exist with barite in such environments. Koch (1888) noted the occurrence of celestite and barite together in bituminous limestones of Transylvania. Bauerman and Foster (1869) had earlier reported celestite in nummulitic limestone of Egypt, and Kraus and Hunt (1906) described disseminated celestite constituting up to 14 per cent of a dolomite in Munroe County, Michigan. Such evidence would seem to suggest that the near-absence of strontium from exhalative ores does not result from solubility factors operating during sedimentation, but reflects a very low abundance in the exhalations themselves.

The naturally occurring carbonate, strontianite, was recognized as a mineral in material from the Strontian mine in Argyllshire by Hope in 1791 and metallic strontium isolated by Davy in 1808, but the first comprehensive account of the geochemistry of strontium did not appear until 1934. Noll (1934), working in V. M. Goldschmidt's laboratory in Göttingen, carried out a detailed examination of determinative methods, and of the occurrence of the element in rocks. His estimates of mean strontium in volcanic lavas (as reported by Goldschmidt 1954) were: basalts, 175 ppm; andesites, 350 ppm; liparites, 87 ppm; trachytes, 612 ppm; nepheline-phonolites, 437 ppm; leucite-phonolites, 1050 ppm Sr.

Since that time there has been a substantial accumulation of data on the incidence of strontium because of its status as the most abundant trace element of igneous rocks and the importance of the ratios of its isotopes (^{86}Sr and ^{87}Sr) in elucidating a variety of geochemical processes. Faure (1978), in a large collation of data, gave a range of 69–190 ppm, mean 124 ppm Sr in 103 mid-ocean ridge tholeiitic basalts; 221–535 ppm, mean 329 ppm Sr in 18 oceanic island tholeiites; and 38–1640 ppm, mean 393 ppm Sr in 632 calc-alkaline basalts. For the other principal categories of the calc-alkaline series he estimated means of 442, 306, 330, and 97 ppm Sr for 361 andesites, 49 dacites, 26 rhyodacite–latite–quartz-latite lavas, and 127 rhyolites respectively. Thus, in broad terms strontium rose rapidly from basalts to a peak in andesites, then to decline fairly regularly through the dacitic spectrum to the rhyolites — a distribution predicted by Gast (1968) and Shaw (1970) on the basis of partition coefficients, and confirmed in the present chapter.

STRONTIUM IN LAVAS OF THE MARINE ENVIRONMENT

There are perhaps more data on the incidence of strontium in lavas than for any other trace element. This is certainly the case for MORBs, and a summary of currently available information is given in Table 6.2. The mean of 566 analyses not including the Icelandic basalts is 202 ppm Sr, and of 707 analyses including the Icelandic material, 200 ppm. Means for Atlantic, Pacific, and Indian MORBs are 157, 207, and 217 ppm respectively and for Iceland, with its partial subaerial emission, 194 ppm Sr. As with copper, zinc, and barium, the Icelandic basalts are thus distinctly higher in strontium than the rest of the mid-Atlantic basalts. While the mean value of strontium in all MORBs is little affected by inclusion of the Icelandic basalts (it is reduced), the mode of the frequency distribution for the latter is slightly higher than that for all MORBs — an apparent anomaly accounted for the greater incidence of high values of strontium in the submerged ridge MORBs (Figs. 10.1(a) and (b)). Mean strontium in 32 Icelandic lavas with SiO_2 >52 per cent is 157 ppm.

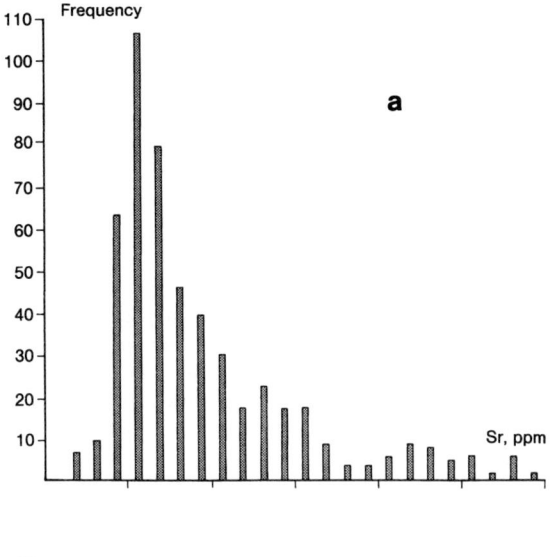

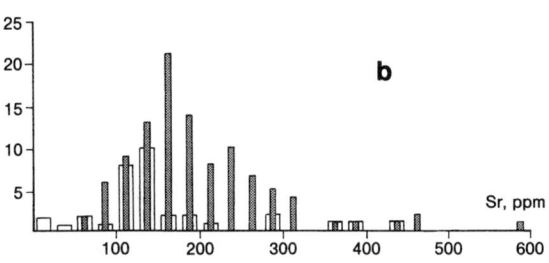

Fig. 10.1. Frequency distributions of strontium abundances in (a) MORBs of all major oceans, (b) in Icelandic basalts (SiO$_2$ <52 per cent; filled bars), and Icelandic lavas more felsic than basalts (SiO$_2$ ≥52 per cent: open bars).

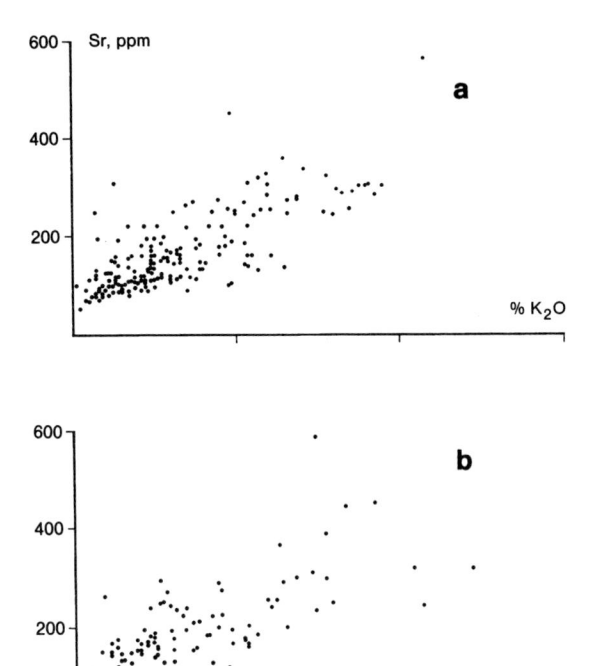

Fig. 10.2. Abundance relations between strontium and K$_2$O in (a) MORBs of the Mid-Atlantic Ridge, and (b) Icelandic basalts.

The principal major element association of strontium in MORBs is that with potassium (Fig. 10.2(a)), and a very similar tie pertains in the Icelandic basalts (Fig. 10.2(b)). Relationships here are very similar to those developed by Ba:K$_2$O (Fig. 9.4). Although, as is shown a little further on, strontium and barium abundances may diverge after the onset of andesite formation in some arc lava series, the behaviour of the two at the basaltic stage is very similar.

The strontium content of arc basalts is, as for barium, conspicuously higher than that of MORBs, with an overall average of 498 ppm Sr for 397 analyses (Table 6.3, Fig. 10.3). Spreads of frequency distributions for strontium are not as broad as those for barium, though there are substantial differences in the general level of strontium between different arcs (e.g. Tonga–Kermadec, mean Sr = 197 ppm, and Vanuatu, mean Sr = 615 ppm; see Table 6.3) and between individual volcanic islands within a given arc (e.g. St Kitts, mean Sr = 290 ppm, and Grenada, mean Sr = 754 ppm in lavas of SiO$_2$ >52 per cent; Fig. 10.4) Mean strontium values for lavas with SiO$_2$ <52 per cent (397 samples, mean = 498 ppm Sr) and for those with SiO$_2$ >52 per cent (620 samples, mean = 472 ppm Sr) are very similar, though the frequency distributions show somewhat different shapes and modes (Fig. 10.3). The reason for the latter is not obvious.

Division of the 1118 lavas into three categories: basalts, SiO$_2$ <52 per cent; andesites, 52 SiO$_2$ <62 per cent; and dacites and more felsic types, SiO$_2$ >52 per cent, gives means of 498, 481, and 463 ppm strontium respectively. This indicates a slight, if doubtfully significant, decrease in strontium with increasingly felsic nature — a tendency that appears to be clearly confirmed by the West Indies lavas, which have mean strontium contents of 679, 429, and 362 ppm respectively. However, precisely the opposite holds for the Solomons lavas, which show a steady increase from 523 ppm Sr in the basalts to 632 ppm in the andesites and to 1082 ppm in the dacites.

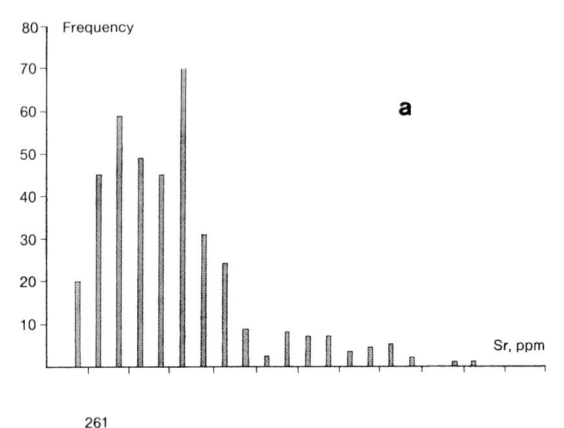

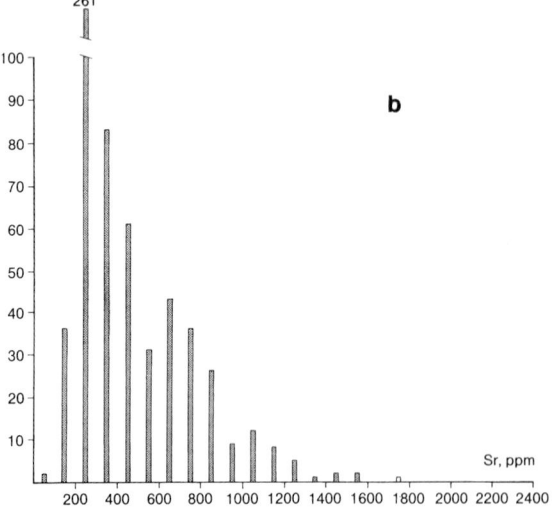

Fig. 10.3. Frequency distributions of strontium abundances in (a) volcanic island-arc basalts (SiO$_2$ <52 per cent), and (b) arc lavas more felsic than basalt (SiO$_2$ ≥52 per cent).

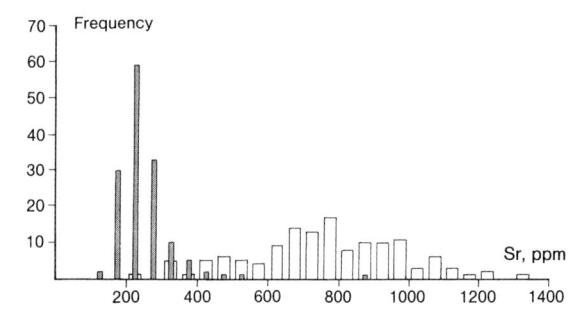

Fig. 10.4. Frequency distributions of strontium abundances in lavas more felsic than basalt (SiO$_2$ >52 per cent) of St Kitts (filled bars) and Grenada (open bars).

On the evidence available the incidence and behaviour of strontium in arc lavas is thus variable. Overall its abundance does not appear to change greatly as the lavas evolve from basalts to dacites, though within the individual arcs it may show pronounced systematic increase or decrease as the relevant lava series becomes more felsic.

CHEMICAL PROPERTIES AND CRYSTAL CHEMISTRY OF STRONTIUM

As a member of the alkaline-earth group strontium is entirely lithophile in character and, stemming from its relatively large ionic size (Sr^{2+}, 0.127 nm), it is not readily accommodated in most common silicate lattices. It is thus, like barium, regarded as a large-ion lithophile, or incompatible, element.

As it happens, the size of the Sr^{2+} ion has been determined differently by the two principal investigators. The early determinations of 0.127 nm by Goldschmidt was followed by one of 0.113 nm by Pauling, and the latter value has been adopted by a number of authors (e.g. Fyfe 1964). Given the similarity of charge and many chemical properties of the alkaline-earth group, Goldschmidt considered that the neatness of the geochemical fit of strontium between calcium (Ca^{2+}, 0.106 nm) and barium (Ba^{2+}, 0.143 nm) resulted from the almost exactly intermediate size of the Sr^{2+} ion. He also noted that the limited range of isomorphous substitutions between calcium and strontium compounds at low and intermediate temperatures, together with the data for the structural dimensions of simple calcium and strontium compounds, favoured a large difference in the ionic radii of these two elements. In addition the close geochemical relations between strontium and potassium (K$^+$, 0.133 nm) indicated that their ionic sizes were probably very similar (Goldschmidt 1954). Goldschmidt's estimate has prevailed and his value is the one now used (Emsley 1989).

The electron configurations and related properties of Sr and the Sr^{2+} ion are given in Table 6.4. As the ionic radius is 0.127 nm, the radius ratio with respect to O^{2-} is 0.907, the coordination number is 8, and the bond approximately 80 per cent ionic. As noted for Ba^{2+} the long inter-atomic distances involved with large ions in complex structures may, however, make coordination polyhedra difficult to define, and there is some variation in coordination number from one crystal structure to another. The difficulty of defining polyhedra is well illustrated by Sr–O relations in Sr(B$_4$O$_7$) (Fischer 1972): in addition to nine oxygen atoms at a distance of 0.284 nm or less, forming an irregular polyhedron, three oxygen atoms occur at distances of 0.304 and 0.305 nm, two at

0.315 nm, and one further oxygen atom at 0.320 nm. As a result of relations such as this, coordination numbers for Sr^{2+} vary from six to twelve, with nine the most stable and most common.

The most common habitat of strontium among the igneous silicates is as a substitution for potassium in feldspars where, in accordance with size–charge considerations, it is incorporated on the 'capture' principle. Such strontium inherits the effective 9-fold coordination of the potassium. Although the calcium of the clinopyroxenes is surrounded by eight oxygens (the ideal 8-fold coordination for strontium), there is little diadochy between calcium and strontium in this situation — perhaps due to the fact that while the calcium has eight nearest neighbour oxygens, it is bonded to only six of them. Sr_2SiO_4 (like Ba_2SiO_4) is isostructural with the olivines, but as the radius of the Sr^{2+} ion is 0.127 nm, some 59 per cent larger than those of Fe^{2+} (0.082 nm) and Mg^{2+} (0.078 nm), no significant substitution is possible. The incidence of strontium in the amphiboles has generally been considered on the basis of substitution for calcium (Moxham 1965; Tauson 1965; Nagasawa and Schnetzler 1971); relations with potassium are considered further on in this chapter. It has been suggested (Taylor 1965) that the relatively low incidence of strontium in the micas is probably due to the 12-fold coordination of potassium here, the site apparently being too large for Sr^{2+}.

THE INCIDENCE OF STRONTIUM IN THE PRINCIPAL MINERAL SPECIES OF THE SOLOMON ISLANDS YOUNGER VOLCANIC SUITE

General

Although of little or no quantitative importance in the Solomons Suite, K-feldspar is in principle the most important captor of strontium among the common volcanic silicates. Most volcanic K-feldspars fall in the range 200–3000 ppm, with a mean of ≈1200–1500 ppm Sr. Overall the volcanic plagioclases appear to contain about half this amount, with a range of 200–2000 ppm, and a mean of ≈500–600 ppm Sr. In a manner reminiscent of Puchelt's (1972) observation that barium tended to achieve a maximum in the intermediate members of the plagioclase series, Ewart and Taylor noted a strong tendency for maximum strontium abundance in plagioclases of the compositional range An_{40}–An_{55}, with no evidence of a simple correlation with calcium, in some New Zealand rhyolitic lavas. The hornblendes — principally of hornblende-andesites — generally contain strontium in the range 10–60 ppm with a mean of ≈ 40–50 ppm.

In general the volcanic clinopyroxenes contain less strontium than the amphiboles (see later in this chapter for hornblende:clinopyroxene partition coefficients) and most fall in the range 10–100 ppm, with a mean of 30–40 ppm Sr. Orthopyroxenes, presumably correlating with their very low calcium content, contain even less: a range of 0–20 ppm with a mean of ≈5–8 ppm Sr is a close approximation on the basis of current data. In spite of the isostructural relations referred to above, strontium is simply too large to fit into the Fe–Mg olivine crystal and in consequence its abundance here is almost vanishingly small, with range of ≈0–15 ppm, and a mean of ≈4 ppm Sr. In spite of their high K_2O contents (8–10 per cent) the biotites are also very ineffective strontium receptors, in this case (in contrast to the olivines) because the available site is too large. What little data are available on strontium in volcanic biotites indicate an abundance of ≈0–50 ppm.

Solomons suite

Mean abundances of strontium in the phenocrystic silicates and magnetite are given in Table 6.5, and spreads of abundances within each mineral group are indicated by the frequency distribution of Fig. 10.5. The plagioclases are clearly by far the principal strontium receptors.

Fe–Mg silicates. Strontium contents here are generally in keeping with those observed in other provinces. That of olivine, at an average of 2 ppm for 50 samples, is almost vanishingly low. Somewhat surprisingly, in view of ionic sizes, its abundance is significantly lower than that of barium. The strontium content of the clinopyroxene is within the generally observed range for this mineral, though the mean for the Solomons material may be a little higher than usual. Ewart and his co-workers (Ewart and Bryan 1972; Ewart *et al.* 1973) found a mean of 26 ppm Sr in 18 separations of Tonga–Kermadec clinopyroxenes, which compares with ≈60 ppm for those of the Solomons. Although the Solomons hornblendes contain only approximately half the CaO content of the clinopyroxenes (hornblende mean CaO = 11.8 per cent; clinopyroxene mean CaO = 20.9 per cent), mean strontium abundance in the hornblendes (176 ppm) is almost exactly three times that of the clinopyroxenes. This clearly indicates a lack of any direct Sr:Ca tie as between these mineral groups, and plotting of individual mineral analyses shows this also to be the case within each group. On the other hand there is a loose correlation between strontium and K_2O in the hornblendes, and the difference in strontium content between the Solomons amphiboles and clinopyroxenes seems likely to reflect, albeit loosely, their differences in K_2O content (hornblende mean K_2O = 0.45 per cent, clinopyroxene mean K_2O = 0.04 per cent).

124 *Ore Elements in Arc Lavas*

Fig. 10.5. Frequency distributions of strontium abundances in (a) clinopyroxenes, (b) hornblendes, and (c) feldspars of the Solomons suite.

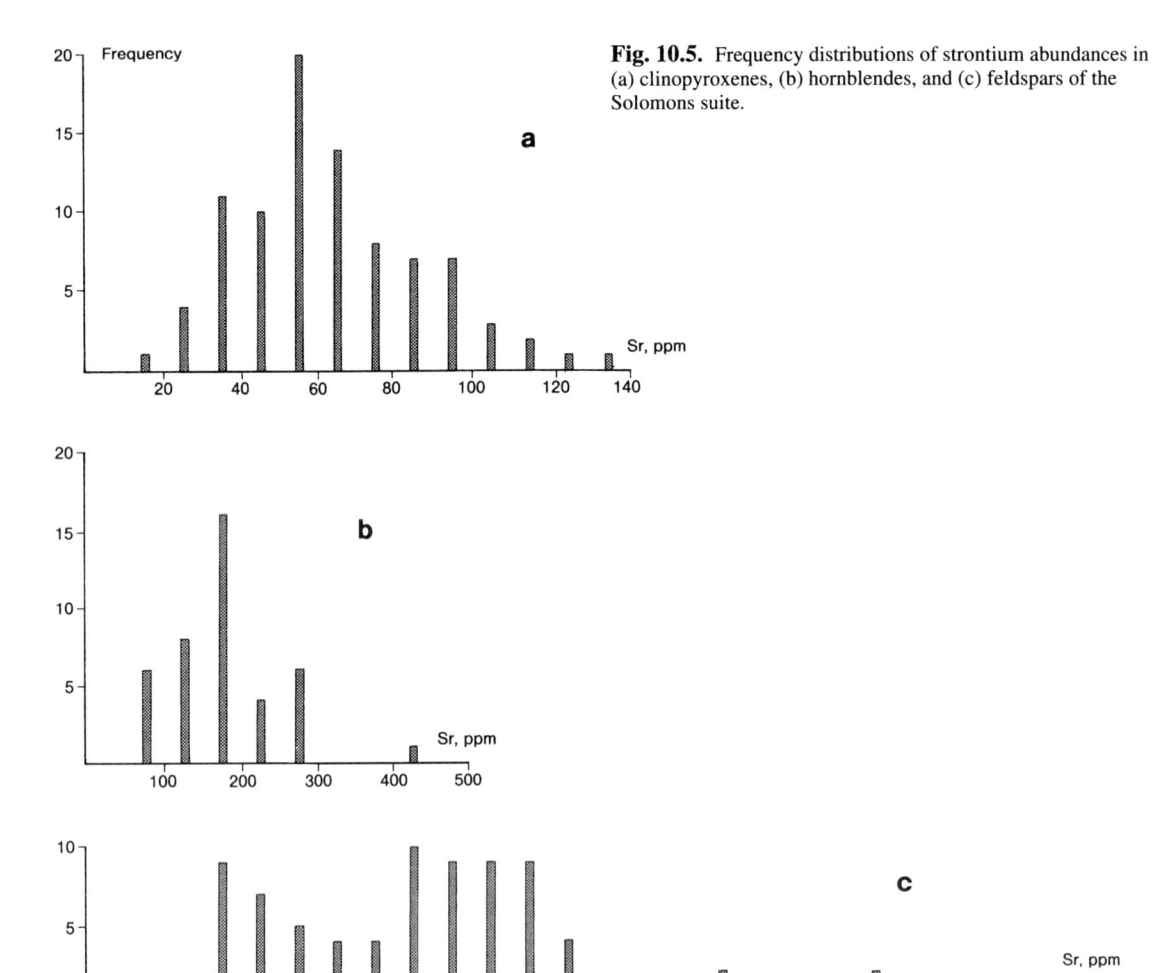

Feldspars. Two features of Table 6.5 stand out: (1) for a given K$_2$O content (0.45 per cent) the feldspars contain almost eight times as much strontium as do the coexisting amphiboles; and (2) overall the feldspars contain almost ten times as much strontium (mean, 1276 ppm) than barium (mean, 130 ppm). The first appears to reflect an expected greater propensity of Sr^{2+} to substitute in the larger available 9-coordinated site in the feldspars than in the smaller 6- to 8-coordinated site available in the amphibole. The second stems from the fact that whereas strontium is a good fit in the 9-coordinated K$^+$ site in the feldspars, barium requires a 12-coordinated site and is therefore largely excluded from the K$^+$ position.

On an individual sample basis Sr:K$_2$O correlation, like that of Ba:K$_2$O, is far from remarkable, with a correlation coefficient of 0.56 for the 78 'ultimate purity' felspar separations. When, however, mean Sr and K$_2$O for each of the three principal lava groups is plotted (Fig. 10.6; compare Fig. 9.7) correlation is good and $r = 0.98$. This *general* tendency for close strontium:potassium ties in the plagioclases is confirmed by the plot of mean Sr for 0.1 per cent intervals of K$_2$O for the 78 separations (Fig. 10.6(b)).

As for barium there is also a good general positive correlation between strontium and sodium in the Solomons plagioclases (Fig. 10.6(a)). It seems likely, however, that this strontium:sodium tie is in fact largely incidental: the real substitutional relationship is between strontium and potassium, and as potassium is more abundant in the more sodium-rich plagioclases, a secondary strontium:sodium correlation appears.

Strontium 125

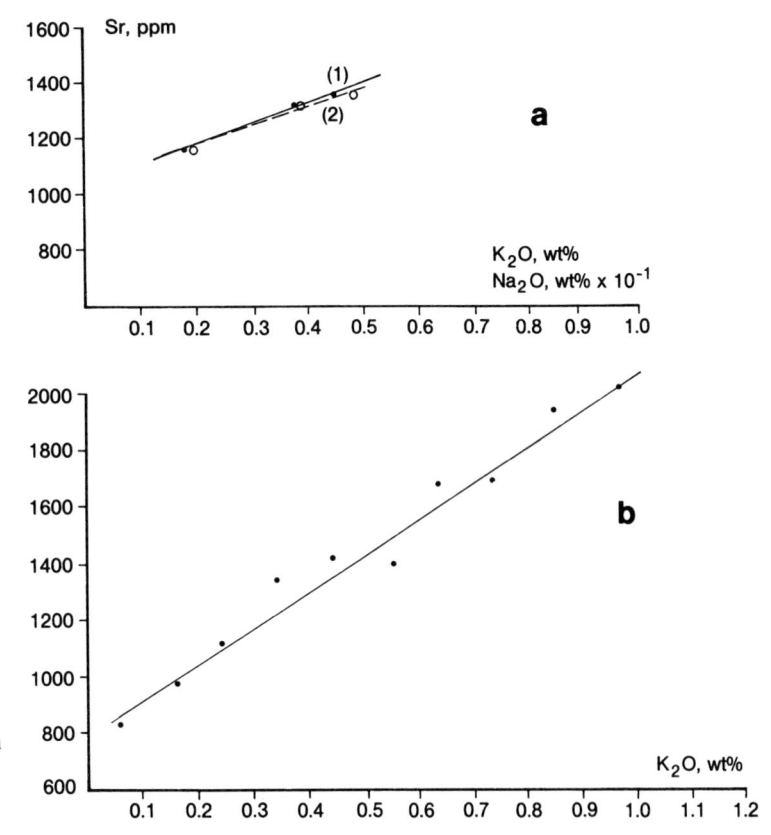

Fig. 10.6. (a) Relations between mean strontium and (1) mean K_2O (filled circles, full line) and (2) mean Na_2O (open circles, broken line) in the feldspars of the three principal lava groups; (b) variations in mean abundances of strontium with respect to 0.1 per cent intervals of K_2O in Solomons plagioclase feldspars.

There remains, however, the question as to why Sr:K_2O correlation is relatively poor on a sample-to-sample basis, and very much better when samples are grouped according, for example, to intervals of K_2O abundance (Fig. 10.6(b)).

An answer may lie in the development of a mosaic structure (Martin 1976) in the plagioclase crystals, and the lodgement of Sr^{2+} ions in mosaic boundaries rather than as simple random substitutions of minor, randomly distributed, K^+.

Suppose that the potassium of the plagioclases occurs not as randomly distributed K atoms but as a component of discrete micro-domains of K-feldspar in a mosaic dominated by micro-domains of plagioclase. Suppose further that the size of the K-feldspar micro-domains varies from one sample, i.e. group of plagioclase crystals, to another, so that for a given plagioclase K_2O content the total area of K-feldspar:plagioclase micro-domain interface might vary substantially between samples. If the strontium tended to lodge in the thin zones of disorder along the K-feldspar:plagioclase interfaces rather than as substitutions for K atoms within the K-feldspar itself, the incidence of strontium would be most closely proportional to the area of interface and less closely proportional to the amount of K-feldspar, i.e. apparent 'plagioclase K_2O'.

On a sample-to-sample basis this would lead to substantial variation in Sr:K_2O abundance relations and hence to the poor correlation referred to. However, where samples were grouped (as in Fig. 10.6(b)), the ratios of K_2O abundance to total area of K-feldspar micro-domain interface would tend to reduce to an overall mean, giving relatively constant relations between K_2O wt per cent in plagioclase, area of interface, and strontium abundance, and hence the high degree of Sr:K_2O correlation displayed in Fig. 10.6(b).

It follows from the latter and the close potassium:lead relation (see Fig. 8.2) that strontium and lead should also be closely correlated in the feldspars, and this is demonstrated in Fig. 10.7. The number of samples in the high lead:high strontium region of the group is small and the reliability is thus poor, but if the true relation is assumed to be linear the correlation coefficient, $r = 0.96$.

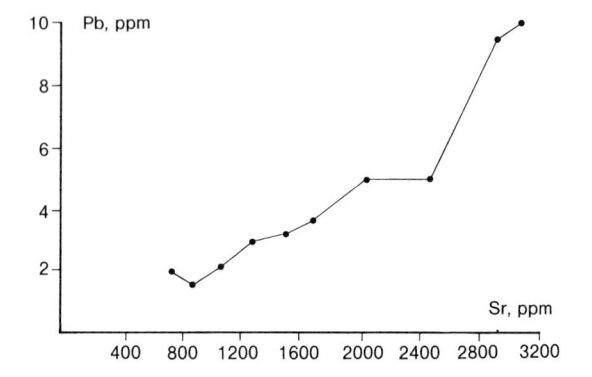

Fig. 10.7. Variation in mean abundance of lead with respect to 200-ppm intervals of strontium in the Solomons plagioclase feldspars (see also Fig. 8.4).

Spinel. Mean values of strontium in spinels of the three principal lava types are given in Table 6.5, and the overall spread is indicated in Fig. 10.8. The full range of values is from <0.05 to 77 ppm Sr, and the mean 23 ppm: lower than that of barium, at 33 ppm Ba. As with the latter, strontium contents of the titanomagnetites of the big-feldspar basalts and basaltic andesites are distinctly lower than those of the high-magnesium basalts and hornblende-andesites. There is no ready explanation for this, but when looked at in terms of progressive increase in whole-rock SiO_2 the pattern of strontium occurrences in the magnetites takes on at least a more systematic appearance: it is at its highest (60–80 ppm Sr) in those of the most mafic lavas, drops rapidly to about 15 ppm in the basalts of 48–50 per cent whole-rock SiO_2, and then maintains a fairly consistent 10–20 ppm level at least as far as the dacites.

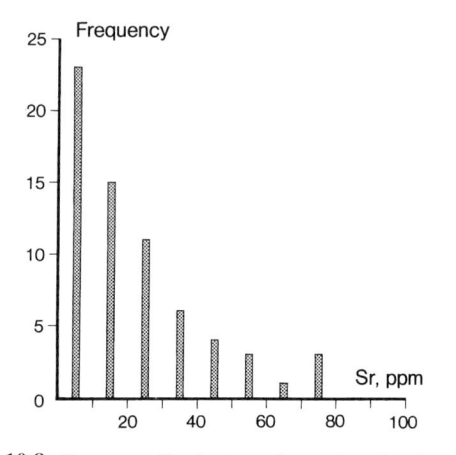

Fig. 10.8. Frequency distributions of strontium abundances in the Solomons spinels.

Perhaps the most conspicuous feature of strontium abundance in spinel is its erratic nature; values vary greatly between spinels that appear to come from similar contexts. As far as can be seen from the limited evidence, strontium occurs as a random substitution in the spinel structure and, like Al^{3+}, Mg^{2+}, and others, is simply incorporated in greatest abundance in the earlier, high-temperature spinels.

CRYSTAL:MELT PARTITIONING

General

Stueber (1978), in a major survey of the literature on crystal:melt partitioning of strontium by the principal volcanic silicates in lavas ranging from basalts to rhyolites, found that the mean partition coefficients for K-feldspar and plagioclase were almost identical at 5.92 and 5.78 respectively. Amphiboles, with their small but significant K_2O content, showed a mean K_D^{Sr} of 0.327 and biotite one of 0.291 (see the previous section for an explanation of low strontium acceptance by biotite in spite of its high K_2O content). Clinopyroxene, orthopyroxene, and olivine, with their very low capacities for accepting strontium, gave mean partition coefficients of 0.128, 0.027, and 0.006 respectively.

Following the pioneering work of Wager and Mitchell (1951) on the Skaergaard intrusion, a large literature has accumulated on the partitioning of strontium between plagioclase and a wide variety of igneous melts. For their six stages A–F, Wager and Mitchell determined a range of K_D^{Sr} from 2.0 to 11.6. Ewart and Taylor (1969) found a range of 2.6 to 11.5 in a group of basalts–andesites– rhyolites from the North Island of New Zealand, with the higher values developed in the more felsic members. Philpotts and Schnetzler (1970) found rather lower values, 1.27–2.84, in a varied group of basalts, andesites, and dacites, and no clear trend related to compositions. Korringa and Noble (1971) calculated K_D^{Sr} plagioclase: melt ranging from 1.5 for An_{90} to 7.0 for An_{30} on the basis of Rayleigh fractionation. Kovalenko *et al.* (1986), experimenting with a series of melts from andesite to 'trachyrhyolites', obtained a range of K_D^{Sr} values ranging from 9.4 at 936 °C to 2.1 at 1240 °C. Slightly earlier work on synthetic materials by Drake and Weill (1975) had yeilded values from 3.06 at 1150 °C to 1.23 at 1400 °C — results to which those of Kovalenko *et al.* (1986) appear to be substantially complementary.

Values for K_D^{Sr} hornblende:melt vary from 0.0224 in a dacite (Nagasawa and Schnetzler 1971) to 0.641 in a camptonite (Philpotts and Schnetzler 1970). The most comprehensive data appear to be those of Ewart and Taylor (1969) which give a range of 0.26 in an andesite to 0.54 in a rhyolite, with a mean of 0.36. Data for biotite are sparse and range from 0.0812 in a mafic phonolite

(Philpotts and Schnetzler 1970) to 0.7 in an alkali basalt (Villemant et al. 1981).

Although the incorporation of strontium in the clinopyroxenes is slight, there is quite extensive information on partition coefficients. Wager and Mitchell (1951) determined a range of 0.03 to 0.25 in the Skaergaard layers; Onuma et al. (1968) obtained 0.11 for clinopyroxene in an alkali olivine-basalt; Ewart and Taylor (1969) 0.07– 0.09 in New Zealand andesites; Philpotts and Schnetzler (1970) 0.02–0.52 in the group of andesites, dacites, and rhyolites already referred to; Kovalenko et al. (1985) 0.05–0.43, mean = 0.23 in a sample of nine lavas ranging from basalt to rhyolite.

The data for olivine are few. Philpotts and Schnetzler (1970) obtained values of 0.0185–0.0094 for oceanite–ankaramite; Hart and Brooks (1974) 0.0002 for an ankaramite; McKay and Weill (1977) 0.003 ± 0.002 for a synthetic Fra Mauro basalt at 1240 °C and 1 atm; and Villemant et al. (1981) 0.02 for the olivine of an alkali basalt (and 0.68 for coexisting magnetite).

Thus the Sr^{2+} ion, at 0.127 nm, is, like barium, too large to fit any crystal sites other than those of K^+. Where the sites involve coordinations of ≈8–9, as in the feldspars, substitution of strontium reaches its maximum and clearly exceeds that of barium, as reflected in the relevant K_D^{Sr} feldspar:melt values. Where, however, K^+ is 12-fold coordinated, as in biotite, the site is too large for strontium, and K_D^{Sr} biotite:melt values are much higher for barium than for strontium.

Solomons suite

Partition coefficients for the principal Solomons minerals are given in Table 10.1. As for previous elements, results yielded by the two sets of calculations are generally similar.

K_D^{Sr} values for olivine: and clinopyroxene:melt are low and, at means of 0.004 and 0.08 respectively, are comparable with, but perhaps a little lower than, those indicated by Stueber (1978) and others. The mean value of 0.32 for the Solomons hornblendes is well in keeping with values in the literature, particularly those of Ewart and Taylor (1969), involving similar material. Values for magnetite are similar to those for the clinopyroxenes, and almost an order of magnitude less than the one value obtained by Villemant et al. (1981). K_D^{Sr} values for magnetite:melt are clearly lower than those of K_D^{Ba}. Once again, however, partitioning into magnetites is at its lowest in those of the big-feldspar lava group.

Clearly the plagioclases are the principal receptors for strontium, with an overall mean K_D^{Sr} plagioclase:melt of 2.01. The slight increase in K_D^{Sr} from the olivine-rich basalts to the andesitic lavas presumably results from a combination of fall in temperature and rise in potassium content of the plagioclase. However, given that these two influences should reinforce each other, and in the light of the results of Korringa and Noble (1971) and Kovalenko et al. (1986), it is perhaps surprising that the observed increase in K_D^{Sr} is so small.

Table 10.1 Crystal:melt distribution coefficients (K_D) with respect to strontium

Mineral species	Lava group	n_1	n_2	m_1	m_2	$K_{D(1)}$	N	M_1	M_2	$K_{D(2)}$
Olivine	OPB	40	44	2	726	0.0028	39	2	704	0.0028
	BFL	7	23	2	624	0.0032	6	3	635	0.0047
Clinopyroxene	OPB	43	44	71	726	0.0978	42	72	721	0.0999
	BFL	26	23	50	624	0.0801	17	50	628	0.0796
	HbA	27	47	54	639	0.0845	20	52	767	0.0678
Hornblende	HbA	25	47	176	639	0.2754	22	180	565	0.3186
Feldspar	OPB	10	44	1160	726	1.5978	10	1160	665	1.7444
	BFL	23	23	1311	624	2.1010	22	1311	619	2.1179
	HbA	46	47	1357	639	2.1236	44	1320	639	2.0657
Spinel	OPB	10	44	51	726	0.0702	8	55	656	0.0838
	BFL	15	23	8	624	0.0128	15	8	627	0.0128
	HbA	37	47	22	639	0.0344	36	21	678	0.0310

OPB = olivine–clinopyroxene-basalt group; BFL = big-feldspar lavas: HbA = hornblende-andesite group; n_1 = total number of mineral separations analysed; n_2 = total number of groundmass separations analysed; m_1 = mean strontium content of mineral of each lava group; m_2 = mean strontium content of groundmass of each lava group; $K_{D(1)}$ = mean distribution coefficient determined as m_1/m_2; N = number of coexisting mineral:groundmass pairs analysed; M_1 and M_2 = mean strontium contents of coexisting mineral and groundmass respectively; $K_{D(2)}$ = distribution coefficient determined as M_1/M_2, for comparison with $K_{D(1)}$. All concentrations as parts per million by mass.

The figures of Table 10.1 are of course averages: strontium values as given are means for feldspars and for melt, the latter as principally represented by the groundmass, i.e. the 'residual melt'. The results derived from this information (78 individual partition coefficients for coexisting pairings) constitute consistent and valuable information, but give no insight into the finer-scale, evolutionary aspects of partitioning in individual melts. Some information of this kind can be obtained by investigating the strontium content not only of the plagioclase crystals and enclosing groundmass, but also of the glassy inclusions within some of the feldspars.

An initial indication that plagioclase has impoverished early formed melt with respect to strontium is provided by Fig. 10.9, which shows the frequencies of strontium values in 78 ultimate purity feldspar separations (solid bars) and in 61 penultimate purity separations — i.e. those containing small quantities of glass included within the feldspar crystals (open bars, Fig. 10.9). A slight but perceptible shift to the left of the latter (mean, 1239 ppm Sr) with respect to the former (mean, 1319 ppm Sr) is evident. As the glass inclusions in the penultimate purity feldspars constitute only a minute percentage of the separations, it follows that the strontium contents of the inclusions must be very much less than those of the containing feldspars, and the relevant partition coefficients thus large.

This has been verified by electron-microprobe analysis of 180 inclusion glass:feldspar crystal:adjacent groundmass glass groupings in 10 of the Solomons lavas. Mean strontium contents were 104:921:331 ppm respectively, giving mean K_D^{Sr} feldspar:inclusion glass of 8.86 and feldspar:groundmass glass of 2.78. In all cases K_D^{Sr} with respect to inclusion glass was higher than that with respect to the enclosing groundmass glass. In some instances the inclusion glass contained strontium at less than the detection limit of the analysis, indicating $K_D^{Sr} \to \infty$ in these cases.

However, from theoretical considerations (temperature; plagioclase composition) and in view of Fig. 10.9, it is — unfortunately — doubtful whether the glass of the inclusion gives a genuine indication of the strontium content of the residual melt at the relevant early stage of lava evolution. It seems more likely that it reflects continuing partitioning to a late stage and at low subsolidus temperatures, with consequent extreme impoverishment of the fine inclusions with respect to strontium. It simply indicates that any minor glass impurity within individual grains of feldspar separations will tend to reduce the apparent strontium content of the feldspar and hence the values of apparent partition coefficients. On the other hand, if the strontium content partly represents late-stage diffusion from glass inclusion to containing plagioclase, the combined analysis may in fact yield a better indication of overall partitioning. The best guide to progressive change in feldspar:melt partitioning thus still lies in the analysis of feldspar:groundmass separations through the spectrum of lava types.

Strontium and barium thus partition essentially similarly between melt and all principal mineral groups, with one notable exception: the plagioclases. With a mean K_D^{Ba} of 0.43 the latter tend to exclude barium such that, overall, the melt contains about 2.3 times the amount of barium as the crystallizing feldspars. Copious plagioclase development thus enriches the melt in barium. With a mean K_D^{Sr} of 2.01, the feldspar tends to capture strontium so that it contains, overall, about twice as much as the melt. Thus copious development of feldspar impoverishes the melt in strontium. A frequency-distribution diagram of K_D^{Ba} and K_D^{Sr} values for each of the 78 ultimate-purity feldspars and their corresponding groundmasses (Fig. 10.10) indicates the contrast clearly. Of the 78 values for K_D^{Sr} none is below 1.0. Of the corresponding values for K_D^{Ba} only four (5 per cent) are above 1.0. The crystallization of the feldspars thus appears to be a very effective way of fractionating barium and strontium in an evolving volcanic melt.

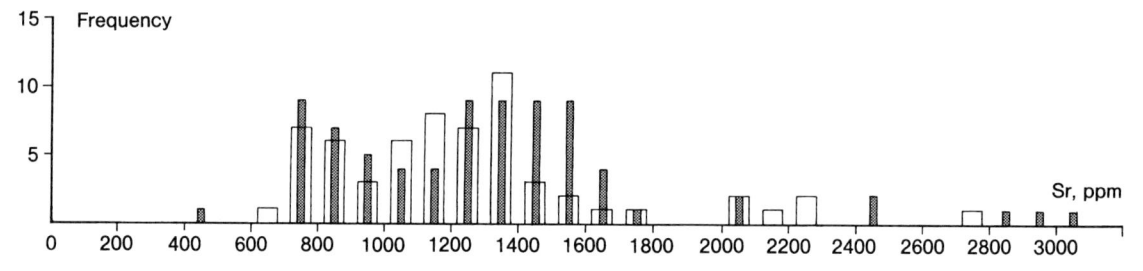

Fig. 10.9. Frequency distributions of strontium abundances in 78 'ultimate purity' feldspar separations (filled bars, mean = 1319 Sr) and in 61 'penultimate purity' separations (open bars, mean = 1239 ppm Sr).

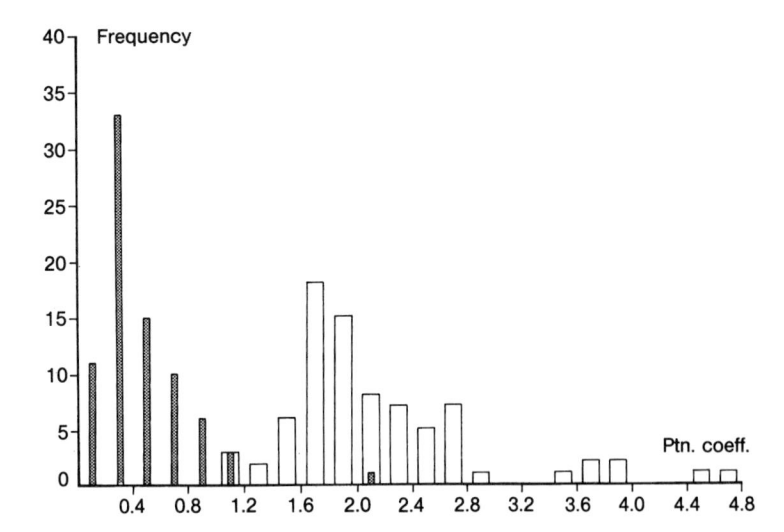

Fig. 10.10. Frequency distributions of values of K_D^{Ba} plagioclase:melt (filled bars) and K_D^{Sr} plagioclase:melt (open bars) developed by the Solomons feldspars as represented by the 78 ultimate purity separations.

THE INCIDENCE OF WHOLE-ROCK STRONTIUM IN THE SOLOMONS YOUNGER VOLCANIC SUITE

Mean strontium in the basaltic members of the suite (SiO_2 <52 per cent) is 523 ppm, which compares with 202 ppm for MORBs, and 584, 679, 615, and 197 ppm for the Aleutian, Lesser Antilles, Vanuatu, and Tonga–Kermadec arcs respectively. Frequency distributions of strontium abundances in the Solomons basalts, and in the associated more felsic lavas (SiO_2 <52 per cent) are given in Fig. 10.11.

Frequency distributions of strontium in the three main lava groups (Fig. 10.12) show features somewhat reminiscent of those of copper (Fig. 6.9), though the similar patterns have probably developed through rather different mechanisms. In the olivine–pyroxene-basalts the copious crystallization of olivine and clinopyroxene has led to a substantial partitioning of strontium into the groundmass (Fig. 10.12(c): mean whole-rock = 434 ppm Sr; mean groundmass = 722 ppm Sr). Development of the big-feldspar lavas is accompanied by an overall increase in whole-rock strontium (mean = 650 ppm) and a slight decrease in that of the groundmass (mean = 624 ppm) to the point where the latter contains slightly less than the whole rock (Fig. 10.12(b)) — a development clearly related to the sharp increase of plagioclase crystallization at this point (cf. Fig 5.6). Although the hornblende-andesites contain slightly higher mean strontium (689 ppm) than the big-feldspar lavas (650), this is largely influenced by a very few lavas from Savo with conspicuously high strontium contents. If the seven whole-rock analyses (all from Savo) with >1000 ppm Sr are momentarily ignored it may be seen (Fig. 10.12(a)) that the major

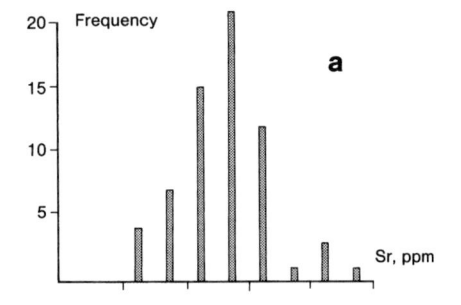

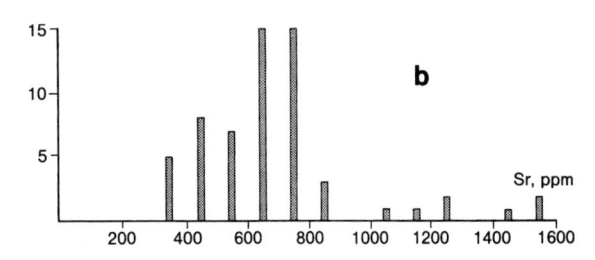

Fig. 10.11. Frequency distributions of strontium abundances in (a) Solomons basalts (SiO_2 <52 per cent) and (b) lavas more felsic than basalts ($SiO_2 \geqslant 52$ per cent).

part of the hornblende-andesite frequency distribution has shifted to the left of that of the big-feldspar lavas, and the mean of the hornblende-andesite whole-rock values is now 586 ppm Sr. The similarity of abundance of strontium as between whole rock and groundmasses of the hornblende-andesites stems from the general abundance — dominance — of feldspar in virtually all lavas of this group.

The close relationship between strontium and K_2O in the feldspars appears in muted form in the whole rocks

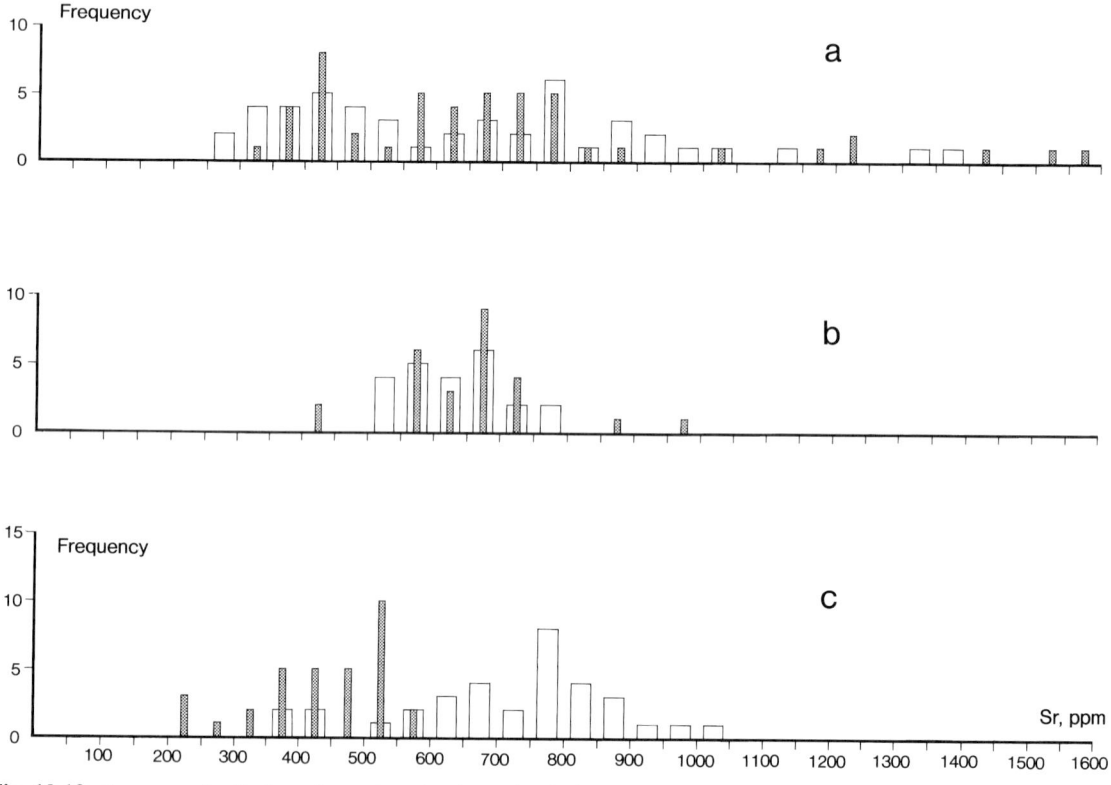

Fig. 10.12. Frequency distributions of strontium abundances in whole rocks (filled bars) and groundmasses (open bars) of (a) lavas of the hornblende-andesite group, (b) big-feldspar lavas, and (c) olivine–clinopyroxene-basalts of the Solomons suite.

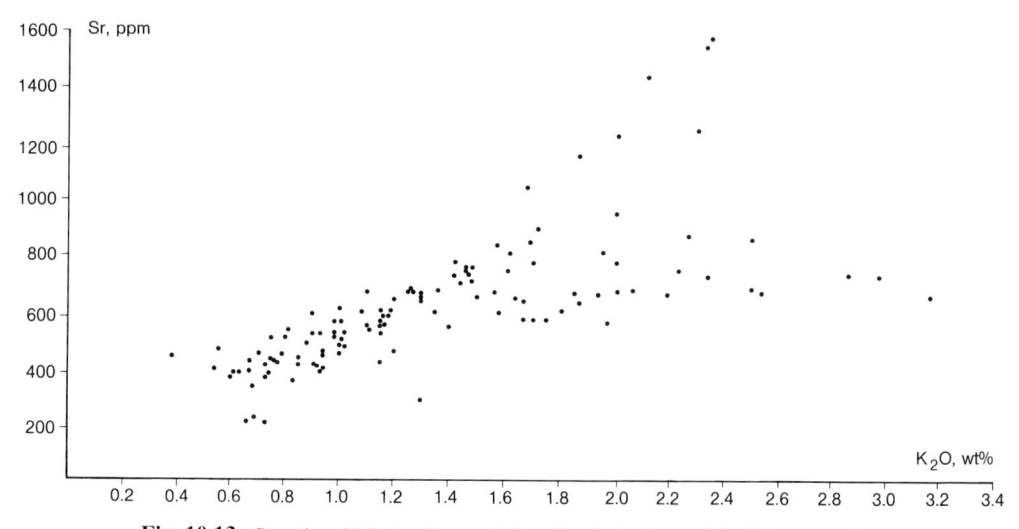

Fig. 10.13. Strontium:K$_2$O abundance relationships in the lavas of the Solomons suite.

and groundmasses. Figure 10.13, showing Sr:K$_2$O relations in individual lavas, indicates a modest degree of correlation to approximately 750 ppm Sr:1.6 per cent K$_2$O, at which point the field splits into two diverging parts. The steeper of this is generated by the andesites and dacites of Savo. The other, relatively flat, component represents some of the big-feldspar lavas, and some hornblende-andesites.The strontium:potassium abundance relationship is therefore not a simple one but presumably it reflects, at least in part, the significance of K$^+$ sites, or K-feldspar domains, in plagioclases as acceptors of Sr^{2+}.

Plotting the three principal lava types and their groundmasses on 'SrMA' diagrams (Fig. 10.14) yields an ordered sequence of fields highly reminiscent of those for barium. With its higher abundance, strontium shows a rather steeper trend than that of barium (Fig. 9.11), but all other features are similar, including a general decrease in the 'orderedness' of the fields proceeding from the picritic basalt whole rocks to hornblende-andesite groundmasses. Once again, although the sequence of lava types and their groundmasses constitutes an easily perceived pattern, the groundmasses of one group do not constitute the whole rocks of the next with respect to the group of elements represented in the figure.

Plotting of mean Sr–MgO–(Na$_2$ + K$_2$O) relations corresponding to 2 per cent whole-rock SiO$_2$ intervals (Fig. 10.15(a)) reveals a tholeiitic trend — perhaps with a slight tendency to hypertholeiitic character — with respect to strontium. Similar plotting of the corresponding groundmasses, however, shows these to develop a hypotholeiitic pattern, with the inflection occurring in the 50–4 per cent whole-rock SiO$_2$ interval. To an immediate first approximation this might be attributed to copious feldspar and associated hornblende, crystallization accompanying the onset of andesite formation — an effect considered in greater detail in Chapter 21. Inclusion of the 119 Bougainville analyses (Rogersen *et al.* 1989) in the relevant calculations yield an essentially straight-line array of points (Fig. 10.15(b)), indicating that the Solomons suite as a whole is probably tholeiitic rather than hypertholeiitic with respect to strontium.

Analogous plots for the three major islands of the Lesser Antilles indicate a less well correlated straight-line (tholeiitic) pattern in the case of Grenada, and almost identical, well-correlated, flat, tholeiitic patterns developed by the Dominica and St Kitts lava assemblages (Fig. 10.15(c)). The Tonga–Kermadec suite also displays a rather flat tholeiitic pattern, but that of Vanuatu is distinctly hypotholeiitic (Fig. 10.15(d)). As has been recognized for a very long time for iron (as FeO), the lava series of different arcs may exhibit quite distinct and different patterns — hypotholeiitic, tholeiitic, or even tending to hypotholeiitic — with respect to strontium.

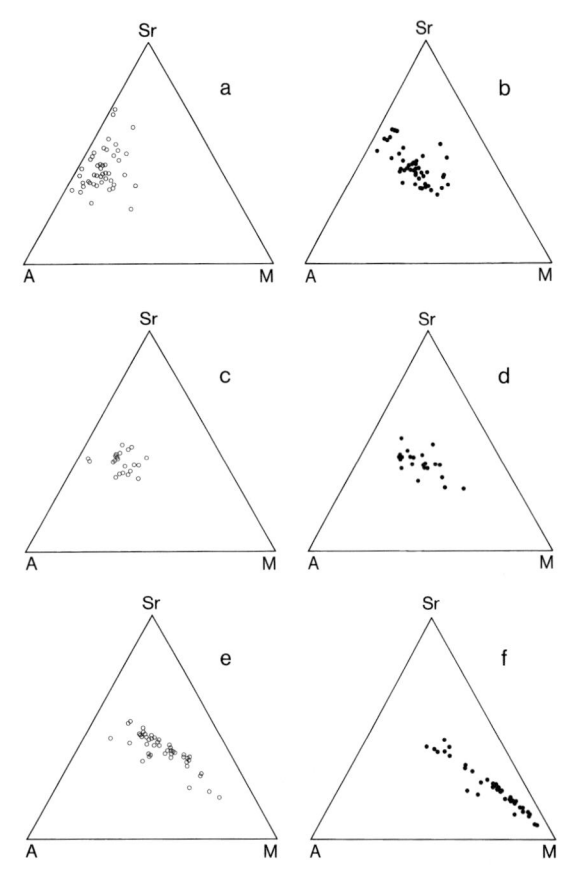

Fig. 10.14. Sr–MgO–(Na$_2$O + K$_2$O) ternary relations (Sr calculated using ppm Sr × 10^{-2}) developed in olivine–clinopyroxene-basalt, big-feldspar lava, and hornblende-andesite whole rocks (filled circles: (f), (d), and (b) respectively), and their corresponding groundmasses (open circles: (e), (c), and (a) respectively).

MORBs as represented by the Mid-Atlantic Ridge basalts display a somewhat diffuse, and linearly restricted, tholeiitic trend (Fig. 10.15(e)). The combination of Icelandic basalts and more felsic lavas (Fig. 10.15(f)) generates a somewhat flat, but quite distinct, hypotholeiitic pattern with respect to strontium.

The graph of strontium v. 2 per cent intervals of whole-rock SiO$_2$ (Fig. 10.16(a)) indicates a general rise in strontium with evolution of the Solomons lava series until the 62–4 per cent SiO$_2$ interval, at which stage (in contrast to barium, but perhaps similarly to lead) strontium looks as if it may decrease or form a plateau. The point for the whole rocks at the 64–6 per cent SiO$_2$ interval of Fig. 10.16(a) represents only five samples (and more data are clearly desirable) but reference to the behaviour of

132 *Ore Elements in Arc Lavas*

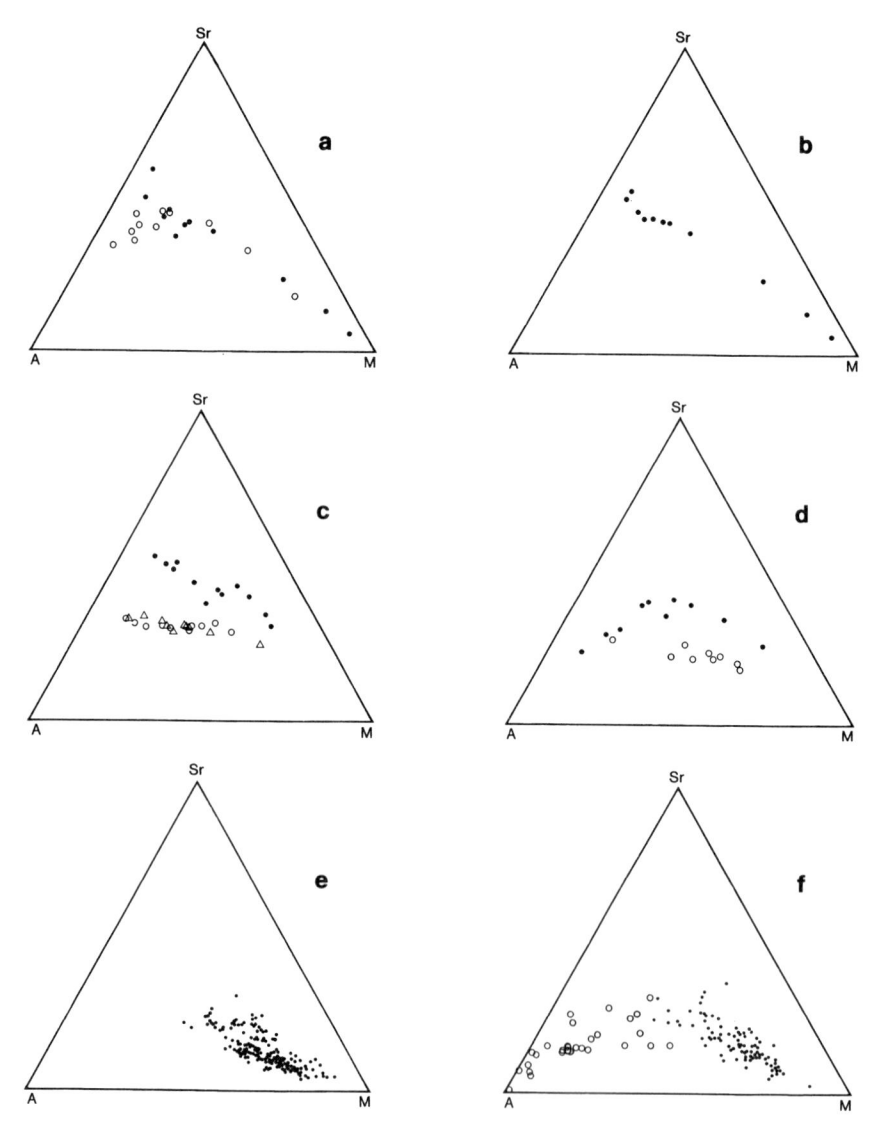

Fig. 10.15. Mean Sr–MgO–(Na$_2$O + K$_2$O) ternary relations corresponding to 2 per cent whole-rock SiO$_2$ intervals in (a) whole rocks (filled circles) and corresponding groundmasses (open circles) of the Solomons suite excluding Bougainville; (b) whole rocks of the Solomons suite including Bougainville (241 analyses); (c) lavas of Grenada (filled circles), Dominica (open circles), and St Kitts (open triangles); (d) Vanuatu (filled circles) and Tonga–Kermadec (open circles). Ternary relations developed in MORBs as represented by individual Mid-Atlantic Ridge basalts are shown in (e), and in individual samplings of Icelandic basalts (SiO$_2$ <52 per cent; filled circles) and Icelandic lavas more felsic than basalts (SiO$_2$ ⩾52 per cent; open circles) in (f).

strontium in other volcanic suites, and to the observations of Gast (1968), Shaw (1970), and Faure (1972), suggests that the break at this point may well be genuine. It is, it is presumably due to copious incorporation of strontium in plagioclase, with consequent gross depletion in the remaining melt.

The corresponding groundmasses show a sharp rise in strontium from 46 to 50 per cent whole-rock SiO$_2$, due to copious crystallization of olivine and clinopyroxenes, and then a general decrease reflecting the crystallization of plagioclase. However, and in contrast to the behaviour of lead and barium, the abundance of strontium in the groundmasses does not remain higher than in the corresponding whole rocks. At the more mafic compositions (Fig. 10.16(a), open circles), strontium in the groundmasses is clearly higher than in the whole rocks, but the trends cross at about 54 per cent whole-rock SiO_2 and thus at the onset of formation of the andesites the abundance of strontium in the groundmass has fallen below that in the whole rocks — a state of affairs that persists for the remainder of the evolutionary sequence.

The behaviour of strontium with respect to whole-rock SiO_2 in the Grenada lava series is very irregular (Fig. 10.16(b)), and the abundance of this element in the most mafic lavas (SiO_2 <44 per cent; mean Sr = 816 ppm) is virtually identical with that in the most felsic (SiO_2 >64 per cent; mean Sr = 827 ppm). The Dominica and St Kitts assemblages, in contrast, exhibit very constant strontium abundances, once again showing a near-equality of strontium at both ends of the relevant lava spectra. The Tonga–Kermadec suite behaves similarly (Fig. 10.16(c)), but that of Vanuatu is characterized by an almost steady decline from a stage just prior to the onset of andesite formation. Apart from the early stage of fractionation (whole rock SiO_2 ≈46–50 per cent) the Solomons and Vanuatu 'curves' of Figs. 10.16(a) and (c) respectively are almost mirror-images of each other. The Aleutians suite also shows a fairly constant decline in strontium abundance with evolution of the lava series (Fig. 10.16(d)). Barium abundances are also plotted in this latter figure (open circles, Fig. 10.16(d)) to demonstrate the remarkable difference in abundance behaviour of these two 'incompatible' elements in some lava series — a result of the capacity of plagioclase to act as a geochemical sieve with respect to these two elements.

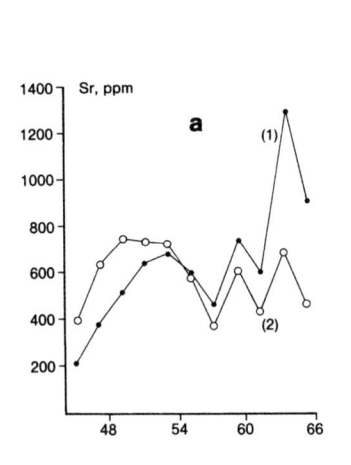

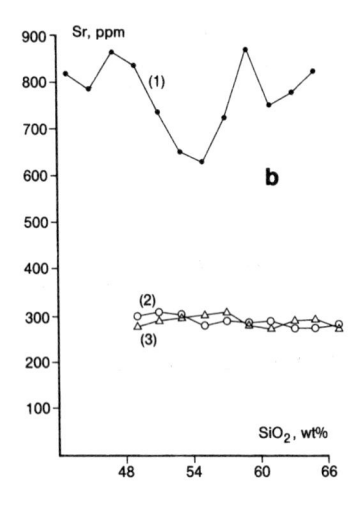

Fig. 10.16. Mean Sr:SiO_2 relations corresponding to 2 per cent whole-rock SiO_2 intervals in (a) whole rocks [(1) filled circles] and corresponding groundmasses [(2) open circles] of the Solomons suite; (b) lavas of (1) Grenada, (2) Dominica, and (3) St Kitts; (c) lavas of (1) the Vanuatu arc and (2) the Tonga–Kermadec arc; and (d) the Aleutians, showing the contrast between the behaviour of strontium (1) and barium (2) accompanying the evolution of this lava series.

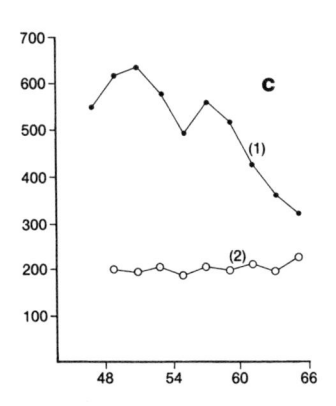

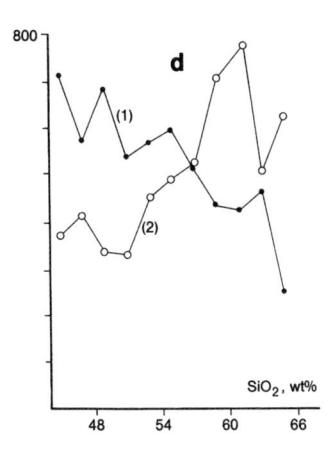

CONCLUDING STATEMENT

Like barium, strontium is about 2.5 times more abundant in arc basalts than in MORBs: on the basis of extensive data, the mean strontium content of MORBs is approximately 200 ppm; that of arc basalts, just under 500 ppm. Overall there appears to be little difference in strontium content between arc lavas with SiO_2 <52 per cent and those with SiO_2 >52 per cent, though there are clear differences between these categories in some individual arcs. The Solomons basalts contain mean Sr = 523 ppm, compared with 702 ppm Sr in lavas more felsic than basalts. Strontium abundance varies greatly, not only from one arc to another, but also between volcanic centres within individual arcs: the Lesser Antilles, particularly St Kitts and Grenada, provide a spectacular example of this.

With its higher charge and ionic radius very similar to, but slightly smaller than, that of K^+, Sr^{2+} tends to be captured into all K^+ sites involving coordination numbers of less than about 10 with respect to O^{2+}. Thus, whereas K_D^{Ba} is <1 for all the major mineral species, Sr^{2+} readily enters the 9-fold coordinated K^+ site in the plagioclases with a resulting mean K_D^{Sr} for the later of ≈ 2. The crystallization of K-bearing plagioclase — and this alone — therefore tends both to remove strontium from the melt and to separate strontium from barium. Thus whereas the olivine–pyroxene-basalts have mean Sr = 434, mean Ba = 96 ppm (Sr/Ba = 4.52), the most 'evolved' lava of the present study — a rhyolite, SiO_2, = 74.75 per cent, from Mt Gallego — contains Sr = 318, Ba = 2480 (Sr/Ba: 0.13).

Strontium displays a somewhat diffuse tholeiitic behaviour in MORBs, and the lava series of the Solomons, Lesser Antilles, and Tonga–Kermadec arc are clearly tholeiitic with respect to this element. On the other hand, the Icelandic suite, those of the Vanuatu and Aleutians arcs, and the groundmasses of the Solomons lavas are unequivocally hypotholeiitic with respect to strontium. Thus, as has long been recognized for FeO, different lava suites may be tholeiitic or hypotholeiitic with respect to strontium.

11

Phosphorus

Although phosphorus is little thought of, and only infrequently referred to, as a constituent of exhalative ores, it is in fact a substantial component — as apatite — of some very important deposits of this class. It is conspicuous in some of the major Proterozoic stratiform Pb–Zn deposits, such as those of Broken Hill, Gamsberg, and Aggenys; occurs in abundance in some iron-rich exhalative ores, such as those of Kiruna; appears as a noteworthy minor component of some Palaeozoic and younger exhalative base-metal deposits; and is conspicuous in some of the exhalites associated with them.

One of the best-known examples of the prominent association of apatite with stratiform sulphide ores is that of Broken Hill, New South Wales, where it occurs at levels of 2 to 20 per cent in the stratigraphically lower layers of sulphide and in the associated iron-rich exhalite ('banded iron formation') (see Fig. 12.1). The incidence of phosphate here is closely tied to that of the metallic sulphides (and iron oxides in the case of the banded iron formation) both across and along bedding, indicating that all probably had a common source.

The early recognition of phosphorus as a ubiquitous minor component of igneous rocks led to its general, if variably accurate, determination in all standard igneous rock analyses. Vogt (1931) in an early estimate of the average phosphorus content of igneous rocks, proposed a figure of 760 ppm. Conway (1945) gave a considerably higher estimate of 1200 ppm. On the basis of 645 analyses Goldschmidt (1954) estimated mean phosphorus in basalts to be 2400 ppm, andesites 1230 ppm and rhyolites 550 ppm. After considering some of the complexities of frequency distributions of phosphorus in the full range

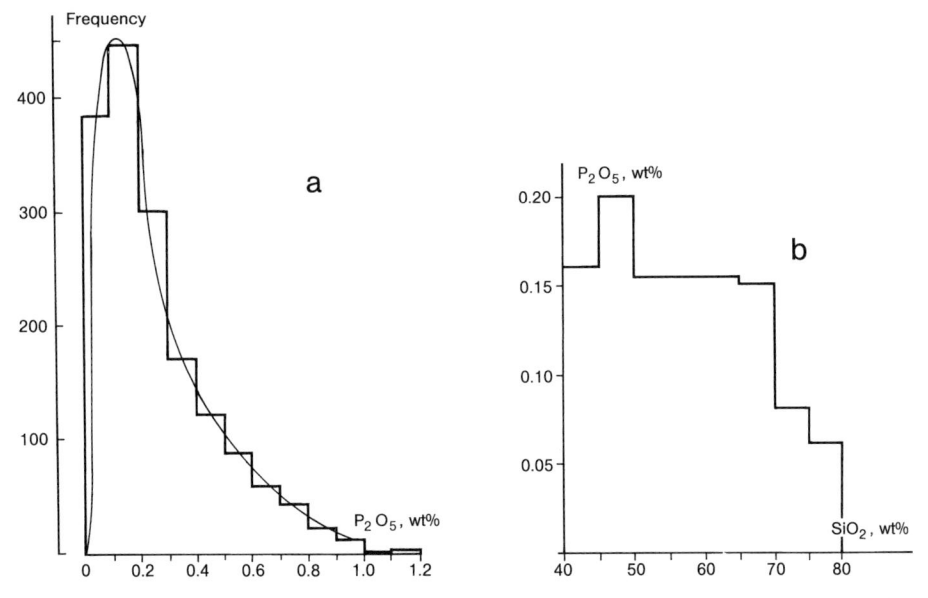

Fig. 11.1. (a) Frequency distribution of P_2O_5 abundances, and (b) relations between mean P_2O_5 and 5 per cent SiO_2 intervals, in 1812 eruptive rocks. After Koritnig (1978).

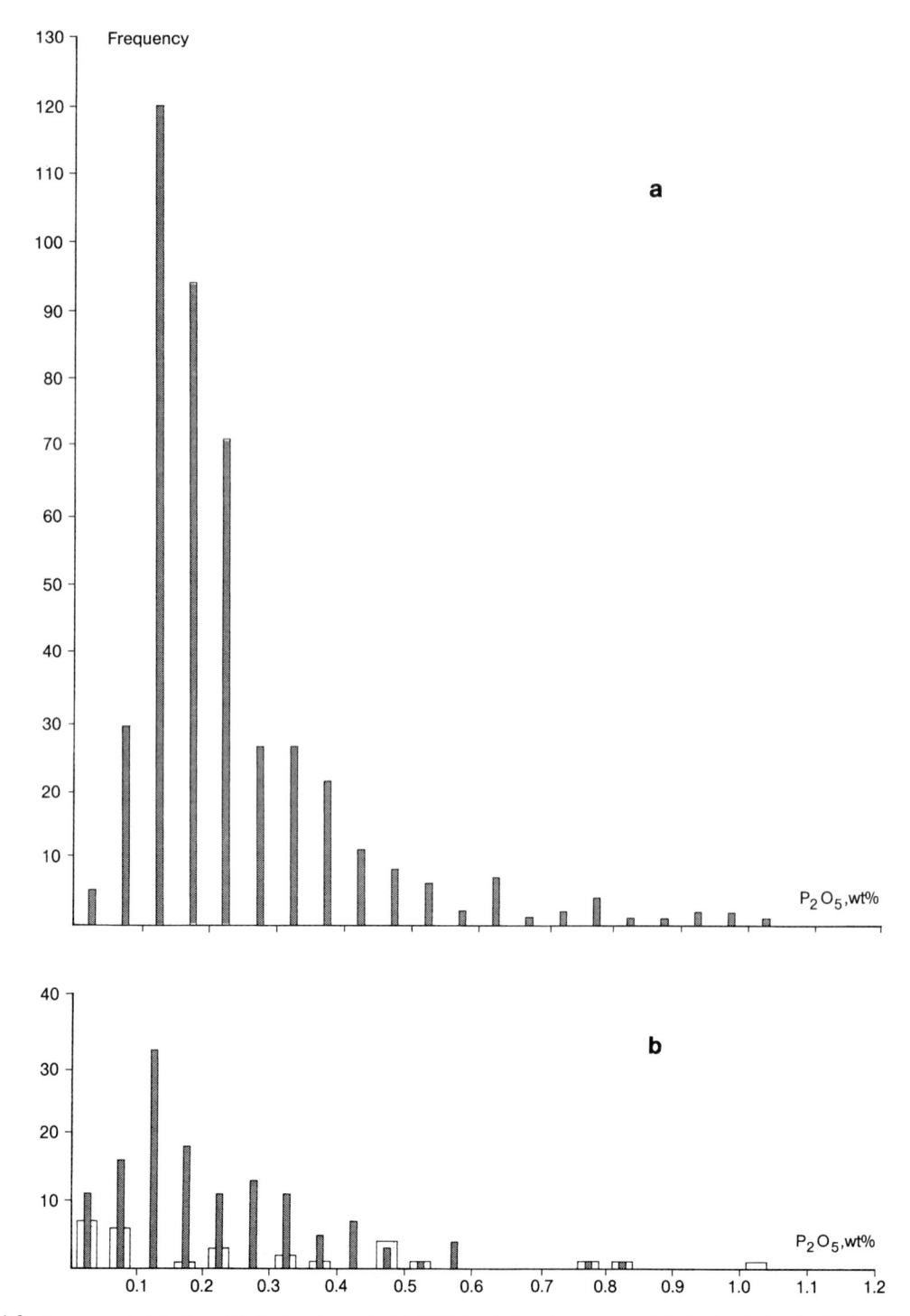

Fig. 11.2. Frequency distribution of P_2O_5 abundances in (a) MORBs of all major oceans and (b) Icelandic basalts (SiO_2 <52 per cent: filled bars) and Icelandic lavas more felsic than basalts ($SiO_2 \geqslant 52$ per cent; open bars).

of igneous rock types, Goldschmidt (1954) came to the general conclusion that 'basalt is the principal "carrier" of phosphorus. In the intermediate rocks the distribution is rather uniform. The most acid rocks show a deficiency of P in comparison with the other rocks.' This has been confirmed more recently by Koritnig (1978) on the basis of 1812 samples (Fig. 11.1).

PHOSPHORUS IN LAVAS OF THE MARINE ENVIRONMENT

A summary of available data on phosphorus in MORBs is set out in Table 6.2. Mean P_2O_5 for 429 analyses not including Iceland is 0.24 per cent (P equivalent ≈1050 ppm), and for 565 analyses including the partly subaerial basalts of Iceland is 0.23 per cent (P equivalent ≈1000 ppm). This compares with a mean of 0.20 per cent (≈870 ppm P) in basaltic rocks overall.

The frequency distribution of non-Icelandic analyses (Fig. 11.2(a)) shows a clear mode in the 0.1–0.15 per cent P_2O_5 range and a pronounced positive skew, with 65 per cent of analyses falling in the 0.1–0.25 per cent (440–1090 ppm P) interval. The distribution for Iceland also has a distinct mode at 0.1–0.15 per cent and a (less pronounced) positive skew; 45 per cent of the analyses fall in the 0.1–0.25 per cent P_2O_5 interval. On the basis of present data, mean values for Iceland and for the rest of the Mid-Atlantic–Indian Ridge are the same at 0.20 per cent P_2O_5 (870 ppm P). Mean P_2O_5 for the Atlantic–Indian Ridge and the Indian and Pacific MORBs are 0.27, 0.24, and 0.34 per cent P_2O_5 (1180, 1050, and 1480 ppm P) respectively. Proportional variation from one area to another of a given ridge may be substantial; e.g. at 36° 50′ N on the Mid-Atlantic Ridge mean P_2O_5 is 0.13 per cent (P ≈570 ppm) (29 analyses, Bougault and Hekinian 1974), and at 43° N on the same ridge it is 0.28 per cent (P ≈1220 ppm) (14 analyses, Shibata et al. 1979). Although the numbers of analyses are too small for reliable comparison, the P_2O_5 content of Icelandic lavas with SiO_2 >52 per cent appears to be slightly lower an average than that of the MORBs themselves (Fig. 11.2(b)).

The data for arcs are summarized in Table 6.3 and Fig. 11.3. The mode for arc basalts is slightly higher than that of MORBs, occurring in the 0.15–0.20 per cent P_2O_5 interval; 53 per cent of all analyses fall in the 0.1–0.25 per cent range, and the population as a whole has a mean of 0.20 per cent P_2O_5 (P ≈873 ppm). Arc lavas with SiO_2 >52 per cent exhibit a prominent mode in the 0.1–0.15 per cent interval; 64 per cent of all analyses are in the 0.1–0.25 per cent range, and the population mean is 0.19 per cent P_2O_5 (P ≈830 ppm). There is also considerable variation in mean P_2O_5 between and within arcs. The

Aleutians basalts exhibit a mean of 0.27 per cent, compared with 0.11 per cent for those of Tonga–Kermadec. Lavas in the range 50–66 per cent SiO_2 on Grenada (151 analyses) have mean P_2O_5 = 0.168 per cent (P ≈730 ppm), whereas lavas of the same silica range on St Kitts (161 analyses) have mean P_2O_5 = 0.104 per cent (P ≈460 ppm) (Fig. 11.4). The hornblende-andesites of the volcano of Mt Gallego in the Solomons have mean P_2O_5 = 0.156 per cent (Fig. 11.4(b)), compared with 0.193 per cent for those of the adjacent volcano of Savo. (The broad pattern of phosphorus abundance in the lavas of these two centres is thus analogous to those of lead and potassium.)

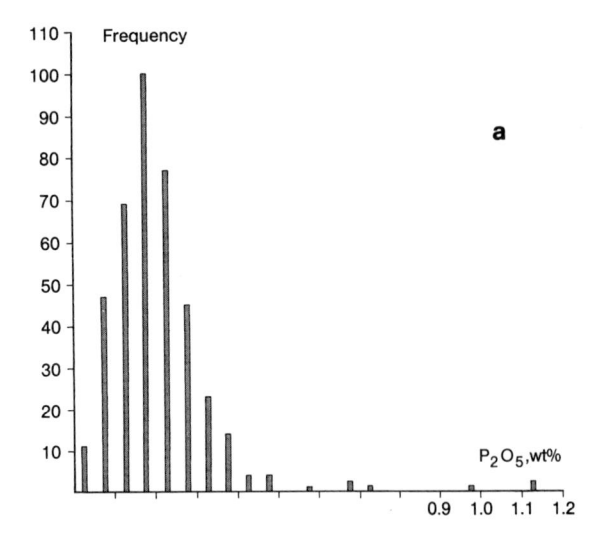

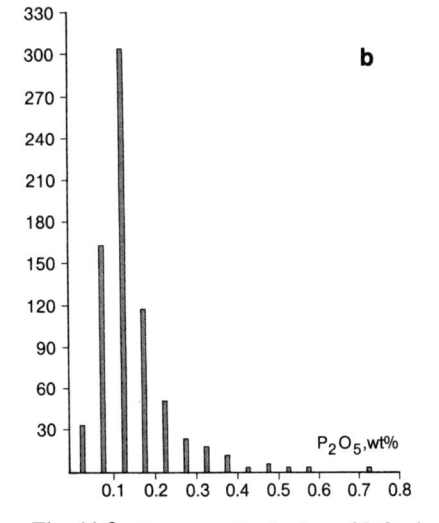

Fig. 11.3. Frequency distributions of P_2O_5 abundances in (a) volcanic island-arc basalts (SiO_2 <52 per cent) and (b) arc lavas more felsic than basalt (SiO_2 ⩾52 per cent).

Mean phosphorus thus appears to be slightly higher in mid-ocean ridge lavas than in those of arcs, though the potential for variation in phosphorus within and between both ridges and arcs is by a factor of 2 to 3, making distinction between the two categories equivocal.

CHEMICAL PROPERTIES AND CRYSTAL CHEMISTRY OF PHOSPHORUS

Like barium and strontium, phosphorus is substantially lithophile and it is one of the more abundant of the minor constituents of the lithosphere. In igneous rocks, and indeed in all crustal materials, it occurs largely in apatite, $Ca_5F(PO_4)_3$. It has, however, been recognized since quite early in the history of mineral chemistry that apatite is by no means the sole repository of phosphorus in igneous rocks. Galkin (1910) found notable amounts of phosphorus in hornblende and pyroxenes, indicating that it might substitute significantly in silicates. Beginning with the work of Kimura (1925) a series of Japanese investigators determined that 10–25 per cent of the silicon of some Japanese zircons was replaced by phosphorus. Machatschki (1931) found that P^{5+} may replace almost 20 per cent of the Si^{4+} in certain zircons, the charge difference being compensated for by the replacement of some Zr^{4+} by trivalent rare earth ions. Mason and Berggren (1941) described a spessartite garnet in which about 8.5 mole per cent of the silicon was replaced by phosphorus. All this gave early analytical indication that PO_4 groups might substitute significantly for SiO_4 groups in common silicates; indeed, Taborszky (1962) went so far as to suggest that most of the P_2O_5 shown in analyses of igneous rocks might be bound in silicates rather than apatite.

That such a state of affairs might have been expected from theoretical considerations was pointed out by Goldschmidt (1954), who noted that the then estimated ionic radii of Si^{4+} and P^{5+} were 0.39 and 0.35 Å respectively and that 'Consequently a diadochic replacement of Si by P can be expected' (Goldschmidt 1954, p. 456). Almost simultaneously Ringwood (1955) deduced that because of the high ionic potential of P^{5+} in igneous melts phosphorus forms PO_4^{3-} complexes which may then replace SiO_4 tetrahedra in the developing silicate minerals — or, at sufficient concentration, are precipitated as a separate phase such as apatite.

Koritnig (1965), in a study that had as its primary concern the replacement of Si^{4+} by P^{5+} in rock-forming silicates, found a general range of 0.2–220 ppm P after removal of fine apatite by HNO_3 dissolution — the equivalent of 1–10 P ions per 10^4 SiO_4 tetrahedra. Koritnig concluded that there was a semi-quantitative relation

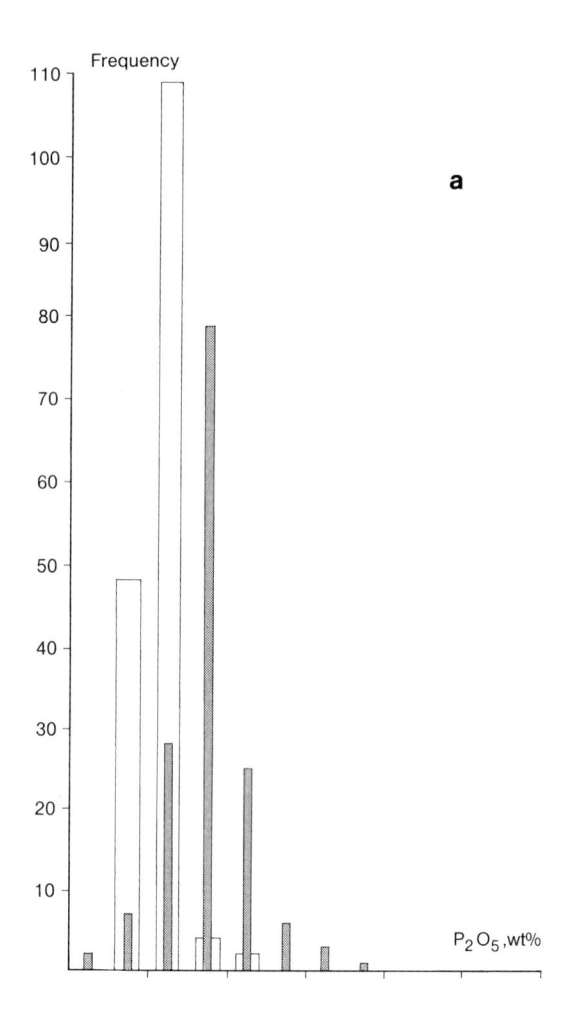

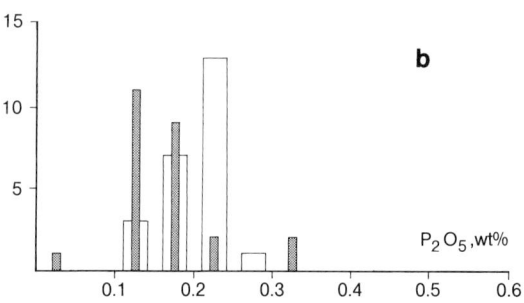

Fig. 11.4. Frequency distributions of P_2O_5 abundances in lavas containing SiO_2 >50 per cent of (a) the Lesser Antilles: Grenada (filled bars) and St Kitts (open bars) and (b) the Solomon Islands: Gallego (filled bars) and Savo (open bars).

between the extent of phosphorus substitution and the structure of the host mineral: the more condensed the linkage of the SiO$_4$ tetrahedra, the smaller the amount of phosphorus admitted. Thus he found a regular diminution in the amount of phosphorus proceeding from neso- to soro-, ino-, phyllo-, and finally to the tektosilicates. He attributed this to the stronger covalent bond character of the P ion as compared with that of Si: the more condensed the SiO$_4$ linkage, the more difficult it is for the P^{5+} to adjust to the bond directions of the SiO$_4$ tetrahedron. He noted that the other important factor influencing substitution is that of chemical equilibrium between the ubiquitous apatite and the coexisting silicates. Koritnig estimated that an average of 2–10 per cent of the total phosphorus of igneous rocks is present as a replacement of Si^{4+} in the silicates, and that such substituted phosphorus may in some cases be as high as 25–30 per cent of the total.

Henderson (1968), in investigating the behaviour of phosphorus in the early to middle stages of fractionation of the layered basic intrusions of Skaergaard, Bushveld, and Rhum found that a number of the relevant gabbros had a large part (up to 80 per cent) of their total phosphorus contained in the cumulus silicates. He questioned whether in fact Koritnig's leaching procedure had removed only phosphate from the rock crushes and observed that until it was established that Koritnig's method removed only the phosphate minerals and not any phosphorus from silicate lattice sites, no further attempt at accurate assessment of the amount of phosphorus camouflaged in the silicates could usefully be made. Any significant removal of such phosphorus by the leaching process would reduce the estimation of the element in the silicates and thus underestimate the role of the latter in removing phosphorus from the melt. Henderson suspected that such removal might be considerably greater than was indicated by Koritnig's measurements.

THE INCIDENCE OF PHOSPHORUS IN THE PRINCIPAL MINERAL SPECIES OF THE SOLOMON ISLANDS YOUNGER VOLCANIC SUITE

General

The incidence of phosphorus in the silicates of four igneous rocks as determined by Koritnig (1965) is shown in Table 11.1. Analogous determinations of phosphorus in minerals of the Skaergaard, Bushveld, and Rhum intrusions as given by Henderson (1968) are shown in Table 11.2.

Table 11.1 Amounts of phosphorus occurring in the structures of silicates of four igneous rocks as determined by Koritnig (1965)

Containing rock	Mineral	Average P conc. (ppm) in mineral
Granite (Konigskopf, Harz, Germany)	Orthoclase	53
	Plagioclase	27
	Quartz	0.2
	Biotite	58
Granodiorite (Bergell, Germany)	K-feldspar	53
	Plagioclase	27
	Quartz	0.2
	Biotite	58
Basalt (Brackenberg, Germany)	Plagioclase	27
	Pyroxene	89
	Olivine	220
Basalt (oceanite, Kilauea, Hawaii)	Alkali feldspar	53
	Plagioclase	27
	Pyroxene	89
	Olivine	220

Table 11.2 Phosphorus abundances in rocks and minerals of the Skaergaard, Bushveld, and Rhum layered gabbroic intrusions, after Henderson (1968)

Igneous mass	Rock sample no.	Mean phosphorus in whole rock P_2O_5, wt%	P, ppm	Mineral species	Mean phosphorus in mineral P_2O_5 (wt%)	P(ppm)	$K_D^{P\S}$
Skaergaard	5181	0.094	410*	plagioclase	0.010	45.5	0.111
				pyroxene	0.007	29.8	0.073
				magnetite	0.002	8.7	0.021
				ilmenite	0.005	20.9	0.051
	5112	0.30 est.	1310	plagioclase	0.012	53	0.040
	4389	0.27 est.	1180	plagioclase	0.025	110	0.0093
	5093	0.078	340	plagioclase	0.321	1400	4.118
				plagioclase	0.064	277	0.815
	5086	0.12	520	plagioclase	0.087	381	0.733
				pyroxene	0.018	77.2	0.148
				olivine	0.018	80.6	0.155
	5109	0.21	920	plagioclase	0.126	550	0.598
				pyroxene	0.018	76.2	0.083
	5107	0.12	520	plagioclase	0.260	1130	2.173
Bushveld	740			pyroxene, Ca-rich	0.006	24.2	
	733	0.010	44.4	plagioclase	0.008	34.0	0.766
				pyroxene, Ca-rich	0.004	19.8	0.446
				pyroxene, Ca-poor	0.006	25.2	0.568
	660	0.011	49.0	plagioclase	0.008	36.8	0.751
				pyroxene, Ca-rich	0.005	20.6	0.420
				pyroxene, Ca-poor	0.003	12.3	0.251
	681	0.010	46.0	olivine (Fo_{88})	0.009	38.2	0.830
Rhum	17127	0.013	57	plagioclase	0.008	33.0	0.579
				pyroxene	0.006	28.0	0.491
				olivine	0.009	40.6	0.712
	17126	0.014	61	plagioclase	0.010	41.6	0.682
				pyroxene	0.004	15.6	0.256
				olivine	0.007	28.7	0.470
	17125	0.014	62	plagioclase	0.006	27.1	0.437
				pyroxene	0.005	21.9	0.353
				olivine	0.008	36.6	0.590
				spinel, Cr-rich	0.043	190	3.065
	17123	0.016	70	plagioclase	0.010	43.2	0.617
	17122	0.028	122	plagioclase	0.027	120	0.984
				olivine	0.016	69	0.566
	17120	0.022	94	plagioclase	0.018	79	0.840
				pyroxene	0.005	24	0.255
				olivine	0.020	86	0.915

*All phosphorus (P) abundances in Skaergaard rocks obtained from the P_2O_5 values by calculation. § All values of K_D^P obtained as the ratio P (mineral)/P (whole rock) by the author using the data of Henderson. This is contrary to the convention used elsewhere in this book, i.e. K_D = concentration in minerals/concentration in groundmass; est = in estimated magma.

Of the minerals represented in the Solomons lavas, Koritnig found olivine to contain up to 517 ppm P, with an average of 220 ppm; pyroxene a maximum and average of 630 and 89 ppm respectively; amphibole 180 and 77 ppm; K-feldspar 106 and 53 ppm; plagioclase a mean of 27; and biotite a mean of 58 ppm P. Henderson found plagioclase to contain 44–1400 ppm, pyroxene 29–82 ppm, olivine 77–84 ppm, and magnetite ≈9 ppm P in the Skaergaard intrusion; plagioclase 34–9 ppm, pyroxene 11–30 ppm, and olivine 38 ppm P in the Bushveld intrusion; and

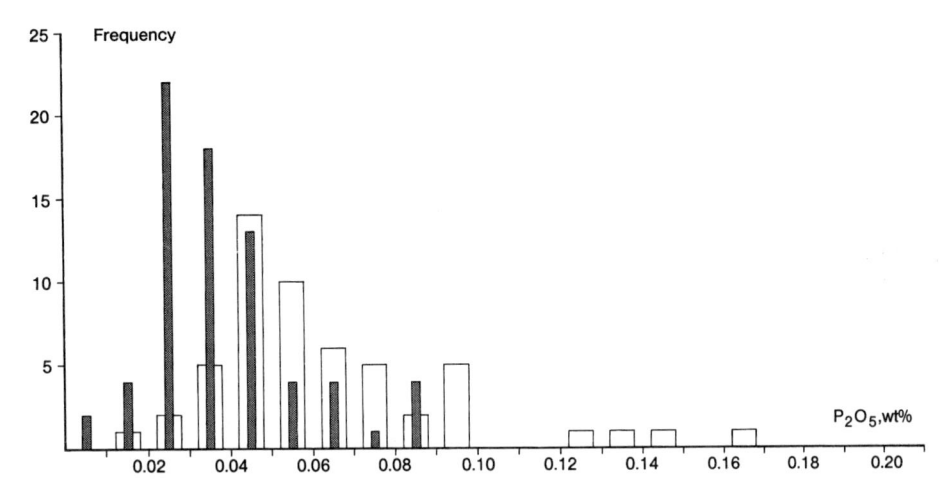

Fig. 11.5. Frequency distributions of P_2O_5 abundances in the two sets of Solomons feldspar separations: ultimate purity, filled bars; penultimate purity, open bars.

plagioclase 27–121 ppm, pyroxene 15–24 ppm, olivine 36–86 ppm, and chrome spinel 186 ppm in Rhum peridotites and allivalites. Henderson concluded that the degree of entry of phosphorus into the three principal minerals was olivine > plagioclase feldspar > pyroxene in all three intrusions — a result deviating slightly from that of Koritnig, whose analyses had indicated olivine > pyroxene > plagioclase feldspar.

Solomons suite

Determination of phosphorus in the Solomons minerals was not undertaken with the order of accuracy of Koritnig and Henderson in mind. P_2O_5 was estimated as a major constituent, as part of the standard XRF programme, and is given to ± 0.01 per cent, i.e. ± 100 ppm (Table 11.4).

Analysis of the two groups of plagioclase separations — 'ultimate purity' and 'penultimate purity' — indicated distinctly higher phosphorus abundances in the latter (Fig. 11.5). For the plagioclases of the olivine–pyroxene lavas, the former gave 695 ppm, whereas the penultimate purity material gave 1386 ppm P. Twenty-three similarly paired separations from the big-feldspar basalts and basaltic andesites gave 150 and 220 ppm P respectively, and 30 paired separations from the hornblende-andesites gave 180 and 317 ppm P. As the very fine glassy inclusions are an extremely minor component of the penultimate purity fractions, the partitioning of phosphorus into the glass is clearly substantial. This is considered further in the next section.

Although the analytical figures are relatively crude, the total amount of data on phosphorus abundances involved in Table 6.5 is large and is likely to give at least a reasonable indication of the relative uptake of this element by the different minerals. Overall means are hornblende (528 ppm) > spinel (440 ppm) > plagioclase (242 ppm)

Table 11.3 Crystal:melt partition coefficients developed by olivine, clinopyroxene, orthopyroxene, plagioclase, and ilmenite in some Hawaiian lavas. After Anderson and Greenland (1969)

Rock type	K_D^{pl}	K_D^{cpx}	K_D^{opx}	K_D^{ol}	K_D^{ilm}
Sub-alkaline basalt	0.024	0.009	0.014	0.043	
Sub-alkaline andesite	0.014	0.009	0.009	0.055	0.05
Alkaline basalt		0.017		0.019	

Superscript pl, cpx, opx, ol, and ilm refer to plagioclase, augite, orthopyroxene, olivine, and ilmenite respectively; K_D are the ratios (P_2O_5 in mineral)/(P_2O_5 in groundmass).

> olivine (150 ppm) = clinopyroxene (151 ppm). Although the number of samples of orthopyroxene (3) was insufficient to give a reliable comparison, all appeared to contain relatively high phosphorus, with a mean of 1628 ppm.

Thirty-nine coexisting olivine:clinopyroxene pairs from the olivine-pyroxene basalts, 11 hornblende:clinopyroxene pairs from the hornblende-andesites and 17 feldspar:clinopyroxene pairs from the big-feldspar basalts and basaltic andesites gave crystal:crystal partition coefficients as follows:

$K_D^{ol:cpx} = 1.3410$; $K_D^{hb:cpx} = 1.2236$; $K_D^{fel:cpx} = 1.2745$.

These groups were selected because they gave the largest number of pairings — and presumably the most reliable mean values for the relevant partition coefficients. They do, however, involve different melt compositions, and comparison of K_D values must be made with this in mind. Although the values are very similar, they indicate relative uptake in the order olivine > feldspar > hornblende > clinopyroxene.

CRYSTAL:MELT PARTITIONING

Apart from its variable capacity to substitute for Si^{4+} in SiO_4 tetrahedra, P^{5+}, with its very small ionic radius of 0.035 nm, is not systematically camouflaged by any of the major or minor components of the principal volcanic silicates, and hence silicate:melt partition coefficients are generally low.

Anderson and Greenland (1969), using the leaching technique of Koritnig (1965) and analysing by neutron activation on separations and by electron probe on individual grains, investigated phosphorus partitioning as between olivine, clino- and orthopyroxene, plagioclase, and ilmenite and melt in five Hawaiian basalts. Their results are given in Table 11.3. They noted that, contrary to the finding of Koritnig (1965), but in agreement with Henderson (1968), their results indicated that acceptance of phosphorus by the pyroxene structure might be slightly lower than by plagioclase. With K_D^P values of the order of 10^{-2} to 10^{-3} throughout, Anderson and Greenland's results indicate that, at the basaltic stage at least, crystallization of the common volcanic silicates must have a strong and consistent tendency to concentrate phosphorus in the remaining melt. It was of course on the assumption that this was the case that Wager (1960, 1963) based his use of P_2O_5 abundances to estimate progressive fractionation of the Skaergaard layers, and that Anderson and Greenland developed their 'phosphorus fractionation diagram' for the estimation of relative proportions of minerals crystallizing prior to the onset of formation of apatite and other phosphorus-rich material.

Determination of P_2O_5 as a major element in the Solomons mineral analyses means that estimation of partition coefficients here must be crude at best (Table 11.4). Clearly they cannot be compared realistically with those of Henderson and Anderson and Greenland, whose estimations they invariably — and in most cases, grossly — exceed. However, in all cases except for those of orthopyroxene and magnetite (both quantitatively very minor components) in the hornblende-andesites $K_D^P <1$. Subtraction of all mineral species of the Solomons lavas thus tends to concentrate phosphorus in the remaining melt.

Table 11.4 Crystal:melt partition coefficients with respect to phosphorus developed by the principal minerals of the Solomons lava suite

Mineral	Lava group	Mineral P	Groundmass P	K_D^P
Olivine	OPB	132	1702	0.0776
Olivine	BFL	255	1833	0.1391
Clinopyroxene	OPB	88	1702	0.0517
Clinopyroxene	BFL	123	1833	0.0671
Clinopyroxene	HbA	264	742	0.3558
Orthopyroxene	HbA	1628	742	2.1941
Hornblende	HbA	528	742	0.7116
Feldspar	OPB	695	1702	0.4083
Feldspar	BFL	150	1833	0.0818
Feldspar	HbA	180	742	0.2426
Spinel	OPB	176	1702	0.1034
Spinel	BFL	308	1833	0.1680
Spinel	HbA	792	742	1.0674

OPB = olivine–clinopyroxene-basalts; BFL = big-feldspar lavas; HbA = hornblende-andesite group. All quantities in ppm by mass. K_D^P are the ratios (P in mineral)/(P in groundmass).

THE INCIDENCE OF WHOLE-ROCK PHOSPHORUS IN THE SOLOMON ISLANDS YOUNGER VOLCANIC SUITE

Mean P_2O_5 for the suite as a whole is 0.206 per cent (900 ppm P), and for the basaltic members 0.215 per cent (940 ppm P). The latter value compares with 0.241 per cent (1050 ppm P) for all MORBs and 0.195 per cent for all arc basalts. Mean P_2O_5 for Solomons lavas with mean SiO_2 >52 per cent is 0.194 per cent, which compares with 0.19 per cent for all arc lavas with SiO_2 >52 per cent. The Solomons lavas thus have an overall P_2O_5 content slightly lower than that for mean MORBs (though the considerable variability of the phosphorus content of MORBs must be kept in mind), and very slightly higher than those of arc lavas in general.

Frequency distributions of P_2O_5 abundances in the Solomons basalts (SiO_2 <52 per cent) and more felsic numbers (SiO_2 >52 per cent) are given in Fig. 11.6 (see also Table 6.3). The modes of both groups are in the 0.20–0.25 per cent P_2O_5 interval but, as already noted for several other elements, the frequency distribution for the lavas more felsic than basalts is less sharp and rather broader than that for the basalts themselves.

The pattern of distribution of phosphorus in the three Solomons lava types (Fig. 11.7) is somewhat similar to those exhibited by copper and zinc: more or less in accordance with the observations of Landergren (1954), Gmelin (1965), and others, the basalts and basaltic andesites of the big-feldspar lava group have the highest mean phosphorus (0.35 per cent P_2O_5); the more mafic olivine–clinopyroxene-basalts and the hornblende-andesites contain significantly less at 0.22 per cent and 0.17 per cent P_2O_5 respectively. Phosphorus is clearly partitioned into the groundmass in the most mafic lavas (Fig. 11.7(c)), becoming proportionately less so in the big-feldspar lavas and substantially evenly distributed in the hornblende-andesites.

This broad pattern of incidence of phosphorus in the three major lava groups is reflected in the P_2O_5 v. whole-rock SiO_2 diagram for the Solomons Younger Volcanic

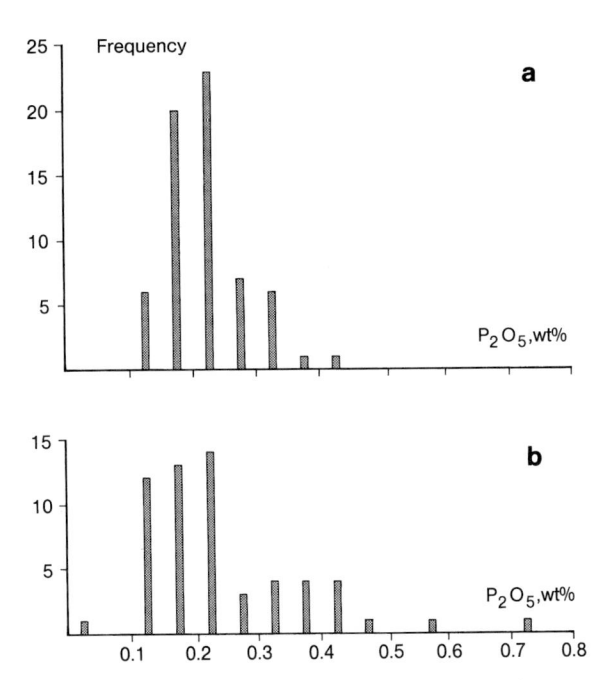

Fig. 11.6. Frequency distributions of P_2O_5 abundances in (a) Solomons basalts (SiO_2 <52 per cent) and (b) Solomons lavas more felsic than basalts (SiO_2 ⩾52 per cent).

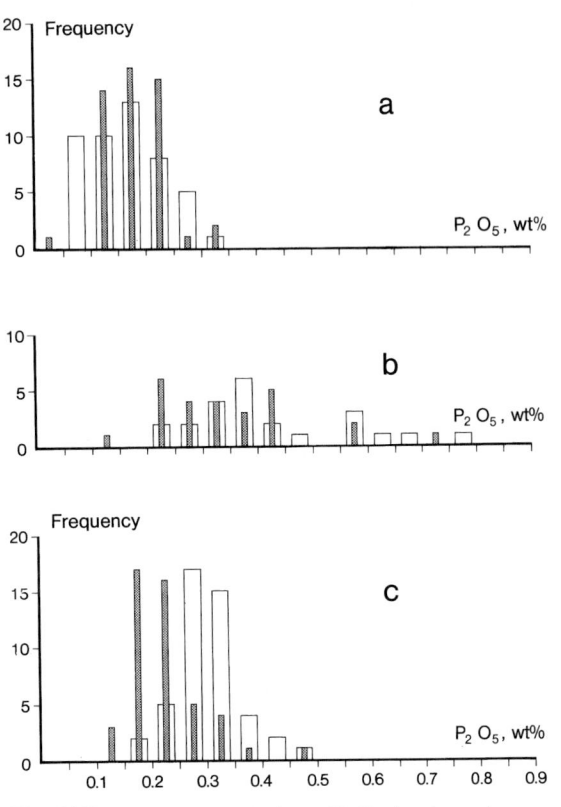

Fig. 11.7. Frequency distributions of P_2O_5 abundances in whole rocks (filled bars) and corresponding groundmasses of (a) lavas of the hornblende-andesite group, (b) big-feldspar lavas, and (c) the olivine–clinopyroxene-basalts of the Solomons suite.

Suite (Fig. 11.8(a)). Phosphorus rises from ≈0.15 per cent (as P_2O_5) in lavas of 44–6 per cent SiO_2 range to a maximum of ≈0.26 per cent in those of 50–2 per cent SiO_2 range. It then decreases fairly steadily to ≈0.10 per cent at whole-rock SiO_2 >64 per cent. It reaches a maximum of ≈0.35 per cent in the groundmass at whole-rock SiO_2 52–4 per cent and then decreases rapidly to 0.1 per cent in the groundmasses of rocks with whole-rock SiO_2 = 56–8 per cent. From here on whole-rock and groundmass P_2O_5 are essentially the same, with P_2O_5 = 0.09 per cent in the groundmass of lavas with SiO_2 >64 per cent.

This rather tidy pattern (made even tidier by the improved statistics stemming from the inclusion of the Bougainville analyses: see Fig. 21.2(d)) of increase and then decrease in phosphorus with increase in whole-rock SiO_2 shown by the Solomons lavas is by no means general, however. The Aleutians data (Fig. 11.8(b)) indicates highest phosphorus in the most mafic lavas (mean P_2O_5 = 0.7 per cent at SiO_2 <46 per cent), a sharp decrease to ≈0.25 per cent at SiO_2 = 48–50 per cent, a fairly consistent maintenance of this level to SiO_2 = 62–4 per cent, and then a general decrease as the lavas become more felsic. A somewhat similar pattern is displayed by the Grenada lava suite (Fig. 11.8(c) (1)). Those of Dominica (Fig. 11.8(c) (2)) and St Kitts, however, show little variation in P_2O_5 over their full range from SiO_2 <50 per cent to >66 per cent. Phosphorus exhibits a slight but consistent increase from 46–60 per cent SiO_2, with a sharp decrease at 60–2 per cent SiO_2, in the Vanuatu lava series (Fig. 11.8(d)), but in the Tonga–Kermadec suite there is a slight decrease then increase in P_2O_5 with increase in whole-rock SiO2 (Fig. 11.8(d) (2)).

Consideration of the abundance behaviour of P_2O_5 in the evolution of the three principal lava groups of the Solomons reveals a very ordered pattern, and again one remarkably similar to those shown by copper and zinc. Ternary plots of PMA (P_2O_5–MgO–(Na_2 + K_2O)) are given in Fig. 11.9. The 'olivine subtraction line' with respect to P_2O_5 shown in Fig. 11.9(f) is remarkably linear — even more sharply defined than that for zinc (Fig. 7.21(f)). The field for olivine–pyroxene-basalt groundmasses is, as would be expected, somewhat less ordered and on the higher P_2O_5 side. Maximum P_2O_5_MgO–(Na_2O + K_2O) is achieved in the big-feldspar lavas and their groundmasses (Fig. 11.9(d) and (c)), after which the field becomes less ordered and descends, fairly systematically, towards the high alkali metal corner of the triangle with the hornblende-andesite groundmasses.

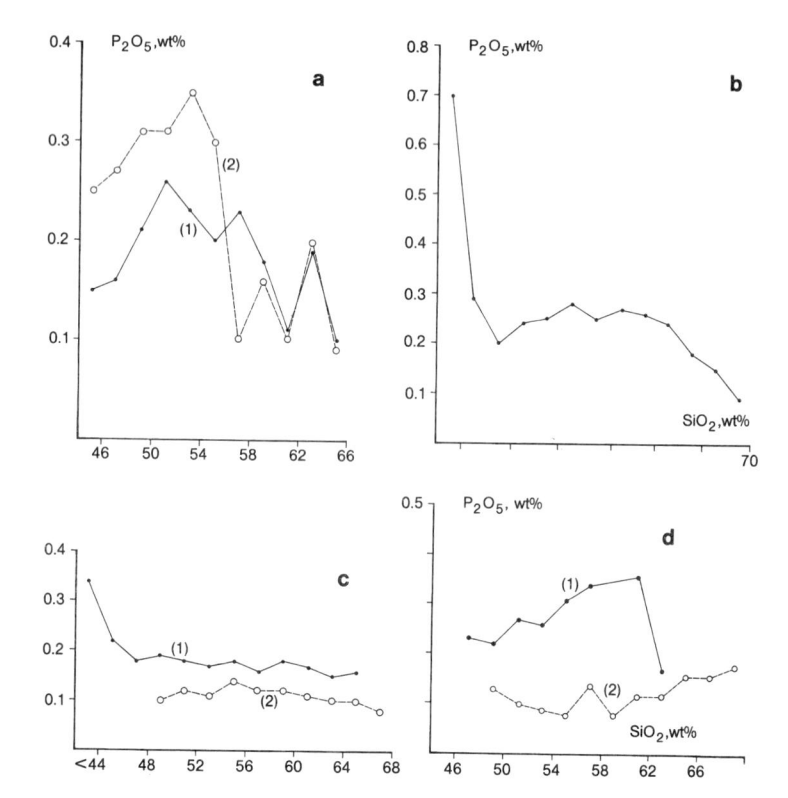

Fig. 11.8. Mean P_2O_5:SiO_2 relations corresponding to 2 per cent whole-rock SiO_2 intervals for (a) whole rocks [(1) filled circles] and corresponding groundmasses [(2) open circles] of the Solomons suite; (b) lavas of the Aleutians; (c) Grenada [(1) filled circles] and Dominica [(2) open circles]; distribution of points for St Kitts almost identical with that for Dominica] (d) Vanuatu [(1) filled circles] and Tonga–Kermadec [(2) open circles].

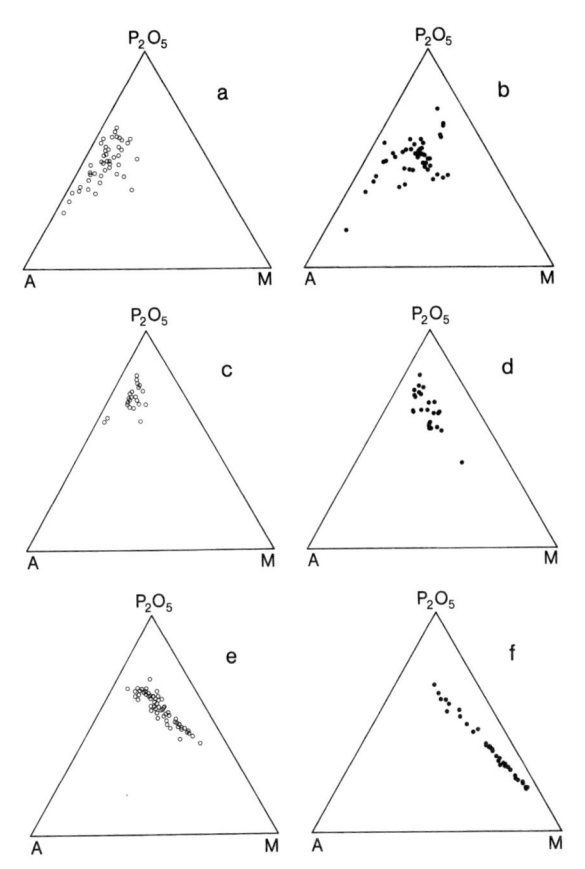

Fig. 11.9. P_2O_5–MgO–($Na_2O + K_2O$) ternary relations (P_2O_5 calculated using weight per cent $P_2O_5 \times 50$) developed in olivine–clinopyroxene-basalt, big-feldspar lava and hornblende-andesite whole rocks [filled circles: (f), (d), and (b) respectively], and their corresponding groundmasses [open circles: (e), (c), and (a) respectively].

This evolutionary pattern is shown in sharper focus in Fig. 11.10(a), which depicts mean PMA relations in whole rocks and groundmasses for 2 per cent whole-rock SiO_2 intervals in the Solomons suite. In the most mafic rocks the proportion of phosphorus is low and P:M:A ≈20:75:5. The proportion of phosphorus rises to a peak in the big-feldspar basalts, with mean proportions of the three components ≈60:20:20 in the whole rocks and 65:10:25 in the groundmasses. With development of the hornblende-andesites phosphorus then decreases again to ≈33:7:60 in the most felsic groundmass. The overall pattern is quite clearly hypotholeiitic, and this is fully confirmed when the full 241 analyses, including those of the 119 Bougainville lavas, are plotted as shown in Fig. 11.10(b)).

Phosphorus   145

This hypotholeiitic pattern is not, however, shown by the Lesser Antilles lavas (Fig. 11.10(c)), which exhibit an essentially constant proportion (in the P_2O_5–MgO–($Na_2O + K_2O$) system) of P_2O_5. Both Grenada, and Grenada–Dominica–St Kitts as a group, develop straight-line distributions, perhaps best described as flatly tholeiitic. The Vanuatu suite (Fig. 11.10(d), filled circles), on the other hand, displays a well-developed hypotholeiitic pattern — at the more felsic end, a slightly truncated version of the pattern for the Solomons suite, with which the Vanuatu lavas have such a close petrological connection. (Indeed, plotting of individual analyses of the high-magnesium lavas of Aoba (see Fig. 23.2) — with their striking petrological and geochemical similarities to those of New Georgia — yields a sharply defined 'olivine subtraction line' with respect to P_2O_5 almost identical with that of Fig. 11.9(f)). The Aleutians lavas are also clearly hypotholeiitic with respect to phosphorus (Fig. 11.10(d), open circles), the distribution of points in this case appearing as a version of the Solomons pattern truncated at the mafic end of the series. The Tonga–Kermadec suite (not shown in Fig. 11.10) is essentially tholeiitic, with perhaps a very slight tendency to be hypertholeiitic.

MORBs as represented by the Mid-Atlantic Ridge basalts (Fig. 11.10(e)) generate a long, clearly defined, tholeiitic pattern, giving the impression of extensive fractionation with respect to phosphorus. The Icelandic basalts behave similarly, but when combined with the Icelandic lavas more felsic than basalt, generate a remarkably well-formed and sharply angled hypotholeiitic distribution (Fig. 11.10(f)). It would indeed be hard to find a greater contrast in ternary patterns generated by a single element than, as in this case, the flatly tholeiitic distribution with respect to phosphorus in the Lesser Antilles lavas, compared with the sharply angled hypotholeiitic one of the Icelandic assemblage.

Thus, while phosphorus certainly does not show the very consistent pattern of behaviour of copper as the various lava series become more felsic, it does display some similarities — particularly in the lavas of the Solomons and perhaps those of Vanuatu and the Aleutians — in the ternary plots. As with copper (and zinc), the proportionately low phosphorus in the more mafic basalts of the Solomons and Vanuatu (New Georgia and Aoba respectively) can readily be explained by the subtraction and accumulation of low-P olivine and clinopyroxene crystals. As for the two metals, however, there appear to be no phosphorus-rich accumulates developed from the intermediate members of the lava series. The proportionate decrease in phosphorus in the more felsic members cannot therefore be accounted for on the basis of fractional crystallization alone.

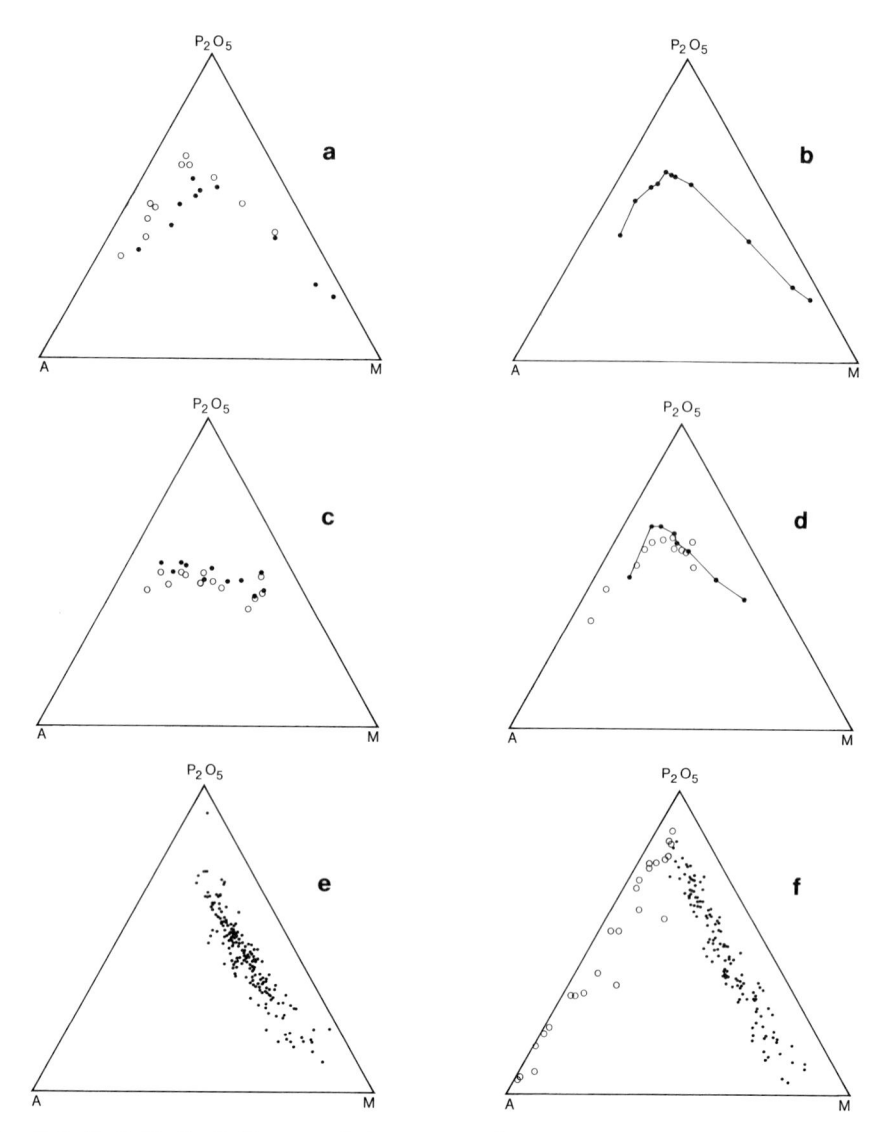

Fig. 11.10. Mean P_2O_5–MgO–($Na_2O + K_2O$) ternary relations corresponding to 2 per cent whole-rock SiO_2 intervals in (a) whole rocks (filled circles) and corresponding groundmasses (open circles) of the Solomons suite excluding Bougainville; (b) whole rocks of the Solomons suite including Bougainville (241 analyses); (c) lavas of Grenada (filled circles) and Dominica–St Kitts (open circles); and (d) Vanuatu (filled circles) and the Aleutians (open circles). Ternary relations developed in MORBs as represented by individual Mid-Atlantic Ridge basalts are given in (e), and in individual samplings of Icelandic basalts ($SiO_2 < 52$ per cent; filled circles), and Icelandic lavas more felsic than basalts ($SiO_2 \geqslant 52$ per cent; open circles) in (f).

CONCLUDING STATEMENT

The mean P_2O_5 content of MORBs is ≈ 0.24 per cent, or ≈ 1050 ppm equivalent phosphorus. That of arc basalts is 0.20 per cent P_2O_5 or ≈ 870 ppm equivalent phosphorus. There is, however, considerable variation both within and between the different mid-ocean ridges, and within and between the various island arcs. For the latter there is, perhaps surprisingly, little difference on the basis of present data between the phosphorus content of basalts (SiO_2 <52 per cent; 365 analyses; 870 ppm P) and that of more felsic lavas (SiO_2 >52 per cent; 722 analyses; 830 ppm

P). For the Solomons suite, the basalts (SiO$_2$ <52 per cent) have mean P$_2$O$_5$ of 0.22 per cent (960 ppm P) and the more felsic types one of 0.19 per cent P$_2$O$_5$ (850 ppm P).

Although phosphorus has not been determined as a trace element in the Solomons rocks and minerals, and the data for the minerals are thus somewhat crude, it is clear that its incidence in most of the silicates is very low. Exceptions are orthopyroxene (mean P$_2$O$_5$ = 0.37 per cent) and the feldspars of the olivine–clinopyroxene-basalts (mean P$_2$O$_5$ = 0.16 per cent). For all silicates except orthopyroxene, crystal:melt partition coefficients are <1. In only one other instance, that of the magnetites of the hornblende-andesite group of lavas, does $K^{\text{cryst:melt}}$ exceed 1. As a result the effect of essentially all crystallization has been to concentrate phosphorus in the melt.

Relations between whole-rock phosphorus and whole-rock SiO$_2$ (as a measure of progressive crystallization) indicate that progressive crystallization has indeed been significant in the earlier stages of lava evolution (to 52–4 per cent SiO$_2$) in the Solomons, Aleutians, and Grenada suites. Among the other arc suites the relationship is, however, variable and there is not a clear general pattern. Ternary P$_2$O$_5$–MgO–(Na$_2$O–K$_2$O) diagrams indicate that the Solomons, Vanuatu, and the Aleutians assemblages are hypotholeiitic; those of Dominica, St Kitts, and Tonga–Kermadec tholeiitic, the last-named perhaps tending to very slight hypertholeiiticity. MORBs as represented by those of the Mid-Atlantic Ridge appear to be an extensively fractionated tholeiitic series with respect to phosphorus, and the Icelandic lava assemblages as a whole is spectacularly hypotholeiitic.

It therefore appears that there are at least some ties between lava evolution and phosphorus abundance, but that these are variable and in some cases erratic. A relationship between crystallization, fractionation, and the incidence of phosphorus seems particularly clear for the more mafic basalts of the Solomons and of Aoba in Vanuatu: in both cases phosphorus conforms to a sharply delineated, highly linear, olivine–pyroxene subtraction –accumulation line. However, while this concentration of phosphorus from mafic to intermediate compositions can be reasonably attributed to crystallization processes, its decrease from intermediate to felsic compositions cannot. Phosphorus is one of the readily volatile components of the melt, and it may be suspected that its relative decrease in abundance in the more felsic lavas — without the development of a complementary phosphorus-rich accumulate — and its somewhat erratic incidence in these rocks may be due, at least in part, to some irregular volatile loss. This is considered in more detail in Chapters 21 and 24.

12

Calcium

Although the presence of 'carbonate' is commonly noted among the non-sulphide components of many exhalative orebodies, comparatively little attention has been paid to the incidence of calcium in deposits of this class. Where the carbonate is dolomitic or ankeritic this is usually referred to, but where it is essentially calcite most investigators do little more than note its presence. In some cases, e.g. Mount Isa and Macarthur River in Australia, carbonate finely interbedded with sulphides may represent a part of the wider, 'normal' depositional context, ranging from fine pelagic limestone to carbonate deposited through sabka–evaporitic processes. In some exhalative ore environments however the carbonate, and its contained calcium, are fairly well limited to the area of ore occurrence. This and other features indicate that the calcium, with the sulphides and other associated elements, is of exhalative origin.

In other instances the calcium occurs in silicates, but the possible significance of its presence tends to be lost sight of in the investigators' concern with the conditions of metamorphism that those (calcium-bearing) silicates might indicate. In yet others it occurs as fluorite, anhydrite or apatite. Attention is then often concentrated on the fluoride, sulphate, or phosphate and the significance that these anions may have had concerning derivation, conditions of sedimentation, and processes of subsequent metamorphism. Again, the presence of the *calcium* is largely overlooked.

Studies of Broken Hill provide a good example of preoccupation with anions to the near exclusion from scientific consideration of the common cation, calcium. It has been recognized for some hundred years that the different ore 'lenses', or layers, at Broken Hill are characterized by conspicuous quantities of rhodonite–bustamite, fluorite, calcite, apatite, etc., and several were named accordingly in the early days of mining: 'rhodonitic zinc lode', 'fluorite lode', and so on. Thus the gangue (non-ore) materials of the 'lodes' (layers), and the lodes themselves, were denoted according to whether silicate, fluoride, carbonate, or phosphate were prominent components of the non-sulphide fraction. What was not emphasized in observers' minds was that these minerals were wholly or largely compounds of *calcium*, and that calcium constituted a constant compositional thread that ran right through the huge stack of sulphide-rich layers that made up the Broken Hill orebody. Fig. 12.1 represents the incidence of sulphide sulphur (and hence a close approximation to lead + zinc) *vis-à-vis* calcium in analysed drill cores through 'B-lode' (one of the major ore layers) at Broken Hill. The figure shows how close the relationships between sulphide and calcium are in this case. Similar ties hold in the associated banded iron formations (ironstones, bifs), in which the incidence of iron — chiefly as magnetite — is closely tied to the incidence of calcium as apatite (Stanton, 1976; Stanton, Roberts, and Chant 1978).

Broken Hill is only one example of the often hidden, but commonly important, incidence of calcium in exhalative bodies: there are many others. Among them are the Aggenys and Gamsberg orebodies of Namaqualand, notable for their complex assemblages of calcium-rich silicates, the Balmat–Edwards deposits of Upper New York State, with their abundant dolomite, calcite, anhydrite, and gypsum, together with diopsidic to tremolitic silicates, and the dolomitic deposits of Mt Isa, Hilton, Macarthur River, Century, and others in Northern Australia.

It has been pointed out (Stanton, 1983) that the exhalative association of calcium and iron, together with variation in sulphur and carbonate availability, may have profound consequences in the mineralogical development of exhalative sulphide ores and their non-sulphide-bearing 'exhalite' associates.

High sulphide and carbonate (HS^-; HCO_3^-) availability in a reducing depositional environment leads to the formation of pyrite, ferrous silicates such as chlorites, and calcium-rich carbonates, all as accompaniments of Cu, Zn, and Pb sulphides as the case may be. Low sulphide availability with abundant carbonate yields low-sulphur iron sulphide (pyrrhotite), ferrous silicates, and calcium-iron (-magnesium) carbonates. A combination of low sulphur and low carbonate in a reducing environment leads to the formation of low-sulphur sulphide together

Calcium 149

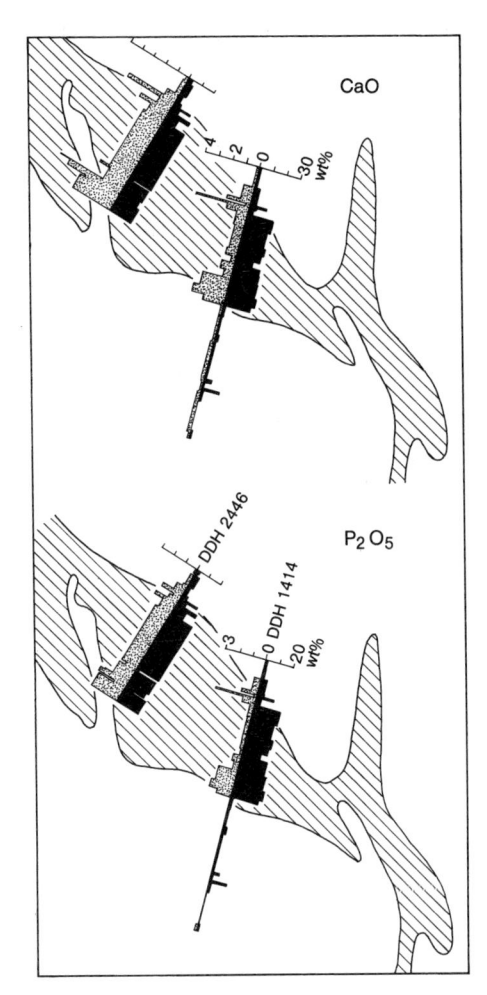

Fig. 12.1. Abundance relations (weight per cent) between total CaO and P$_2$O$_5$ (both indicated by stippled bars) and total sulphide sulphur (filled bars) in two diamond drill cores through Broken Hill B lode (shaded area of irregular outline) on Section 84, New Broken Hill Consolidated Ltd. Bar widths indicate lengths of intersection of individual lithological units (and hence samplings), and weight percentages are indicated by the scales normal to the drill holes. After Stanton, Roberts, and Chant (1978).

with a range of iron–calcium silicates which, with ageing and metamorphism, may yield minerals such as hedenbergite, grossularitic garnet, Ca-amphiboles, and related calcium-bearing species such as wollastonite, diopside, bustamite, and so on. Under a slightly less reducing regime magnetite may appear together with the ferric iron-calcium silicates andradite and/or epidote.

The result is a stratiform exhalative skarn assemblage — a 'stratiform skarn' — having no tie with the skarns generated by the contact metamorphism of carbonate rocks. Broken Hill, the silicate facies of the Gamsberg iron–zinc formation, a number of orebodies in Finland, the Geco orebody of Ontario, and many others are of this kind. Most of these deposits have been recognized as being closely related to iron formations: many of them could equally well be referred to as 'exhalative calcium formations'.

CHEMICAL PROPERTIES AND CRYSTAL CHEMISTRY OF CALCIUM

Like the other members of the alkaline-earth group, calcium is strongly lithophile in character (the sulphide, oldhamite (CaS), is known but rare in nature) and is an important component of both ferromagnesian and plagioclase mineral groups. It constitutes ≈20, 1–2, and 12 wt per cent of the Solomons clinopyroxenes, orthopyroxenes, and hornblendes respectively, and ≈16, 13, and 11 wt per cent of the plagioclases of the olivine–pyroxene, bigfeldspar, and hornblende-andesite lava groups in that order.

With an ionic radius of 0.106 nm and radius ratio 0.803 with respect to O^{2-}, Ca^{2+} favours 8-fold coordinated sites — or close approximations to this — in the rock-forming silicates. As a result of this 'intermediate' ionic size and hence preferred coordination, calcium is little involved in diadochical relationships with the more common igneous trace elements. On the one hand Ba^{2+}, Sr^{2+}, and Pb^{2+} (0.143, 0.127, 0.132 nm) are too large and, as already noted, tend to develop substitutional relationships with K^+ (0.133 nm). On the other hand Cu^{2+}, Zn^{2+}, Co^{2+}, Ni^{2+} (0.072, 0.083, 0.082, 0.078 nm) are too small and tend to have substitutional relationships with the other alkaline earth ion Mg^{2+} (0.078 nm). The only common trace-to-minor ion of appropriate size and charge for ready substitutional ties with Ca^{2+} is Mn^{2+}, at 0.091 nm.

What initially appears as an anomaly arises when the behaviour of strontium relative to calcium is examined in the series of groundmass fractions referred to in Chap. 7. In Fig. 12.2(a), barium is, as expected (see Chapter 9), linearly and directly related to the abundance of K_2O. In a manner apparently contrary to the considerations of Chapter 10, strontium is linearly but *inversely* related to K_2O. In keeping with this apparently anomalous behaviour with respect to potassium, strontium is directly related to CaO in the groundmass fractions, whereas (in accordance with the prognostications of Chapter 9) barium exhibits an inverse relationship (Fig. 12.2(b)). It

150 *Ore Elements in Arc Lavas*

might have been expected that *both* barium and strontium (Ba^{2+} = 0.143; Sr^{2+} = 0.127 nm) would be directly related to K^+ (0.133 nm) and hence inversely related to Ca^{2+} (0.106 nm). Fig. 12.2 shows this to be, quite unequivocally, not the case.

The answer lies in petrography. The groundmass concerned — separated into five fractions according to magmatic susceptibility — consists of glass, fine plagioclase feldspar, and a very fine magnetite dust located in the glass. Electromagnetic separation using progressively increasing field strengths thus tends to remove early fractions rich in glass + magnetite dust and later fractions richer in plagioclase. The early fractions are the repository for most of the potassium of the groundmass (the residual melt); the fine plagioclase is dominated by calcium and contains only very minor potassium. Barium is fixed stably in K^+ sites, separates with glass + magnetite

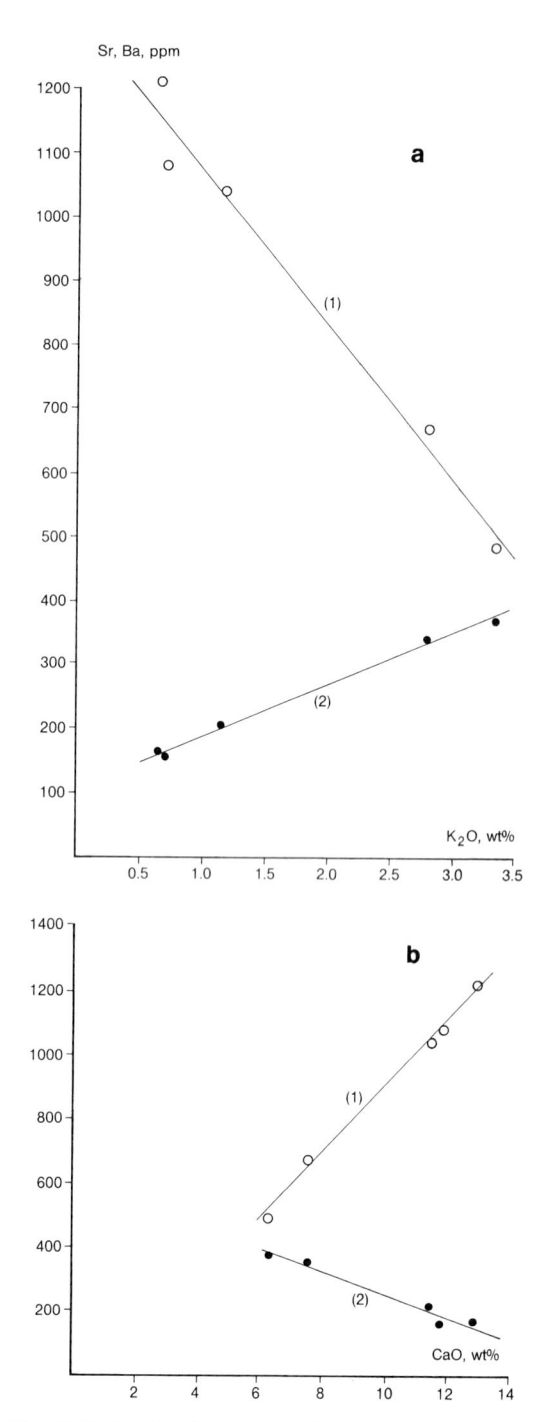

Fig. 12.2. Relations between strontium [(1), open circles] and barium [(2), filled circles] and (a) K_2O and (b) CaO in the five magnetic fractions of the big-feldspar lava groundmass of Fig. 7.18.

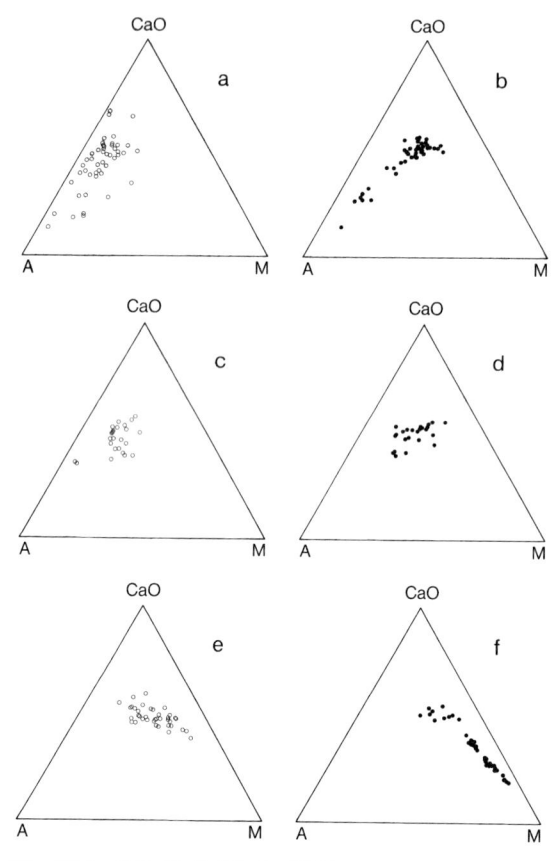

Fig. 12.3. CaO–MgO–($Na_2O + K_2O$) ternary relations developed in olivine–clinopyroxene-basalt, big-feldspar lava, and hornblende-andesite whole rocks [filled circles; (f), (d), and (b) respectively], and their corresponding groundmasses [open circles: (e), (c), and (a) respectively].

dust, and correlates with K$_2$O in the series of groundmass fractions. Strontium on the other hand is, as considered in Chapter 10, localized by some feature of K$^+$ occurrence *in the plagioclase feldspars*. It is the plagioclase structure, or some KAlSi$_3$O$_8$–CaAl$_2$Si$_2$O$_3$ mosaic structure within it, that appears to be important in capturing Sr^{2+}. The relevant plagioclase compositions are dominated by calcium and contain only very minor potassium. Strontium therefore appears to follow calcium in the groundmass. The tie is, however, indirect and no significant Ca^{2+}–Sr^{2+} diadochy is involved.

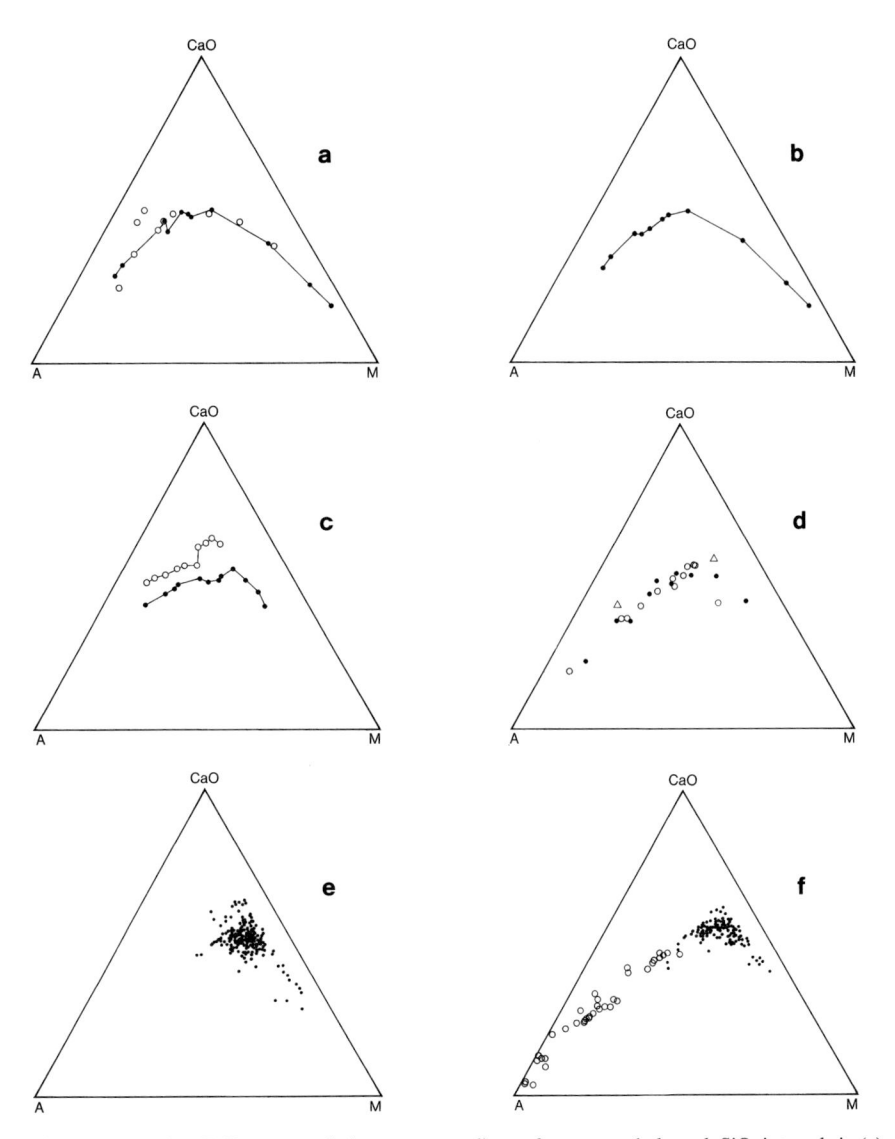

Fig. 12.4. Mean CaO–MgO–(Na$_2$O + K$_2$O) ternary relations corresponding to 2 per cent whole-rock SiO$_2$ intervals in (a) whole rocks (filled circles) and corresponding groundmasses (open circles) of the Solomons suite excluding Bougainville; (b) whole rocks of the Solomons suite including Bougainville (241 analyses); (c) lavas of Grenada (filled circles) and Dominica (open circles); (d) Vanuatu (filled circles), Aleutians (open circles), and Tonga–Kermadec (limits of linear field indicated by two open triangles). Ternary relations developed in MORBs as represented by individual Mid-Atlantic Ridge basalts are given in (e), and in individual samplings of Icelandic basalts (SiO$_2$ <52 per cent; filled circles), and Icelandic lavas more felsic than basalt (SiO$_2$ ⩾52 per cent; open circles) in (f).

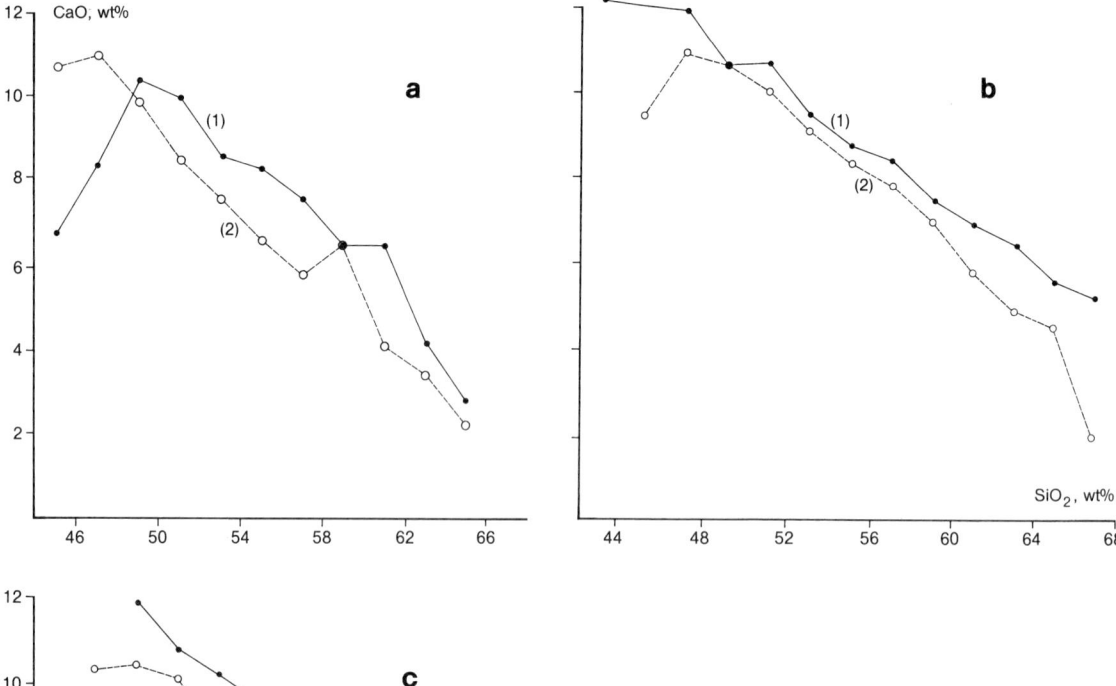

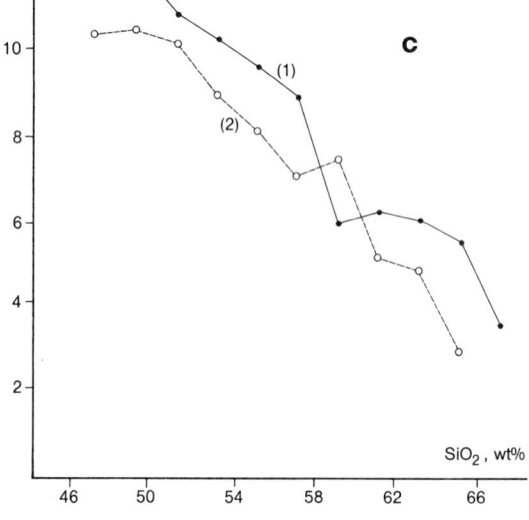

Fig. 12.5. Mean CaO:SiO$_2$ relations corresponding to 2 per cent whole-rock SiO$_2$ intervals in (a) whole rocks [(1) filled circles] and corresponding groundmasses [(2) open circles] of the Solomons suite; (b) lavas of the Lesser Antilles (Grenada, Dominica, and St Kitts combined [(1) filled circles], and the Aleutians [(2) open circles]; and (c) Tonga–Kermadec [(1) filled circles] and Vanuatu [(2) open circles].

THE INCIDENCE OF WHOLE-ROCK CALCIUM IN THE SOLOMON ISLANDS YOUNGER VOLCANIC SUITE

The broad incidence of calcium as a major element of the Solomons lavas has been considered in Chapter 5, and is given in Table 6.3.

When relative abundances are plotted on CaO–MgO–(Na$_2$O + K$_2$O) i.e. 'CaMA' diagrams (Fig. 12.3) ternary relations are seen to be very similar to those developed by total iron as FeO (Stanton 1967), manganese, and zinc (see Chapters 17, 16, and 7 respectively). The mafic basalts and picrites are represented by a well-defined olivine–clinopyroxene subtraction line (Fig. 12.3(f)), which gives way to an inflexion generated by the big-feldspar lavas and their groundmasses (Figs. 12.3(d) and (c)) and then to the 'calc-alkaline' (i.e. hypotholeiitic) slope of the hornblende-andesites and their groundmasses (Figs. 12.3(b) and (a)).

The hypotholeiitic nature of the Solomons suite with respect to calcium is confirmed unambiguously when mean values corresponding to 2 per cent whole-rock SiO_2 intervals are plotted on the CaMA diagram (Fig. 12.4). Their hypotholeiitic character shows with even greater clarity when the Bougainville analyses are included (12.4(b)). The Grenada assemblage is also clearly hypotholeiitic (Fig. 12.4(c), filled circles); Dominica on the other hand — and also St Kitts — appears to represent a hypotholeiitic series truncated at the mafic end (Fig. 12.4(c), open circles). The Tonga–Kermadec suite, the limits of which are defined by the two open triangles in Fig. 12.4(d), is similar to those of Dominica and St Kitts in this respect, as for the most part is that of the Aleutians. As might have been expected on the basis of its analogies with the Solomons and Grenada assemblages, the Vanuatu suite also displays a well-developed hypotholeiitic distribution.

The MORBs pattern is substantially tholeiitic, but there is also a fairly clear indication of incipient hypotholeiiticity (Fig. 12.4(e)). The latter appears to be extended (see also the Solomons olivine–clinopyroxene-basalts in Fig. 12.4(f)) in the Icelandic suite, which, in its more felsic members, exhibits a distinctive, extended hypotholeiitic pattern with respect to calcium.

CaO–SiO_2 relations in the Solomons suite (as expressed by the plot of CaO v. 2 per cent intervals of whole-rock SiO_2: Fig. 12.5) are somewhat reminiscent of those of copper. The whole-rock curve rises from the low calcium contents of the picrites and picritic basalts to a peak in the 48–52 per cent whole-rock SiO_2 range, then decreases steadily with increase in SiO_2 through the andesites and into the dacite range. The Aleutians and, to a lesser extent, the Vanuatu assemblages (Fig. 12.5(b) and (c)) behave somewhat similarly, but of the Lesser Antilles, and Tonga–Kermadec show little other than simple decline in CaO with increasing whole-rock SiO_2.

CONCLUDING STATEMENT

Judging from its incidence in exhalative ores, calcium is an important exhalative element — perhaps considerably more important from the quantitative point of view than has been generally recognized. It develops hypotholeiitic abundance patterns very similar to those of iron (Stanton 1967), manganese, zinc, and copper, reflecting early crystallization and subtraction of highly magnesian olivines (which contain little of these elements) followed by subtraction of another set of compounds in which, in contrast, the four elements are relatively abundant.

13

Titanium

Strontium has been considered among the present group of elements (Chapter 10) because, in spite of its close chemical similarity to barium (a commonly abundant component of exhalative ores) and its substantially greater incidence than barium as a trace element in volcanic rocks, it is virtually absent from exhalative ore deposits. Barium (and calcium) are abundant components of many orebodies of the volcanic regime. It seemed, therefore, that the conspicuous absence of the closely related alkaline-earth element strontium might well provide a clue to the identity of some important fractionating process — and hence an indication of mechanisms of derivation and transport of the materials of ores overall.

A somewhat similar line of reasoning leads to a consideration of titanium in volcanic processes. Systematic chemical analysis of diamond drill cores intersecting a variety of exhalative ore types shows that, almost invariably, titanium is depressed — virtually to nil — in the orebodies relative to the sedimentary and metasedimentary rocks enclosing them. Examples are Broken Hill (Fig. 13.1) and Pegmont in the Lower Proterozoic, and Woodlawn and Captain's Flat in the Palaeozoic, of Eastern Australia. There are many others. In view of the high abundance of titanium in most volcanic magnetites ($TiO_2 \geq 5.0$ per cent) and hornblendes ($TiO_2 \geq 1.0$ per cent), and the likelihood that Ti^{3+} might be transported as a chloride complex in reduced brines, it is perhaps surprising that titanium is characteristically so low in exhalative ores.

If the titanium of the ore locale is no more than external detrital and non-exhalative (it always bears a very close and direct abundance relationship with aluminium: see Fig. 13.1), its paucity in the ores may be attributed quite simply to dilution of the detrital component by local copious precipitation of hydrothermal products. If on the other hand the low titanium of the orebodies results from a combination of such dilution together with the inhibition of precipitation of hydrothermal Ti^{4+} by the preservation of highly reducing conditions at the site of sulphide deposition, it might be expected that the sulphide lenses would commonly be surrounded by a high-Ti rim repres-

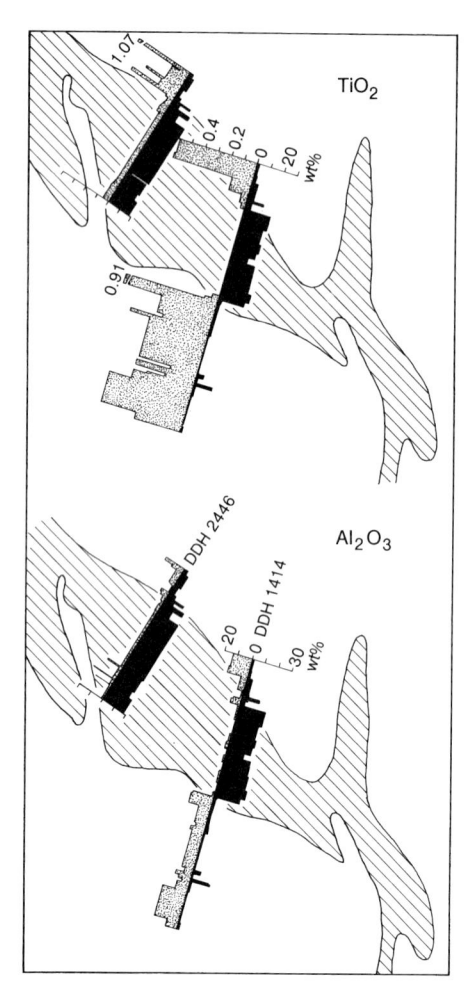

Fig. 13.1. Abundance relations between TiO_2 and Al_2O_3 (both shown as stippled bars) and total sulphide sulphur (filled bars) in two diamond drill cores through Broken Hill B lode (shaded area of irregular outline) on Section 84, New Broken Hill Consolidated Ltd. Bar widths indicate lengths of intersection of individual lithological units (and hence samplings), and weight percentages are indicated by the scales normal to the drill holes. After Stanton, Roberts, and Chant (1978).

enting the rapid, localized precipitation of hydrothermal titanium as soon as this encountered the surrounding well-oxygenated sea water. The writer is unaware of any description of such high-Ti selvages.

This seems to indicate that the ore solutions, whatever their nature and derivation, carried negligible titanium. In view of the abundance of this element in basalts and related volcanic rocks its conspicuously low incidence in the hydrothermal locale may thus provide some valuable clues concerning the derivation of exhalative ores.

Titanium was first detected and recognized as an element in ilmenite by the Revd W. Gregor at Creed in Cornwall in 1791 (Emsley 1989; given as 1789 by Partington 1943). Since the earliest days of rock analysis titanium has been recognized as an important minor component of igneous and other rocks. More recently it has been used as a petrological and petrotectonic discriminator element, and a variety of possible applications have been suggested.

Early estimates of the amounts of titanium in igneous rocks were made by Clarke and Washington (1924) of 0.64 per cent Ti (1.07 per cent TiO_2) and by von Hevesy *et al.* (1930) at 0.61 per cent Ti (1.02 per cent TiO_2). Goldschmidt (1954) considered, however, that these estimates were probably somewhat high, since the averaging had been unweighted and included a number of analyses of rare, notably Ti-rich, rock types. A modern estimate (Emsley 1989) of the crustal abundance of titanium is 6320 ppm Ti, or 1.05 per cent TiO_2.

TITANIUM IN LAVAS OF THE MARINE ENVIRONMENT

Data on the incidence of titanium in MORBs are summarized in Table 6.2. The mean value for 566 basalts (SiO_2 <52 per cent) is 1.60 per cent TiO_2 or 9592 ppm Ti, which compares with the recent estimate quoted above of 6320 ppm Ti for the Earth's crust (Emsley 1989). If the Icelandic basalts are included, the mean for 707 samples is 1.68 per cent TiO_2 (10 072 ppm Ti). Weighted means for the Atlantic, Pacific, and Indian Ocean MORBs are 1.34, 1.59, and 1.65 per cent TiO_2 respectively, and for Iceland, 2.01 per cent. Individual samplings vary greatly (Fig. 13.2). The non-Icelandic basalts range from 0.32 to 3.96 per cent TiO_2; those of Iceland from 0.43 to 5.22 per cent TiO_2. The frequency distribution of values for the former (Fig. 13.2(a)) is fairly regular, with a probable mode at 1.3–1.4 per cent TiO_2; 26 per cent of analyses fall in the 1.2–1.5 per cent range and 58 per cent in the 1.0–1.8 TiO_2 range. There are too few analyses of materials with SiO_2 >52 per cent to yield a clear pattern of frequency distribution, but values are generally lower than in the basalts. With the much smaller number of analyses available the pattern of values in the Icelandic basalts (Fig. 13.2(b)) lacks the clarity of that for the materials of the ridges proper, but a mode in the 1.5–2.1 per cent TiO_2 range is indicated. Again, the sparse information on the more felsic rocks (SiO_2 >52 per cent) points to lower overall TiO_2 than in the basaltic members of the assemblage.

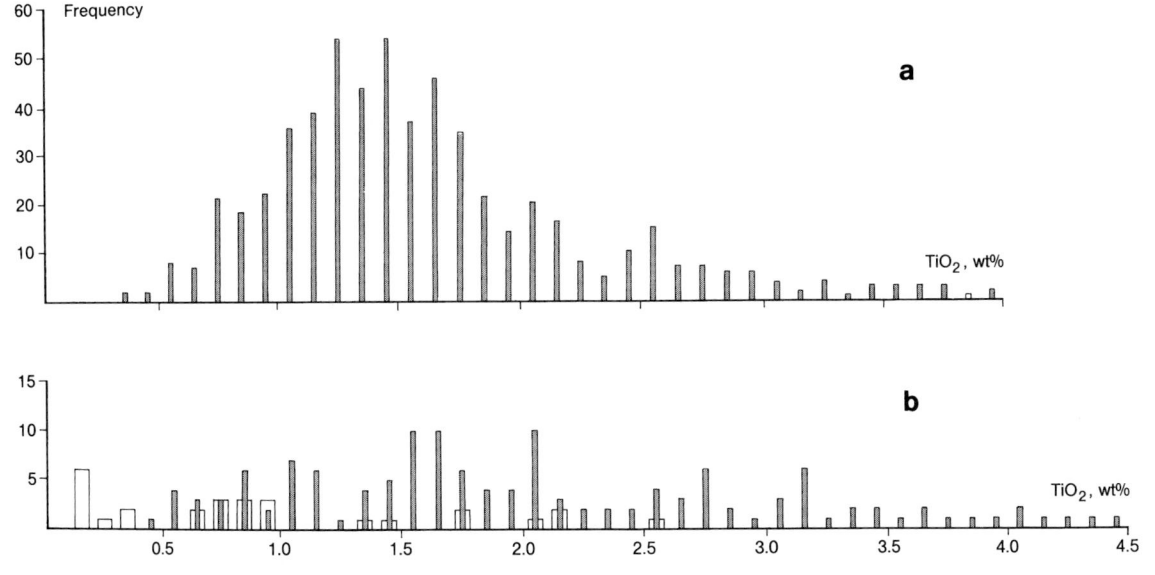

Fig. 13.2. Frequency distributions of TiO_2 abundances in (a) MORBs of all major oceans, and (b) Icelandic basalts (SiO_2 <52 per cent, filled bars; for reasons of space, a few higher values to 5.22 per cent are omitted) and Icelandic lavas more felsic than basalt (SiO_2 ≥52 per cent, open bars).

TiO$_2$ contents of arc lavas are given in Table 6.3 and Fig. 13.3. Values are distinctly lower than for MORBs. The overall mean for 494 arc basalts (SiO$_2$ <52 per cent) is 0.89 per cent TiO$_2$: approximately half that for MORBs. There is a clear mode in the 0.8–0.9 per cent interval, 62 per cent of all analyses falling in the 0.7–1.0 per cent and 97 per cent in the 0.2–1.5 per cent TiO$_2$ ranges.

Among the basaltic members there is substantial variation in TiO$_2$ between arcs. The mean value for Aleutians basalts is 1.22 per cent, whereas that for the Solomons is half this at 0.62 per cent. The values for the Lesser Antilles (mean ≈0.92 per cent for Grenada, Dominica, and St Kitts combined), Tonga–Kermadec (0.86 per cent), and Vanuatu (0.79 per cent) are essentially comparable with each other.

Arc lavas more felsic than basalts (SiO$_2$ >52 per cent), with an overall mean of 0.68 per cent, exhibit distinctly lower TiO$_2$ than the basalts (Fig. 13.3(b)). The frequency distribution for 739 analyses shows a clear mode at

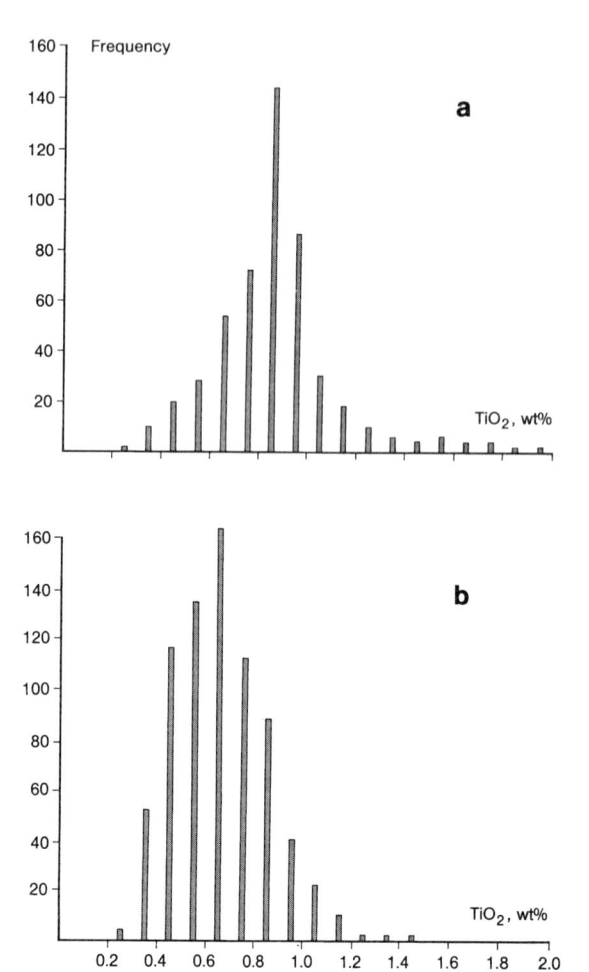

Fig. 13.3. Frequency distributions of TiO$_2$ abundances in (a) volcanic island-arc basalts (SiO$_2$ <52 per cent) and (b) arc lavas more felsic than basalt (SiO$_2$ ⩾52 per cent).

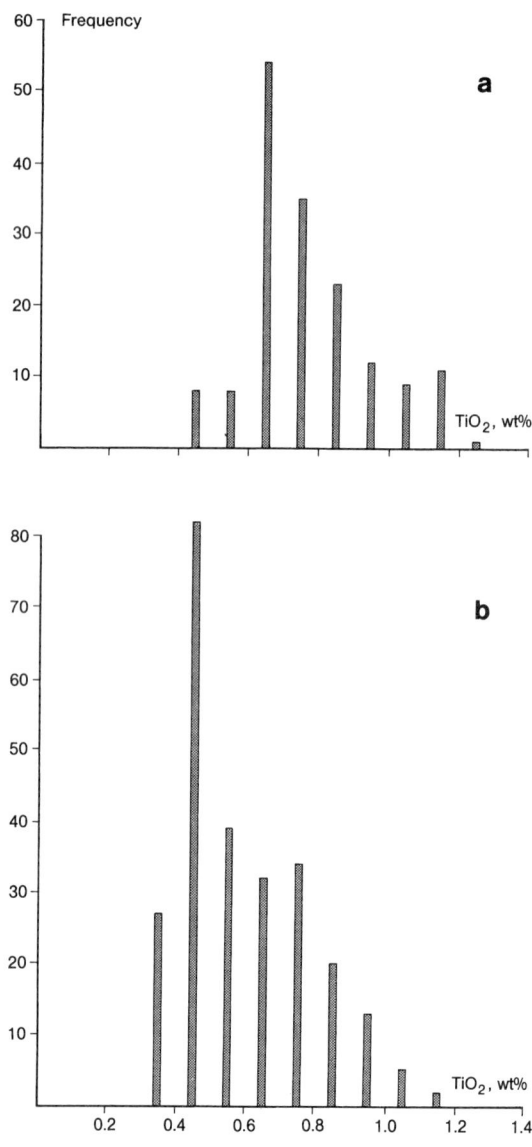

Fig. 13.4. Frequency distributions of TiO$_2$ abundances in lavas containing 50–66 per cent SiO$_2$ of (a) St Kitts, and (b) Dominica.

0.6–0.7 per cent, 55 per cent of analyses lying between 0.5 and 0.8 per cent and 83 per cent between 0.4 and 0.9 per cent TiO_2. There is minor intra-arc variation in some cases, a feature of which St Kitts (0.72 per cent) and Dominica (0.65 per cent) provide a good example (Fig. 13.4). Inter-arc variation among the more felsic lavas is notably less than among the basalts. In spite of the overall tendency for titanium to be lower in the more felsic group, this is not, however, pronouncedly so for some arcs. The Solomons lavas with SiO_2 <52 per cent have mean TiO_2 = 0.62 per cent and those with SiO_2 >52 per cent have mean TiO_2 = 0.65 per cent, a general similarity emphasized by the frequency distributions of Fig. 13.5. The corresponding TiO_2 contents of the Vanuatu lavas are 0.79 per cent and 0.76 per cent. For those arcs for which there are relatively comprehensive samplings of all of basalts, andesites (52 <SiO_2 <62 per cent), and dacites (SiO_2 >62 per cent) there is a steady decrease in TiO_2 with increasingly felsic nature. The West Indies basalts, andesites, and dacites contain 0.92, 0.74 and 0.46 per cent mean TiO_2 respectively, and for Tonga–Kermadec the corresponding figures are 0.86, 0.74 and 0.56 per cent TiO_2.

CHEMICAL PROPERTIES AND CRYSTAL CHEMISTRY OF TITANIUM

Although it is well known to form stable sulphides (TiS and TiS_2) and the nitride TiN and carbide TiC, titanium is very strongly lithophile in nature. It does not enter notably into the sulphides of igneous sulphide melts, and it is not known as a significant component of the sulphides of exhalative ores. It occurs in crustal rocks virtually entirely as Ti^{4+}, though Ti^{3+} may occur to a very limited extent in some specialized, highly reduced, environments.

The ionic radius of Ti^{4+} is 0.068 nm (Table 6.4), giving a radius ratio with respect to O^{2-} of 0.486. This is too large for tetrahedral coordination, and titanium therefore normally coordinates six oxygens in approximately octahedral arrangement. Si^{4+} has an ionic radius of 0.041 nm, and a radius ratio with respect to O^{2-} of 0.29, readily permitting 4-fold coordination and the development of the ubiquitous SiO_4 tetrahedron. For these reasons, in spite of similarity of valence and close similarity of chemical properties, titanium does not readily substitute for Si^{4+} in SiO^4 tetrahedra, but does so for other ions in octahedral sites, charge differences being compensated for by coupled substitution or vacancy formation. Most commonly Ti^{4+} substitutes for Fe^{3+} (ionic radius 0.067 nm) or Al^{3+} (0.057 nm) in octahedral positions. Hartmann (1969) proposed that where Ti^{4+} does substitute for Si^{4+} in the tetrahedral position it follows Al^{3+} and Fe^{3+} in order of preference; i.e. Ti^{4+} enters tetrahedral positions with respect to O^{2-} only if there is insufficient Al^{3+} and Fe^{3+} present to compensate for any deficiency in the tetrahedral site. Schröpfer (1968) found that Ti^{4+} could partly replace Si^{4+} in synthetic diopside by the coupled substitution of $Ti^{4+} + Al^{3+}$ for $Mg^{2+} + Si^{4+}$. He also deduced some substitution of Si^{4+} by Ti^{4+} even where there was sufficient Al^{3+} to occupy all vacant tetrahedral positions. Moore and White (1971) concluded that most of the titanium of andradite garnet occurs as a replacement of Fe^{3+} in the octahedral sites, and that an unknown proportion of the titanium occurs as Ti^{3+} (see, however, Goldschmidt 1954, p. 412). Tilmanns (1972) suggested that the silica deficiency of high-titanium garnets is probably compensated for by the substitution of Al^{3+} and Fe^{3+} into the tetrahedral sites.

The ground-state electron configuration, atomic and ionic radii, and electronegativity of titanium are given in Table 6.4.

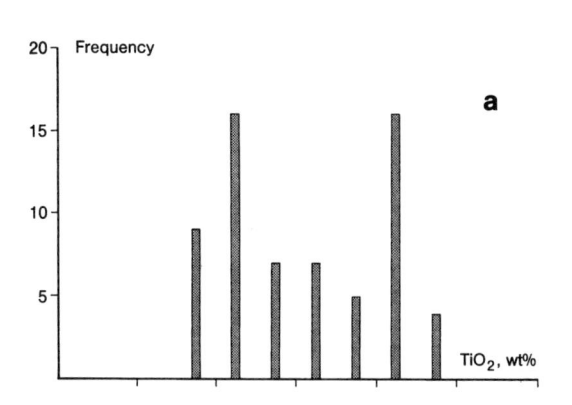

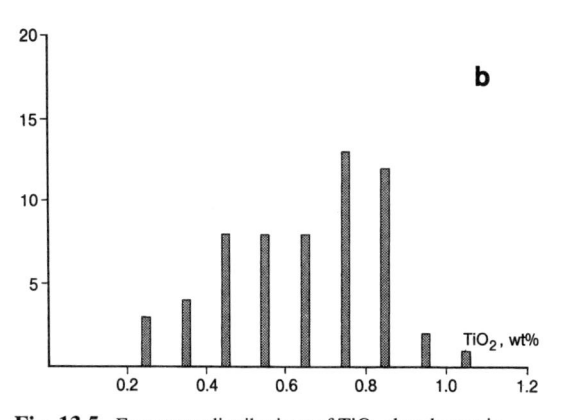

Fig. 13.5. Frequency distributions of TiO_2 abundances in (a) Solomons basalts (SiO_2 <52 per cent), and (b) Solomons lavas more felsic than basalt (SiO_2 ≥52 per cent).

THE INCIDENCE OF TITANIUM IN THE PRINCIPAL MINERAL SPECIES OF THE SOLOMONS YOUNGER VOLCANIC SUITE

General

As a 'major' element of standard igneous rock analyses, titanium has been determined a in a large number of igneous minerals, and there is now a substantial body of data on its incidence in the more common oxides and silicates of the volcanic regime.

Substitution of Ti in olivines is slight, presumably because of the large charge difference with respect to Fe^{2+} and Mg^{2+}, the principal ions of the octahedral site, and the paucity of Al^{3+} and Fe^{3+} to facilitate coupled substitution. As a result olivines generally contain very low titanium and TiO_2 falls in the range 0–0.4 per cent.

The 'non-titaniferous' clinopyroxenes contain slightly more titanium than the olivines, ≈0.10–1.0 per cent TiO_2, presumably largely because of the common presence of small quantities of aluminium and the opportunity this provides for the coupled substitution $Ti^{4+} + Al^{3+}$ for $Mg^{2+} + Si^{4+}$ referred to above. Correns (1978) points out that at low oxygen activities titanium may enter the structure as Ti^{3+}: a favoured substitution because the introduction of the lower charged iron requires less substitution of Al^{3+} for Si^{4+}. Apart from this, acceptance of Ti^{4+} by the clinopyroxene structure depends on the ease with which Si^{4+} can be replaced by Al^{3+}, and hence, at least in part, on the activity of silica in the melt. The degree of entry of Ti^{4+} appears in fact to depend both on silica activity (Kushiro 1960) and temperature (Verhoogen 1962). The incorporation of titanium into the orthopyroxenes is rather less, ≈0.1–0.5 per cent TiO_2, and the incidence of aluminium in the structure appears to be concomitantly lower.

The common hornblendes, with their abundant octahedral sites, complex compositions involving mono-, di-, and trivalent elements, and abundant aluminium, make them much better acceptors of Ti^{4+} than the pyroxenes. Most fall in the range 0–4.0 per cent TiO_2, with 1.0–2.0 per cent most common. The biotites are, for similar reasons, relatively ready acceptors of titanium and have been found to contain ≈1.0–6.0 per cent TiO_2, with 2.0–4.0 per cent the most common range.

The feldspars do not offer appropriate structural sites for the accommodation of Ti^{4+}, and their titanium content is accordingly very low: generally of the order of 0.01–0.10 per cent TiO_2. Even much of this is likely to be found — where analysis has been carried out on mineral separations rather than by microprobe — in minute glass inclusions in volcanic feldspars rather than in the feldspar structure itself (see below).

The principal acceptors of titanium are of course the oxides ilmenite and titanomagnetite, a subject much too extensive for treatment here. Perhaps 90 per cent of all volcanic magnetites contain TiO_2 in the range 5–15 per cent TiO_2, with 8–10 per cent the most common.

Solomons suite

Mean values for TiO_2 in the principal titanium-bearing silicate minerals and spinel solid solutions are given in Table 6.5. Spreads of values of TiO_2 in three of the principal titanium-bearing mineral groups, clinopyroxene, hornblende, and spinel, are shown in Fig. 13.6. Data for

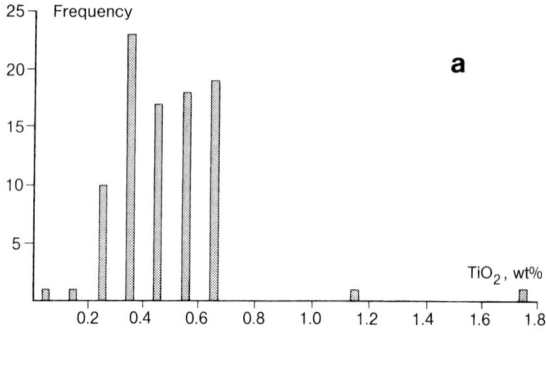

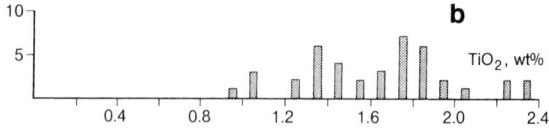

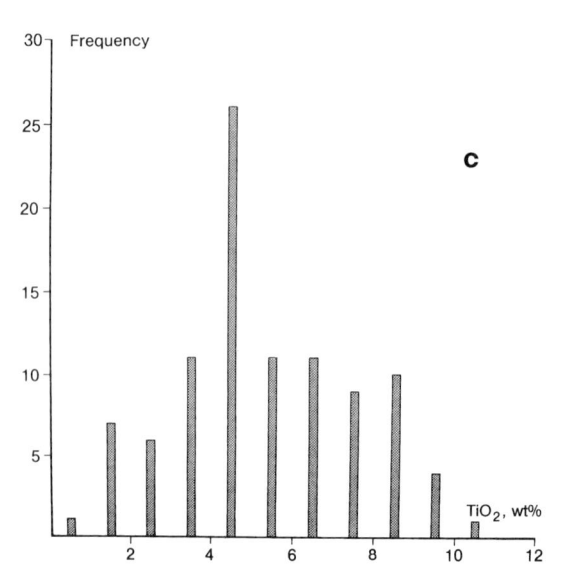

Fig. 13.6. Frequency distributions of TiO_2 abundances in (a) clinopyroxenes, (b) hornblende, and (c) spinels of the Solomons suite, as determined by X-ray fluorescence analysis of mineral separations.

Titanium 159

biotite are given in Fig. 13.9(c); for the reason already given, the values for this mineral represent microprobe analyses of individual crystals in the two Savo dacites; the other three diagrams (Fig. 13.6) represent analyses of bulk mineral separations.

Fe–Mg silicates. As expected from the earlier investigations referred to above, and the low entry of potential coupling ions (chiefly Al^{3+}) into its structure, the amount of titanium in the Solomons olivines is almost vanishingly small. The mean Ti content of 85 olivine crystals of a range of olivine–pyroxene- and big-feldspar lavas (as determined by the Cameca microprobe) is 52 ppm. The full spread of values obtained is 14–210 ppm, with 92 per cent of all analyses <100 ppm Ti (Fig. 13.7(a); note that the lowest class interval here is <30 ppm Ti). Fig. 13.7(b) depicts the Ti content of 63 clinopyroxenes coexisting with the 85 olivines. It shows the pronounced partitioning of titanium into the (Al-bearing) pyroxene (the calculated mean $K_D^{ol:cpx}$ based on this grouping is 0.019).

Mean TiO_2 in the clinopyroxenes overall is 0.48 per cent, with a range from mean = 0.38 per cent in those of the olivine–pyroxene-lavas to mean = 0.57 per cent in the big-feldspar group (Table 6.5; Figs. 13.6(a), 13.8). The slightly higher titanium of the big-feldspar lava pyroxenes is in keeping with the somewhat elevated titanium values in this rock group overall. Table 13.1 indicates that within

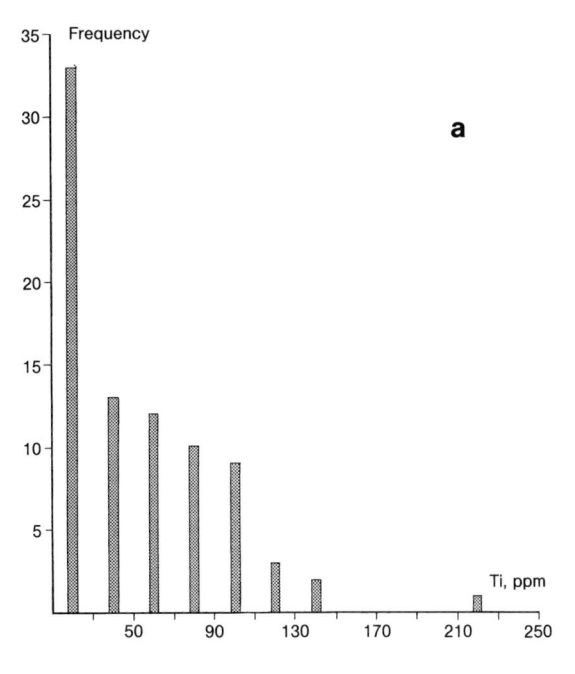

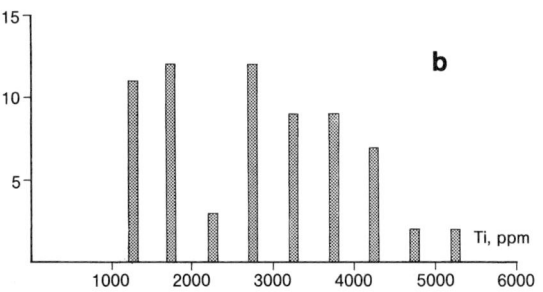

Fig. 13.7. Frequency distributions of titanium abundances in (a) 85 olivines, and (b) 63 clinopyroxenes coexisting in an olivine–clinopyroxene-basalt of the Solomons suite. The lowest class interval of (a) is <30 ppm. All analyses by Cameca electron-microprobe.

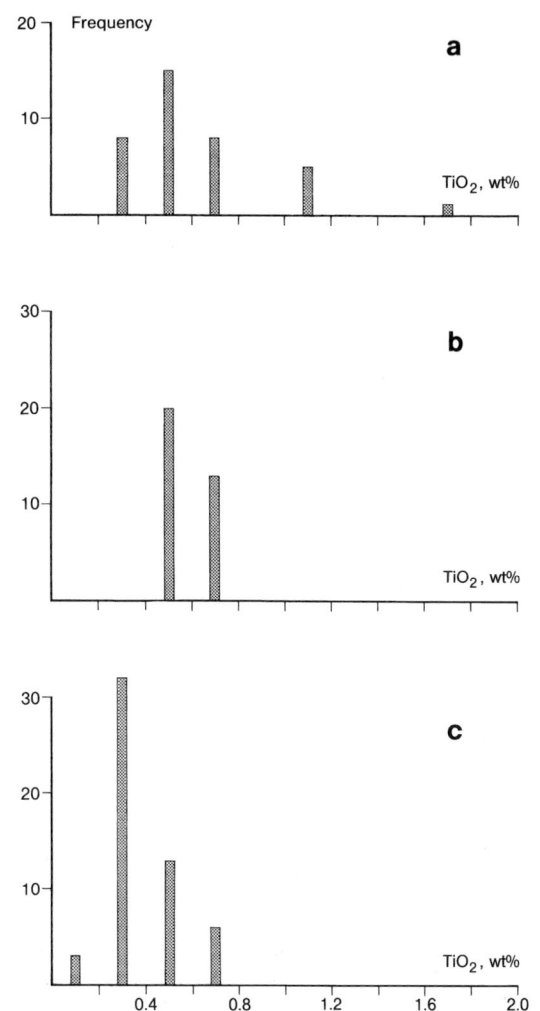

Fig. 13.8. Frequency distributions of TiO_2 abundances separations of clinopyroxenes of (a) the hornblende-andesite group, (b) big-feldspar lavas, and (c) the olivine–clinopyroxene-basalts of the Solomons suite.

the clinopyroxenes the incidence of titanium is certainly not related functionally to the incidence of aluminium: indeed, on this evidence it might appear that the relation between the two elements is an inverse one. However

Table 13.1 Mean TiO_2 and Al_2O_3 in clinopyroxenes of the three principal lava groups of the Solomons suite

Lava group	TiO_2 (wt %)	Al_2O_3 (wt %)
Hornblende-andesites	0.55	3.34
Big-feldspar lavas	0.57	2.90
Olivine–clinopyroxene-basalts	0.38	3.84

All determinations by X-ray fluorescence analysis of mineral separations.

apparently in keeping with its lower mean Al_2O_3, the mean titanium content of the orthopyroxenes is clearly lower than that of the clinopyroxenes — though, as indicated by the 85 coexisting pairs shown in Fig. 13.9(a), there is considerable overlap. TiO_2:Al_2O_3 relations in coexisting ortho- and clinopyroxenes taken together (Fig. 13.10) indicate a direct tie with a moderate degree of linearity. It may be suggested that the apparent lack of order in TiO_2:Al_2O_3 relationships as between the three groups of clinopyroxenes of Table 13.1 reflects the complicating effects of different concentrations of Ti in the melt and of the resulting different crystal: melt partition coefficients. For the eight two-pyroxene lavas of Fig. 13.10, mean $K_D^{cpx:opx} = 1.7790$.

The mean TiO_2 of 41 hornblendes separated from hornblende-andesites and dacites is 1.63 per cent, with a spread of values between 0.99 and 2.32 per cent (Fig. 13.6(b)). There is no indication that the variation is due to variation in Al_2O_3, SiO_2 or any other major component of the hornblende. The greater tendency for titanium to enter hornblende rather than pyroxene is illustrated by Fig. 13.9(b), which shows the incidence of TiO_2 in 81 hornblendes and 60 clinopyroxenes coexisting

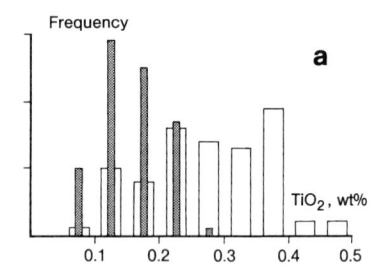

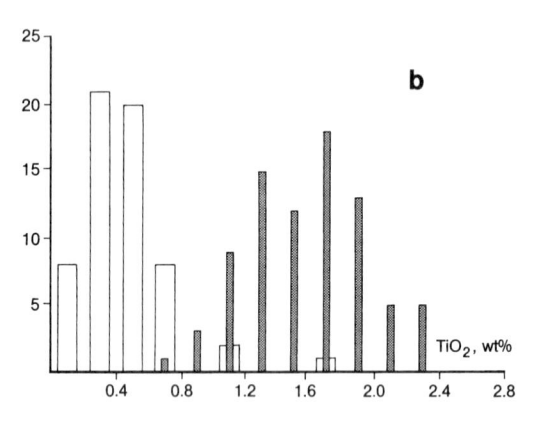

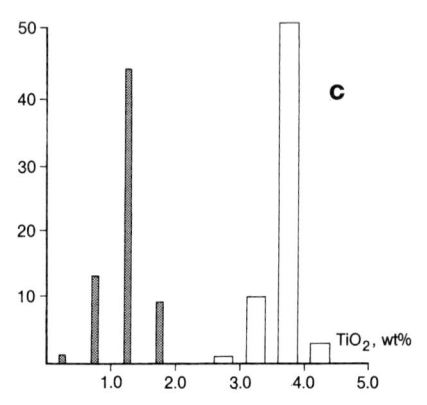

Fig. 13.9. Frequency distributions of TiO_2 abundances in (a) 85 coexisting orthopyroxene–clinopyroxene pairs (orthopyroxenes, filled bars, mean = 0.15; clinopyroxenes, open bars, mean = 0.27 per cent TiO_2; mean $K_D^{cpx:opx} = 1.78$); (b) coexisting clinopyroxene–hornblende (hornblende, 81 analyses, filled bars, mean = 1.58; clinopyroxene, 60 analyses, open bars, mean = 0.44 per cent TiO_2; $K_D^{Hb:cpx} = 3.59$); and (c) 66 coexisting hornblende–biotite pairs (hornblende, filled bars, mean = 1.19; biotite, open bars, mean = 3.67 per cent TiO_2; $K_D^{biot:Hb} = 3.08$). All analyses by Cameca electron-microprobe).

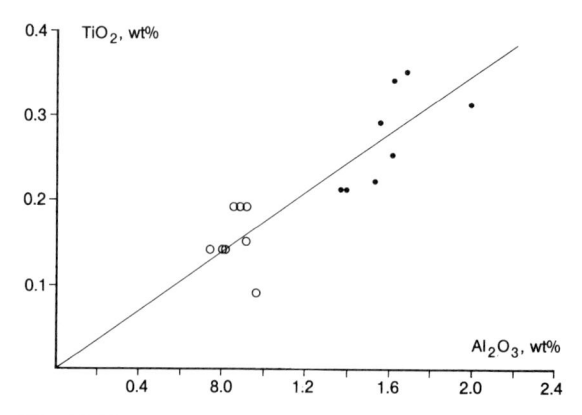

Fig. 13.10. Relations between TiO_2 and Al_2O_3 in coexisting orthopyroxene–clinopyroxene pairs in eight two-pyroxene-andesites of the Solomons suite. Orthopyroxene, open circles; clinopyroxene, filled circles; each point represents the mean of 12 electron-microprobe analyses.

in a hornblende-andesite. For this sampling, the mean TiO$_2$ of the hornblende is 1.58 per cent; that of the clinopyroxene is 0.44 per cent. The mean $K_D^{Hb:cpx}$ is 3.5909. Titanium tends to concentrate in the hornblende dark rims relative to the hornblende itself, although the association is somewhat equivocal overall (Fig. 13.11(b); see also Tables 5.4 and 5.5 and Fig. 5.12).

In the absence of bulk separations, data on biotite are limited to results obtained on the Cameca microprobe. Mean TiO$_2$ for 66 biotite crystals of a single dacite (Fig. 13.9(c) is 3.67 per cent and the mean for 66 hornblende crystals coexisting with them (a number of the pairs were in fact intergrown) is 1.19 per cent TiO$_2$. On this basis, mean $K_D^{biot:Hb}$ = 3.0840. Thus in the Solomons lava suite biotite concentrates titanium relative to hornblende to about the same degree as hornblende concentrates it relative to clinopyroxene.

Feldspars. With their lack of structural sites appropriate for the incorporation of Ti^{4+} the feldspars accommodate negligible titanium. Mean TiO$_2$ for 73 ultimate purity feldspar separations is 0.04 per cent; values for the three lava groups are given in Table 6.5. Plotting of frequency distributions of TiO$_2$ contents of the 73 ultimate purity, and 56 penultimate purity, feldspar separations indicates that significant titanium is located in the very fine glassy inclusions in the feldspars (Fig. 13.12). Correction for such impurity indicates a general level of TiO$_2$ at about 0.02 (120 ppm Ti) and this is confirmed by probe analyses. Fifty Cameca microprobe analyses of feldspars, covering the full spectrum of lava types, gave mean feldspar Ti ≈180 ppm (0.03 per cent TiO$_2$).

Spinels. Mean TiO$_2$ for 84 separations of magnetite is 6.0 per cent, and the corresponding spread of values from 0.74 to 10.14 per cent (Fig. 13.6(c)). Mean TiO$_2$ in the spinels of the olivine–clinopyroxene-basalts, big-feldspar lavas, and hornblende-andesites is 6.32, 6.46, and 5.09 per cent respectively (Table 6.5). Frequency distributions of values in individual crystals of the three lava groups, as

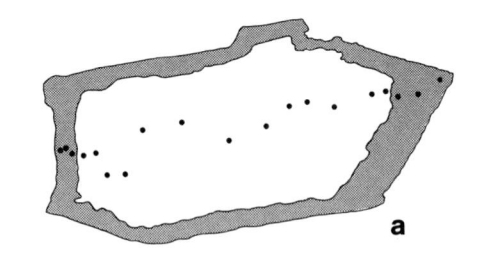

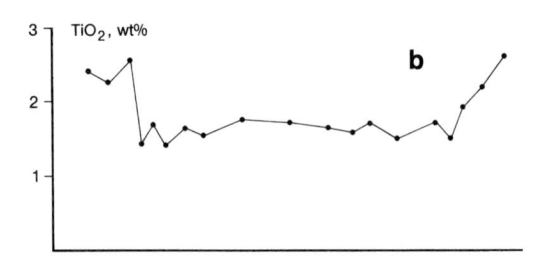

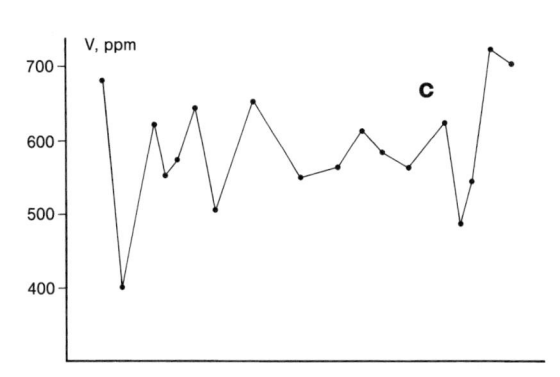

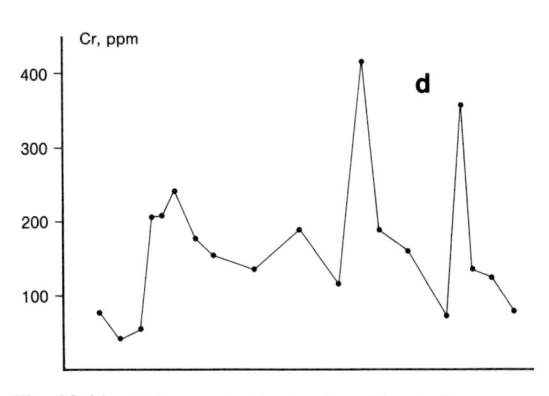

Fig. 13.11. (a) Camera lucida drawing of the single hornblende crystal and rim of Plate 3; (b) TiO$_2$, (c) V, and (d) Cr profiles along the length of the composite entity as indicated by the black dots in (a).

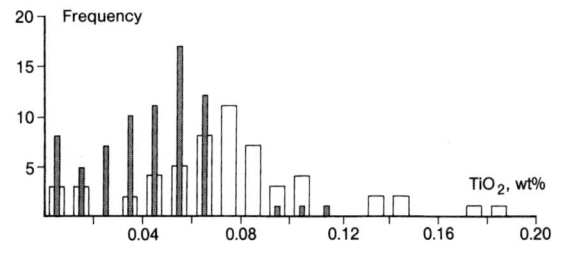

Fig. 13.12. Frequency distributions of TiO$_2$ abundances in ultimate purity (filled bars) and penultimate purity (open bars) separations of Solomons feldspars.

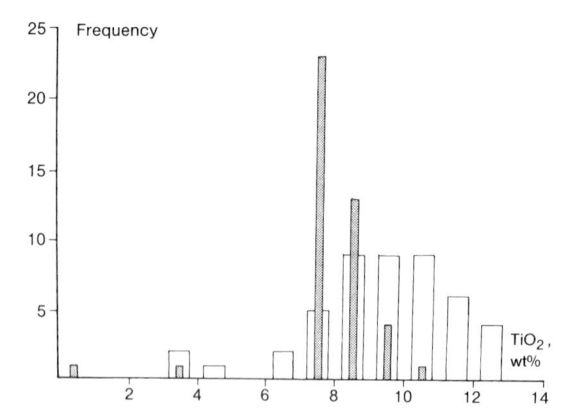

Fig. 13.14. Frequency distributions of TiO$_2$ abundances in spinels (magnetites) included within large clinopyroxene phenocrysts (filled bars; mean = 7.82) and within the groundmass (open bars; mean = 9.36 per cent TiO$_2$) of the Rendova ankaramitic lava referred to in Chapters 5 and 7 (Figs. 5.17 and 7.15).

determined by electron microprobe, are given in Fig. 13.13. Electron-probe analysis indicates a general increase in Ti from core to rim in individual magnetite crystals. The relatively high Ti content of the opaque rims of many hornblendes doubtless reflects the magnetite content of such rims. As noted in Chapter 5 (Fig. 5.16) magnetite TiO$_2$ rises rapidly from a mean of ≈3.00 per cent in lavas of 46–8 per cent SiO$_2$ to one of ≈7.5 per cent in lavas of 50–2 per cent SiO$_2$ (largely the big-feldspar basalts) and then decreases steadily to ≈4.0 per cent. Microprobe analysis of the magnetites included in the clinopyroxenes and the groundmass of the ankaramitic basalt already referred to in Chapters 5 and 7 indicates (Fig. 13.14) that the crystallization of the earlier (clinopyroxene-included) magnetites occurred at a stage when titanium was still being partitioned into the melt. As may be seen from Fig. 5.16, there is a pronounced antipathetic relationship between magnetite Cr$_2$O$_3$ and TiO$_2$ in the more mafic lavas (SiO$_2$ <52 per cent).

The spinels and biotite are clearly the principal receptors of titanium among the major minerals of the Solomons suite. Having in mind the limited data for biotite, a partition coefficient for coexisting magnetite:biotite of $K_D^{Ti} \approx 1.50$ is indicated.

CRYSTAL:MELT PARTITIONING

General

As a ubiquitous, and petrologically interesting, minor component of igneous rocks, titanium has been the subject of fairly extensive partitioning studies.

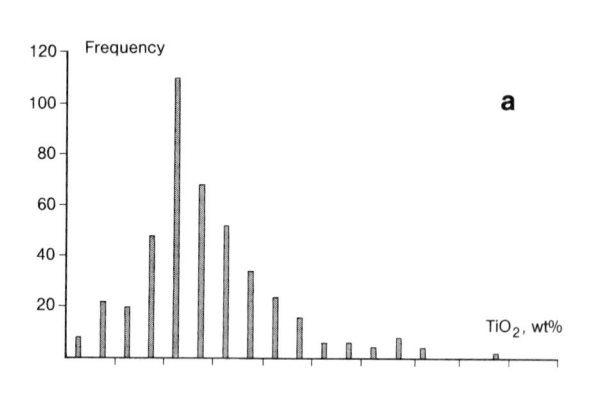

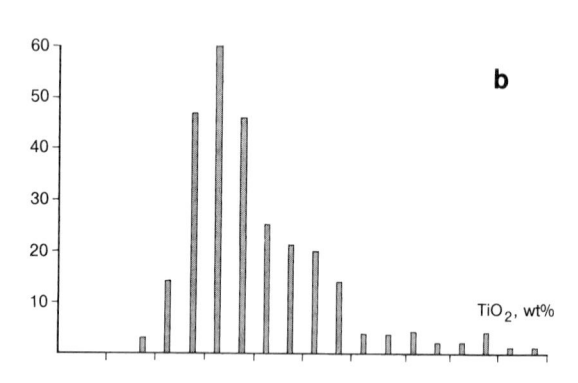

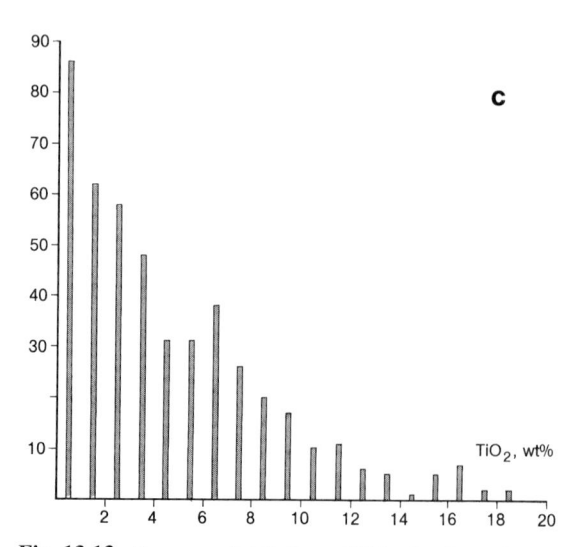

Fig. 13.13. Frequency distributions of TiO$_2$ abundances in individual spinel crystals within (a) lavas of the hornblende-andesite group, (b) big-feldspar lavas, and (c) olivine–clinopyroxene-basalts, as determined by electron-microprobe.

Bougault and Hekinian (1974) determined a $K_D^{ol:melt}$ value of 0.037 in a mid-Atlantic ridge basalt. Lindstrom (1976) obtained a value of 0.024 in a natural alkalic basalt at 1112–34°C at 1 atm; Duke (1976), values ranging from 0.03 to 0.11 (mean = 0.07) between 1125–50°C at 1 atmos in the system $SiO_2–Al_2O_3–FeO–MgO–CaO–Na_2O$; and Dunn (1987), values of 0.015–0.018 in a natural low-K tholeiitic MORB between 1150 and 1177°C at 1 bar.

Bougault and Hekinian (1974) calculated a $K_D^{cxp:melt}$ value of 0.2 for the above mid-Atlantic ridge basalt, and Grove and Bryan (1983) one of 0.27 in another mid-Atlantic ridge sampling. Pearce and Norry (1979) estimated approximate values of 0.3 in mafic rocks, 0.4 in intermediate, and 0.7 in felsic types. Duke (1976) obtained values ranging from 0.23 to 0.50 at 1125–50°C at 1 atmos. in the $SiO_2–Al_2O_3–FeO–MgO–CaO–Na_2O$ system, and Dunn (1987) values of 0.35–0.43 at temperatures of 1250–1290°C and 10–20 kb.

Information on the other important igneous silicates is relatively sparse. In experiments on mid-Atlantic ridge basalt (from the FAMOUS area) at 1310°C and 15 kb, Bender et al. (1978) found $K_D^{opx:melt} \approx 0.32$. Pearce and Norry (1979) estimated $K_D^{opx:melt} \approx 0.1$ in mafic rocks, ≈ 0.25 in intermediate, and ≈ 0.4 in felsic materials. Their estimates of $K_D^{Hb:melt}$ in the three categories were 1.5, 3.0, and 7.0 respectively, and of K_D^{Ti} for phlogopite:melt, 0.9, 1.5, and 2.5 respectively. Bougault and Hekinian (1974) estimated $K_D^{plag:melt} \approx 0.038$ in their mid-Atlantic ridge basalt, and Lindstrom (1976) obtained a value of ≈ 0.045 at 1135°C and 1 atm pressure using the natural alkalic basalt mentioned above as starting material. Pearce and Norry (1979) estimated values of 0.04, 0.05, and 0.05 for mafic, intermediate, and felsic rocks respectively.

With its propensity for accepting relatively large amounts of titanium into its structure, partition coefficients developed by magnetite are much larger than those characteristic of the silicates. Dudas et al. (1973) obtained a value of 16.2 for magnetite in dacitic tephra, Leeman et al. (1978a,b) values from 7.4 to 19.4 for magnetites in lavas ranging from ferrobasalts to ferrolatite, and Pearce and Norry values of 7.5, 9.0, and 12.5 in mafic, intermediate, and felsic rocks respectively.

Solomons suite

The order of acceptance of titanium into the structures of the principal Solomons minerals is clearly magnetite > biotite > hornblende > clinopyroxene > orthopyroxene > feldspar ≈ olivine.

From the data of Table 13.2, approximate mean partition coefficients for feldspar and olivine vis-à-vis groundmass are ≈0.07 and 0.05–0.08 respectively, and crystallization of these two minerals concentrates titanium strongly into the melt.

With its abundant octahedral sites and minor Al content the clinopyroxene accepts small but significant amounts of titanium and $K_D^{cpx:melt}$ increases nearly linearly with evolution of the containing lavas from mafic to intermediate (Fig. 13.15(a)). This may not be the equilibrium effect it appears to be at first sight, however: the clinopyroxenes of the hornblende-andesites, while exhibiting an apparent $K_D^{cpx:melt}$ of 1.25, in fact contain no more Ti

Table 13.2 Crystal:melt distribution coefficients (K_D) with respect to titanium

Mineral species	Lava group	n_1	n_2	m_1	m_2	$K_{D(1)}$	N	M_1	M_2	$K_{D(2)}$
Olivine	OPB	40	44	0.03	0.38	0.079	39	0.03	0.38	0.079
	BFL	7	23	0.02	0.44	0.045	6	0.02	0.49	0.041
Clinopyroxene	OPB	43	44	0.23	0.38	0.605	42	0.22	0.39	0.564
	BFL	26	23	0.34	0.44	0.773	17	0.36	0.44	0.818
	HbA	27	47	0.33	0.20	1.650	20	0.30	0.24	1.250
Hornblende	HbA	41	47	0.98	0.20	4.900	22	0.95	0.22	4.318
Feldspar	OPB	10	44	0.02	0.38	0.053	10	0.02	0.43	0.047
	BFL	23	23	0.03	0.44	0.068	22	0.03	0.44	0.068
	HbA	46	47	0.02	0.20	0.100	44	0.02	0.21	0.095
Spinel	OPB	20	44	3.05	0.38	8.026	20	2.73	0.37	7.378
	BFL	23	23	3.87	0.44	8.795	22	4.02	0.45	8.933
	HbA	37	47	3.79	0.20	18.950	36	3.34	0.20	16.700

OPB = olivine–clinopyroxene-basalt group; BFL = big-feldspar lavas; HbA = hornblende-andesite group; n_1 = total number of mineral separations analysed; n_2 = total number of groundmass separations analysed; m_1 = mean titanium content of mineral of each lava group; m_2 = mean titanium content of groundmass of each lava group; $K_{D(1)}$ = mean distribution coefficient determined as m_1/m_2; N = number of coexisting mineral:groundmass pairs analysed; M_1 and M_2 = mean titanium contents of coexisting mineral and groundmass respectively; $K_{D(2)}$ = distribution coefficient determined as M_1/M_2, for comparison with $K_{D(1)}$. All concentrations as per cent by mass of the element.

than do the pyroxenes of the other two lava groups, which exhibit much lower K_D values. The rise in $K_D^{\text{cpx:melt}}$ in the andesites reflects, rather, a sharp decrease in titanium content of the residual melt, presumably due to the rapid crystallization of copious quantities of magnetite at this state of lava evolution. If this principle does indeed apply, the relatively high $K_D^{\text{cpx:melt}}$ of the hornblende-andesite reflects in fact a *disequilibrium* phenomenon, already formed clinopyroxene crystals simply constituting passive inheritors of a feature of melt composition induced by the rapid formation of high-titanium magnetite.

Crystallization of hornblende, with mean $K_D^{\text{Hb:melt}} \approx 4.32$ leads to impoverishment of the melt in titanium, as does the formation of biotite, which, on the basis of very limited data, develops a $K_D^{\text{biot:melt}} \approx 8.4$. Hornblende:melt partition coefficients vary greatly from one lava to another, with values as high as 16.25 in one of the dacites. As with the pyroxenes, however this reflects low titanium in the groundmass rather than any increase in the hornblende and may be reasonably suspected of being a magnetite subtraction effect rather than, say, a temperature-dependent hornblende:melt equilibrium state.

As the principal acceptor of titanium, magnetite establishes very large partition coefficients, and its crystallization must exert a profound influence on the titanium content of the residual melt. On the basis of mean $K_D^{\text{mag:melt}}$ for each of the three lava groups, partitioning increases as the lava series evolves (Fig. 13.15b). However although $K_D^{\text{mag:melt}}$ is highest in the hornblende andesites, the absolute titanium content of both magnetite and melt are lower here than in the big feldspar lava group. The development of the latter, and the concomitant crystallization of relatively large amounts of high titanium magnetite, clearly marks a turning point in the abundance pattern of titanium in the Solomons lava series. This is referred to again in the following section.

THE INCIDENCE OF WHOLE-ROCK TIO₂ IN THE SOLOMON ISLANDS YOUNGER VOLCANIC SUITE

As pointed out earlier (Stanton and Bell, 1969), the general abundance of titanium in the Solomons basalts (SiO_2 <52 per cent) is notably low. The mean of 0.62 per cent determined in the present study compares with those of 1.60 per cent for MORBS (Table 6.2) and 0.79, 0.86, 0.92 and 1.22 per cent for the basaltic fraction of Vanuatu, Tonga-Kermadec, the Lesser Antilles, and the Aleutians respectively (Table 6.3). The essentially similar mean of 0.65 per cent for the Solomons lavas more felsic than basalts (SiO_2 >52 per cent) is also clearly lower than for this category on other arcs and the mid-ocean ridges.

The distribution of TiO_2 in the three principal Solomons lava groups and their groundmasses is given in Fig. 13.16. In the olivine-pyroxene basalts, representing the earlier stages of crystallization and the concentration of low-Ti olivine and clinopyroxene in the solid products, the titanium has tended strongly to partition into the melt — a phenomenon now reflected in the relatively higher incidence of TiO_2 in the groundmasses (mean = 0.66 per cent) than in the whole rocks (mean = 0.47 per cent) (Fig. 13.16c). The development of the big feldspar basalts and basaltic andesites has involved the onset of copious crystallization of the titanium-rich oxides resulting in an approximate equivalence of whole-rock (mean 0.79 per cent) and groundmass (mean 0.74 per cent) TiO_2 (Fig. 13.16(b)). This general trend has persisted with the formation of the hornblende-andesites and more felsic lavas, characterized as this was by the continued copious crystallization of the titaniferous

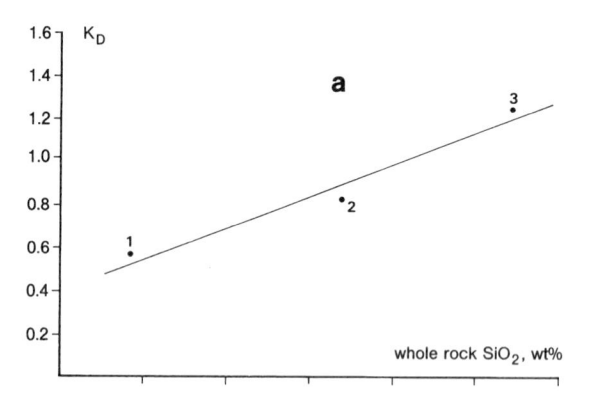

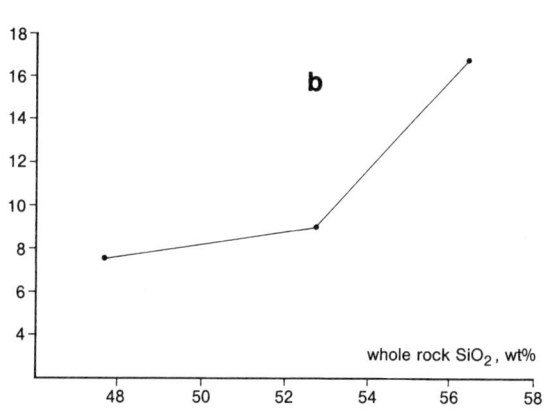

Fig. 13.15. Relations between (a) mean K_D^{Ti} clinopyroxene:melt and (b) mean K_D^{Ti} magnetite:melt and whole-rock SiO_2 in (1) olivine–clinopyroxene-basalts, (2) big-feldspar lavas, and (3) the hornblende-andesite group of the Solomons suite.

Titanium 165

relations in the olivine–pyroxene-basalt whole rocks, shows a highly linear olivine–pyroxene subtraction line, with TiO$_2$ increasing with decrease in MgO. The groundmasses, as complements of the olivine and clinopyroxene in the whole rock, extend this apparently evolutionary trend into the higher TiO$_2$ values (Fig. 13.16(e)). These groundmasses, together with the big-feldspar basalts and basaltic andesites and the less felsic hornblende-andesites, represent a maximum in the proportionate abundance of titanium in the lava series (Fig. 13.17(e), (d), and (b)). To the left of this somewhat diffuse inflexion the big-feldspar lava groundmasses, the more felsic hornblende-andesites, and the hornblende-andesite groundmasses yield a downward slope of proportionately decreasing MgO and TiO$_2$ and increasing (Na$_2$O + K$_2$O) (Figs. 13.17 (c), (b), and (a)). Clearly the suite as a whole is hypotholeiitic.

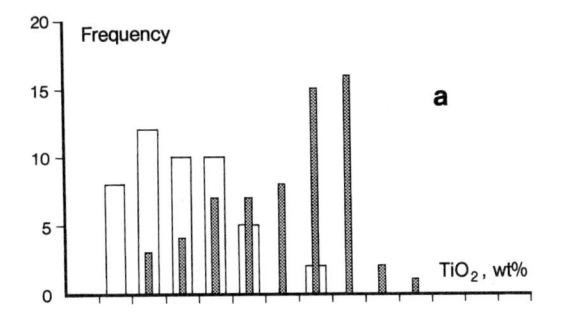

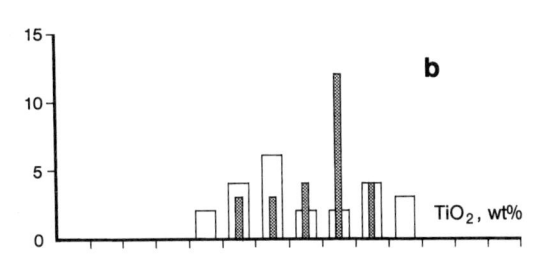

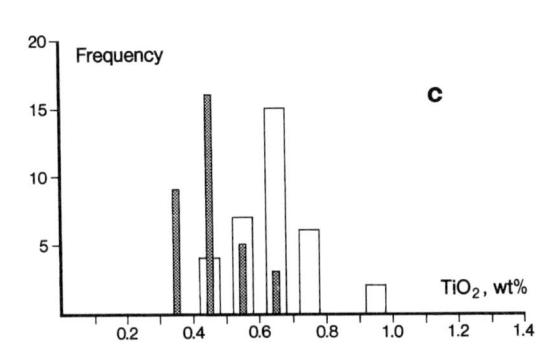

Fig. 13.16. Frequency distributions of TiO$_2$ abundances in whole rocks (filled bars) and groundmasses (open bars) of (a) the hornblende-andesite group, (b) the big-feldspar lavas, and (c) the olivine–clinopyroxene-basalts of the Solomons suite.

oxides and the onset of crystallization of the high-Ti silicates hornblende and minor biotite. At this stage mean whole-rock TiO$_2$ at 0.63 per cent is almost twice that of the groundmasses at 0.34 per cent, and the earlier solid:melt relations with respect to TiO$_2$ have been substantially inverted (compare Figs. 13.16(a) and (c)).

As with most of the elements already examined, this suggests the possibility of an ordered relationship between titanium abundance and lava type. This is confirmed by the series of TiMA diagrams of Fig. 13.17 (constructed on the basis of Ti = weight per cent TiO$_2$ × 5). Fig. 13.17(f), depicting TiO$_2$–MgO–(Na$_2$O + K$_2$O)

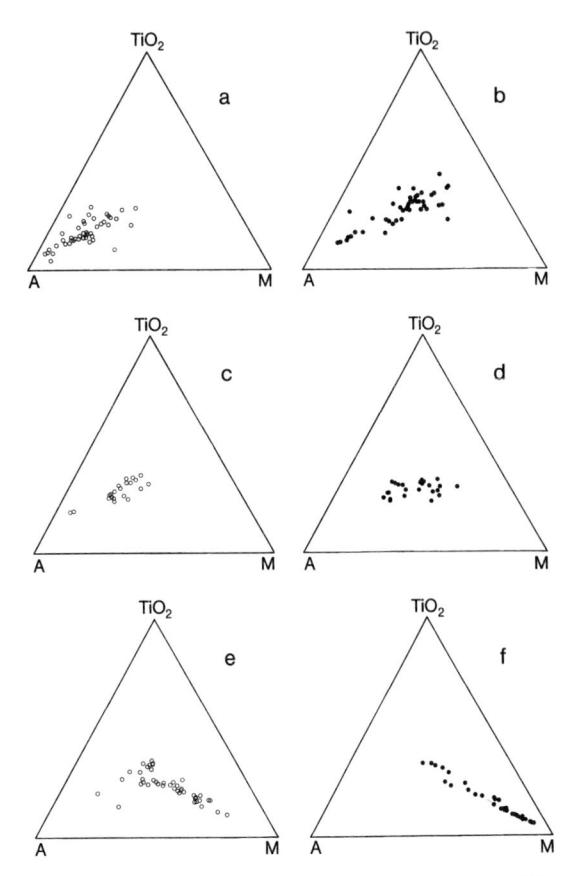

Fig. 13.17. TiO$_2$–MgO–(Na$_2$O + K$_2$O) ternary relations (TiO$_2$ calculated as weight per cent TiO$_2$ × 5) in the Solomons olivine–clinopyroxene-basalts, big-feldspar lavas, and hornblende-andesites [filled circles: (f), (d), and (b) respectively] and their groundmasses [open circles: (e), (c), and (a) respectively].

This is demonstrated with even greater clarity when mean TiO$_2$–MgO–(Na$_2$O + K$_2$O) values corresponding to two per cent intervals of whole-rock SiO$_2$ are plotted (Fig. 13.18); and the curve is smoothed further when the Bougainville analyses are included (Fig. 13.18(b)). The Grenada (Fig. 13.18(c)) and Vanuatu (Fig. 13.18(d)) assemblages exhibit similar, rather flat, hypotholeiitic patterns. Those of Dominica and St Kitts (Fig. 13.18(c)) and the Aleutians (Fig. 13.18(d)), on the other hand, show what are presumably hypotholeiitic distributions devoid of their more mafic component. As was pointed out in Chapter 5, truncated hypotholeiitic and negatively tholeiitic patterns may at first sight appear very similar: the series is negatively tholeiitic *only* if a high magnesium

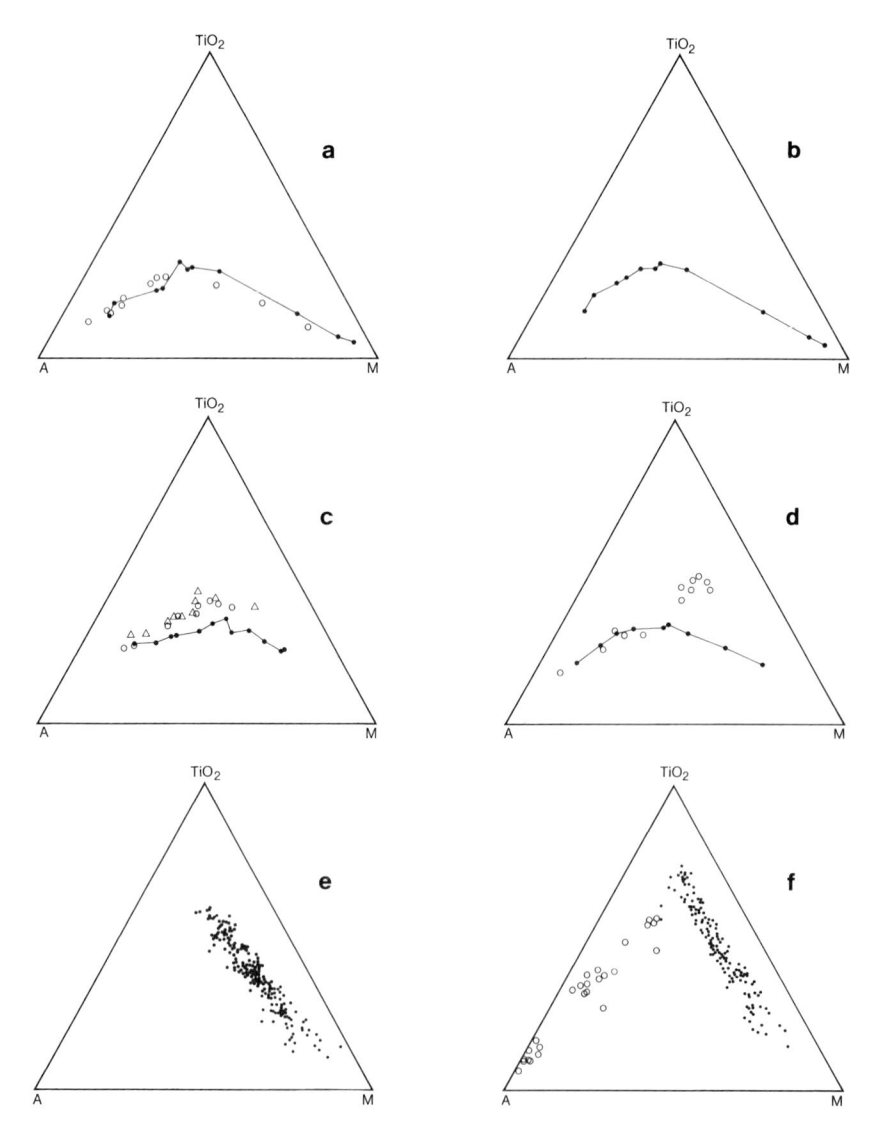

Fig. 13.18. (a) Mean TiO$_2$–MgO–(Na$_2$O + K$_2$O) ternary relations corresponding to 2 per cent whole-rock SiO$_2$ intervals in (a) whole rocks (filled circles) and corresponding groundmasses (open circles) of the Solomons suite excluding Bougainville; (b) whole rocks of the Solomons suite including Bougainville (241 analyses); (c) lavas of Grenada (filled circles), Dominica (open circles), and St Kitts (open triangles); (d) Vanuatu (filled circles) and the Aleutians (open circles). Ternary relations developed in MORBs as represented by Mid-Atlantic Ridge basalts are shown in (e), and in individual samplings of Icelandic basalts (SiO$_2$ <52 per cent; filled circles) and Icelandic lavas more felsic than basalts (SiO$_2$ ⩾52 per cent; open circles) in (f).

component is included as an intrinsic part of the linear array of points. If on the other hand the upper right of the field is constituted of felsic basalts and basaltic andesites, the array as a whole almost certainly represents a truncated hypotholeiitic distribution.

The MORBs array (Fig. 13.18(e)) is tholeiitic, the marked attenuation of the field presumably reflecting extensive fractionation with respect to titanium. The Icelandic basalts display a similar attenuation, in this case almost parallel to the TiO$_2$–MgO edge of the triangle, but develop an incipient hypotholeiiticity in the high TiO$_2$ region. This is then fully developed by the more felsic members of the suite, leading to a quite unambiguous hypotholeiitic distribution overall (Fig. 13.18(f)).

When mean whole-rock TiO$_2$ is plotted against 2 per cent intervals of whole-rock SiO$_2$, the Solomons lava series shows a sharp rise from 0.39 per cent TiO$_2$ in rocks with SiO$_2$ <48 per cent to a peak of 0.80–0.81 per cent in those for which SiO$_2$ is 50–4 per cent. It then decreases fairly constantly to a mean of 0.30 per cent TiO$_2$ in lavas with SiO$_2$ >64 per cent (Fig. 13.19(a)). Reflecting the state of affairs indicated by Fig. 13.16, mean TiO$_2$ for the groundmasses corresponding to the same 2 per cent whole-rock SiO$_2$ intervals is higher than for the whole rocks up to 50 per cent SiO$_2$, peaks (at 0.67 per cent) at 48–50 per cent SiO$_2$, and then decreases, always at a lower level than the whole rocks, to 0.25 per cent TiO$_2$ at SiO$_2$ >64 per cent. The Tonga–Kermadec lavas show a rather similar whole-rock pattern (Fig. 13.19(b), filled

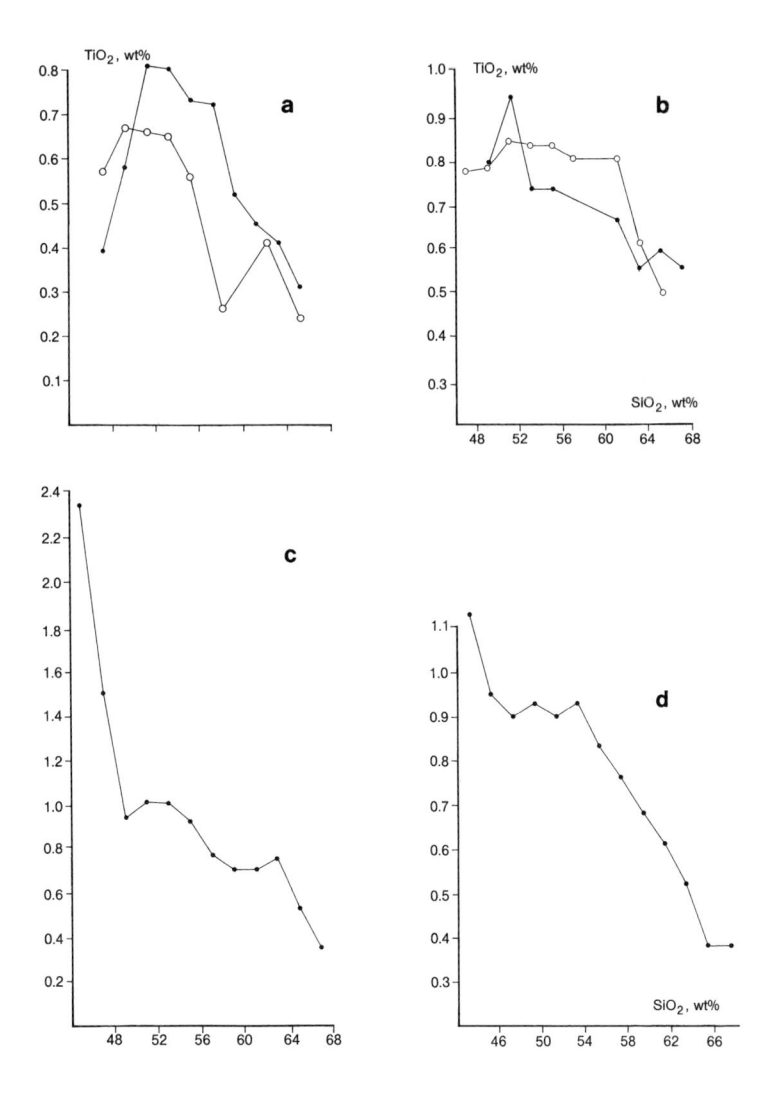

Fig. 13.19. Mean TiO$_2$:SiO$_2$ relations corresponding to 2 per cent whole-rock SiO$_2$ intervals in (a) whole rocks (filled circles), and corresponding groundmasses (open circles) of the Solomons suite; (b) lavas of Tonga–Kermadec (filled circles) and Vanuatu (open circles); and (c) the Aleutians; (d) the Lesser Antilles (Grenada, Dominica, and St Kitts combined).

circles). Those of Vanuatu (Fig. 13.19(b), open circles) are a little different in that they maintain a fairly constant level of TiO$_2$ (≈0.8 per cent) to ≈60–2 per cent SiO$_2$, at which point TiO$_2$ decreases sharply. The lava suites of the Aleutians (Fig. 13.19(c)) and the Lesser Antilles (Fig. 13.19(d)) show no sign of an initial increase in TiO$_2$, and decrease fairly consistently from low- to high-SiO$_2$ lava types.

CONCLUDING STATEMENT

On the basis of abundant data, mean TiO$_2$ in MORBs is ≈1.6 per cent, in basalts of the principal modern volcanic arcs it is 0.89 per cent, and in those of the Solomon Islands suite, 0.62 per cent. Mean TiO$_2$ of arc lavas more felsic than basalt (i.e. SiO$_2$ >52 per cent) is ≈0.68 per cent, and for the Solomons is 0.65 per cent. There is thus a general tendency for titanium to be impoverished in more felsic lavas, though this is not the case with those of the Solomons suite, which are notably low in this element overall. In some suites, including the Solomons, titanium exhibits a maximum in lavas of ≈50–4 per cent SiO$_2$; in others it displays a fairly simple, steady decrease from more mafic to more felsic members of the relevant lava series.

Olivine, pyroxene, and feldspars of the Solomons lavas contain very little titanium, and their crystallization induces strong partitioning into the melt. Copious precipitation and accumulation of olivine and clinopyroxene undoubtedly account for the relatively low TiO$_2$ of the more mafic members. Hornblende, biotite, and magnetite–particularly the latter two–incorporate substantial titanium in their structures and therefore tend to partition titanium out of the melt and into the crystalline phase. Titanomagnetite is a powerful abstractor of titanium and hornblende–magnetite crystallization must make a substantial contribution to the conspicuous decrease in titanium in the progressively more felsic fractions of the melts in question. The onset of copious crystallization of the plagioclase feldspars, with their minute titanium content, would tend to cancel this trend if the crystals were subtracted. Their low density, however, precludes this; they remain in the later fractions, and the effect of hornblende–magnetite subtraction on titanium contents continues to dominate. This is considered again in a more quantitative way in Chapter 21.

The Solomons suite is clearly hypotholeiitic with respect to titanium, and this appears to hold for island-arc assemblages generally. In contrast MORBs are tholeiitic, and appear to have developed by extensive closed-system fractionation with respect to titanium. The Icelandic basalts also appear to reflect extensive fractionation, but this has involved a slightly different subtractant, and the late-stage development of an open system. As a result the Icelandic lava assemblage as a whole displays a pronounced — it might almost be said an exaggerated — hypotholeiitic character.

14

Vanadium

Like titanium, vanadium is conspicuous by its absence or near-absence from exhalative ore deposits. Although I. and W. Noddack (1931) recorded vanadium in quantities up to several hundred parts per million in a variety of primary sulphides and although the vanadium sulphide patronite [VS_4] and the copper vanadium sulphide sulvanite [$3Cu_2S.V_2S_5$] are well known, vanadium is barely recorded in, or in association with, exhalative ores. A spectacular range of vanadiferous minerals is a conspicuous feature of the Hemlo gold deposit of Ontario, but this is the only such occurrence known to the author. Although vanadium occurs in relatively large amounts (2000–10 000 ppm) in some igneous magnetites and titanomagnetites (*vide* the Bushveld intrusion), and indeed although these are important commercial sources of the metal, it does not seem to be present in amounts of more than a few parts per million in exhalative sulphide or oxide ores. The fact that weathered cappings and gossans developed on such deposits do not contain conspicuous vanadate minerals appears to provide eloquent evidence of its paucity in the primary exhalative accumulations.

Although vanadium was discovered in 1801 by A. M. del Rio in Mexico, this was not well recorded and the element was recovered in 1831 by N. G. Selfstrom at Falun in Sweden as a minor component of the igneous titanomagnetite ore of Taberg. Reliable determination of amounts of vanadium in natural materials began at the end of the nineteenth century. W. F. Hillebrand (1900) detected measurable amounts of vanadium in all of a group of 57 igneous rocks: the lowest contents were found in the more felsic rocks, but in more mafic varieties vanadium as V_2O_3 was found to be as high as 0.05 per cent. One sample of biotite contained 0.127 per cent V_2O_3. Clarke and Washington (1924) estimated an average of 170 ppm V in igneous rocks and von Hevesy *et al.* (1930) estimated an average of 200 ppm. Goldschmidt (1938; 1954) suggested a rather lower figure (150 ppm V) for the upper lithosphere and Taylor (1964) and Mason (1966) proposed a mean value of 135 ppm for the crust. The modern estimate of mean crustal abundance (Emsley 1989) is 136 ppm.

Extensive investigation of vanadium in igneous rocks during the past forty years has shown substantial variation, even in materials of quite closely defined type. Prinz (1967) has drawn attention to the fact that, for example, a large sample of 'tholeiitic basalts' exhibited a range of 10–600 ppm V and that other analogous categories showed similarly substantial variation. However major groupings of particular categories of basaltic rocks display rather similar mean vanadium values, generally in the 200–50 ppm range. Ultrabasic rocks show a somewhat lower mean value, as do more felsic igneous rocks, which exhibit an overall progressive decline in vanadium with increase in whole-rock SiO_2. This is demonstrated in Fig. 14.12(e), constructed from the large number of analyses gathered by Gmelin (1968).

While orthomagmatic titaniferous iron ores contain relatively abundant vanadium (commonly in the range 1000–4000 ppm V) exhalative iron ores, as noted above, generally contain conspicuously little, and vanadium contents less than the crustal average seem typical for this class of deposit.

It is now recognized that there are three principle geological settings in which vanadium concentrates:

1. As a minor component (as V^{3+}) of igneous titano–magnetite deposits, chiefly in association with gabbroic and anorthitic rocks.
2. As vanadates (V^{5+}) in sedimentary rocks and weathering products in environments such as those of the sandstone-type uranium–vanadium deposits.
3. As reduced vanadium (V^{2+}), including sulphides, in reduced and bituminous sediments.

Thus although vanadium is an important trace element of basaltic igneous systems and although it is readily precipitated and adsorbed in reducing sedimentary environments, with only a very few possible exceptions it does not appear as a significant component of exhalative sediments in the volcanic environment. The great Hemlo gold deposit of the superior Province of Ontario — earlier identified as an exhalative deposit but now the subject of divided opinion — is one such exception, and exhibits a spectacular range of vanadium-bearing minerals. Harris

(1989) identified abundant vanadiferous green muscovite containing up to 8.5 weight per cent V_2O_5 and sphene, epidote–clinozoisite, and garnet containing ⩽14.2, 6.6, and 13.5 per cent V_2O_3 respectively. Other vanadium-bearing minerals found associated with the Helmo deposit were rutile (⩽5.6 weight per cent V_2O_3, barian tomichite (⩽44.8), haematite (⩽40.6), karelianite (⩽50.7), chromite (⩽2.1), hemloite (⩽20.4), and cafarsite (⩽4.7 weight per cent V_2O_3). Pan and Fleet (1992 and a series of earlier contributions on Hemlo mineralogy) analysed green muscovite containing ⩽17.6 per cent V_2O_3, phlogopite (⩽10.1), pumpellyite (⩽25.7), garnet (⩽18.5), epidote-group minerals (⩽9.1), antimonian vesuvianite (⩽4.3), and sphene (⩽18.5 weight per cent V_2O_3). Bernier (1990) found a vanadiferous zincian–chromian hercynite containing up to 5.72 per cent V_2O_3 in the basaltic alteration zone and overlying auriferous chert and associated quartz–grunerite–magnetite iron formation of the Atik Lake Fe, As, Zn, Cu, Au, Ag mineralization in Manitoba. Vanadiferous zinc–chromium-bearing spinels have also been found associated with the Outokumpu deposits (Treloar et al. 1981; Treloar 1987). However, occurrences such as these appear to be rare, and — given that the Helmo deposit may not be exhalative — vanadium is, overall, a very minor to insignificant component of exhalative ores and their immediate environments.

VANADIUM IN LAVAS OF THE MARINE ENVIRONMENT

Data on the incidence of vanadium in MORBs and Icelandic basalts is given in Table 6.2 and Fig. 14.1. The mean value for all ridge basalts (406 analyses) excluding those of Iceland is 263 ppm. As clearly indicated by Fig. 14.1(b), that for the Icelandic basalts (76 analyses) is distinctly higher, at 323 ppm V. Some 90 per cent of all MORB samples contain between 150 and 350 ppm V, and 66 per cent contain between 200 and 300 ppm V. Within individual ridges vanadium contents are fairly constant, but there appears to be significant variation between the principal ridge systems: mean vanadium for the mid-Atlantic, Atlantic–Indian, Indian, and Pacific ocean ridge systems are 244, 256, 302, and 289 ppm respectively. None of these attains the mean value exhibited by the Icelandic basalts.

There are insufficient analyses available for MORBs and Icelandic lavas with SiO_2 >52 per cent for any reliable quantitative comparison to be made at present, but the limited data shown by the open bars in Fig. 14.1(b) indicate that the more felsic lavas contain less vanadium than the basalts.

Data for the arc basalts (SiO_2 <52 per cent) and more felsic lavas are given in Table 6.3 and Fig 14.2. The frequency of vanadium values in MORBs and arc basalts is remarkably similar, as are their mean values at 263 ppm and 289 ppm V respectively. Of the 350 arc samples represented by Fig. 14.2(a), 71 per cent contain between 200 and 300 ppm V. The accuracy of Goldschmidt's (1954) estimate 'In gabbroid and basaltic rocks the amount of vanadium is often about 150– 300 ppm' is confirmed by the fact that of the 756 basalts represented in Figs 14.1(a) and 14.2(a) 582, or 77 per cent, are between these limits.

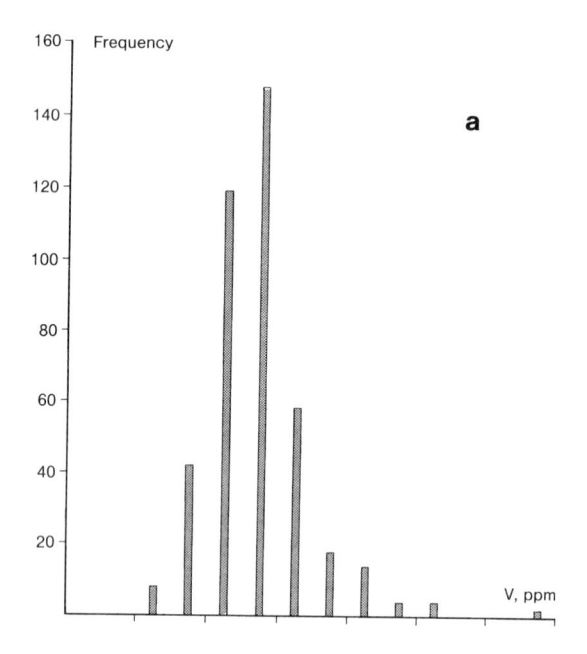

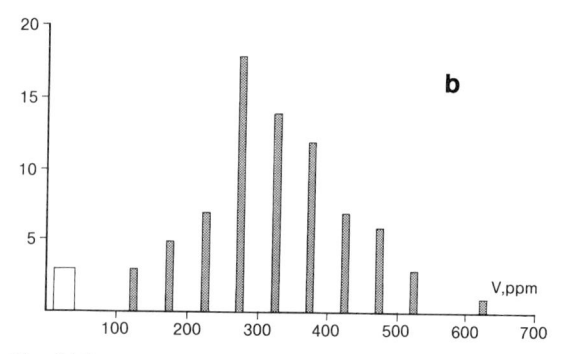

Fig. 14.1. Frequency distributions of vanadium abundances in (a) MORBs of all major oceans, and (b) Icelandic basalts (SiO_2 < 52 per cent: filled bars) and Icelandic lavas more felsic than basalt (SiO_2 ⩾ 52 per cent: open bars).

There is significant inter-arc variation: mean values for the basalts of the Lesser Antilles, Solomons, Tonga–Kermadec, Vanuatu, and the Aleutians are 244, 235, 297, 355 and 270 ppm V respectively.

As a group arc lavas more felsic than basalts (SiO_2 >52 per cent) contain less vanadium than the latter (Table 6.3; Fig. 14.2(b)). Of the 437 samples represented in Fig. 14.2(b), 314, or 72 per cent, have vanadium contents between 50 and 200 ppm, and the overall mean value is 151 ppm. Lavas in the SiO_2 >52 per cent category from the Lesser Antilles, Solomons, Tonga–Kermadec, Vanuatu, and the Aleutians exhibit means of 129, 187, 206, 155, and 154 ppm V respectively. For those arcs for which sampling has given reliable representation of all of basalts, andesites (52 <SiO_2 <62 per cent), and dacites (62 <SiO_2 <70 per cent) there is a general decline in vanadium with increase in SiO_2: for the Lesser Antilles means are 244, 141, and 83 ppm V respectively, and for Tonga–Kermadec they are 297, 294, and 71. It may be noted, incidentally, that there is little difference in vanadium content of the Tonga–Kermadec basalts and andesites (297 and 294 ppm) and when the three categories of basalts, andesites, and dacites are considered it is only in the most felsic group, the dacites, that the vanadium decreases markedly.

Abundance relations between vanadium and titanium in MORBs (as represented by those of the mid-Atlantic ridge) and in Icelandic basalts are shown in Fig. 14.3(a) and (b) respectively.

CHEMICAL PROPERTIES AND CRYSTAL CHEMISTRY OF VANADIUM

As indicated above, vanadium does not have a dominating lithophile or chalcophile nature and occurs as oxide, silicate, or sulphide in a manner not dissimilar from that of iron. In its trivalent state its ionic radius of 0.065 nm is almost identical with that of Fe^{3+} (0.067 nm), and it thus tends to occur with Fe^{3+}, particularly in magnetite, in igneous systems. The large amounts of vanadium occurring in the variety of muscovite (a dioctahedral mica) known as roscoelite substitute not, of course for Fe^{3+}, but for Al^{3+}, together with any minor Fe in the octahedral layer of the mica. Although, for reasons of charge and ionic size, vanadium substitutes very much less for Fe^{2+} (ionic radius 0.082 nm), it is none the less commonly a significant trace in Fe^{2+}-bearing minerals such as the amphiboles, micas, and pyroxenes. Among the common igneous minerals the spinels, notably the titanomagnetites, are its principal host. Ilmenite, with virtually all its Fe as Fe^{2+}, accommodates notably less.

The principal features of the vanadium atom and ions are given in Table 6.4.

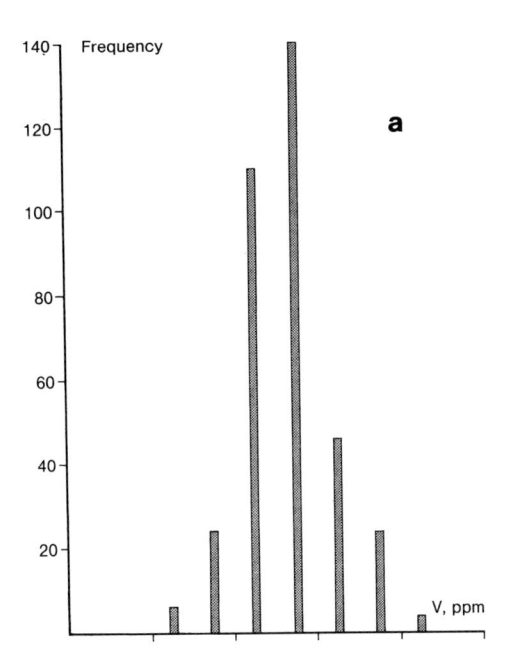

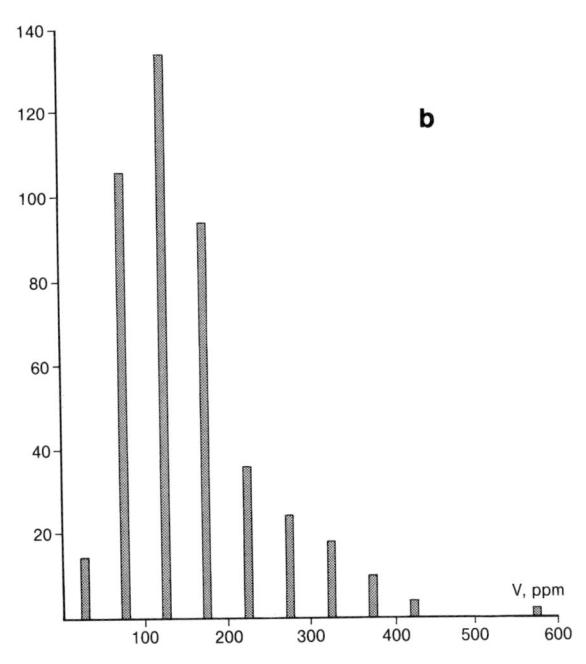

Fig. 14.2. Frequency distributions of vanadium abundances in (a) volcanic island-arc basalts (SiO_2 <52 per cent) and (b) arc lavas more felsic than basalt (SiO_2 ≥52 per cent).

172 *Ore Elements in Arc Lavas*

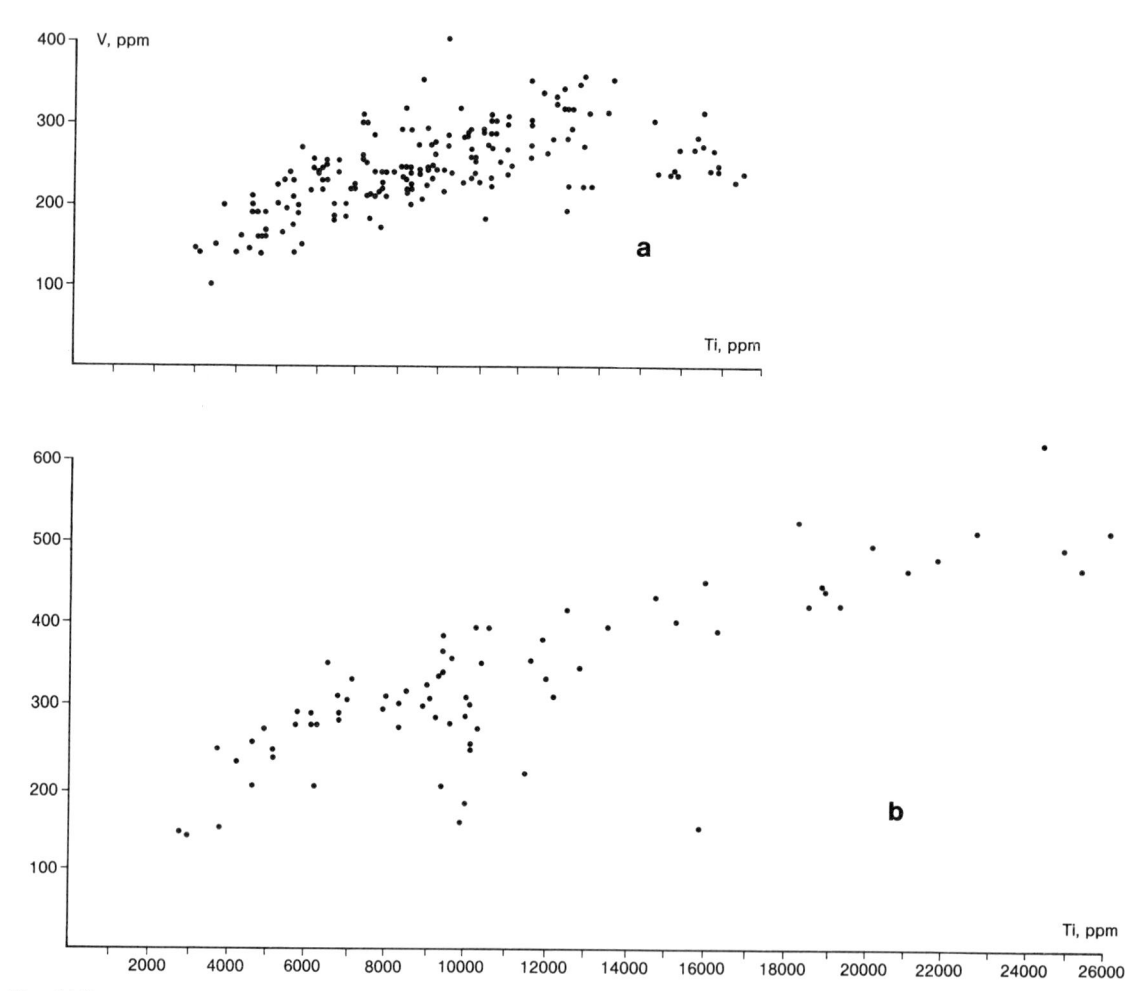

Fig. 14.3. Abundance relations between vanadium and titanium in (a) MORBs as represented by the basalts of the Mid-Atlantic Ridge and (b) Icelandic basalts.

THE INCIDENCE OF VANADIUM IN THE PRINCIPAL MINERAL SPECIES OF THE SOLOMONS YOUNGER VOLCANIC SUITE

General

As the principal ionic species in igneous system is V^{3+}, there is little substitution of vanadium in the olivines of igneous rocks. Minute amounts of V^{2+} in the melt would have appropriate size and charge for substitution in either Fe^{2+} or Mg^{2+} sites, and this may account for very small quantities of vanadium in olivines. While the V^{3+} ion is of appropriate size to fit into the distorted M–O coordination octahedra of olivine, there are no monovalent ions in the structure to provide an opportunity for coupled substitution $M_2^{2+} = V^{3+} + M^{1+}$. The relatively few available determinations of vanadium are therefore very low: generally <10 and many <5 ppm V.

The plagioclase feldspars, although always very low in vanadium, appear to contain genuinely slightly greater amounts than do the olivines. Wager and Mitchell (1951), in their pioneering work on the Skaergaard minerals, found <5–10 ppm in plagioclases, and 10 ppm appears now to be a common level of vanadium in members of the plagioclase series. The V^{3+} may be present as a substitution for the small quantities of Fe^{3+} that almost invariably appear in the igneous plagioclases.

With greater opportunity for coupling and minor Fe^{3+} substitution, vanadium appears at distinctly higher levels in the pyroxenes, amphiboles, and micas. Pyroxenes and amphiboles commonly contain 100–300 ppm and biotite 100–400 ppm V.

By far the major acceptors of vanadium among the igneous minerals are, as noted above, the iron oxides, the amounts accommodated being related to the proportion of Fe^{3+} in the structure. Ilmenite, with its relatively low incidence of Fe^{3+}, accepts least vanadium, values of 300–1000 ppm V being common. Magnetite, particularly the spectrum of titano-magnetites, almost characteristically contains 1000–4000 ppm V, and where magnetite and ilmenite coexist $K_D^{mag:ilm}$ values generally range from about 2 to 6. Abundant haematite is not so common in igneous systems, but where this mineral and magnetite coexist, $K_D^{haem:mag}$ values appear to lie generally in the range. 1.0–1.5.

Solomons suite

Mean values of vanadium in the principal silicates and in the magnetites of the Solomons suite are given in Table 6.5, and spreads of values in the relevant minerals are shown in Figs. 14.4–14.7.

Fe–Mg silicates. Vanadium contents of the Solomons olivines appear to be slightly higher that those previously determined in other localities. Figure 14.4(a) shows a clear mode in the 6–8 ppm interval, and the overall weighted mean is 8.6 ppm V. It might be suggested that the comparatively high vanadium here (one olivine from an olivine–pyroxene-basalt contained 43 ppm V) could result from minor spinel contamination. This, however, is extremely unlikely; as already noted (Chapter 5) the olivines of these lavas characteristically include little or no opaque material, and all separations were repeatedly examined under oil for such contamination. The slightly high vanadium of these olivines appears genuine.

In keeping with the results of other investigators the vanadium content of the pyroxenes, both clino- and orthopyroxenes, is distinctly higher. The clinopyroxenes exhibit a mode in the 150–250 ppm range (Fig. 14.4(b)) and an overall mean of 204 ppm V. There appears to be no trend in vanadium content of the clinopyroxenes related to, e.g., an increasingly felsic nature of the containing lavas. The clinopyroxenes of the big-feldspar lavas, which as a group possess the highest whole-rock vanadium content of the Solomons suite, are, however, with a mean of 244 ppm V (Table 6.5), distinctly higher in vanadium than the clinopyroxenes of the other two lava groups.

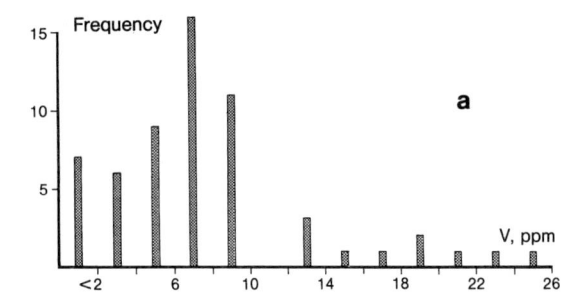

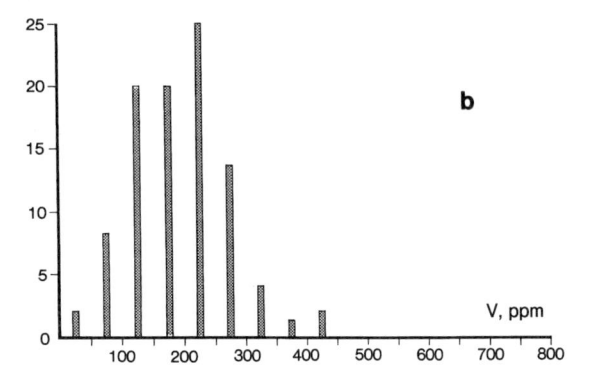

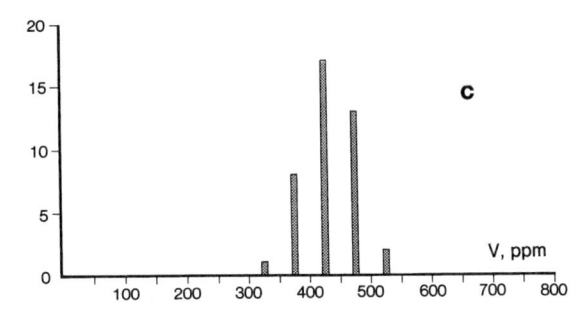

Fig. 14.4. Frequency distributions of vanadium abundances in (a) olivines (one value of 43 ppm V omitted for reasons of space), (b) clinopyroxenes, and (c) hornblendes of the Solomons suite.

Less abundant data (all by probe) for the orthopyroxenes indicate a mean of ≈92 ppm V. This is in general agreement with the result of Ewart and his co-workers on pyroxenes from the Tonga–Kermadec arc. Eighteen clinopyroxenes and fifteen orthopyroxenes from this province gave means of 308 ppm and 144 ppm V respectively. Eight coexisting pairs from Fonualei Island (Ewart *et al.* 1973) gave means of 197 ppm and 105 ppm V for clino- and orthopyroxenes respectively, and hence a mean partition coefficient of $K_D^{cpx:opx} \approx 1.88$. The mean vanadium contents of 67 coexisting clinopyroxenes and orthopyroxenes from a Solomon Islands two-pyroxene andesite (Fig. 14.5(a)) were 266 ppm and 91 ppm respectively, yielding a mean $K_D^{cpx:opx} = 2.92$.

Vanadium contents of the Solomons hornblendes are usually somewhat higher than those of the pyroxenes with a clear mode in the 400–50 ppm range (Fig. 14.4(c); see also Fig. 14.5(b)) and an overall mean of ≈ 434 ppm V (Table 6.5). The development of opaque rims leads to concentration of vanadium in the latter (compare Figs. 14.5(b) and (c)), the means for parent hornblende and daughter rims in this particular section being 568 ppm and 611 ppm V respectively. In this instance the transformation has thus generated a 'partition coefficient' $K_D^{Hb:rim} = 0.93$. This, however, is probably spurious: as suggested earlier (Chapter 5), the rim is almost certainly developed by reaction between hornblende crystal and residual melt of suddenly increased O_2 content, and the relevant equilibrium is therefore between hornblende:melt and then hornblende:rim:melt, not simply between hornblende and rim.

Limited investigation (65 grains of each mineral in a single lava sample) of partitioning between coexisting clinopyroxenes and hornblende indicates, as might be expected from Fig. 14.4 and Table 6.5, that vanadium tends to concentrate in the hornblende (Fig. 14.5(b)). Means for hornblende and clinopyroxenes in this lava were 568 ppm and 172 ppm V respectively, giving $K_D^{Hb:cpx} = 3.30$.

Electron-probe determinations of 28 grains of each of coexisting biotite and hornblende in a single dacite gave mean values of 445 ppm and 377 ppm respectively, and hence $K_D^{biot:Hb} = 1.18$. As noted in earlier chapters, however, this biotite has crystallized at a stage of lava evolution at which vanadium has become grossly impoverished in the melt: the dacite concerned contains 64.42 per cent SiO_2 and 62 ppm vanadium. Had the biotite crystallized in an environment containing, say, 200 ppm V, its capacity of accommodating vanadium relative to that of pyroxene and hornblende may have appeared as much more marked.

Feldspars. The vanadium content of the 'ultimate purity' separations of Solomons plagioclase feldspars is very low (Fig. 14.6; Table 6.5), with an overall mean of 3.53 ppm. The vanadium content of the feldspars of the big feldspar lavas is surprisingly less than that of the other two groups, in contrast to the state of affairs with the respective whole rocks, pyroxenes, and magnetites. Consideration of the paired 'penultimate purity' feldspar separations (Fig. 14.6(b)) indicates relative concentration of

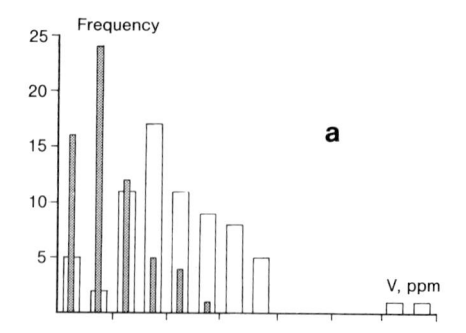

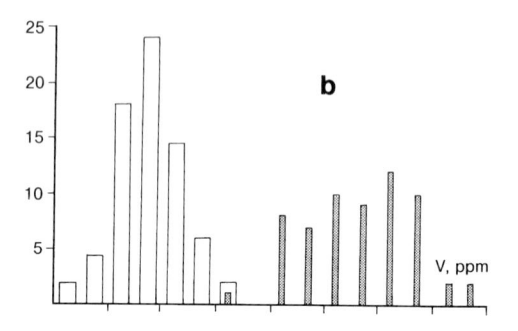

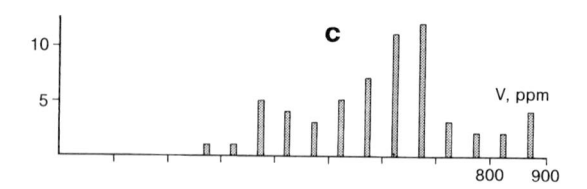

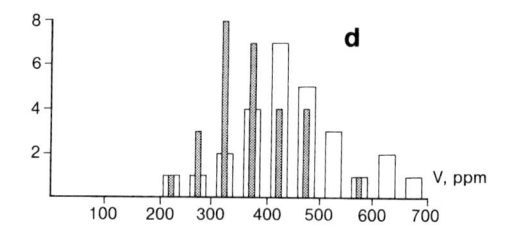

Fig. 14.5. Frequency distributions of vanadium abundances in (a) orthopyroxene (filled bars; mean = 91 ppm V), and clinopyroxene (open bars; mean = 266 ppm V) of 66 coexisting pairs in a Solomons two-pyroxene andesite; (b) hornblende (filled bars; 61 grains; mean = 568 ppm V) and clinopyroxene (open bars; 70 grains; mean 172 ppm V) coexisting in a hornblende–clinopyroxene-andesite of the Solomons suite; (c) opaque rims to the 61 hornblende grains of (b); and (d) hornblende (filled bars; 28 grains; mean = 377 ppm V) and biotite (open bars; 28 grains; mean 445 V) in a hornblende–biotite-dacite of Savo.

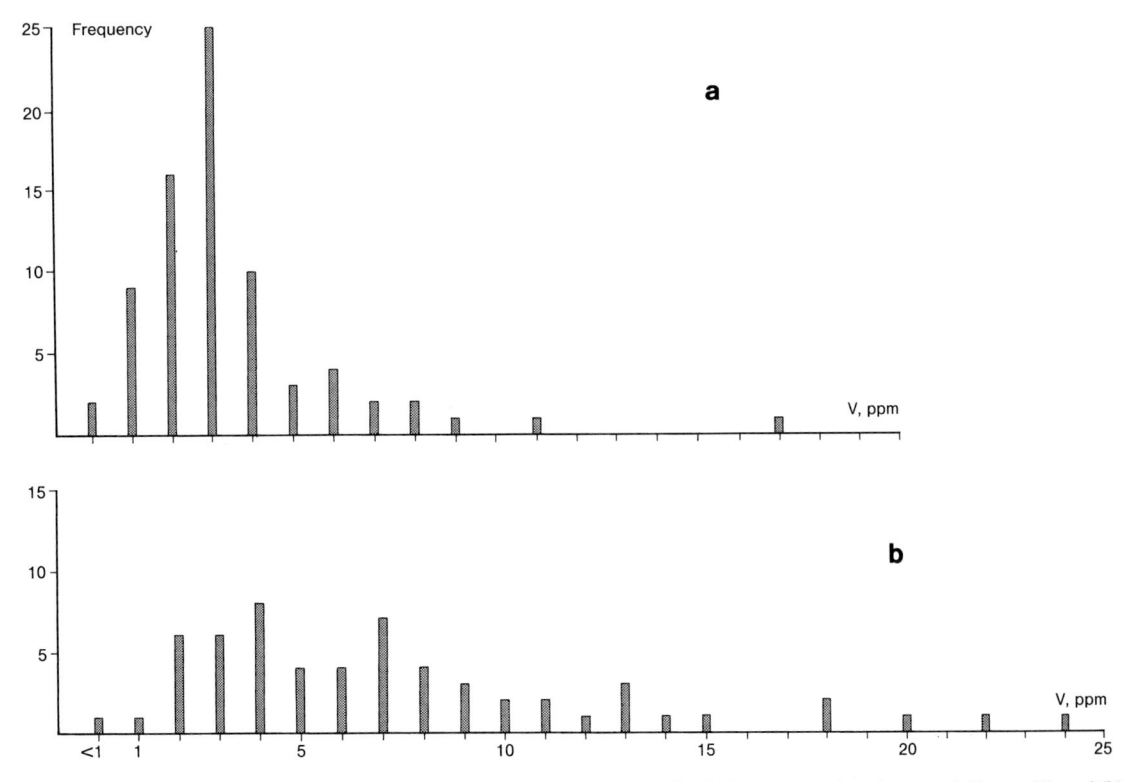

Fig. 14.6. Frequency distributions of vanadium abundances in (a) ultimate purity feldspar separations (mean = 3.53 ppm V), and (b) penultimate purity separations (mean 8.79 ppm V).

vanadium into included glass. The mean value for this fraction, 8.79 ppm V, is 2.5 times that for the 'ultimate purity' material indicating, with the very small amount of glass involved, that there must have been quite substantial concentration of vanadium in the fine glass inclusions. Recognizing that there must be at least a minute amount of glass in the ultimate purity fraction, a real mean value of ≈3 ppm V for the feldspars is probably a reasonable estimate.

Spinel. As elsewhere, spinel is by far the major receptor of vanadium in the Solomons lavas, with a mean value of ≈3650 ppm V for 62 separations. The titanomagnetites of the big-feldspar lavas are highest in vanadium (mean = 4271 ppm, Table 6.5) and those of the hornblende-andesites the lowest (mean = 3029 ppm). The mean of 3029 ppm is likely to be in part a reflection of the relatively low vanadium content of the melt at the stage at which the hornblende-andesites crystallized, though it will be recalled (see Chapter 7) that the zinc content of magnetites *increased* under analogous circumstances. Vanadium values range overall from 633 to 5184 ppm, with 70 per cent between 2500 and 4000 ppm. The frequency distribution (Fig. 14.7) is, however, ambiguous; it may be interpreted as an insufficiently sampled single population with a mode in the 3000–3500 ppm interval, or as two heavily overlapping populations with modes in the 2500–3000 and 3500–4000 ppm V intervals. As the other trace components of magnetite tend to unimodal frequency distributions, it is suspected that the apparent bimodality of Fig. 14.7 reflects limitations of sampling.

Several investigators (e.g. Goldschmidt 1954) have proposed that whereas Cr^{3+} abundance is directly related to Al^{3+} in naturally occurring spinels, V^{3+} is related to Fe^{3+}. Consideration of Fig. 5.19 with Table 6.5 indicates that at least this cannot be entirely so in the Solomons spinel; whereas increase in spinel vanadium does appear to be related to increase in spinel Fe_2O_3 in the olivine–pyroxene-basalt:big-feldspar lava grouping, further increase in Fe_2O_3 in the spinel of the hornblende-andesites is accompanied by a clear decrease in vanadium. Plotting of V_2O_3 vs Fe_2O_3 confirms this: in those lavas in which there is an abundance of vanadium in the

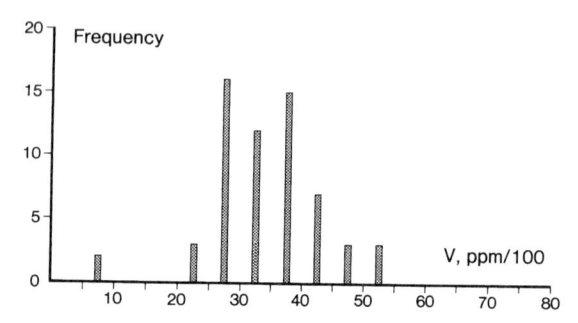

Fig. 14.7. Frequency distribution of vanadium abundances in Solomons spinel separations. The full range of values is 633–5184 ppm V.

melt, here is a crude but discernible correlation between V^{3+} and Fe^{3+} in the magnetites. With the development of the hornblende-andesites and of spinels of high Fe^{3+} content (Fig. 5.19) the vanadium content of the melt declines rapidly, however (Fig. 14.12). The amount of vanadium available thus, presumably, becomes much less than the increasingly ferric spinel is capable of accommodating: concomitantly less vanadium is incorporated, and the V^{3+}:Fe^+ correlation in the spinel is lost.

In a manner generally analogous with TiO_2, vanadium in magnetites increases from a low point in the most mafic lavas to a peak in those containing 50–2 per cent whole-rock SiO_2, and then decreases fairly uniformly with further increase in whole-rock SiO_2 (Fig. 14.8). This apparent decrease in the vanadium content of the magnetite as the lava series evolves from basalt to dacite is, as

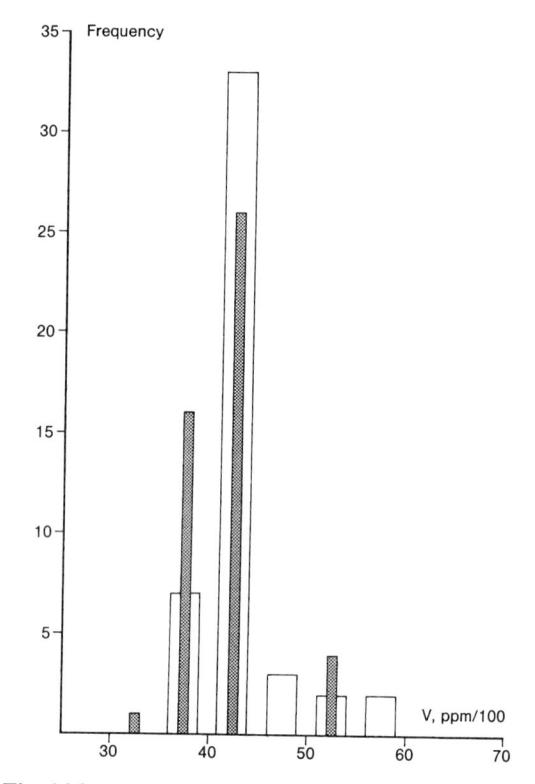

Fig. 14.9. Frequency distributions of vanadium abundances in magnetites included within large pyroxene phenocrysts (filled bars) and within the groundmass (open bars) of the Rendova ankaramitic basalt.

already noted in the case of Cu, Zn, and Ti, reflected in the differing vanadium contents of inclusion vis-à-vis groundmass magnetite of a single lava. In a manner entirely analogous with that of titanium (Fig. 5.16 and Chapter 13), the lava shown here as an example, with whole-rock SiO_2 = 49.01 per cent, falls to the left of the peak of Fig. 14.8, i.e. in that part of the lava sequence in which the vanadium content of the magnetite is rising. As would be expected from this, the groundmass magnetites contain slightly more vanadium (mean = 4284 ppm) than do the earlier-formed magnetites included in the pyroxenes (mean = 3990 ppm). Frequency distributions are shown in Fig. 14.9.

CRYSTAL:MELT PARTITIONING

There appears to be little information on crystal:melt partitioning of vanadium in volcanic systems in spite of the fact that it is widely recognized as an important trace element and is relatively abundant in all of the important

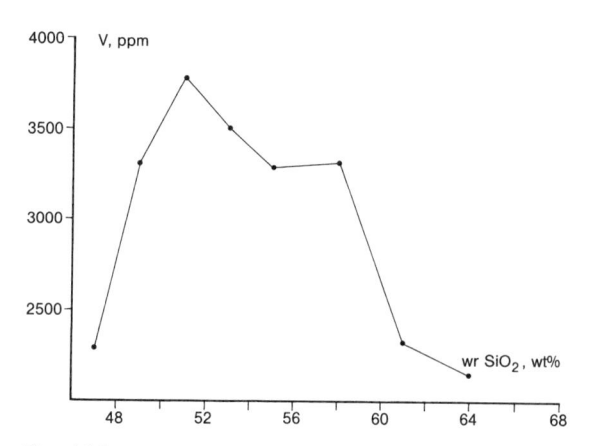

Fig. 14.8. Variation in vanadium content of spinel with respect to variation in whole-rock SiO_2 abundance in the containing lavas. Points represent mean vanadium in spinel corresponding to 2 per cent whole-rock SiO_2 intervals.

lava types. There is sufficient data to give a clear indication of the relative propensities of the principal volcanic minerals for accommodating vanadium, but virtually none concerning crystal:melt relations.

Data for the Solomons lavas are presented in Table 14.1.

$K_D^{xst:melt}$ for olivine and feldspar are very low, both for the most part in the range 0.02–0.03, and their crystallization has led to strong partitioning of vanadium into the melt. The mean value for the clinopyroxenes (Table 14.1, col. 11) is slightly greater than unity so that overall it affects the vanadium content of the melt only slightly. Hornblende, biotite, and magnetite, with mean $K_D^{xst:melt}$ of 7.02, 5.71, and 29.95 respectively, are substantial 'sinks' for vanadium and their crystallization impoverishes the melt accordingly.

The data may not, however, justify such simple application. It is noticeable (Table 14.1) that for all minerals $K_D^{xst:melt}$ values increase systematically with increase in the felsic nature of the containing rock. This might be taken to indicate that, for example, as the various Fe–Mg silicates, and perhaps the plagioclase, become more iron-rich, and as the spinel becomes more Fe^{3+}-rich, each mixed-crystal series becomes capable of accommodating greater amounts of vanadium and the relevant partition coefficients therefore increase. On the other hand, it may reflect declining vanadium content of the melt and crystal:melt disequilibrium. That is, a given crystal may form in a relatively early, relatively vanadium-rich environment and then be preserved, without re-equilibration, in later, vanadium-impoverished, residual melt (Fig. 14.12). The systematic, apparently remarkably orderly, increase in apparent K_D values would thus represent disequilibrium rather than changing equilibrium concentrations and the progressive increase in K_D with, say, decline in melt temperature.

THE INCIDENCE OF WHOLE-ROCK VANADIUM IN THE SOLOMON ISLANDS YOUNGER VOLCANIC SUITE

Mean vanadium for the full spectrum of Solomons basalts and more felsic lavas is 215 ppm, for the basalt fraction, 235 ppm, and for the more felsic group (SiO_2 >52 per cent), 187 ppm. Frequency distributions of vanadium values in the two categories (Fig. 14.10) indicate clear mode at 200–50 ppm for the basalts and 150–200 ppm for the more felsic rocks. The mean for the basalts compares with the overall mean for MORBs of 263 ppm, and for arc basalts, 289 ppm. The mean value for the more felsic Solomons lavas (SiO_2 >52 per cent), at 187 ppm V, is rather higher than the mean of 151 ppm V for this rock group in arcs overall. As a province the Solomons basalts might be seen as somewhat low in vanadium, though certainly not markedly so as in the case of titanium.

Frequency distributions of vanadium contents of the three principal Solomons lava groupings and their groundmasses (Fig. 14.11) yield a pattern generally very similar to that of titanium. The mafic basalts — olivine–

Table 14.1 Crystal:melt distribution coefficients (K_D) with respect to vanadium

Mineral species	Lava group	n_1	n_2	m_1	m_2	$K_{D(1)}$	N	M_1	M_2	$K_{D(2)}$
Olivine	OPB	40	44	8	256	0.031	39	7	259	0.027
	BFL	7	23	10	188	0.053	6	9	229	0.039
Clinopyroxene	OPB	43	44	152	256	0.594	42	143	265	0.540
	BFL	26	23	244	188	1.298	17	263	197	1.335
	HbA	27	47	215	64	3.359	20	211	75	2.813
Hornblende	HbA	25	47	434	64	6.781	22	428	61	7.016
Feldspar	OPB	10	44	3	256	0.012	10	3	204	0.015
	BFL	23	23	3	188	0.016	22	3	187	0.016
	HbA	46	47	4	64	0.063	44	3	64	0.047
Spinel	OPB	9	44	3707	256	14.480	9	3707	151	24.55
	BFL	15	23	4271	188	22.718	15	4271	151	28.28
	HbA	39	47	3029	64	47.328	36	2993	60	49.88

OPB = olivine–clinopyroxene-basalt group; BFL = big-feldspar lavas; HbA = hornblende-andesite group; n_1 = total number of mineral separations analysed; n_2 = total number of groundmass separations analysed; m_1 = mean vanadium content of mineral of each lava group; m_2 = mean vanadium content of groundmass of each lava group; $K_{D(1)}$ = mean distribution coefficient determined as m_1/m_2; N = number of coexisting mineral:groundmass pairs analysed; M_1 and M_2 = mean vanadium contents of coexisting mineral and groundmass respectively; $K_{D(2)}$ = distribution coefficient determined as M_1/M_2, for comparison with $K_{D(1)}$. All concentrations as parts per million by mass.

178 *Ore Elements in Arc Lavas*

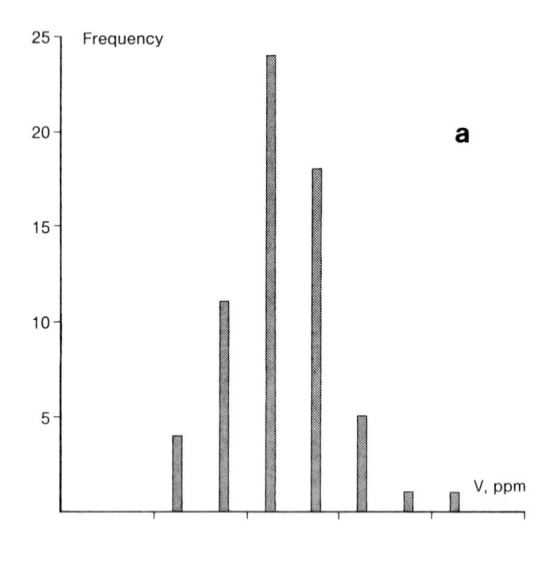

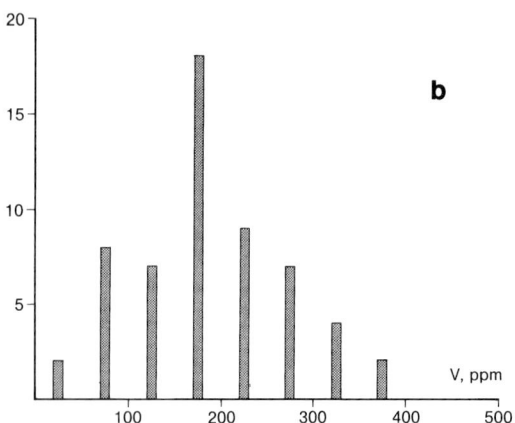

Fig. 14.10. Frequency distributions of vanadium abundances in (a) Solomons basalts (SiO$_2$ < 52 per cent) and (b) Solomons lavas more felsic than basalts (SiO$_2$ ⩾ 52 per cent).

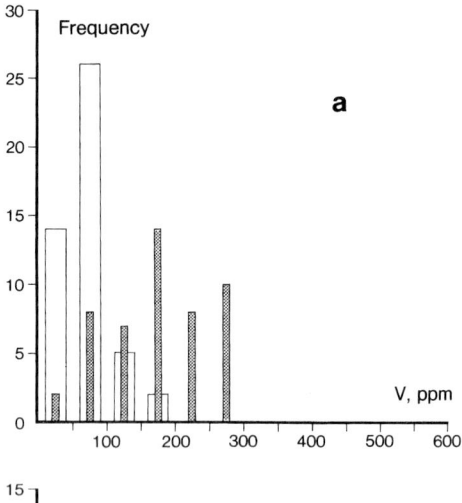

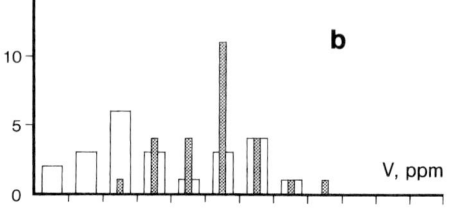

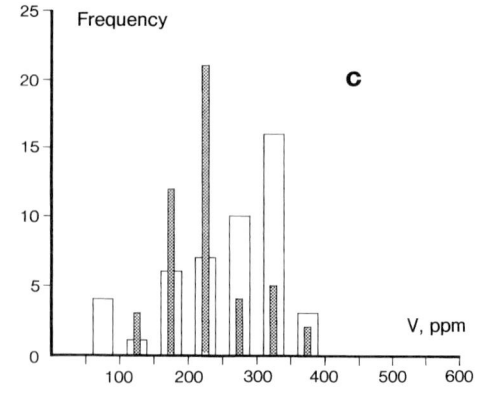

Fig. 14.11. Frequency distributions of vanadium abundances in whole rocks (filled bars) and groundmasses (open bars) of (a) lavas of the hornblende-andesite group, (b) big-feldspar lavas, and (c) the olivine–clinopyroxene-basalts of the Solomons suite.

clinopyroxene lavas — show a tendency to concentration of vanadium in the groundmass (Fig. 14.11(c); whole-rock mean = 221 ppm; groundmass mean = 258 ppm V). Development of the somewhat less mafic big-feldspar basalts and basaltic andesites induces a reversal of this relationship (Fig. 14.11(b): whole-rock mean = 265 ppm; groundmass mean = 188 ppm V) and this is accentuated in the formation of the hornblende-andesites (Fig. 14.11(a): whole-rock mean = 172 ppm; groundmass mean = 64 ppm V). Fig 14.11 and the whole-rock/groundmass means thus indicate the whole-rock vanadium achieves a maximum in the big-feldspar lavas whereas groundmass vanadium decreases systematically from olivine–clinopyroxene- basalts to andesites, and that the principal concentration of vanadium shifts from groundmass to whole-rock going from the mafic basalts to andesites and dacites.

This muted indication of a link between the incidence of vanadium and the evolution of the lava series is delineated much more sharply in vanadium:whole rock SiO$_2$

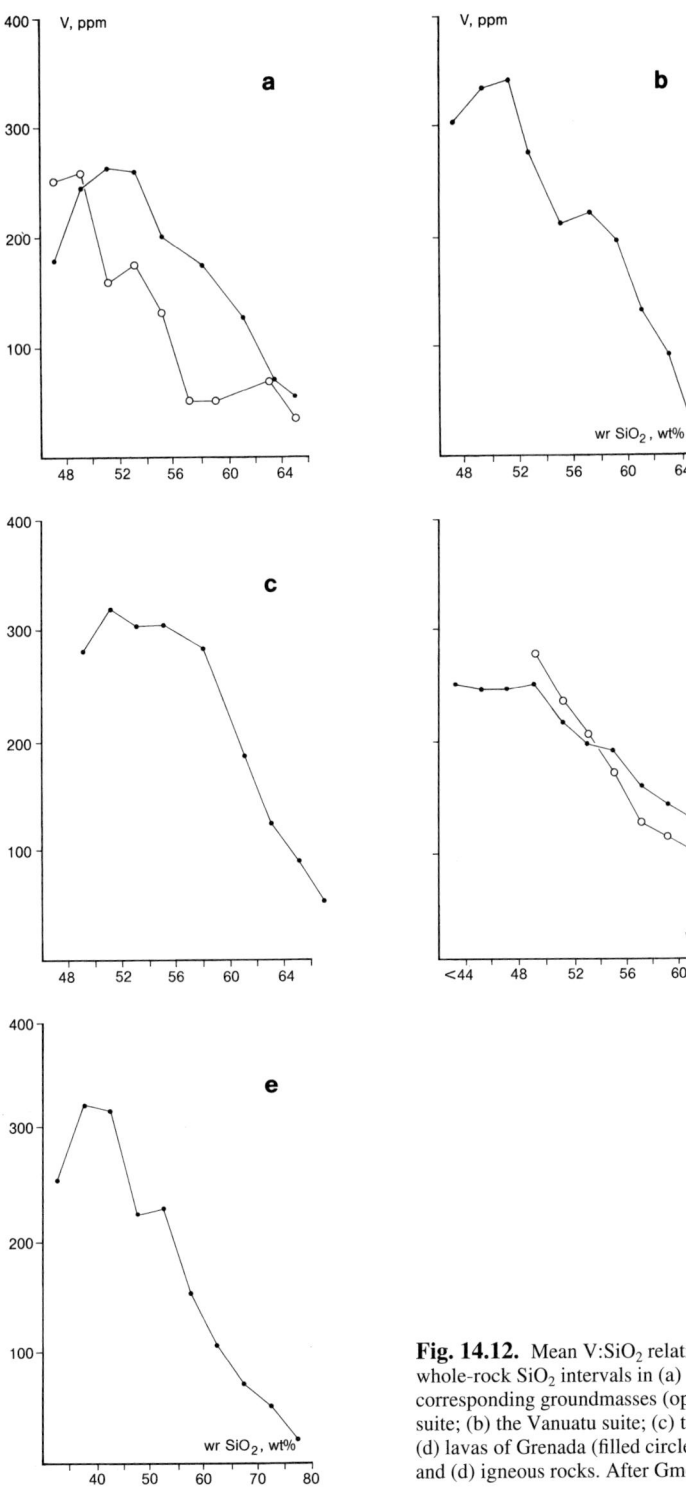

Fig. 14.12. Mean V:SiO$_2$ relations corresponding to 2 per cent whole-rock SiO$_2$ intervals in (a) whole rocks (filled circles) and corresponding groundmasses (open circles) of the Solomons suite; (b) the Vanuatu suite; (c) the Tonga–Kermadec suite; (d) lavas of Grenada (filled circles) and St Kitts (open circles); and (d) igneous rocks. After Gmelin (1968).

180   Ore Elements in Arc Lavas

relations and in the behaviour of the ternary grouping V–MgO–(Na$_2$O+K$_2$O).

A V–SiO$_2$ diagram for the Solomons lavas (Fig. 14.12(a)) shows a clear rise in vanadium as the lavas change from mafic basalts to basalts, with a vanadium maximum ≈260 ppm in the 50–4 per cent whole-rock SiO$_2$ interval: the SiO$_2$ content of many of the big feldspar basalts and basaltic andesites. Vanadium then decreases steadily, to about 50–60 ppm, in the dacites. Groundmass vanadium, higher than that in the whole rock at lower SiO$_2$ levels, quickly falls below that of the whole rocks as SiO$_2$ increases, and continues to decline to a level of about 30 ppm in the dacites. The two south-west Pacific arcs, Vanuatu and Tonga–Kermadec, show V–SiO$_2$ patterns essentially similar to the Solomons (Figs. 14.12(b) and (c)). Of the two representative islands of the Lesser Antilles, St Kitts exhibits an essentially even decrease in vanadium with increase in SiO$_2$; Grenada a flat pattern from <44 to 50 per cent SiO$_2$ and then a steady decrease almost identical to that of the St Kitts lavas (Figs. 14.12(d)). The behaviour of vanadium with respect to increasing whole-rock SiO$_2$ in all of these arc suites is clearly very similar to that of titanium. Although the SiO$_2$ values do not quite correspond, this general *pattern* of V–SiO$_2$ in arc lavas is also very similar to that demonstrated by Gmelin (1968) for a very large sampling of igneous rocks of all types (Fig. 14.12e).

As might have been expected, the pattern of V–MgO–(Na$_2$O+K$_2$O)) relations is also very similar to that for titanium. Beginning with the picritic lavas in the low-vanadium region of the lower right of the triangle (Fig. 14.13(f)), vanadium rises to a maximum in the big-feldspar and hornblende-andesite groups and then declines — MgO decreasing and (Na$_2$O+K$_2$O) increasing — in the big-feldspar lava groundmasses, felsic andesites, and andesite–dacite groundmasses (Fig. 14.13(c) to (a)).

These trends, and their analogies with those for titanium, appear in sharper focus in Fig 14.14, which shows mean VMA relations corresponding to 2 per cent whole-rock SiO$_2$ intervals. The Solomons whole rock and groundmasses describe a nicely symmetrical arc (Fig 14.14(a)) very similar to that for titanium (Fig. 13.18(a)) and are thus 'flatly hypotholeiitic'. This is brought out with slightly greater crispness by the inclusion of the Bougainville lavas. (Fig 14.14(b)). Whole-rock trends for Grenada and Vanuatu (Figs. 14.14(c) and (d)) appear to be similar in principle whereas St Kitts (Fig. 14.14(c), open circles) and the Tonga–Kermadec suite (Fig. 14.14(d), open circles), with their relative lack of high-Mg–low-Ti members, are apparently truncated at the mafic end.

MORBs are tholeiitic with respect to vanadium (Fig. 14.14(e)), though the field of points is not nearly so

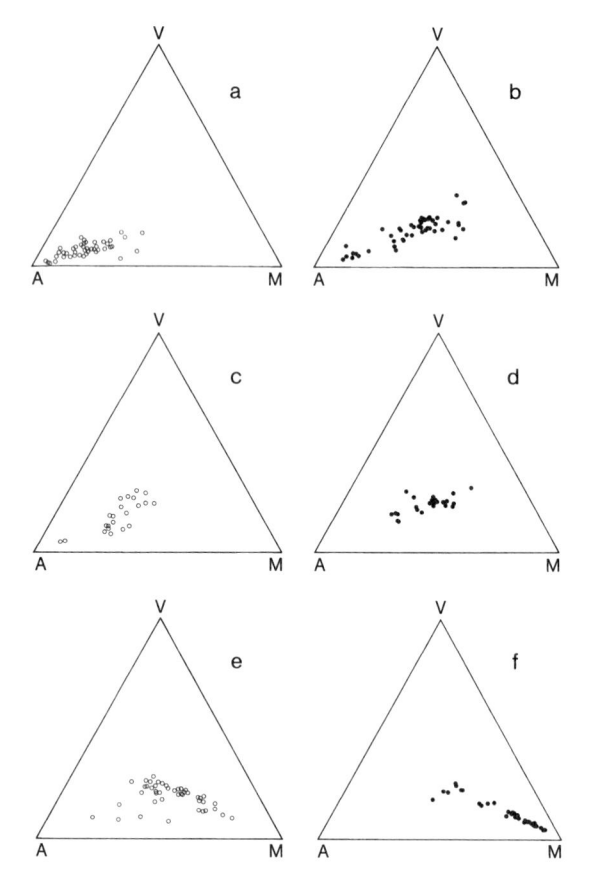

Fig. 14.13. V–MgO–(Na$_2$O + K$_2$O) ternary relations (V calculated using ppm V × 10^{-2}) developed in olivine–clinopyroxene-basalt, big-feldspar lava, and hornblende-andesite whole rocks [filled circles: (f), (d), and (b) respectively], and their corresponding groundmasses [open circles: (e), (c), and (a) respectively].

attenuated as that for titanium (Fig. 13.18(e)). As with Fig 13.18(e), Fig. 14.14(e) shows signs of incipient hypotholeiiticity. The Icelandic basalts (Fig. 14.14(f)) show a rather more pronounced tholeiitic tendency than do MORBs, but the very sparse data for vanadium in Icelandic lavas more felsic than basalts (the four open circles at the lower left of the diagram) indicate that this suite is probably hypotholeiitic overall.

Qualitatively these trends are attributable to early subtraction of low-V olivine and clinopyroxenes leading to enrichment of vanadium in the melt to the basalt–basaltic andesite stage, followed by subtraction of high-V magnetite and hornblende with the concomitant generation of andesites and dacites and the progressive reduction of vanadium in the melt.

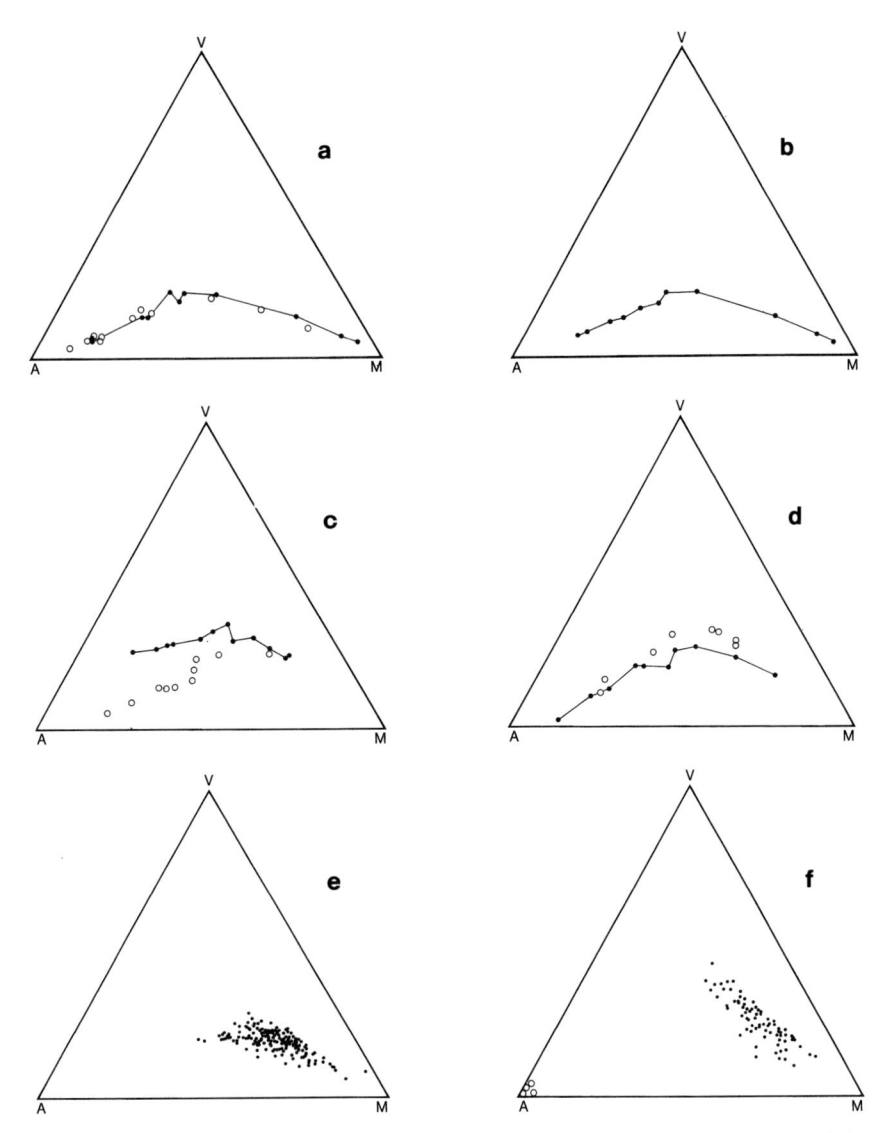

Fig. 14.14. Mean V–MgO–(Na$_2$O + K$_2$O) ternary relations corresponding to 2 per cent whole-rock SiO$_2$ intervals in (a) whole rocks (filled circles) and corresponding groundmasses (open circles) of the Solomons suite excluding Bougainville; (b) whole rocks of the Solomons suite including Bougainville (241 analyses); (c) lavas of Grenada (filled circles) and St Kitts (open circles); (d) lavas of Vanuatu (filled circles) and Tonga–Kermadec (open circles). Ternary relations as developed in MORBs as represented by individual Mid-Atlantic Ridge basalts are given in (e); and in individual samplings of Icelandic basalts (SiO$_2$ < 52 per cent; filled circles), and Icelandic lavas more felsic than basalt (SiO$_2$ ⩾ 52 per cent; open circles) in (f).

CONCLUDING STATEMENT

The mean vanadium content of a large sampling of MORBs is 263 ppm, and of a similarly large sampling of arc basalts it is 289 ppm. That of the Solomons basalts is 235 ppm. The mean vanadium content of the Solomons lavas more felsic than basalts (i.e. SiO$_2$ >52 per cent) is 187 ppm, which compares with a mean of 151 ppm for such rocks in the five major volcanic arcs considered.

Olivine and feldspar generally contain less than 10 ppm V and their crystallization tends to fractionate vanadium into the remaining melt. The pyroxenes most commonly contain 150–250 ppm, develop $K_D^{xst:melt}$ values of about unity, and their crystallization therefore does not have a pronounced effect on the vanadium content of the melt. Hornblende and biotite accommodate 400–50 ppm V, develop $K_D^{xst:melt}$ values of ≈6–7, and hence tend to impoverish the remaining melt with respect to vanadium. Magnetite and the titano-magnetite range of spinels are the principal acceptors of vanadium, commonly containing 2500–3500 ppm V. They generate $K_D^{xst:melt}$ values in the 20–50 range and are thus, pro rata, important in removing vanadium from the melt. In terms of vanadium abstraction, 2 per cent magnetite removed is, on average, the equivalent of approximately 14 per cent hornblende subtraction.

Early subtraction of olivine and low-vanadium pyroxene thus leads to a brief phase of vanadium enrichment in the melt. With the development of felsic basalts and the relatively copious crystallization of magnetite, followed by the crystallization of hornblende and continuing precipitation of magnetite, the trend is reversed and vanadium steadily decreases in the melt. The overall pattern then becomes hypotholeiitic with respect to this element. In its general behaviour and pattern of abundance in the Solomons and other arc lava series vanadium is thus very similar to titanium.

15

Chromium

Like titanium and vanadium, its close neighbours in series 4 of the periodic table of the elements, chromium is essentially absent from exhalative ores and their associated chemical sediments of the volcanic regime. Although chromium has long been recognized as a significant component of the troilite phase of iron meteorites, it has not been found as other than minute traces in the iron sulphides of volcanogenic orebodies. Further eloquent evidence for its essential absence from exhalative ores lies in its extremely low incidence in what are well-recognized chromium acceptors among associated oxides and silicates. Thus, in spite of its propensity for entering the spinel structure (see chromite and related spinels), chromium is not found in the magnetite of accompanying iron oxide-rich exhalites and their weathering products. Similarly, where the relevant ores have been metamorphosed, chromium does not commonly appear in significant amounts in Ca-rich garnet, hedenbergitic pyroxenes, muscovitic micas, or the chlorites — all ready acceptors of Cr^{3+} in crustal environments. Exceptions are the ore units of Outokumpu in Finland, which contain an average of ≈ 0.2 per cent whole rock Cr_2O_3 and a group of chromium-rich silicates and chromite (Treloar 1987a, b), and the calc-silicate rocks of the Cadi Zone of the Hemlo deposit in Ontario (Pan and Fleet 1989), which also contain a variety of Cr-rich silicates and chromite. (As noted in Chapter 14, however, the Hemlo orebody may not be of exhalative origin.) Chromium-bearing silicates have been reported from a number of exhalative ores, but the overall abundance of whole-rock chromium has been low. Chromium-bearing mica (fuchite) is a well-known associate of some of the exhalative sulphide orebodies of Western Tasmania (Que River, Hellyer, etc.) and indeed is used here as an indicator in mineral exploration. This association is, however, of a secondary nature* Perhaps the most sensitive indicator of the lack of chromium in exhalative ore deposits is the absence of chromates from their gossans. Although the lead chromate crocoite ($PbCrO_4$) occurs as spectacular crystals in the alteration zones of some of the Tasmanian orebodies, chromates do not appear in the weathered cappings or secondary zones of any of the major exhalative deposits, even where these have been exposed to the most appropriate conditions.

As chromium is a relatively abundant trace element of basaltic rocks — particularly of the more magnesium-rich, mafic varieties such as those associated with the Cyprus orebodies and many of their kind elsewhere — it appears that the general absence of this element, like that of strontium, titanium, and vanadium, may constitute evidence concerning the derivation and transport of exhalative ores.

Perhaps the first estimate of the average chromium content of igneous rocks was that of Clarke and Washington (1924), who proposed a figure of 370 ppm. This was soon recognized as too high: the result of the inclusion in the sample of an unrealistically large number of unusual (high-magnesium) rock types. Goldschmidt (1938; see Goldschmidt 1954, p. 545) estimated an overall average of 200 ppm for igneous rocks, and a similar figure for the Earth's crust. The modern estimate of the crustal abundance of chromium is 122 ppm (Emsley 1989).

It was pointed out by Goldschmidt (1937) that in the igneous regime the pattern of incidence of chromium 'is clearly correlated with the sequence of crystallization in

---

* It has been proposed by Jack (1989) that the chromium of the mica is derived from the late-stage hydrothermal alteration of chromium-rich spinel and clinopyroxene occurring as components of a mafic lava overlying the relevant orebody. Persistence of the hydrothermal plume not only after the formation of the exhalative sulphides but also for some time following the eruption of the overlying mafic basalt led to leaching of spinel and clinopyroxene from the basalt, the development of secondary muscovite, the transfer of chromium from spinel and clinopyroxene to mica, and the localized formation of fuchite. While fuchite and ore are thus related to each other through their different relationships with the plume, the chromium is primarily orthomagmatic, not exhalative, and its incidence in the mica results from alteration-transfer by, not deposition from, the hydrothermal solutions. The chromium occurs *above* the orebody, not as a component of it.

from all tectonic environments and thus, like the estimates of Goldschmidt (1954), included the continental as well as the oceanic milieu. Jakes and White (1972) gave 50 ppm and 40 ppm as the mean Cr abundances in island-arc tholeiites and calc-alkaline rocks respectively, and 15 ppm and 25 ppm respectively for the andesites of island-arc tholeiitic and calc-alkaline association. Shiraki (1978) constructed histograms of chromium contents of 125 ocean ridge and ocean floor basalts, and of 258 basaltic rocks from oceanic islands (Figs. 15.1(a) and (b) respectively), both showing modes, the first very sharp, the second somewhat broader, in the 200–800 ppm Cr range. Shiraki's (1978) compilation of chromium values in lavas of the marine environment is reproduced in Table 15.1.

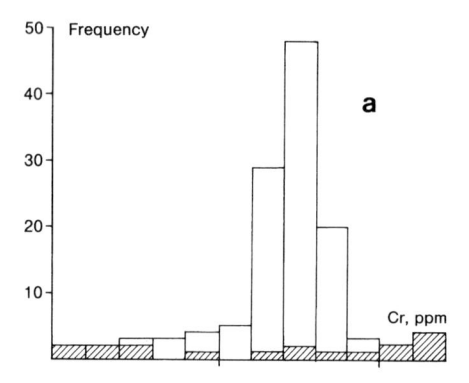

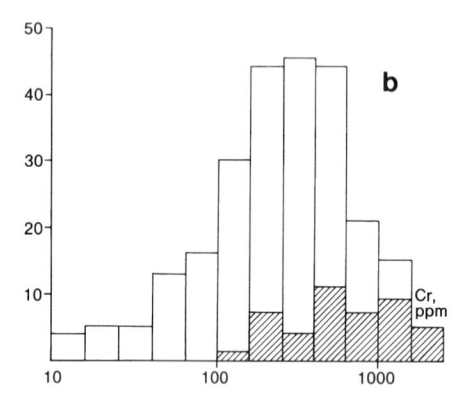

Fig. 15.1. Frequency distributions of chromium abundances in (a) 125 basaltic rocks from oceanic ridges and ocean floors (hatched bars, associated gabbroic rocks), and (b) 258 basaltic lavas from oceanic islands (hatched bars, associated oceanites, picrites, and ankaramites). Chromium abundances shown on logarithmic scale. After Shiraki (1978).

CHROMIUM IN LAVAS OF THE MARINE ENVIRONMENT

The incidence of chromium in MORBs is summarized in Table 6.2. The mean for 474 analyses of basalts (SiO_2 <52 per cent) from the Atlantic, Atlantic–Indian, Indian, and Pacific Ocean ridges is 268 ppm Cr, and for 599 analyses including those of Iceland, 266 ppm Cr. This compares with an estimate of 307 ppm by Shiraki (1978). Means for the four ridge systems are 355, 163, 214, and 283 ppm Cr respectively. This broad variability between the major ridge systems is also apparent within individual ridges and sampling locations. For example, samplings of the mid-Atlantic ridge at 30° S (7 samples; Frey *et al.* (1974) gave a mean of 566 ppm Cr, whereas others at 51°–54.5° S (40 samples; le Roex *et al.* 1987) gave 297 ppm Cr. Among the latter group chromium varied from 52 ppm (whole-rock SiO_2 = 50.64 per cent; MgO = 6.7 per cent) to 1077 ppm (whole-rock SiO_2 = 47.55 per cent; MgO = 15.63 per cent). Large factors are involved — a state of affairs doubtless due to the rapid incorporation of chromium into earliest-formed crystals, and its consequent sensitivity to even minor differences in degree of magmatic fractionation.

The full range of variation of chromium in MORBs is indicated in Fig. 15.2. The frequency distribution is conspicuously bimodal, with frequency maxima corresponding to the intervals 50–100 and 250–300 ppm Cr. This may well reflect fractionation: a widespread parent melt containing 100–150 ppm Cr undergoes early crystallization yielding accumulates most frequently containing 250–300 ppm Cr, and complementary residua most frequently containing 50–100 ppm Cr. The same feature appears in muted form in the Icelandic basalts (Fig. 15.2(b)) and may represent the process at a barely more than incipient stage.

Chromium contents of the relatively few Icelandic samplings with SiO_2 >52 per cent are shown by open bars

such a way that the element appears in the earliest crystallizates, rich in silicates of magnesium, such as olivine rocks and, to a somewhat less extent, pyroxenites' (in Goldschmidt 1954, p. 548). Goldschmidt also noted that the abundance of chromium decreased sharply, with magnesium, in the less mafic rock types. He observed that in ultramafic rocks chromium contents may range from about 1000 to 4000 ppm Cr; in gabbros and basalts it is only about one-tenth of that: 100–400 ppm; in andesites 25–80 ppm; and in rhyolites 2–10 ppm.

Prinz (1967) collated 236 superior analyses of basalt from the literature and determined the following arithmetical means for seven principal types: quartz-normative tholeiite, 153 ppm Cr; olivine-normative tholeiite, 218 ppm Cr; olivine-normative alkali basalt, 185 ppm Cr; nepheline-normative alkali basalt, 190 ppm Cr; olivine-normative basalt, 199 ppm Cr; all tholeiites, 162 ppm Cr; all alkali basalts, 187 ppm Cr. These included basalts

Chromium

Table 15.1 Chromium abundances in the principal categories of marine lavas, after Shiraki (1978)

Geological and tectonic affiliation, and lava type	Number of analyses	Mean chromium abundance (ppm)
Mid-ocean ridges and ocean floors		
Tholeiitic basalt	110	307
Oceanic islands		
Oceanite, picrite	18	1270
Ankaramite	15	448
Basalt, basanite	195	245
Trachybasalt, hawaiite	66	30
Trachyandesite, mugearite	33	11
Trachyte, phonolite, rhyolite	109	5
Island arcs		
Basalt	238	191
Andesite	308	55
Dacite	62	16
Rhyolite	36	4
Tholeiitic basalt	45	100
Tholeiitic andesite	53	16
Tholeiitic dacite	23	4
Calc-alkali basalt	50	179

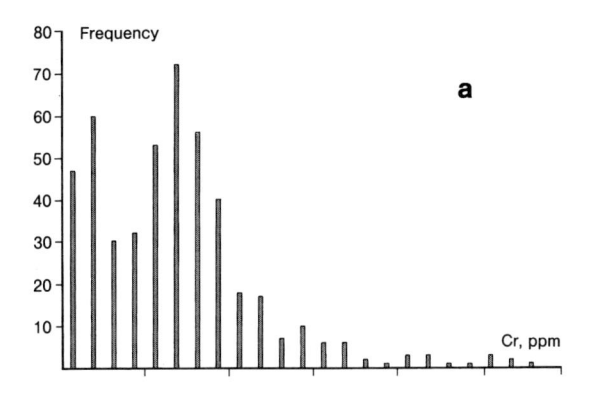

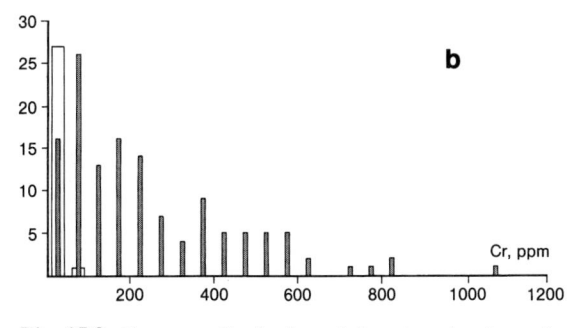

Fig. 15.2. Frequency distributions of chromium abundances in (a) 476 analyses of MORBs of all major oceans, and (b) 136 analyses of Icelandic basalts (SiO_2 <52 per cent). Chromium abundances in Icelandic lavas more felsic than basalt ($SiO_2 \geqslant 52$ per cent) shown by open bars in (b).

in Fig. 15.2(b). They reflect the ubiquitous sharp decrease in chromium accompanying even the slightest development of more felsic lavas.

Data on the incidence of chromium in arc rocks are also extensive (Table 6.3). Means for 358 analyses of basalts (SiO_2 <52 per cent) and 428 analyses of lavas more felsic than basalts (SiO_2 >52 per cent) are 284 ppm and 35 ppm Cr respectively. Frequency distributions (Fig. 15.3) for both categories are strongly J-shaped, that for basalts (Fig. 15.3(a)) extending out to 1800–50 ppm and that for the more felsic rocks (Fig. 15.3(b)) to 850–900 ppm Cr. For the two arcs — the Lesser Antilles (Grenada) and Tonga–Kermadec — for which there is a sufficient number of analyses for estimation of chromium in all of basalts, andesites, and dacites, the abundances are 551, 204, and 120 ppm and 104, 28, and 3 ppm Cr respectively.

Variation in chromium content of comparable lavas (e.g. basalts) both between and within arcs is considerable. Mean chromium in the basalts of the Lesser Antilles, Aleutians, Solomons, Vanuatu, and Tonga–Kermadec is 419, 95, 606, 303, and 104 ppm respectively, and for lavas more felsic than basalts, 112, 45, 29, 17, and 19 ppm respectively (Table 6.3). Such differences probably represent some differences in source compositions, accentuated to varying degrees by the vagaries of differentiation. There may similarly be considerable variation within individual arcs: e.g. the lavas of the three main islands of the Lesser Antilles. Here the overall mean chromium abundances for Grenada, Dominica, and St Kitts are 551, 133, and 11 ppm respectively. The chromium

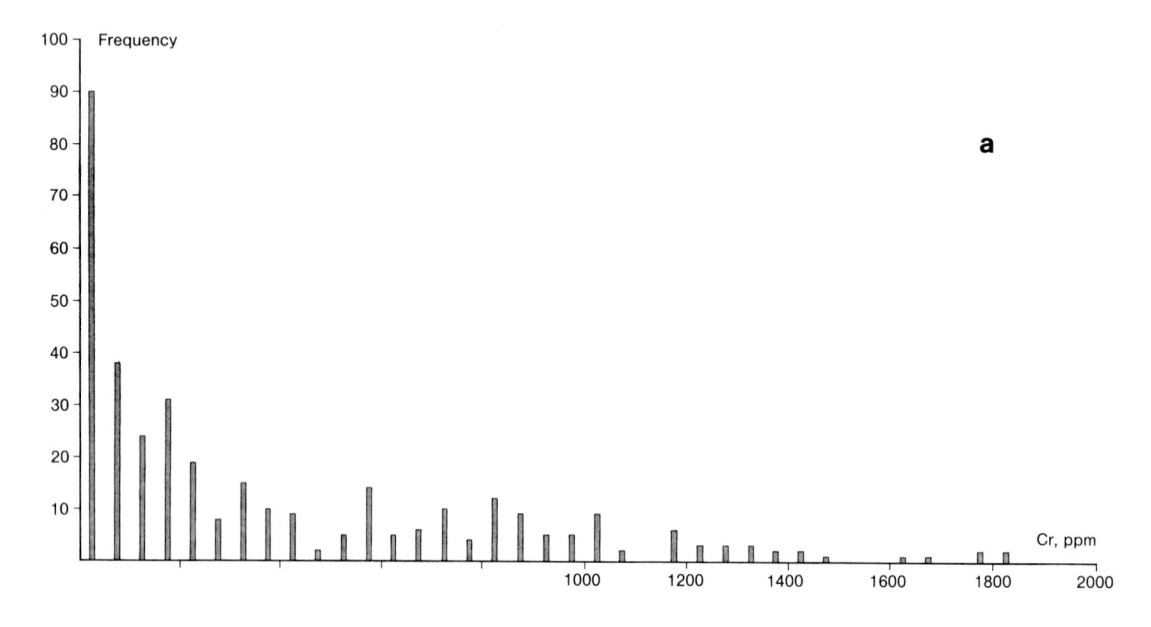

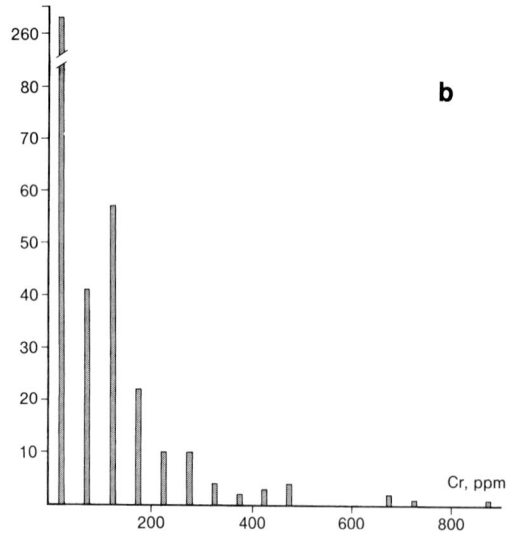

Fig. 15.3. Frequency distributions of chromium abundances in (a) 363 analyses of volcanic island-arc basalts (SiO$_2$<52 per cent), and (b) 434 analyses of arc lavas more felsic than basalt (SiO$_2 \geqslant$ 52 per cent).

content of the Grenada lavas is thus 4.1 times that of Dominica and 50 times that of St Kitts, leading, somewhat fortuitously, to the remarkably high chromium content of the West Indies dacites relative to the andesites indicated above. These remarkable differences in the chromium content of lavas of different volcanic centres of a single arc are further highlighted by a consideration of a single compositional category of volcanic products. If, of 'basalts', 'andesites', and 'dacites' the completely defined group, 'andesites' (52 per cent <SiO$_2$ <62 per cent) is taken as the index, the mean chromium contents for lavas of this category on Grenada, Dominica, and St Kitts are 204, 130, and 8.6 ppm respectively. The ratios are 1:0.64:0.04. At the same time mean chromium contents of basalts, andesites, and dacites on Grenada are 551, 204, and 120 ppm respectively, the ratios of these being 1:0.37:0.22. That is, variation in chromium content of the andesites alone — a single well-defined category of lavas — between islands is of the same order as the variation between all of basalts, andesites, and dacites of Grenada.

CHEMICAL PROPERTIES AND CRYSTAL CHEMISTRY OF CHROMIUM

Although chromium shows strong chalcophile characteristics in meteorites it appears to be entirely lithophile in the terrestrial crustal environment. It occurs to very limited extent in the hexavalent Cr^{6+} state in minerals such as crocoite, where it is in tetrahedral coordination with oxygen and Cr–O bond lengths are in the range 0.160–0.167 nm. In the great majority of cases — in the chrome spinels, pyroxenes, amphiboles, etc. — it occurs in the trivalent Cr^{3+} state, where it is in octahedral co-ordination with oxygen (usually slightly distorted octahedra) and develops Cr–O bond lengths in the 0.197–0.200 nm range.

The principal features of the chromium atom and ions are given in Table 6.4.

THE INCIDENCE OF CHROMIUM IN THE PRINCIPAL MINERAL SPECIES OF THE SOLOMON ISLANDS YOUNGER VOLCANIC SUITE

General

As it is one of the important minor or trace elements of mafic and ultramafic igneous rocks, there is now quite a large amount of data on chromium in igneous minerals of the mafic to ultramafic environment. These include, pre-eminently, chromite and its related chromium-bearing spinels, but also clino- and orthopyroxenes, olivine, and, to a lesser extent, kammererite and fuchite (chromian chlorite and muscovite respectively), both of which are found occasionally as metamorphic and alteration products in ultramafic intrusions and picritic lavas.

There is little to say about chromium in chromite and related spinels other than that the Cr content of ideal chromite $(MgO.Cr_2O_3)$ is 54 per cent, that commercial 'chromite' of the ultramafic environment generally contains 45–60 per cent Cr_2O_3 (23–31 per cent Cr), and that the Cr content of the spinels of mafic to picritic lavas falls in the range of a few hundreds of ppm to ≈55 per cent Cr_2O_3 (≈29 per cent Cr).

As remarked by Shiraki (1978), although the environments in which many olivines crystallize are rich in chromium, most terrestrial olivines are very poor in it. High chromium determined in some early analyses of olivine separations has been found to be due to minute inclusions of spinel, and most recent electron-probe analyses show little chromium in the olivines themselves. This is not surprising; while Cr^{3+} and Mg^{2+} have somewhat similar ionic radii (0.064 and 0.078 nm respectively), the compositional simplicity of olivine provides little opportunity for compensation of the charge difference between the two ions.

Among the common volcanic silicates clinopyroxene appears to be the principal host for chromium, perhaps because of its relatively high Al^{3+} and compensatory Na^+ content. Chromium contents of lava clinopyroxenes have been found to range from a few hundred parts per million in mafic basalts to a few thousand parts per million (c. 3000–4000 ppm) in picritic types. When orthopyroxene coexists with clinopyroxene it contains less chromium than the latter. As in the case of olivine, this is presumably due to a lack of Al^{3+}–Na^+ or analogous groupings that might facilitate charge compensation. $K_D^{cpx:opx}$ is generally in the range 2 to 8. Acceptance of chromium into the structures, and resultant partition coefficients, may be pressure- as well as temperature-sensitive.

There appears to be little information on the incidence of chromium in the hornblendes of calc-alkaline and related lavas, though Ewart and Taylor (1969) give a range of 1.5–125 ppm Cr in those of the andesites–rhyolites of the Taupo area of New Zealand. Information on chromium in volcanic biotites is even more sparse. Shiraki (1978) observes that chromium contents of both minerals (hornblende and mica) decrease gradually with decrease in the host melt; and in some plutonic rocks Cr is more concentrated in biotite than in hornblende, but that in volcanic rocks and most plutonic rocks the reverse is the case, perhaps indicating that the hornblende has crystallized earlier than the biotite.

Data on chromium in chromian spinels and magnetite are extensive.

Solomons suite

Mean values of chromium in the principal silicates and the range of spinels of the Solomons suite are given in Table 6.5.

Olivine. The range of chromium values found in the spectrum of olivine separations, chiefly from the olivine–pyroxene-basalts, is <1–724 ppm, and the nature of the distribution is shown in Fig. 15.4. These figures appear high in the light of current views on the incidence of chromium in olivine (see previous section). Chromium abundances were therefore checked by probe analysis of olivines in two of the lavas for which XRF analyses of olivine separations had already been made. Chromium determinations on the separated materials were 724 and 195 ppm, and the averages of probe analyses of 20 microscopically clear grains in each case were 317 and 163 ppm respectively.

188 *Ore Elements in Arc Lavas*

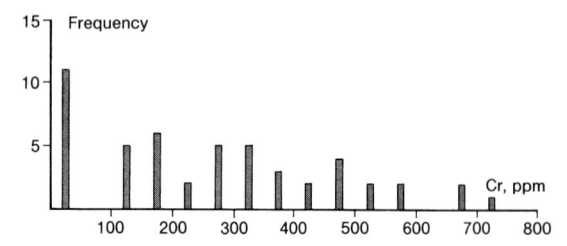

Fig. 15.4. Frequency distribution of chromium abundances in olivines of the Solomons suite.

This appeared to indicate that at least a substantial part of the apparent chromium content of these olivines occurred in inclusion material of some kind — presumably chrome-rich spinel. It has, however, already been pointed out (Chapter 5) that the olivine of the Solomons lavas, in contrast to much of the associated clinopyroxene, contains little if any included spinel. In addition, all separations were repeatedly checked in oil and recycled through the separation process until all microscopically visible impurities had been eliminated. The only explanation apparent to the writer is that at least some of the cloudiness present in a minority of olivine grains represents ultrafine, submicroscopic, chromium-rich spinel. The problem clearly calls for systematic electron microscopical examination.

Plotting of Cr *v*. MgO (or FeO) for all of the olivine separations reveals a high degree of correlation (*c.* 0.80) but also a very high slope of the regression line (Fig. 15.5(a)). What this indicates is that MgO is principally confined to the 40–50 per cent range, but that within this chromium may have values anywhere in the range $\approx$100–700 ppm, i.e. there is essentially no relationship between Cr and MgO. While this might reflect essentially unsystematic incorporation of chromium into the olivine structure related to chromium concentration in the co-existing melt, it is perhaps more likely that it reflects a more-or-less random inclusion of ultrafine, almost pure, chromite — material so high in chromium that it could have a pronounced effect on the apparent abundance of this element in the olivine at the parts per million level while having little effect on Mg or Fe at the per cent level.

The electron-probe results on the clear olivines do, however, indicate that the olivine structure itself may incorporate quite significant amounts of chromium (in the case of the Solomons olivine, up to $\approx$670 ppm) and that this has some relation with the Mg number. Fig. 15.5b shows Cr:MgO relations in individual probe analyses of a range of Solomons olivines (all completely clear at × 450). Extrapolation of the curve of best fit indicates that the olivine begins to accommodate some chromium con-

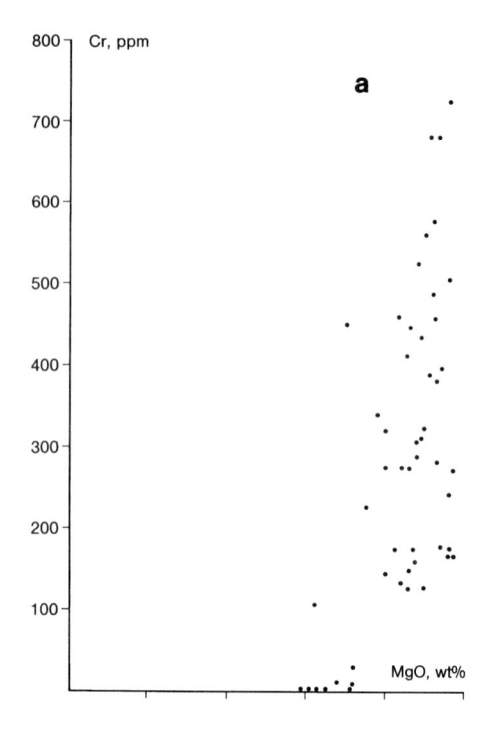

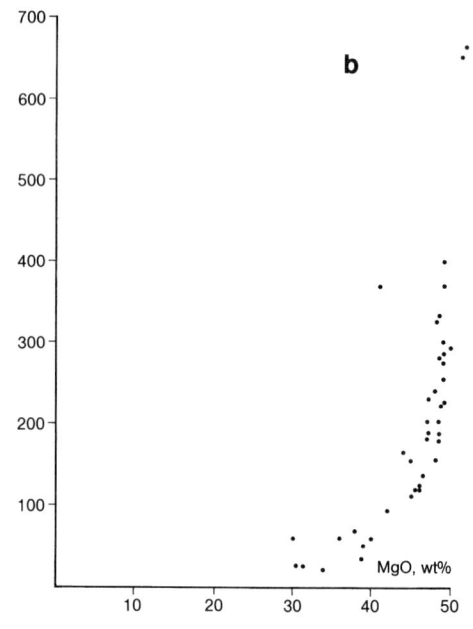

Fig. 15.5. Abundance relations between chromium and MgO in (a) olivine separations analysed by X-ray fluorescence, and (b) clear, inclusion-free olivine (each point represents the mean of $\approx$10 to 15 individual olivine grains) analysed by electron microprobe.

tent at MgO ≈25 per cent. Cr then rises with MgO until the latter reaches ≈45 per cent, at which point the curve steepens; at MgO >50 per cent olivine may contain Cr >600 ppm.

Thus in spite of the current view that the olivine structure itself is unlikely to accommodate significant chromium, and despite the unpromising nature of the olivine structure for any such substitution, it does appear that chromium may substitute in this mineral to at least ≈650 ppm, and that chromium tends to increase with increase in MgO. In the light of this, Fig. 15.5(a) may be seen as a muted form of Fig. 15.5(b), the sharpness of the former having been blurred by the effects of submicroscopic impurities in the olivines of the separations.

Pyroxenes. Despite the slightness of change in their Mg number with change in host rock type (see Chapter 5), the clinopyroxenes exhibit gross variation in chromium content from those of the olivine–pyroxene-basalts to those of the hornblende-andesites. The frequency distributions of Cr values in clinopyroxenes separated from members of the three principal lava types are accordingly plotted separately in Fig. 15.6. Mean values in the clinopyroxenes of the olivine–pyroxene-basalts, big-feldspar basalts–basaltic andesites, and hornblende-andesites are 2435, 270, and 241 ppm Cr respectively (Table 6.5). Although it has not been practicable to repeat all these analyses using the electron probe, strategic checking by this means gives results fully compatible with those obtained on the corresponding separated material. This general trend in relationships between chromium content of clinopyroxene and composition of host lava is illustrated more clearly by Fig. 15.7, which shows Cr content of clinopyroxene v. SiO_2 content of the host rock.

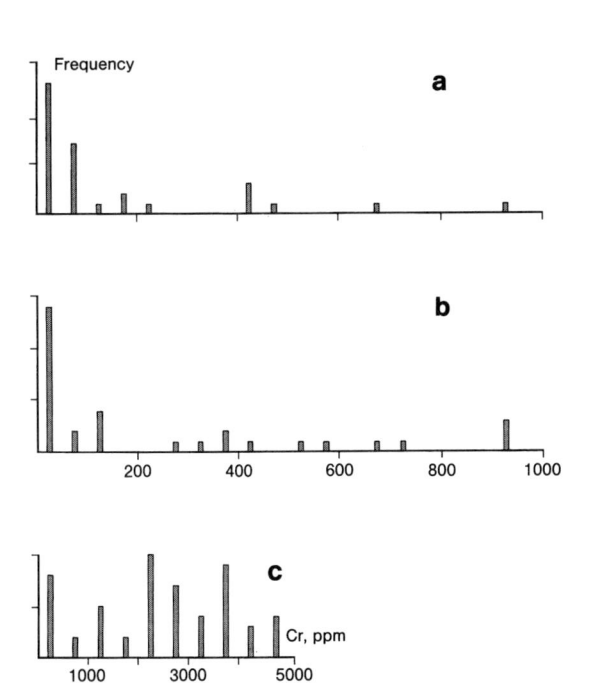

Fig. 15.6. Frequency distributions of chromium abundances in clinopyroxene separations from lavas of (a) the hornblende-andesite group, (b) the big-feldspar lava group, and (c) the olivine–clinopyroxene-basalts.

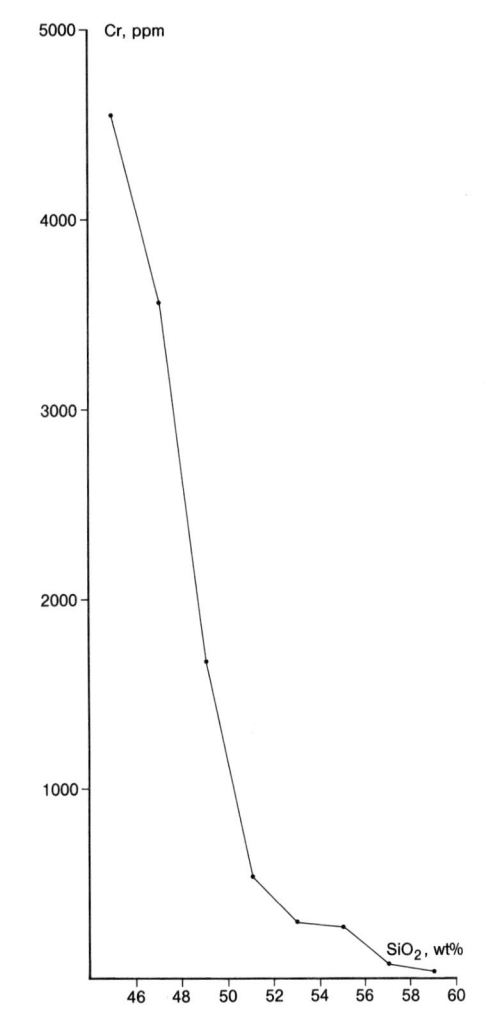

Fig. 15.7. Mean abundance of chromium in clinopyroxenes with respect to 2 per cent whole-rock SiO_2 intervals in the host lavas. The latter range from picrites to felsic andesites.

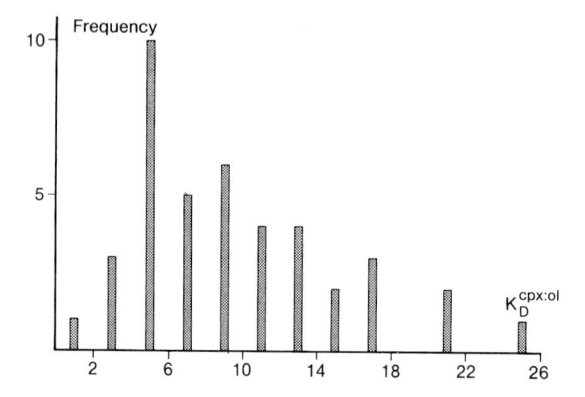

Fig. 15.8. Frequency distribution of values of $K_D^{cpx:ol}$ with respect to chromium in the Solomons lavas.

As clinopyroxene coexists with olivine and hornblende respectively at the two ends of the Solomon Islands lava spectrum, it is possible to calculate partition coefficients for the two pairings. $K_D^{ol:cpx}$ in 44 coexisting pyroxene:olivine pairs ranges from 1.18 to ∞, the latter pertaining to three of the more felsic lavas, in which the olivine contains <1 ppm Cr. Apart from these, i.e. for 41 pairings, K_D values fall in the range 1.18–24.43 (Fig. 15.8) and average 7.76. $K_D^{cpx:Hb}$ values for nine pyroxene:hornblende pairings are in the range 0.26–4.78 and average 1.63.

Chromium concentrates in the clinopyroxenes relative to the orthopyroxenes. A total of 166 coexisting grains (85 clinopyroxene, 81 orthopyroxene) in eight two-pyroxene andesites (i.e. approximately 10 grains of each pyroxene per section) were analysed with the microprobe (Table 15.2). The grand mean for the 85 clinopyroxenes is 788 ppm, and for the 81 orthopyroxenes, 211 ppm Cr, giving an overall $K_d^{cpx:opx} = 3.73$. Apart from the fact that the clinopyroxenes also contain the greater amount of aluminium, there appears to be no close abundance relationship between aluminium and chromium in either of these groups of pyroxenes.

A striking feature of the incidence of chromium in the Solomons pyroxenes — both clino- and orthopyroxenes, in whatever context they occur — is its propensity for variation from one crystal to another within a single section. For example, chromium in clinopyroxene has been found to range, in just one thin section of a single two-pyroxene-andesite, from 30 ppm to 4676 ppm and from 25 to 2625 ppm in the coexisting orthopyroxene of the same section. This is not attributable to compositional zoning of individual grains: while increase in Al and Cr from core to rim is common, it has not been found at this magnitude. It is suspected that the variation may reflect turbulent mixing of pyroxenes of two generations: a phenomenon of which, with the tendency of chromium to partition very rapidly into the earliest formed crystals, the Cr content of pyroxenes (particularly clinopyroxenes), may be a delicate indicator. In this connection, both the clino- and orthopyroxenes of the two-pyroxene-andesites tend to fall into two groups on the basis of their chromium content. The division occurs at about 16 per cent MgO in the clinopyroxenes. Of the 85 crystals referred to above, 59 have MgO <16 per cent; their mean MgO = 14.97 per cent and mean Cr = 100 ppm. The remaining 26 crystals have MgO >16 per cent; their mean MgO = 18.34 per cent and mean Cr = 2246 ppm. The analogous division among the orthopyroxenes occurs at MgO ≈27 per cent. Of the 81 crystals, 67 have MgO <27 per cent; their mean MgO = 25.13 per cent and mean Cr = 51 ppm. The remaining 14 crystals have MgO >27 per cent; their mean MgO = 31.25 per cent and mean Cr = 1052 ppm. There appears to have been some kind of hiatus between the two pairings: perhaps a mixing of two batches of melt that had crystallized to slightly different stages, a state of affairs 'finger-printed' with great

Table 15.2 Chromium and aluminium abundances in coexisting clino- and orthopyroxenes of eight two-pyroxene-bearing andesites of the Solomons suite

	Al^{cpx}	Al^{opx}	Al$^{cpx/opx}$	Cr^{cpx}	Cr^{opx}	Cr$^{cpx/opx}$
1	1.52	0.81	1.88	1018	41	24.83
2	1.60	0.80	2.00	668	192	3.48
3	1.36	0.96	1.42	1619	1122	1.44
4	1.38	0.91	1.52	1031	84	12.27
5	1.98	0.74	2.68	173	43	4.02
6	1.66	0.85	1.95	424	55	7.71
7	1.54	0.88	1.75	827	60	13.78
8	1.46	0.91	1.60	542	89	6.09
Mean	1.56	0.86	1.81	788	211	3.73

Aluminium abundances given as per cent by mass; chromium abundances as parts per million by mass.

sensitivity by the chromium contents of the relevant pyroxenes.

Similar indicators are provided by the chromium of the spinels and this is referred to again below.

Hornblende. Mean chromium for 25 separations from the hornblende andesites — the sole host for hornblende in the Younger Volcanic Suite — is 120 ppm, which compares with a mean of 241 ppm Cr in the clinopyroxenes of the same lava group. The spread of values in the hornblendes (25 separations together with averaged probe analyses for a further 16 lavas) is shown in Fig. 15.9, which may be compared with that of the coexisting clinopyroxenes in Fig. 15.6(a). No correlation between Cr and Al, Fe, or Mg in the hornblendes is apparent, and variation in the amount of chromium in the mineral may thus simply reflect the amount of chromium available in the melt at the time of crystallization. There is no consistent crystal:groundmass relation with respect to chromium, however, and the systematics of the incidence of this element in the hornblende are therefore obscure. The only hint of a correlation is one between chromium and nickel in hornblende: this is clearly discernible but poor.

Probe analysis of the single crystal of hornblende and its well-developed rim already referred to (see Figs. 5.13, 7.9, and 13.11) indicates that rim formation is accompanied by a *reduction* of chromium; mean Cr for the 13 analyses of the hornblende crystal is 204 ppm, and for the 6 analyses of rim material, 87 ppm. Analyses of hornblende crystals and opaque rims in sections of other Solomons hornblende-andesites gave generally similar results (Table 5.4 and 5.5). Chromium contents of 100 crystals (mean = 200 ppm Cr) and 100 rims (mean = 89 ppm Cr) in the same probe section as the large crystal are indicated in Fig. 15.10. The mean value of 9 ppm Cr

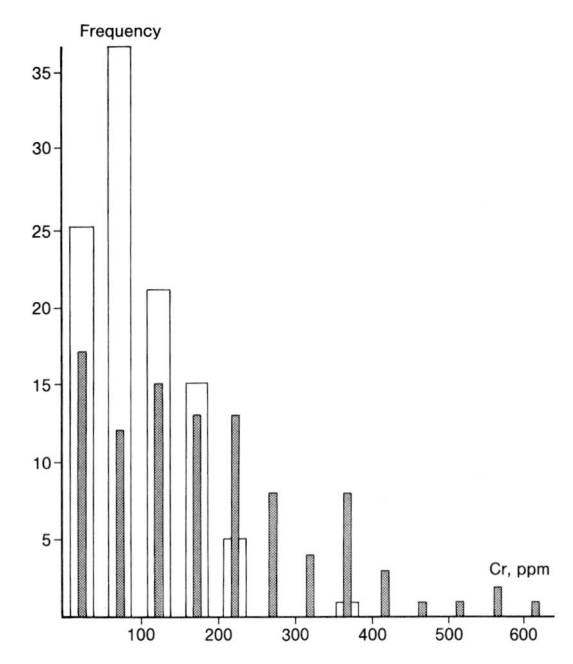

Fig. 15.10. Frequency distributions of chromium abundances in 100 hornblende crystals (filled bars; mean = 200 ppm Cr) and in the corresponding opaque rims (open bars; mean = 89 ppm Cr) in a single hornblende (–clinopyroxene)-andesite of the Solomons suite.

in seven 'best separations' of rim material compares with the mean of 120 ppm Cr in the 25 separations of hornblende (Table 5.4).

That the rim material should contain *less* chromium than the pristine hornblende is surprising given that magnetite and clinopyroxene, the two most effective abstractors of chromium from the melt, are two of the three principal products of rim formation. Elements such as Zn, Mn, and Ti, of which magnetite is a notable host and which would therefore be expected to concentrate in the rims, are indeed enriched conspicuously in the latter. However, chromium, which might have been expected to behave in a similar — or even more accentuated — way, clearly does not do so.

Biotite. The mean value of chromium for 99 biotite crystals occurring in the two Solomons dacites is 52 ppm, and the spread of values as determined by the electron probe is shown in Fig. 15.11. This low incidence of chromium in biotite may be attributable to its very low abundance in the melt from which the biotite crystallized: ≈7 ppm Cr on the basis of groundmass analyses.

In keeping with Shiraki's (1978) allusion, hornblende contains more chromium than coexisting biotite in the

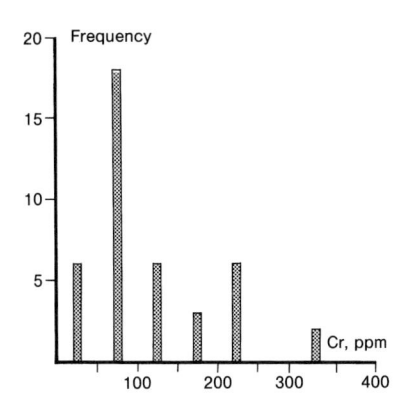

Fig. 15.9. Frequency distribution of chromium in hornblendes separated from lavas of the hornblende-andesite group.

192 Ore Elements in Arc Lavas

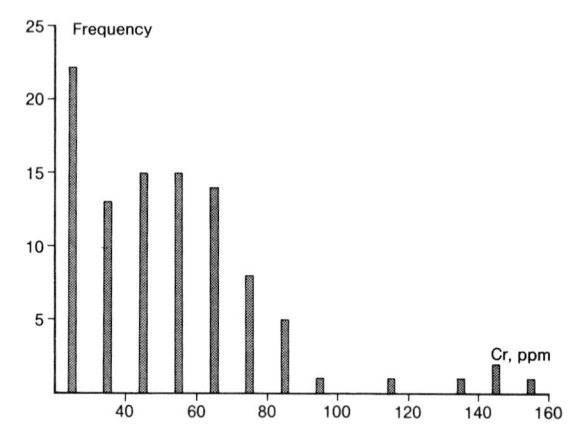

Fig. 15.11. Frequency distribution of chromium abundances in 99 individual grains of biotite in the two Savo dacites; analysis by electron microprobe.

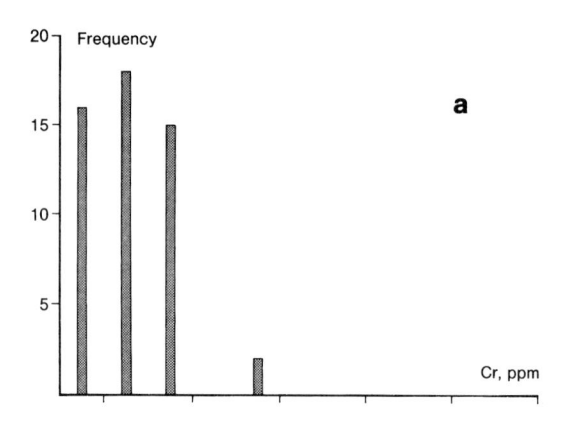

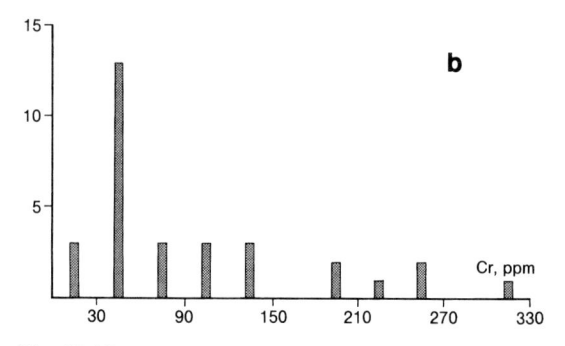

Fig. 15.12. Frequency distributions of chromium abundances in (a) biotite (51 grains; mean = 46 ppm Cr) and (b) hornblende (33 grains; mean = 94 ppm Cr) coexisting in a single Savo dacite.

Solomons dacites (Fig. 15.12). For a single probe section of the lava concerned, 51 biotites gave a mean of 45.83 ppm Cr, and 33 hornblendes a mean of 94.35 ppm Cr, indicating a mean $K^{Hb:biot} = 2.06$ on the assumption of essentially contemporaneous crystallization.

Feldspar. Most X-ray fluorescence analyses of feldspars indicated Cr <1.0 ppm, and the overall average for the 76 ultimate purity feldspars is 0.3 ppm Cr. While part of this may occur as a genuine component of the feldspar lattice (as in the case of the ubiquitous small quantities of Fe^{3+} in the feldspars) some of it is likely to be present in the very fine glass inclusions referred to earlier.

Spinel. As indicated in Chapter 5 (see Fig. 5.16), the chromium content of the Solomons spinels varies grossly, and does this substantially in sympathy with paragenesis, or evolutionary development, of the lavas in which it occurs. The mean value of chromium in magnetites separated from members of the olivine–pyroxene-basalt group is 6.31 per cent. Spinel is not the first mineral to crystallize from the more mafic fractions of the melt (olivine is the first to appear: Chapter 5); it is thus present only as scattered grains, and hence in amounts insufficient for normal separation and X-ray fluorescence analysis, in many of the most mafic lavas. Analysis of such material requires the use of the microprobe. As would be expected these spinels tend to be the most chromium-rich, and thus when probe analyses are combined with X-ray fluorescence determinations the mean chromium content for the group as a whole increases to 9.63 per cent (Table 6.5). The spread of mean values for the spinels of individual olivine–pyroxene-basalts ranges from ≈1000 ppm to 21.05 per cent Cr (40.48 per cent Cr_2O_3) (Fig. 15.13(c)). Mean chromium in the spinels decreases sharply proceeding from the olivine–pyroxene-basalts (9.63 per cent) to the big-feldspar lavas (2933 ppm Cr) and hornblende-andesites (475 ppm), the spreads for which are shown in Figs. 15.13 (c),(b), and (a) respectively.

Values obtained from X-ray fluorescence analysis of mineral separations, or by averaging probe analyses of numbers of individual grains of a single probe section, commonly mask substantial between-grain differences in chromite composition within an individual lava sample. Table 15.3 lists electron-probe determinations of Cr_2O_3 in 10 individual spinel grains in each of 6 olivine–pyroxene-basalts. Analyses are listed simply in the order in which they were carried out. Little of the variation can be attributed to composition zoning: in general grains were too small to warrant more than a single analysis, and this was usually carried out close to the geometric centre in section.

All the listed variation occurred within single probe sections, and many of the grossly different grains oc-

Table 15.3 Chromium abundances in ten individual spinel crystals in each of six Solomons lavas

No.	1	2	3	4	5	6
1	38.83	35.31	29.19	56.57	0	33.45
2	22.56	11.55	27.62	13.26	0.23	54.08
3	55.13	9.69	22.81	22.75	13.18	49.25
4	30.37	19.41	33.28	27.69	1.66	53.28
5	9.58	25.91	35.40	5.38	0.47	55.98
6	48.50	8.17	29.90	30.38	10.46	34.46
7	24.30	20.81	22.38	18.43	16.29	47.26
8	9.95	12.75	43.72	3.70	2.96	15.18
9	29.80	2.90	56.80	6.59	6.35	39.50
10	46.73	37.21	11.88	3.08	0.27	19.28
Mean	31.58	18.37	31.30	18.78	5.19	40.17

Chromium abundances are given as per cent by mass Cr_2O_3; in each of the six lavas, all ten spinel crystals occur within an area of ≈ 1.0 cm^2 or less of the relevant probe section

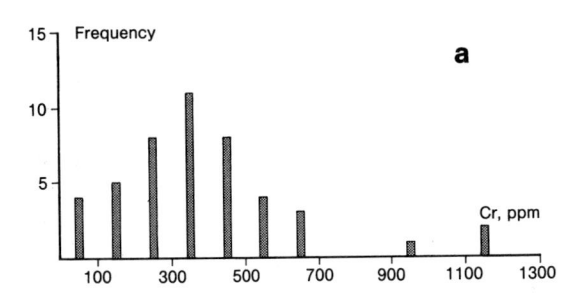

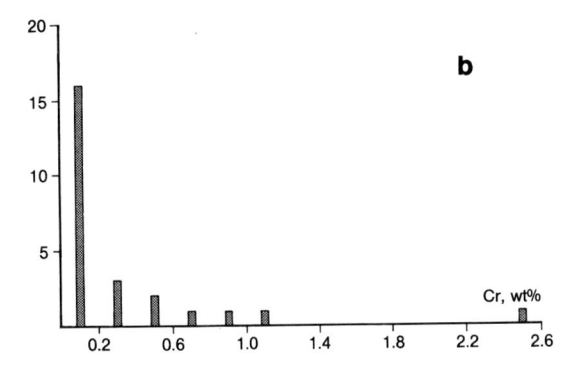

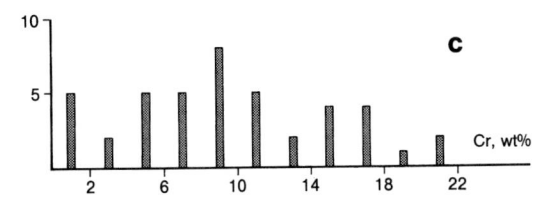

Fig. 15.13. Frequency distributions of chromium abundances in the spinels of (a) the hornblende-andesite group, (b) the big-feldspar lavas, and (c) the olivine–clinopyroxene-basalts. The scale of (a) is in parts per million by mass, those of (b) and (c) in per cent by mass.

curred within a fraction of a millimetre of each other: e.g. grains 3 and 5 of lava No. 1 (collection no. Sol. 16/81; whole-rock MgO = 22.32 per cent; whole-rock Mg no. = 80) were virtually juxtaposed. This, again, may indicate substantial turbulence in a rapidly rising — and rapidly cooling — melt. It certainly does not give the impression of a resting, slowly cooling melt undergoing equilibrium crystallization.

CRYSTAL:MELT PARTITIONING

K_D^{Cr} values, set out as in the preceding chapters, are given in Table 15.4.

One of the earlier investigations of the partitioning of chromium between olivine and basaltic melt was that of Ringwood (1970), who determined $K_D^{ol:melt}$ 0.9 for a synthetic lunar basalt at ~1140 °C at 1 atm. Bougault and Hekinian (1974) determined a value of 1.1 in a mid-Atlantic Ridge basalt. Duke (1976) obtained values of 1.1 to 5.2 in a series of experiments on the SiO_2–Al_2O_3–FeO–MgO–CaO–Na_2O system at 1125–250 °C and 1 atm. and Longhi et al. (1978) determined a range of 0.7 to 1.53 in experimental runs on lunar basalts between 1124 and 1375 °C and 1 atm. The impression gained from this experimental work and analysis of natural material is that values for the partition coefficient generally vary around 1.0–2.5, and within this range are probably influenced significantly by minor compositional variation of the melt.

Mean $K_D^{ol:melt}$ for the Solomon Islands olivine–pyroxene lavas is 1.54, and the range of values is 0–3.65. This appears to be in accord with earlier investigations. The value for the olivines of the big-feldspar basalts, at $K_D^{ol:melt} = 10.00$ is, however, rather high. Two considerations may be important here.

1. Sample size: the K_D estimation is based on only six coexisting olivine:melt pairs. Values range from 0–∞, and the three intermediate coefficients are 16.14, 7.50, and 6.85. The estimation of a mean $K_D \approx 10$ is therefore highly tentative and of a reliability not to be compared with that of the value obtained for the more mafic lavas, which is based on 39 coexisting pairs.
2. Inheritance: at least two of the olivines may have been inherited, metastably, from a more mafic environment of olivine crystallization. Reference has been made above to the possibility of melt turbulence in bringing together members of a given mineral species of greatly differing composition (e.g. chromian spinels), and in the incorporation of earlier-formed minerals in melts with which they are not in equilibrium. It is suspected that these two olivines (Cr = 339 and 226 ppm) formed in a relatively highly mafic batch of melt which was then rapidly and turbulently mixed with a larger volume of more evolved melt from which most of the chromium had been stripped by clinopyroxene and spinel. The result was the non-equilibrium incorporation of high-Cr olivine in low-Cr melt and the development of a high, but spurious, apparent partition coefficient.

Partitioning of chromium by clinopyroxenes has received much attention, presumably because of the propensity of this mineral group for incorporating chromium in its structure. Wager and Mitchell (1951) determined $K_D^{cpx:melt}$ of 17.6 and 7.0 for their Skaergaard A and B stages; Ewart and Taylor (1969) determined values of 70 and '>200' for the clinopyroxenes of two New Zealand andesites; and Ringwood (1970) obtained a value of 3.5 in the experiments on synthetic lunar basalt referred to above. Bougault and Hekinian (1974) obtained 13.0 for their mid-Atlantic Ridge basalt; Duke (1976) determined values in the range 8.0–36.0 in the experiments already referred to; and Dostal et al. (1983) found values of 28 and 40 for clinopyroxenes of a plutonic cumulate of a Lesser Antilles lava.

Mean $K_D^{cpx:melt}$ for the olivine–pyroxene lavas is 12.26, with a range from 3.38 to 54.78. The values bear a crude but discernible inverse relation with whole-rock MgO (Fig. 15.14). Thus if MgO be taken as a qualitative indication of crystallization temperatures, partition coefficients are, as might be expected, generally inversely related to temperature. The values of 41.22 and 42.89 for clinopyroxenes in big-feldspar lavas and hornblende-andesites respectively appear inordinately high, and it is suspected that, at least in part, they reflect the inheritance factor already referred to in connection with olivine. Chromium contents of clinopyroxenes of the big-feldspar lavas range from 0 to 912 ppm; those of the hornblende-andesites from 10 to 1440 ppm. It seems probable that those with the higher Cr values have been inherited, by turbulent melt mixing, from more mafic environments of crystallization.

While there is insufficient data on the relevant groundmasses for the calculation of a reliable value for $K_D^{opx:melt}$, probe analysis of coexisting clinopyroxenes and

Table 15.4 Crystal:melt distribution coefficients (K_D) with respect to chromium

Mineral species	Lava group	n_1	n_2	m_1	m_2	$K_{D(1)}$	N	M_1	M_2	$K_{D(2)}$
Olivine	OPB	40	44	346	201	1.72	39	338	219	1.54
	BFL	7	23	83	4.52	18.36	6	120	12	10.00
Clinopyroxene	OPB	43	44	2435	201	12.11	42	2685	219	12.26
	BFL	26	23	270	4.52	59.73	17	209	5.07	41.22
	HbA	27	47	241	4.66	51.72	20	190	4.43	42.89
Hornblende	HbA	25	47	120	4.66	25.75	22	110	6.00	18.33
Feldspar	OPB	10	44	0.50	201	0.002	10	3.30	204	0.016
	BFL	23	23	0.65	4.52	0.144	22	0.68	3.23	0.211
	HbA	46	47	0.07	4.66	0.015	44	0.15	4.66	0.032
Spinel	OPB	36	44	96306	201	479	20	84941	212	401
	BFL	23	23	2933	4.52	649	22	3318	3.23	1027
	HbA	37	47	475	4.66	102	36	481	5.03	96

OPB = olivine–clinopyroxene-basalt group; BFL = big-feldspar lavas; HbA = hornblende-andesite group; n_1 = total number of mineral separations analysed; n_2 = total number of groundmass separations analysed; m_1 = mean chromium content of mineral of each lava group; m_2 = mean chromium content of groundmass of each lava group; $K_{D(1)}$ = mean distribution coefficient determined as m_1/m_2; N = number of coexisting mineral:groundmass pairs analysed; M_1 and M_2 = mean chromium contents of coexisting mineral and groundmass respectively; $K_{D(2)}$ = distribution coefficient determined as M_1/M_2, for comparison with $K_{D(1)}$. All concentrations as parts per million by mass of the element.

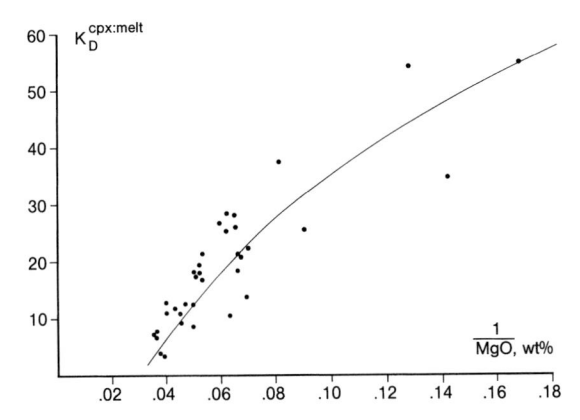

Fig. 15.14. Relation between value of $K_D^{cpx:melt}$ and the reciprocal of whole-rock MgO in the corresponding host lava.

orthopyroxenes (see earlier section on pyroxenes) indicates that within a given crystal:melt environment clinopyroxenes accommodate 3–4 times as much chromium as do the orthopyroxenes. This implies a $K_D^{opx:melt}$ of ≈10–15 in an andesitic melt.

Partitioning of chromium by volcanic hornblendes is intermediate between that induced by the two pyroxene types. Ewart and Taylor (1969) obtained values from 0.52 to 11.0 (mean = 4.08) in rhyolitic matrices and 23 in an andesitic matrix in their study of New Zealand lavas. Dudas et al. (1973) found values from 5.5 to 43.6 (mean = 23.3) in samples of dacitic tephra and Villement et al. (1981) obtained a value of 2.9 for hornblende in alkali basalt. Dostal et al. (1983) calculated a value of 12.5 for hornblende of the plutonic cumulate contained in the Lesser Antilles lava.

$K_D^{Hb:melt}$ values in the Solomons lavas are much more consistent than those calculated for $K_D^{cpx:melt}$, and the mean of 18.33 for 22 coexisting pairs (Table 15.4) is essentially in keeping with values for individual pairings. It is likely that the inheritance factor is less significant for the hornblendes than the clinopyroxenes. Mean $K_D^{cpx:Hb}$ in the andesites, as calculated from the results of column 11, Table 15.4, is 2.34. If some inheritance of clinopyroxenes is taken into account (clearly such inheritance tends to elevate the apparent K_D^{Cr}) this value appears in reasonable agreement with that of 1.63 obtained from probe analysis of the nine clinopyroxene:hornblende pairings referred to earlier.

There appears to be little information on the partitioning of chromium by biotite in the volcanic regime: Higuchi and Nagasawa (1969) obtained a value of 12.6 ± 4.8 for biotite in a dacite and Villement et al. (1981) one of 5.4 in an alkali basalt series. The mean Cr for the biotites of the two Solomon Islands dacites is ≈50 ppm (Table 6.5) and 7 ppm for their groundmasses, giving a mean $K_D^{biot:melt}$ ≈7.

As chromium is one of the principal elements of the spinels, and as the latter have such a pronounced capacity for abstracting chromium from the melt, this element appears in the spinels over a very large range: from major constituent to minor trace. Thus $K_D^{mag:melt}$ values recorded in the literature range from about 4 (e.g. Villement et al. 1981) to >600 (Lindstrom 1976). The very large range of chromium in the Solomons lavas has already been alluded to, and this gives rise to a range of K_D values from 0 to 10^3. The mean value of 401 for $K_D^{mag:melt}$ in the olivine–pyroxene-basalts represents a range from 31 to 6504, the latter relating to a magnetite containing 6504 ppm Cr in a groundmass containing 1 ppm Cr. Magnetite separations from the big-feldspar lavas exhibit chromium contents from 0 to 24 000 ppm, giving a correspondingly large spread of apparent partition coefficients and a mean $K_D^{mag:melt}$ = 1027 (Table 15.4). Of the 22 magnetite: groundmass pairings of Table 15.4, 11 are for samples with <1 ppm Cr in the groundmass. Chromium is more consistent in both magnetite and groundmass in the hornblende-andesites, and the mean of 96 is much more representative of individual pairings than is the case with that for the big-feldspar lavas. Once again the relative abundances of chromium in magnetites and groundmasses, and the resulting apparently exaggerated K_D values, seem to indicate rapid and turbulent mixing of melts and the consequent inheritance of earlier-formed spinels by later-formed hybrid lavas.

THE INCIDENCE OF WHOLE-ROCK CHROMIUM IN THE SOLOMON ISLANDS YOUNGER VOLCANIC SUITE

Mean chromium for the suite as a whole is 324 ppm, and for the basalts, 606 ppm (Fig. 15.15(a)). The latter compares with 268 ppm for MORBs, and 419, 95, 104, and 303 ppm Cr for the basalts of the Lesser Antilles, Aleutians, Tonga–Kermadec, and Vanuatu respectively. The mean chromium content of the Solomons lavas more felsic than basalt (SiO_2 >52 per cent; Fig. 15.15(b)) is, at 29 ppm, comparable with those of 45, 19, and 17 ppm Cr for materials of this category from the Aleutians, Tonga–Kermadec, and Vanuatu respectively, but less than that for the Lesser Antilles, for which the mean value is 112 ppm. As noted in an earlier section, however, this figure is heavily influenced by the very high-Cr andesites and dacites of Grenada. If only Dominica and St Kitts are considered, the relevant mean value is 68 ppm Cr. In comparison with those of other major volcanic arcs the Solomons lavas are thus distinctly high in chromium at

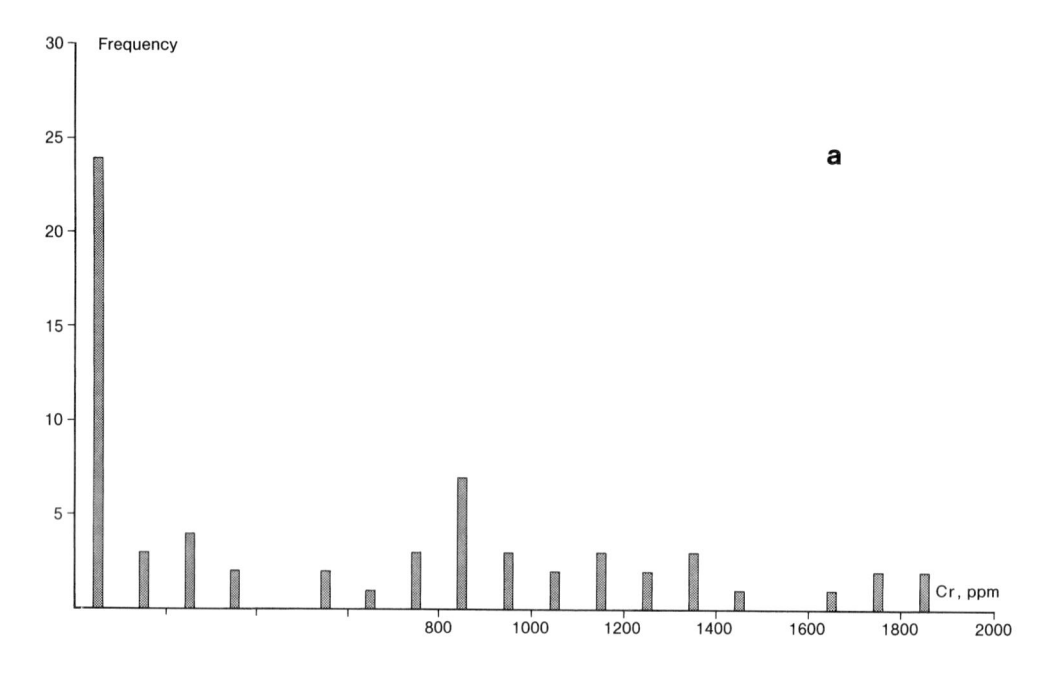

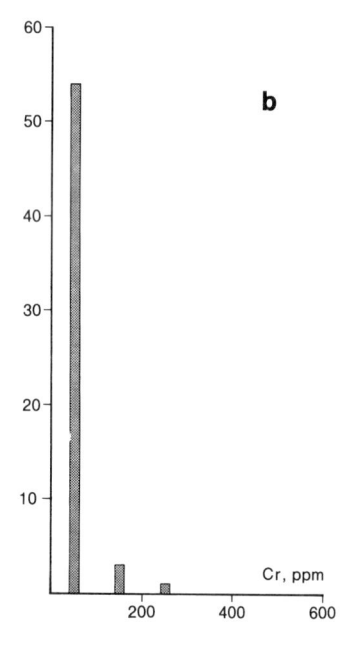

Fig. 15.15. Frequency distributions of chromium abundances in (a) Solomons basalts (SiO_2 <52 per cent), and (b) Solomons lavas more felsic than basalt ($SiO_2 \geqslant 52$ per cent).

the basaltic end of the spectrum, and a little below average in the more felsic members.

The distribution of chromium in the three principal Solomons lava types and their groundmasses (Fig. 15.16) is rather different from those of titanium and vanadium. Whereas these two develop abundance maxima in the big-feldspar basalts and basaltic andesites (Figs. 13.16 and 14.11), chromium simply decreases rapidly from the most mafic olivine–pyroxene-basalts (up to 1820 ppm Cr) through basalts, basaltic andesites, and andesites to

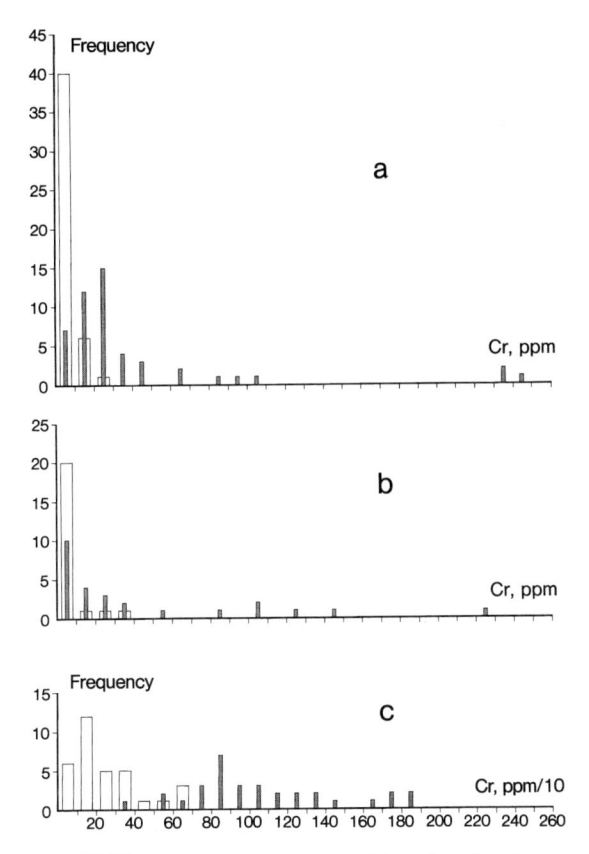

Fig. 15.16. Frequency distributions of chromium abundances in whole rocks (filled bars) and groundmasses (open bars) of (a) lavas of the hornblende-andesite group, (b) big-feldspar lavas, and (c) the olivine–clinopyroxene-basalts of the Solomons suite.

dacites (≈10 ppm Cr). Further, while Ti and V of the mafic basalts tend to be enriched in the groundmass, and that of the andesites and dacites in the crystalline phases, chromium tends to enrichment in the crystalline phases throughout. This stems chiefly from the incorporation of large amounts of chromium in early formed clinopyroxene (mean Cr = 2435 ppm; maximum determined Cr = 4844 ppm). Subtraction of large quantities of such pyroxene, closely followed by the subtraction of very high-chromium spinels, from a parent melt containing, say, ≈800 ppm Cr, rapidly impoverishes the residual melt in this element. The result is a rapid decrease of chromium in the resultant groundmasses and more felsic differentiates. Titanium and vanadium, on the other hand, do not enter any of the early crystal phases in high abundance, and thus initially concentrate in the melt until high-Ti and high-V spinels and associated hornblende begin to crystallize at the basic andesite stage. They therefore tend to maximum abundance in the intermediate members of the lava series, and impoverishment in the groundmasses does not become apparent until the andesites and dacites develop.

A striking feature of chromium abundance in the Solomons lavas — in particular those with MgO >6 per cent — is its close, positive correlation with MgO (Fig. 15.17). It has already been noted that chromium does not show a close correlation with magnesium, or Mg numbers, in any of the principal minerals, and its abundance in olivine (the principal repository of magnesium in many of the lavas) is relatively low. There is, however, a very clear correlation between the two elements in the whole rocks and also, if in somewhat muted form, in the groundmasses. The relationship appears again in the whole-rock analyses of material from Grenada, and Aoba, Vanuatu in the lavas (SiO_2 <48 per cent, MgO >6 per cent) Fig. 15.17(b) and (c)), with a notably steeper slope in the case of the former. It would appear that in its incorporation in spinel, clinopyroxene, and olivine, the combined effect is the removal of chromium from the melt at a rate very closely related to the rate of removal of magnesium; and, as is shown later (Chapter 19), that of nickel.

Ternary plots of CrMA (Cr–MgO–($Na_2O + K_2O$)) (chromium calculated as ppm Cr × 10^{-1}) (Fig. 15.18) show chromium abundances to have a fairly systematic relationship with the evolution of the lava series. The behaviour of chromium in this system is, however, somewhat different from that of the other elements considered so far. First, although the fields for the olivine–pyroxene whole rocks and groundmasses are relatively orderly (Fig. 15.18(f) and (e)), those for the other categories, particularly the big-feldspar lava and hornblende-andesite whole rocks, show much greater scatter than do the analogous plots for the other elements. Second, while the latter either increase with increase in ($Na_2O + K_2O$)/MgO, or increase to a maximum and then decrease, chromium decreases in both whole rocks and groundmasses from the very inception of crystallization. This is almost linear from Cr:M:A ≈85:15:0 to ≈45:35:20, at which stage the field as a whole tends to flatten and trend towards the alkali apex.

The general nature of the trend is shown more clearly in Fig. 15.19, in which mean CrMA points for 2 per cent whole-rock SiO_2 intervals (46–8 per cent — >66 per cent SiO_2) are shown for whole rocks and groundmasses. When the three rock groups are treated together in this way the points for the whole rocks and their groundmasses trace out a relatively smooth curve of Cr (and MgO) reduction to Cr:M:A ≈0:10:90. To facilitate ready comparison with the corresponding figures for titanium

198 *Ore Elements in Arc Lavas*

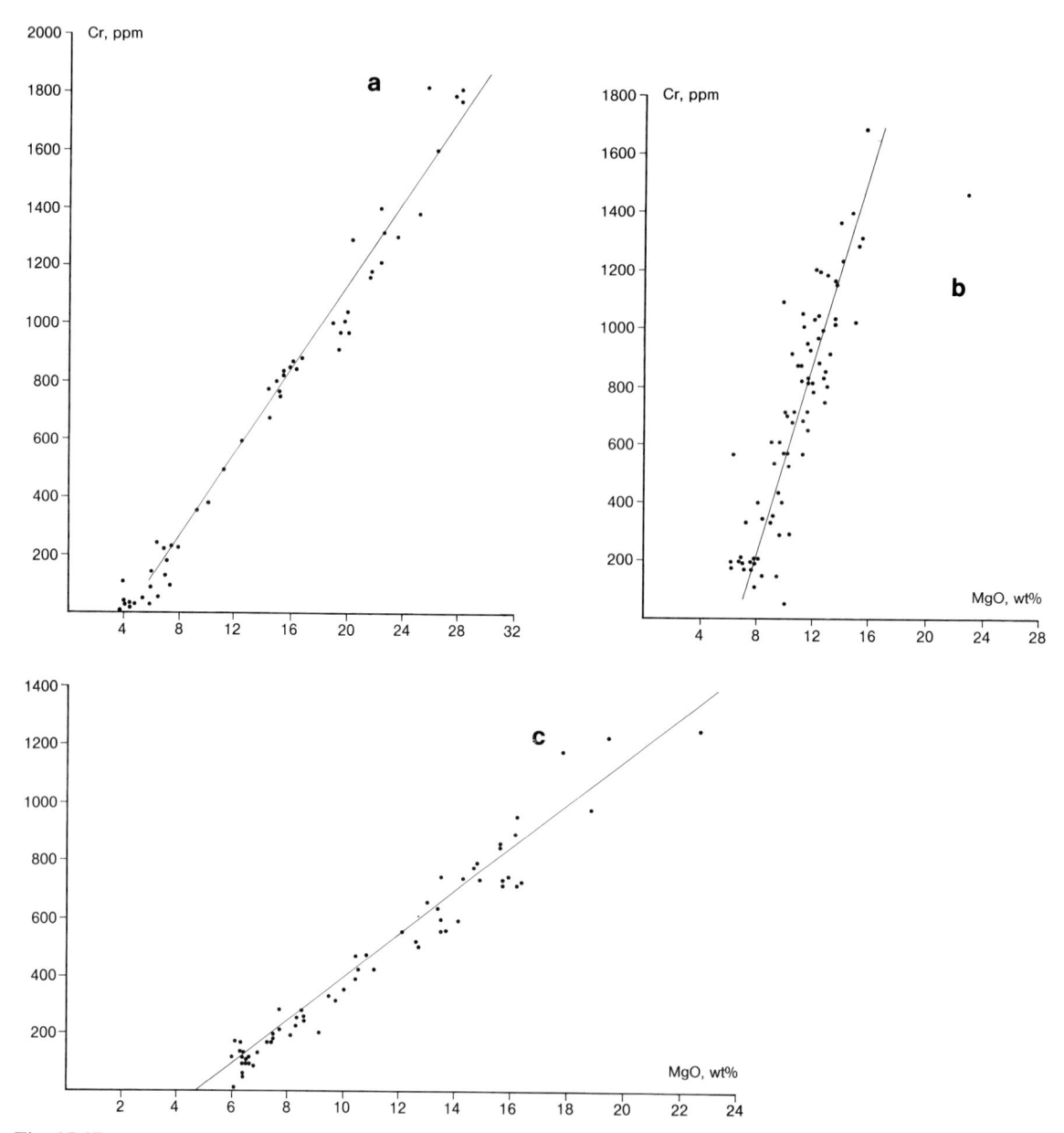

Fig. 15.17. Abundance relations between chromium and MgO in (a) the spectrum of New Georgia basalts, (b) Grenada (Lesser Antilles) basalts, and (c) the basalts of the island of Aoba, Vanuatu. The Grenada basalts contain higher chromium relative to MgO than do those of New Georgia and Aoba, yielding a conspicuously steeper line of correlation.

(Fig. 13.18) and vanadium (Fig. 14.14), the ternary plots of Fig. 15.20 have been constructed using Cr as ppm × 10^{-2} rather than ppm × 10^{-1} as in Figs. 15.18 and 15.19. The points for the Solomons lavas and groundmasses of Fig. 15.20 follow a smooth path of chromium decrease to MgO ≈40 per cent at which stage the curve flattens and Cr → 0 at MgO ≈10 per cent. As in previous cases, the curve is rendered smoother by the inclusion of the

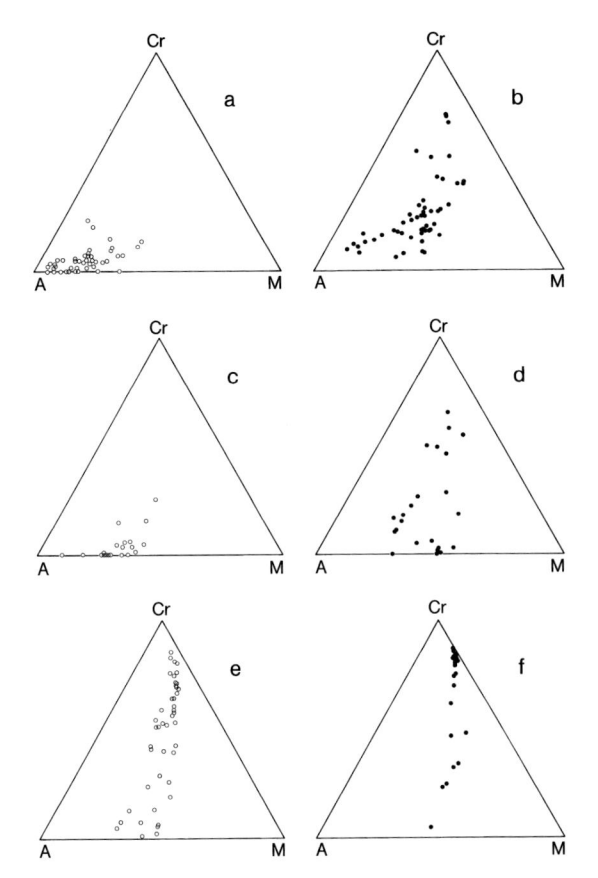

Fig. 15.18. Cr–MgO–(Na$_2$O + K$_2$O) ternary relations (Cr calculated using ppm Cr $\times$ 10^{-1}) developed in olivine–clinopyroxene-basalt, big-feldspar lava and hornblende-andesite whole rocks [filled circles; (f), (d), and (b) respectively], and their corresponding groundmasses [open circles; (e), (c), and (a) respectively].

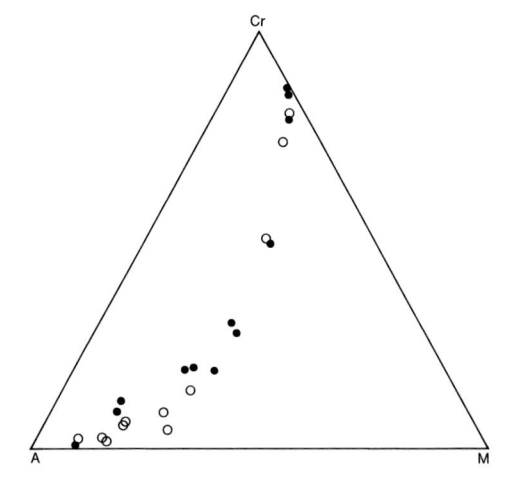

Fig. 15.19. Mean Cr–MgO–(Na$_2$O + K$_2$O) ternary relations corresponding to 2 per cent whole-rock SiO$_2$ intervals in whole rocks (filled circles) and corresponding groundmasses (open circles) of the Solomons suite, excluding Bougainville, with Cr calculated using parts per million Cr $\times$ 10^{-1}. (This figure facilitates comparison of Figs. 15.18, 15.20, and 19.21).

Bougainville lavas (Fig. 15.20(b)). Somewhat similar patterns (at least in principle) are generated by the lavas of Grenada, Vanuatu, and Tonga–Kermadec (Figs. 15.20(c) and (d)). The Solomons suite clearly constitutes a beautiful example of a 'nichrome' pattern with respect to chromium. The Grenada, Vanuatu, and Tonga–Kermadec distributions are less spectacular examples, but conform to the pattern none the less. The lavas of Dominica and St Kitts (Fig. 15.20(c) show little proportionate change in chromium and almost certainly represent nichrome patterns truncated at their mafic ends.

MORBs as represented by the basalts of the mid-Atlantic Ridge display a diffuse, but none the less clearly apparent, nichrome pattern truncated at its more felsic end. The spectrum of Icelandic lavas on the other hand develops a nichrome pattern with almost as broad a span as that of the Solomons.

The general tendency for chromium to decrease in proportion throughout the development of a lava series is, as already noted, due to its subtraction from the melt from the very onset of crystallization. The contrasting tendency of Ti and V to rise to a peak at SiO$_2$ $\approx$52–4 per cent and then decrease stems from the fact that they are not incorporated in crystalline phases to any substantial extent until the point of andesite development is approached and titaniferous spinel and hornblende begin to crystallize in substantial amount.

The broad relation between the abundance of chromium and the development of a lava series is portrayed in a slightly different way in the Cr:SiO$_2$ diagrams of Fig. 15.21. By the time the SiO$_2$ content has reached 52–4 per cent most of the chromium of the Solomons and Vanuatu suites has been incorporated in crystalline phases and little (generally <50 ppm) is left in the melt (Figs 15.21(a) and (c) respectively). The curve representing the Grenada suite (Fig. 15.21(b) also begins to flatten at $\approx$52 per cent SiO$_2$, but does so at a perceptibly higher chromium level: at $\approx$250 ppm Cr as compared with $\approx$50 ppm in the Solomons and Vanuatu suites. The Dominica and St Kitts (Fig. 15.21(b)) and Tonga–Kermadec (Fig. 15.21(d)) curves appear, once again, to represent the more felsic (low chromium) tails of lava assemblages from which the more mafic — accumulative — components are missing.

200 *Ore Elements in Arc Lavas*

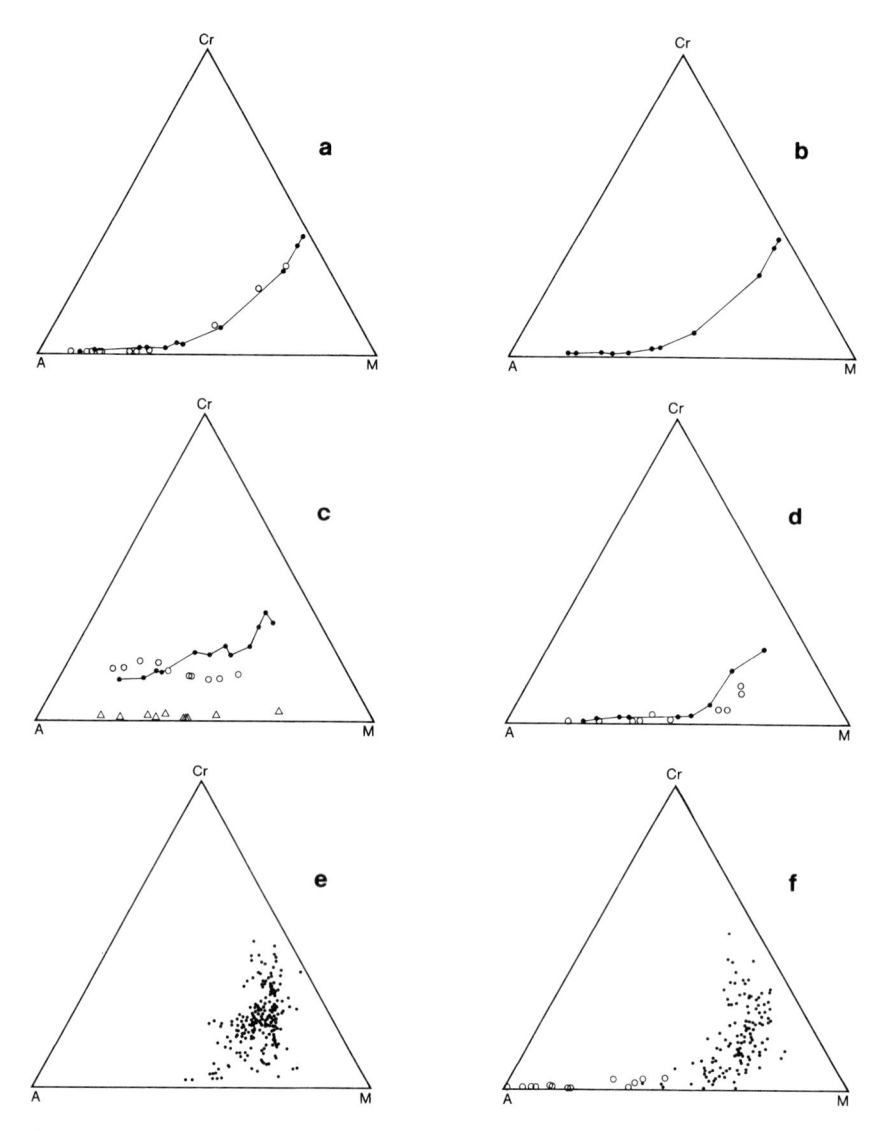

Fig. 15.20. Mean Cr–MgO–(Na$_2$O + K$_2$O) ternary relations (Cr calculated using ppm Cr × 10^{-2}) corresponding to 2 per cent whole-rock SiO$_2$ intervals in (a) whole rocks (filled circles) and corresponding groundmasses (open circles) of the Solomons suite excluding Bougainville; (b) whole rocks of the Solomons suite including Bougainville (241 analyses); (c) lavas of Grenada (filled circles), Dominica (open circles), and St Kitts (open triangles); (d) lavas of Vanuatu (filled circles) and Tonga–Kermadec (open circles). Ternary relations developed in MORBs as represented by Mid-Atlantic Ridge basalts are shown in (e), and in individual samplings of Icelandic basalts (SiO$_2$ <52 per cent; filled circles) and Icelandic lavas more felsic than basalt (SiO$_2$ ⩾52 per cent; open circles) in (f).

CONCLUDING STATEMENT

The approximate abundance of chromium in MORBs is 268 ppm, and in arc basalts, 284 ppm. The mean chromium content of the basalts of the Solomon Islands Younger Volcanic suite is 606 ppm, making them the most Cr-rich of the principal documented arc basalts. As a general phenomenon chromium decreases rapidly from its maximum in picritic basalts to much lower levels in basaltic andesites, andesites, and more felsic lavas. This reflects a very close, linear relationship with whole-rock magnesium.

The principal acceptors of chromium are spinels and clinopyroxenes, both of which commence formation in

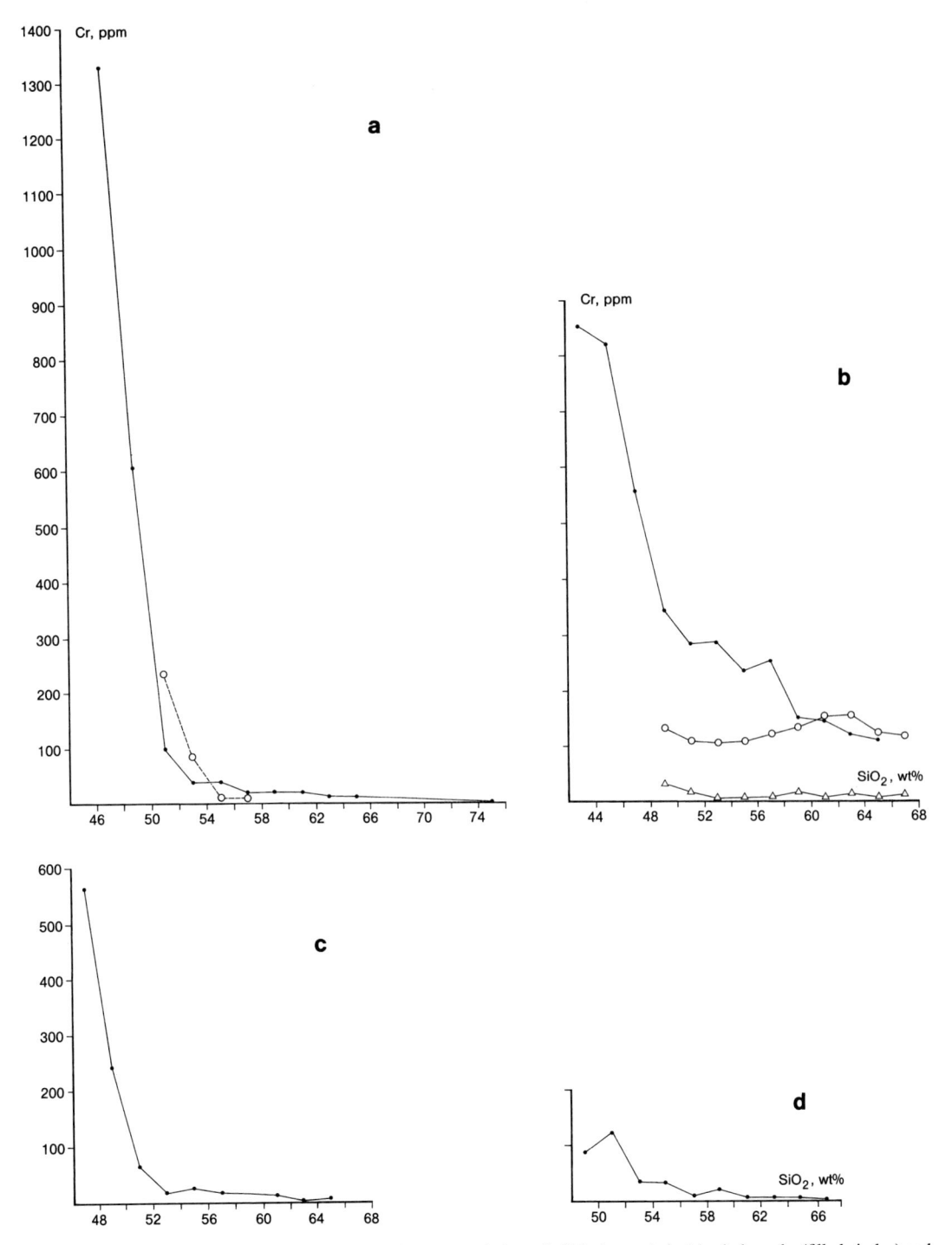

Fig. 15.21. Mean Cr:SiO$_2$ relations corresponding to 2 per cent whole-rock SiO$_2$ intervals in (a) whole rocks (filled circles) and corresponding groundmasses (open circles) of the Solomons suite; (b) lavas of Grenada (filled circles), Dominica (open circles), and St Kitts (open triangles), (c) Vanuatu, and (d) Tonga–Kermadec.

the early stages of lava crystallization. This leads to the very rapid removal of chromium from the melt, a marked sensitivity of chromium abundance to degree of fractionation, and to Cr–MgO–(Na$_2$O + K$_2$O) relations in sharp contrast to those of V–MgO–(Na$_2$O + K$_2$O). Whereas vanadium tends not to be accepted into earliest-formed crystals and hence reaches a peak in abundance in lavas in the 52–8 per cent SiO$_2$ range before declining in the more felsic residual melts, chromium is abstracted strongly from the beginning of crystallization, and as a result exhibits a rapid initial decrease to ≈52 per cent whole-rock SiO$_2$, after which the rate of subtraction declines as Cr is virtually eliminated from the residual melt.

Such a sequence of crystallization leads to the development of a 'nichrome' pattern of Cr–MgO–(Na$_2$O + K$_2$O) relations in both arc and mid-ocean ridge (including Icelandic) lava assemblages.

16

Manganese

Manganese is a commonly prominent, but erratically occurring, component of exhalative ores. It may appear (usually as complex mixed oxides, but also as silicate and carbonate and as a component of some sulphides) as the dominating element of exhalative sediments, and a number of important manganese ore deposits are of this nature. It also occurs, as a constituent of iron:manganese-rich exhalites, as an associate of, but separate from, some base metal sulphide ores; and in other instances constitutes an important component of the sulphide orebody itself.

Exhalative manganese oxide–silicate lenses and beds of manganiferous chert and jasper are common features of the 'Steinmann Trinity' or 'Steinmann Association' (Stanton 1972). As such they occur in general association with Cyprus-type Cu and Cu–Zn ores of the appropriate provinces in various parts of the world. Cyprus itself, together with related provinces in nearby Greece and Turkey, the Sandbagawa schist belt in Japan, the Bay of Islands complex of Newfoundland, the Franciscan of California, and the Woolomin terrain of New South Wales, all provide examples of the prominent association of manganiferous cherts and argillites with oceanic basaltic rock (ranging from picritic basalts to basaltic andesites) and Cyprus-type sulphide ores.

Similarly many of the zinc and zinc:lead-rich exhalative ores of volcanic arc provinces such as those of the Northern Appalachians, Norwegian Caledonides, southeastern Australia, and the Iberian Peninsular are also accompanied by notable levels of exhalative–sedimentary manganese. Here the volcanic association tends to be with andesites, dacites, and rhyolites rather than with the more basaltic rocks of the Cyprus-type environment. The incidence of manganese in these occurrences ranges from abundant and conspicuous ($\approx$5–10 per cent) to amounts that are significantly, but only slightly, above the normal background level of the enclosing sedimentary rocks.

Some exhalative base metal sulphide deposits contain very large amounts of manganese as an intrinsic part of the ore. Outstanding examples are the deposits of Broken Hill, New South Wales, Pegmont, Queensland, and those of Namaqualand in southern Africa. Manganese contents of routine samplings may exceed 20 per cent, and some segments of the Broken Hill deposits contain thousands of tons of almost pure manganese silicate. The manganese occurs as rhodonite, manganhedenbergite and related pyroxenoids, tephroitic olivine, spessatine-rich garnet, manganiferous amphiboles, as a component of sphalerite and of any associated magnetite. In most deposits of this kind the incidence of the manganese is closely tied to that of the sulphide metals yielding a general spatial coincidence very similar to that developed by phosphate, referred to in Chapter 11. Fig. 16.1 shows this manganese–sulphide metal relationship as it appears in three drill cores through A-lode of the Broken Hill orebody. Not surprisingly, many of the deposits of this general kind are characterized by conspicuously manganiferous gossans.

In spite of this abundance of manganese, and its quite intimate association with the sulphides in these deposits, there are other Lower Proterozoic occurrences, of very similar age, size and metal abundances, from which manganese is conspicuously low or absent. Examples are the Mount Isa-Hilton and Macarthur River deposits of Australia and the Sullivan and related deposits of British Columbia. Similarly most of the Archaean volcanogenic copper and copper–zinc orebodies of Canada and the Yilgarn province of Western Australia contain little manganese. Whether this reflects a difference in the nature of the materials contributed to the ore-forming environment, i.e. differences in the ore solutions as derived from their source or as modified by transport factors, or whether it stems from differences in the sedimentary conditions of ore deposition, is not known. With its solubility so sensitively dependent on small changes in Eh and pH in the aquatic environment, the absence of manganese from some exhalative orebodies may well be a reflection of depositional environment rather than of variation in source material.

Be this as it may, manganese is a common and commonly abundant component of exhalative stratiform and other ores of volcanic association, and the spatial tie between the manganese and other ore elements is often so close that a fundamental genetic relationship seems certain.

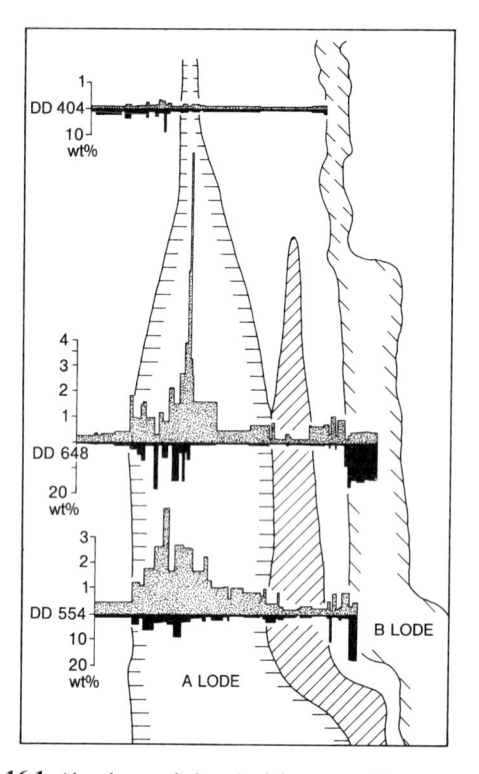

Fig. 16.1 Abundance relations (weight per cent) between MnO (stippled bars) and sulphide sulphur (filled bars) in three diamond drill cores through Broken Hill A-lode (horizontal hatchuring) on Section 68, New Broken Hill Consolidated Ltd. Bar widths indicate length of intersection of individual lithological units (and hence samplings); weight percentages are indicated by the scales normal to the drill holes. Partial right-hand diagonal hatchuring, B-lode; full left hand diagonal hatchuring, Separation Pegmatite. Graduated bars at left-hand extremities of drill holes indicate 1.0 per cent intervals of MnO, 10 per cent intervals of sulphide sulphur. After Stanton, Roberts, and Chant (1978).

That manganese constituted a 'minor major' component of virtually all igneous rocks was established from the earliest days of rock analysis. Clarke and Washington (1924) estimated an average of 0.086 per cent Mn (0.11 per cent MnO) for all igneous rocks. von Hevesy et al. (1934) proposed an average of 0.098 per cent Mn (0.13 per cent MnO) and Goldschmidt (1954) estimated the average content in the upper lithosphere at close to 0.09 per cent Mn or 0.12 per cent MnO.

Goldschmidt (1954) observed that the geochemistry of manganese in igneous rocks is dominated by the manganous (Mn^{2+}) ion and that he knew of 'not a single instance' of appreciable amounts of trivalent or quadrivalent manganese in any igneous rock or mineral. He pointed out that igneous melts scarcely attain an oxidation potential high enough for the formation of trivalent manganese. He noted that in general the highest absolute amounts of manganese occur in basaltic rocks and their relatives, and that abundances decrease with increase in felsic index: a pattern of behaviour shared with the other members of 'the magnesium–iron group of doubly-charged cations (Mg, Fe, Ni, Co, Mn)'. He also drew attention to the fact that, owing to its larger ionic size, Mn^{2+} tended to enter mineral lattice sites after Mg^{2+} and Fe^{2+}. Thus Mn^{2+} decreased as the whole-rock felsic index increased, but in doing so it lagged behind Mg^{2+} and Fe^{2+}. The result was that decrease in Mn was accompanied by an increase in Mn/Mg and Mn/Fe as crystallization of the melt proceeded. Goldschmidt suggested from this that the ratio MnO:MgO might be used as a crystallization, or differentiation, index.

In a very comprehensive compilation of analyses of Cenozoic volcanic rocks Wedepohl (1978) confirmed Goldschmidt's deduction that manganese decreased progressively from basalt to rhyolite and demonstrated that the overall factor was close to 2 (basalts, 1318 ppm; rhyolites, 620 ppm Mn). In addition he estimated an average of 1320 ppm Mn (0.17 per cent MnO) in 'abyssal oceanic basalts', 1317 ppm Mn (0.17 per cent MnO) in tholeiitic basalts of island arcs, and 1239 ppm Mn (0.16 per cent MnO) in oceanic island alkali olivine basalts.

MANGANESE IN LAVAS OF THE MARINE ENVIRONMENT

There are few modern igneous rock analyses that do not include an estimation of MnO, and the data on manganese in MORBs are thus extensive (Table 6.2). The overall average for 507 samplings of the ridges themselves is 0.171 per cent MnO (0.132 per cent Mn) and for samplings including Icelandic basalts 0.177 per cent MnO (0.137 per cent Mn). Averages for the four ridge systems considered — mid-Atlantic, Atlantic–Indian, Indian, and Pacific — are 0.18, 0.18, 0.16, and 0.16 per cent MnO respectively. The spread of values for all individual samplings from all ridge systems is given in Fig. 16.2(a) which shows a clear mode in the 0.16–0.20 per cent MnO range.

Manganese is distinctly higher in the Icelandic basalts than in sea-floor MORBs (Fig. 16.2(b)), with a mean of 0.201 per cent MnO (0.156 per cent Mn) and a clear mode in the 0.18–0.22 per cent range. The Icelandic lavas more felsic than basalts ($SiO_2 \geqslant 52$ per cent; open bars at twice the vertical scale for the filled bars) have a larger spread but a lower mean (0.14 per cent MnO) than the basalts. The reason for this clearly higher level of manganese in the Icelandic basalts relative to MORBs is not obvious.

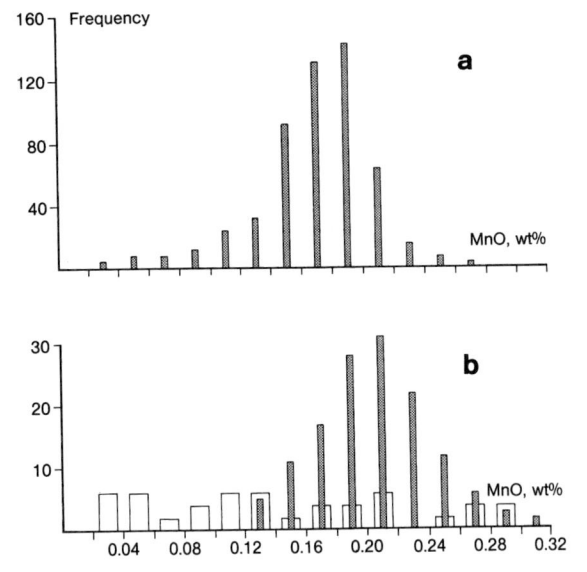

Fig. 16.2. Frequency distributions of MnO abundances in (a) MORBs of all major oceans, and (b) Icelandic basalts (SiO_2 <52 per cent; solid bars) and Icelandic lavas more felsic than basalts ($SiO_2 \geq 52$ per cent; open bars).

Abundance relations between MnO and total iron as FeO in MORBs (as represented by mid-Atlantic ridge basalts) and in Icelandic basalts are shown in Fig. 16.3.

The frequency distribution of MnO in 489 arc basalts (Fig. 16.4(a)) is generally similar to that of MORBs in Fig. 16.2(a) in that both have similar modes (0.18–0.20 per cent MnO) and approximately similar spreads. The distribution of values for the arcs is, however, slightly positively skewed and that for MORBs slightly negatively. This is reflected in the relevant mean values: 0.19–0.20 per cent for the arcs; 0.17 per cent MnO for MORBs. Average MnO for individual arcs ranges from 0.17 in the lesser Antilles to 0.20 per cent for Vanuatu (Table 6.3). The three Melanesian arcs, the Solomons, Vanuatu, and Tonga–Kermadec, show remarkable similarity in the MnO content of their basalts at 0.196, 0.197, and 0.195 per cent respectively.

The frequency distribution for 746 arc lavas with SiO_2 ≥ 52 per cent exhibits a substantially symmetrical bell shape (Fig. 16.4(b)) with a mode to the left (0.14–0.18 per cent MnO) of that for the basalts and a correspondingly lower mean value of 0.15–0.16 per cent MnO. This mean for the broad category of lavas of SiO_2 >52 per cent does, however, mask the relatively high MnO in the andesite category (52 <SiO_2 <62 per cent) of the Lesser Antilles. Mean MnO contents for basalts, andesites, and dacites of this province are 0.17, 0.17, and 0.14 per cent respectively. Apart from this the arc suites show

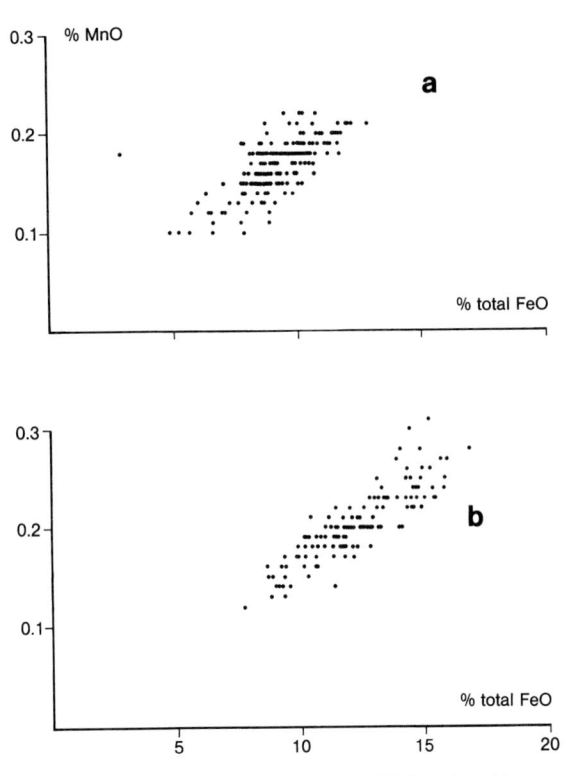

Fig. 16.3. Abundance relations between MnO and total iron as FeO in (a) MORBs of the Mid-Atlantic Ridge, and (b) Icelandic basalts.

a progressive reduction in manganese from basalts to dacites and more felsic types. On the basis of the data of Table 6.3, mean values for basalts, andesites, and dacites for the five arcs taken together are 0.20, 0.17, and 0.13 per cent MnO respectively.

CHEMICAL PROPERTIES AND CRYSTAL CHEMISTRY OF MANGANESE

Stemming from its relatively common incidence in any one of three oxidation states, Mn^{2+}, Mn^{3+}, and Mn^{4+}, the crystal chemistry and hence mineralogical affinities of manganese are complex. It is dominantly lithophile in the crustal environment, but it is also clearly chalcophile to a degree. It is well known as occurring as the monosulphide alabandite (MnS) and the disulphide hauerite (MnS_2; pyrite structure), but perhaps its best-known form of sulphide occurrence is as a minor component of sphalerite, of which it may constitute 1 per cent or more. It is a component of rhodochrosite and mixed (Ca, Fe^{2+}, Mn^{2+}) carbonates, rhodonite, and mixed (Ca, Fe^{2+}, Mn^{2+}) SiO_3

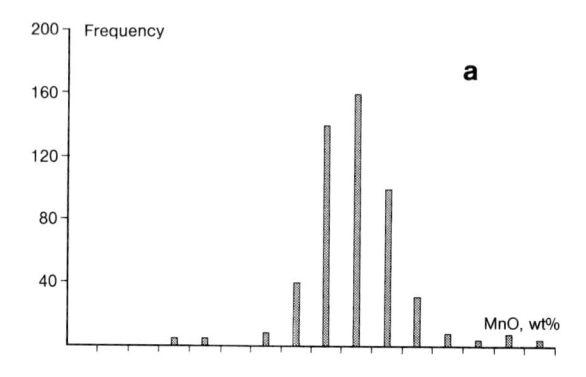

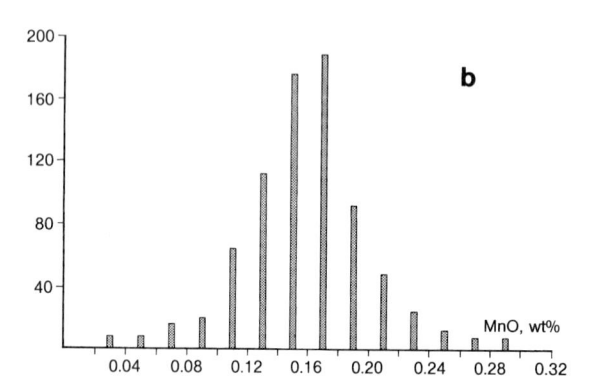

Fig. 16.4. Frequency distribution of MnO abundances in (a) volcanic island-arc basalts ($SiO_2 < 52$ per cent), and (b) arc lavas more felsic than basalt ($SiO_2 \geq 52$ per cent).

pyroxenoids, and of a large number of other silicates in the olivine, pyroxene, amphibole, and mica groups. With its propensity for entering the spinel structure manganese is a common component of magnetite, and where exhalative sulphide ores are accompanied by Fe:Mn-rich exhalites, the associated manganese may be present in all of sulphide, silicate, and oxide combinations.

The incidence of manganese in igneous, including volcanic, rocks is, however, virtually entirely dominated by the Mn^{2+} ion in the silicate Mn–O context. Its ionic radius of 0.091 nm yields a radius ratio with respect to O^{2-} of 0.650, almost exactly half-way between 0.414 and 0.732, the limiting ratios for octahedral coordination with oxygen atoms. Mn^{2+} therefore occurs dominantly (though not exclusively) in divalent octahedral positions in silicate lattices, with an Mn–O bond length ≈0.116 nm. As pointed out by Goldschmidt (1954) and earlier in this chapter, the size of the Mn^{2+} ion relative to Mg^{2+} (0.078 nm) and Fe^{2+} (0.082 nm) leads to a tendency — now recognized as a consequence of octahedral site-preference energies — for the acceptance of magnesium, iron, and manganese into these sites in that order, so that Mn:Mg and Mn:Fe ratios tend to increase with progress in crystallization of igneous silicates. Manganese does not have any sympathetic abundance relationship with magnesium, however: its direct relationship — which in many of the common silicates is very close — is with ferrous iron, and any relationship with magnesium thus tends to be negative. As well as being the closer to Mn^{2+} in size, Fe^{2+} also forms as M–O bonding very similar to that of manganese.

As the size difference between Mn^{2+} and Ca^{2+} (0.106 nm) is 14.15 per cent of the latter, manganous manganese may substitute extensively for calcium in appropriate sites. Little manganese substitutes for Ca in the plagioclases, however — a state of affairs considered in more detail a little further on. Almost all the manganese of volcanic rocks occurs in the octahedral Mg:Fe:Mn sites of ferromagnesian silicates and in the tetrahedrally coordinated divalent sites of the spinels.

The principal features of the manganese atom and ions are given in Table 6.4.

THE INCIDENCE OF MANGANESE IN THE PRINCIPAL MINERAL SPECIES OF THE SOLOMON ISLANDS YOUNGER VOLCANIC SUITE

General

Stemming from the pronounced affinity of Mn^{2+} for Fe^{2+} lattice positions referred to above, manganese shows a marked positive correlation with ferrous iron in most ferromagnesian silicates. It may therefore vary greatly in abundance in a single mineral species. Statements of abundances of manganese in the principal volcanic silicates thus have little meaning unless the relevant Fe:Mg ratios are given.

Fig. 16.5 shows Mn:FeO relations in analysed olivines, ortho- and clinopyroxenes, and hornblendes from the Lesser Antilles, Aleutians, Tonga–Kermadec, and Vanuatu arcs. Correlation of the two in the volcanic olivines (Fig. 16.5(a)) is moderate, with a possible indication that the distribution is curved rather than linear, or in the form of two linear distributions with a slight inflexion at ≈25 per cent FeO:4000 ppm Mn (cf. FeO:Zn relations of Fig. 7.5). The latter pattern is similar in principle to that noted between Mg and Mn (and hence, in a complementary way, between Fe and Mn) in a wide range of igneous olivines by Simpkin and Smith (1970). The overall Fe:Mn ratio for the 105 olivines considered is 43.74.

The pyroxenes show lower Fe:Mn ratios and a somewhat lesser degree of correlation between the two

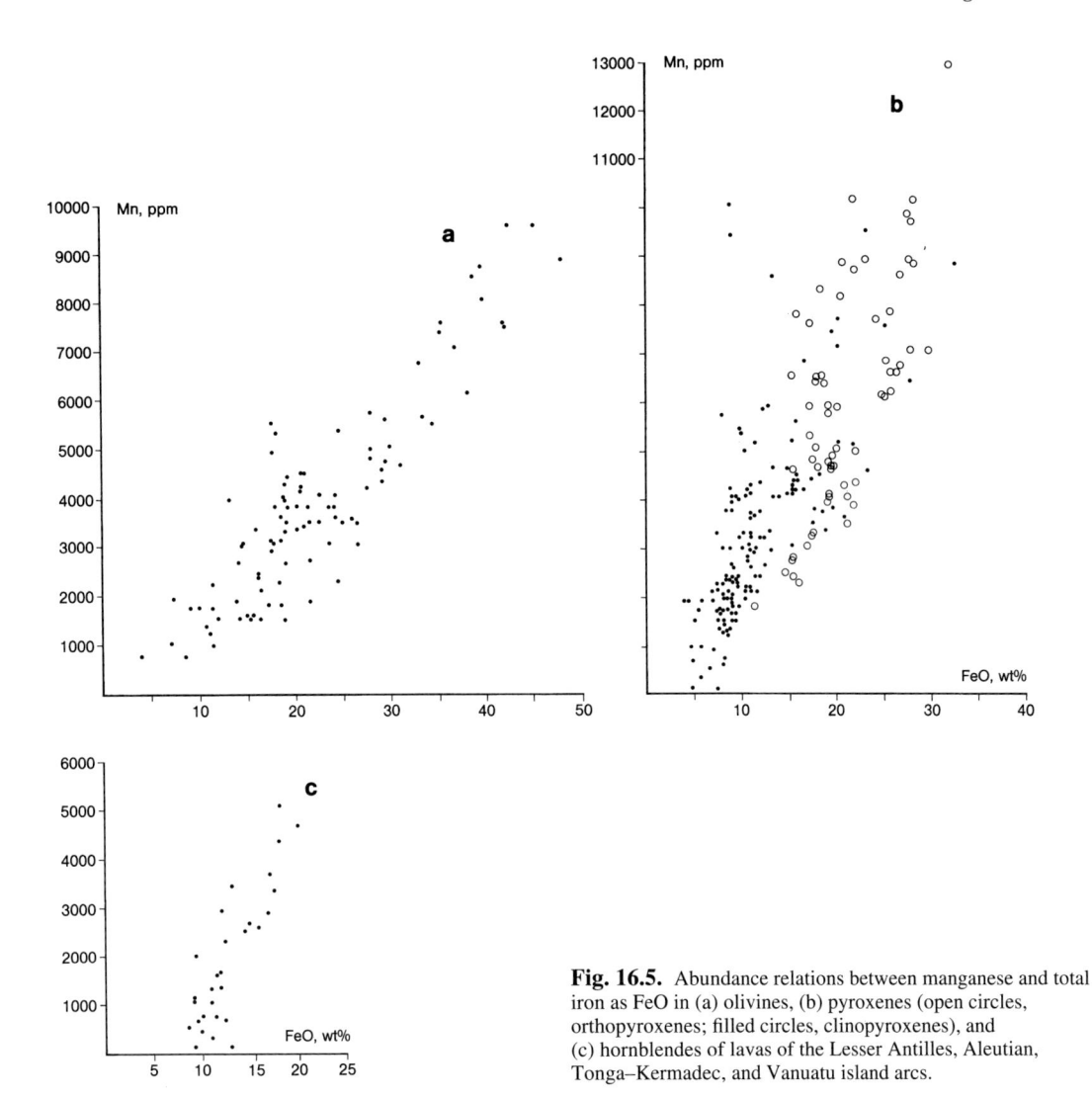

Fig. 16.5. Abundance relations between manganese and total iron as FeO in (a) olivines, (b) pyroxenes (open circles, orthopyroxenes; filled circles, clinopyroxenes), and (c) hornblendes of lavas of the Lesser Antilles, Aleutian, Tonga–Kermadec, and Vanuatu island arcs.

(Fig. 16.5(b)). It must, however, be kept in mind that the 232 analyses (165 clinopyroxenes; 67 orthopyroxenes) cover a very wide geographical spread; when pyroxenes from individual arcs are considered, the degree of correlation is higher. The mean Fe/Mn ratio for the 165 clinopyroxenes of Fig. 16.5(b) is 26.87, and for the 67 orthopyroxenes 27.21. The remarkable similarity of the two seems to suggest that it is the iron content, rather than differences in structure or other compositional features, that constitutes the principal influence in the incorporation of manganese in the pyroxene lattice.

The available sampling of volcanic-arc hornblendes in the literature is small, but the 46 analyses of Fig. 16.5(c) indicate quite a good correlation between FeO and Mn, and a mean Fe:Mn ratio of 52.62.

Mean Mn contents of the olivines, ortho- and clinopyroxenes, and hornblendes of Fig. 16.5 are 3923, 6050, 3227, and 1965 ppm respectively. Very sparse information on biotites indicates values generally in the 2000–3000 ppm range, for plagioclases 50–120 ppm Mn, and for magnetites 2000–7000 ppm — though some individual volcanic magnetites may contain more than 20 000 ppm Mn.

Solomons suite

Mean values of manganese (as MnO) in the principal silicates and in the spinels are given in Table 6.5. The ferromagnesian minerals contain remarkably similar amounts of manganese, with an overall arithmetical mean of 2640 ppm Mn (0.34 per cent MnO). In spite of the potential for substitution of Mn^{2+} for Ca^{2+} the feldspars show the normal very low manganese contents and a mean (which is almost certainly an over-estimate) of 107 ppm. As with the other transition elements, the spinel structure is the principal Mn acceptor, and the titano-magnetites have a mean of ≈0.6 per cent MnO (≈4600 ppm Mn).

Olivine. For the full range of Fe:Mg ratios the Solomons olivines contain an average of ≈2600 ppm Mn and an overall spread of values as shown in Fig. 16.6(a). Whereas the volcanic-arc olivines of Fig. 16.5(a) showed no more than a muted inflexion in their FeO: Mn relations, those of the Solomons show a quite unambiguous change of slope at ≈23 per cent FeO: 3500 ppm Mn (Fig. 16.7(a)). Whether, however, this represents a curve or two distinct trends remains unclear: the writer suspects the latter, though further data is required to resolve the ambiguity. The mean Fe:Mn ratio for the Solomons olivines is 60.48: distinctly higher than that of 43.74 developed in the olivines of the other arcs. This is due to the heavy weighting of highly forsteritic olivines, with their low Mn contents, in the Solomons sampling. The mean Fe:Mn ratio of the lower trend of Fig. 16.7(a) is ≈66 but that of the upper, steeper slope is ≈43: very similar to that for the olivines of the other arcs.

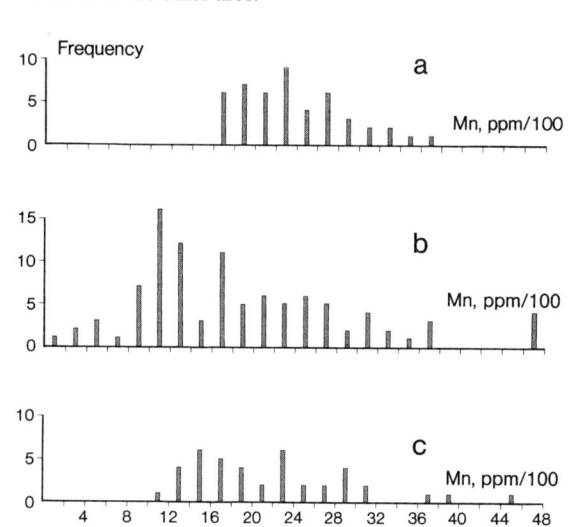

Fig. 16.6. Frequency distributions of manganese abundances in (a) olivines, (b) clinopyroxenes, and (c) hornblendes of the Solomons suite, as determined by X-ray fluorescence analysis of mineral separations. (See Fig. 16.8 (a) for clinopyroxenes containing >4800 ppm Mn).

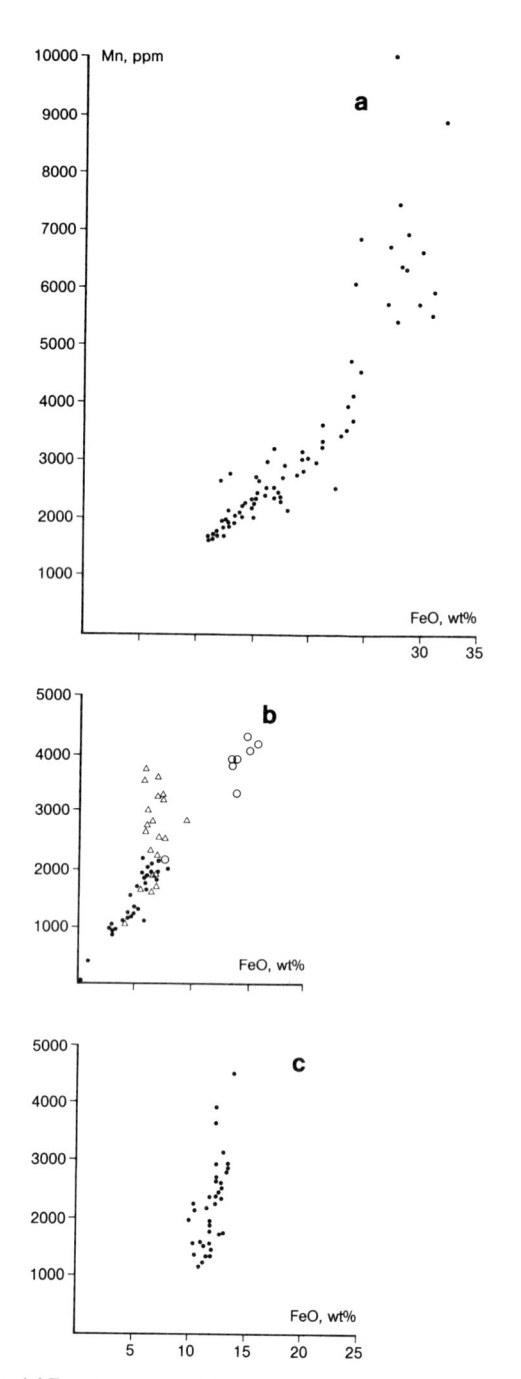

Fig. 16.7. Abundance relations between manganese and total iron as FeO in (a) olivines; (b) clinopyroxenes of the olivine–clinopyroxene-basalts (filled circles), clinopyroxenes of the big-feldspar lavas and hornblende-andesite group (open triangles), and orthopyroxenes (open circles); and (c) hornblendes of the Solomons suite. Olivines, clinopyroxenes, and hornblendes analysed by X-ray fluorescence; orthopyroxenes by electron microprobe.

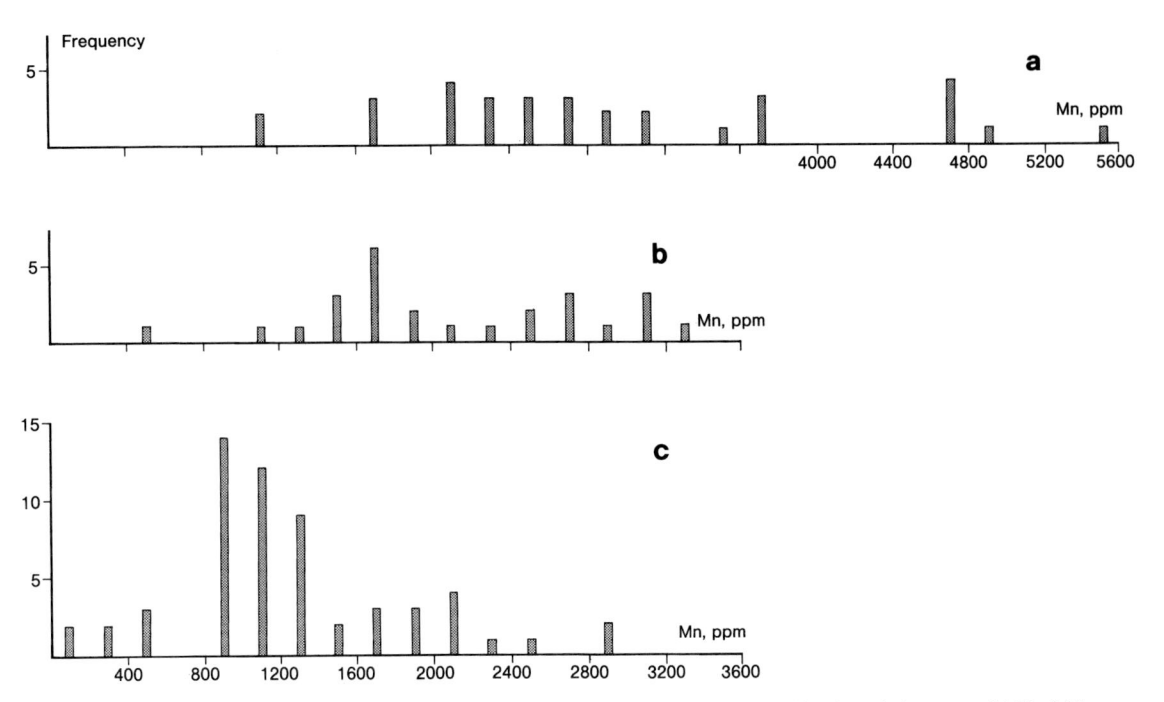

Fig. 16.8. Frequency distributions of manganese abundances in clinopyroxenes of (a) hornblende-andesite group, (b) big-feldspar lavas, and (c) olivine–clinopyroxene-basalts of the Solomons suite.

Pyroxenes. The manganese content of the clinopyroxenes varies greatly in accordance with petrological context (Table 6.5; Fig. 16.8). Frequency distributions of manganese values in the clinopyroxenes of the three principal lava types overlap substantially, but there is a clear trend towards increasing manganese in the pyroxenes as the host rocks become more felsic. Mean Mn in the clinopyroxenes of the hornblende-andesites (2757 ppm) is 2.3 times that in the olivine–pyroxene-basalts (1223). As in all other volcanic two-pyroxene suites, manganese tends to partition into the orthopyroxenes: Fig. 16.9 shows the distribution of Mn between coexisting clinopyroxene (85 analyses) and orthopyroxene (81 analyses) in eight Solomon Islands two-pyroxene andesites. The small group of low-Mn orthopyroxenes of Fig. 16.9 are the pigeonites already referred to. Mean Mn for orthopyroxene and clinopyroxene in these eight lavas is 4022 and 1855 ppm respectively, giving $K_D^{opx:cpx} = 2.17$, in good agreement with the range 1.5–3.0 indicated by Wedepohl (1978).

Although the correlation between FeO and Mn in individual pyroxenes can at best be described as 'clear but not close' (Fig. 16.7(b)), the relevant mean values for the eight two-pyroxene-andesites show a high level of correlation and a strikingly close tie between the FeO: Mn relations in the two groups of pyroxenes (Fig. 16.10).

Hornblende. At ≈2000 ppm (0.26 per cent MnO) the mean manganese content of the Solomons hornblendes is similar to that of the Solomons clinopyroxenes overall (≈1950 ppm Mn; 0.25 per cent MnO) (Table 6.5; Fig. 16.6(c)). At first sight Fig. 16.7(c) appears to indicate, as in the case of the hornblendes of the other arcs, a good linear correlation of Mn with FeO. Closer inspection reveals, however, that the very steep slope of the 'cor-

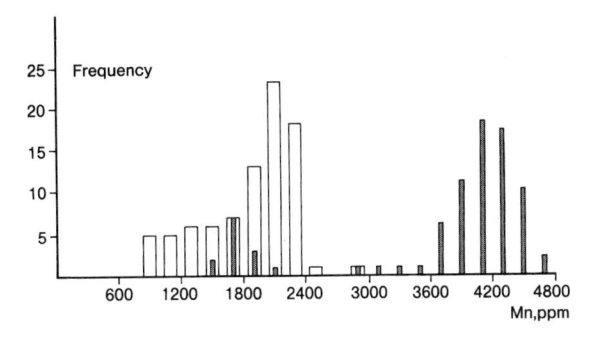

Fig. 16.9. Frequency distributions of manganese abundances in clinopyroxenes (open bars, 85 analyses, mean = 1855 ppm Mn) and orthopyroxenes (filled bars, 81, analyses, mean = 4022 Mn) in eight Solomon Islands two-pyroxene-bearing andesites.

210 *Ore Elements in Arc Lavas*

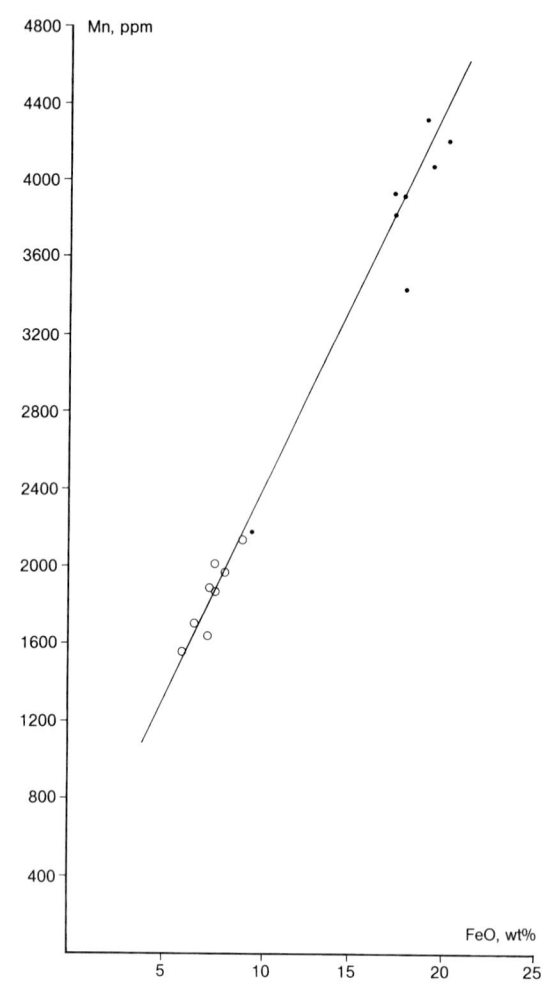

Fig. 16.10. Abundance relations between manganese and total iron as FeO in the coexisting orthopyroxenes (filled circles) and clinopyroxenes (open circles) of eight two-pyroxene-andesites of the Solomons suite. Each point represents the mean of 12 electron-microprobe analyses.

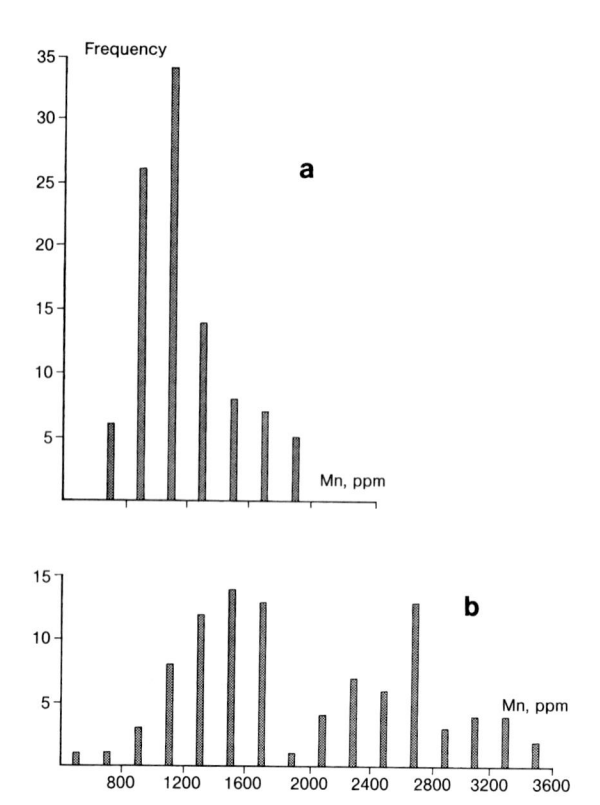

Fig. 16.11. Frequency distributions of manganese abundances in (a) 100 hornblende crystals, and (b) the 100 corresponding opaque rims in a single Solomon Islands hornblende-andesite.

relation' may reflect no more than that the FeO contents of the Solomons are confined to a narrow range 10–14 per cent and that within this Mn may vary anywhere between 1000 and 4500 ppm, i.e. that there is no correlation between iron and manganese in these hornblendes.

The development of opaque rims to the hornblendes is accompanied by a clear and ubiquitous concentration of manganese in the rims (Tables 5.4 and 5.5; Fig. 7.9). The enrichment factor ranges from ≈1.2 to 4.5 in the pairing analysed, with a mean of ≈2.0. Fig. 16.11 shows the results of paired analysis of 100 hornblende crystals and their rims in the same section in which the single crystal of Fig. 7.9 occurs. In all cases the relevant rim contained higher Mn than the parent crystal, though variation from one crystal to another leads to a substantial overlap in the two frequency distributions of Fig. 16.11. In this case mean Mn for crystal and rim are 1162 and 2014 ppm respectively, giving a mean enrichment factor of 1.73. (The reason for the bimodality of the frequency distribution of Fig. 16.11(b) is not known, but probably lies in the finely intergrown polyphase nature of the rim material. The feature calls for further investigation.) Although the pattern of Fig. 7.9 is not entirely unambiguous on this point, it seems most likely that at least the major part of the manganese added to the rim as it forms is derived from the melt as crystal and melt react. The enrichment factor for this single crystal:rim pairing is 1.81. Coexisting clinopyroxene is moderately abundant in this lava, and $K_D^{Hb:cpx} = 0.67$.

Biotite. To the order of accuracy of sampling, the manganese content of biotite (mean = 2556 ppm Mn (0.33 per cent MnO); Table 6.5) is generally similar to that of clinopyroxene and hornblende, and less than that of orthopyroxene. Detailed microprobe analysis of 66 co-

Manganese 211

mate and penultimate purity plagioclase separations (Fig. 16.13) indicates manganese to be slightly more abundant in the latter, and hence the likelihood that there has been significant partitioning of this element into the minute inclusions of glass referred to in preceding chapters.

Spinel. Spinel is by far the principal host of manganese in the Solomons, as in other lavas. Mean manganese for

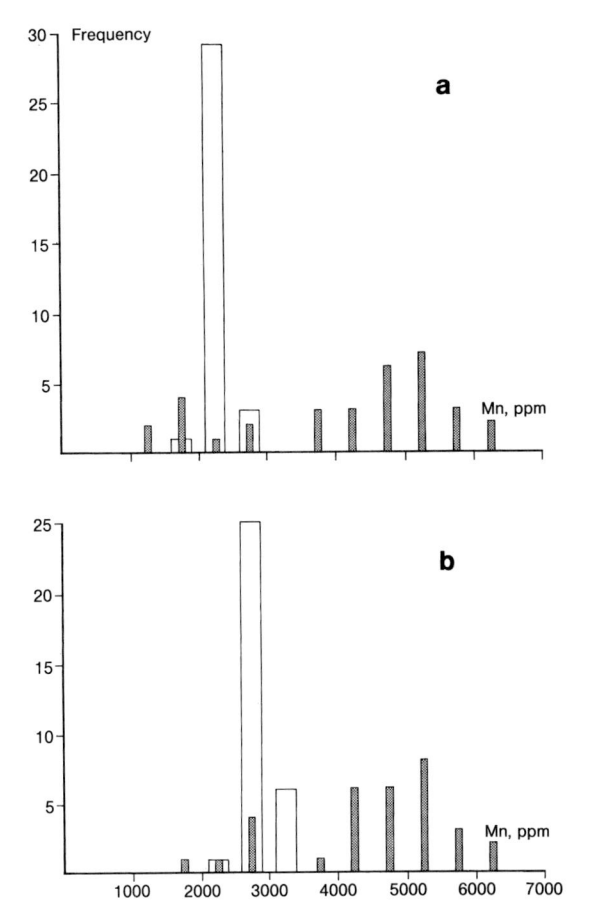

Fig. 16.12. Frequency distributions of manganese abundances in coexisting biotites (open bars), and hornblendes (filled bars) in the two Savo dacites 41/80 and 6/81, designated (a) and (b) respectively.

existing hornblende–biotite pairs (i.e. a total of 66 hornblendes and 66 biotites) in each of the two Savo dacites (Fig. 16.12) indicates, however, that, at this point in lava evolution at least, hornblende accepts more manganese than does biotite. Mean manganese contents in hornblende and biotite of the relevant two dacites are 4162 and 2251 ppm and 4660 and 2834 ppm respectively. The two pairs of histograms also show that the biotites are much more consistent in their manganese content than are the hornblendes.

Plagioclase. The abundances of Mn in the Solomons feldspars (Table 6.5) are generally in keeping with the previously observed general range of 50–150 ppm. Whether the manganese occurs as a substitute for Ca^{2+} or as minute quantities of Mn^{3+} in $Al^{3+}:Fe^{3+}$ sites is not known. Plotting of frequency distributions of Mn in ulti-

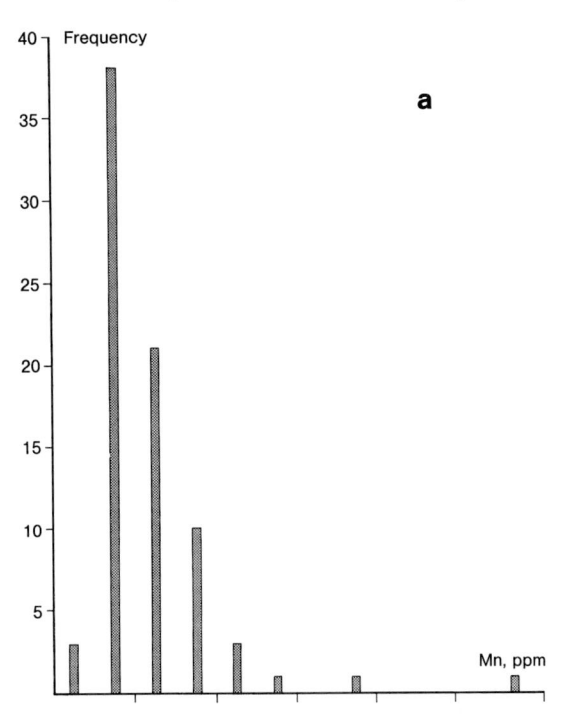

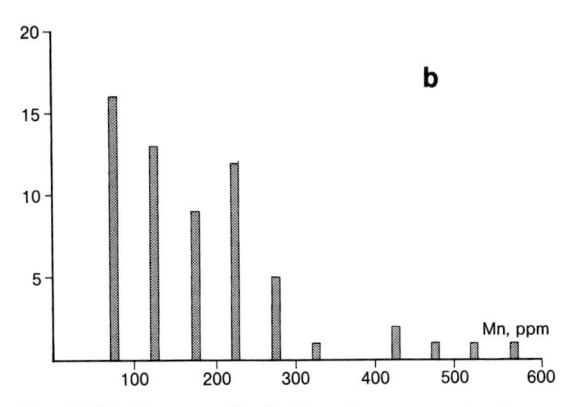

Fig. 16.13. Frequency distributions of manganese abundances in (a) ultimate purity feldspar separations (mean = 117 ppm Mn), and (b) penultimate purity feldspar separations (mean = 198 ppm Mn).

212 *Ore Elements in Arc Lavas*

to the hornblende-andesites (Table 6.5; Fig. 16.14) and, as might have been anticipated from this, increases more or less systematically with increase in whole-rock SiO_2 (Fig. 7.14(b)).

Manganese in magnetites thus behaves antipathetically with respect to chromium, magnesium, and aluminium, and to titanium and vanadium at higher SiO_2 values, and sympathetically with respect to zinc. Coupled with the fact that FeO in the magnetites tends to increase as the lavas become more felsic (Fig. 5.19), this might be taken to indicate that manganese is accepted into the magnetite structure as Mn^{2+} essentially in proportion to the abundance of Fe^{2+}. As indicated in Chapter 7 the mechanism is probably not, however, as simple as this.

Electron microprobe analysis of cores and rims of individual magnetite grains indicates a common, but certainly not universal, tendency for manganese and Fe^{2+} to vary sympathetically and to be more abundant in the rims: a feature in general conformity with the relationship between lava type and magnetite FeO and manganese. This tendency for magnetite FeO and manganese to increase as crystallization progresses is nicely illustrated by the two groups of magnetites in the clinopyroxene-rich lava already referred to. The magnetites included in the pyroxene crystals, and hence formed earlier in the

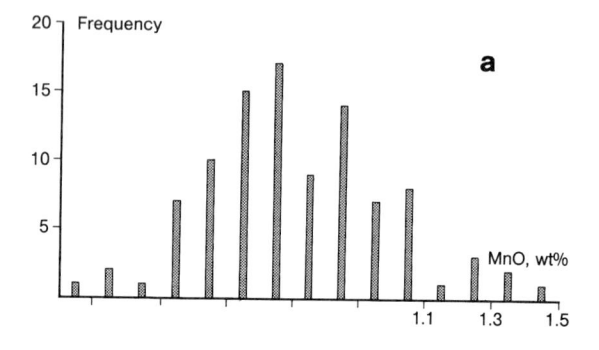

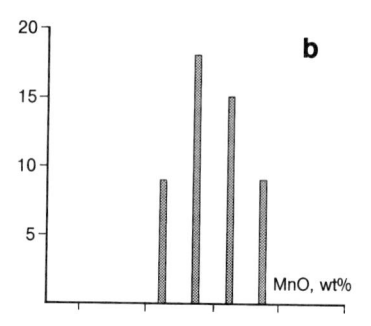

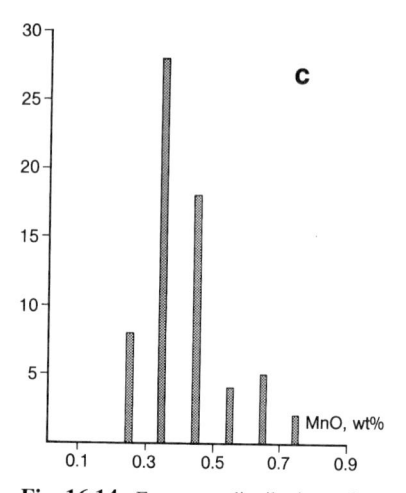

Fig. 16.14. Frequency distributions of manganese abundances in spinels of (a) the hornblende-andesite group, (b) big-feldspar lavas, (c) olivine–clinopyroxene-basalts of the Solomons suite (X-ray fluorescence analyses of 80 separations; average of probe analyses of 20 magnetite grains in each of the remaining lavas).

separated material from all lava groups is ≈4000 ppm (Table 6.5), and values for individual lavas range from <1000 to >1200 ppm (0.13 <MnO <1.60 per cent). Mean magnetite manganese for the three principal lava groups shows a clear increase from the olivine–pyroxene-basalts

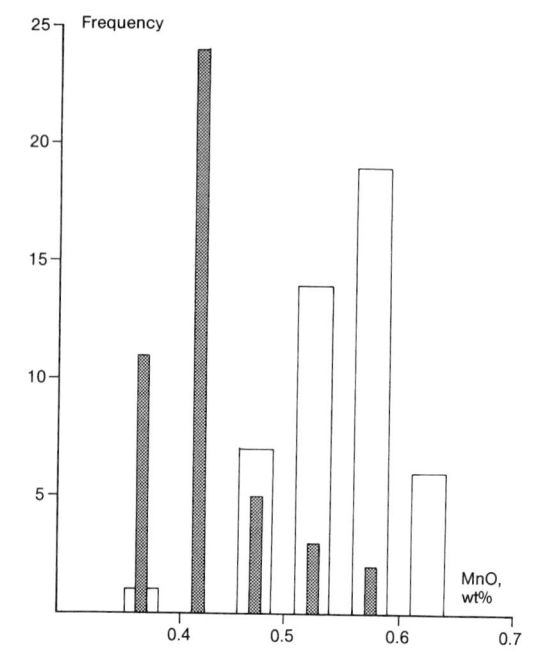

Fig. 16.15. Frequency distributions of manganese abundances in magnetites included within the clinopyroxenes (filled bars, mean MnO = 0.42), and within the groundmass (open bars, mean MnO = 0.54) of the ankaramitic basalt referred to in Chapter 5.

crystallization history of the rock, contain slightly, but significantly, less Mn and Fe than those formed later and now included in the groundmass (Fig. 16.15).

CRYSTAL:MELT PARTITIONING

K_D^{Mn} values for the principal silicates and magnetite are given in Table 16.1.

The propensity of olivine to accept manganese into its structure (and the well-known existence of tephroitic olivines) has led to extensive investigation of the partitioning of manganese between olivine and igneous melts. Early investigations by Wager and Mitchell (1951) on Skaergaard materials indicated $K_D^{ol:melt}$ values between 1.4 (hypersthene–olivine-gabbro) and 4.3 (fayalite-ferrogabbro) with a mean of ≈2.7. Henderson and Dale (1969) determined values in the range 1.1–1.3 for samples of basaltic material from oceanic islands, and a value of 1.8 for one of Stanton and Bell's (1969) samples from the Solomon Islands. Gunn (1971) deduced a value of 1.0 for a Hawaiian olivine-tholeiite, and Bougault and Hekinian (1974) and Bender et al. (1978) determined values of 0.92 and 0.77 respectively in two mid-Atlantic Ridge basalts. Roeder (1974), in a series of experiments, mostly on natural materials, between 1154 and 1306 °C obtained values between 0.32 and 0.73 (mean from 19 experimental runs = 0.48); Duke (1976) determined values of 0.99 to 1.5 (mean = 1.26) on six synthetic melts between 1150 and 1250 °C; Watson (1977), also using synthetic material (in this case from 1250 to 1450 °C) obtained values from 0.625 to 1.789 in MgO–CaO–Na$_2$O–Al$_2$O$_3$–SiO$_2$ charges ranging from 44.45 to 63.33 per cent SiO$_2$ respectively; and Longhi et al. (1978) obtained values of 0.80–1.32 (mean of 10 experiments = 1.06) in a series of experiments on lunar basalts between 1135 and 1328 °C. Dunn (1987) determined values from 1.38 to 1.52 (mean of three = 1.44) in experiments on natural low-K mid-ocean ridge basalts.

Investigations on olivines in basaltic–andesitic liquids during the past 20 years thus indicate $K_D^{ol:melt}$ values generally in the range 0.5–1.8 and that, in addition to temperature effects, partition coefficients tend to decrease with decrease in SiO$_2$, and increase in manganese, in the melt.

Mean $K_D^{ol:melt}$ for the Solomons lavas is, at ≈2–4, somewhat on the high side of generally observed values. The fact that the values are higher in the big-feldspar lavas than in the olivine–pyroxene types indicates that partition coefficients may increase with increase in SiO$_2$ content of the melt, and this appears to be confirmed, in agreement with the results of Watson (1977), by a plot of $K_D^{ol:melt}$ v. whole-rock SiO$_2$ (Fig. 16.16(a)). The crystallization of olivine thus impoverishes the coexisting melt in manganese, and does this increasingly as melt SiO$_2$ increases (and as the olivine becomes more fayalitic) and melt manganese decreases.

The partitioning of manganese between pyroxenes — particularly the clinopyroxenes — and coexisting melts has also received considerable attention. Wager and Mitchell (1951) obtained values ranging from 0.97 to 3.3 (mean = 1.89) for the clinopyroxenes of the various Skaergaard layers; Onuma et al. (1968) determined $K_D^{cpx:melt}$ = 1.27 in an alkali olivine-basalt; and Bougault and Hekinian (1974)

Table 16.1 Crystal:melt distribution coefficients (K_D) with respect to manganese

Mineral species	Lava group	n_1	n_2	m_1	m_2	$K_{D(1)}$	N	M_1	M_2	$K_{D(2)}$
Olivine	OPB	40	44	2619	1212	2.16	39	2613	1194	2.18
	BFL	7	23	5838	1212	4.82	7	5838	1490	3.92
Clinopyroxene	OPB	43	44	1223	1212	1.01	42	1179	1221	0.97
	BFL	26	23	2410	1212	1.99	17	2585	1295	2.00
	HbA	27	47	2757	675	4.08	20	2823	840	3.36
Hornblende	HbA	25	47	2207	675	3.27	22	2217	793	2.80
Feldspar	OPB	10	44	147	1212	0.121	10	147	1259	0.117
	BFL	23	23	121	1212	0.100	22	121	1208	0.100
	HbA	46	47	108	675	0.160	44	107	689	0.155
Spinel	OPB	20	44	3689	1212	3.04	20	3692	1218	3.03
	BFL	23	23	4250	1212	3.51	22	4359	1212	3.60
	HbA	37	47	6003	675	8.89	36	6947	733	9.48

OPB = olivine–clinopyroxene-basalt group; BFL = big-feldspar lavas; HbA = hornblende-andesite group; n_1 = total number of mineral separations analysed; n_2 = total number of groundmass separations analysed; m_1 = mean manganese content of mineral of each lava group; m_2 = mean manganese content of groundmass of each lava group; $K_{D(1)}$ = mean distribution coefficient determined as m_1/m_2; N = number of coexisting mineral:groundmass pairs analysed; M_1 and M_2 = mean manganese contents of coexisting mineral and groundmass respectively; $K_{D(2)}$ = distribution coefficient determined as M_1/M_2, for comparison with $K_{D(1)}$. All concentrations as parts per million by mass of the element.

experimental runs referred to above determined values from 0.55 to 0.75 (mean = 0.63); and Dunn (1987), experimenting on the natural materials already mentioned, a narrow range from 0.81 to 0.91 (mean = 0.86).

The mean partition coefficients obtained for the 79 Solomons clinopyroxenes (Table 16.1) are thus in keeping with the early results of Wager and Mitchell but are rather high compared with more recent determinations — particularly those obtained in experiment. The Solomons values are, however, based on a large amount of data and appear systematic, and probably constitute the most reliable observational information available. Mean K_D values rise progressively from the olivine–pyroxene-basalts to the hornblende-andesites, reflecting a simultaneous increase in manganese in the clinopyroxene, and decrease in manganese in the melt. Calculation of mean $K_D^{cpx:melt}$ values corresponding to 1 per cent whole-rock SiO_2 intervals (Fig. 16.16) indicates the development of relationships generally analogous to those found for olivine.

Information on $K_D^{opx:melt}$ is sparse. Onuma *et al.* (1968), investigating a two-pyroxene basalt, found a value of 1.44: only slightly higher than that for the coexisting clinopyroxene (1.27). Bender *et al.* (1978) obtained a value of 1.75 for a Mid-Atlantic Ridge basalt at 1310 °C, 15 kbar; Dudas *et al.* (1973) values of 10.7 and 9.77 in orthopyroxenes in dacitic tephra; and Nagasawa and Schnetzler (1971) values of 29.2 and 33.7 in two dacites. In keeping with their higher iron content, the orthopyroxenes clearly capture more manganese than do the clinopyroxenes, a difference accentuated as the containing melt becomes richer in SiO_2.

No direct data are yet available on Solomons $K_D^{opx:melt}$ values, but a rough approximation may be obtained from $K_D^{opx:cpx}$ and $K_D^{cpx:melt}$ values. As already noted, mean $K_D^{opx:cpx}$ for the 85 coexisting pairings of Fig. 16.9 is 2.17. If the mean $K_D^{cpx:melt}$ for the Solomons andesites is estimated at 3.4, the imputed value for $K_D^{opx:melt}$ is 7.38.

Although it would appear that many clinopyroxenes, particularly Mg-rich varieties, develop crystal:melt partition coefficients <1 and hence tend to concentrate manganese in the melt as they crystallize, those of the Solomons develop K_D values >1, and hence have tended to impoverish the relevant melts in this element — and to do so at accelerated rates as the relevant melts became more SiO_2-rich. Orthopyroxene develops K_D values $\gg 1$, and thus always tends to impoverish the melt in manganese.

As noted in the preceding section, hornblende accommodates manganese to about the same extent as clinopyroxene. This combined with the fact that hornblende tends to crystallize later, when the melt has become depleted in manganese, leads to the development of somewhat higher K_D values for this mineral than for the clinopyroxenes of the same lava series.

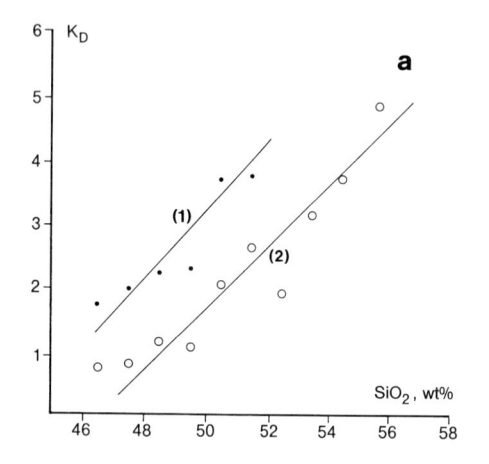

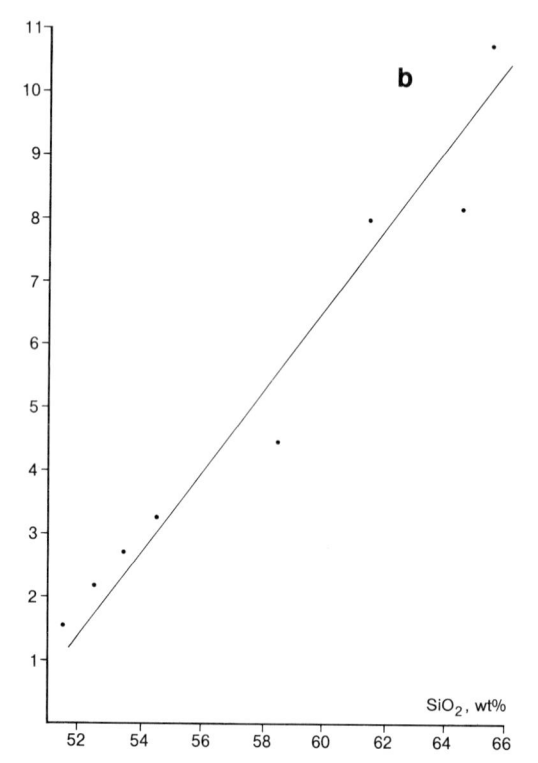

Fig. 16.16. Relations between mineral:melt partition coefficients (K_D) for manganese and two per cent intervals of whole-rock SiO_2 in the host lavas, as developed by (a) olivine [(1) filled circles], and clinopyroxene [(2) open circles], and (b) hornblende.

obtained 0.81 for clinopyroxene in a Mid-Atlantic Ridge tholeiitic basalt. Seward (1971), in an extensive series of experiments using synthetic mixes, obtained values from 0.36 to 1.41 and averaging 0.60; Duke (1976) in the

Higuchi and Nagasawa (1969) obtained a value of 0.94 for hornblende in basalt; Nagasawa and Schnetzler (1971) found values of 11.0 and 11.9 in the two dacites mentioned above; and Dudas et al. (1973) found values ranging from 3.04 to 5.10 (mean = 3.35) in the dacitic tephra already referred to. There is thus an indication that, as with olivine and pyroxene, the effectiveness of the hornblende in abstracting manganese from the melt increases with increasing SiO$_2$ in the melt.

$K_D^{Hb:melt}$ values exhibited by the Solomons andesites (Table 16.1) appear to be in reasonable accord with these observations by other investigators and, in keeping with the behaviour of the associated olivines and pyroxenes, increase with increase in SiO$_2$ in the melt (Fig. 16.16b). With mean $K_D^{Hb:melt}$ >1, hornblende thus tends to impoverish the melt in manganese, and to do so increasingly as the latter evolves to higher SiO$_2$ contents.

The only data available to the writer on the partitioning of manganese by biotite are those of Higuchi and Nagasawa (1969), who determined a value of 6.00 ± 0.13 for a biotite in a dacite. The biotites of the two Savo dacites gave $K_D^{biot:melt}$ values of 2.24 and 6.50. Biotite has therefore also been a net abstractor of manganese from the Solomons lavas, though in view of the very limited incidence of this mineral the absolute amount of manganese involved must have been infinitesimally small.

Stemming from their very low capacity for accommodating Mn^{2+}, the plagioclases show a strong tendency to partition manganese into the melt. Wager and Mitchell (1951) determined values between 0.03 and 0.13 (mean = 0.07) for plagioclases of the Skaergaard layers; Higuchi and Nagasawa (1969), one of 0.05 for the plagioclase of an alkali olivine-basalt; and Bougault and Hekinian (1974), 0.07 for the plagioclase of a Mid-Atlantic Ridge basalt. Lindstrom (1976) obtained an experimental value (1135 °C at 1 atm) of 0.05 on a natural alkali basalt, and Dudas et al. (1973) determined a value of 0.18 for plagioclase in dacitic tephra.

The Solomons plagioclases are in reasonable conformity with these results, mean values ranging from 0.10 to 0.12 in the basalts to 0.16 in the andesites and minor dacites. Copious crystallization of plagioclases during the solidification of the more felsic members of the Solomons suite would therefore have tended to concentrate manganese in the residual melt.

In contrast to the plagioclases the spinels — particularly titanomagnetite — with their marked propensity for accepting manganese, develop high K_D values with respect to the melt from which they crystallize. There is, however, comparatively little precise information in the literature on magnetite partitioning of manganese. Wager and Mitchell (1951) determined a value of 4.0 for the magnetite of Skaergaard Stage B; Dudas et al. (1973) obtained values of 5.6 and 6.7 for magnetite of the dacitic tephra referred to above; and Lindstrom (1976) obtained surprisingly low values of 1.70–1.81 in experimental runs (1111–67 °C, 1 atm) using his natural alkalic basalt.

The Solomons magnetites give mean $K_D^{mag:melt}$ ranging from 3.03 in the olivine–pyroxene-basalts to 9.48 in the hornblende-andesites and dacites. As with the clinopyroxenes, this stems from concomitant increase in magnetite manganese with decrease in melt manganese. Magnetite is therefore an important abstractor of manganese from the melt: a rôle that is, again, progressively accentuated as the melt becomes increasingly felsic.

MANGANESE–ZINC RELATIONS IN THE SOLOMONS MINERALS

The clear if sometimes loose tie between iron and manganese in the ferromagnesian minerals and magnetite, coupled with the equally clear correlation of zinc with iron noted in Chapter 7, points to some kind of systematic relationship between manganese and zinc in the Solomons minerals.

This is confirmed by Mn:Zn plots for olivine, clinopyroxene, hornblende, and magnetite (XRF analyses of mineral separations) shown in Fig. 16.17. As might have been expected from Zn:Fe and Mn:Fe relations, the Zn:Mn field for olivine (Fig. 16.17(a)) represents either a curve or an inflection at ≈250 ppm Zn:4000 ppm Mn. The distribution for clinopyroxene (Fig. 16.17(b)) appears to be linear with good correlation, and that for hornblende (Fig. 16.17(c)) also linear, with a high degree of correlation (coefficient of correlation >0.90). Spinel (Fig. 16.17(d)) also shows a linear relationship, but with a somewhat lower degree of correlation. This may stem from the incidence of some Mn^{3+}, which would enable manganese to substitute in both di- and trivalent positions in the spinel structure, whereas zinc would be restricted to the divalent site. Analogous plotting of Zn:Mn relations in the eight orthopyroxene:clinopyroxene pairings referred to in previous sections shows a good linear relationship involving both mineral groups, with the points for the orthopyroxenes occupying, as expected, the higher end of the field.

Biotite on the other hand shows no clear, systematic Zn:Mn relationship. Electron microprobe data for individual biotite crystals of the two Savo dacites (Fig. 16.18(a)) indicate that the biotites of one (Sol. 41/80; filled circles) *may* have a Zn–Mn correlation with a very steep slope: however, the field for the other (Sol. 6/81; open circles), although similarly linear, is vertical, indicating that zinc here may vary in abundance from 240

216 *Ore Elements in Arc Lavas*

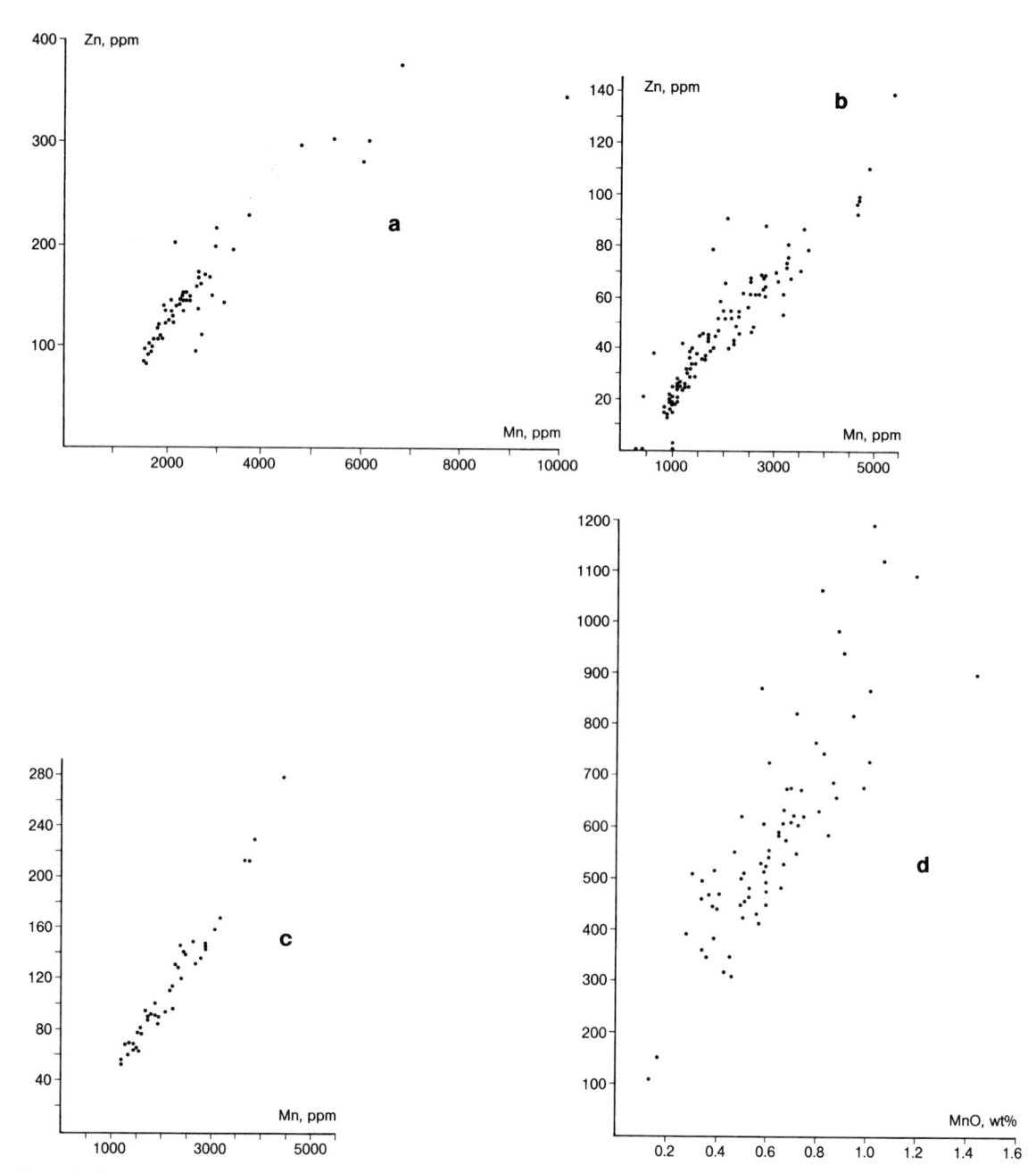

Fig. 16.17. Abundance relations between manganese and zinc in (a) olivines, (b) clinopyroxenes, (c) hornblendes, and (d) spinels of the Solomons suite.

to 570 ppm independently of the abundance of associated manganese. It seems likely that the field for Sol. 41/80 is in fact also essentially vertical, indicating that the abundance of zinc is again independent of the abundance of manganese. Fig. 16.18(b) shows, for comparison, the nature of Zn–Mn relations in individual hornblende crystals coexisting with the biotites of the two dacites. This may also be compared with Fig. 16.17(c) which

Manganese 217

basalts, and in the associated lavas more felsic than basalts (SiO$_2$ >52 per cent) are given in Figs. 16.19(a) and (b) respectively. The mean for the more felsic rocks is 0.132 per cent MnO (1022 ppm Mn) and the frequency distribution is, accordingly, distinctly to the left of that for the basalts.

Frequency distributions of manganese in the three principal lava types (Fig. 16.20) show, for the Solomons lavas, the steady decrease in manganese from basaltic to progressively more felsic lavas commented upon as a general phenomenon by Goldschmidt (1954) and Wedepohl (1978). Means for the olivine–pyroxene-basalts, big-feldspar lavas, and hornblende-andesites are 0.171, 0.161, and 0.128 per cent MnO (1324, 1247, 991 ppm Mn) respectively. When grouped as basalts, andesites, and dacites the averages are 0.196, 0.143, and 0.072 per cent MnO respectively (Table 6.3). Fig. 16.20 also indicates that manganese has a consistent tendency to partition into the solid phases (see previous section); groundmass means are

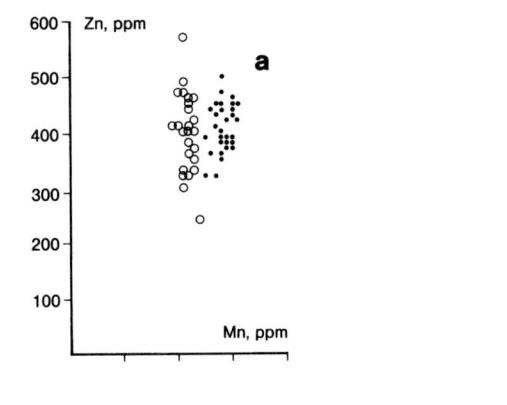

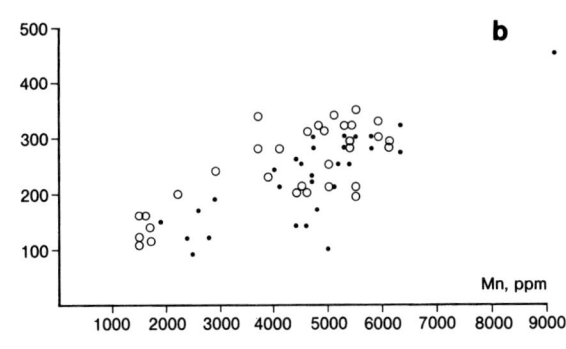

Fig. 16.18. Abundance relations between manganese and zinc in (a) biotites and (b) hornblendes of the two Savo dacites, one lava being designated by filled circles in (a) and (b), the other by open circles.

shows how much closer the Zn–Mn correlation is made to appear by the 'averaging' process of analysing bulk separations. The parallel X-ray fluorescence analysis of bulk mineral separations and electron-microprobe analysis of individual mineral grains in all of the writer's Solomons material has shown — starkly, in many cases — just how effectively the analysis of separations may conceal substantial within- and between-grain variation among individual crystals of a single lava sample.

THE INCIDENCE OF WHOLE-ROCK MANGANESE IN THE SOLOMONS YOUNGER VOLCANIC SUITE

Mean manganese as MnO in the Solomons basaltic rocks is 0.196 per cent (0.1518 ppm Mn), which is distinctly higher than that of MORBs (0.171 per cent MnO) and very similar to that of the basalts of the other four arcs considered in the present study: 0.193 per cent MnO. Frequency distributions of MnO values in the Solomons

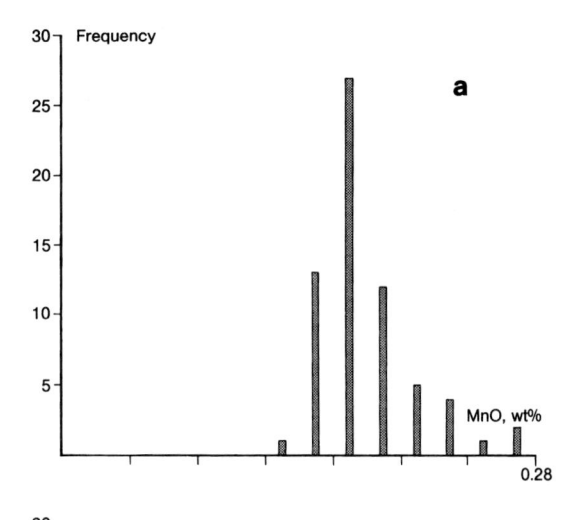

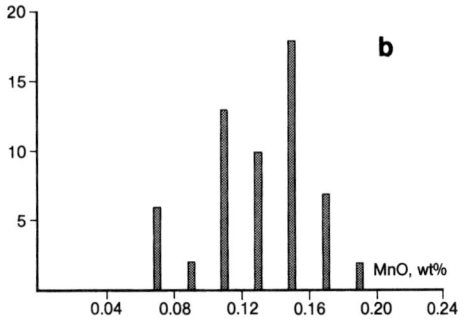

Fig. 16.19. Frequency distributions of MnO abundances in (a) basalts (SiO$_2$ <52 per cent), and (b) lavas more felsic than basalt (SiO$_2$ ≥52 per cent) of the Solomons suite.

218 *Ore Elements in Arc Lavas*

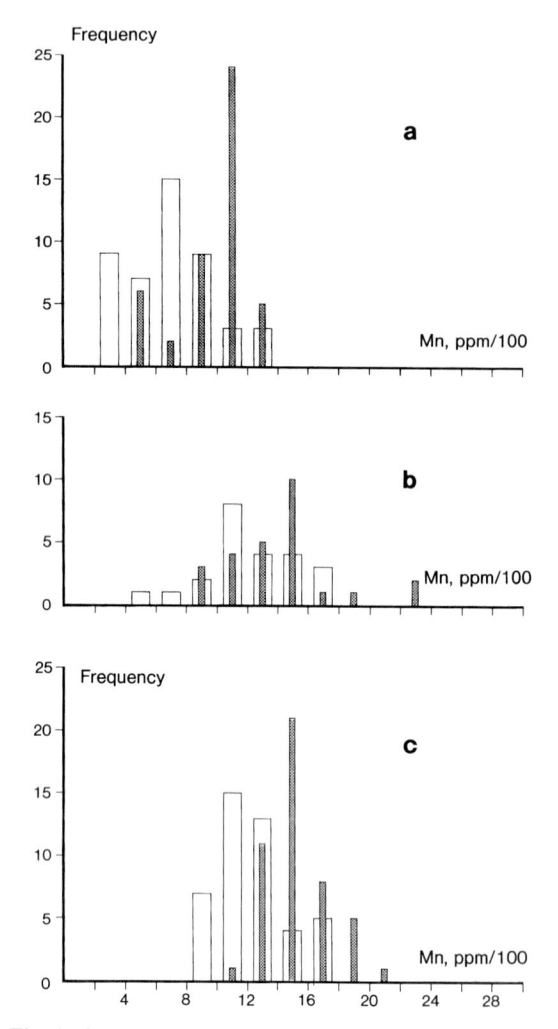

Fig. 16.20. Frequency distributions of MnO abundances in whole rocks (filled bars) and groundmasses (open bars) of (a) the hornblende-andesite group, (b) big-feldspar lavas, and (c) the olivine–clinopyroxene-basalts of the Solomons suite.

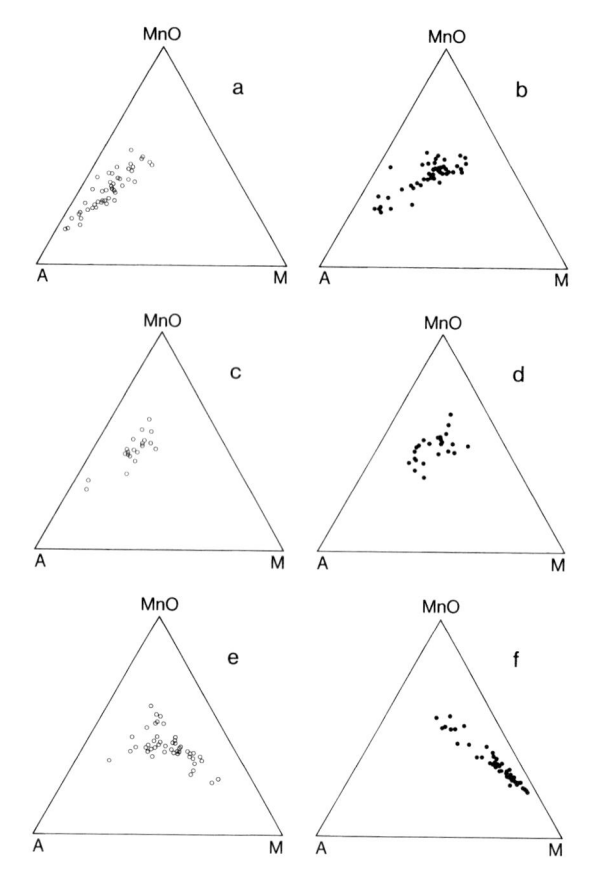

Fig. 16.21. MnO–MgO–(Na$_2$O + K$_2$O) ternary relations (MnO calculated using weight per cent MnO × 50) developed in olivine–clinopyroxene-basalts, big-feldspar lavas and hornblende-andesite whole rocks [filled circles: (f), (d), and (b) respectively], and their corresponding groundmasses [(e), (c), and (a) respectively].

0.148, 0.146, and 0.081 per cent MnO (1146, 1131, 627 ppm Mn) for the three categories of Fig. 16.20.

This relationship between manganese abundance and lava type is reflected in the (MnO–MgO–(Na$_2$O + K$_2$O)) diagram for whole rocks and groundmasses of Fig. 16.21. As already shown for several of the preceding elements, the sequence begins (at the mafic end of the spectrum, Fig. 16.21(f)) with a highly linear, well-defined 'olivine subtraction' line, or field. There is then a succession of whole-rock:groundmass fields tracing out a hypotholeiitic sequence very similar to those developed by zinc (Chapter 7) and iron (Chapters 5 and 17). The proportionate abundance of MnO reaches a peak in the big-feldspar lavas, falling away in the groundmasses of this group and in the hornblende-andesites and dacites. This well-developed hypotholeiitic pattern is once again demonstrated more sharply by plots of MnMA relations corresponding to 2 per cent whole-rock SiO$_2$ intervals (SiO$_2$ <46 per cent to SiO$_2$ >66 per cent). This indicates (Fig. 16.22(a)) that the peak in proportionate MnO corresponds to the quite broad range of 50–8 per cent whole-rock SiO$_2$ content: i.e. from felsic basalts to 'intermediate' andesites. The curve is confirmed by the inclusion of the Bougainville data in Fig. 16.22(b). For comparison, a similar pattern is developed in the Grenada lavas (filled circles in Fig. 16.22(c), peak at 52–6 per cent whole-rock SiO$_2$), and muted ones in the Dominica, St Kitts, Vanuatu, and Aleutians suites (Figs. 16.22(c) and (d)). The Tonga–Kermadec suite exhibits little proportionate variation in MnO. MORBs appear to be essentially tholeiitic with respect to manganese (Fig. 16.22(e)),

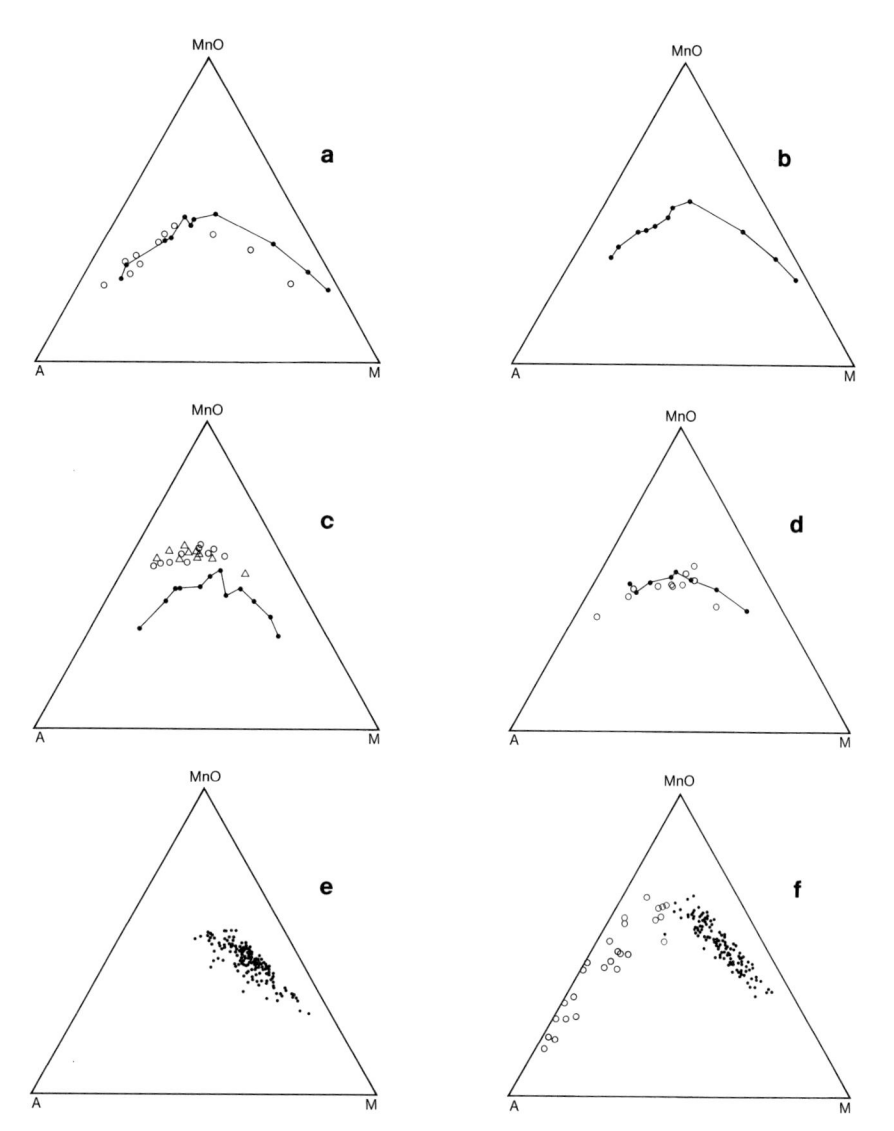

Fig. 16.22. Mean MnO–MgO–(Na$_2$O + K$_2$O) ternary relations corresponding to 2 per cent whole-rock SiO$_2$ intervals in (a) whole rocks [(1) filled circles] and corresponding groundmasses [(2), open circles] of the Solomons suite, excluding Bougainville; (b) whole rocks of the Solomons suite including Bougainville (241 analyses); (c) lavas of Grenada (filled circles), Dominica (open circles), and St Kitts (open triangles); (d) Vanuatu (filled circles) and the Aleutians (open circles). Ternary relations developed in MORBs as represented by Mid-Atlantic Ridge basalts are shown in (e), and in individual samplings of Icelandic basalts (SiO$_2$ <52 per cent: filled circles) and Icelandic lavas more felsic than basalt (SiO$_2$ ⩾52 per cent: open circles) in (f).

whereas the Icelandic suite as a whole develops a beautiful hypotholeiitic pattern (Fig. 16.22(f)).

Fig. 16.23 shows variation in mean absolute abundance of MnO as related to 2 per cent intervals of whole-rock SiO$_2$. The Solomons suite exhibits an initial rise in MnO to a plateau of 0.20 per cent in the whole-rock SiO$_2$ interval 46–50 per cent. It then declines steadily to a mean of ≈0.065 per cent for lavas with SiO$_2$ >64 per cent.

The groundmasses behave similarly at a level of ≈0.06 per cent MnO below the whole rocks as far as 60–2 per cent whole-rock SiO$_2$. Its irregular behaviour at SiO$_2$ >62 per cent may reflect unreliability stemming from the small number of samples, already referred to, in these higher SiO$_2$ intervals. Lava suites from the Lesser Antilles, Aleutians, and Tonga–Kermadec (Figs. 16.23(b) and (c)) behave rather similarly to those of the Solomons: MnO

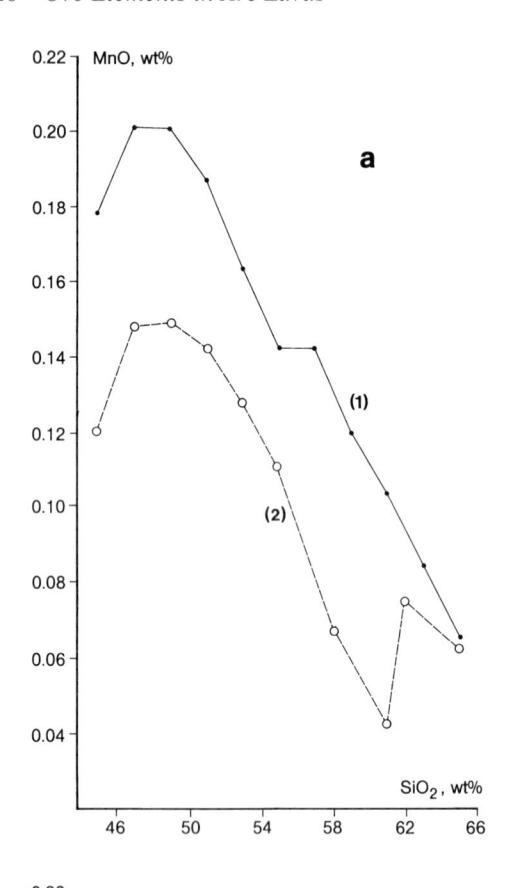

Fig. 16.23. Mean MnO:SiO$_2$ relations corresponding to 2 per cent whole-rock SiO$_2$ intervals in (a) whole rocks [(1) filled circles] and corresponding groundmasses [(2), open circles] of the Solomons suite; (b) the lava series of (1) Grenada and (2) Dominica (to which MnO:SiO$_2$ relations in the St Kitts series are very similar); and (c) the lava series of (1) the Aleutians, (2) Vanuatu, and (3) Tonga–Kermadec.

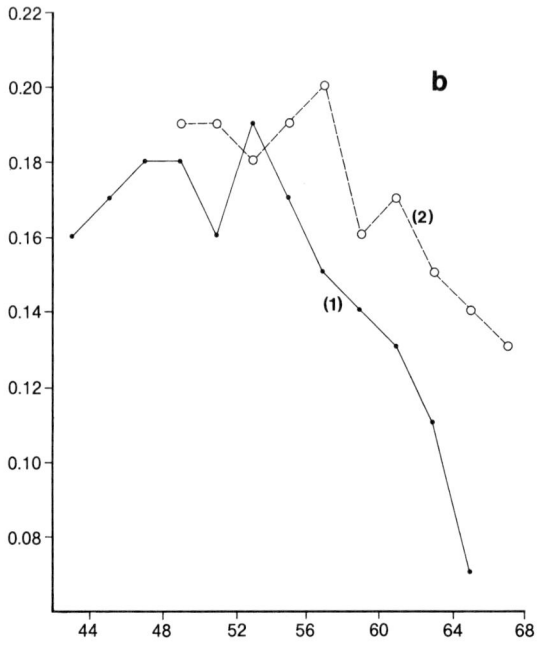

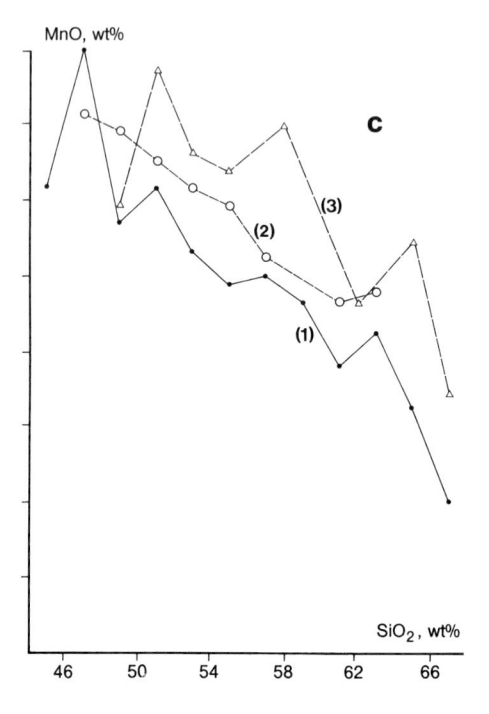

shows an initial small rise, or plateau, at lower whole-rock SiO$_2$ levels, and then decreases fairly steadily as the rocks become more felsic. The pattern for Vanuatu (Fig. 16.23(c)) is, however, one of simple steady decrease in MnO with increase in SiO$_2$.

The rather close relationships between manganese and iron in most of the Solomons minerals (see previous section) point to the possibility of correlations between these elements in the whole rocks: correlations perhaps similar to those between whole-rock chromium and MgO (Chapter 15, Fig. 15.17) and whole-rock nickel and MgO (Fig. 19.16). Such a likelihood is further emphasized by the similarities in the patterns of ternary relations of Figs. 5.4, 17.1, and 16.21 and 16.22.

The relationships are not, however, as close as might have been anticipated. Fig. 16.24(a) shows whole-rock Fe^{2+} (as FeO) v. MnO — a pairing that might have been expected to yield a correlation as close as those between Cr–MgO (Fig. 15.17) and Ni–MgO (Fig. 19.16). Although a general MnO–FeO correlation is apparent, the

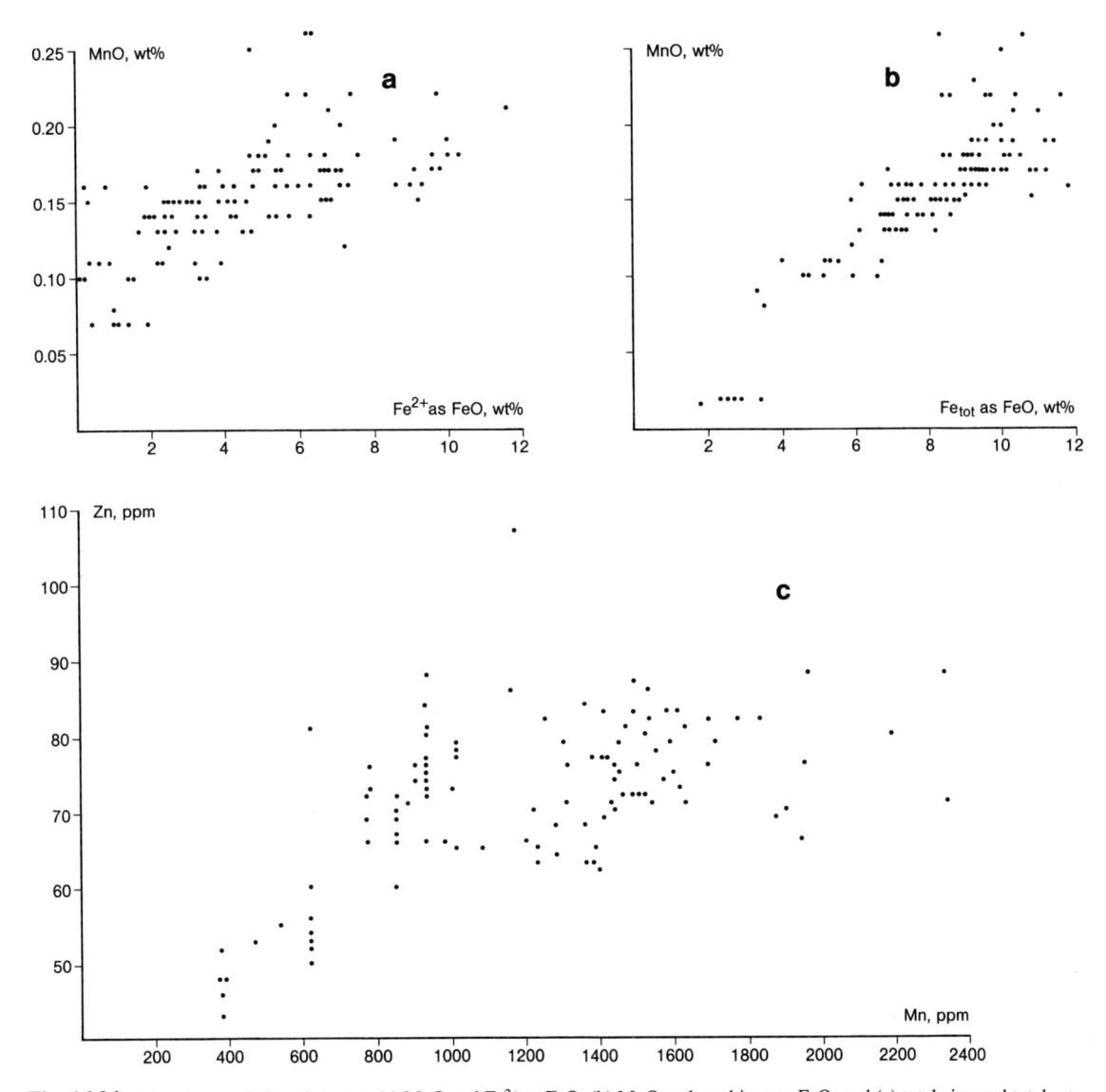

Fig. 16.24. Abundance relations between (a) MnO and Fe^{2+} as FeO; (b) MnO and total iron as FeO; and (c) total zinc and total manganese in the spectrum of lavas of the Solomons suite.

field of points is diffuse and the coefficient of correlation low. Somewhat surprisingly the tie between MnO and *total* iron (again expressed as FeO; Fig. 16.24(b)) is closer, with a better-defined field of points and a higher coefficient of correlation. It was noted in Chapter 15 that although chromium tends to enter the earliest-formed crystals of magmatic systems and hence, indirectly, tends to parallel magnesium in abundance, the very high degree of correlation between the two found in the Solomons lavas was unanticipated. The Ni–Mg tie is well known and to be expected given the similar size (0.078 nm), charge, and properties of these two ions. The chromium and magnesium ions, however, have different charges and somewhat different sizes (0.064 and 0.078 nm) respectively; chromium forms a much more covalent M–O bond than does magnesium; and chromium does not tend to substitute for magnesium in igneous crystals (*vide* olivine). It would therefore appear that for the development of the Cr–MgO correlation the limited mineralogical (and groundmass) correlations must have combined with the relative abundances of the relevant minerals to yield a high degree of whole-rock correlation. For manganese and iron the relatively good mineralogical correlations must have combined with the relative abundances of the minerals concerned to yield poor whole-rock correlations. Similarly, and presumably for similar reasons, whole-rock Mn–Zn correlation is low (Fig. 16.24(c)). That such a hypothesis may be close to the truth is indicated by the series of magnetic fractions of the three rocks used previously to examine whole-rock correlations of Zn–Fe, Ba–K, Sr–Ca, etc. Fig. 16.25 shows clearly that there are close correlations between whole-rock manganese and total iron where the *single* mineralogical groupings of *individual* lavas are considered.

CONCLUDING STATEMENT

Mean MnO for arc basalts at 0.20 per cent is slightly higher than that for MORBs at 0.17 per cent, though frequency distributions of MnO values in the two groups are very similar. There is a strong tendency for MnO contents to decrease as lava compositions become more felsic. The Lesser Antilles suites are, however, an exception to this in that in these manganese values tend to peak in the andesites.

Manganese occurs virtually entirely as Mn^{2+} in volcanic systems and hence tends to correlate with Fe^{2+} and total iron. It shows variable but generally close ties with iron in the ferromagnesian silicates, and obscure — if any — relationship in the plagioclase feldspars. Although manganese tends to increase in the magnetites in parallel with Fe^{2+} and total iron as the host lavas become more felsic, correlation of Mn with Fe^{2+} and Fe_{tot} from one individual magnetite to another is poor to non-existent. $K_D^{cryst:melt}$ for the ferromagnesian minerals ranges ≈ 1.0–3.5, plagioclase 0.10–0.16, and magnetite 3–10, the value of the partition coefficient increasing in all cases as the host rock becomes more felsic and poorer in manganese. Apart from plagioclase and the most magnesian of the clinopyroxenes, crystallization of all silicates and magnetite thus tends to impoverish the melt in manganese. Crystallization of copious quantities of plagioclase should enrich the residual melt in this element.

The Solomons lavas exhibit a slight rise in manganese from the most mafic types to peak in the interval 46–50 per cent SiO_2 (presumably the result of subtraction of high-Mg:low-Mn clinopyroxene) and then a steady decline as the lavas become more felsic. The other arc suites behave substantially similarly in principle. MnO–MgO–$(Na_2O + K_2O)$ ternary relations are systematic and develop a hypotholeiitic pattern very similar to those of Zn–MgO–$(Na_2O + K_2O)$ and FeO–MgO–$(Na_2O + K_2O)$.

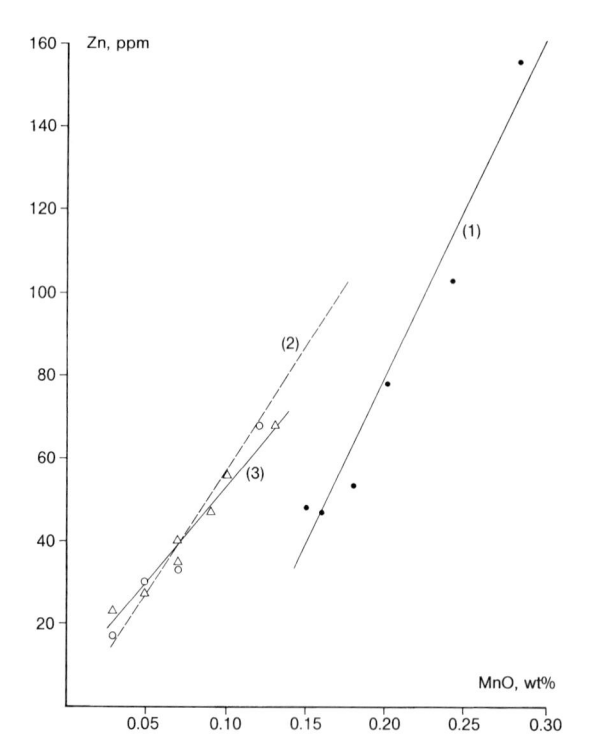

Fig. 16.25. Relations between total manganese as MnO and zinc in groundmass fractions separated, according to magnetic susceptibility, from the total groundmass of the representatives of each of (1) olivine–clinopyroxene-basalt (filled circles); (2) big-feldspar lava (open circles); and (3) hornblende-andesite (open triangles) referred to in Chapters 7 and 12. The high degree of correlation in each case stems from the incidence of most of both zinc and manganese of the groundmass in the fine magnetite of the latter.

17

Iron

From the quantitative point of view iron is by far the most important exhalative sulphide metal. Many exhalative deposits contain little sulphide other than that of iron: as pyrite, pyrrhotite, or both. Most Cyprus-type deposits are dominated by pyrite, with chalcopyrite and sphalerite appearing as no more than subordinate components. The Archaean deposits of Canada, the felsic-associated exhalative deposits of the Appalachians, the Caledonides, the Urals, and the Eastern Highlands of Australia, and many others, contain pyrite and/or pyrrhotite as their principal sulphide. In some cases, such as Mt Isa in Queensland, deposits may be most accurately described as large sulphide iron formations containing sufficient 'impurities' of chalcopyrite, sphalerite, and galena to constitute base-metal orebodies. The pyrite of such deposits is sometimes used as a source of sulphur and, less commonly, of iron itself. In addition to the iron of these simple iron sulphides, chalcopyrite ($CuFeS_2$) contains some 30 per cent Fe, and most sphalerite is $(Zn, Fe)S$ rather than ZnS and commonly contains 5–10 per cent Fe.

The iron component of exhalative orebodies is, however, by no means all, or necessarily, sulphide. As well as pyrite and/or pyrrhotite and the Fe-bearing base-metal sulphides, most deposits contain greater or lesser quantities of iron as silicate and oxide, and occasionally as sideritic carbonate. Much iron is commonly bound in chlorite, and it may also occur in silicates such as garnet, hedenbergite, and iron-rich amphiboles. Some deposits may contain large quantities of iron as magnetite, particularly where a deposit straddles several sedimentary facies of iron formation. The deposits of Aggenys and Gamsberg in Namaqualand (see Chapter 12) exhibit well-developed oxide facies consisting largely of well-bedded exhalative magnetite. This gives way to silicate facies consisting of a variety of iron-rich silicates, and then to sulphide–silicate facies in which iron sulphides and sphalerite (with minor galena) predominate.

The role of calcium and iron in the development of skarn mineralogies in sulphur- and carbonate-poor exhalative environments has already been alluded to (Chapter 12). Lack of sulphide sulphur in the case of iron and of carbonate in the case of calcium leads to the precipitation of the two as silicates: usually mixed Ca–Fe and Ca–Fe–Al silicates similar to those commonly developed in contact metamorphic skarns. The result is the development of stratiform concentrations of skarn-like mineral assemblages: the 'stratiform skarns' or 'reaction skarns' of Scandinavian usage. In part these may provide delicate indicators of oxygen activity in the silicate facies of the exhalative environment: in reducing environments such as that in which the Broken Hill layers appear to have deposited the garnet tends to be grossularitic and epidote is not conspicuous, whereas in slightly more oxygenated environments such as that in which the Gamsberg iron formation was laid down the garnet tends to be andraditic, epidote may be prominent, and other silicates may have a small Fe^{3+} component. In such sulphur-poor environments some zinc may be 'forced' into silicate and oxide compounds, and if notable manganese is available highly complex Fe–Ca–Mn–Zn silicate (-oxide) mineralogies result. Stripped of its sulphides, Broken Hill is a spectacular example of such stratiform skarns.

CHEMICAL PROPERTIES AND CRYSTAL CHEMISTRY OF IRON

Iron exhibits both chalcophile and lithophile characteristics. In exhalative–sedimentary environments it partitions between sulphide, silicate, oxide, and carbonate, establishing equilibrium concentrations in the relevant phases depending on sulphide and carbonate ion activities and prevailing P–T–Eh–pH conditions. A good example of this is the effect of sulphide activity on Fe:Mg ratios in sedimentary–diagenetic chlorites of the exhalative milieu. At low sulphide activities iron has a greater tendency (opportunity) to enter the silicate phase and resulting chlorite Fe/Mg ratios are relatively high. At high sulphide activities iron has a greater opportunity to enter the sulphides, and the resulting chlorites are conspicuously magnesian — a prominent feature of many chlorite-rich exhalative sulphide bodies.

With its ionic radius of 0.082 nm, Fe^{2+} sites provide good camouflage for divalent metals such as Co^{2+}

(0.082), Ni^{2+} (0.078), Mn^{2+} (0.091), and Zn^{2+} (0.083 nm) in most ferromagnesian silicates, and in the divalent position in magnetites. The ferric iron Fe^{3+} (0.067 nm) is not important in most volcanic silicates (though small quantities of Fe^{3+} are present in most pyroxenes and hornblendes) and is thus not a major provider of sites for trace elements in these minerals. Its trivalent site in magnetite is, however, the basis for the development of the isomorphous chromite and spinel series, and for the camouflaging (at lower concentrations) of Cr^{3+} (0.064), V^{3+} (0.065), and Mn^{3+} (0.070 nm).

Mean iron contents (as FeO) of the major Solomon Islands volcanic silicates are: olivines 21 per cent; clinopyroxenes 7 per cent; orthopyroxenes 19 per cent; hornblendes 12 per cent; and biotites 15 per cent. The two minor elements exhibiting the closest affinity with iron are zinc (Chapter 7) and manganese (Chapter 16).

THE INCIDENCE OF WHOLE-ROCK IRON IN THE SOLOMON ISLANDS YOUNGER VOLCANIC SUITE

The general incidence of iron as a major element in the Solomons lava series has been considered in Chapter 5.

The behaviour of total Fe as FeO in FeO–MgO–(Na_2O + K_2O) diagrams for lava suites of this kind is well known, and (as shown in Chapter 5) the members of the Solomons series follow the usual pattern (Fig. 5.4). The picrites and related olivine–pyroxene-basalts (Fig. 5.4(f)) generate a well-defined linear array which, at its high FeO extremity, merges into the field of the big-feldspar lavas (Fig. 5.4(d)). This constitutes an area of inflexion, which in turn gives way to the field of the hornblende-andesites (Fig. 5.4(b)). The fields for the groundmasses develop in sympathy with those of the whole rocks.

The Solomons Younger Volcanic suite thus traces out, with respect to iron in this ternary system, an extended evolutionary path from ultra-high magnesium lavas to a typical 'calc-alkali' trend of andesites and dacites. For many volcano–tectonic provinces, and those of arcs in particular, the basalt–andesite–dacite portion of this hypotholeiitic pattern is a familiar one.

The curve as developed by the Solomons lavas is sharpened when mean FMA relations corresponding to two per cent intervals of whole-rock SiO_2 are plotted as in Fig. 17.1(a). The improved statistics resulting from the inclusion of the Bougainville lavas yield the even smoother curve of Fig. 17.1(b). The Grenada suite also shows a well-developed hypotholeiiticity, though the pattern is truncated at the mafic end of the Dominica and St Kitts assemblages (Fig. 17.1(c); see also Brown et al. 1977, Fig. 6). The Vanuatu suite, including as it does the picritic lavas of Aoba, also exhibits a well-defined hypotholeiitic pattern; that of the Aleutians is truncated at the mafic end and extended at the felsic end of the spectrum (Fig. 17.1(d)).

The diagram depicting MORBs as represented by the mid-Atlantic Ridge basalts (Fig. 17.1(e)) re-affirms the well-known tholeiitic nature of ocean-ridge basalts (the development of extreme tholeiiticity with respect not only to iron but also to zinc by some ocean-ridge basalts is referred to in Chapter 24). The full assemblage of Icelandic basalts and the more felsic lavas constitutes a quite spectacular 'high iron' hypotholeiitic suite (Fig. 17.1(f)). Although similar in principle, the sharply angled hypotholeiitic pattern of the Icelandic lavas is in marked contrast to the flatter, lower-Fe curves of the arc suites as exemplified by that of the Solomons in Fig. 17.1(b).

Total Fe as FeO–whole rock SiO_2 relations in the Solomons lavas and their groundmasses are shown in Fig. 17.2(a). Iron tends to be partitioned into the solid phases throughout, leading to a groundmass curve consistently beneath that for the whole rocks. The initial increase in iron in the groundmasses (44–50 per cent SiO_2) is presumably due to early crystallization of high Mg–low Fe clinopyroxenes (FeO ≈5 per cent) with a concomitant brief partitioning of iron into the melt. The whole rocks exhibit a small plateau at 9.5–9.6 per cent FeO between 44 and 50 per cent SiO_2, followed by a steady decline reflecting, at least in part, the progressive crystallization of more Fe-rich olivines and pyroxenes, together with hornblende and magnetite. Variation in modal magnetite with evolution of the lava series is indicated by curve (3) in Fig. 17.2(a), and it may be seen that the onset of decline in iron in both whole rocks and groundmasses coincides, in the 48–52 per cent SiO_2 range, with an abrupt rise in the incidence of magnetite.

It has been shown that, contrary to assumptions made in some comprehensive studies of lava suites elsewhere (e.g. Brown et al. 1977 on the Lesser Antilles; Gorton 1974 on Vanuatu), the Solomons series exhibits substantial, and apparently systematic, variation in oxidation index as expressed by $Fe_2O_3/FeO + FeO_3$ (Fig. 5.7). This increases steadily in the whole rocks from 0.27 in the picrites (SiO_4 <46 per cent) to a peak of 0.85 in the most felsic andesites (SiO_2 = 60–2 per cent). In the materials represented in Fig. 17.2(a) it then decreases to 0.73 in the dacites in the 64–6 per cent SiO_2 range. The oxidation indices of the corresponding groundmasses behave similarly beginning at a slightly higher base of 0.42 in the picrites, peaking at 0.89 in the felsic andesites and then decreasing to 0.66 in the dacites. However, as was demonstrated in Chapter 5, these apparent reductions are almost certainly spurious, and a result of the relatively small number of samples in the >62 per cent SiO_2 category. Improved statistics concomitant with the in-

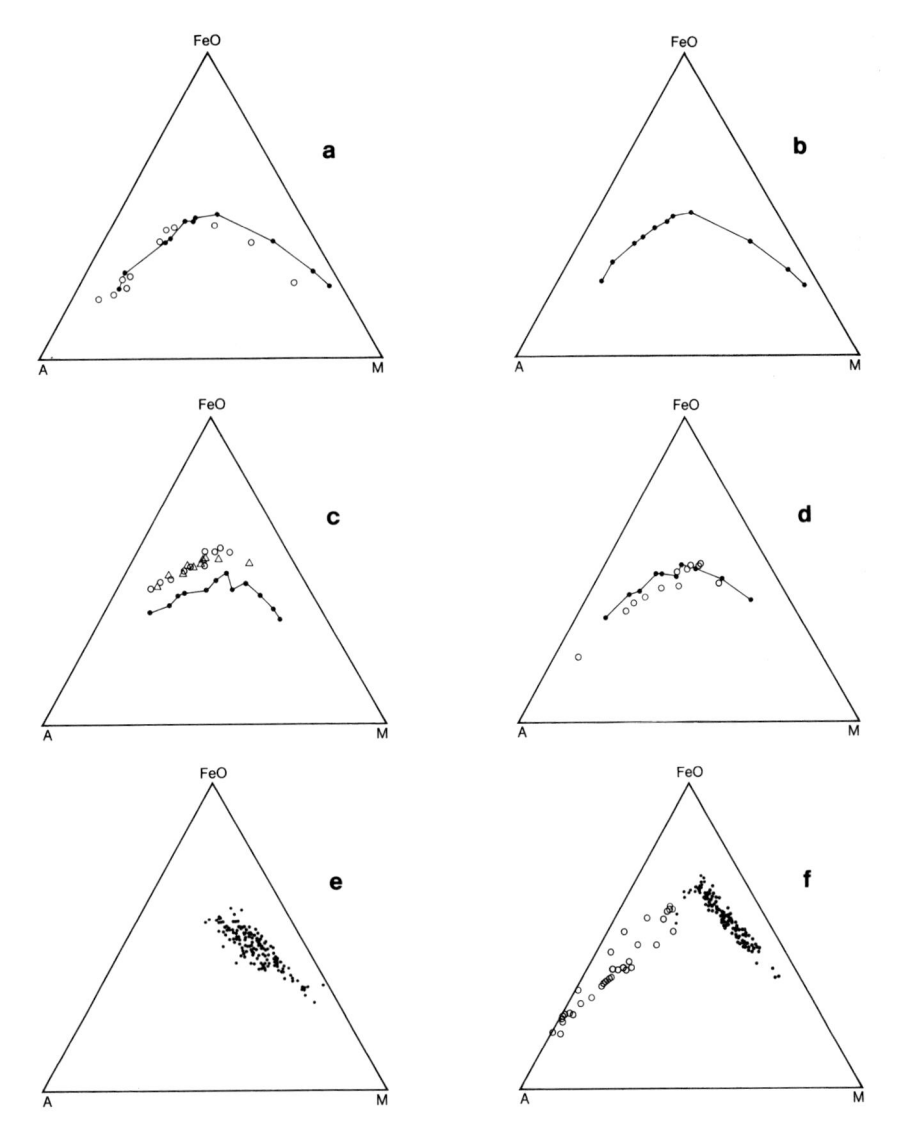

Fig. 17.1. Mean FeO–MgO–(Na$_2$O + K$_2$O) ternary relations corresponding to 2 per cent whole-rock SiO$_2$ intervals (<46, 46–8, 48–50 → 64 per cent SiO$_2$) in (a) whole rocks (filled circles) and corresponding groundmasses (open circles) of the Solomons suite excluding Bougainville; (b) whole rocks of the Solomons suite including Bougainville (241 analyses); (c) lavas of Grenada (filled circles), Dominica (open circles), and St Kitts (open triangles); (d) Vanuatu (filled circles), and the Aleutians (open circles). Ternary relations developed in MORBs as represented by Mid-Atlantic Ridge basalts are given in (e), and in individual samplings of Icelandic basalts (SiO$_2$ < 52 per cent: filled circles) and Icelandic lavas more felsic than basalt (SiO$_2$ ⩾ 52 per cent: open circles) in (f).

clusion of the Bougainville lavas yields a smooth and apparently continuing increase in whole-rock oxidation index with increase in SiO$_2$, as shown in Fig. 5.7(b). Comparison of the latter with curve (2) in Fig. 17.2(b) (variation of oxidation indices of 'average' basalts, andesites, dacites, and rhyolites involving large numbers of analyses from the literature, with variation in SiO$_2$) shows the Solomons curve to be generally similar in slope to that for the averages, but to reflect a generally higher oxidation level in the Solomons lavas.

The behaviour of total iron as FeO in the lavas of the Lesser Antilles (Fig. 17.2(c)), the Aleutian Islands, and Vanuatu (Fig. 17.2(d)) is generally similar to that displayed in the Solomons suite.

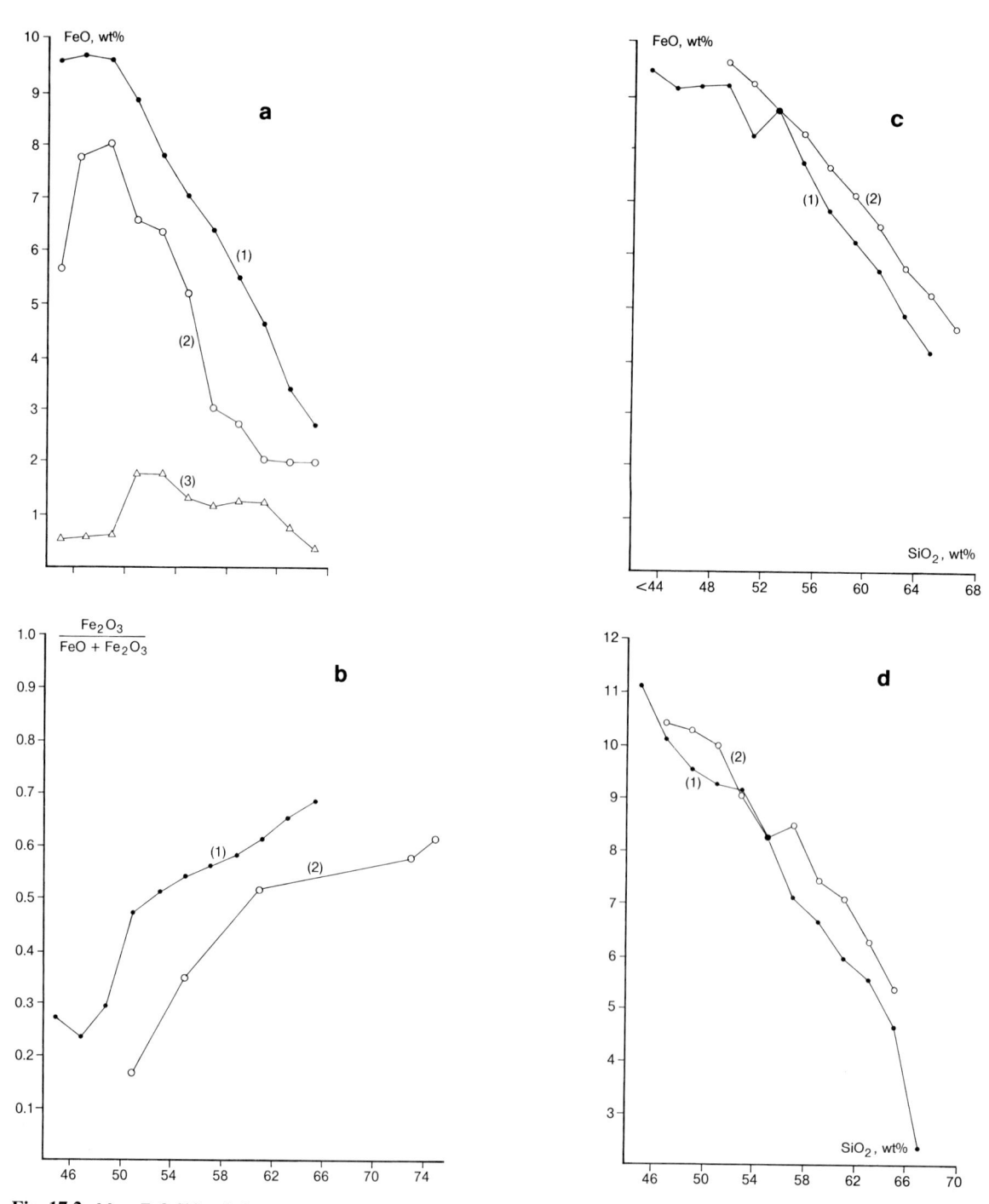

Fig. 17.2. Mean FeO:SiO$_2$ relations corresponding to 2 per cent whole-rock SiO$_2$ intervals in (a) whole rocks [(1) filled circles] and groundmasses [(2) open circles] of the Solomons suite; (c) lavas of Grenada [(1) filled circles] and Dominica [(2) open circles; the 'curve' for St Kitts is virtually identical with that of Dominica]; and (d) lavas of Aleutians [(1) filled circles] and of Vanuatu [(2) open circles]. Curve (3) of (a) portrays variation of modal magnetite with SiO$_2$ in the Solomons suite. In (b) variation of mean oxidation indices corresponding to 2 per cent whole-rock SiO$_2$ intervals in the lavas of the Solomons suite is compared with the corresponding variation in basalts, andesites, dacites, and rhyolites as represented by means of large numbers of analyses from the literature (compare with Fig. 5.7).

CONCLUDING STATEMENT

Owing to its comparatively low monetary value, little attention is paid to the incidence of iron in exhalative ores. Iron is, however, by far the most important exhalative sulphide metal by mass, and it may also occur in substantial quantities as oxide, silicate, and, occasionally, as carbonate.

Most basalts contain iron in the 9–11 per cent total Fe-as-FeO range, andesites 6–8 per cent, dacites 3–5 per cent, and rhyolites 1–2 per cent. The lavas of the Solomons suite conform with this pattern, showing a steady decrease from 9.6 per cent at whole-rock SiO_2 = 48–50 per cent to 2.7 per cent at SiO_2 = 64–6 per cent.

For most of the crystallization sequence iron is partitioned strongly into the crystalline phases and hence is higher throughout in the whole rocks than their groundmasses. The FMA diagram is characterized by a well-defined, highly linear olivine–clinopyroxene subtraction line in the basalt–picrite portion of the field, and displays the general hypotholeiitic pattern common in arc suites.

The oxidation index ($Fe_2O_3/(FeO + Fe_2O_3)$) of the Solomons lavas is ≈0.25 in the mafic basalts, increases quite steeply to 0.47 in the 50–2 per cent SiO_2 range, and then follows a remarkably even slope to 0.68 at SiO >64 per cent. Overall the oxidation index of the Solomons lavas is 0.1–0.2 higher than that of 'average' basalt–andesite–dacite–rhyolite at any given SiO_2 level.

Cobalt

Cobalt is virtually ubiquitous as a trace to very minor constituent of exhalative sulphide ores, but rarely if ever appears as a major component. Its average level of abundance in such deposits is difficult to estimate but is probably no more than about 0.05 per cent. The cobalt may occur as a major element of sulpharsenides and related minerals, which themselves occur in very small amount, or as a trace to minor element of the more common and abundant sulphides such as pyrite, pyrrhotite, and chalcopyrite.

Cobaltite, CoAsS, is a common trace mineral in exhalative deposits, and arsenides such as safflorite (Co, Fe)As$_2$ and loellingite (Fe, CoAs$_2$) appear occasionally. By far the major amount of exhalative cobalt occurs as a minor component of pyrite, pyrrohotite, chalcopyrite, and sphalerite, in which it generally appears at the 100–2000 ppm level. The pentlandite of the somewhat questionable small group of exhalative nickel deposits (see Chapter 19) may contain cobalt to ≈1.0 per cent.

The earliest reliable estimate of cobalt in igneous rocks seems to be that of Clarke and Washington (1924), who proposed an average of 0.001 per cent. A little later Goldschmidt (with H. Witte and H. Hörman; unpublished, reported in Goldschmidt 1954) estimated an average of 0.004 per cent for the upper lithosphere. Goldschmidt (1937) estimated averages of 240 ppm Co for peridotite, 80 ppm for gabbro, 30 ppm for diorite, and 8 ppm for granite. Sandell and Goldich (1943), in their landmark paper on the rarer metallic constituents of some American igneous rocks, found means of 0.0003 per cent Co in 'silicic rocks' (SiO$_2$ >63 per cent) and 0.0032 per cent in 'subsilicic rocks' (SiO$_2$ <63 per cent). They observed that a 'relation between cobalt (0.83Å) and magnesium (0.78Å) is more apparent than one between cobalt and ferrous iron (0.83Å). This relation between cobalt and magnesium, in contrast to that between nickel and magnesium, is linear over a wide magnesia range' (Sandell and Goldich 1943, p. 178) and they used this to calculate expected cobalt averages from known MgO averages for the igneous rock groups considered earlier by Goldschmidt (1937). The values they obtained, 0.026 per cent Co for peridotites, 0.005 per cent for gabbros, 0.002 per cent for diorites, and 0.0007 per cent for granites, are in fact remarkably close to those earlier determined by direct chemical analysis by Goldschmidt. On the assumption that the Co/MgO ratio is constant and equal to about 0.00066, Sandell and Goldich calculated an average cobalt value of 0.0023 per cent for the igneous rocks of the Earth's crust. The modern estimate of cobalt in the crust is 29 ppm or 0.0029 per cent (Emsley 1989): a figure to which that of Sandell and Goldich is remarkably close.

In their investigation of the Skaergaard Intrusion, Wager and Mitchell (1951) determined averages of 55, 40, and 50 ppm, Co for hypersthene–olivine-gabbro, hortonolite-ferrogabbro, and chilled marginal gabbro respectively. For the six stages A, B, C, D, E, and F of the layered series they estimated 53, 30, 25, 18, 13, and 6 ppm Co respectively. More modern information on the cobalt content of the major volcanic rock groups is sparse, perhaps at least partly due to problems with contamination of analytical crushes during preparation. Stueber and Goles (1967) gave averages of 102 ppm Co for alpine-type ultramafic rocks, 119 ppm for dunites, and 88 ppm for pyroxenites. Taylor et al. (1969) estimated averages of 50 ppm Co for alkali and tholeiitic basalts, 40 ppm for high-Al basalts, 28 ppm for 'low-silica' andesites (Mean SiO$_2$ = 54.9 per cent), 24 ppm for andesites (mean SiO$_2$ = 59.5 per cent), 20 ppm for 'low-K' andesites (mean K$_2$O = 0.7 per cent), and 13 ppm for 'high-K' andesites (mean K$_2$O = 3.3 per cent). Wedepohl (1975) estimated 37 ppm for oceanic tholeiites and 43 ppm for alkali olivine-basalts.

The rather sparse data thus indicate that most basaltic lavas probably contain ≈35 to 50 ppm Co, andesites 15–30 ppm, and more felsic types progressively less.

Concerning the broad petrochemical affinities of cobalt, Goldschmidt (1954) considered that it generally followed ferrous iron more closely than magnesium in magmatic processes. He cited the work of Wager and Mitchell (1951), who had shown (see above) that in the Skaergaard layered series cobalt was more evenly distributed than nickel and, following Fe^{2+}, was relatively more abundant in the middle phases of the intrusion. Goldschmidt, however, noted the contrary findings of Sandell and Goldich (1943) — that cobalt tended to correlate with whole-rock MgO rather than with FeO — and the compatible results of Nockolds and Mitchell (1948). Carr and Turekian (1961) also demonstrated a close relationship between cobalt and magnesium in granitic rocks, and between Co and (Fe + Mg) in basalts.

The results of the present investigation confirm the cobalt–magnesium relation, and show that, for island-arc lavas, this is much closer than that between cobalt and ferrous iron.

COBALT IN LAVAS OF THE MARINE ENVIRONMENT

Perhaps because of the contamination problems inherent in some preparation procedures, and perhaps also because of its perceived lack of substantial petrochemical significance, the literature database on the incidence of cobalt in marine lavas it not as large as those for elements such as vanadium, chromium, and nickel. There is, however, sufficient information for quite reliable evaluation of its incidence and behaviour in the principal lava groups of the marine environment.

The abundances of cobalt in samplings of MORBs are summarized in Table 6.2 The mean for 376 analyses of ridge material is 44 ppm and for 429 analyses including basalts from Iceland is 45 ppm. Averages for the Mid-Atlantic, Atlantic–Indian, Indian, and Pacific Ocean ridges are 44, 44, 47, and 37 ppm Co respectively: a remarkable consistency emphasized by Fig. 18.1 The Icelandic basalts contain an average of 53 ppm: slightly, but probably significantly, higher than those of the ridges proper. This

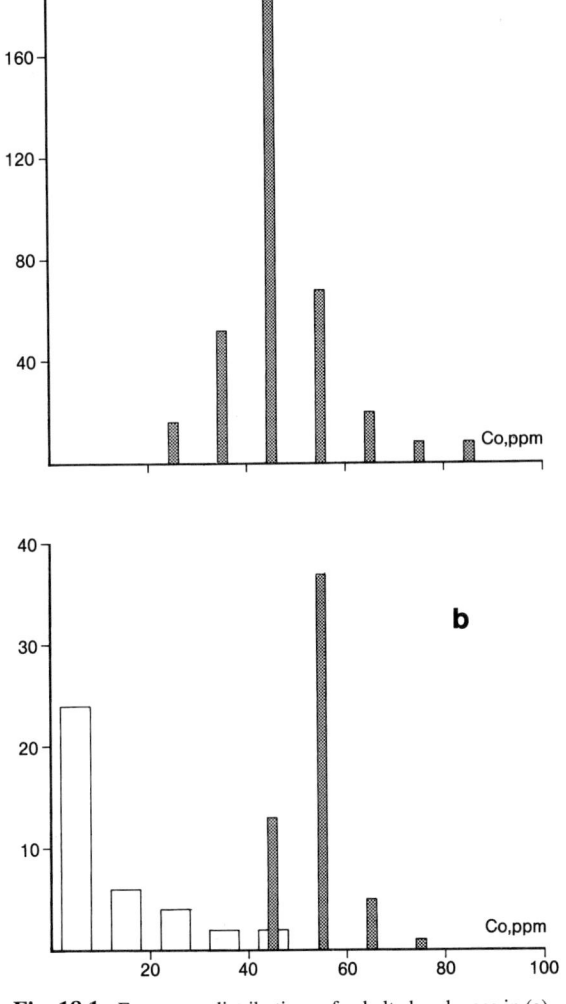

Fig. 18.1. Frequency distributions of cobalt abundances in (a) MORBs of all major oceans, and (b) Icelandic basalts (SiO$_2$ < 52 per cent: solid bars) and Icelandic lavas more felsic than basalts (SiO$_2$ ≥ 52 per cent: open bars).

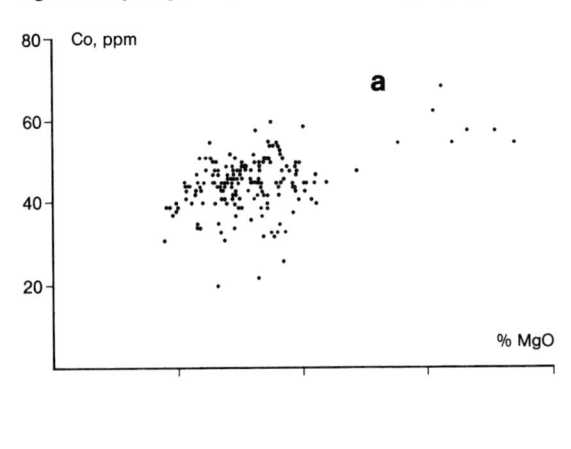

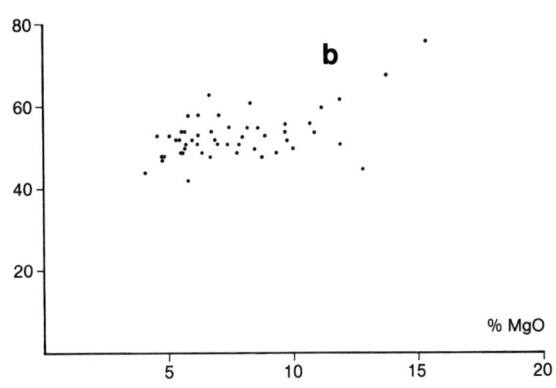

Fig. 18.2. Abundance relations between cobalt and MgO in (a) MORBs as represented by those of the Mid-Atlantic Ridge, and (b) Icelandic basalts.

230 *Ore Elements in Arc Lavas*

difference is shown clearly in Fig. 18.1. Cobalt in MORBs exhibits a discernible but poor correlation with whole-rock MgO (Fig. 18.2) and FeO. In agreement with the observation of Carr and Turekian (1961) its abundance is, however, related to Mg + Fe: presumably a reflection of its strong octahedral site-preference (see below).

The striking tendency to uniformity of cobalt in MORBs is presumably due to its relative insensitivity to olivine crystallization and removal, and is in sharp contrast to the abundance behaviour of nickel, which is highly sensitive to olivine fractionation. Figure 18.3, based on cobalt and nickel analyses of 188 MORBs samplings from ridge areas surrounding the South Atlantic triple junction, shows clearly the difference in abundance behaviour of the two metals.

The generally lower values of cobalt in the more felsic members (SiO_2 >52 per cent) of the Icelandic suite are indicated by the open bars in Fig. 18.1(b).

Data on cobalt in arc basalts (Table 6.3) are less extensive than for MORBs. Frequency distributions of cobalt abundances in arc basalts other than those of the Solomons for which analyses are available (Aleutians, Tonga–Kermadec, Vanuatu) are given in Fig. 18.4. Mean cobalt for the 85 arc basalts is 37 ppm; if the author's data for the Solomons are included, the mean for the four arcs (149 analyses) is 43 ppm: almost precisely that for the 376 analyses of MORBs. Mean cobalt for the 74 arc lavas more felsic than basalts (SiO_2 >52 per cent; Fig. 18.3(b)) is 22 ppm. Inclusion of the Solomons data yields an array of 132 analyses, which again yield a mean of 22 ppm. In the only case — Tonga–Kermadec — in which significant data are available for all three lava groups, mean cobalt contents are: basalts (SiO_2 <52 per cent), 35 ppm; andesites (52 per cent <SiO_2 <62 per cent), 30 ppm; dacites (SiO_2 >62 per cent), 12 ppm.

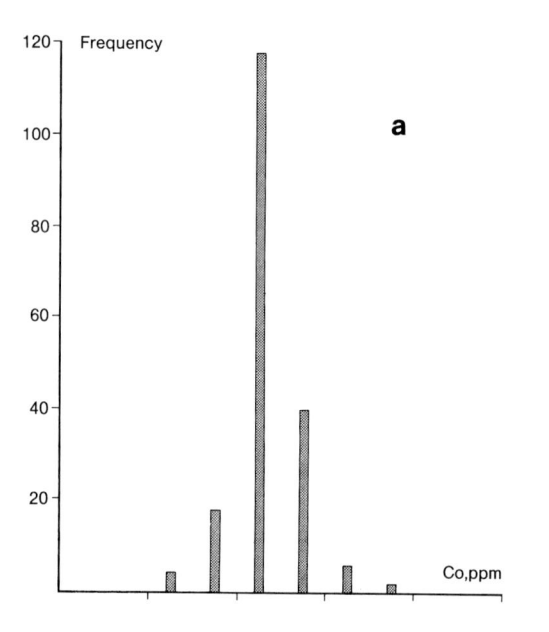

Fig. 18.3. Frequency distributions of (a) cobalt and (b) nickel abundances in a set of samples of MORBs from the vicinity of the South Atlantic triple junction. Data chiefly from le Roex *et al.* (1983).

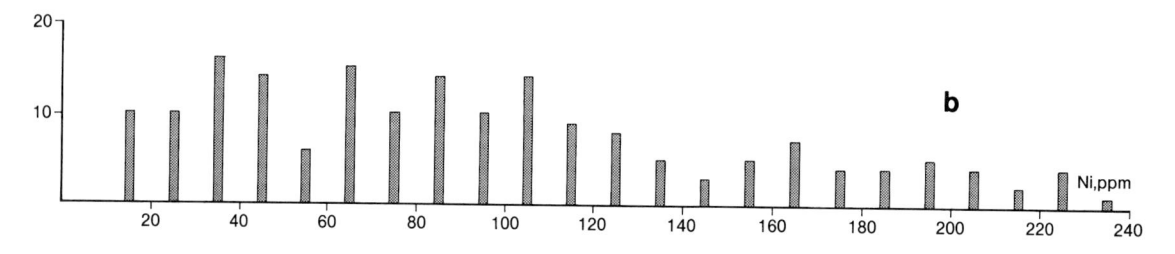

Cobalt 231

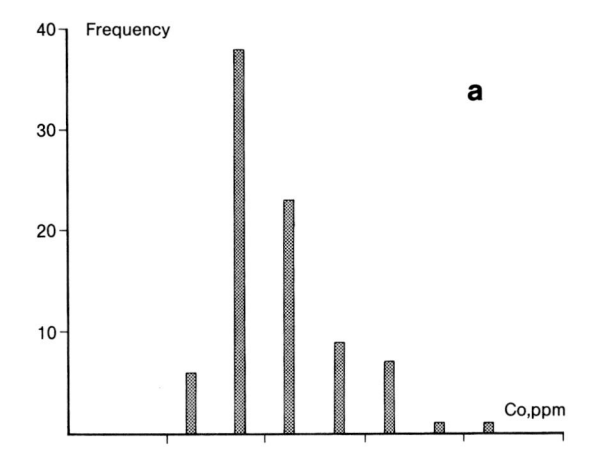

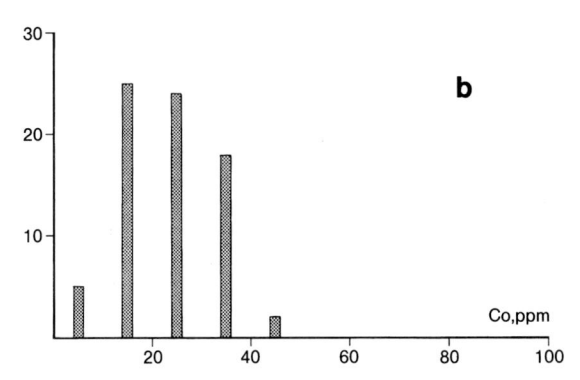

Fig. 18.4. Frequency distributions of cobalt abundances in (a) volcanic island-arc basalts ($SiO_2 < 52$ per cent), (b) arc lavas more felsic than basalt ($SiO_2 \geq 52$ per cent).

CHEMICAL PROPERTIES AND CRYSTAL CHEMISTRY OF COBALT

Cobalt occurs in nature as the Co^{2+} and Co^{3+} ions, the former greatly dominant. The electron configuration of the Co^{2+} ion, [Ar] $3d^7$, is such that in ionic compounds it is stabilized in octahedral and tetrahedral coordinations (Burns 1970). On purely geometrical grounds its radius ratio of 0.586 with respect to oxygen indicates a preference for 6-fold (octahedral) coordination in silicates. Goldschmidt (1954) listed the divalent ions of nickel, magnesium, cobalt, and iron and their respective ionic radii (in Å), thus:

Ni^{2+}	Mg^{2+}	Co^{2+}	Fe^{2+}
0.78	0.78	0.82	0.83

On this basis he suggested (as noted above) that whereas Ni^{2+} tends to substitute for Mg^{2+}, Co^{2+} should show a preference for Fe^{2+} sites. He considered that this principle underlay the tendency for trace nickel to occur in the earliest crystallizing silicates, and for trace cobalt to occur in the somewhat later (intermediate) crystallizing members of a progressively solidifying magmatic system. Burns and Burns (1974), however, give the ionic radius of Co^{2+} in octahedral coordination as 0.735Å, 'intermediate between Mg^{2+} (0.72Å) and Fe^{2+} (0.77Å) — so that it substitutes for these cations in several silicates' (1974, p. 27-A-1). The most recent standard estimate of the radius of the Co^{2+} ion remains at 0.082 nm (Emsley 1989), but the indication by Burns and Burns that Co^{2+} substituted for both Mg^{2+} and Fe^{2+} in silicates appears to be borne out by what may be described as the somewhat ambiguous behaviour found in the present study.

The principal properties of cobalt and Co^{2+} are given in Table 6.4.

THE INCIDENCE OF COBALT IN THE PRINCIPAL MINERAL SPECIES OF THE SOLOMON ISLANDS YOUNGER VOLCANIC SUITE

General

In their investigation of the Skaergaard intrusion, Wager and Mitchell (1951, p. 151) noted that cobalt was present in pyroxene, olivine, ilmenite, and magnetite 'in amounts roughly proportional to the ferrous iron content of these minerals'. Cobalt content remained steady at about 50 ppm through most of the pyroxene series, falling to 15 ppm in the most iron-rich members. The olivines contained 100–50 ppm Co and also showed a decline in this element in the later — presumably more iron-rich — members. The ilmenite and magnetite of the hortonolite ferrogabbro contained 70 and 60 ppm Co respectively. Carr and Turekian (1961) determined cobalt values of 73, 85, 164, and 256 ppm in clinopyroxene, orthopyroxene, olivine, and magnetite respectively in a Mexican basalt, and Taylor *et al.* (1969) found 38, 31, 48, 30, 45 and 43 ppm Co (mean = 39 ppm) in the clinopyroxenes of six eclogite inclusions obtained from kimberlitic pipes in Australia, Africa, and Norway.

Solomons suite

Mean values of cobalt in the principal silicates and the spinels of the Solomons suite are given in Table 6.5.

Ferromagnesian minerals. Olivine is by far the principal receptor of cobalt, with an overall average abundance of 185 ppm Co and a range of values as indicated in

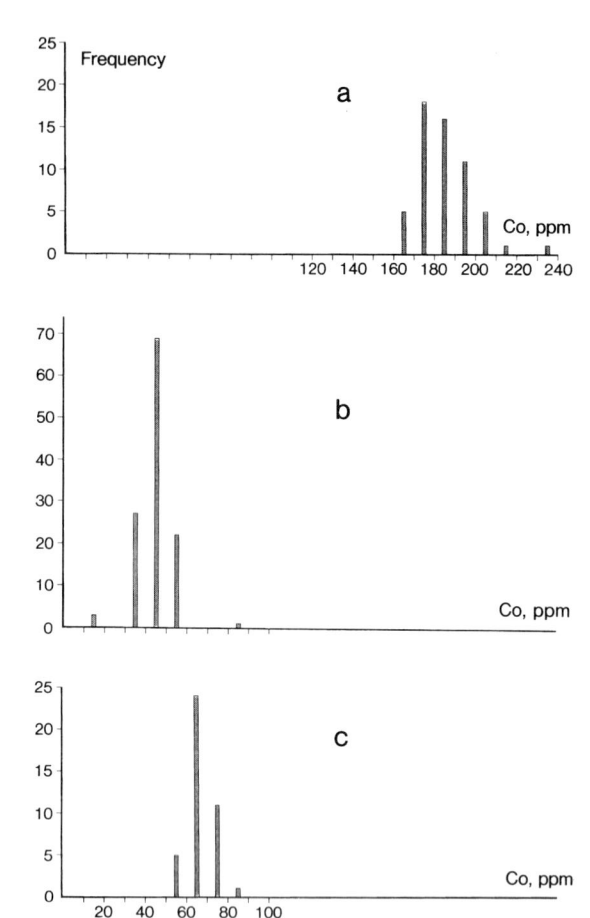

Fig. 18.5. Frequency distributions of cobalt abundances in (a) olivines, (b) clinopyroxenes, and (c) hornblendes of the Solomons suite.

Fig. 18.5(a). It contains some three times as much as the next most cobalt-rich silicate, hornblende, and about 1.6 times as much the overall average in magnetite. Although Table 6.5 indicates a slightly higher mean abundance of cobalt in the more iron-rich olivines of the big-feldspar lavas — and hence perhaps some general systematic relationship between cobalt and iron in the olivines — Fig. 18.6 indicates that large changes in olivine FeO:MgO are accompanied by little variation in cobalt. The combination of greatly varying FeO and substantially constant Co leads to the conspicuously linear field and apparently high degree of correlation of Fig. 18.6(a) but the positive slope of the regression is very slight. By necessity Co:MgO relations (Fig. 18.6(b)) are generally similar but the regression line has a slightly negative slope. Thus it may be said that cobalt shows *almost* no preference for one or the other of Fe^{2+} or Mg^{2+} sites in olivine: what very slight preference there is, is in favour of the former.

The cobalt content of the clinopyroxenes is even more consistent (Fig. 18.5(b)) than that of the olivines, with an overall average of 43 ppm. Table 6.5 indicates that those of the big-feldspar lavas contain slightly but significantly more cobalt than do those of the other two major lava groups, and this appears to be borne out by Fig. 18.7. Because the clinopyroxenes as a group do not exhibit the large variation in FeO:MgO shown by the olivines, Co:FeO and Co:MgO relations in the former do not develop linear arrays similar to those of Fig. 18.6(a) and (b): the relationship is the substantially random one of Fig. 18.6(c).

The orthopyroxenes contain a mean of 98 ppm cobalt (Table 6.5). At first sight this appears to be in keeping with the higher iron content of the orthopyroxenes, but again plotting of individual Co:Fe contents reveals no systematic relation between the two elements within the orthopyroxene group. However, mean FeO_{opx}/mean FeO_{cpx} = 18.55/7.37 = 2.5, and mean Co_{opx}/mean Co_{cpx} = 98/43 = 2.3, indicating a very close Co:FeO correlation on a broader scale.

This appears to be confirmed when the other two ferromagnesian minerals, hornblende and biotite, are brought into consideration, as shown in Fig. 18.8. Hornblende, like the clinopyroxenes and orthopyroxenes, is remarkably consistent in its cobalt content (Fig. 18.5(c)) but shows no 'within-group' Co:FeO correlation. Data for the biotites are insufficient to indicate uniformity or otherwise. However, when mean Co:FeO for each of the three groups of clinopyroxenes, the orthopyroxenes, hornblendes, and biotites are plotted together as in Fig. 18.8 there develops a marked linear relationship, with a correlation coefficient of 0.93. All this seems to indicate that while there is a high degree of randomness of Co:FeO relations on a between-sample basis, there is nevertheless a broad but strong *tendency* for the two elements to be correlated over the full spectrum of pyroxene–amphibole–mica — a correlation doubtless reflecting the crystal structural affinities of these three mineral groups.

As indicated in Fig. 18.8 there is a sharp increase in Co:FeO ratios from orthopyroxene to olivine: the latter clearly does not conform with the linear pyroxene–amphibole–mica Co:FeO relationship, presumably reflecting the greater strength of the bonding of the cobalt ion in the olivine structure.

The impression gained from the limited biotite data of Table 6.5 is that the cobalt content of hornblende is equal to or greater than that of biotite. However, in the two instances in which it has been possible to investigate the

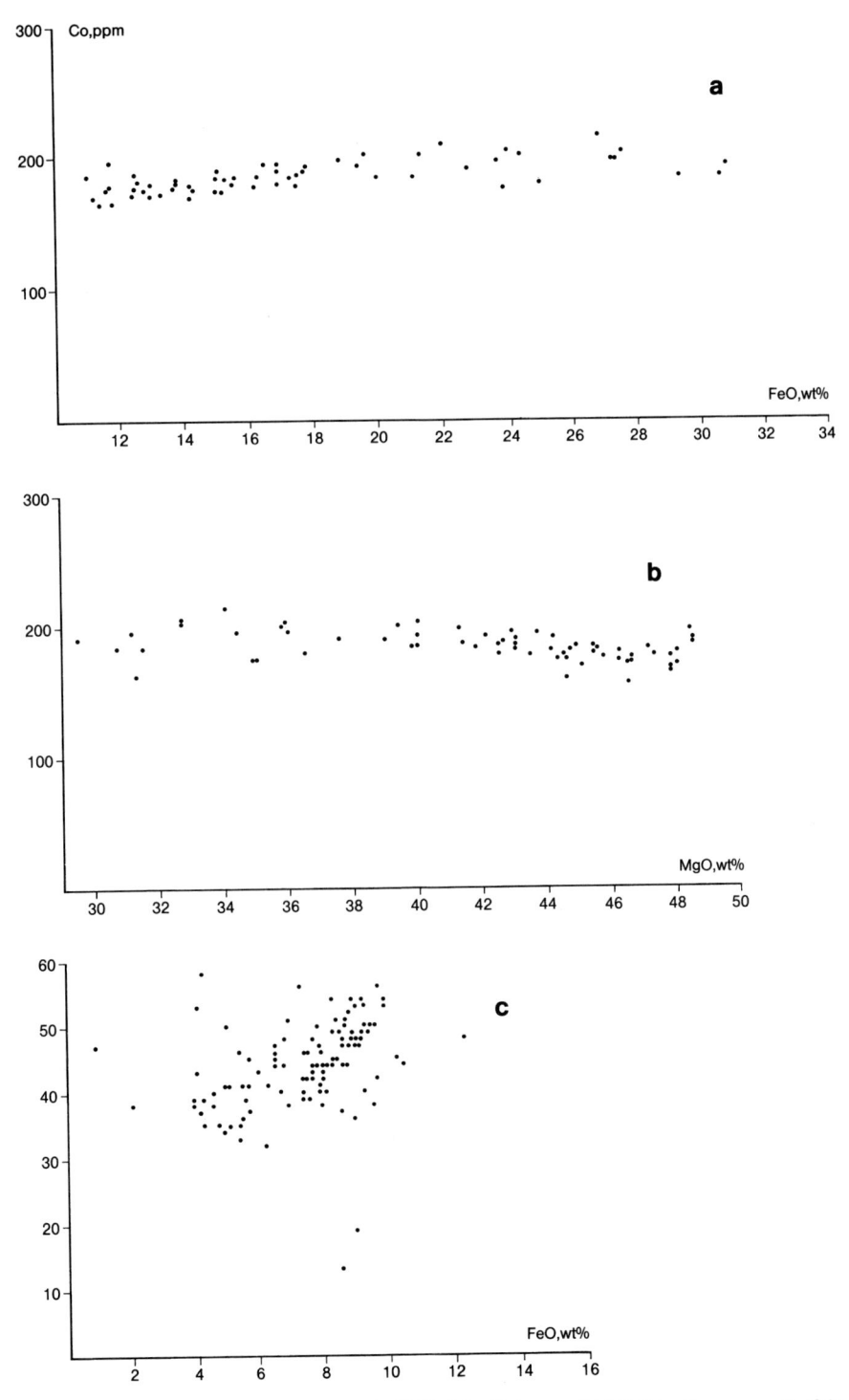

Fig. 18.6. Abundance relations between cobalt and (a) FeO, (b) MgO in olivines, and (c) FeO in clinopyroxenes of the Solomons suite.

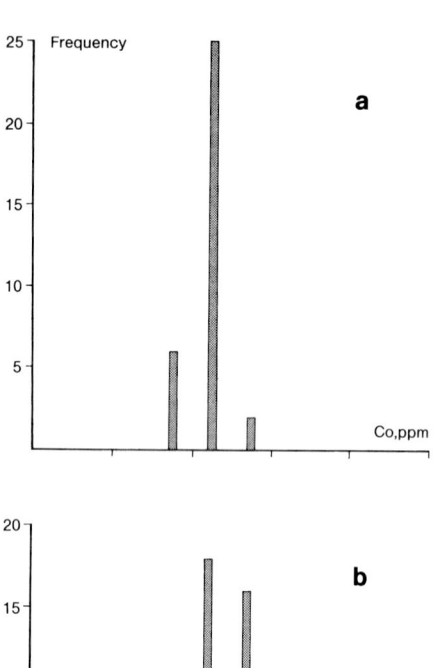

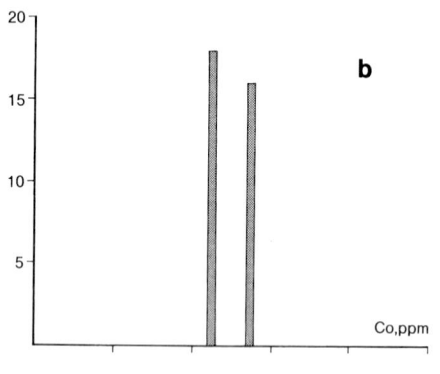

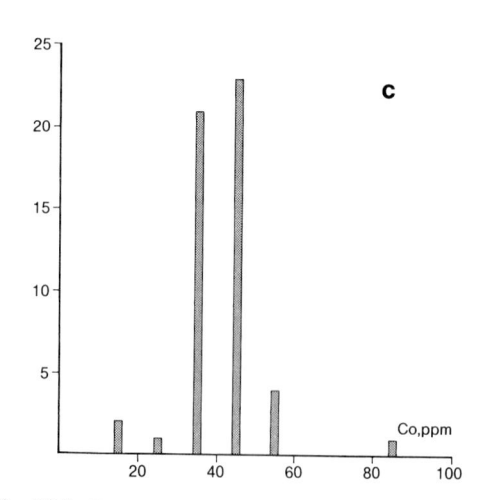

Fig. 18.7. Frequency distributions of cobalt abundances in clinopyroxenes of (a) the hornblende-andesite group, (b) big-feldspar lavas, and (c) the olivine–clinopyroxene-basalts of the Solomons suite.

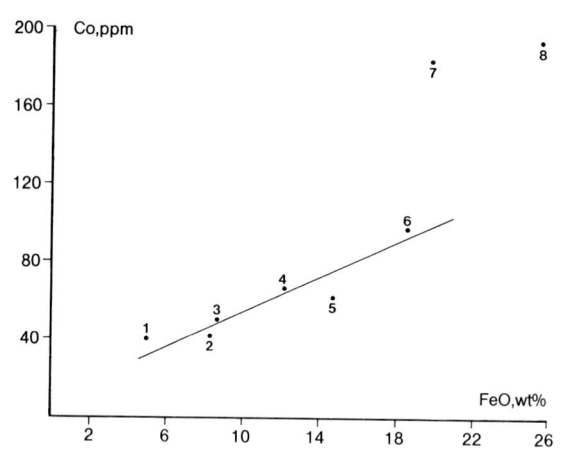

Fig. 18.8. Abundance relations between mean cobalt and mean FeO in (1) clinopyroxenes of the olivine–clinopyroxene-basalts, (2) clinopyroxenes of the hornblende-andesite group (3) clinopyroxenes of the big-feldspar lavas, (4) hornblendes, (5) biotites, (6) orthopyroxenes, (7) olivines of the olivine–clinopyroxene-basalts and (8) olivines of the big-feldspar lavas.

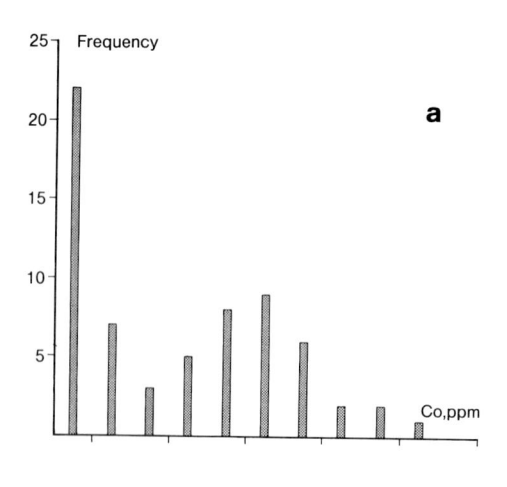

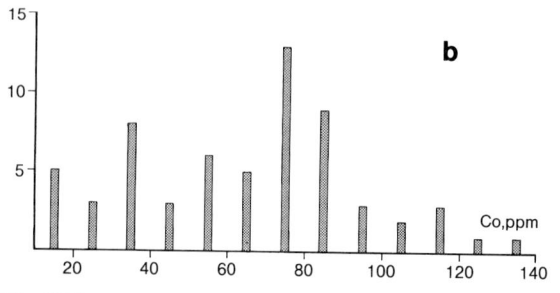

Fig. 18.9. Frequency distributions of cobalt abundances in (a) hornblende, and (b) biotite coexisting in two Savo dacites.

abundance of cobalt in *coexisting* biotite and hornblende (in the two Savo biotite-bearing dacites), biotite exhibits the greater cobalt content. Thirty-two biotite:hornblende pairs in one dacite gave 62 and 33 ppm Co respectively, and 33 pairs in the other gave 62 and 46 ppm. The full 65 pairings are plotted together in Fig. 18.9: for these, mean Co in biotite = 62 ppm; mean cobalt in hornblende = 40 ppm. Mean Co_{biot}/Co_{Hb} is thus 1.55.

Electron-probe determination of cobalt across the large single crystal of hornblende of Fig. 7.9 showed, as for chromium and in contrast to such elements as zinc, manganese, and potassium, a completely random pattern of cobalt abundance, and no discernible difference in the amounts of cobalt as between hornblende crystal and opaque rim (Fig. 18.10). This was confirmed by XRF analysis of separations of rim material, which gave cobalt values of ≈60–62 ppm: essentially the same as the hornblendes themselves.

Feldspars. The feldspars contain an almost vanishingly small amount of cobalt (Table 6.5). Comparison of the

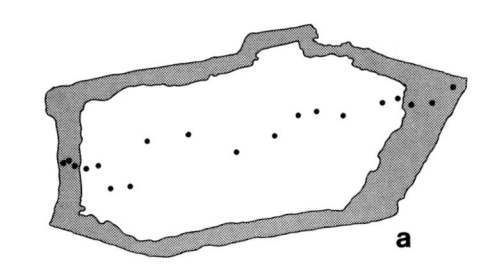

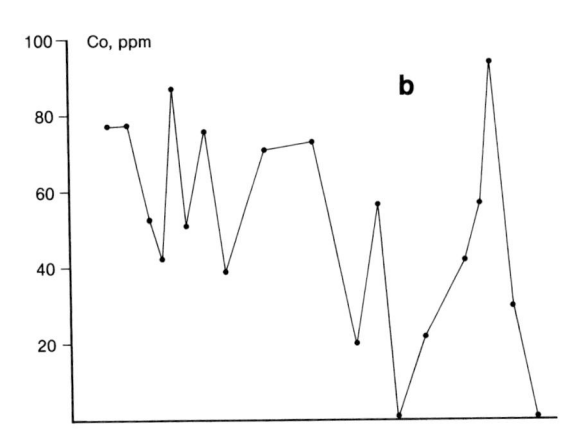

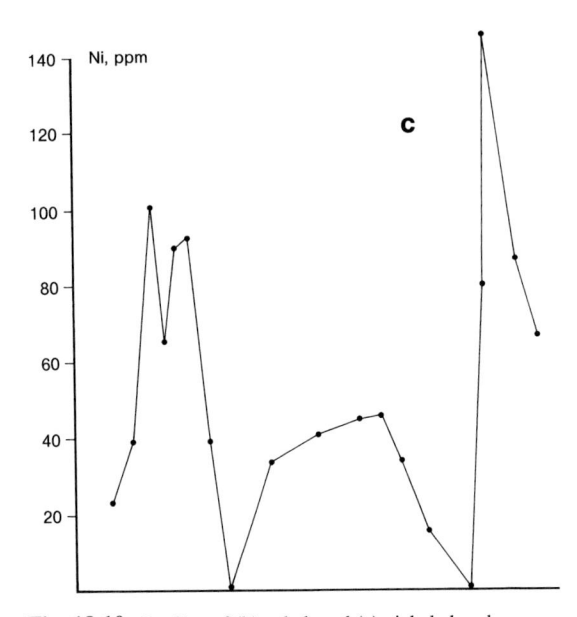

Fig. 18.10. Profiles of (b) cobalt and (c) nickel abundances along the length of the single crystal of hornblende (a) and its opaque rim (shown stippled). See also Chapters 5, 7, and 13 for other constitutional features of this crystal and its opaque rim.

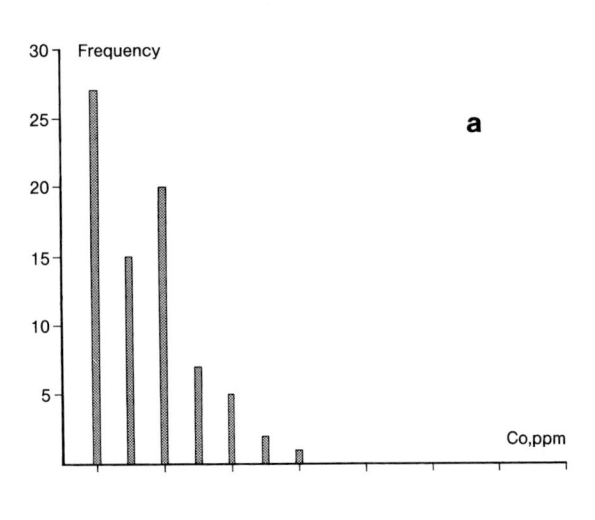

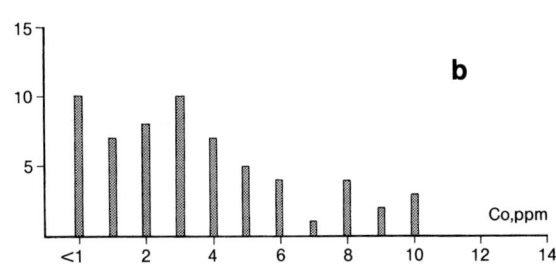

Fig. 18.11. Frequency distributions of cobalt abundances in (a) ultimate purity feldspar separations and (b) penultimate purity feldspar separations.

'ultimate purity' and 'penultimate purity' plagioclase fractions previously referred to (Fig. 18.11) indicates a slightly higher incidence of cobalt in the latter: a mean of 3.56 ppm Co compared with 1.64 ppm in the 'ultimate purity' fraction (Fig. 18.11). This indicates that there is a relative concentration of cobalt in the fine glassy inclusions in the feldspars. Appropriate correction for this yields intrinsic abundances of cobalt in the plagioclases of 1–2 ppm in almost all cases.

Spinel. Spinel follows olivine as a cobalt acceptor, with an overall mean of 116 ppm for the 80 magnetites analysed. In contrast to the broad groups of olivines, clino- and orthopyroxenes, and hornblendes, magnetite exhibits considerable variation in uptake of cobalt: from 13 to 183 ppm Co (Fig. 18.12). This is not random, however: as indicated in Fig. 18.12 (and Table 6.5), there is a progressive and quite ordered decrease in magnetite cobalt from the olivine–pyroxene-basalts (mean = 144 ppm Co) through the big-feldspar lavas (mean = 123 ppm Co) to the magnetites of the hornblende-andesites and dacites (mean = 102 ppm Co).

CRYSTAL:MELT PARTITIONING

K_D^{Co} values for the principal mineral:melt pairings are given in Table 18.1

The earliest reliable determinations of partitioning of cobalt between olivine and basaltic melt are those of Wager and Mitchell (1951) who obtained values of $K_D^{ol:melt}$ of 3.0, 4.17, 5.56, 3.8, and 3.3 for the Skaergaard A, B, D, E, and

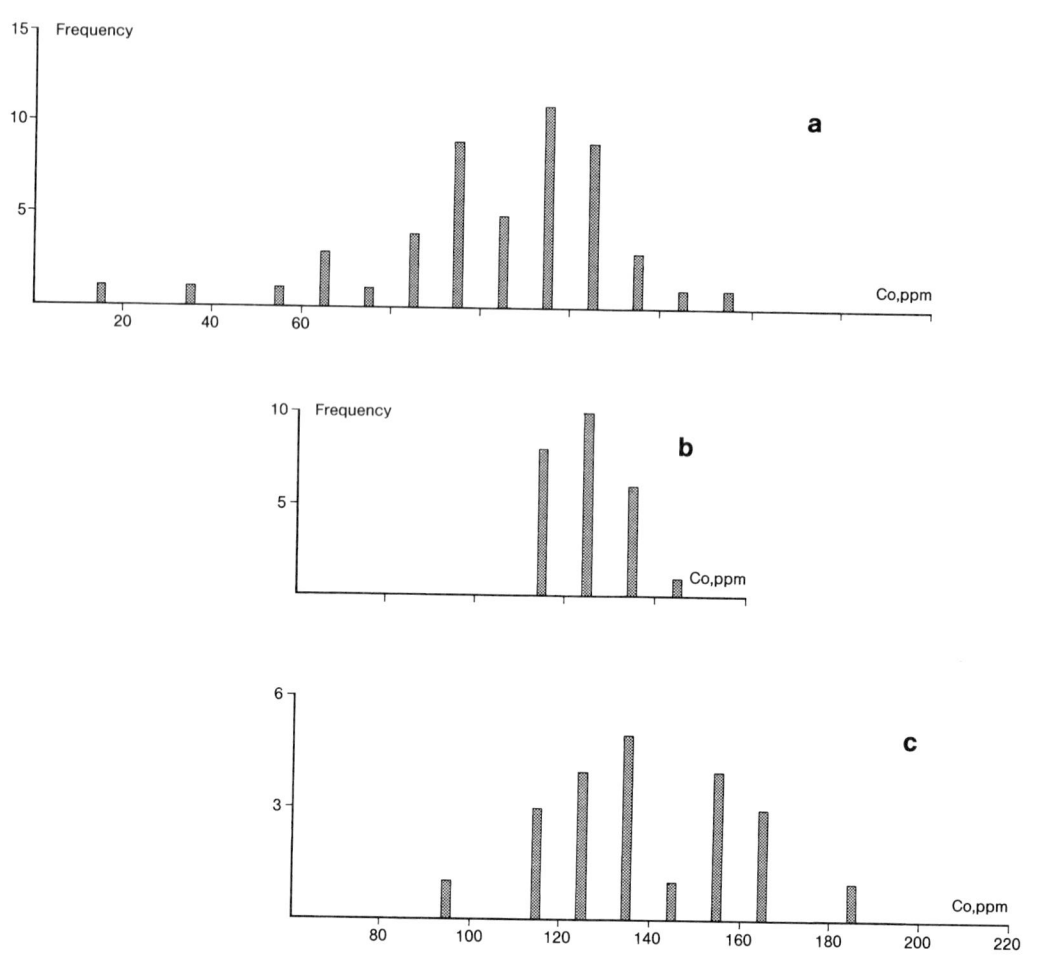

Fig. 18.12. Frequency distributions of cobalt abundances in spinels of (a) the hornblende-andesite group, (b) the big-feldspar lavas, and (c) the olivine–clinopyroxene-basalts.

Table 18.1 Crystal:melt distribution coefficients (K_D) with respect to cobalt

Mineral species	Lava group	n_1	n_2	m_1	m_2	$K_{D(1)}$	N	M_1	M_2	$K_{D(2)}$
Olivine	OPB	40	44	184	40	4.60	39	182	38	4.79
	BFL	7	23	193	25	7.72	6	198	35	5.66
Clinopyroxene	OPB	43	44	40	40	1.00	42	38	39	0.97
	BFL	26	23	50	25	2.00	17	50	28	1.79
	HbA	27	47	42	10	4.20	20	45	12	3.75
Hornblende	HbA	25	47	66	10	6.60	22	66	9.26	7.13
Feldspar	OPB	10	44	2.10	40	0.053	10	2.10	28	0.075
	BFL	23	23	2.41	25	0.096	22	3.18	26	0.122
	HbA	46	47	0.83	10	0.083	44	0.81	9.96	0.081
Spinel	OPB	20	44	144	40	3.60	20	144	36	4.00
	BFL	23	23	123	25	4.92	22	124	27	4.59
	HbA	37	47	102	10	10.20	36	104	10	10.40

OPB = olivine–clinopyroxene-basalt group; BFL = big-feldspar lavas; HbA = hornblende-andesite group; n_1 = total number of mineral separations analysed; n_2 = total number of groundmass separations analysed; m_1 = mean cobalt content of mineral of each lava group; m_2 = mean cobalt content of groundmass of each lava group; $K_{D(1)}$ = mean distribution coefficient determined as m_1/m_2; N = number of coexisting mineral:groundmass pairs analysed; M_1 and M_2 = mean cobalt contents of coexisting mineral and groundmass respectively; $K_{D(2)}$ = distribution coefficient determined as M_1/M_2, for comparison with $K_{D(1)}$. All concentrations as parts per million by mass.

F stages respectively. Henderson and Dale (1969) found values of 1.9 to 3.4 in a group of oceanic basalts; Bougault and Hekinian (1974) a value of 3.0 in a mid-Atlantic ridge basalt; Villemant et al. (1981) a value of 5.1 in alkali basalts; and Lemarchand et al. (1987) values of 4.51, 6.43, and 14.60 in a basalt, mugearite, and trachyte respectively of an alkaline lava series. Experimental results by Duke (1976) in the SiO_2–Al_2O_3–FeO–MgO–CaO–Na_2O system between 1125° and 1250 °C at 1 atm, anhydrous, gave Nernst partition coefficients ranging from 5.0 to 6.3. Seifert et al. (1988), experimenting with lunar basaltic liquid in equilibrium with cobalt metal, obtained $K_D^{ol:melt}$ values ranging from 1.6 to 2.22 at temperatures from 1420 to 1260 °C.

$K_D^{ol:melt}$ for cobalt in the Solomons lavas is in keeping with the above determinations on natural materials (Table 18.1). Partition coefficients developed by the slightly more iron-rich olivines of the big-feldspar lavas are a little higher than for the olivines of the olivine–pyroxene-basalts, though the sampling is too small for certainty. Overall values among samplings are relatively constant within the limits 2 to 8. Fig. 18.13, based on 39 olivine:groundmass pairings for the olivine–pyroxene-basalts, shows the marked consistency of cobalt K_D values when compared with the spread of nickel K_D values for precisely the same olivine:groundmass pairs.

Wager and Mitchell (1951) also provide the earliest reliable data on the partitioning of cobalt between clinopyroxene and melt in basaltic materials. They give values of 1.1, 1.7, 2.4, 2.2, 2.3, and 2.5 for the A–F stages respectively of the Skaergaard intrusion. Onuma et al. (1968) determined values of 1.12 and 2.08 for clinopyroxene and orthopyroxene respectively in an alkali olivine-basalt; Ewart and Taylor (1969) give rather higher values: 2.3 and 8.0 for clinopyroxenes in two New Zealand andesites, and 3.9, 9.5, and 12 for associated

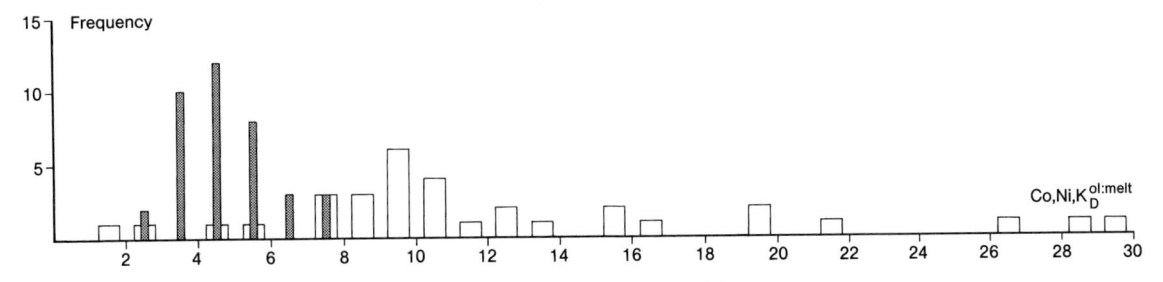

Fig. 18.13. Ranges of variation of values of K_D^{Co} olivine:melt (filled bars) and K_D^{Ni} olivine:melt (open bars) in 39 olivine–clinopyroxene-basalts of the Solomons suite.

orthopyroxenes. Bougault and Hekinian (1974) determined a value of 1.32 for clinopyroxene in the mid-Atlantic ridge basalt referred to above; Villemant et al. (1981) a value of 1.02 for alkali basalt; and Lemarchand et al. (1987) values of 0.95, 1.55, and 5.30 for clinopyroxenes in hawaiite, mugearite, and trachyte respectively. The experimental work of Duke (1976) gave values ranging from 1.0 to 1.7, and that of Lindstrom and Weill (1978) a range from 0.54 to 1.33.

Although variable from one individual crystal: groundmass pairing to another, the results for the Solomons lavas are in general agreement with these values. Mean values increase from 0.97 for the clinopyroxenes of the olivine–pyroxene-basalts to 3.75 for those of the andesites. This, however, does not indicate an increasing tendency for the clinopyroxenes to capture cobalt as the host melt becomes more felsic: it simply reflects a more or less constant uptake of cobalt by the crystals as the melt becomes progressively poorer in it (Table 18.1). At first sight columns 7 and 11 of Table 18.1 seem to indicate that the clinopyroxenes of the andesites are some three to four times more effective than those of the olivine–pyroxene-basalts in abstracting cobalt from the melt. This, however, is not so: all three of the clinopyroxene groups accommodate ≈40–50 ppm Co, and the apparent progressive increase in $K_D^{cpx:melt}$ is due to the progressive decrease of cobalt in the melt.

The limited amount of electron microprobe data available indicate that in the Solomons two-pyroxene andesites, coexisting clino- and orthopyroxenes contain 40–50 and 90–100 ppm Co respectively. $K_D^{opx:cpx}$ is thus ≈2 and by inference $K_D^{opx:melt}$ is therefore ≈6–8 in the andesitic members of the Solomons suite.

Values for $K_D^{Hb:melt}$ given in the literature vary greatly. Ewart and Taylor (1969) give a value of 7.5 for hornblende in a New Zealand andesite; Dudas et al. (1973) values ranging from 5.2 to 17.9 for hornblendes of dacitic tephra; and Lemarchand et al. (1987) values of 1.3, 2.1, and 16.7 for amphiboles in hawaiite, mugearite, and trachyte respectively.

$K_D^{Hb:melt}$ for the Solomons andesites and dacites is fairly constant, with a mode in the 6–7 range. The mean value of 7.1 for 22 coexisting pairs is very close to that of 7.5 obtained by Ewart and Taylor for hornblende in the New Zealand andesite. As noted earlier, although mean cobalt for the Solomons hornblendes and biotites, at 66 and 62 ppm respectively, indicates a very similar uptake of this element by the two minerals, this is gainsaid by consideration of coexisting pairs (vide Fig. 18.9). In each of the two cases investigated in detail by electron microprobe cobalt was higher in the biotite (61.7, 62.1 ppm: mean = 62 ppm) than the hornblende (32.6, 46.4 ppm: mean = 40 ppm). Mean $K_D^{biot:Hb}$ for these two lavas is thus 1.55. Mean $K_D^{biot.:melt}$ for the same two are 20.6 and 6.9, as compared with a value of 23 obtained by Villemant et al. (1981) for biotite in an alkali basalt.

A range of 0.03–0.10 for $K_D^{plag:melt}$ in basaltic lavas has been obtained by various investigators. Dudas et al. (1973) determined a value of 0.38 for the plagioclase of a sample of dacitic tephra. Such estimations must be accepted with doubt, however; it has already been noted that cobalt may occur at relatively elevated levels in very fine glassy inclusions in many volcanic plagioclases, and this may have a substantial proportional effect on the apparent cobalt content of plagioclase separations. The writer would prefer to accept a close approximation to nil cobalt *in the plagioclase lattice* — and hence that partition coefficients approach zero.

As might be expected, there is a considerable body of data in the literature on the partitioning of cobalt by magnetite in basaltic and related rocks. Wager and Mitchell (1951) obtained values of 2.7, 3.2, 3.3, and 2.3 for $K_D^{mag:melt}$ in the Skaergaard B, C, D, and E stages respectively. Dudas et al. (1973) found very high values, 19.5–36, developed by magnetites in dacitic tephra; Lindstrom (1976) determined values from 5.8 to 17 in experiments using a natural alkali basalt at ≈1100–70 °C at 1 atm; Villemant et al. (1981) obtained a value of 4.3 for magnetite in an alkali basalt; and Lemarchand et al. found values of 4.61, 8.52, and 41.70 in hawaiite, mugearite, and alkaline trachyte respectively.

Mean values for $K_D^{mag:gmass}$ in the three Solomons lava groups are generally similar to the values and trends found by these other investigators. Values vary substantially from one sample to another within a given lava group, and some individual K_D values are quite as large as those determined by Dudas et al. and Lemarchand et al. The increase in partition coefficients with increase in the felsic nature of the relevant lavas (Fig. 18.14), noted also with the clinopyroxenes (Table 18.1), results, in the case of magnetite, from a progressive decrease in cobalt in *both* magnetite and melt. Its decrease in the melt is, however, much the more rapid, leading to a progressive increase in mean partition coefficients.

All the major Solomons minerals other than plagioclase thus develop mean partition coefficients ($K_D^{cryst:melt}$) greater than one, and hence tend to impoverish the melt in cobalt as they crystallize. Copious early crystallization of olivine, with mean K_D ≈5.0, leads to rapid diminution of cobalt in the melt, and this continues at later stages by the crystallization of hornblende and magnetite. The feldspars incorporate essentially no cobalt, and their copious formation from the onset of andesite formation must have tended to buffer the effects of hornblende and magnetite crystallization.

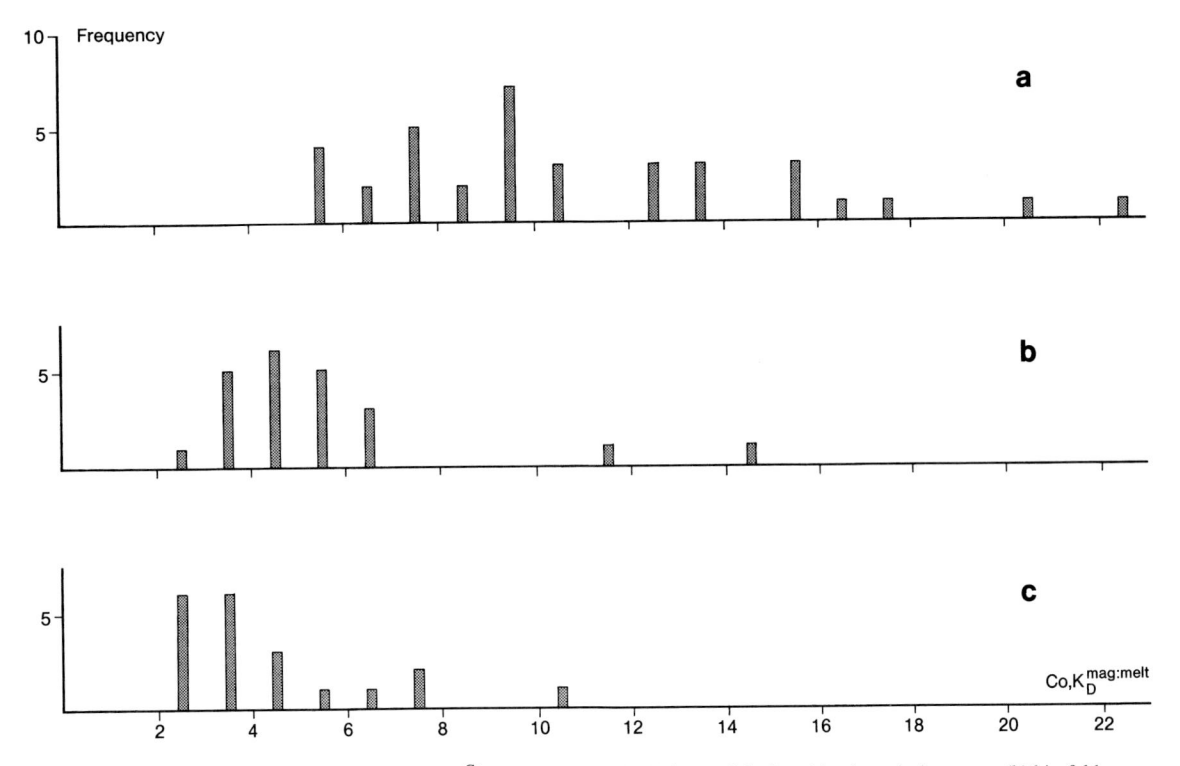

Fig. 18.14. Frequency distributions of values of K_D^{Co} magnetite:melt in (a) lavas of the hornblende-andesite group, (b) big-feldspar lavas, and (c) olivine–clinopyroxene-basalts of the Solomons suite.

THE INCIDENCE OF WHOLE-ROCK COBALT IN THE SOLOMON ISLANDS YOUNGER VOLCANIC SUITE

Mean cobalt for the suite as a whole is 42 ppm and, for the basalts (SiO$_2$ <52 per cent), 59 ppm.

Inspection of Fig. 18.15(a) appears to indicate that the frequency distribution of cobalt values is bimodal with peaks in the 30–40 ppm and 80–90 ppm ranges. This is almost certainly an artefact of sampling stemming from the necessity of including the olivine–pyroxene- and, particularly, the picritic lavas within a manageable total number of samples. A completely random sampling would doubtless have yielded a frequency distribution with a pronounced mode in the 30–50 ppm Co range, the picrites simply constituting a high-cobalt 'tail'. If, therefore, the 20–60 ppm Co portion of the frequency distribution is taken as the probable major component of a full random sampling, the mean value obtained is 39 ppm. This then compares with 44 ppm Co in MORBs, and 35, 33, and 44 ppm in the basalts of Tonga–Kermadec, the Aleutians, and Vanuatu respectively.

Mean cobalt abundance in Solomons lavas with SiO$_2$ >52 per cent is 22 ppm, and the frequency distribution of values is as shown in Fig. 18.15(b). This compares with a mean of 21 ppm Co for all arcs other than the Solomons, and means of 24, 19, and 20 ppm for Tonga–Kermadec, the Aleutians, and Vanuatu respectively. It appears that arc lavas, like MORBs, are notably consistent in their cobalt content.

The distribution of cobalt in the three principal Solomons lava types and their groundmasses is marked by a general decrease in whole-rock cobalt from olivine–pyroxene-basalts through to hornblende-andesites and dacites, and by a consistent tendency for the groundmasses to be impoverished in it relative to the whole rocks (Fig. 18.16). The latter effect is gross in the case of the olivine–pyroxene-basalts, owing to the pronounced sequestering of cobalt in the abundant phenocrystic olivine. Mean whole-rock cobalt is 79, 30, and 22 ppm for olivine–pyroxene-basalts, big-feldspar basalts, and hornblende-andesites respectively, and 40, 25, and 10 ppm for the corresponding groundmasses.

240 *Ore Elements in Arc Lavas*

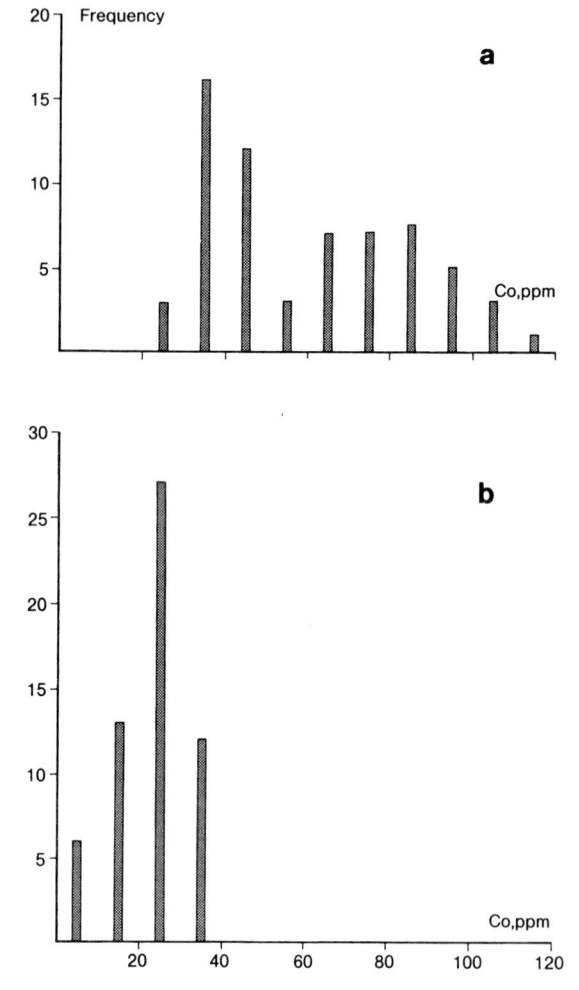

Fig. 18.15. Frequency distributions of cobalt abundances in (a) Solomons basalts (SiO$_2$ < 52 per cent), and (b) Solomons lavas more felsic than basalts (SiO$_2$ ⩾ 52 per cent).

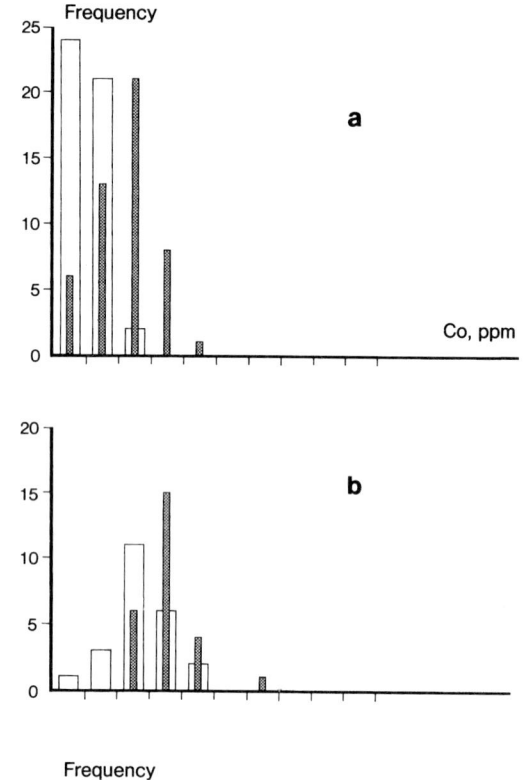

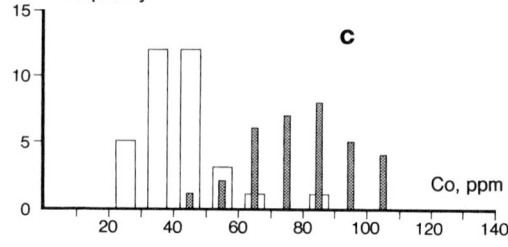

Fig. 18.16. Frequency distributions of cobalt abundances in whole rocks (filled bars) and corresponding groundmasses (open bars) of (a) lavas of the hornblende-andesite group, (b) big-feldspar lavas, and (c) the olivine–clinopyroxene-basalts of the Solomons suite.

As much of the early removal of cobalt from the melt occurs via copious early crystallization of olivine, it is not surprising that in the basaltic rocks with MgO ⩾ 6 per cent, cobalt shows a marked correlation with MgO in the whole rocks (Fig. 18.17(a)). This Co:MgO tie is much closer than those between Co:FeO, and Co:total iron (Figs. 18.17(b) and (c)), and is reminiscent of the Cr:MgO relationship (Fig. 15.17). It must, however, be remembered that the Co:MgO correlation is not due to an affinity of Co^{2+} for Mg^{2+} sites, but of Co^{2+} for olivine.

As shown above (see Fig. 18.6), Co^{2+} has a very slight Fe^{2+} site-preference in olivine, but a strong tendency to be incorporated in the olivine lattice. Thus cobalt is incorporated, and subtracted, in olivine, whatever the Mg:Fe proportion in the latter. The olivines of the olivine– pyroxene-basalts, however, happen to be magnesium-rich, and so the cobalt that is in fact correlated with olivine develops an indirect, or secondary, correlation with MgO in this particular series of rocks.

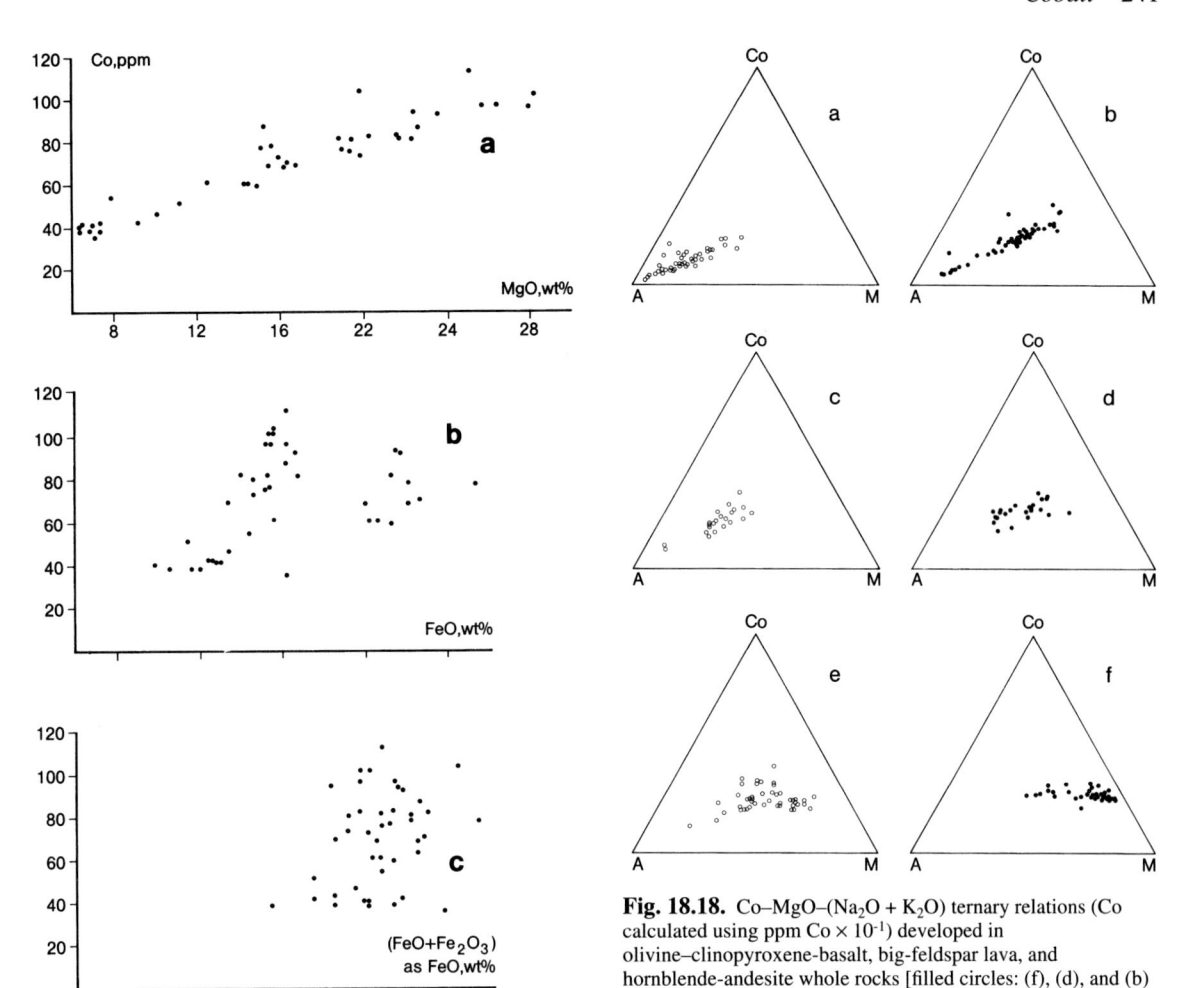

Fig. 18.17. Abundance relations between cobalt and (a) MgO, (b) FeO, and (c) total iron as FeO in the olivine–clinopyroxene-basalts of the Solomons suite.

Fig. 18.18. Co–MgO–(Na$_2$O + K$_2$O) ternary relations (Co calculated using ppm Co × 10^{-1}) developed in olivine–clinopyroxene-basalt, big-feldspar lava, and hornblende-andesite whole rocks [filled circles: (f), (d), and (b) respectively] and their corresponding groundmasses [open circles: (e), (c), and (a) respectively].

Ternary Co–MgO–(Na$_2$O + K$_2$O) relationships are distinctive, yielding a pattern (Fig. 18.18) that, as in the case of vanadium, appears to be highly muted form of those of iron, zinc, and manganese. The trend generated by the olivine–pyroxene-basalts (Fig. 18.18(f)) indicates no more than a very slight increase in proportionate cobalt with decrease in MgO, in contrast to the steep slopes of the analogous trends for the other three metals (Figs. 5.4, 7.21, and 16.21 respectively). Highest proportionate cobalt occurs in the big-feldspar lavas, and this then falls away in a very regular, well-correlated trend generated by the andesitic lavas. The Solomons suite may thus be said to exhibit a 'flat' hypotholeiitic pattern with respect to cobalt. The near-horizontal olivine–clinopyroxene subtraction line for cobalt represents in fact a state of affairs midway between that for nickel — for which olivine crystallization leads to subtraction from the melt of a high proportion of nickel relative to MgO (Figs. 19.20, 19.21) — and those for manganese, iron, and zinc, for which olivine crystallization leads to the subtraction of *low* proportions of these metals relative to MgO. As a result of this nickel is relatively impoverished in the remaining melt, whereas iron, manganese, and zinc are relatively enriched. That crystallization of olivine and clinopyroxene should have so little effect on proportionate cobalt in the Co–MgO–(Na$_2$O + K$_2$O) system is at first sight somewhat surprising: olivine (mean Co = 185 ppm) and clinopyroxene (mean Co = 40 ppm) crystal-

242 *Ore Elements in Arc Lavas*

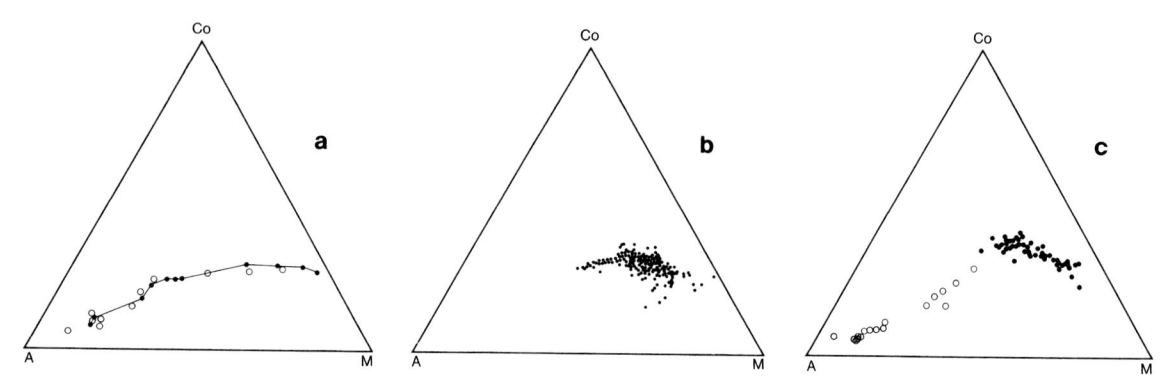

Fig. 18.19. Mean Co–MgO–(Na$_2$O + K$_2$O) ternary relations corresponding to 2 per cent whole-rock SiO$_2$ intervals in (a) whole rocks (filled circles) and corresponding groundmasses (open circles) of the Solomons suite excluding Bougainville; (b) MORBs as represented by the Mid-Atlantic Ridge basalts; and (c) samplings of Icelandic basalts (SiO$_2$ < 52 per cent: filled circles) and Icelandic lavas more felsic than basalt (SiO$_2$ ≥ 52 per cent: open circles).

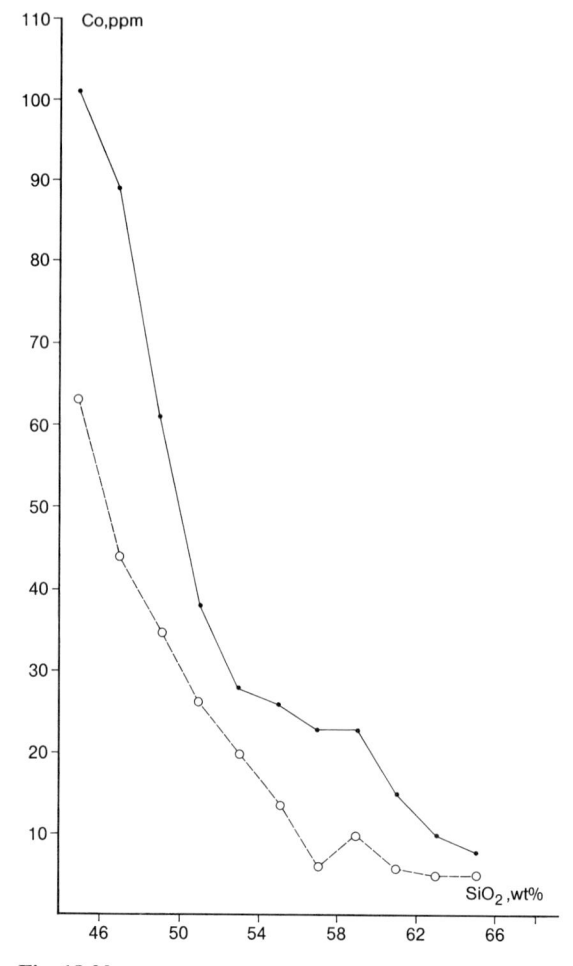

Fig. 18.20. Mean CO:SiO$_2$ relations corresponding to 2 per cent whole-rock SiO$_2$ intervals in whole rocks (filled circles) and corresponding groundmasses (open circles) of the Solomons suite.

lizing in an overall ratio of ≈1:1 would incorporate ≈112 ppm Co: about 2.5 times the mean cobalt content of MORBs and almost certainly more than would be contained in a basalt parental to the Solomons suite. However in the case of the olivine–pyroxene-lavas it just so happens that the abundance behaviour of the three components is such that proportionate cobalt remains almost constant: a feature demonstrated with greater clarity in Fig. 18.19, which shows mean Co–MgO–(Na$_2$O + K$_2$O) proportions corresponding to two per cent intervals (<46, 46–8, 48–50 — >64 per cent) of whole-rock SiO$_2$. The distribution of points comes very close to being negatively tholeiitic.

MORBs, as represented by the mid-Atlantic ridge basalts (Fig. 18.19(b)), generate what looks like a truncated version — the more mafic half — of the Solomons distribution (In this connection it may be noted that the fields of Figs. 18.18(f) and 18.19(b) are quite similar. The Icelandic assemblage, on the other hand, is made up of two essentially linear trends (Fig. 18.19(c)): one generated by the basalts, the other by the more felsic lavas, the two linear fields combining to form a hypotholeiitic pattern overall.

The effect of olivine subtraction on the *absolute* abundance (as distinct from 3-component proportionate abundance) of cobalt is however shown quite unambiguously in Fig. 18.20. The change from 46 to 54 per cent whole-rock SiO$_2$ is accompanied by a decrease of some 75 per cent (≈100 to 25 ppm) in cobalt content. At >64 per cent SiO$_2$ only 5–6 ppm Co remain in the melt. Whereas the steep decrease in cobalt to 52–4 per cent SiO$_2$ results principally from the aforementioned olivine crystallization, the more gradual reduction from this point and to increasingly felsic rocks is due to hornblende and lesser magnetite formation. This combination of factors results in residual melts, now represented by groundmasses, always impoverished in cobalt with respect to the contemporary total of crystals-plus-melt, now represented by the whole rocks (Figs. 18.16, 18.20).

CONCLUDING STATEMENT

The mean cobalt contents of MORBs and arc basalts are virtually identical at 43–4 ppm. Cobalt decreases steadily with increase in the felsic nature of the containing lava; on the basis of limited data arc andesites contain ≈24 ppm, dacites ≈12 ppm and rhyolites ≤5 ppm Co. Mean cobalt for the Solomons basalts is 59 ppm: a figure rendered artificially high by the inclusion of a large amount of picritic material in the basalt sampling. If the latter is excluded the Solomons basalt mean is ≈40 ppm Co, and that for the andesites–dacites is 25 ppm.

Cobalt shows virtually no site-preference for Fe^{2+} vis-à-vis Mg^{2+} octahedral sites; a very slight preference only for Fe^{2+} sites in olivines has been detected in the Solomons material. Olivine is the principal acceptor of cobalt, followed by magnetite, in the basaltic–andesitic segment of the lava spectrum. All the principal mineral species of the Solomons lavas apart from the feldspars incorporate cobalt at higher levels than those at which it occurs in the melt, and thus tend to impoverish the latter in cobalt. Plagioclase incorporates virtually none, and its crystallization must therefore tend to enrich the melt in cobalt.

On the CoMA diagram cobalt develops only the slightest inflexion of the kind generated so markedly by iron, zinc, manganese, and calcium. The field overall is well correlated, and while it is certainly not linear the change of slope from the 'olivine subtraction line' of the olivine–pyroxene-basalts to the trend of the hornblende-andesites is relatively muted. While the distribution as generated by the Solomons suite is probably most accurately referred to as 'flatly hypotholeiitic', it may also be seen as a close approach to a single, negatively tholeiitic pattern. A possible reason for this is considered in Chapter 21.

19

Nickel

Although nickel is a highly chalcophile element, and indeed has a stronger affinity for sulphur than has iron, and although its average abundance in sea-floor basalts is some 60 per cent greater than that of copper, it is conspicuous for its scarcity in exhalative ores. It has been suggested by Lusk (1976a,b) that some of the stratiform nickel deposits associated with mafic to ultramafic lavas of the Archaean of Australia and Canada may be exhalative, and a firm case for such an origin for the nickel–cobalt-bearing copper deposits of Outokumpu, in the Lower Proterozoic Karelian terrane of Finland, has been made by Peltola (1978), Koistinen (1981), and others. Apart from this, however, nickel appears as no more than a trace in exhalative deposits. As far as the writer is aware, exhalative ores containing commercial quantities of nickel are unknown in the two great epochs of exhalative ore formation: the Lower Proterozoic, c. 1500–2000 Ma, and the Lower to Middle Palaeozoic, 350–550 Ma.

The nickel of Outokumpu, the only commercial nickel deposit for which such an origin seems firmly established, occurs chiefly as a component of abundant nickeliferous pyrrhotite, and as disseminated pentlandite. Unlike cobaltite (CoAsS), the nickel sulpharsenide gersdorffite (NiAsS) is not a common trace mineral of exhalative copper, zinc, and lead sulphide ores. This near-absence of nickel from exhalative deposits is almost certainly not a depositional phenomenon. The propensity for nickel to precipitate as a sulphide in reduced sedimentary environments is well known, and the common occurrence of trace quantities of millerite (NiS) in coal measures has been recognized since Miller (1842) noted its presence in samples from the South Wales coalfield and Des Cloiseaux (1880) reported it from some of the coal-bearing sediments of Belgium. The conspicuously low abundance of nickel, and of cobalt, in exhalative sulphide ores appears to stem from some factor of supply, and is considered again in Chapters 21 and 24.

The earliest reliable estimate of nickel in igneous rocks appears to be that of Clarke and Washington (1924), who proposed an average of 0.02 per cent. In 262 analyses of igneous rocks made in the laboratory of the United States Geological Survey to that time, an average of 0.0274 per cent of nickel oxide (0.0215 per cent Ni) was found. Clarke commented 'Had it been sought for in all cases, this figure might have been slightly reduced' (Clarke and Washington 1924, p. 708). Both J. H. L. Vogt (1931) and Goldschmidt (1954) considered this value too high and proposed one of 0.01 per cent. In their investigation of the trace metal content of North American igneous rocks Sandell and Goldich (1943) obtained averages of 0.0097 per cent Ni and 0.0032 per cent Co (Co:Ni = 0.33:1) in sub-silicic rocks (SiO_2 <63 per cent) and 0.00058 per cent Ni and 0.00030 per cent Co (Co:Ni = 0.52:1) for silicic rocks (SiO_2 >63 per cent). As noted in the preceding chapter, they found a moderately correlated but consistently linear Co:MgO relationship. Nickel, on the other hand, exhibited less simple relationships (Fig. 19.1): a somewhat doubtful correlation of small positive slope from the origin to ≈0.0035 per cent Ni — 4.5 per cent MgO, and then a sharp inflexion with a fairly high degree of correlation and much greater positive slope above these values.

Apart from Vogt (1923), Goldschmidt (1954) seems to have been the first to recognize the tie between magmatic processes and the relative abundance of cobalt and nickel. He observed that whereas early investigators had obtained Co:Ni averages for ≈1:12 for igneous rocks and the lithosphere, Sandell and Goldich (1943) had obtained one of 3:10 and Goldschmidt and his co-workers a value of ≈4:10. Goldschmidt (1954) suggested that the discrepancy arose from excessive, and incorrect, extrapolation of results obtained from an unrepresentative sampling of the spectrum of igneous rock types. He pointed out that when purely chemical methods were employed quantitative analyses of cobalt and nickel were limited to rocks particularly rich in nickel, e.g. dunites, harzburgites, pyroxenites, etc. In these rocks there is indeed a great preponderance of nickel over cobalt, the Co:Ni ratio being ≈1:12. Early assumptions that this applied to all igneous rocks — a generalization that was soon shown to be incorrect — then led to much too low an estimate of average Co:Ni ratios.

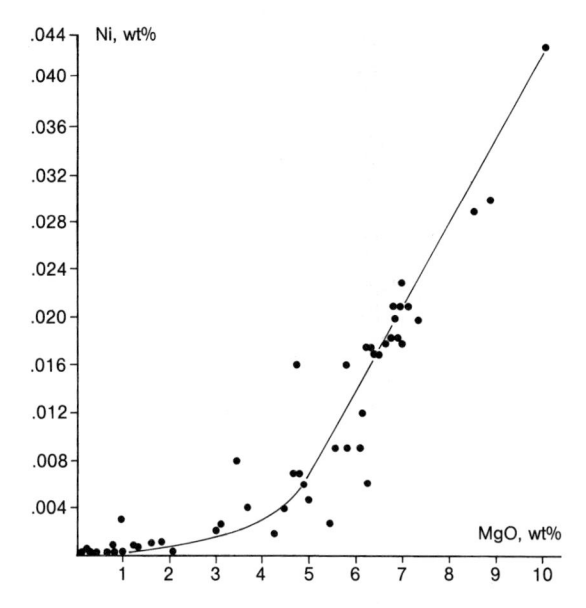

Fig. 19.1. Abundance relations between nickel and MgO in a range of North American igneous rocks, as found by Sandell and Goldich (1943).

The work of Sandell and Goldich provided an early indication that cobalt and nickel values, and in particular Co:Ni ratios, were different for igneous rocks of different SiO_2 levels. Goldschmidt (1954) tabulated the results of various investigators who had determined cobalt and nickel in samplings representing *ranges* in igneous rock composition (Table 19.1) and showed that (1) there was a general decrease in cobalt and nickel going from more mafic to more felsic rock types and (2) nickel decreased more rapidly than cobalt, leading to an increase in Co:Ni ratios from about 1:7.8 in ultramafic rocks, 1:2.9 in mafic, 1:1.5 in intermediate, and 1:0.9 in felsic igneous rocks, i.e. an inversion in Co:Ni ratios from ≪1 to >1 as whole-rock SiO_2 increased. This of course results from the greater incorporation of nickel *vis-à-vis* cobalt in the earlier products of igneous crystallization, a phenomenon illustrated quite unambiguously by the Solomons and other volcanic-arc lava series.

The modern estimate of the crustal abundance of nickel is 99 ppm (0.0099 per cent: Emsley 1989): remarkably close to that of 0.01 per cent (100 ppm) by Vogt (1931) and Goldschmidt (1954).

NICKEL IN LAVAS OF THE MARINE ENVIRONMENT

Nickel is one of the metallic trace elements recognized by igneous petrologists as being — in contrast to copper, for example — of petrogenetic significance, and most superior analyses of lavas and lava series now include it. The data bank is therefore quite substantial and, perhaps beginning with the work of Sandell and Goldich (1943) and Goldschmidt (1954), the behaviour of nickel in the principal volcanic minerals and lava types is now fairly well known.

The incidence of nickel in MORBs is summarized in Table 6.2. The mean of 500 analyses of basalts (SiO_2 <52 per cent) representing all of the principal mid-ocean ridges is 114 ppm Ni; that of 631 basalts including those of Iceland, 112 ppm Ni. Averages for the Mid-Atlantic, Atlantic–Indian, Indian, and Pacific ocean ridges are 154, 79, 90, and 101 ppm respectively. The average for 131 Icelandic basalts is 105 ppm. Thus there is less variation in nickel as between ridge systems than in chromium (see Chapter 15), though there is a general correlation of nickel and chromium averages from one system to another (Fig. 19.2).

Table 19.1 Nickel and cobalt abundances in igneous rocks, after Goldschmidt (1954)

	Ultrabasic		Basic		Intermediate		Acid	
	Ni	Co	Ni	Co	Ni	Co	Ni	Co
Germany (Goldschmidt)	3600	240	160	80	40	31	2.5	8
Scotland (Nockolds and Mitchell)	650	200	230	70	80	30	20	15
Finland (Sahama)	>800	240	47	24	—	—	2.5	—
Greenland (Wager and Mitchell)	1000	90	150	50	<2	20	—	5
USA (Minn.) (Sandell and Goldich)	—	—	160	30	15	6	10	4

Early estimation of average abundances of nickel and cobalt in the principal categories of igneous rocks, as collated by Goldschmidt.

246 Ore Elements in Arc Lavas

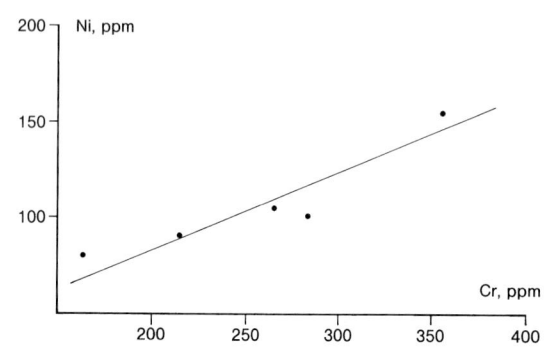

Fig. 19.2. Co-variation of mean nickel and chromium in the five ridges and ridge systems: Mid-Atlantic, American–Antarctic, Atlantic-South-west Indian, Pacific, and Indian Ocean ridges.

The full range of variation of nickel in MORBs is shown in Fig. 19.3(a). Inspection of the frequency distribution indicates the possibility, as for chromium, of bimodality, with a principal mode at 80–90 ppm Ni and a subsidiary one at 40–50 ppm Ni. This feature appears also in the frequency distribution of nickel in Icelandic basalts (Fig. 19.3(b)), leading to the suspicion that it is a 'real' feature and that it may result from early crystallization in a parent melt containing 50–60 ppm Ni. The resulting crystal settling and differentiation yields an accumulate containing slightly higher nickel than the parent (mode ≈80–90 ppm) and a complementary derivative containing slightly less (mode ≈40–50 ppm).

As noted in the preceding chapter, the very rapid incorporation of nickel in early crystallizing high-Mg olivines leads to an extreme sensitivity of nickel abundance to differentiation, and hence wide variation in the nickel content of relatively similar mid-ocean ridge basalts. In spite of its close physical and chemical similarity to nickel, cobalt has a marked insensitivity to early differentiation processes, leading to the contrast in patterns of abundance of the two elements in MORBs (Fig. 18.3; and Fig. 18.1 *vis-à-vis* Fig. 19.3).

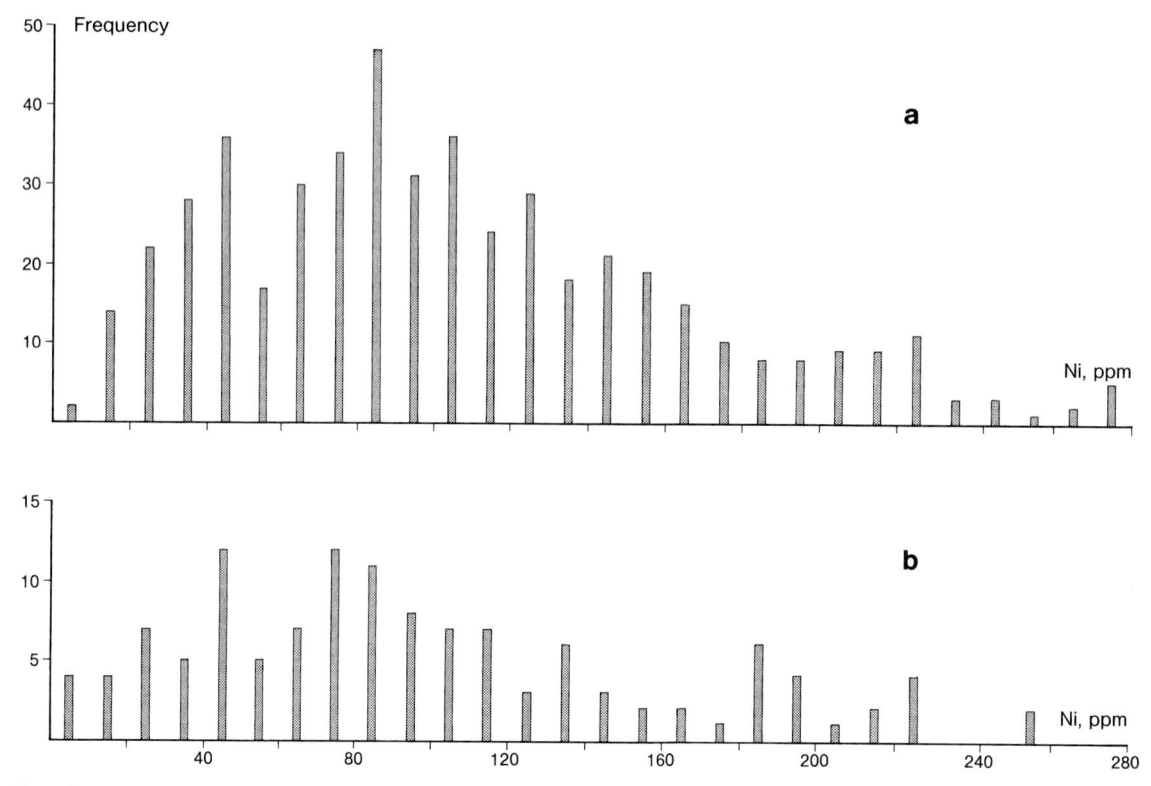

Fig. 19.3. Frequency distributions of nickel abundances in (a) MORBs of all major oceans, and (b) Icelandic basalts (SiO$_2$ <52 per cent).

Relations between Ni and MgO in MORBs (Fig. 19.4) are reminiscent of those found in the continental igneous rocks of North America by Sandell and Goldich (1943) shown in Fig. 19.1.

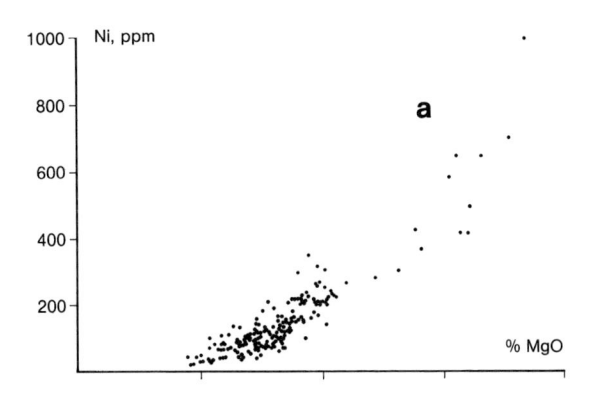

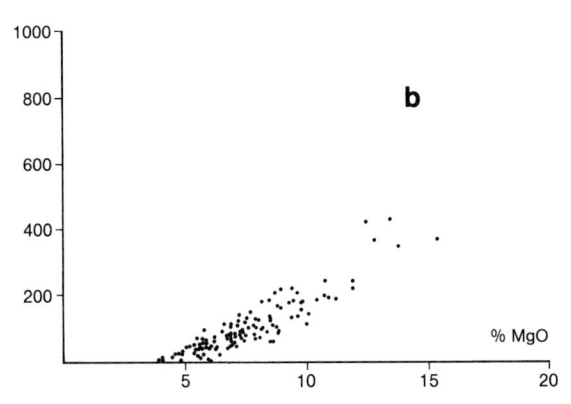

Fig. 19.4. Abundance relations between nickel and MgO in (a) MORBs as represented by basalts of the Mid-Atlantic Ridge, and (b) Icelandic basalts (SiO$_2$ <52 per cent).

The incidence of nickel in arc lavas is now very well documented; Fig. 19.5 represents 1030 analyses. The spread of values in the basaltic rocks is extensive, and in order to represent the major part of it in a single diagram, Fig. 19.5(a) has been constructed using the large class interval of 50 ppm Ni. Fig. 19.5(b), with a class interval of 10 ppm Ni, shows the distribution of values in the range 0–300 ppm Ni, which includes 75 per cent of the samples with SiO$_2$ <52 per cent (340 out of 453 analyses of arc basalts). Like the distribution of nickel in MORBs and Icelandic basalts, Fig. 19.5(b) shows a hint of bimodality, with an intermediate minimum of 40–50 ppm Ni. Fig. 19.5(c), depicting the frequency distribution of nickel abundances in arc lavas more felsic than basalts (SiO$_2$ >52 per cent), shows — as in the case of the mid-ocean ridge lavas — the sharpness of the decline in nickel accompanying the development of more felsic rock types.

Mean nickel for 453 analyses of arc basalts and for 587 analyses of arc lavas more felsic than basalts are 134 ppm and 19 ppm respectively. For the two arcs, the Lesser Antilles and Tonga–Kermadec, for which there are sufficient analyses for reliable calculation of averages for all of basalts, andesites, and dacites, the average abundances are 183, 41, and 35 ppm and 35, 15, and 3 ppm respectively. Variation in average nickel both within and between arcs may be considerable. To illustrate potential within-arc variation, the basalts of Grenada contain an average of 242 ppm Ni, whereas those of St Kitts contain 14 ppm. Potential between-arc variation is indicated above: the basalts of the three major west Indies islands, Grenada, Dominica, and St Kitts, contain an average of 183 ppm Ni, whereas those of Tonga–Kermadec contain only 35 ppm, a value less than that of 41 ppm found in the andesites of the West Indies.

CHEMICAL PROPERTIES AND CRYSTAL CHEMISTRY OF NICKEL

In spite of its strongly chalcophile nature, nickel commonly occurs as an important trace in most of the igneous ferromagnesian silicates, particularly olivine. With its tendency to occur in higher amounts in more magnesium-rich rocks, and in the more magnesian members of ferromagnesian mixed-crystal series such as olivine, it was suspected from the earliest days of geochemistry (i.e. the late nineteenth century) that nickel had a strong 'affinity' for magnesium in magmatic processes. With the recognition and determination of ionic size and the formulation of the 'Goldschmidt Rules' (Goldschmidt 1937) the correlation of nickel with magnesium in olivines came to be regarded as the example *par excellence* of the diadochic substitution of a trace for a major element in a common igneous mineral (Vogt 1923).

As Ni^{2+} and Mg^{2+} had almost identical ionic radii (that of Ni^{2+} being perhaps the slightly smaller of the two) and similar charges, this seemed a satisfactory explanation for the observed abundance relationships. Such a view appeared to be confirmed by J. H. L. Vogt's (1923) very thorough documentation of the abundances of the two elements in a wide variety of igneous rocks and minerals, and his conclusion from this that nickel was camouflaged by magnesium in silicate crystal lattices.

248   *Ore Elements in Arc Lavas*

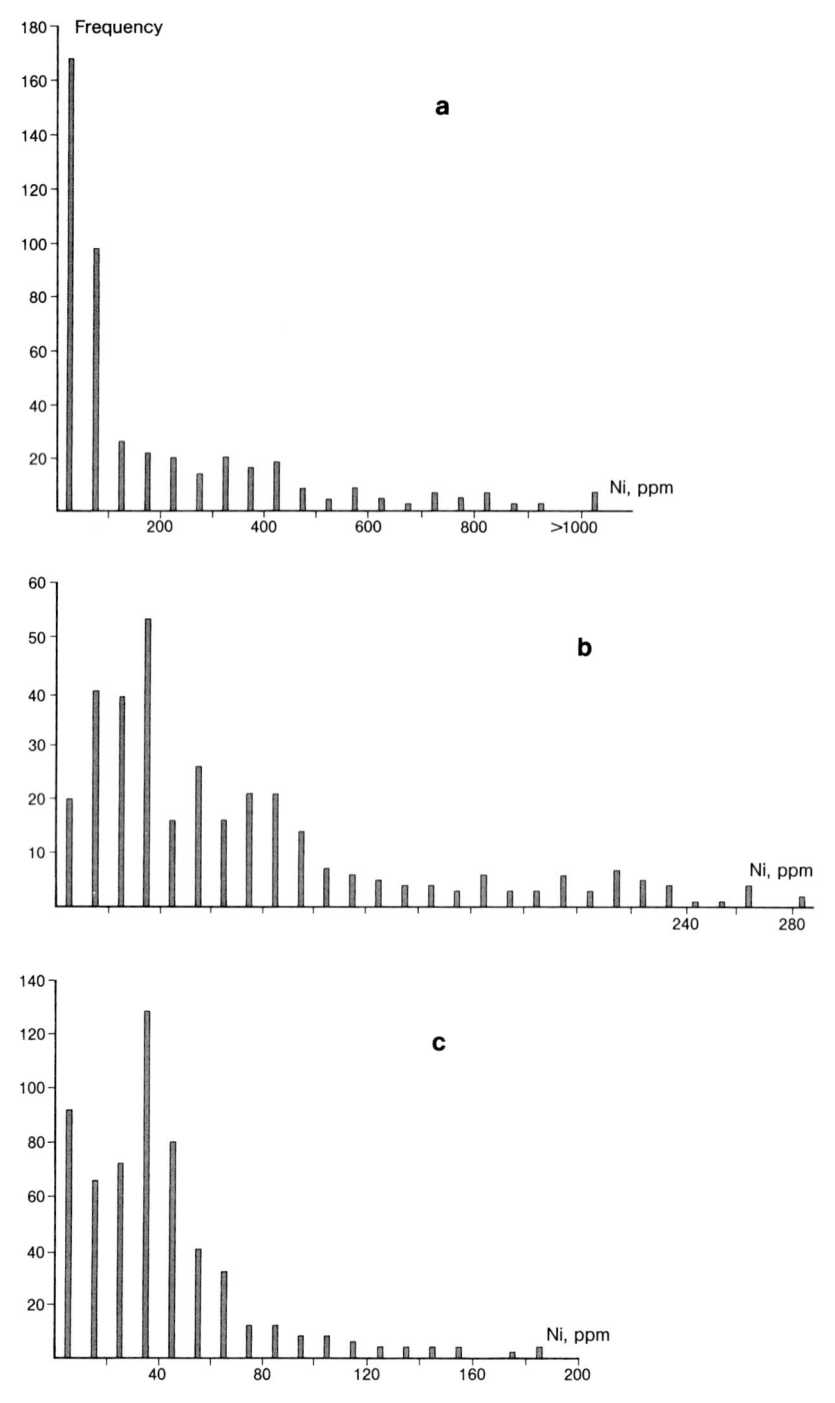

Fig. 19.5. Frequency distributions of nickel abundances in (a) volcanic island-arc basalts (SiO$_2$ <52 per cent); (b) horizontal expansion of (a) in the interval 0–300 ppm Ni; and (c) arc lavas more felsic than basalt (SiO$_2$ ⩾52 per cent).

However, Ringwood (1955), on the basis of considerations of bond type and electronegativity, pointed out that it is far more likely that Ni^{2+} enters crystal lattices at the expense of Fe^{2+} rather than Mg^{2+}. Ringwood drew attention to the fact that, as indicated by melting points, the Mg–O bond was stronger than the Ni–O bond. He also noted the results of Wager and Mitchell (1951) who had shown that, for the Skaergaard layers, the later-crystallizing olivines and pyroxenes were lower in nickel, not because they were less Mg-rich, but because the residual melt from which they formed was very much poorer in nickel. He also noted that although Vogt (1923) had observed that when olivine and orthopyroxene of the same rock are compared the olivines are always richer in nickel than the orthopyroxenes, Ramberg and de Vore (1951) had demonstrated that in such cases the olivines are usually richer in Fe relative to Mg than the orthopyroxenes. Had the nickel been camouflaged by the magnesium it might have been expected that the orthopyroxene would be richer in nickel than the co-existing olivine. The reverse was the case, indicating that nickel was accommodated — i.e. camouflaged — by iron rather than magnesium. Thus, Ringwood suggested, the incorporation of nickel did not have a *direct* relationship with the abundance of magnesium in the crystal: the apparent Ni:Mg tie was due to (1) high $K_D^{crystal:melt}$ values at the early stages of crystallization, leading to the incorporation of a large proportion of total nickel in the early (largely olivine) crystals and (2) the fact that the earlier-formed members of the ferromagnesian mixed-crystal series were those highest in magnesium. In this way the Ni:Mg abundance tie, though close, was induced indirectly rather than directly.

More recently Burns (1970) and Burns and Burns (1974) have pointed out that two features dominate the crystal chemistry of nickel: (1) the readiness of nickel to form strong metallic bonds, leading to its occurrence in crystal structures with short metal–metal distances and a variety of coordination polyhedra and (2) the strong preference of nickel for octahedral coordination in crystal structures *prohibiting* the close proximity of neighbouring metal atoms. Burns considered that the preference of Ni^{2+} for octahedral coordination and its early incorporation in olivines with the onset of magmatic crystallization results from the high crystal field octahedral site-preference energy of Ni^{2+}, combined with its small ionic radius.

Estimations of the size of the Ni^{2+} ion, the dominant nickel ion of the magmatic milieu, have varied considerably, from 0.070 to 0.078 nm. The most modern estimate (0.078 nm; Emsley 1989) is given in Table 6.4, together with the other principal properties of the nickel atom and Ni^{2+} ion.

THE INCIDENCE OF NICKEL IN THE PRINCIPAL MINERAL SPECIES OF THE SOLOMON ISLANDS YOUNGER VOLCANIC SUITE

General

Vogt (1923) appears to have been the first to attempt a systematic synthesis of the incidence of nickel in igneous minerals as well as igneous rocks. He noted that thirteen analyses of magnesium-rich olivines (MgO from 47 to 51 per cent) contained 0.19–0.50 per cent NiO (1493–3929 ppm Ni) and that for 12 of these NiO/MgO × 100 = 1.01, 0.97, 0.89, 0.82, 0.74, 0.71, 0.67, 0.65, 0.57, 0.51, 0.46, and 0.40. Four iron-rich olivines (FeO = 18–30 per cent) on the other hand contained 0.15–0.32 per cent NiO (1179–2515 ppm Ni), indicating *inter alia*, that 'It appears that the iron-poor olivines as a whole contain more nickel than the iron-rich olivines' (Vogt 1923, p. 317). An analysis of a Massachusetts enstatite (quoted from F. W. Clarke 1910) gave 34.4 per cent MgO and 0.23 per cent NiO (1807 ppm Ni). Six clinopyroxenes of widely differing geological context and geographical location gave means of 13.39 per cent MgO and 0.04 NiO (314 ppm Ni). Vogt also noted analyses of amphibole and biotite by Clarke (1910) giving 12.47 per cent MgO:0.10 per cent (786 ppm) Ni and 0.34 per cent NiO (2672 ppm Ni) respectively.

In considering the incidence of nickel in the principal minerals of the Skaergaard layers, Wager and Mitchell (1951, p. 153) observed, as had Vogt, that it 'enters early olivine in abundance'. They found 250 ppm in the olivine of the chilled marginal gabbro, 325 ppm in that of the hypersthene-olivine-gabbro (Fo_{64}), and 10 ppm in the olivine (Fo_{41}) of the hortonolite-gabbro. They also noted a marked reduction in Ni/Mg ratios in the olivines as crystallization progressed, i.e. although the Mg content of the olivines decreased as crystallization proceeded, the Ni content decreased even more rapidly. Nickel occurred in the early clinopyroxenes (200 ppm Ni in the pyroxene of the chilled marginal gabbros and 140 in that (En_{34}) of the hypersthene-olivine-gabbro) but fell to <1 ppm in the pyroxene of the hortonolite-ferrogabbro. Again, nickel decreased rapidly with respect to magnesium as crystallization proceeded: the relation as expressed by Ni/Mg × 1000 was 2.1 for the earliest pyroxene, changing successively with fractionation to 1.6, 0.7, <0.03. Nickel was found to enter magnetite in amounts 'roughly similar to the amounts found in corresponding pyroxenes and olivines' (Wager and Mitchell 1951, p. 158) and was thus abundant in early formed magnetites, but absent from the later ones.

More recently Mercy and O'Hara (1967), in an investigation of 23 assorted ultramafic rocks, found 15 olivines to contain an average of 3127 ppm Ni, 21 orthopyroxenes 765 ppm, and 15 clinopyroxenes 481 ppm Ni. For six specimens in which olivine, orthopyroxene, and clinopyroxene coexisted in these samples, the mean nickel contents were 3336, 698, and 560 ppm respectively. Four coexisting olivine:orthopyroxene pairs gave 3160 and 755 ppm Ni; and thirteen orthopyroxene: clinopyroxene pairs gave 763 and 472 ppm Ni respectively.

In their investigation of the minor elements in the Keweenawan lavas of Michigan, Cornwall and Rose (1957) found nickel most abundant in the olivines (max. = 2000 ppm; mean = 1660 ppm), but also to occur in significant quantities in the clinopyroxenes (max. = 380 ppm; mean = 258 ppm) and spinel (max. = 1000 ppm; mean = 675 ppm). For two coexisting olivine: clinopyroxene pairs, mean nickel abundances were 1660 ppm and 270 ppm; for four spinel:clinopyroxene pairs, 675 ppm and 243 ppm. Ewart et al. (1973) found 0.34 per cent NiO (2672 ppm Ni) in the highly magnesian olivine ($\approx$Fo$_{90}$) of the Metis Shoal in the Tonga–Kermadec arc. Nickel abundances in the pyroxenes of this island group, while showing systematic clinopyroxene:orthopyroxene relations, are apparently also related to geographical province: eight clinopyroxene:orthopyroxene pairs from Fonualei Island contained 3.18 and 5.03 ppm Ni and 13.38 and 18.88 per cent MgO respectively, whereas four such pairs from Late Island contained 97 and 136 ppm Ni and 16.36 and 23.70 per cent MgO respectively. The clinopyroxene:orthopyroxene pairing from the Metis Shoal contained 135 and 200 ppm Ni and 14.96 and 22.67 per cent MgO respectively. Ewart and his co-workers found comparatively little nickel in the titanomagnetites, five samplings giving an average of 63 ppm. Estimations of nickel in olivines of the Aleutians arc by various investigators range from 100 ppm Ni in olivines $\approx$Fo$_{73}$ to 3900 ppm Ni in olivines $\approx$Fo$_{76}$; as in the case of Tonga–Kermadec, the variation in nickel cannot be simply related to variation in MgO content, and a geographical component is indicated.

Solomons suite

Mean values of nickel in the principal mineral groups are given in Table 6.5.

Olivine. The full range of variation in nickel found by XRF analysis of mineral separations is 123–2194 ppm (Fig. 19.6(a)), corresponding to olivine MgO contents from 32.74 to 47.80 per cent, i.e. Fo$_{66}$ to Fo$_{89}$. The range obtained by electron-probe analysis of individual crystals is 25–3337 ppm Ni, corresponding to olivine MgO = 29.84 per cent (Fo$_{63}$) to 51.46 per cent (Fo$_{94}$).

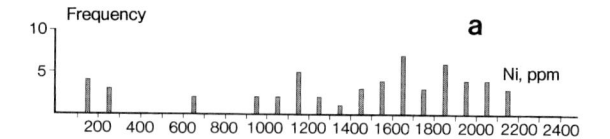

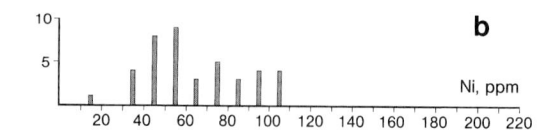

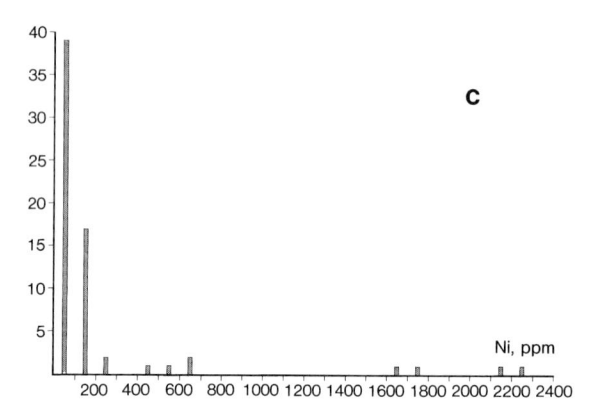

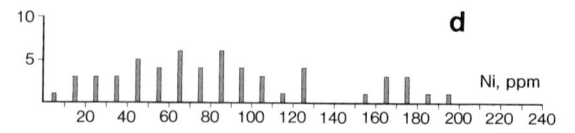

Fig. 19.6. Frequency distributions of nickel abundances in separated (a) olivines, (b) hornblende, (c) magnetite, and (d) magnetite, expanded in the 0–200 ppm Ni range of (c), of the Solomons suite.

Plotting of Ni v. MgO for all of the separations, and for the olivines for those other samples for which sufficient probe analyses are available for reliable averaging ($\approx$20 analyses/section), yields the type of close, linear relationship (Fig. 19.7) found by earlier investigators. The regression line intersects the x axis at MgO $\approx$32 per cent, indicating that significant amounts of nickel enter these olivines only at Mg levels >Fo$_{65}$. The coefficient of correlation is 0.904, so that the abundance relationship between Ni and MgO in this set of olivines is virtually functional (regression equation: Ni = 112.21 MgO– 3366.5). This seems more likely to reflect the strong octahedral site-preference of Ni^{2+} than the vagaries of nickel abundance in the coexisting melts.

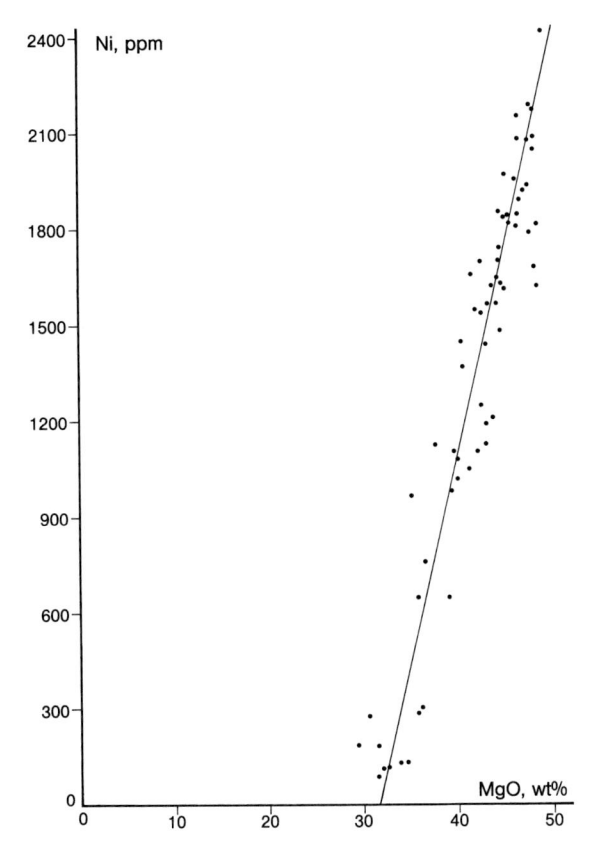

Fig. 19.7. Abundance relations between nickel and MgO in olivines of the Solomons suite.

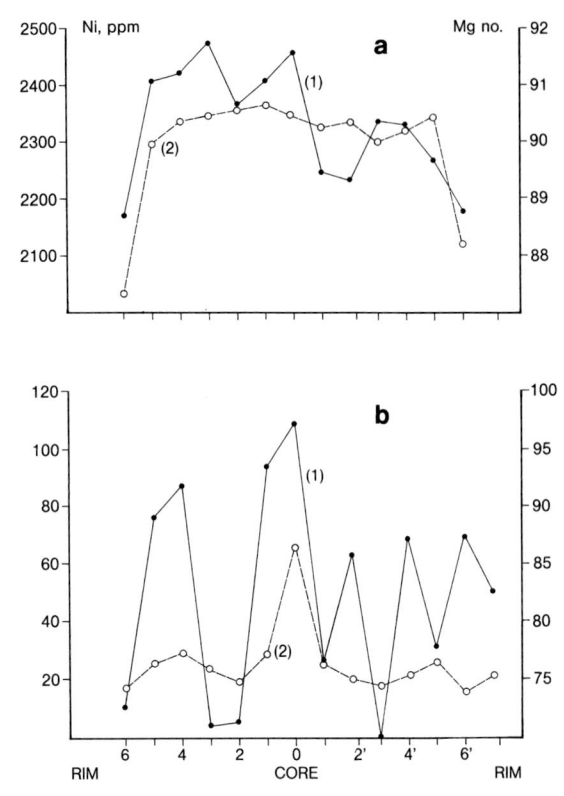

Fig. 19.8. Core-to rim variation in nickel (ppm) and MgO (as Mg number) in (a) a large olivine phenocryst ($\approx$ 7.0 mm diameter) of a Kohinggo picrite, and (b) a clinopyroxene phenocryst ($\approx$ 8.0 mm diameter) of a Rendova ankaramitic basalt. Nickel abundances indicated by filled circles; Mg numbers by open circles.

Most of the Solomons olivines are zoned with respect to nickel, in general sympathy with magnesium. Thus most individual crystals exhibit higher Ni–MgO cores and lower Ni–MgO rims. Such zoning may, however, be absent or very slight, is very occasionally reversed and, where larger crystals are involved, may display considerable irregularity between core and rim (Fig. 19.8(a)). Although in such cases nickel and magnesium tend to vary in sympathy, variation in MgO is generally a somewhat subdued version of that of nickel. Occasionally Ni–MgO abundance relations within single crystals are slightly antipathetic, but this is very much the exception. Nickel contents of groundmass olivines are not only distinctly lower than those of the associated phenocrystic grains, but are almost invariably much lower than those of the rims of the latter. Electron microprobe analyses of ten each of phenocryst cores and rims, and ten groundmass olivines (30 analyses in all) of one of the picritic basalts gave mean nickel values of 2732, 2346, and 1594 ppm respectively.

Pyroxenes. Although far less effective as a nickel acceptor than olivine, clinopyroxene accommodates significant quantities of nickel where this is available in the melt, and material separated from one of the Solomons picrite basalts contained 1982 ppm Ni. Nickel abundance in the clinopyroxenes is, however, generally much less than this, and mean values for the three principal lava groups are 370, 80, and 36 ppm Ni respectively (Table 6.5; Fig. 19.9). In spite of the fact that the Mg numbers of the clinopyroxenes vary little from olivine-basalt to hornblende-andesite (Chapter 5), their mean nickel contents clearly do vary, perhaps indicating that, to a greater extent than with the olivines, clinopyroxene nickel largely reflects availability in the coexisting melt. This may be corroborated by what is a very clear relationship between the nickel content of clinopyroxene and whole-rock SiO_2 (as an indicator of magmatic evolution) of the containing

252   *Ore Elements in Arc Lavas*

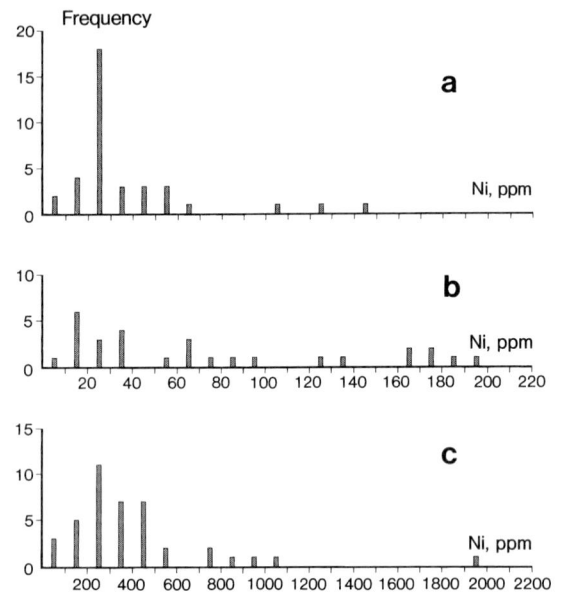

Fig. 19.9. Frequency distributions of nickel abundances in clinopyroxenes of (a) lavas of the hornblende-andesite group, (b) the big-feldspar lavas, and (c) the olivine–clinopyroxene-basalts of the Solomons suite.

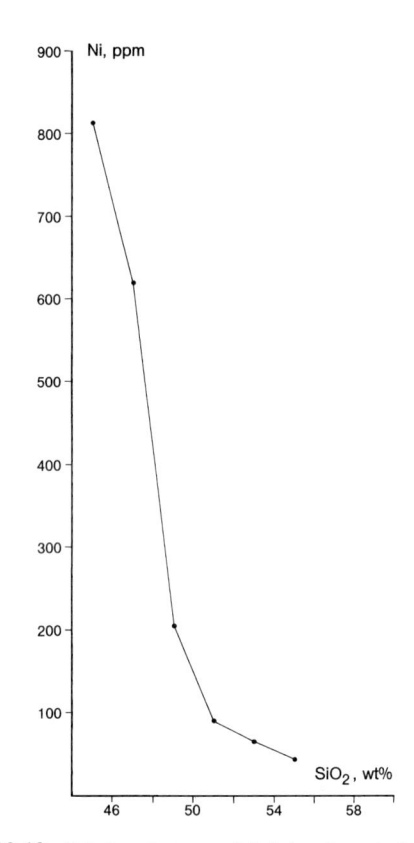

Fig. 19.10. Relations between nickel abundance in the clinopyroxenes of the Solomons basalts and basaltic andesites, and 2 per cent intervals of whole-rock SiO_2 in the containing lavas.

lavas (Fig. 19.10). As with the olivines, zonation of nickel content in individual clinopyroxene crystals is common, and tends to develop as a somewhat exaggerated reflection of accompanying Mg:Fe ratios (Fig. 19.8(b)).

Partitioning of nickel between olivine and clinopyroxene has been investigated in detail by probe analysis of crystal margins where the two minerals lie close together or in contact. Ten to twenty such pairs of analyses per section gave K_D^{Mg} olivine:clinopyroxene generally in the range 2.5–3.0, and K_D^{Ni} olivine: clinopyroxene principally in the range 2.0–7.0 (Fig. 19.11). There is, however, no systematic relation between magnesium and nickel distribution coefficients in the two minerals. Two examples, a coexisting olivine-clinopyroxene pair from a

Table 19.2 MgO and nickel abundances in coexisting olivine–clinopyroxene pairs

No.	Olivine MgO	Olivine Ni	Clinopyroxene MgO	Clinopyroxene Ni	K_D^{Mg}	K_D^{Ni}
1	48.89	2101	19.31	269	2.53	7.81
2	35.06	289	16.96	73	2.07	3.96

MgO and nickel abundances in coexisting olivine–clinopyroxene pairs of two Solomon Islands basalts, and consequent values of $K_D^{ol:cpx}$ with respect to magnesium and nickel. All analyses carried out by electron microprobe and represent margins of crystals where olivine and clinopyroxene were in contact or within ≈0.01 mm. 1 = picrite basalt host rock; 2 = big-feldspar basalt. MgO given as per cent by mass; Ni as parts per million by mass.

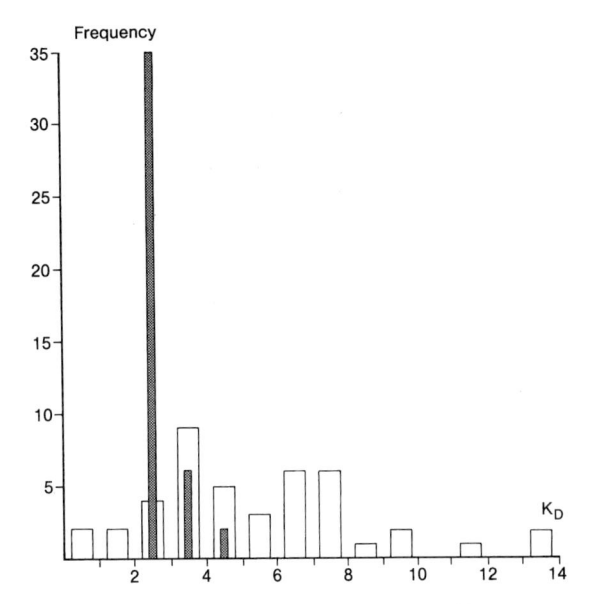

Fig. 19.11. Frequency distribution of values of K_D^{Mg} (filled bars) and K_D^{Ni} (open bars) between phenocrystic olivine and clinopyroxene in 43 Solomon Islands basalts.

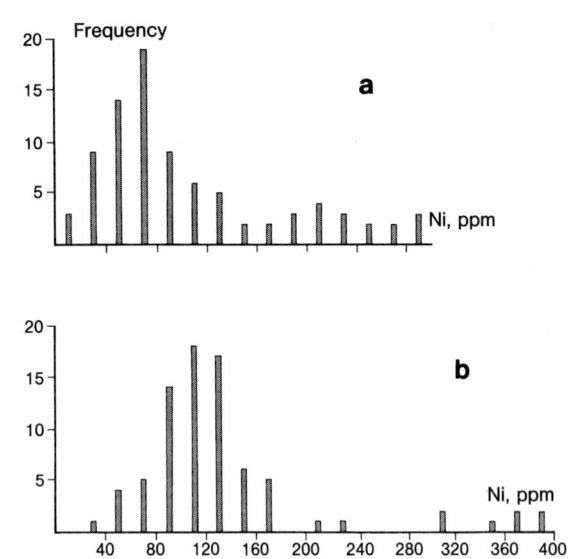

Fig. 19.12. Frequency distributions of nickel abundances in (a) clinopyroxenes (86 analyses; mean = 106 ppm Ni) and (b) orthopyroxenes (83 analyses; mean = 146 ppm Ni) coexisting in a Solomon Islands two-pyroxene-andesite.

picritic basalt and another from a big-feldspar basalt, yielded the results given in Table 19.2

The orthopyroxenes (which occur only in andesites) contain distinctly higher nickel than the clinopyroxenes of the basaltic andesites and andesites (Table 6.5). Where the two pyroxenes coexist the orthopyroxenes usually, but not quite invariably, contain the higher nickel. Eight coexisting pairings gave averages of 26.08 per cent and 16.05 per cent MgO ($K_D^{opx:cpx} = 1.62$) and 146 and 106 ppm Ni ($K_D^{opx:cpx} = 1.38$) respectively. A total of 85 pairings (170 grains) analysed by electron probe gave nickel values as indicated by the frequency distributions of Fig. 19.12

Hornblende. The clinopyroxenes and hornblendes of the hornblende-andesites contain comparable mean amounts of MgO: 14.58 and 13.65 per cent respectively. The mean nickel abundance in the hornblendes is, however, 64 ppm compared with 36 ppm in the corresponding clinopyroxenes (Table 6.5), indicating that, at least at the low melt levels of nickel pertaining at this stage of lava evolution, hornblende is the more effective acceptor of nickel and incorporation of the latter in the relevant crystal structures is related to factors more complex than magnesium content alone. This latter deduction appears to be corroborated by a complete lack of correlation of Ni and MgO in the two minerals as these occur in the Solomons hornblende-andesites.

The frequency distribution of nickel abundances in the hornblendes is shown in Fig. 19.6(b). The indicated overall partition coefficient $K_D^{Hb:cpx} = 1.78$.

The behaviour of nickel in hornblende rim formation is not altogether clear (Table 19.3). One hundred and twenty-two, largely paired, hornblende:opaque rim analyses (55 analysis of hornblende; 67 of opaque rims) in three of the hornblende-andesites gave the results in Table 19.3, indicating a consistent slight to moderate decrease in nickel accompanying the development of the rims. This is a little surprising, given that magnetite is one of the products of rim formation, and that it is a somewhat more receptive host to nickel than is hornblende (Table 19.3). The profile of Fig. 18.10(c) seems to indicate a concentration of nickel in the rim close to the rim: hornblende interface, with an ordered decrease towards the outer margin of the rim. (Mean values derived from this profile indicate that the rim material (78 ppm) contains more nickel than the hornblende (45 ppm) in this particular instance. The crystal concerned is, however, just one of the 26 crystal-rim pairs analysed in this particular lava (21/80, Table 19.3), the mean values for which indicate slightly lower nickel in the rims overall).

254 Ore Elements in Arc Lavas

Table 19.3 Nickel abundances in hornblende

Sample No.	Hornblende		Rims	
	n	Ni	n	Ni
21/80	26	100	29	92
24/80	14	23	19	20
92/81	15	128	19	79
Total/Mean	55	84	67	64

Mean nickel abundances in hornblende (ppm) and its opaque rims in three Guadalcanal hornblende-andesites. n = number of grains analysed.

Biotite. The limited data provided by probe analysis of the biotites of the two Savo dacites indicates a mean nickel of 110 ppm: some three times that of the clinopyroxenes and almost twice that of the hornblendes of the Solomons andesite series. Analysis of 62 biotite: hornblende pairings (a total of 124 coexisting grains) yielded the frequency distributions of Fig. 19.13. Mean nickel in the biotites is 110 ppm (as above) and in the hornblendes is 80 ppm: rather higher than the mean of 64 ppm Ni determined for the 41 hornblendes given in Table 6.5. Mean $K_D^{biot:Hb}$ for the 62 pairings is 1.375.

Feldspar. The mean abundance of nickel in the ultra-pure feldspar separations is 3 ppm, and in about 25 per cent of these it is <1 ppm. Comparison of ultimate and penultimate purity fractions indicates that virtually all feldspar nickel is present in minute inclusions: perhaps ultra-fine magnetite within the minute glassy inclusions in the feldspars already referred to in earlier chapters. The abundance of nickel in the feldspar structure is therefore estimated to be close to zero.

Spinel. In view of the existence of the nickel spinel, trevorite ($NiO.Fe_2O_3$), it might be expected that the spinels of the Solomons picrites would contain abundant nickel. This is so: the chrome-magnetites of a picrite from

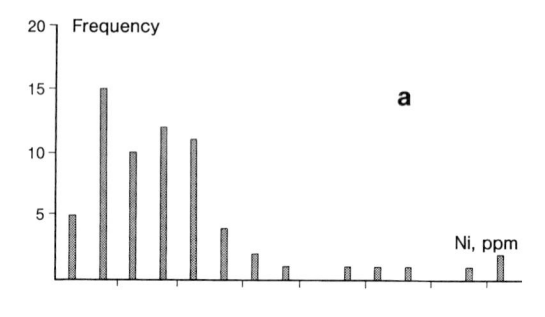

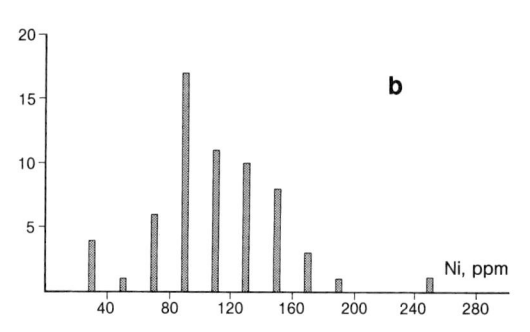

Fig. 19.13. Frequency distributions of nickel abundances in (a) hornblende (62 analyses; mean = 80 ppm Ni), and (b) biotite (62 analyses; mean = 110 ppm Ni) coexisting in the two Savo dacites.

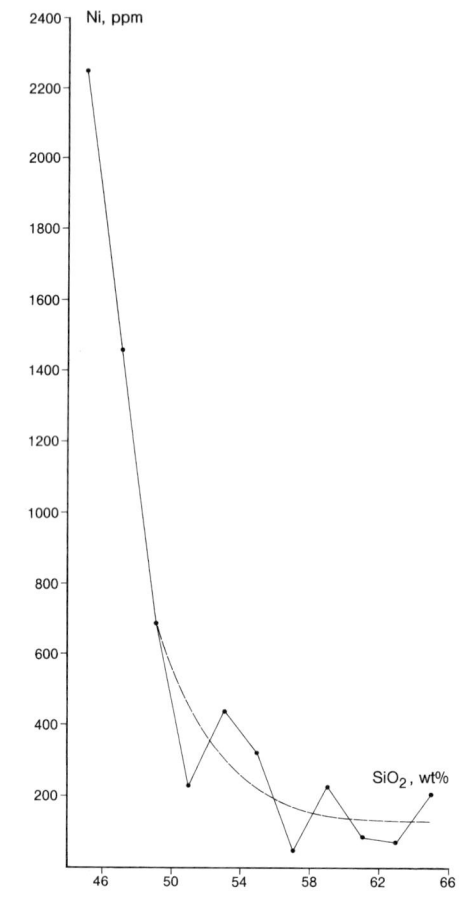

Fig. 19.14. Relations between nickel abundances in the spinels of the Solomons Younger Volcanic Suite, and 2 per cent whole-rock SiO_2 intervals in the containing lavas. The finer broken line represents only a smoothing of the curve at higher SiO_2 abundances: the trend that might have been expected to develop with large sample populations.

Kohingo Island contain 2227 ppm Ni (whole-rock MgO = 28.06 per cent; whole rock Ni = 1410 ppm; coexisting olivine MgO = 46.64 per cent (Fo$_{87}$), Ni = 2161 ppm). The frequency distribution of nickel abundances in magnetite is shown on two scales (Fig. 19.6(c) and (d)) to indicate the full range (19.6c) and also the nature of the distribution for values <200 ppm Ni.

Abundances decrease from a mean of 943 ppm in the magnetites of the olivine–pyroxene-basalts to 103 ppm Ni in those of the hornblende-andesites (Table 6.5): a trend closely related to magmatic evolution (Fig. 19.14) and hence to the availability of nickel in the melt (Fig. 19.19).

In spite of the slightly higher nickel in the magnetite (2227 ppm) compared with that of the olivine (2161 ppm) of the Kohingo Island coexisting pair, olivine is the more effective acceptor of nickel overall. Mean $K_D^{ol:mag}$ for the olivine–pyroxene lavas is 1.60, and for the big-feldspar lavas, 2.31. As these values indicate, $K_D^{ol:mag}$ with respect to Ni tends to increase with decrease in the concentration of nickel in the coexisting melt, though the correlation — which of course is negative — is not close. Some lack of clarity in this relationship may, however, have been induced by the inheritance of olivine and, particularly, magnetite by turbulent mixing of the kind referred to in Chapter 15.

The spinels are thus overall less effective than olivine in sequestering nickel from the melt, but are more effective than all of the other ferromagnesian minerals and feldspars.

CRYSTAL:MELT PARTITIONING

K_D^{Ni} values for all of the principal Solomons silicates, and for magnetite, are set out in Table 19.4. Plagioclase has been included although (as noted in the previous section) it seems likely that much of the very small amounts of nickel detected here are located in minute glassy inclusions rather than in the crystal structure itself. If this is the case, the relevant partition coefficients must be vanishingly small.

Nickel is the chalcophile trace element of greatest interest to igneous petrology — a situation arising from its very early recognized abundance relationships with magnesium in igneous rocks and, particularly, in their olivines. As a result the literature on partitioning of nickel in igneous systems is now extensive. That for olivine and clinopyroxene is too large for detailed treatment here, and the following is no more than a brief note on what is a very extensive analytical and experimental literature.

As with most of the preceding trace elements, the history of partitioning studies more or less begins with the work of Wager and Mitchell (1951) on the Skaergaard Intrusion: they determined $K_D^{ol:melt}$ values of 12.0, 9.3, and 10.0 for the olivines of their A, B, and D stages respectively. Hakli and Wright (1967) measured coefficients in the interval 13.5–16.8 in basalts of Makaopuhi lava lake in Hawaii; Henderson and Dale (1969) obtained values ranging from 4.9 to 18.6 in oceanic basalts; Gunn (1971) deduced a value of 10 in a basalt from the 1959 Kilauea

Table 19.4 Crystal:melt distribution coefficients (K_D) with respect to nickel

Mineral species	Lava group	n_1	n_2	m_1	m_2	$K_{D(1)}$	N	M_1	M_2	$K_{D(2)}$
Olivine	OPB	40	44	1511	144	10.49	39	1493	155	9.63
	BFL	7	23	545	18	30.28	6	552	29	19.03
Clinopyroxene	OPB	43	44	370	144	2.57	42	414	158	2.62
	BFL	26	23	80	18	4.44	17	68	19	3.58
	HbA	27	47	36	5.77	6.24	20	33	5.67	5.82
Hornblende	HbA	25	47	64	5.77	11.09	22	61	5.96	10.23
Feldspar	OPB	10	44	7.10	144	0.049	10	7.10	38	0.187
	BFL	23	23	2.17	18	0.121	22	2.27	17	0.134
	HbA	46	47	1.98	5.77	0.343	44	2.00	5.77	0.347
Spinel	OPB	20	44	943	144	6.55	20	943	163	5.79
	BFL	23	23	236	18	13.11	22	244	18	13.56
	HbA	37	47	103	5.77	17.85	36	83	5.80	14.31

OPB = olivine–clinopyroxene-basalt group; BFL = big-feldspar lavas; HbA = hornblende-andesite group; n_1 = total number of mineral separations analysed; n_2 = total number of groundmass separations analysed; m_1 = mean nickel content of mineral of each lava group; m_2 = mean nickel content of groundmass of each lava group; $K_{D(1)}$ = mean distribution coefficient determined as m_1/m_2; N = number of coexisting mineral:groundmass pairs analysed; M_1 and M_2 = mean nickel contents of coexisting mineral and groundmass respectively; $K_{D(2)}$ = distribution coefficient determined as M_1/M_2, for comparison with $K_{D(1)}$. All concentrations as parts per million by mass.

Iki eruption, and Bougault and Hekinean (1974) determined a value of 12.2 for olivine in a mid-Atlantic ridge basalt. Among others, Duke (1976), Mysen and Kushiro (1976), Irvine and Kushiro (1976), Arndt (1977), Leeman and Lindstrom (1978), Mysen (1978), and Seifert et al. (1988) have produced a very large amount of experimental data indicating, for a wide range of temperatures, total and partial pressures, and olivine and melt compositions, potential olivine–melt K_D^{Ni} from ≈1.0 to ≈36.05, with values most commonly in the range 6–15 at lower confining pressures.

Mean $K_D^{ol:melt}$ for the Solomon Islands olivine–pyroxene lavas is 9.63, and for the fewer number of pairs in the bigfeldspar lavas, 19.03 (Table 19.4). The overall mean is 10.88, which is very close to the value of 12 obtained by Wager and Mitchell (1951) for stage A of the Skaergaard gabbros.

Clinopyroxene:melt partition coefficients as determined both by analysis of natural materials and by experiment are much lower and less variable than those developed by olivine. Wager and Mitchell determined 1.2, 4.0, and 5 for the clinopyroxenes of the Skaergaard A, B, and C stages respectively; Hakli and Wright (1967) found a range of 2.22 to 4.40 in the basalts of Makaopuhi lava lake; Ewart and Taylor (1969) determined values of 3.5 and 5.2 in two New Zealand andesites, and Bougault and Hekinian (1974) found a value of 4.4 in the mid-Atlantic ridge basalt referred to above. Seward (1971), Duke (1976), Lindstrom and Weill (1978), Mysen (1978), and Dostal et al. (1983) in experiments involving a wide variety of starting materials and conditions, obtained values in the range 1–11, with the majority in the interval $K_D^{cpx:melt} = 1.5$–4.0.

As for the Solomons olivines, the mean values of partition coefficients for clinopyroxene mask very large variations from one lava to another. The mean of 2.62 for 42 crystal:melt pairings within the olivine–pyroxene-basalt group (Table 19.4) involves a range of individual values from 0.89 to 20.42. There is, however, a strong clustering in the 1.5–3.0 interval. Again, mean K_D increases systematically with increase in the felsic nature of the lavas: nickel decreases in both crystals and melt proceeding from mafic basalt to andesite and dacite, but does so more rapidly in the melt than in the crystals.

Somewhat surprisingly there is little information on partitioning of nickel between orthopyroxenes and melt. Ewart and Taylor (1969) determined values of 4.8, 5.2, and 6.5 for three New Zealand andesites and Mysen (1978), in a series of experiments on andesitic liquids at 1025–1075 °C and 10 and 20 atmospheres, obtained values from 0.45 to 1.15. These experiments indicated a decrease in $K_D^{opx:melt}$ with decrease in the availability of nickel in the system. Taking the value of 1.38 for $K_D^{opx:cpx}$ in the Solomons two-pyroxene-andesites referred to above and the mean value of 5.82 for $K_D^{cpx:melt}$ in the Solomons andesites overall (Table 19.4), a very approximate general value of $K_D^{opx:melt} = 8$ is indicated. This is somewhat higher than, but in reasonable agreement with, the determinations of Ewart and Taylor (1969) on the New Zealand andesites.

These authors have also obtained values for $K_D^{Hb:melt}$ in the New Zealand lavas: 1.5 for hornblende in a rhyolite and 7.2 for a hornblende in an andesite. Mysen (1978) obtained a range of values from 1.41 to 2.87 in a series of experiments involving H_2O-saturated andesitic liquids at 1000 °C and 15 kbar, $K_D^{Hb:melt}$ decreasing with decrease in concentration of nickel in the system. Dostal et al. (1983) found a value of 6.8 developed by the hornblende of a plutonic cumulate in a Lesser Antilles lava. As in the case of olivine and the pyroxenes, there is substantial variation in $K_D^{Hb:melt}$ in the Solomons andesites. The mean, 10.23, is somewhat higher than values obtained by others on natural materials. It is, however, derived from 22 individual pairings in closely related lavas, and is thus likely to be quite as reliable as any previous determinations.

The only estimation of $K_D^{biot:melt}$ in volcanic materials known to the writer is that of Villemant et al. (1981), who obtained a value of 1.3 for biotite in an alkali basalt series. For the Solomons, the two Savo dacites give $K_D^{biot:melt} = 19.59$ and 10.19, with a mean of 14.89: a value well above that of Villemant et al.

As for the orthopyroxenes, there is remarkably little available data on the partitioning of nickel between magnetite and co-existing melts. Wager and Mitchell (1951) obtained values of 15 and 10 for the magnetites of the Skaergaard B and C stages; Leeman (1974) and Lindstrom (1976) found values between 12.2 and 77 in experiments using a natural picritic tholeiite; Villemant et al. (1981) determined a value of 3.5 for the magnetite of an alkali basalt, and Dostal et al. (1983) a value of 29 for the plutonic cumulate of the Lesser Antilles lava.

Again, the Solomons material is highly variable from one individual pairing to another, but mean values for the principal lava groups show a general increase in $K_D^{mag:melt}$ as the relevant lavas become more felsic: 5.79 for the olivine–pyroxene-basalts rising to 14.31 for the hornblende-andesites, with an overall average of 11.91 for 78 magnetite–melt pairings.

All the ferromagnesian minerals and magnetite thus develop partition coefficients $K_D^{cryst:melt} > 1$, and hence tend to impoverish the melt with respect to nickel. The earliest-formed crystals, olivine and magnetite, immediately establish high coefficients (≈10 and ≈6 respectively) and these increase progressively with decrease in the amount of nickel remaining in the melt. The result is rapid depletion of nickel in the evolving melt.

THE INCIDENCE OF WHOLE-ROCK NICKEL IN THE SOLOMON ISLANDS YOUNGER VOLCANIC SUITE

Mean nickel for the suite as a whole is 190 ppm, for the basalts (SiO_2 <52 per cent) 351 ppm, and for the lavas more felsic than basalts (SiO_2 >52 per cent), 19 ppm (Table 6.3; Fig. 19.15). The value for basalts appears grossly higher than that for MORBs (114 ppm Ni) but this is due (as noted in connection with chromium) to the large number of picritic lavas included in the Solomons sample. The nickel content of Solomons basalts for which 48 <SiO_2 <52 per cent — the SiO_2 range into which most MORBs fall — is 181 ppm; for the range 49 <SiO_2 <52 per cent it is 94 ppm. The mean nickel content of the 'non-picritic' Solomons basalts is thus probably fairly close to that of mean MORBs. This compares with 183 ppm Ni in the West Indies basalts (also weighted towards high nickel by the highly mafic basalts of Grenada), and 50, 35, and 97 ppm Ni in the basalts of the Aleutians, Tonga–Kermadec, and Vanuatu arcs respectively. Mean nickel for the Solomons lavas more felsic than basalts (SiO_2 >52 per cent), at 19 ppm, compares with 39, 8, 10, and 8 ppm Ni for the West Indies, Aleutians, Tonga–Kermadec, and Vanuatu respectively. Again the mean nickel of the West Indies lavas is rendered relatively high by the notably high nickel of the Grenada suite. The mean nickel abundance in the lavas with SiO_2 >52 per cent of Dominica and St Kitts is 20 ppm: almost identical with that of their Solomons analogues.

The relationship between nickel and MgO in the Solomons whole rocks is close. Fig. 19.16(a) shows Ni v. MgO is all Solomons lavas with MgO >6.0 per cent; the coefficient of correlation is 0.943 and the regression line intersects the x-axis at MgO ≈6.0 per cent. The tie is not quite so close in the corresponding groundmasses (coefficient of correlation = 0.888), but is clear none the less. For comparison, the basalts of the island of Aoba in the Vanuatu arc show Ni:MgO relations very similar to those of the Solomons (Fig. 19.16(b), with a coefficient of correlation of 0.971 and the regression line intersecting the x-axis at 5.0 per cent MgO. The Grenada basalts, as represented by mean nickel and MgO corresponding to 2 per cent whole-rock SiO_2 intervals (<44–52 per cent) as calculated by Brown et al. (1977), yield a coefficient of correlation of 0.982 and intersect the x-axis at ≈3.5 per cent MgO (Fig. 19.16(c)). The relatively good correlation between Ni and MgO in the Solomons groundmasses leads to the suspicion that nickel moves to magnesium positions in the developing olivine structure while this is still part of the overall structure of the liquid.

It follows from the closeness of Cr:MgO relations in these lavas (vide Chapter 15), that there should be a close tie between chromium and nickel, and this is indeed the case (Fig. 19.17). The distribution of points may constitute a curve, or two linear arrays intersecting at ≈400 ppm Ni, 800 ppm Cr.. The upper array has a coefficient of correlation of 0.966 and may be expressed by the regression equation Cr = 1.1251 Ni + 359.1; the lower a coefficient of correlation of 0.964 and the regression equation Cr = 2.1121 Ni–11.265. The cluster of points round the postulated intersection represents lavas containing c. 14–15 per cent MgO: a state of affairs referred to again in Chapter 23. The Grenada basalts show a clear Ni:Cr correlation (Fig. 19.17(b)), but no apparent inflection around Ni = 400 ppm, Cr = 800 ppm. The slope

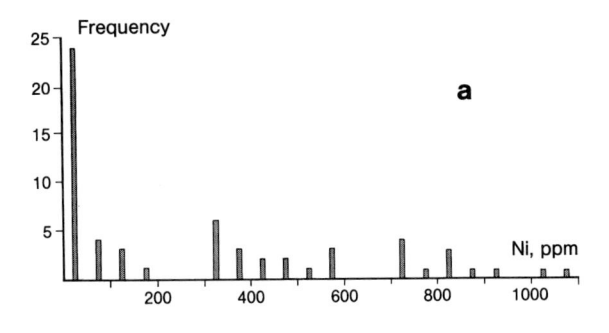

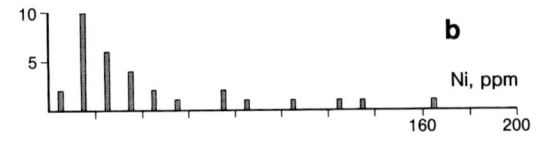

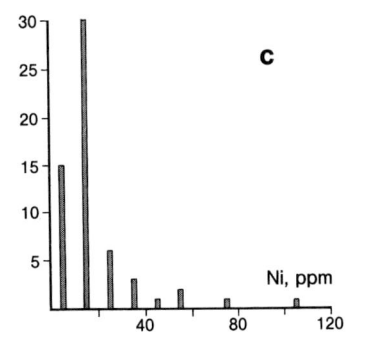

Fig. 19.15. Frequency distributions of nickel abundances in (a) Solomons basalts (SiO_2 <52 per cent); (b) Solomons basalts in the 0–200 ppm Ni interval of (a); and (c) Solomons lavas more felsic than basalts (SiO_2 ⩾52 per cent).

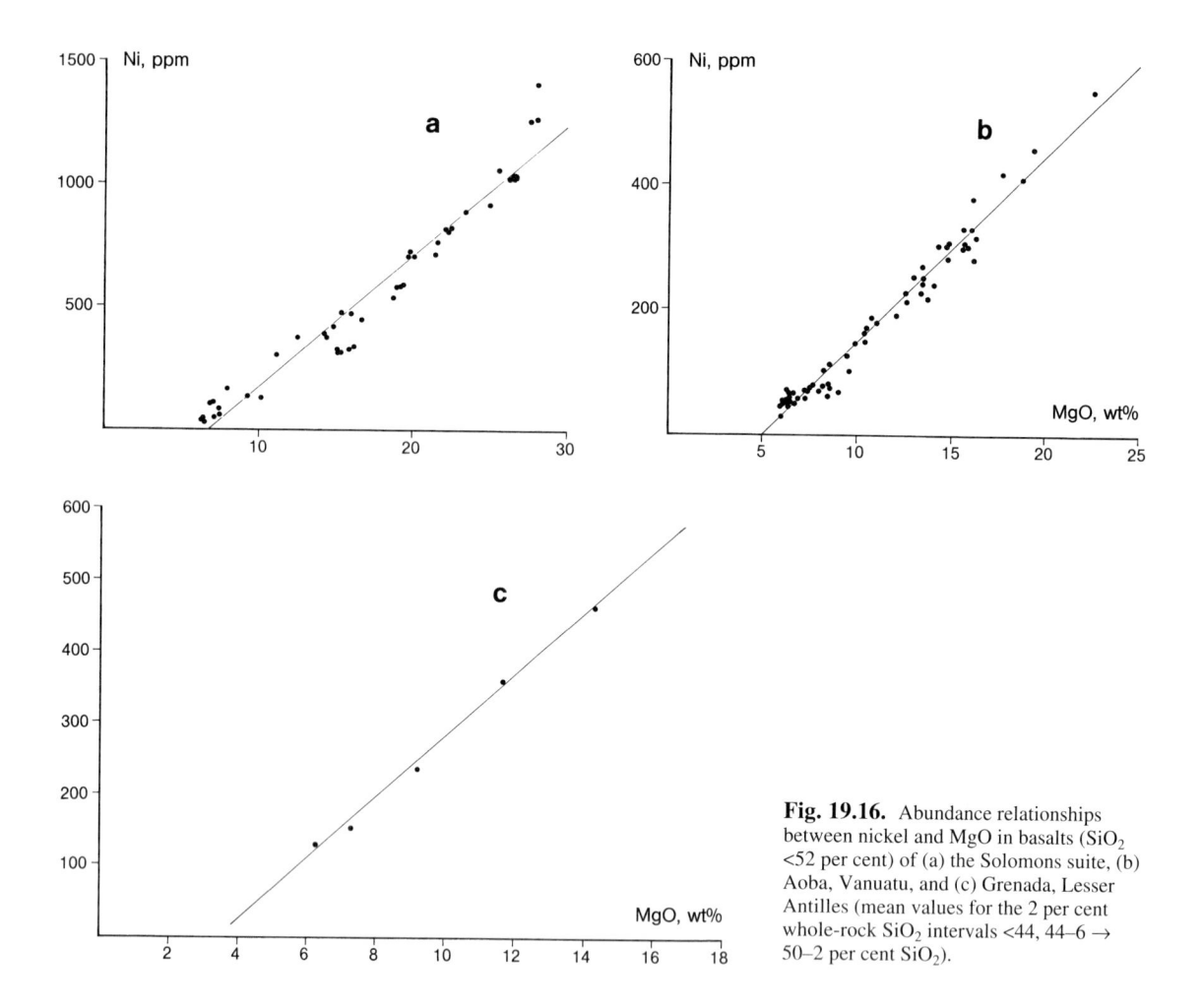

Fig. 19.16. Abundance relationships between nickel and MgO in basalts (SiO$_2$ <52 per cent) of (a) the Solomons suite, (b) Aoba, Vanuatu, and (c) Grenada, Lesser Antilles (mean values for the 2 per cent whole-rock SiO$_2$ intervals <44, 44–6 → 50–2 per cent SiO$_2$).

of the lower point of the array (Cr = 2.175 Ni + 32.499; r = 0.957)) is, however, very similar to that generated by the Solomons lavas.

The distribution of nickel in the three principal Solomons lava types and their groundmasses (Fig. 19.18) emphasizes the strong partitioning of nickel into crystalline phases in the olivine–pyroxene-basalts, a progressive diminution in degree of such partitioning as the rocks become more felsic, and a sharp decrease in whole-rock and groundmass nickel accompanying the change from mafic to felsic basalts and basaltic andesites. Mean nickel in the olivine–pyroxene-basalt, big-feldspar lava, and hornblende-andesite whole rocks is 642, 20, and 15 ppm respectively, and in the corresponding groundmasses is 178, 18, and 6 ppm.

The rapidity of the decrease in whole-rock nickel from the very earliest stage of crystallization is indicated in Fig. 19.19, in which mean nickel corresponding to 2 per cent intervals of whole-rock SiO$_2$ is plotted against the latter. The 'curve' descends very steeply from 1410 ppm Ni at <46 per cent SiO$_2$ to 46 ppm Ni at 50–2 per cent SiO$_2$: the stage at which olivine, the principal abstractor of nickel, virtually ceases to crystallize (see Fig. 5.6). Patterns similar in principle are developed by the lavas of the West Indies (Fig. 19.19(b)) and Vanuatu (Fig. 19.19(e)), which reach their points of 'flattening' at 48–50

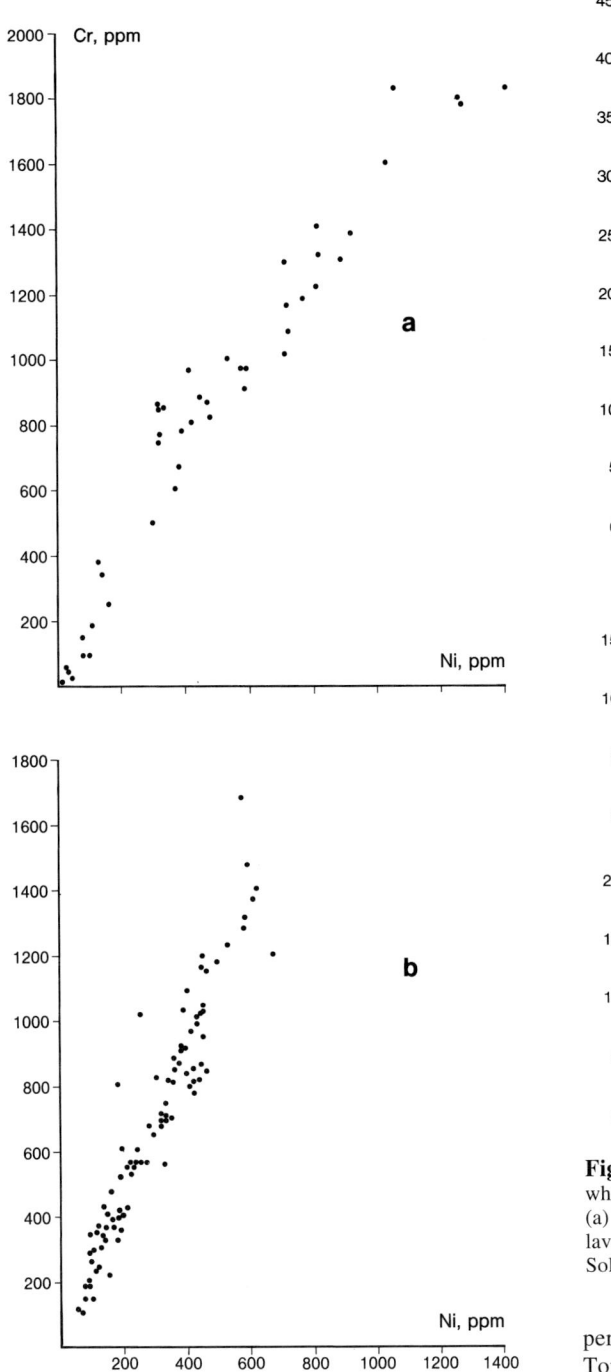

Fig. 19.17. Abundance relations between nickel and chromium in basalts (SiO$_2$ <52 per cent) of (a) the Solomons suite, and (b) Grenada.

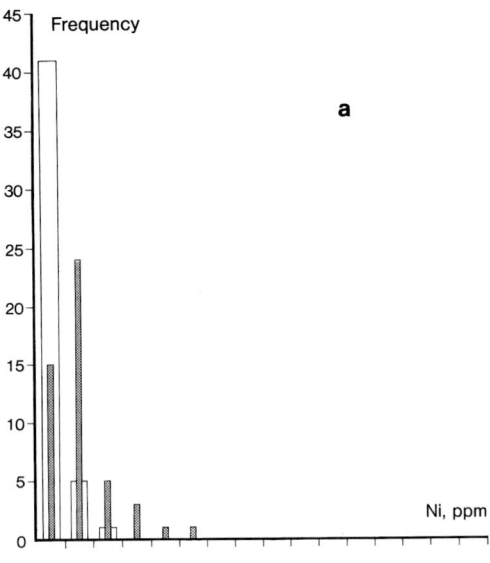

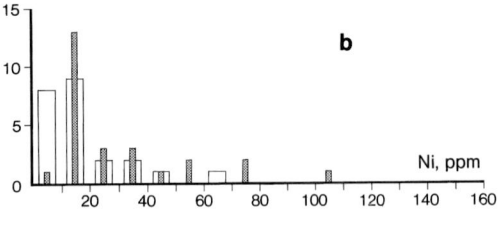

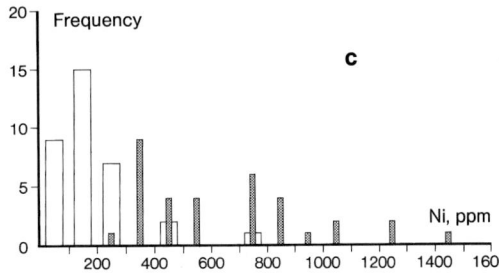

Fig. 19.18. Frequency distributions of nickel abundances in whole rocks (filled bars) and groundmasses (open bars) of (a) lavas of the hornblende-andesite group, (b) big-feldspar lavas, and (c) the olivine–clinopyroxene-basalts of the Solomons suite.

per cent and 52–4 per cent SiO$_2$ respectively. The Tonga–Kermadec suite (Fig. 19.19(d)) is almost certainly truncated, with its more mafic (and more Ni-rich) segment missing. Flattening has probably occurred at 48–50 per cent SiO$_2$.

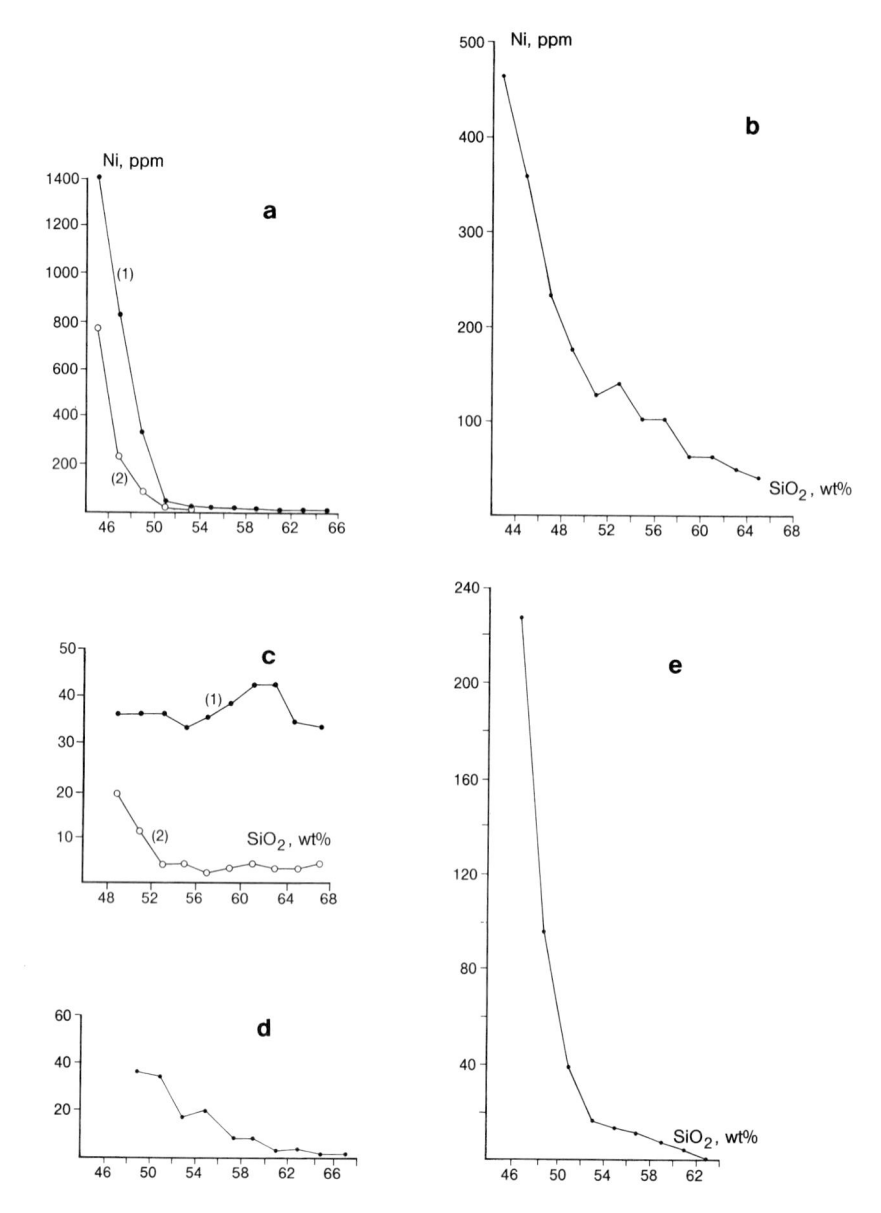

Fig. 19.19. Mean Ni:SiO$_2$ relations corresponding to 2 per cent whole-rock SiO$_2$ intervals in (a) whole rocks [(1) filled circles] and corresponding groundmasses [(2) open circles] of the Solomons suite; (b) the lavas of Grenada; (c) Dominica [(1) filled circles]; and St Kitts [(2) open circles]; (d) Tonga–Kermadec; and (e) Vanuatu.

Ternary plots of Ni–MgO–(Na$_2$O + K$_2$O) (Fig. 19.20) demonstrate orderly trends from olivine–pyroxene-basalt whole rocks to hornblende-andesite groundmasses with an overall pattern reminiscent of that of chromium, and hence of 'nichrome' type. The plots for the olivine–pyroxene-basalt groundmasses and the big-feldspar lavas are, however, very diffuse, perhaps reflecting the extreme sensitivity of nickel occurrence to the frequency of nucleation of olivine as this approaches the end of its crystallization path. The distinctive nichrome pattern appears more clearly when NiMA points corresponding to 2 per cent intervals of whole-rock SiO$_2$ are plotted.

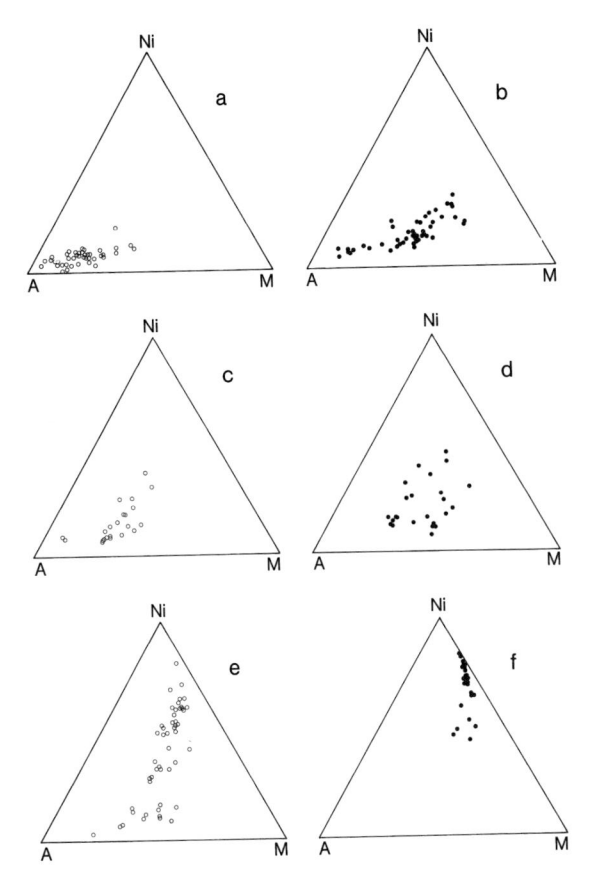

Fig. 19.20. Ni–MgO–(Na$_2$O + K$_2$O) ternary relations (Ni calculated using ppm Ni × 10^{-1}) developed in olivine–clinopyroxene-basalt, big-feldspar lava, and hornblende-andesite whole rocks [filled circles: (f), (d), and (b) respectively], and their corresponding groundmasses [open circles: (e), (c), and (a) respectively].

Figure 19.21 shows a fairly orderly progression of points for Solomons whole rocks and groundmasses, with a break in slope occurring between 50–2 and 52–4 per cent SiO$_2$, the point at which conspicuous flattening of the Ni:SiO$_2$ curve develops (Fig. 19.19(a)). The curve is smoothed considerably in the region of the more felsic lavas when analyses of Bougainville material are included (Fig. 19.21(b)). The array of points generated by the Grenada lavas (Fig. 19.21(c)) appears similar in principle to that of the Solomons; it is, however, located in a much higher nickel region of the diagram and extends over a much shorter compositional span than the latter. The Dominica and St Kitts arrays in contrast are very flat; indeed, the three lesser Antilles suites show remarkable differences in NiMA behaviour. The Vanuatu curve is rather similar to that of the Solomons, that of the Tonga–Kermadec suite (Fig. 19.21(d)) distinctly flatter. A minimum of licence allows one to suggest that Grenada, the Solomons, Vanuatu, Tonga–Kermadec, and St Kitts constitute a series, the curve shapes of which reflect a progressive decrease in nickel on a provincial scale. The Dominica series, however, fits the pattern very poorly.

The pattern generated by MORBs as represented by those of the mid-Atlantic ridge (Fig. 19.21(e)) is remarkably similar to that of the more mafic portion of the Solomons curve, and the array representing the full Icelandic assemblage (Fig. 19.21(f)) even more so.

A brief digression on the contrariness of Nature, and the pitfalls that this may create. Throughout most of the foregoing geochemical chapters the propensity of the different minerals for accepting various elements into their crystal structures, combined with the general position of each mineral in the sequence of crystallization, has been sufficient to account for much of the trace-element abundance behaviour observed in the Solomons lavas as these have evolved from picrites to dacites. However, for those who would place all of their faith in such mechanisms, the spinels of the Rendova ankaramitic basalt bring a cautionary note.

It will be recalled that, with its SiO$_2$ content of 49.09 per cent, this lava falls to the left of the peaks in the metal:SiO$_2$ variation diagrams for copper (Fig. 6.12), zinc (Fig. 7.23), titanium (Fig. 13.19), and vanadium (Fig. 14.12). Thus the earlier set of spinels — those included within the large clinopyroxene crystals of this lava — contained lower concentrations of each of these trace elements than did the later-formed set: those included in the groundmass.

The incidence of nickel is in good agreement with this pattern in terms of the above 'controls': in accordance with the relationship of Fig. 19.19, the concentration of nickel in the earlier-formed spinels — those included in the clinopyroxenes — is higher (mean = 402 ppm Ni) than in the later, groundmass, spinels (mean = 234 ppm Ni) as shown in Fig. 19.22(a).

The relative abundances of chromium in the two sets of spinels are, however, quite out of keeping with the principle with which the other five elements appear to conform so well. In the light of the relationships of Fig. 15.21 it might have been expected that, as with nickel, the concentration of chromium would have been higher in the earlier-formed, clinopyroxene-included spinels. Exactly the opposite is the case: the spinels within the

262 *Ore Elements in Arc Lavas*

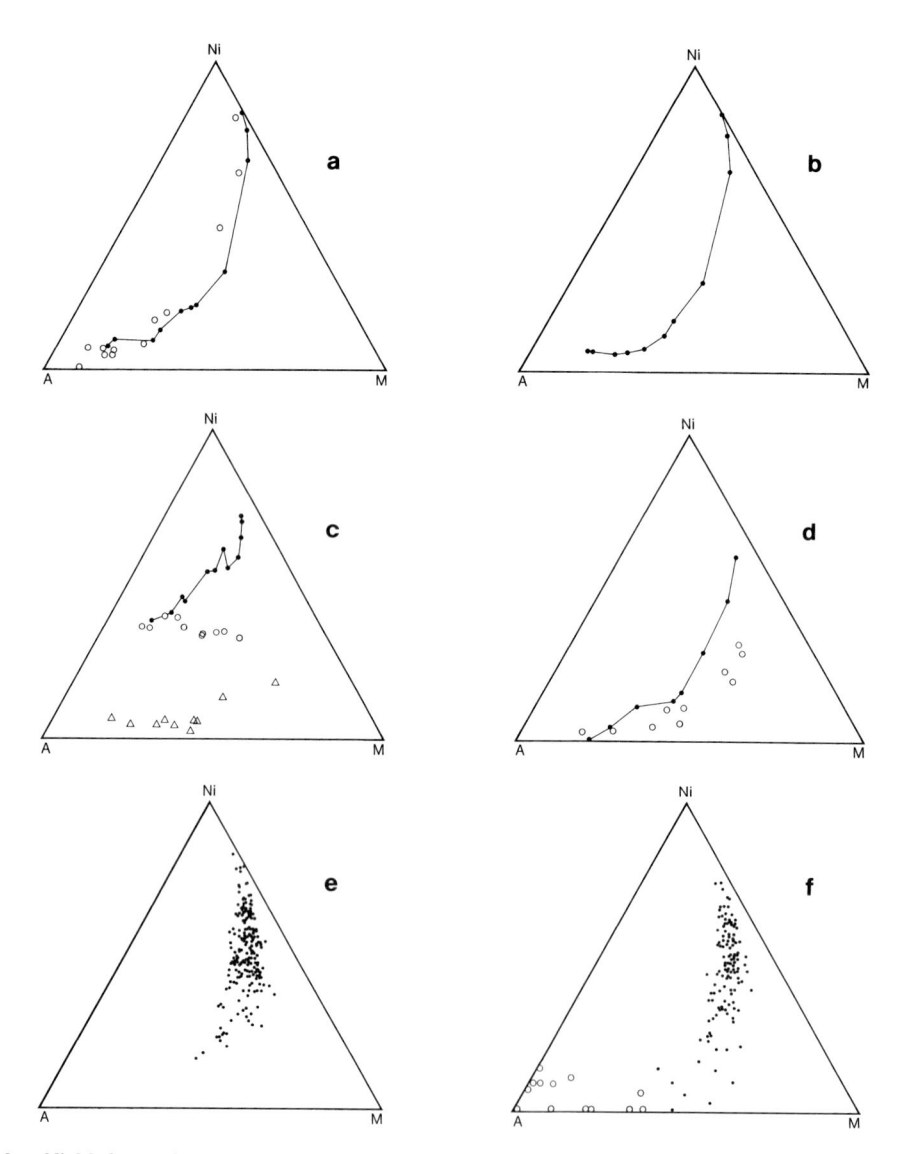

Fig. 19.21. Mean Ni–MgO–(Na$_2$O + K$_2$O) ternary relations corresponding to 2 per cent whole-rock SiO$_2$ intervals in (a) whole rocks (filled circles) and corresponding groundmasses (open circles) of the Solomons suite excluding Bougainville; (b) whole rocks of Solomons suite including Bougainville (241 analyses); (c) lavas of Grenada (filled circles), Dominica (open circles), and St Kitts (open triangles); (d) Vanuatu (filled circles) and Tonga–Kermadec (open circles). Ternary relations developed in MORBs as represented by Mid-Atlantic Ridge basalts are shown in (e), and in Icelandic basalts (SiO$_2$ <52 per cent: filled circles) and Icelandic lavas more felsic than basalt (SiO$_2$ ≥52 per cent: open circles) in (f).

clinopyroxene contain distinctly less chromium (mean = 731 ppm Cr) than do those of the groundmass (mean = 1030 ppm Cr): a state of affairs shown very clearly by the frequency distributions of Fig. 19.22(b).

The reason for this 'discordance' on the part of chromium is not obvious. It indicates, however, that the crystallization histories of some of these arc lavas may be far from simple.

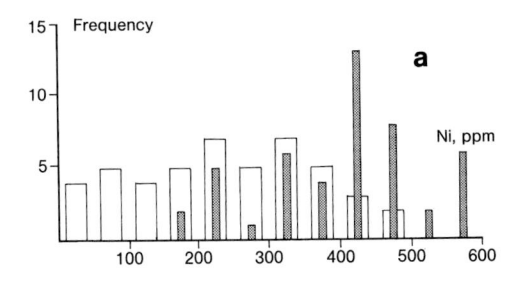

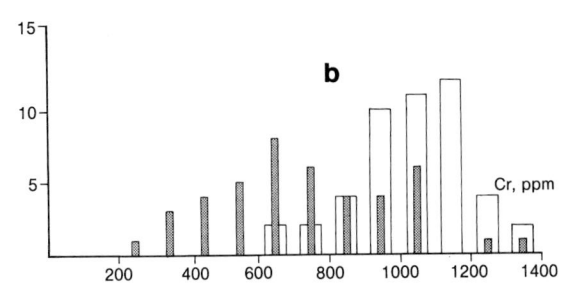

Fig. 19.22. Frequency distributions of (a) nickel abundances in magnetites included within the clinopyroxenes (filled bars; mean = 402 ppm) and within the groundmass (open bars; mean = 234 ppm Ni), and (b) chromium abundances (means = 731 ppm and 1030 ppm Cr respectively) in the same two groups of magnetites of the Rendova ankaramitic basalt referred to in Chapter 5.

CONCLUDING STATEMENT

The approximate abundance of nickel in MORBs is 114 ppm, and in arc basalts, 134 ppm. If the picritic accumulates of the Solomons, Vanuatu, and Grenada were excluded from the calculation, the nickel content of arc basalts would be significantly less than that of MORBs: perhaps between 60 and 70 ppm. The mean nickel content of the basalts of the Solomon Islands Younger Volcanic suite is 351, making them the most Ni-rich of the five arc suites considered. If the SiO_2 interval 48–52 per cent only is used for calculation (the SiO_2 interval that includes most MORBs), the mean value for the Solomons basalts is 181 ppm. In all the arc suites for which adequate data is available nickel decreases rapidly from its maximum in the most picritic basalts to much lower levels in basaltic andesites and more felsic lavas. The point of inflexion, or pronounced flattening, of the 'curves' is ≈52 per cent SiO_2: the stage in basalt–andesite formation at which copious olivine crystallization appears to taper off. In all this the abundance of nickel closely follows that of magnesium, as observed in other provinces and contexts by earlier investigators. Thus the whole pattern of abundance of whole-rock nickel, particularly in its relationship with magnesium, parallels that of chromium, with which it is closely correlated.

This parallelism is, however, largely fortuitous and stems from the substantially coeval crystallization of clinopyroxene and spinel on the one hand and of olivine on the other, rather than from any diadochical relationship between nickel, chromium, and magnesium. Nickel and magnesium occur in the melt as divalent ions of radius 0.078 nm; chromium as the trivalent ion Cr^{3+} with a radius of 0.064 nm. Nickel and chromium thus carry different charges and the radii of Ni^{2+} and Mg^{2+} are 18 per cent greater than that of Cr^{3+}: differences unconducive to mutual substitution. The close correlation of the two is thus due to a near-coincidence in the timing of crystallization of their respective host minerals. Nickel and chromium are both abstracted strongly from the melt from the beginning of crystallization — of olivine and clinopyroxene + spinel respectively — and as a result exhibit rapid initial decrease to ≈52 per cent whole-rock SiO_2, after which rate of subtraction declines sharply as both are virtually eliminated from the residual melt.

On three-component Ni–MgO–($Na_2 + K_2O$) diagrams, the Solomons suite exhibits a well-developed nichrome patten, and this is reproduced in large part by the Vanuatu assemblage. The Grenada lavas generate a steeper, higher-nickel curve; those of St Kitts and Tonga–Kermadec flatter, lower-nickel arrays. All, however, appear to reflect rapid, early removal of most of the nickel of the melt in high-Ni phases, notably olivine, followed by much slower removal in the later stages of fractionation.

20

Sundry elements

All the Solomons lavas of present concern contain, in addition to the trace elements considered in Chapters 6–19, very small quantities of elements such as cadmium, gallium, gold, silver, tungsten, uranium, and others that also commonly occur as minor to trace components of exhalative ores. They also contain varying, very small, amounts of sulphur. In addition they contain traces of elements such as rubidium, zirconium, and thorium, which, through their often highly systematic abundance relations with the traces of the 'ore elements' serve to emphasize the intimacy of the tie between the incidence of the latter and the geochemical and mineralogical evolution of the relevant lava systems.

Together with the elements considered in the preceding chapters, rubidium, zirconium, and gallium were determined in all 122 samplings of the Solomons Younger Volcanic Suite. In addition a 28-sample subset (a 'condensed series' representing the full SiO_2 span of the parent set) was analysed for lithium, sulphur, chlorine, fluorine, molybdenum, silver, cadmium, indium, caesium, arsenic, antimony, bismuth, the rare earths, hafnium, tantalum, tungsten, gold, thorium, and uranium (Table 20.1).

Sulphur as S^{2-} is of course an abundant and universal component of the exhalative ores (even the dominantly silicate–oxide–carbonate ores of Franklin and Stirling Hill, New Jersey, contain minor sulphide) and of their closely related porphyry and epithermal deposits. Fluorine and chlorine are common components of ore-transporting systems, are present in virtually all volcanic discharges, and occur in many exhalative ores as components of fluorite, apatite, and the micas. Cadmium, and to a lesser extent gallium and indium, are found as components of sphalerite and hence as associates of zinc in many exhalative and related ores. Antimony, arsenic, and bismuth have been detected in many volcanic sublimates. Antimony and arsenic are components of tetrahedrite–tennantite and arsenic is in the arsenopyrite that occur in minor quantities in most exhalative stratiform ores. Bismuth as native bismuth and bismuthinite occurs in trace quantities in many orebodies of the exhalative regime. Silver and gold are associated with copper in almost the full range of deposits of volcanic affiliation, and molybdenum is associated with copper, silver, and gold in many porphyry deposits. Some stratiform tungsten-bearing ores are now considered to have a volcanic (possibly basaltic) affiliation, and gold and silver are the principal components of many young volcanic–epithermal deposits. The list is long and while a minute suite of 28 samples could not be expected to yield a definitive insight into the petrological basis of these associations, it does at least provide some indications.

Sulphur. A recent estimate (Emsley 1989) of the abundance of sulphur in the Earth's crust is 340 ppm. Most estimates of average abundances of this element in the spectrum of volcanic rocks range around 250–300 ppm. Sulphur is, however, one of the most conspicuous components of volcanic gases, and the amounts found in any particular solidified lava must largely reflect the degassing history of that lava (see Chapter 22).

The central emphasis of the present contribution is on the incidence of the 'ore elements' in the silicates and oxides. As mentioned in the Preface, collection of samples was carried out with this very much in mind. Any sample showing the slightest microscopical ($\times$ 430) indication of sulphide was discarded. The only sulphide observed under the microscope was pyrite, but as this may have contained traces of Cu, Zn, and other chalcophile metals, the relevant sample was eliminated from the investigation.

The possibility that significant sulphur occurred in forms undetectable by the optical microscope remained, however, and this was investigated by X-ray fluorescence analysis of the condensed suite (Table 20.1). For these samples mean total sulphur is 27.5 ppm, and the range is 0 < S < 90 ppm. Six samples contained S < 5 ppm and three contained S > 70 ppm.

As the mass ratios of copper and zinc to sulphur in chalcopyrite and sphalerite are $\approx$1:1 and 2:1 respectively, a mean sulphur content of 27.5 ppm gives scope for the occurrence of significant quantities of the metals as submicroscopic sulphides where, say, Cu $\approx$80 ppm and Zn $\approx$100 ppm. Copper being the chalcophile metal with the

Table 20.1 SiO_2, MgO, and 16 trace elements in the Solomon Islands 'condensed suite'

Rock number	Lava group	SiO_2	MgO	S	Cl	F	Rb	Zr	Hf	As	Mo	Au	Ag	Cd	Zn	Ga	W	Th	U
Sol 16/81-RLS 387	OPL	46.20	22.32	5	75	140	13.0	5	0.4	<0.4	0.14	1.5	0.18	0.22	70	8.0	2.14	0.31	0.11
Sol 64/80-RLS 378	OPL	46.29	28.07	10	100	270	8.5	8	0.2	0.8	0.18	1.4	0.33	<0.10	62	6.0	1.88	0.15	<0.05
Sol 35/81-RLS 407	OPL	46.49	26.29	10	120	130	7.5	6	0.3	0.4	0.19	<1.0	0.23	0.14	71	7.5	2.79	0.27	0.09
Sol 29/81-RLS 400	OPL	46.88	23.50	<5	130	150	10.0	6	0.4	<0.4	0.41	2.4	0.30	<0.10	72	8.0	1.27	0.30	0.08
Sol 40/81-RLS 412	OPL	48.03	21.51	10	95	180	9.5	4	0.4	<0.4	0.16	2.6	0.29	0.24	71	9.0	1.75	0.31	0.11
Sol 59/80-RLS 374	OPL	48.15	11.16	5	80	200	8.5	9	0.5	0.4	0.51	1.0	0.26	<0.10	56	12.5	3.59	0.78	0.15
Sol 34/81-RLS 406	OPL	48.55	18.94	<5	85	190	11.5	9	0.6	<0.4	0.42	2.5	0.27	0.18	76	10.0	0.93	0.42	0.13
Sol 54/81-RLS 580	BFL	48.75	7.11	25	290	110	11.0	32	0.7	0.4	nd	4.3	nd	nd	80	16.4	nd	nd	nd
Sol 19/81-RLS 390	OPL	49.20	16.06	<5	135	360	12.0	8	0.5	0.8	0.60	2.9	0.29	0.13	75	11.5	0.81	0.58	0.08
Sol 78/80-RLS 385	OPL	49.28	9.23	30	380	330	19.0	21	0.8	1.0	0.42	2.3	0.18	0.21	66	14.0	1.69	1.00	0.49
Sol 60/80-RLS 375	OPL	49.74	7.93	<5	115	260	11.0	30	1.1	0.4	0.43	2.9	0.28	0.24	66	16.0	2.81	0.88	0.22
Sol 23/81-RLS 395	OPL	50.22	14.29	<5	175	220	12.5	27	0.9	1.0	0.78	1.6	0.20	<0.10	74	11.5	1.21	0.49	0.16
GZ 32-RLS 651	BFL	50.38	5.55	45	200	550	22.5	50	1.0	1.6	0.69	5.2	0.47	0.19	84	16.0	nd	1.00	0.47
Sol 15/81-RLS 386	BFL	51.35	3.79	10	115	560	23.0	53	1.7	1.6	1.08	3.2	0.44	0.18	71	18.0	1.80	1.50	0.44
Sol 24/80-RLS 22	HbA	51.63	5.35	15	70	210	11.0	83	1.8	0.6	1.28	2.0	0.96	0.33	77	18.0	4.39	7.80	0.35
Sol 6/80-RLS 6	HbA	51.69	6.37	30	115	420	45.0	94	2.0	1.0	1.21	2.0	0.86	0.25	80	16.5	1.78	10.30	1.50
NG 789-RLS 652	BFL	54.31	3.04	10	250	740	44.5	129	2.7	2.2	2.32	8.5	0.43	0.23	107	18.4	nd	2.30	0.77
TP 089-RLS 591	BFL	54.36	4.23	75	295	400	46.0	81	1.9	2.0	1.73	4.7	0.35	<0.10	71	17.2	nd	2.00	0.71
Sol 18/80-RLS 18	HbA	54.87	3.22	25	160	190	14.5	97	2.2	0.8	1.36	<1.0	0.67	0.26	88	20.0	1.73	6.80	0.61
Sol 26/80-RLS 24	HbA	54.90	4.73	90	390	340	13.5	72	1.5	1.4	0.80	1.0	0.36	0.18	70	16.0	2.60	2.30	0.41
Sol 49/81-RLS 575	BFL	55.31	3.54	5	175	450	25.0	74	1.6	0.6	0.66	2.1	0.36	0.15	77	18.6	nd	1.30	0.42
Sol 32/80-RLS 32	HbA	55.42	3.64	55	580	540	22.5	85	2.0	1.0	0.54	2.0	0.40	0.12	67	20.5	2.60	0.92	0.23
Sol 94/81-RLS 55	HbA	57.29	2.99	30	80	190	15.5	49	1.2	1.0	0.45	1.0	0.32	0.17	60	16.5	2.58	2.10	0.28
Sol 7/81-RLS 52	HbA	58.97	2.55	70	150	280	29.0	108	2.4	0.6	0.61	1.0	0.48	0.17	60	21.5	1.32	1.80	0.74
Sol 104/81-RLS 57	HbA	60.94	2.83	50	100	170	11.5	74	0.7	0.8	0.48	1.0	0.38	<0.10	54	18.0	1.65	1.20	0.27
Sol 46/80-RLS 45	HbA	62.80	1.56	70	430	610	38.5	141	3.3	1.8	0.45	1.0	0.32	0.16	55	23.0	nd	2.70	0.96
Sol 31/80-RLS 31	HbA	64.17	1.27	85	1740	390	39.0	138	3.3	2.4	0.68	<1.0	0.12	<0.10	52	22.5	2.14	1.50	0.68
Sol 41/80-RLS 41	HbA	65.36	1.06	35	275	440	38.5	129	3.1	1.0	0.45	1.0	0.32	0.18	46	22.5	2.11	1.30	0.42

OPL = olivine–clinopyroxene-basalt; BFL = big-feldspar lava; HbA = member of hornblende-andesite group. SiO_2 and MgO given as per cent by mass; all trace elements except Au given as parts per million by mass; Au given as parts per billion (10^{-9}) by mass; nd = not determined.

greatest propensity for combining with sulphur, Cu:S abundance relations should indicate whether or not there is any systematic metal:sulphur tie, and hence whether the metals in question occur as sulphides rather than as traces in silicates. In fact Cu:S relations in the 28 samples appear to be completely random, indicating that at least the major part of the copper does not occur as sulphide and hence that it occurs principally as a trace in silicate or in oxide, or both: a state of affairs confirmed by electron microprobe analysis.

Sulphur does not show a clear-cut relationship with whole-rock SiO_2, but it may be seen that there is a general tendency for its abundance to be greater in the more felsic members of the suite. This presumably reflects its progressive concentration in the melt as the latter evolved. It is slightly but significantly higher in the lavas of Savo (mean S = 63 ppm) than in those of the closely related Gallego volcano (mean S = 42 ppm).

Chlorine. Although chlorine is abundant in most high-temperature volcanic discharges it is not, because of the relatively high solubility of most of its common compounds, a conspicuous component of most exhalative ores.

As for sulphur, the chlorine contents of crystallized lavas are largely reflections of degassing histories. Volcanic glasses tend to contain more chlorine than their crystalline equivalents, presumably the result of rapid quenching and entrapment of chlorine in the glass, as compared with the progressive degassing of the less rapidly quenched, crystallized materials. At least partly as a result of such volatile loss, analytical data give no clear indication of any relation between chlorine content and extent of differentiation of lavas: what is little more than 'informed estimation' suggests however that basalts contain, on average, 100–200 ppm, rhyolites 200–300 mean ppm Cl.

For the Solomons condensed suite as a whole, mean Cl = 247 ppm. The full range is 75–1740 ppm, the latter figure representing an unusually Cl-rich lava from Savo. If this is excluded, the range is 75–580 ppm. Chlorine content has no obvious relationship with whole-rock SiO_2 or with any other index of evolution of the series. On the basis of 11 samples there does, however, appear to be a genuine difference in chlorine content between the closely related Gallego and Savo lavas: those from Gallego have mean Cl = 167 ppm; those from Savo 549 ppm. If the Savo sample containing the unusually high Cl content of 1740 ppm be excluded, the mean for the Savo sampling is 311 ppm.

Fluorine. Fluorine is a common constituent of both volcanic exhalations and exhalative ores. Its principal habitat in the latter is fluor-apatite, a conspicuous component of many exhalative sulphide deposits and of the iron formations associated with them. In some, e.g. Broken Hill, significant quantities of fluorine occurs also in fluorite, and total fluorine combined in apatite and/or fluorite in such orebodies may exceed one per cent by mass. Smaller quantities may occur in associated biotite.

Like sulphur and chlorine, fluorine is an important component of many volcanic gases, and its abundance in any given lava depends in part on the degassing history of the lava. This is emphasized by broad comparisons of plutonic and volcanic rocks: large numbers of analyses indicate that plutonic rocks generally have higher fluorine contents than volcanic rocks of comparable composition. Once again we may make 'informed estimates' of means of 350–400 ppm in basalts, 450–550 ppm F in rhyolites.

Fluorine is somewhat more abundant than chlorine in the Solomons condensed suite, with an overall mean of 322 ppm and a range from 110 to 740 ppm. While the data of Table 20.1 may be taken to indicate a general increase in fluorine with increase in whole-rock SiO_2, the F v. SiO_2 plot shows substantial scatter and a negligible correlation between the two. As for chlorine, the lavas of the Savo volcano contain distinctly higher fluorine (mean, 452 ppm; range 280–610 ppm F) than do those of the closely related Gallego volcano (mean, 253 ppm; range 170–420 ppm F).

Rubidium. Rubidium is one of the minor trace elements for which data are available for the full Solomons suite and for the relevant groundmass and mineral separations.

The abundance of rubidium in the Earth's crust is ≈78 ppm (Emsley 1989). Stemming from its very close physical and chemical relationships with potassium (see below), rubidium tends to concentrate in the high-K minerals and K-rich residual melts. It commonly occurs at the 200–500 ppm Rb level in K-feldspars, and 400–800 ppm in igneous biotites. Its abundance behaviour in volcanic rocks reflects its identity as a large-ion lithophile, or incompatible, element: mean abundances in oceanic and circum-oceanic basalts are ≈5–15 ppm; andesites, ≈20–40 ppm; rhyolites, ≈50–150 ppm Rb.

The mean Rb content of MORBs as represented by 157 analyses of material from the southern part of the Mid-Atlantic Ridge and associated structures is ≈8.5 ppm, with 99 per cent of values within the 0–15 ppm range. The mean value for all lavas of the Lesser Antilles arc (1518 analyses; Brown *et al.* 1977) is 27 ppm. Average values for basalts and andesites of the three principal Lesser Antilles islands are: Grenada, 23 and 43 ppm Rb respectively; Dominica, 16 and 46 ppm; St Kitts, 5 and 13 ppm. A suite of 72 basaltic lavas of Aoba and Ambrym in Vanuatu contain an average of 24 ppm Rb, and four associated andesites an average of 45 ppm. Mean Rb in the 119, mainly andesitic, lavas of Bougainville is 43 ppm.

Mean Rb values for the Solomons olivine–pyroxene-basalts, big-feldspar lavas, and hornblende-andesites are 13.9, 27.9, and 26.0 ppm respectively, and 23.0, 36.0, and 29.6 ppm for their groundmasses (Table 20.2). Whole-rock and groundmass frequency distributions (Fig. 20.1) are very similar to those of K_2O in the same materials (Fig. 5.9): there is a substantial overlap of whole-rock and groundmass distributions in each case, a clear tendency for partitioning into the melt (groundmass), and the development of maxima in the big-feldspar lavas rather than in the hornblende-andesites as might have been expected.

Table 20.2 Mean values of rubidium and zirconium as compared with lead and K_2O in the Solomons lavas and their groundmasses

	Rb	Zr	Pb	K_2O
OPB: whole rocks	13.9	14.8	2.5	1.05
OPB: groundmasses	23.0	20.0	3.9	1.66
BFL: whole rocks	27.9	73.3	4.2	1.95
BFL: groundmasses	36.0	92.6	5.3	2.47
HbA: whole rocks	26.0	87.9	5.6	1.29
HbA: groundmasses	29.6	87.5	6.8	1.81

OPB = olivine–clinopyroxene-basalts; BFL = big-felspar lavas; HbA = hornblende-andesite group. Rb, Zr, and Pb given as parts per million by mass; K_2O per cent by mass.

With its ionic radius of 0.149 nm, Rb^+ tends to occupy K^+ (0.133 nm) sites (Goldschmidt 1954; Patterson and Swaine 1955) and in doing so develops a correlation not only with potassium but also with Pb^{2+} (0.132 nm) (Fig. 20.2). The $Rb:K_2O$ correlation (Fig. 20.3) is, however, of distinctly higher degree than that between Rb and Pb. In both cases closer correlation appears at the lower levels of the relevant elements. Rb:Pb exhibits a diffuse linear relationship to ≈Rb = 30 ppm: Pb = 5 ppm. $Rb:K_2O$ is better defined, but tends to become diffuse above Rb = 20 ppm: K_2O = 1.5 per cent. This diffuseness resolves, however, into three highly correlated sets when points representing the three lava groups are plotted separately (Fig. 20.3(a), (b), and (c)). Coefficients of

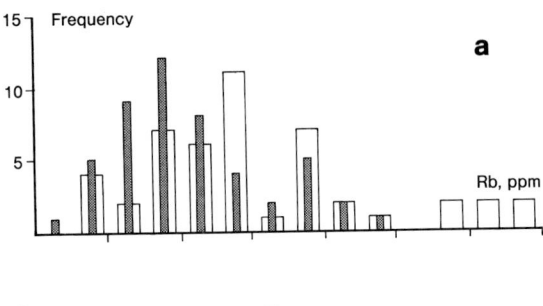

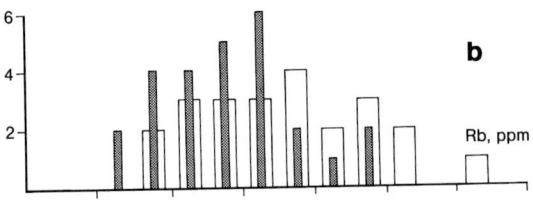

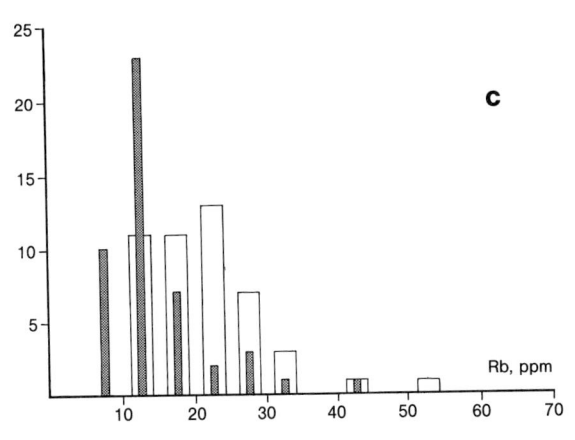

Fig. 20.1. Frequency distributions of rubidium abundances in whole rocks (filled bars) and groundmasses (open bars) of (a) lavas of the hornblende-andesite group, (b) big-feldspar lavas, and (c) olivine–clinopyroxene-basalts.

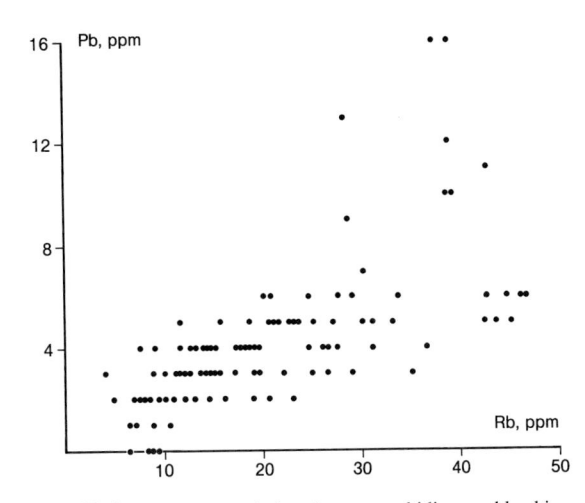

Fig. 20.2. Abundance relations between rubidium and lead in the lavas of the Solomons suite.

correlation of Rb:K$_2$O for the hornblende-andesites, big-feldspar lavas, and olivine–pyroxene-basalts are 0.825, 0.907, and 0.961 respectively. The Savo lavas (Fig. 20.3(a), filled circles), already noted as containing distinctly higher K$_2$O and Pb than those of Gallego (Chapter 8), also contain the higher Rb. The sharp increase in lead relative to rubidium shown by some lavas with Rb >25 ppm might be taken as indicating that Rb:Pb relations conform to a curve. On the other hand, it may simply reflect a breakdown in the correlation at Rb levels above about 25 ppm.

Mean rubidium in the principal potassium-bearing silicates — feldspar and hornblende — and in the clinopyroxenes of the three principal lava groups (Table 20.3) confirm the general diadochic relationships between rubidium and potassium in the Solomons minerals. Comparison of the 'ultimate purity' and 'penultimate purity' feldspar separations indicates that rubidium, like potassium and lead (see Chapters 5 and 8), tends to partition into the glassy inclusions where these occur in feldspar. Mean distribution coefficients developed between the principal minerals and their coexisting melts are given in Table 20.4. All are less than unity (and most are <0.1), leading to strong partitioning of rubidium into the melt throughout the crystallization history of the melt.

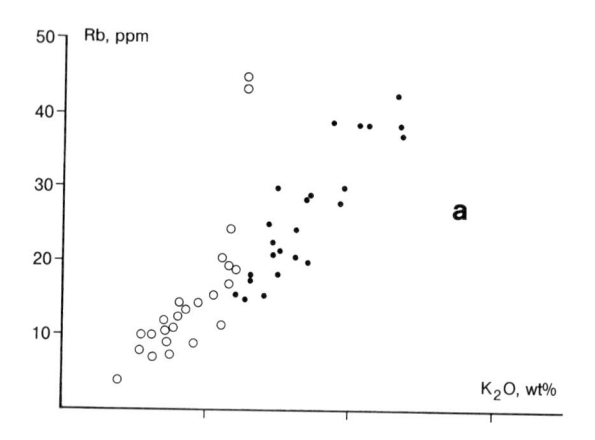

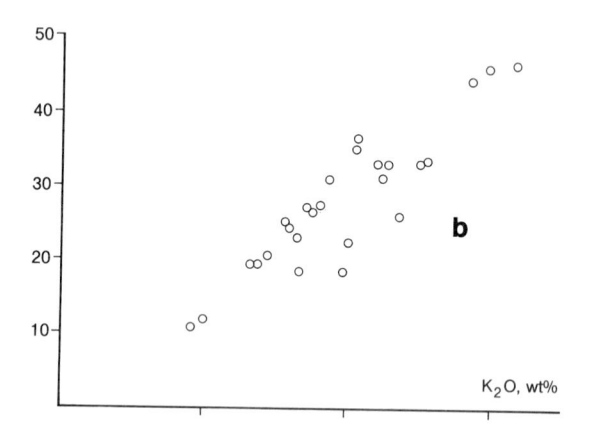

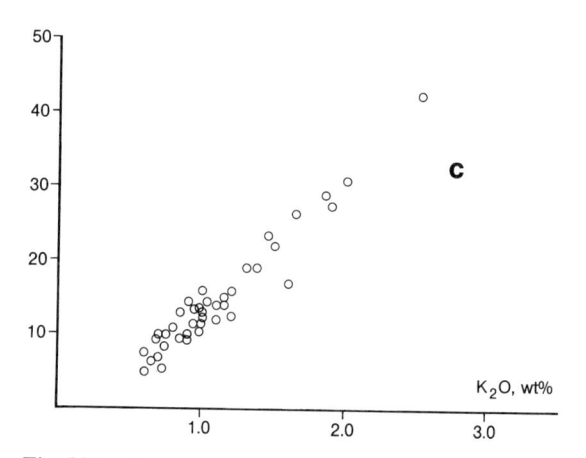

Fig. 20.3. Abundance relations between rubidium and K$_2$O in lavas of (a) the hornblende-andesite group (filled circles, Savo volcano: open circles, Gallego volcano); (b) the big-feldspar lavas; and (c) olivine–clinopyroxene-basalts of the Solomons suite.

Table 20.3 Mean values of rubidium and zirconium as compared with lead and K$_2$O in the principal minerals of the Solomons lavas

	Rb	Zr	Pb	K$_2$O
Olivine (OPL)	0.23	2.63	nd	0.02
Olivine (BFL)	0.56	10.55	nd	0.04
Clinopyroxene (OPL)	0.34	12.78	nd	0.05
Clinopyroxene (BFL)	1.14	26.07	nd	0.04
Clinopyroxene (HbA)	0.74	36.09	nd	0.04
Hornblende (HbA)	2.07	47.85	1.8	0.45
Plagioclase (OPL)				
up	0.50	12.60	1.2	0.07
pup	1.50	17.75	2.3	0.14
Plagioclase (BFL)				
up	1.43	25.86	2.5	0.38
pup	3.47	28.17	2.6	0.49
Plagioclase (HbA)				
up	5.27	15.11	3.6	0.44
pup	7.64	16.82	3.4	0.64
Spinel (OPL)	2.33	43.70	13.0	—
Spinel (BFL)	1.21	36.79	6.0	—
Spinel (HbA)	2.58	142.32	18.0	—

OPL = olivine–clinopyroxene-basalts; BFL = big-felspar lavas; HbA = hornblende-andesite group; up = ultimate purity plagioclase separations; pup = penultimate purity plagioclase separations. Rb, Zr, and Pb given as parts per million by mass; K$_2$O per cent by mass. nd = not determined.

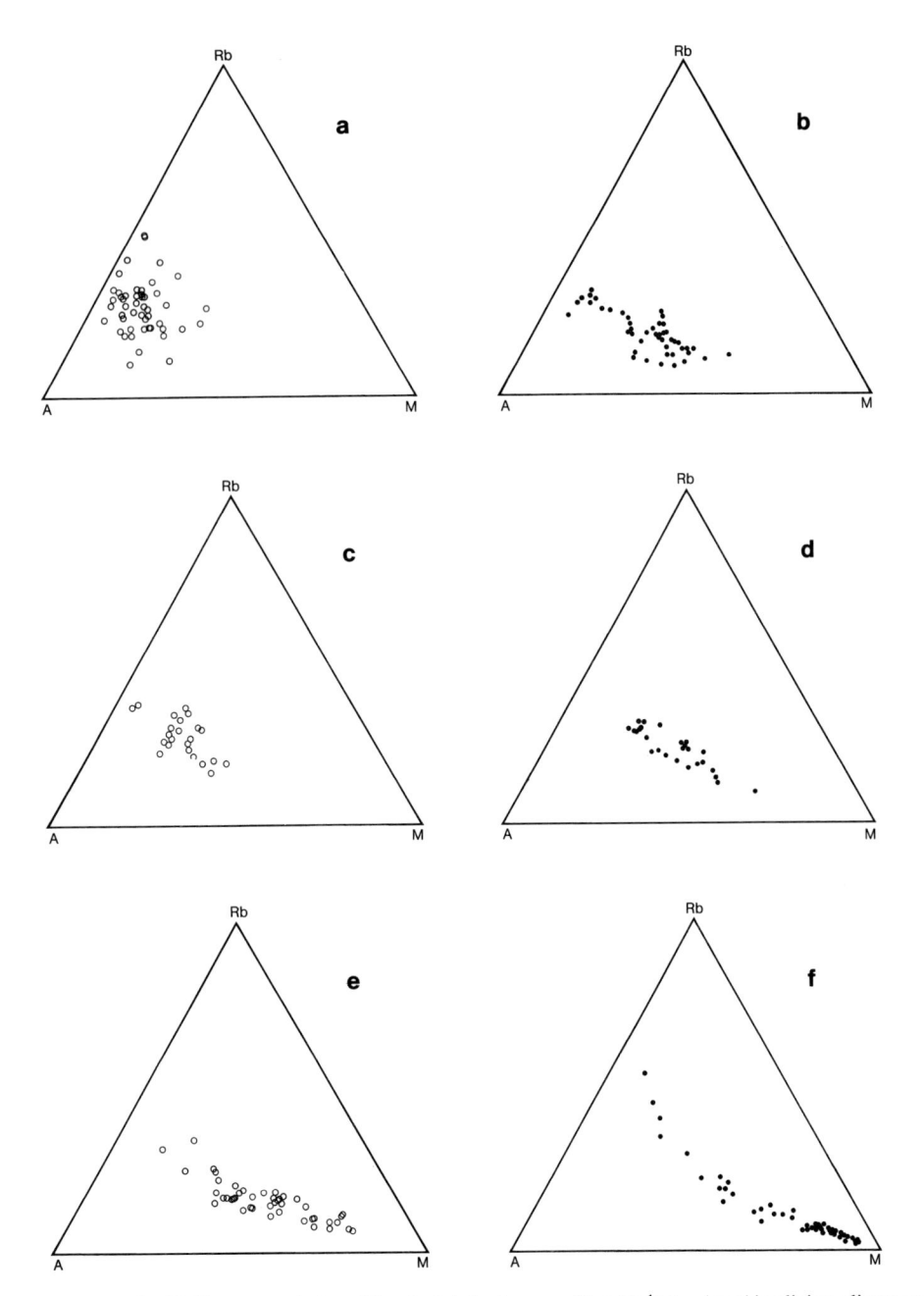

Fig. 20.4. Rb–MgO–(Na$_2$O + K$_2$O) ternary relations (Rb calculated using ppm Rb × 10^{-1}) developed in olivine–clinopyroxene-basalt, big-feldspar lavas, and hornblende-andesite whole rocks [filled circles: (f), (d), and (b) respectively], and their corresponding groundmasses [open circles: (e), (c), and (a) respectively].

Rb–MgO–(Na$_2$O + K$_2$O) relations (Fig. 20.4) are highly systematic, generally similar to those of PbMA (Fig. 8.8), and show a tendency — particularly, though somewhat surprisingly, in the large span generated by the olivine–clinopyroxene-basalts (Fig. 20.4(f)) — to a hypertholeiitic nature. This is confirmed by the plot of

270   Ore Elements in Arc Lavas

Table 20.4 Crystal: melt distribution coefficients (K_D^{Rb} and K_D^{Zr}) with respect to rubidium and zirconium

Mineral species	Lava group	n_1	n_2	m_1^{Rb}	m_2^{Rb}	K_D^{Rb}	m_1^{Zr}	m_2^{Zr}	K_D^{Zr}
Olivine	OPB	40	44	0.23	23	0.0100	2.63	20	0.13
	BFL	7	23	0.56	36	0.0156	10.55	93	0.11
Clinopyroxene	OPB	43	44	0.34	23	0.0148	12.78	20	0.64
	BFL	26	23	1.14	36	0.0317	26.07	93	0.28
	HbA	27	47	0.74	30	0.0247	36.09	87	0.41
Hornblende	HbA	25	47	2.07	30	0.0690	47.85	87	0.55
Feldspar	OPB	10	44	0.50	23	0.0217	12.60	20	0.63
	BFL	23	23	1.43	36	0.0397	25.86	93	0.28
	HbA	46	47	5.27	30	0.1757	15.11	87	0.17
Spinel	OPB	10	44	2.33	23	0.1013	43.70	20	2.19
	BFL	23	23	1.21	36	0.0336	36.79	93	0.40
	HbA	37	47	2.58	30	0.0860	142.32	87	1.64

n_1 = total number of mineral separations analysed; n_2 = total number of groundmass separations analysed; m_1 = mean rubidium/zirconium content of mineral of each lava group; m_2 = mean rubidium/zirconium content of groundmass of each lava group; K_D^{Rb} and K_D^{Zr} = mean distribution coefficients determined as m_1/m_2.

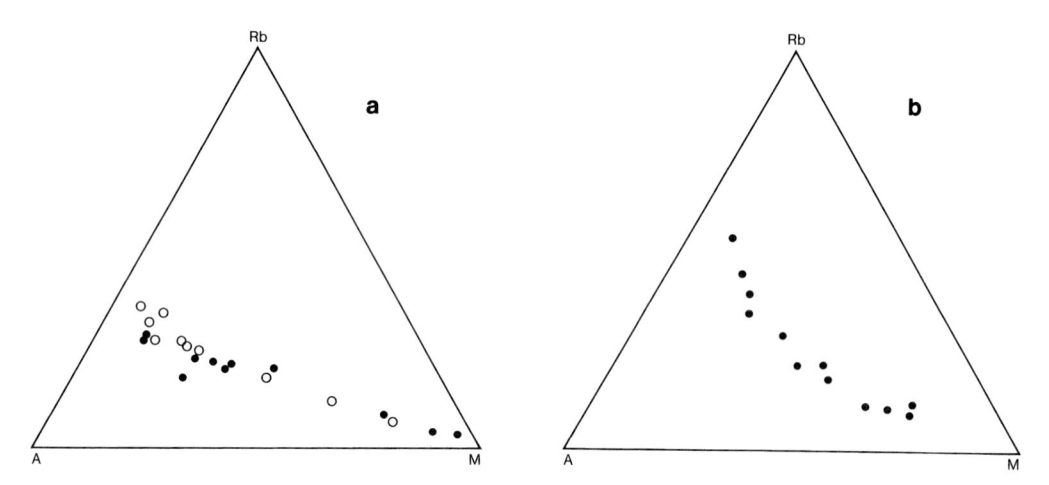

Fig. 20.5. Mean Rb–MgO–(Na$_2$O + K$_2$O) ternary relations corresponding to 2 per cent whole-rock SiO$_2$ intervals in (a) whole rocks (filled circles) and corresponding groundmasses (open circles) of the Solomons suite, and (b) the lavas of Grenada.

Fig. 20.5(a), which shows mean RbMA relations in whole rocks and their groundmasses corresponding to 2 per cent intervals of whole-rock SiO$_2$. The groundmasses (open circles) show a very slight upward inflection at the low-MgO end of the distribution, indicating what might be best referred to as an 'incipient' hypertholeiitic trend. By comparison the lavas of Grenada exhibit a very pronounced hypertholeiitic pattern (Fig. 20.5(b)), which, as with the other incompatible elements Pb, Ba, and Sr, is probably general for Rb in extensively differentiated series of igneous rocks. The increasing abundance of Rb in the melt with increase in SiO$_2$ (the latter as an indicator of progressive differentiation) is shown in Fig. 20.6(a). Rb:SiO$_2$ relationships in the lavas of the three principal Lesser Antilles islands are shown in Fig. 20.6(b) for comparison. The similarity between Rb:SiO$_2$ and the

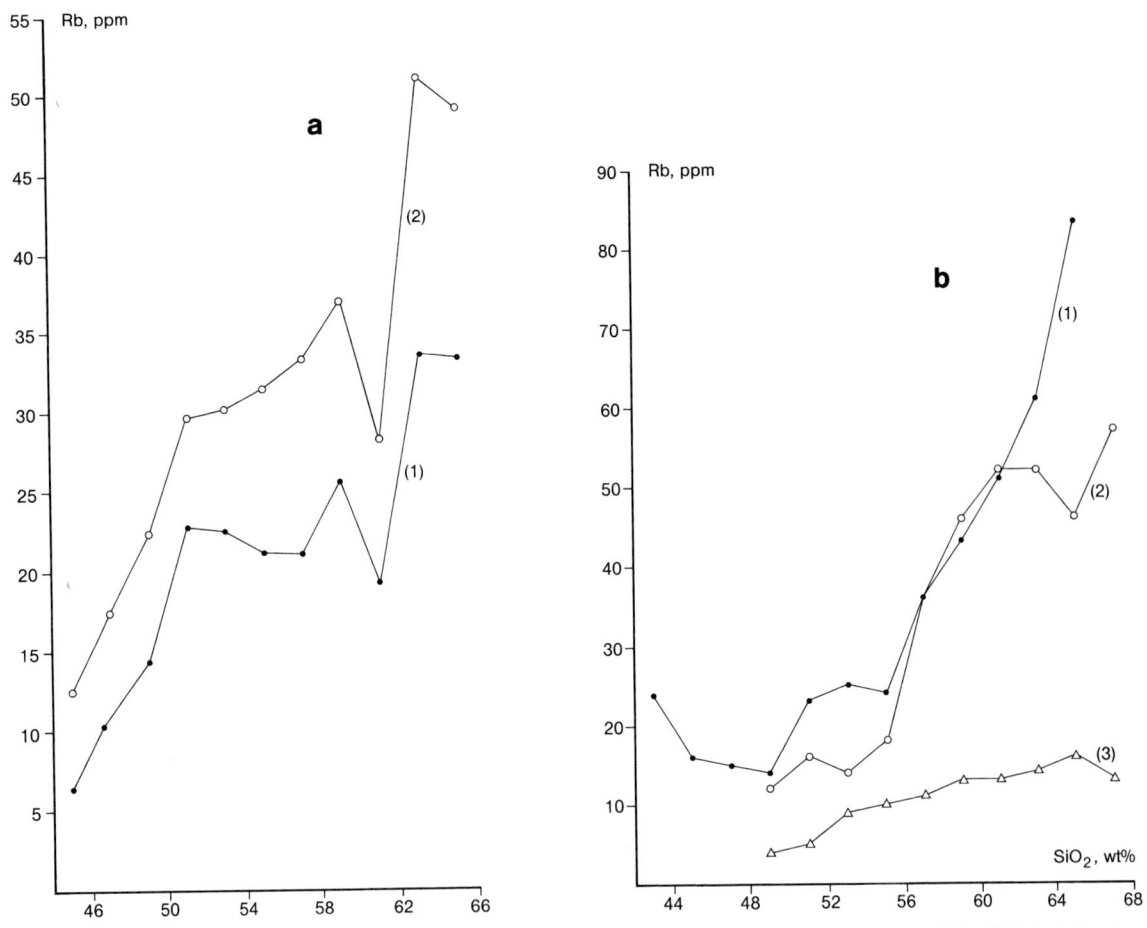

Fig. 20.6. Mean Rb:SiO$_2$ relations corresponding to 2 per cent whole-rock SiO$_2$ intervals in (a) whole rocks [(1) filled circles] and corresponding groundmasses [(2) open circles] of the Solomons suite; and (b) the lava suites of Grenada [(1) filled circles], Dominica [(2) open circles], and St Kitts [(3) open triangles] in Lesser Antilles.

corresponding graphs for K$_2$O and Pb (Fig. 20.7) and Ba and Sr (Figs. 9.13 and 10.16 respectively) is striking, if expected.

Zirconium. The mean abundance of zirconium in the Earth's crust is ≈162 ppm (Emsley 1989), and in igneous rocks ≈155 ppm Zr (Erlank *et al.* 1978). Mean Zr in MORBs as represented by the southern extremity of the Mid-Atlantic Ridge and associated structures is 117 ppm. The mean Zr content of the Lesser Antilles arc lavas (Brown *et al.* 1977) is 107 ppm. Average values for basalts and andesites of the three principal islands are: Grenada, 96 and 169 ppm Zr respectively; Dominica 44 and 102 ppm; St Kitts 65 and 96 ppm. Averages in Vanuatu basalts and andesites are 65 and 134 ppm respectively. Mean zirconium in the mainly andesitic assemblage of Bougainville is 101 ppm.

Average zirconium contents of the three principal Solomons lava groups and their groundmasses are given in Table 20.2, and frequency distributions of individual values in Fig. 20.8. Zirconium increases sharply from the olivine–pyroxene-basalts to the members of the big-feldspar lava group, but rate of increase then tapers off with the development of the hornblende-andesites. This change in the abundance behaviour of zirconium as the lavas became more highly felsic is emphasized by the fact that not only do the groundmasses of the hornblende-andesites contain slightly lower mean Zr than the

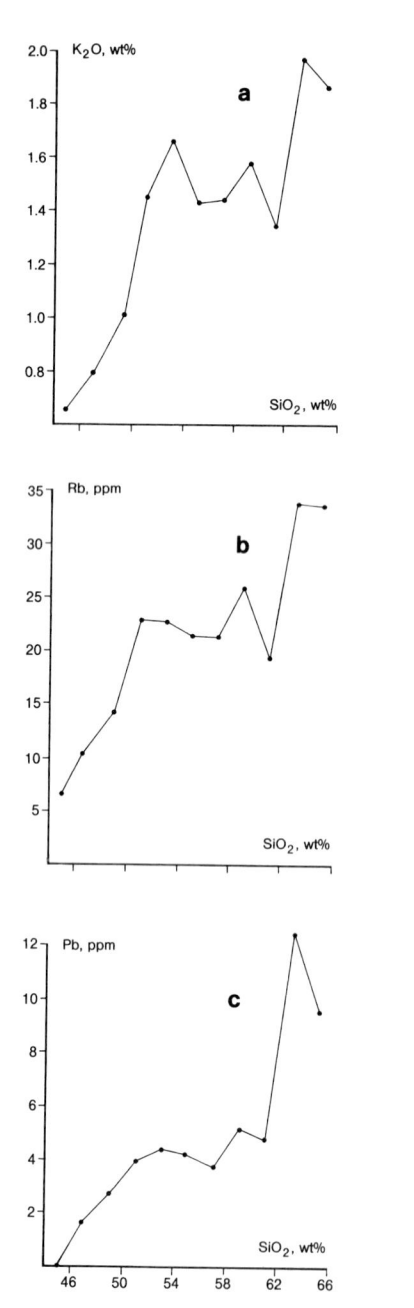

Fig. 20.7. Mean (a) K$_2$O:SiO$_2$, (b) Rb:SiO$_2$, and (c) Pb:SiO$_2$ relations corresponding to 2 per cent whole-rock SiO$_2$ intervals in the 122 samples of the Solomons suite, showing the similarities in abundance patterns generated by these three 'incompatible' elements, two of them lithophile, one chalcophile, as the lava series has evolved. Such evidence emphasizes the close ties between the 'ore' elements and the 'normal' components of the melt during the geochemical evolution of the latter.

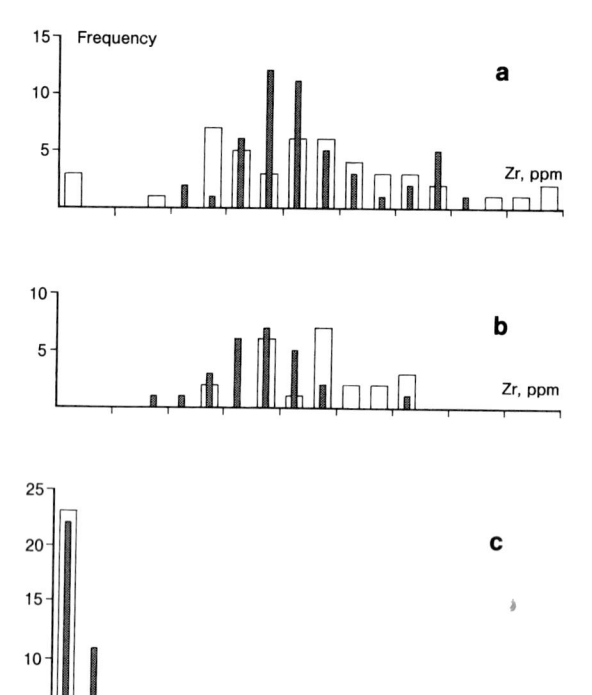

Fig. 20.8. Frequency distributions of zirconium abundances in whole rocks (filled bars) and groundmasses (open bars) of (a) lavas of the hornblende-andesite group, (b) big-feldspar lavas, and (c) the olivine–clinopyroxene-basalts of the Solomons suite.

corresponding whole rocks, but they also contain less Zr overall than the groundmasses of the big-feldspar basalts and basaltic andesites.

Goldschmidt (1954) observed that the two elements zirconium and hafnium, with atomic numbers 40 and 72, constitute a remarkable example of a pair of elements displaying almost complete chemical similarity in spite of very different total electron numbers and atomic weights. With their very similar ionic radii, Zr^{4+} (0.087 nm) and Hf^{4+} (0.084 nm) may develop a high degree of diadochic substitution. This is shown well by the incidence of the two elements in the condensed suite of Solomons lavas (Table 20.1; Fig. 20.9). A conspicuous feature of Zr:Hf relations is that the full condensed series, ranging from 46 <SiO$_2$ <66, and embracing the complete spectrum from picrites to dacites, plots (Fig. 20.9) as a single linear set (coefficient of correlation $r = 0.96$), with no inflexion marking the change from olivine-basalts to hornblende-andesites. This is a striking feature, and is referred to again in Chapter 23.

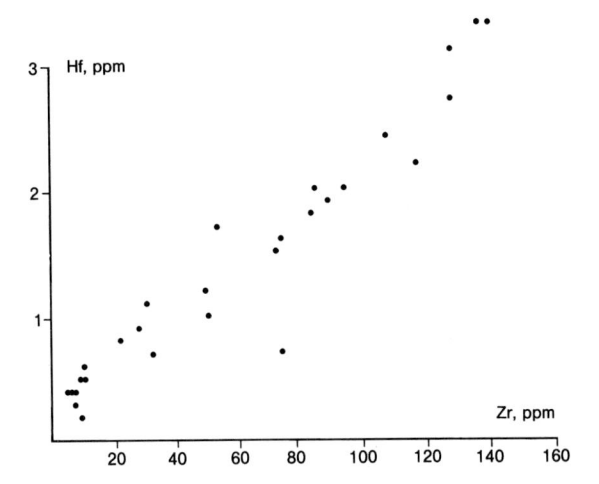

Fig. 20.9. Hafnium:zirconium relations in the 28 members of the Solomons condensed suite. The SiO$_2$ content of the lavas ranges from 46.20 to 65.36 per cent. Coefficient of correlation of the distribution $r = 0.962$.

Of greater concern in the present context, however, is the abundance behaviour of zirconium with respect to lead (Fig. 20.10). The tendency for lead to occur as an abundant trace element in zircons is the basis for the use of this mineral in lead dating of relatively old rocks, and it is clear that lead is quite readily accommodated, perhaps largely as an interstitial ion, in the zircon structure. In older rocks, particularly those of the Palaeozoic and Precambrian eras, it might have been suspected that any correlation between zirconium and lead was derivative: that initially U^{4+} (0.097 nm) substituted for Zr^{4+} (0.087 nm) in the zircon structure, leading to a U:Zr correlation in the whole rock, and that uranium then underwent radioactive decay to lead, generating an 'inherited' Pb:Zr correlation. However, although some of the lead of Solomons zircons must have been derived in this way, the lavas are far too young (Chapter 4) for any significant proportion of their lead to have been generated by radioactive breakdown. The size difference (10 per cent) between U^{4+} and Zr^{4+} is appropriate for extensive substitution; it is such that U^{4+} can substitute with little lattice distortion — and no charge imbalance — and hence to a quite significant extent. The difference between the radii of Pb^{2+} and Zr^{4+} (34 per cent) is too great for simple substitution, and the observed diffuse Pb:Zr correlation therefore must be due either to interstitial solid solution of lead in zircon, or to a casual, loose, association of lead and zirconium, resulting quite simply from a shared incompatibility in most of the common silicate structures, in the residual melt as this evolved.

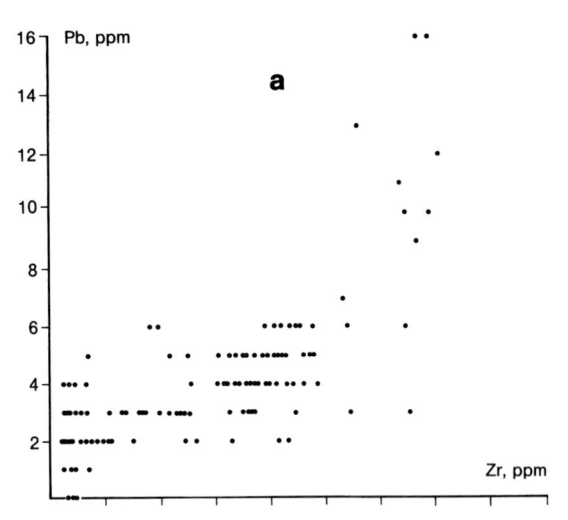

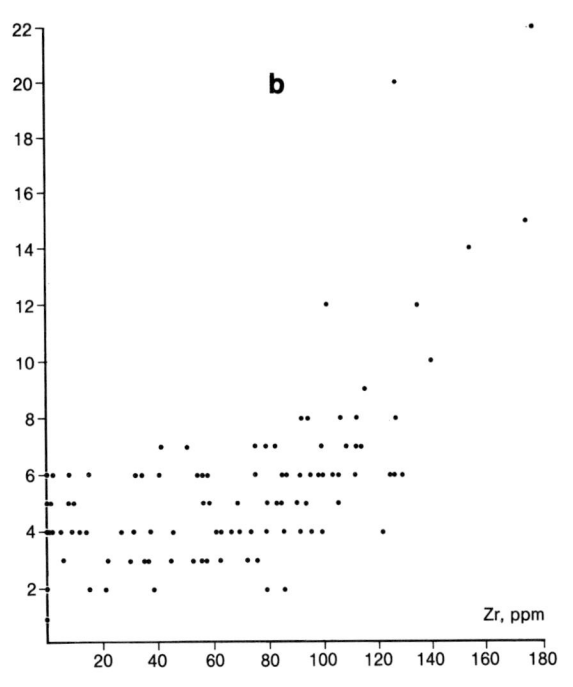

Fig. 20.10. Lead versus zirconium in (a) whole rocks, and (b) corresponding groundmasses of the Solomons suite.

Mean zirconium in the principal minerals ranges from 2.6 ppm in the olivines of the olivine–pyroxene-basalts to

142.3 in the magnetites of the hornblende-andesites (Table 20.3). Among the ferromagnesian silicates, hornblende is the principal zirconium receptor overall. $K_D^{Hb:cpx}$ in the hornblende-andesites is 1.33, and $K_D^{cpx:ol}$ in the olivine–pyroxene-basalts is 4.86. $K_D^{mag:Hb}$ in the hornblende-andesites is 2.97. A notable feature of all mineral groups (other than hornblende, which occurs only in the hornblende-andesites) is that zirconium content increases with increase in the felsic nature of the host rock. Mean distribution coefficients developed between the principal minerals and their coexisting melts are given in Table 20.4. Apart from the spinels of the olivine–clinopyroxene-basalts and the hornblende-andesite group, all are less than unity, so that by far the major part of the crystallization process induces concentration of zirconium in the melt.

The pattern of Zr–MgO–(Na$_2$O + K$_2$O) relations in the Solomons suite (Fig. 20.11) is very similar to those of analogous ternary groupings involving rubidium and strontium (Figs. 20.5 and 10.15 respectively) and, to a lesser extent, those for lead and barium (Figs. 8.9 and 9.12 respectively). Plotting of mean ZrMA values corresponding to 2 per cent whole-rock SiO$_2$ intervals as in Fig. 20.11 indicates quite clearly the approximately linear pattern of the whole rocks, though two groundmasses (corresponding to 46–8 and 48–50 per cent whole-rock SiO$_2$) have comparatively low proportionate zirconium contents. The analogous diagram for Grenada (Fig. 20.11(b)) indicates pronouncedly high comparative zirconium in the picrites, but apart from this a steep and steady increase in the proportion of zirconium with increase in (Na$_2$O + K$_2$O) and hence SiO$_2$. The two lava series may be reasonably described as tholeiitic with respect to zirconium.

In spite of the relatively low zirconium content of some of the Solomons groundmasses in the ZrMA system, Zr:SiO$_2$ relations in whole rocks and groundmasses are remarkably similar (Fig. 20.12) to those exhibited by strontium, barium, lead, potassium, and rubidium, i.e. the five important incompatible elements. Zr:SiO$_2$ relations in the Grenada suite (Fig. 20.12(b), curve (1)) reflect the relatively high zirconium of the more mafic members indicated in the ternary ZrMA diagram. A low-Zr trough occurs at 46–8 per cent SiO$_2$, after which Zr increases more or less uniformly with increase in SiO$_2$. The curves for Dominica (Curve (2)) and St Kitts (curve (3)) indicate a general lower level of zirconium than in the Grenada lavas, and fairly uniform increases of zirconium from SiO$_2 \approx 48$–50 per cent.

Arsenic, antimony, bismuth. Compounds of arsenic, antimony, and bismuth have long been recognized as components of modern volcanic sublimates and, as noted earlier in this chapter, all three commonly appear as traces in exhalative ores.

The mean crustal abundance of arsenic is 1.8 ppm and of igneous rocks ≈ 1.5 ppm. The mean arsenic content of basalts is ≈ 1.5 ppm, of andesites ≈ 2.0 ppm, and of rhyolitic lavas ≈ 2.7 ppm. Present estimates of the mean crustal abundance of antimony are ≈ 0.20 ppm, and the sparse data indicate that mean abundances in the basalts

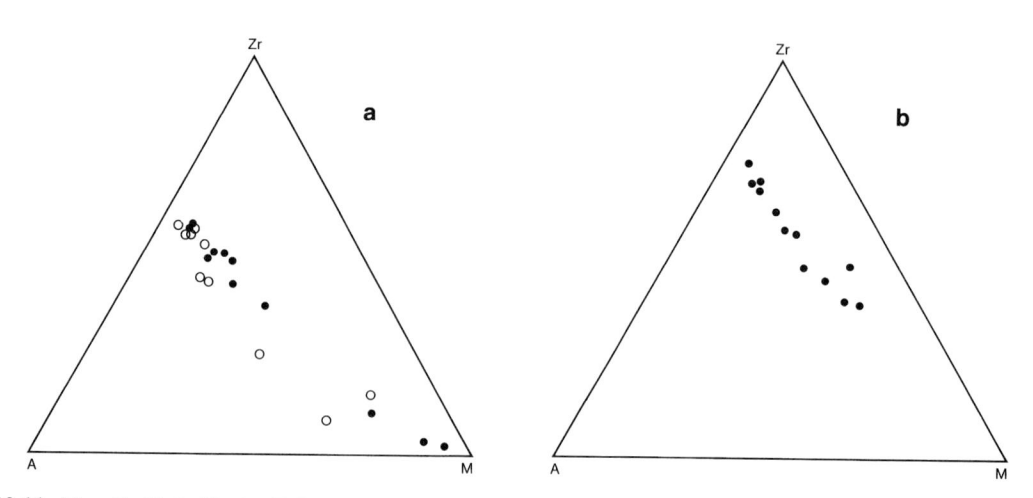

Fig. 20.11. Mean Zr–MgO–(Na$_2$O + K$_2$O) ternary relations (Zr calculated using ppm Zr $\times 10^{-1}$) corresponding to 2 per cent whole-rock SiO$_2$ intervals in (a) whole rocks (filled circles) and corresponding groundmasses (open circles) of the Solomons suite, and (b) the lava series of Grenada.

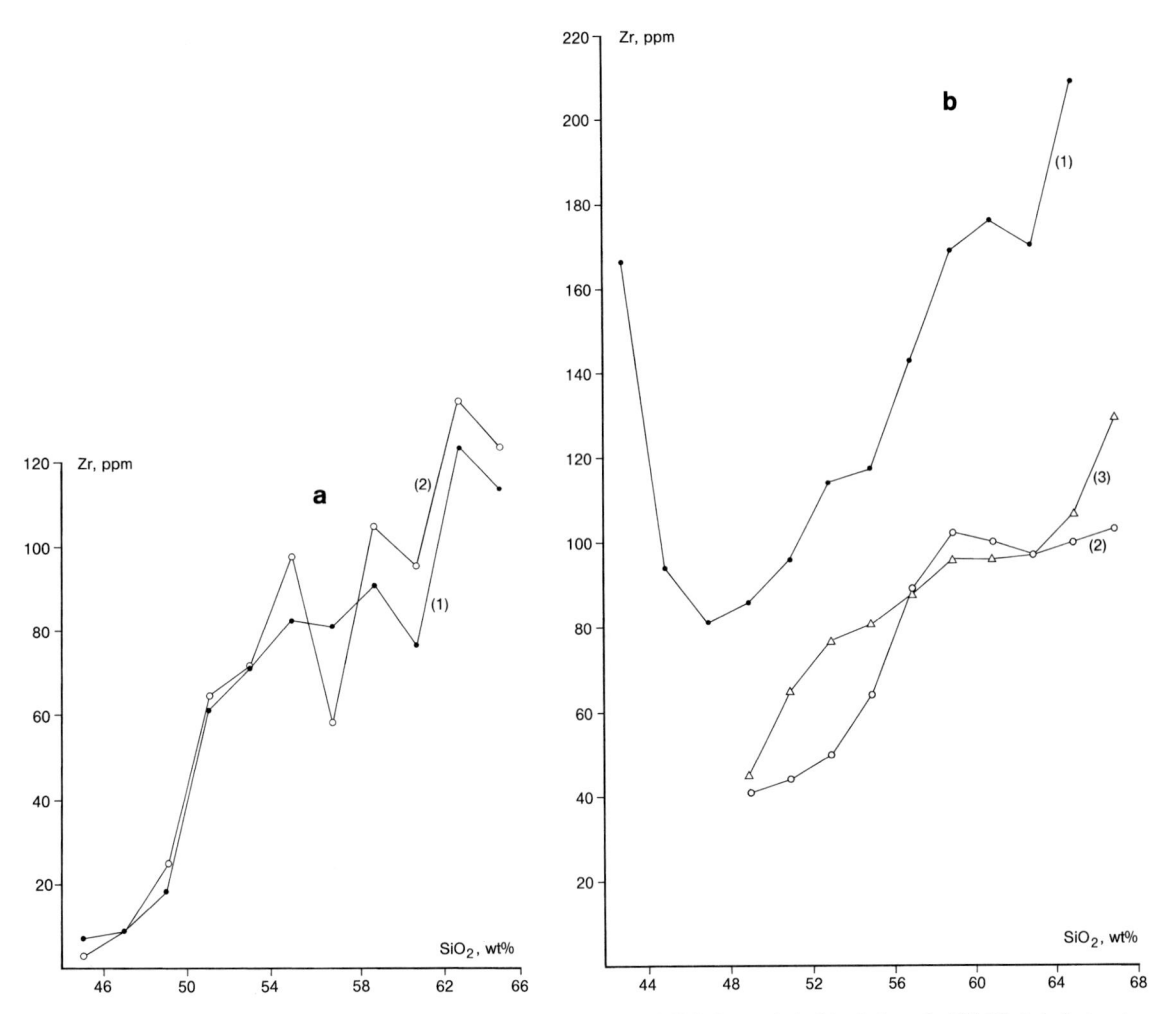

Fig. 20.12. Mean Zr:SiO$_2$ relations corresponding to 2 per cent whole-rock SiO$_2$ intervals in (a) whole rocks [(1) filled circles] and corresponding groundmasses [(2) open circles] of the Solomons suite, and (b) the lava suites of Grenada [(1) filled circles], Dominica [(2) open circles], and St Kitts [(3) open triangles] in the Lesser Antilles.

are ≈0.1–0.2 ppm and in andesites ≈0.2 ppm Sb. Bismuth is even less abundant: the current estimate of its crustal abundance is 0.008 ppm (Emsley 1989). Values of 0.01–0.80 ppm have been reported for basalts and 0.02–0.90 ppm for more felsic lavas.

Any antimony and bismuth present in the lavas of the Solomons condensed suite occurs at levels <0.2 ppm. The elements are thus very low throughout the full spectrum from picrites to dacites. Arsenic is present in measurable quantities in most of the samples, but is very low none the less. The overall mean for the 28 samplings is 0.9 ppm As, and the range is from <0.4 ppm (4 samples) to 2.4 ppm As. As indicated in Table 20.1, there is a general increase in arsenic with increase in whole-rock SiO$_2$, but the field of the As:SiO$_2$ plot is in fact very scattered and any correlation is of very low degree. As with Pb, Rb, K, Zr, S, F, and Cl, arsenic is higher in the lavas of Savo (mean 1.4 ppm As) than in those of the contemporary Gallego volcano (mean = 0.9 ppm As).

Molybdenum. Although common as a trace in volcanic exhalative ores, molybdenum rarely if ever occurs as a significant component of deposits of this type. Its principal habitat in the volcanic regime is that of sub-volcanic porphyries, in which it may occur, in amounts ≈0.05 per cent Mo, as the principal economic element of 'porphyry molybdenum' deposits. There can be little doubt that in such settings the molybdenum is of volcanic derivation and has been emplaced by volcanic processes.

The current estimate of mean molybdenum abundance in the Earth's crust is 1.2 ppm (Emsley 1989). Manheim (1978) noted that 73 per cent of all reliable analyses of molybdenum in igneous rocks gave Mo = 0–1 ppm, and that the overall range was 0–15 ppm Mo. Basalts appear to contain mean Mo ≈1.5 ppm, andesites ≈1.3 ppm. Peralkaline lavas have been found to contain 13–15 ppm (see Manheim 1978).

With its ionic radius of 0.092 nm, Mo^{2+} might be expected to appear in those volcanic silicate lattice sites that also accommodate Mn^{2+} (0.091 nm) and perhaps Zn^{2+} (0.083 nm) and Fe^{2+} (0.082 nm). While the 'condensed suite' of 28 Solomons lavas constitutes a very small sampling, it gives no indication that the abundance of molybdenum is related to the incidence of these elements in the whole rocks.

Within this limited group molybdenum does, however, exhibit a marked relationship with whole-rock SiO_2 (Fig. 20.13). In a manner perhaps analogous in principle to the behaviour of copper and zinc with respect to increasing whole-rock SiO_2 (Figs. 6.12 and 7.23), molybdenum increases from a level of ≈0.1–0.2 ppm in the most mafic lavas to a peak of 2.3 ppm in the mafic andesites (SiO_2 = 54.3 per cent), then falls sharply to ≈0.5 ppm in lavas in the 55–6 per cent SiO_2 interval. This general level is then maintained into the dacite range.

In the absence of molybdenum estimations for the relevant minerals it cannot be determined whether the sudden decrease in molybdenum at 54–5 per cent SiO_2 results from crystal subtraction or from some other mechanism such as volatile loss. It may, however, be significant that this rise and fall of molybdenum relative to whole-rock silica is rather similar to (and indeed even better developed than) the Cu–SiO_2 relationship. The copper of at least many of the hundreds of porphyry copper deposits now known is recognized as a product of sub-volcanic volatile loss, quickly recaptured in the chilled and cracked upper crust of the intrusion concerned. As the principal volcanic minerals accept little copper into their structures (Chapter 6), such volatile loss may explain the rise and then sudden decrease in abundance of copper as the melt differentiates, evolves, and becomes increasingly SiO_2-rich. While the problem clearly requires further investigation via determination of molybdenum in the principal volcanic minerals, analogy suggests that Mo–SiO_2 abundance relations may have been induced partly by volatile loss — and that we may be seeing here a manifestation of a process that has led, elsewhere, to the formation of sub-volcanic porphyry molybdenum deposits. (The writer prefers to follow the course of candidness concerning Fig. 20.13, and acknowledges his suspicion that its sequence of Mo:SiO_2 relations looks almost too good to be true. The capacity for small samples occasionally to yield apparently unambiguous, and pretty, results of this kind is too well known to require emphasis. The Mo:SiO_2 tie in this case clearly requires much more thorough investigation.)

Gold and silver. If similarity of charge and ionic size were the determining factors, it might be expected that gold (Au^+ = 0.137 nm) would tend to occupy potassium sites in the volcanic silicates and residual melt structures, and hence would show a general abundance relationship with K_2O. While there is some indication of a very low degree of correlation of the two in the Solomons condensed suite, there is no indication of the kind of unequivocal relationship such as is developed by each of rubidium and lead with potassium.

On the basis of charge and ionic size there is little reason to expect systematic substitution of Ag^+ (0.113 nm) for any of the common ions of the volcanic silicates. This is borne out by the analytical data. There is some indication of muted correlation with Th^{4+} (0.099 nm), but no apparent correlation of silver with gold, or of either of these with copper.

The abundances of gold and silver, like those of copper and molybdenum, do, however, display ties with the whole-rock SiO_2 contents of the series of host lavas. The variation diagrams for gold and silver show greater scatter than that for molybdenum, but the patterns are reasonably clear none the less: gold peaks at ≈54–5 per cent SiO_2, silver at ≈51–2 per cent SiO_2, and, for comparison, copper in the same 28 samples peaks at ≈54–5 per cent SiO_2.

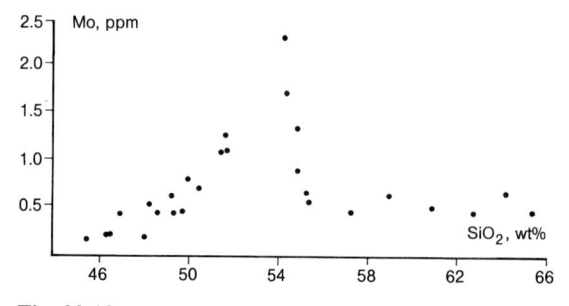

Fig. 20.13. Mean Mo:SiO_2 relations corresponding to 2 per cent whole-rock SiO_2 intervals as shown by the 28 samples of the Solomons condensed suite.

Cadmium. The occurrence of notable quantities of cadmium in natural zinc sulphides is one of the better-known mineral–chemical associations, and such 'camouflaged' cadmium constitutes the principal source of this metal. Abundances up to ≈0.5 per cent Cd in sphalerite are not uncommon, the development of such concentrations being facilitated by the isostructural nature of both cubic and hexagonal close-packed forms of the zinc and cadmium sulphides. This close association of the two elements in sulphide ore deposits — including those of volcanic–exhalative variety — might be taken to indicate the possibility of a systematic association at source. Such a possibility is, of course, reinforced by their membership of the same group of the periodic table and their resulting close similarity of chemical properties.

Their ionic sizes are, however, rather different. Taking Zn^{2+} = 0.083 and Cd^{2+} = 0.103 nm, we find a difference in ionic radius of 19.4 per cent of the larger ion. The difference between Cd^{2+} and Fe^{2+}, the principal camouflage element for Zn^{2+} in the volcanic silicates, is 20.4 per cent of the larger ion. Such size differences are not conducive to the coexistence of trace Zn^{2+} and Cd^{2+} in similar sites within the ferromagnesian minerals, and so the development of close abundance relations during igneous crystallization and differentiation processes is unlikely.

The evidence provided by the Solomons condensed suite does not contradict this: zinc:cadmium abundance relations here are random. The same is the case for indium (In^{3+}, 0.092 nm) and gallium (Ga^{3+}, 0.062 nm), two elements that occasionally occur as noteworthy traces in sphalerite. Within the sampling limits of the 28 Solomons lavas indium and gallium show no abundance relationships either between themselves or with zinc. Nor do abundances of cadmium, indium, or gallium show any tie with whole-rock SiO_2 as an index of lava evolution.

Gallium. The geochemistry of gallium, atomic number 31, is closely related to that of aluminium, atomic number 13, just as the geochemistry of germanium, 32, is closely related to that of silicon, 14 (Goldschmidt 1954). Ga^{3+} (0.062 nm) and Al^{3+} (0.057 nm) are so similar in size that most of the gallium in the lithosphere is camouflaged in aluminium-bearing minerals, particularly the feldspars. Only a very minor amount of gallium is accommodated other than in the aluminosilicates, and most of this occurs in sulphides such as sphalerite, as noted above.

The ratio of Ga:Al_2O_3 in volcanic rocks is commonly close to 1:1 on a ppm/wt per cent basis, and this is true of the Solomons suite overall. The olivine–pyroxene-basalt and big-feldspar lava groups demonstrate the phenomenon almost precisely: in the former, mean Ga = 11.40 ppm, mean Al_2O_3 = 11.41 per cent; in the latter, mean Ga = 17.63 ppm, mean Al_2O_3 = 17.76 per cent. The hornblende-andesite group diverges slightly from the pattern, with mean Ga = 19.20 ppm, mean Al_2O_3 = 17.61 per cent. The lavas containing <11.0 per cent Al_2O_3 (these constitute >60 per cent of the olivine–pyroxene-basalts and ≈22 per cent of the whole suite) exhibit a high degree of Ga–Al_2O_3 correlation (Fig. 20.14), with r = 0.96. In quite marked contrast, those containing >11.0 per cent Al_2O_3 show a much lower degree of correlation, with r = 0.63. This sharp break in the pattern of Ga:Al_2O_3 relationships may have petrogenetic connotations, and is referred to again in Chapter 23.

Tungsten. An important group of tungsten deposits currently receiving increasing recognition are the stratiform, exhalative, scheelite-bearing 'skarns'. As the name indicates, these have a conspicuous calc-silicate component, and they commonly exhibit a marine basaltic association. Although tungsten has a distinct felsic affiliation in granitic–cratonic settings, the association of sedimentary–exhalative tungsten concentrations with basaltic rocks indicates the possibility that this element may be released at a much earlier stage of melt evolution

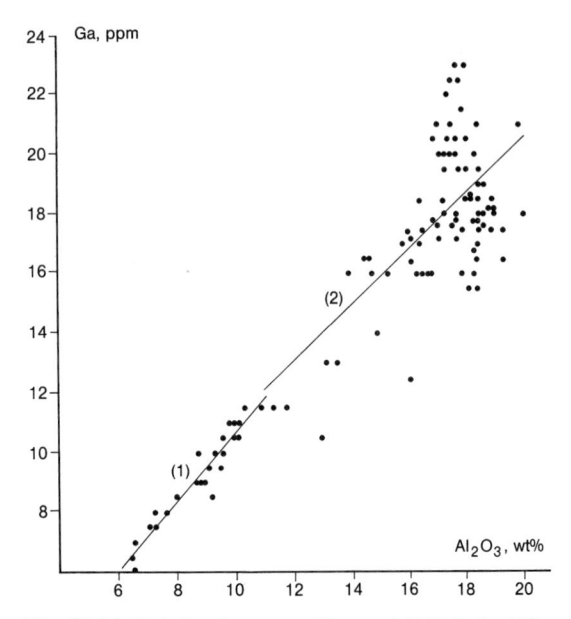

Fig. 20.14. Relations between gallium and Al_2O_3 in the 122 lavas of the Solomons suite, showing calculated regression lines for (1) all lavas (n = 27) in which Al_2O_3 <11 per cent and (2) all lavas (n = 95) in which Al_2O_3 >11 per cent. For population (1) the coefficient of correlation r = 0.962, and the regression line is a good description of the distribution of points; for population (2), r = 0.633, and the regression line is essentially meaningless.

in certain volcanic regimes. (The place of such deposits in leaching systems has not, to the writer's knowledge, received detailed consideration.)

The average crustal abundance of tungsten is low at 1.2 ppm (Emsley 1989). The very sparse data indicate that basalts and andesites probably contain ≈1.0 ppm, rhyolites ≈1.0–2.5 ppm W. Like molybdenum, tungsten achieves highest concentrations in peralkaline lavas, in which abundances may reach 15 ppm. This close similarity in abundance reflects in part the identical size (0.062 nm) and charge of Mo^{6+} and W^{6+} and also their mutual tetrahedral coordination with respect to oxygen.

On the basis of charge and size and its lithophile character, it might be suspected that W^{4+} (0.068 nm) would find principal acceptance in Ti^{4+} (0.068 nm) sites in the oxides and pyroxenes of the more mafic lavas, and hence that tungsten might show a general correlation with titanium in the whole rocks. This is not the case; for the 22 samples of the condensed suite analysed for tungsten, W–Ti abundance relations are quite random. Nor does there appear to be any tie between tungsten and whole-rock SiO_2, such as is exhibited by copper, molybdenum, and some of their associates.

Uranium. Like tungsten, uranium occurs (as uraninite) in some stratiform exhalative skarns, and under these and related circumstances is likely to have been an exhalative product. On the basis of charge, size and (again like tungsten) its lithophile character, it might be expected that U^{4+} (0.097 nm) would tend to share lattice sites with Th^{4+} (0.099 nm) and perhaps Y^{3+} (0.106 nm), and that it might be accepted into Ca^{2+} (0.106 nm) sites in, say, apatite.

The uranium:thorium affinity, and the common U:Th ratio of ≈3:1, are well known and are borne out for the most part by their abundances in the Solomons suite (Fig. 20.15). However, three of the lavas, all of them basalts or basaltic andesites from Mount Gallego on Guadalcanal, have notably high thorium contents relative to uranium and show gross divergence from the 3:1 ratio. The four Gallego lavas included in the condensed suite contain an average of 6.8 ppm Th, which compares with the overall average of 1.94 ppm Th. Gallego may well be a thorium-rich volcano, just as Savo is lead-rich.

Uranium shows no correlation with yttrium or with phosphate, indicating that it neither shares lattice sites with Y^{3+} nor substitutes significantly for Ca^{2+} in apatites. It does, however, show a modest degree of correlation (coefficient of correlation = 0.74) with zirconium, indicating the possibility, discussed above, of admission into zircon and primitive zirconium groupings in the melt structure.

The abundance relation of uranium to whole-rock SiO_2 is far from unambiguous, but there is some indication that

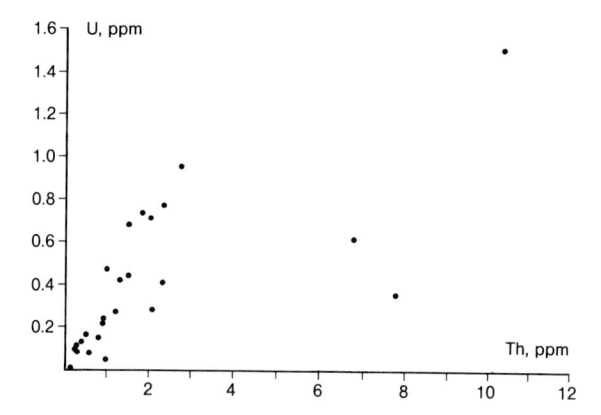

Fig. 20.15. Relations between uranium and thorium in the Solomon Islands condensed series. Apart from the three high-thorium (Th >6 ppm) samples from Mount Gallego, the sampling shows good correlation between the two elements with mean Th: U ≈ 3:1.

it rises from 0.05 ppm on the most mafic lavas to ≈1.0 ppm at SiO_2 = 51–2 per cent, then falls to a level of ≈0.5 ppm that persists for the remainder of the SiO_2 range, i.e. from 54 to 65.36 per cent SiO_2.

CONCLUDING STATEMENT

Consideration of 'non-ore', lithophile trace elements such as rubidium and zirconium, and a group of what may be termed 'minor trace elements' such as molybdenum, gold, silver, uranium, thorium, gallium, and others, yields information of relevance to the petrogenesis of both lavas and exhalative ores. In particular it links (1) lithophile elements that are generally regarded as normal, intrinsic parts of the melt, and the varying abundances of which are widely recognized as indicators of melt processes, with (2) the 'ore elements', most of which have been considered of little relevance to such processes by most igneous petrologists (see, e.g., Brown *et al.* 1977) but which constitute important components of ores associated with volcanic rocks. For example, in the low-sulphur volcanic environment the highly chalcophile large-ion element lead behaves almost precisely similarly to the large ion *lithophile* elements such as potassium and rubidium — and barium and strontium — and shows close correlation with these as the relevant lava suite evolves. Such systematic ties between the 'ore elements' on the one hand, and those such as potassium and rubidium — unquestioningly accepted as normal participants of the igneous process — on the other, serves to emphasize that the chalcophile and associated elements may be just as systematic in their behaviour and abundance in the crystal–melt–vapour milieu as any other component of an evolving lava system.

Determination of trace sulphur in the lavas confirms the microscopical evidence that this element occurs in no more than minute quantities throughout the suite. Furthermore, those traces of sulphur that do occur exhibit quite random abundance relationships with the coexisting traces of copper and zinc. It thus appears confirmed that the metal traces we have considered are indeed components of the silicates and oxides: they are not present simply as fine particles of sulphide. Some of the sulphur may occur as microscopical iron sulphides — FeS or FeS_2 — but it is suspected that most of it occurs within the silicates, as appears to be the case with at least some of the coexisting phosphorus (Chapter 11).

Rubidium displays close abundance relations with potassium and hence with lead. Degrees of correlation are highest in the more mafic lavas, decreasing conspicuously in the andesites and dacites. Correlation of lead and zirconium also breaks down in the more felsic rocks, though the well-developed linear relationship between zirconium and hafnium persists through the full lava suite. Molybdenum, and to a lesser extent gold and silver, show pronounced abundance relationships with whole-rock SiO_2 of a kind earlier shown to be exhibited by copper. Cadmium, indium, and gallium, all of which are known to occur as minor and trace constituents of exhalative sphalerites, show no correlation with zinc in the Solomons lavas. Gallium displays the expected correlation with whole-rock aluminium, though this is conspicuously best developed in those lavas with Al_2O_3 <11.0 per cent. Uranium is well correlated with thorium in the approximate ratio U:Th = 3:1 except in the lavas from Mount Gallego, which appear to be notably thorium-rich.

21

Abundance patterns in the crystallizing melt

In Chapters 6–20 we have considered the geochemical behaviour of a number of elements, some of them well known as important components of exhalative ores, in the lavas, principal minerals, and associated residual melts of the Solomon Islands Younger Volcanic Suite. Some of these elements appear to have been rapidly abstracted from the melt in early formed olivine and spinel, others to have been incorporated in somewhat later phases such as the pyroxenes and hornblende, and yet others have remained principally in the melt even at the latest stages of evolution.

We may now turn to consider the ways in which these different patterns of behaviour influence the enrichment of impoverishment of the relevant elements in the melt, and hence their availability for incorporation — and escape from the melt — in any late-stage volatile phase.

The basic principles, and results, of trace-element partitioning during fractional crystallization were laid down by V. M. Goldschmidt in 1937 and 1954 (to be later refined by Shaw (1953), Ringwood (1955), Curtis (1964), Burns (1970, Burns and Burns 1974), and others). Just prior to the publication of Goldschmidt's great work in 1954, Wager and Mitchell (1951) had presented their seminal contribution on partitioning and trace-element abundance behaviour in the highly differentiated basaltic Skaergaard intrusion. This work was added to, and many other layered intrusions considered, by Wager and Brown (1968), and since that time numerous related studies have been carried out, particularly on the more spectacular layered intrusions such as the Bushveld, Stillwater, Muskox, Great Dyke, and other, generally similar, occurrences.

In their detailed consideration of trace-element abundances in the Skaergaard layering, Wager and Brown (1968) set out in diagrammatic form three principal patterns of trace-element concentration (Fig. 21.1(a)–(c)), to which Vincent (1974) later added a fourth (Fig. 21.1(d)):

1. Elements that readily enter the early formed minerals, leading to a decrease in their amount in both crystal accumulations (the igneous rocks) and residual liquids as fractionation processes (e.g. Ni, Cr; Fig. 21.1(a)).

2. Elements that most readily enter minerals forming in the middle to somewhat later stages for fractionation: these rise to a maximum and then decrease in amount (e.g. Ti, V; Fig. 21.1(b)).
3. Elements that are not accepted, or are accepted to only a very limited extent, by the principal minerals: these elements tend to remain in the residual liquid and may become progressively more concentrated in it, leading to highest abundance in the most felsic rocks of the relevant differentiation series (e.g. Pb, Ba; Fig. 21.1(c)).
4. A minor class of rather 'inert' elements that do not readily form ions under magmatic conditions but tend to be incorporated more or less randomly as uncharged atoms (interstitials) within the crystal lattices of almost any phases crystallizing (e.g. Au, Ag; Fig. 21.1(d)).

PARAGENESIS AND CONTEXT OF CRYSTALLIZATION

A very general paragenetic sequence deduced for the five principal Solomons minerals — olivine, spinel, clinopyroxene, hornblende, and plagioclase — has already been set out in Chapter 5. Olivine is the first mineral to appear, overlapped to a minor extent by spinel and clinopyroxene. Spinel and clinopyroxene have a long span of crystallization together, clinopyroxene existing at whole-rock $SiO_2 \approx 60$ per cent and magnetite continuing on. Feldspar appears shortly after the entrance of clinopyroxene and continues to crystallize throughout the remainder of the lava series. Hornblende first appears at ≈ 49 per cent whole-rock SiO_2 (and thus overlaps extensively with clinopyroxene) and exists at ≈ 65 per cent SiO_2. The minor orthopyroxene of the Solomons suite is more or less confined to lavas in the 52–60 per cent SiO_2 range, and the very minor biotite to those in the $SiO_2 > 62$ per cent category.

The setting out of such a sequence may give the impression that crystallization is seen to have occurred under the sort of quiet, more or less uninterrupted cooling

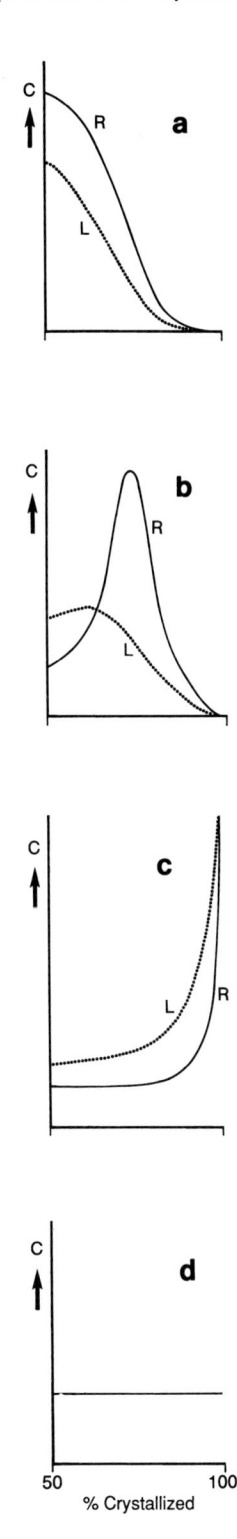

Fig. 21.1. Patterns of abundance developed in crystal accumulations (R) and residual liquids (L) by igneous trace elements (a) that readily enter the early formed minerals, (b) that most readily enter minerals formed in the middle to later stages of fractionation, (c) that are accepted to no more than a very limited extent in the principal minerals and therefore tend to accumulate in the residual melt, and (d) that tend to remain in the atomic state and are incorporated more or less randomly in all phases. After Wager and Brown (1968) and Vincent (1974).

conditions often visualized for prominently layered sequences such as that of the Skaergaard. On the other hand, it might be suspected intuitively that in a seismically highly active area such as the Solomons, in which much of the volcanism has been characterized by explosive degassing, such placid conditions of crystallization are unlikely to have pertained, or at least to have prevailed for any significant length of time. Such a suspicion is supported by the spectacular short-range variation in the compositions of early formed minerals: reference has been made earlier (Chapters 15 and 19) to gross variation of spinel and olivine compositions on the scale of millimetres and the likelihood that this has resulted from melt turbulence and the mixing of crystals formed under different conditions. That some layered intrusion chambers had in fact been sub-volcanic, and that eruption of lava and simultaneous replenishment of the melt from below may have led to substantial variation in physical and chemical conditions of crystallization, was first put forward by Brown (1956), and stated in explicit form by Wager and Brown:

The connection between layered intrusion chambers and surface volcanos was proposed (Brown, 1956) as a general petrogenetic concept, because the changing composition of lavas erupted from central-type volcanos, in contrast to fissure-type eruptions reaching the surface directly, is more than likely due to crystal fractionation and this can only take place in a fairly high-level magma chamber in which slow crystal separation is taking place. It would appear, therefore, that whereas certain layered intrusions, such as the Skaergaard, were formed by a single episode of magma emplacement and provide an example of a closed system, others, such as Rhum, formed in magma chambers lying along the path of magma on its way to the surface. Erupted magma would come from this magma chamber and fresh supplies would flow into it from below. Periods of volcanic eruption would result in periodic changes both in physical and chemical conditions in the magma chamber and, in turn, leave their imprint on the character of the crystallizing layered rocks. Such a volcano as that postulated on Rhum, would have continued to erupt basalt during the whole period represented by the exposed layered series, because the minerals found in these ultrabasic rocks are the characteristic phenocrysts of many basalts, the groundmass of which is still capable of precipitating a wide range of calcic plagioclase, magnesian olivine and magnesian pyroxene. (Wager and Brown 1968, p. 297).

Related references to magma-chamber replenishment and the mixing of new 'parental' melt with a variety of differentiation products in the sub-volcanic regime have since been made by O'Hara (1977; O'Hara and Mathews 1981), Arculus (1991), and others, and it seems reasonable to suspect — indeed to assume — that the Solomons lavas crystallized, and developed at least many of their compositional features, under non-static conditions. In addition to the magma-chamber convection currents proposed for Skaergaard, the Bushveld, and many other layered intrusions, one may visualize in the case of the Solomons such phenomena as:

1. magma surges from depth leading to mixing of new parental magma with older mafic differentiates and to 'shunting' of the upper parts of the magma body into the volcanic conduit;
2. tectonic compression or dilation of the magma chamber leading to various degrees of disruption of the latter, to the expulsion of lava and/or to the acceptance of further batches of parent melt;
3. double-diffusive overturn, with its consequent magma turbulence.

The activity of any or all of such phenomena would clearly lead to a dynamic state in which crystals formed under one set of conditions may be mixed with melt, and any crystals it may contain, generated under another set of conditions. Such turbulent mixing would be expected to lead to the juxtaposition of crystals of a given mineral species of quite different compositions, and to wide variation in partition coefficients as alluded to in the geochemical chapters.

A crude paragenetic sequence can none the less be discerned in the Solomons lavas, particularly in the more mafic members: the olivine–clinopyroxene-rich basalts. Olivine may be found as the sole mineral embedded in glass, and it may occur in such concentration as to indicate accumulation. Spinel does not occur without olivine, and the two are in some cases the only mineral constituents of a given lava. Clinopyroxene does not appear without at least a little olivine and spinel, and so on. Although magma mixing has almost certainly occurred, it is likely that this has involved — at least in the case of the olivine–clinopyroxene-basalts — no more than the mixing of new batches of parental melt with the fractionated products of earlier batches of essentially similar melt. While this would have led to a blurring of the sequence of nucleation, it has not obliterated the paragenetic pattern. Hence the resulting pattern of geochemical evolution, while also somewhat blurred, maintains a clarity quite sufficient to indicate the broad role of crystallization in progressively enriching and impoverishing those elements of present concern.

We may now look at patterns of trace-element abundances in the Solomons lavas from two points of view: (1) the patterns of trace-element acceptance by the principal minerals of the lavas, and (2) the resulting abundance behaviour of the trace elements as those lavas have progressively crystallized and the residual melts evolved.

PRINCIPAL MINERALS AND THEIR PATTERNS OF TRACE-ELEMENT ACCEPTANCE

Evaluation of quantities such as the incidence of trace elements is made easier and more effective if a standard is available for comparison. Table 21.1 (derived from Tables 5.2 and 6.2) sets out mean values of the relevant elements and oxides in average MORBs of the Mid-Atlantic Ridge and the environs of the triple junction at its southern extremity. As noted in Chapter 5, this particular MORBs sampling is used because it represents a relatively clearly defined tectonic feature, the products of which have been analysed in only a small group of laboratories. Mean compositions (summed to 100 per cent on an anhydrous basis) and abundances of the trace elements in average Solomons olivine–clinopyroxene-basalts, big-feldspar lavas, and hornblende-andesites are also given in Table 21.1

For convenience a condensed list of mean trace-element abundances in the principal Solomons minerals, and the related mean partition coefficients developed in the three Solomons lava groups, is given in Table 21.2.

Olivine. Olivine is by far the principal receptor of nickel among the Solomons minerals, particularly in the earlier stages of crystallization when high-Mg members of the olivine series are forming. Although K_D^{Ni} values developed between hornblende and later spinels and the relevant melts are comparable with those developed by olivine (Table 21.2), it must be remembered that such hornblende and spinel represent the later stages of lava evolution when the nickel content of the residual melt has become very low. The absolute amount of nickel removed from the melt by these minerals is therefore much less than that sequestered by the early olivines.

Individual olivine crystals and crystal cores containing ≈52 per cent MgO (Fo 94) contain 3000–500 ppm Ni; mean nickel in the olivines of the Solomons olivine–pyroxene-basalts is 1511 ppm, which compares with averages of 497 ppm in the basalts themselves and 154 ppm in the Mid-Atlantic Ridge MORBs. The capacity of early olivine subtraction to reduce melt nickel is thus substantial. A direct comparison of nickel acceptance by high-Mg olivine and spinel is provided by their respective mean $K_D^{cryst:melt}$ values in the olivine–pyroxene-basalts: here $K_D^{ol:melt} = 1.66 \times K_D^{mag:melt}$. Very much larger differences pertain in some individual picrites and basalts.

Olivine is also — in a manner virtually independent of Mg:Fe ratios — the principal acceptor of cobalt, with a mean of 185 ppm (mean Solomons olivine–pyroxene-basalts = 67 ppm Co; mean MORBs, 44 ppm Co). It also accepts noteworthy amounts of zinc and manganese, with observed maxima and means of Zn max = 410 ppm, mean = 156 ppm and MnO max = 0.89 per cent, mean = 0.31 per cent, in the basaltic olivines compared with mean Zn = 72 ppm, mean MnO = 0.17 in the Solomons basalts and mean Zn = 72 ppm, mean MnO = 0.18 per cent in MORBs. Zinc and manganese abundances are directly related to the iron content of the olivine and therefore bear essentially no relation to cobalt and are inversely related to those of nickel. The mean abundance of chromium in the olivines of the basalts is 346 ppm, a value almost identical with that of MORBs (355 ppm Cr) and substantially less than that of the basalts themselves (777 ppm). A mean $K_D^{ol:melt}$ of 1.54 for the olivines of the basalts indicates that these have little effect — one of slight impoverishment — on chromium in the evolving melt. At a mean value of 36 ppm, copper is very much lower in the Solomons olivines than it is in their containing basalts (89 ppm) and in MORBs (66 ppm Cu). The crystallization of olivines thus tends to concentrate copper in the residual melt, as also it does titanium, vanadium, barium, and strontium.

Table 21.1 Mean bulk compositions of MORBs (Mid-Atlantic Ridge) and of the olivine–clinopyroxene-basalt, big-feldspar lava, and hornblende-andesite lava groups of the Solomon Islands Younger Volcanic Suite

	1	2	3	4
SiO$_2$	50.16	49.21	53.32	57.68
TiO$_2$	1.35	0.57	0.80	0.64
Al$_2$O$_3$	15.73	11.93	17.96	17.94
FeO*	9.40			
FeO		7.21	4.56	2.24
Fe$_2$O$_3$		2.68	4.15	4.52
MnO	0.18	0.17	0.16	0.13
MgO	8.51	15.36	4.25	3.46
CaO	11.80	9.55	8.98	7.73
Na$_2$O	2.38	2.01	3.51	4.17
K$_2$O	0.28	1.10	1.97	1.31
P$_2$O$_5$	0.20	0.22	0.35	0.17
Total§	99.99	100.01	100.01	99.99
Ba	56	121	260	303
Sr	157	488	650	689
Pb	id	2.5	4.2	5.6
V	244	225	265	172
Cr	355	777	41	40
Co	44	67	35	22
Ni	154	497	30	15
Cu	66	89	161	66
Zn	72	72	80	67

FeO* = total Fe as FeO; total§, major components summed to 100 per cent on an anhydrous basis; id, insufficient data for reliable estimate.
1 = MORB; **2** = olivine–clinopyroxene-basalt; **3** = big-feldspar lava; **4** = hornblende andesite.

284 Ore Elements in Arc Lavas

Table 21.2 Condensed tabulation of minor and trace element mean concentrations and distribution coefficients in the principal minerals of the Solomons suite

	Cu Conc.	K_D	Zn Conc.	K_D	Pb Conc.	K_D	Ba Conc.	K_D	Sr Conc.	K_D	Ti Conc.	K_D
Olivine OPB	36	0.33	156	2.64	nd	-	14	0.08	2	0.003	0.03	0.08
Olivine BFL	51	0.35	285	3.86	nd	-	18	0.07	2	0.005	0.02	0.04
Clinopyroxene OPB	31	0.29	30	0.42	nd	-	10	0.06	71	0.10	0.23	0.56
Clinopyroxene BFL	31	0.20	58	0.78	nd	-	11	0.06	50	0.08	0.34	0.82
Clinopyroxene HbA	31	0.53	67	1.42	2.4	0.32	11	0.04	54	0.07	0.33	1.25
Orthopyroxene HbA, incl. 2-pyrox. lavas	35	nd	213	nd	nd	-	nd	-	nd	-	0.09	nd
Hornblende HbA	39	0.69	111	2.98	1.8	0.20	67	0.17	176	0.32	0.98	4.32
Biotite HbA	70	2.92	411	9.00	nd	-	4875	7.12	nd	-	2.20	8.40
Feldspar OPB	9	0.07	7	0.12	1.2	0.33	45	0.24	1160	1.74	0.02	0.05
Feldspar BFL	24	0.15	6	0.09	2.5	0.47	141	0.45	1311	2.12	0.03	0.07
Feldspar HbA	14	0.21	3	0.12	3.6	0.54	205	0.56	1357	2.07	0.02	0.10
Spinel OPB	295	2.84	490	9.07	13.0	2.97	53	0.23	51	0.08	3.05	7.38
Spinel BFL	145	0.87	487	6.47	6.2	1.27	16	0.05	8	0.01	3.87	8.93
Spinel HbA	175	2.41	643	17.05	18.3	2.60	55	0.14	22	0.03	3.79	16.70

OPB = olivine–clinopyroxene-basalt group; BFL = big-feldspar lavas; HbA = hornblende-andesite group; Conc. = mean concentration; $K_{(D)}$ = crystal:melt mean distribution coefficient; nd = not determined; all concentrations given as parts per million by mass of the element.

The principal effect of olivine formation is thus the well-known one: the rapid abstraction of nickel from the melt, particularly in the earliest stages when the high-Mg members are crystallizing. Olivine formation depletes the melt in cobalt at a fairly even rate throughout its crystallization history, and in zinc and manganese in the later stages as it becomes more Fe-rich. It is essentially neutral in its effect on chromium concentration in any MORBs-like parent and, by non-acceptance, tends to concentrate Ti, V, Ba, Sr, and Cu into the remaining melt.

Spinel. Because of their propensity, as a mixed crystal series, for accepting Cr^{3+} into their structure, the spinels are highly effective abstractors of chromium from the melt. It has been noted earlier (Chapter 15) that some individual spinel grains in the more highly picritic lavas contain >55 per cent Cr_2O_3, i.e. they are chromites, and it is thus clear that from the very onset of their crystallization the spinels sequester chromium. The mean chromium contents of the spinels of the three main lava groups are: olivine–pyroxene-basalts, mean spinel Cr = 96 306 ppm, big-feldspar lavas, 2933 ppm Cr; hornblende-andesites, 475 ppm Cr. The principal limitation on the abstraction of chromium by spinel thus appears to be the concentration of Cr in the melt. Although a few individual Solomons spinels contain less chromium than do MORBs (355 ppm Cr) the great majority have chromium contents well above this value and would rapidly deplete any MORB-like parent in this element. The role of the spinels is certainly to abstract chromium from the melt from the onset of their crystallization. The principal limitations on the absolute amount of this element removed is the amount of spinel precipitated and, as noted above, the concentration of chromium in the coexisting melt.

As the majority of the spinels are in fact titanomagnetites (TiO_2 range ≈5–10 per cent; overall mean ≈6.5 per cent; see Fig. 5.16), their crystallization strongly depletes any likely parent (MORBs = 1.35 per cent TiO_2) in titanium. Those of the felsic basalt–basaltic andesite portion of the lava spectrum tend to have highest TiO_2 contents and therefore titanium stripping of the melt by the spinels is at its peak in melts of ≈52 per cent SiO_2. Similarly the overall mean vanadium content of the spinels is ≈3700 ppm compared with 244 ppm V in MORBs, which would lead to the depletion of vanadium in any such parental melt as spinel crystallized. Likewise the overall mean MnO content of the Solomons magnetites is 0.60 per cent compared with 0.17 per cent in the Solomons basalts and 0.18 per cent in MORBs, leading to a strong subtractive effect. This is referred to again below.

The spinels are also very effective receptors of the chalcophile metals in terms of the incidence of these materials such as the Solomons basalts and MORBs. Reference has already been made (Chapter 6) to the remarkably high copper contents of the magnetites of the Kohinggo Island picrites (≈1000 ppm Cu). Copper (mean Solomons spinel Cu = 205 ppm), zinc (548 ppm), lead (15 ppm), cobalt (123 ppm), and nickel (427 ppm) all occur at much higher mean concentrations in the

V		Cr		Mn		Co		Ni	
Conc.	K_D	Conc.	K_D	Conc.	K_D	Conc.	K_D	Conc.	K_D
8	0.03	346	1.54	2619	2.18	184	4.79	1511	9.63
10	0.04	83	10.00	5838	3.92	193	5.66	545	19.03
152	0.54	2435	12.26	1223	0.97	40	0.97	370	2.62
244	1.34	270	41.22	2410	2.00	50	1.79	80	3.58
215	2.81	241	42.89	2757	3.36	42	3.75	36	5.82
92	nd	211	nd	3949	nd	98	nd	146	nd
434	7.02	120	18.33	2207	2.79	66	7.13	64	10.23
445	5.71	52	7.43	2556	4.12	62	10.33	110	12.94
3	0.015	0.50	0.016	148	0.117	2.1	0.08	7	0.19
3	0.016	0.65	0.211	121	0.100	2.4	0.12	2	0.13
4	0.047	0.07	0.032	108	0.155	0.8	0.08	2	0.35
3707	24.55	96306	401	3689	3.03	144	4.00	943	5.79
4271	28.28	2933	1027	4250	3.60	123	4.59	236	13.56
3029	49.88	475	96	6003	9.48	102	10.40	103	14.31

Solomons spinels than in MORBs, which would lead to corresponding depletion of any such parent melt in these elements as spinel crystallization proceeded.

One striking feature of the Solomons spinel trace-element assemblage calls for particular comment. Whereas the concentrations of Cr, Ti, Co, Ni, and Cu in the spinels all decrease in sympathy with decrease in their concentrations in the melt, the concentrations of Mn and Zn in the spinel *increase* as those in the melt decrease (Chapters 7 and 17). The general phenomenon of contrasting trace-element patterns in the mineral concerned has also been noted in the case of olivine, in which — in close relationship with Mg/Fe — manganese and zinc increase in amount as nickel decreases. In the latter case, however, the decrease in olivine nickel accompanies the decrease in melt nickel, and the increases in manganese and zinc in the olivine do no more than mute a slight increase in these in the melt stemming from the copious crystallization of relatively manganese- and zinc-poor clinopyroxene. The crystallization (to some limiting point) of increasingly Mn- and Zn-rich spinel from a *decreasingly* Mn- and Zn-rich melt would, on the other hand, accelerate significantly the decrease of these two elements in the evolving residual melt.

Clinopyroxene. The principal trace element of the Solomons clinopyroxenes is chromium, with an overall mean abundance of ≈980 ppm Cr. As with chromium concentrations in the Solomons spinels, those in the clinopyroxenes fall away rapidly, however, with decrease in this element in the melt. Mean clinopyroxene chromium in the olivine–pyroxene-basalts is 2435 ppm, in the big-feldspar lavas 270 ppm, and in the hornblende-andesites, 241 ppm. As the major element compositions (particularly Mg/Fe ratios) of the clinopyroxenes change little through the evolutionary span of the Solomons lavas, the decrease in clinopyroxene chromium appears likely to be entirely a reflection of the decrease in chromium available in the melt. It is clear, however, that clinopyroxene is a net abstractor of chromium throughout, and particularly in the early stages of crystallization. Nickel behaves similarly, decreasing from a mean of 370 ppm in the clinopyroxenes of the olivine–pyroxene-basalts to 80 ppm and 36 ppm in those of the big-feldspar lavas and hornblende-andesites respectively. Thus, as in the case of chromium, clinopyroxene abstracts relatively large amounts of nickel while this is at high concentration in the melt and then, in spite of increases in $K_D^{cpx:melt}$ (Table 21.2), progressively smaller amounts as melt concentrations decline. The clinopyroxenes similarly tend to deplete the melt in cobalt: cobalt contents of the pyroxenes of the three lava groups remain relatively constant at 40–50 ppm, but $K_D^{cpx:melt}$ increases systematically from 0.97 in the olivine–pyroxene-basalts to 3.75 in the andesites and dacites.

In the earlier stages of crystallization clinopyroxene tends to partition titanium into the melt, and it is only in the hornblende-andesites that mean $K_D^{cpx:melt}$ exceeds 1 (Table 21.2). Overall mean TiO$_2$ in the clinopyroxenes is ≈0.5 per cent, compared with 1.35 per cent in MORBs, indicating that their broad role is one of melt enrichment with respect to titanium. This is accentuated by the crystallization of orthopyroxene. Mean MnO ranges from 0.13 per cent to 0.34 per cent in the clinopyroxenes of the three principal lava groups, with an overall value of 0.25 per cent. Thus, while it tends to partition manganese into the melt at the very earliest stages of crystallization, clinopyroxene is a moderate abstractor of this element for the major part of the overall crystallization history (Table 21.2). It similarly tends to enrich vanadium in the earlier stages of lava evolution, and then becomes a modest abstractor of that element as the lavas evolve into basaltic andesites and andesites. Apart from zinc in the hornblende-andesites ($K_D^{cpx:melt}$ = 1.42), crystallization of clinopyroxene increases the concentration of Cu, Zn, Pb, Ba, Sr, and P in the liquid.

Hornblende. The amphiboles of the Solomons lavas are significant scavengers of all of Zn, Ti, V, Cr, Mn, Co, and Ni. As a mineral that appears relatively late in the paragenesis, at a stage when many of the trace and minor species have already been sequestered by olivine, spinel, and clinopyroxene, hornblende exhibits remarkably high concentrations of these elements: a state of affairs reflected in the relevant partition coefficients (Table 21.2). It therefore has an important role in 'mopping up' much of the small traces of these elements remaining in the melt after the crystallization of olivine and of much of the clinopyroxene and spinel. Like the other major silicates, hornblende accepts only small quantities of copper, and although it is a slightly more receptive host than olivine and the pyroxenes, it, too, partitions copper into the melt. It also partitions Pb, Ba, Sr, and P into the residual liquid.

The development of opaque reaction rims (Plates 1, 2, 3) round hornblende crystals has a significant effect on the concentrations of several of the trace elements (Tables 5.4, 5.5). Abundances of Zn and Mn are substantially higher in the opaque rims as compared with the parent hornblende. Ba and Sr are lower in the rims than in the crystals. Ni, Co, and Cr are generally less abundant in the rims than in the unreacted hornblende, Ti and V are variably affected, and Cu and Pb show little change from crystals to rims. The increase in Zn and Mn with rim formation might be accounted for by the development of abundant magnetite in the reaction intergrowth, though such an hypothesis is not supported by the relatively lower abundances of Ni, Co, and Cr in the rims

Development of the rims thus accentuates the role of hornblende in removing Zn and Mn from, and concentrating Ba and Sr in, the melt. It may also tend to mute the effects of this mineral in reducing Ni, Co, and Cr in the residual liquid, is likely to show little influence on the incidence of Cu and Pb, and may have erratic effects on relative concentrations of Ti and V in crystals and melt.

Hornblende thus plays an important role in *accentuating* the effects of olivine and clinopyroxene in removing Zn, Mn, Co, and Ni from, and concentrating Cu, Pb, Ba, Sr, and P in, the developing residual liquid right into the final stages of its evolution. It also accentuates the subtraction of Ti by the spinels. These effects are variably modified by the development of the opaque reaction rims.

Feldspar. The feldspars, a very important component of the Solomons lavas, accommodate only minute amounts of all of the major trace elements except strontium. Although it contains much higher concentrations of lead — particularly in the more potassium-rich members — than the ferromagnesian minerals, the amounts are small; $K_D^{felspar:melt}$ is <1 (Table 21.2) and the role of feldspar is, overall, one of the concentrating lead in the evolving melt. The degree of acceptance of barium is very similar to that of lead.

This brings us to the principal role of plagioclases in influencing trace-element patterns in the Solomons lava series — and indeed in any arc-lava suite: the well-known tendency towards separation of barium from strontium as the lavas evolve from basalts to andesites and more felsic compositions. As noted in Chapters 9 and 10, strontium is readily accepted into the plagioclase lattice, but the larger Ba^{2+} ion is not, and so the feldspars become mineralogical 'sieves' for sorting strontium from barium and for concentrating the latter in the residual melt. Related to this is the role of the *plagioclase* feldspars in reinforcing the effect of all other silicate crystallization in concentrating lead in the progressively evolving residual melt. The amount of lead accepted by the feldspars is directly related to the amount of potassium the latter contain. The copious crystallization of low-K plagioclases thus continues the accumulation of lead in the melt: a process which, as discussed in Chapter 24, may have substantial implications in exhalative ore petrogenesis and the development of ore type:lava type associations.

Orthopyroxene and biotite. Although orthopyroxene is prominent in the Simbo lavas and occurs in many of the andesites of the Solomons Younger Volcanic Suite, it is generally present in quantities too small for adequate magnetic and heavy liquid separation and examination of crystal:melt equilibria. Biotite has been found, as noted earlier, only in the two Savo dacites (though there is little doubt that it will in due course be found to occur more widely than this), and it was judged that satisfactory separation from these would have been extremely difficult if not impossible.

The electron microprobe analyses of coexisting ortho- and clinopyroxenes do, however, yield some insight into the relative influence of these two minerals on trace-element abundance patterns in the andesitic lavas.

The principal differences in degree of trace-element acceptance by the ortho- as compared with the clinopyroxenes parallel the higher FeO/MgO ratios, and at the same time higher absolute MgO contents, of the orthopyroxenes. With their higher iron and higher FeO/MgO, the orthopyroxenes accept higher Zn, Mn, and Co. The orthopyroxenes also contain higher Ni than the clinopyroxenes. They contain less chromium than the latter, however, and also less titanium. Copper contents are generally too low for satisfactory determination by electron probe, but limited data indicates that the orthopyroxenes contain copper at about the 30–40 ppm level as found in all of olivine, clinopyroxene, and hornblende. Although the orthopyroxenes are present at very low abundance, they must at least accentuate the effect of the late-stage clinopyroxenes in impoverishing the melt in Zn, Mn, and Co and in enriching it in Ti and probably Cu. The formation of orthopyroxene would tend to reduce the effect of clinopyroxene crystallization in subtracting chromium from the melt.

As microprobe analysis of biotite has been restricted to the two dacites from Savo, the degree to which the data can be applied is questionable. The two groups of biotite crystals analysed (65 analyses) show several features quite clearly, however. The most notable is their high zinc content: mean Zn = 411 ppm and mean $K_D^{biot:melt}$ = 9.0. Biotite is therefore, per unit of mass, a significant abstractor of zinc from the melt in the later stages of differentiation. In addition it is the only ferromagnesian silicate to contain more copper than the melt with which it coexists. With mean Cu = 70 ppm and mean $K_D^{biot:melt}$ = 2.9, biotite is an important (per unit mass) abstractor of copper from the late-stage melt. Indeed, biotite constitutes a late-stage 'sink', not only for these two elements but also for Ti, Ba, Mn, Cr, Co, and Ni. As it constitutes only 1–2 modal per cent of the lavas concerned, its actual subtractive effect is slight, but it is clear that the crystallization of large quantities, e.g. 10–15 per cent, of biotite would quickly lead to a stripping from any late-stage residual melt of most of the remaining trace metals.

PRINCIPAL TRACE-ELEMENT ABUNDANCE PATTERNS IN THE SOLOMONS LAVAS

We may now consider the incidence of the ore-element traces from the second perspective. Having reviewed the more prominent habits of trace element acceptance by the Solomons minerals, we may turn to consider the patterns of abundance of these traces in the lavas of which the minerals form a part. In particular we shall examine the ways in which their concentrations vary with variation in silica content of the containing lavas and hence (if we are not basing our procedure too much on hypothesis) with the evolution of the lava series.

It is now clear from Chapters 6–20 that the Solomons 'whole-rock' trace abundances do indeed develop a variety of patterns reminiscent of those deduced by Wager and Brown (1968) and Vincent (1974) to result from igneous crystallization in substantially closed, intrusive, systems. As noted earlier in the present chapter, such patterns stem at least largely from a combination of differences in trace-element acceptance by different minerals, and a tendency for those minerals to crystallize — and hence remove their substance from the melt — in a time sequence. Although eruptions of the different lava types of the Solomons suite do not show any real sequence in time there is, as noted above (see also Chapter 5), an indication of a sequence in the beginning of crystallization, i.e. the first appearance of the various minerals relative to each other, viz. olivine, spinel, clinopyroxene, plagioclase, orthopyroxene, hornblende, biotite (Bowen's reaction series). This conforms with the usual pattern of increasing whole-rock total SiO_2. To a first approximation the latter may thus be taken to have a time connotation, i.e. the evolution of the lava series, and hence to Wager, Brown, and Vincent's progression based on percentage of total melt crystallized. It therefore seems reasonable to consider the Solomons patterns — based on an increase in whole-rock SiO_2 (Fig. 21.2) — in the light of Wager, Brown, and Vincent's deductions based on the fraction of melt crystallized. This is done using the full 241 analyses representing the Solomon Islands and Bougainville, as set out in Chapter 5 and referred to in Chapters 6–20. The element and oxide-versus-SiO_2 diagrams of Fig. 21.2 thus have a greater statistical reliability than their analogues of Chapters 6–19, and this is reflected in most cases by increased smoothness of curves at the higher SiO_2 levels.

Elements readily accepted by, and removed from the melt in, early-crystallizing minerals

Nickel. Nickel, as we have seen, is captured in large amount (3000–500 ppm) by the earliest, high-Mg phases of the first mineral to crystallize: olivine. Such an 'olivine effect' is reinforced by almost simultaneous capture in the earliest clinopyroxenes (≈1000 ppm Ni) and spinels (≈2000 ppm Ni). The identity and hence nickel content of the parent to the olivine–pyroxene-basalt spectrum is of course not known (see Chapter 23). However, in terms of the latter's average nickel concentration (497 ppm) and the average of mid-Atlantic MORBs (154 ppm Ni), the nickel abundances in the early Solomons olivines, pyroxenes, and spinels are high and would have a pronounced nickel-subtraction effect on any likely basalt parent. As indicated by curve (1) of Fig. 21.2(a), a steep trend of high-Ni accumulates and low-Ni residua flattens abruptly in the 50–2 per cent SiO_2 range, lavas of higher SiO_2 contents averaging no more than 18 ppm Ni and tapering down to below 8 ppm at SiO_2 >64 per cent.

Chromium. Although the spinels are the pre-eminent captors of chromium, early formed clinopyroxenes contain >5000 ppm Cr and some olivines >700 ppm Cr. These, with their greater modal abundances, reinforce the effect of spinel in subtracting chromium from the melt. As a result chromium, like nickel, is rapidly stripped from the melt by the earliest-formed crystals of these minerals, yielding a $Cr:SiO_2$ curve virtually parallel to that of $Ni:SiO_2$: a steep trend (curve (2), Fig. 21.2(a)) of high-Cr accumulates and lower-Cr residua. As with the nickel curve, that for chromium flattens at the 50–2 per cent SiO_2 interval: lavas of higher SiO_2 average 29 ppm Cr, and <9 ppm Ni at SiO_2 >64 per cent. As noted earlier (Chapter 15), nickel and chromium abundances behave similarly, not because they are captured by the same mineral (one is incorporated principally in olivine, the other in spinel), but because they are incorporated in the same group of minerals at the same stage of crystallization history.

Cobalt. While cobalt is certainly one of the elements incorporated in the earlier-formed crystals, its removal from the melt is not quite so rapid as that of nickel and chromium, and hence $Co:SiO_2$ relations (Fig. 21.2(b)) are rather less dramatic. As was pointed out in Chapter 18, the incidence of cobalt within any given ferromagnesian silicate is substantially independent of Mg:Fe ratios. It clearly favours the olivine and spinel structures, however, occurring in somewhat similar mean abundances, 184 and 144 ppm respectively, in the two minerals as these occur in the Solomons olivine–clinopyroxene-basalts. The steepest part of the curve extends to the SiO_2 interval 52–4 per cent — slightly higher than for nickel and chromium — and ≈30 ppm Co. The rate of decrease in cobalt at higher SiO_2 levels changes relatively evenly, not abruptly as in Fig. 21.2(a), as crystallization of olivine gives way to that of clinopyroxene and hornblende, and of spinel of progressively lower cobalt content.

While these abundance patterns of nickel, chromium, and cobalt all conform with Wager and Brown's (1968) 'early crystal subtraction' pattern and may be attributed entirely — if qualitatively — to fractional crystallization, there is a clear difference between the behaviour of cobalt on the one hand and of nickel and chromium on the other. Nickel and chromium are removed from the melt in large proportions at the very earliest stages of crystallization. Removal of cobalt, on the other hand, begins early, but in smaller proportional amount, and continues further along the crystallization paths of the lavas concerned. This difference may be significant in exhalative ore formation, and is referred to again in Chapter 24.

Elements most readily entering minerals formed at middle to later stages of fractionation

Almost inevitably this category represents a range. Some elements are readily accepted by crystals, or members of mixed-crystal series, that begin to form after, but closely following, the earliest-nucleated crystals; such elements do not simply undergo uninterrupted decline as do nickel, chromium, and cobalt, but develop a small peak at an early stage of lava evolution, which is then followed by steady decrease with increase in whole-rock SiO_2. Other elements exhibit peak whole-rock abundances at later stages of the evolutionary process.

Manganese and iron. Manganese and iron appear to represent the earlier-formed of this general group. As a result of early crystallization of the highest-Mg, lowest-Fe clinopyroxenes and olivines — the members of the relevant mixed-crystal series that also contain lowest manganese — iron and manganese are slightly enriched in the melt now represented by lavas in the 46–50 per cent SiO_2 range. This is, however, quickly followed by the crystallization of more Fe:Mn-rich olivines and clinopyroxenes, spinel and, somewhat later, orthopyroxene, hornblende, and biotite. All these establish partition coefficients $K_D^{Mn,\ Fe} >1$ with respect to the coexisting melts, and there develop the lines of apparent liquid descent shown in Fig. 21.2(c). While such an attribution is entirely qualitative, the abundance behaviour of manganese and iron might be accounted for substantially by fractional crystallization. This is examined quantitatively later in this chapter.

Calcium and phosphorus. Calcium and phosphorus also develop quite early maxima, closely following manganese and iron in the achievement of their highest whole-rock values (Fig. 21.2 (d)). Calcium peaks in the 48–50 per cent SiO_2 range, P_2O_5 in the 52–8 per cent SiO_2 interval. The initial increase may be attributed to olivine subtraction in both cases; the following decline to the crystallization of clinopyroxene and apatite. The effects of clinopyroxene subtraction on calcium abundance are

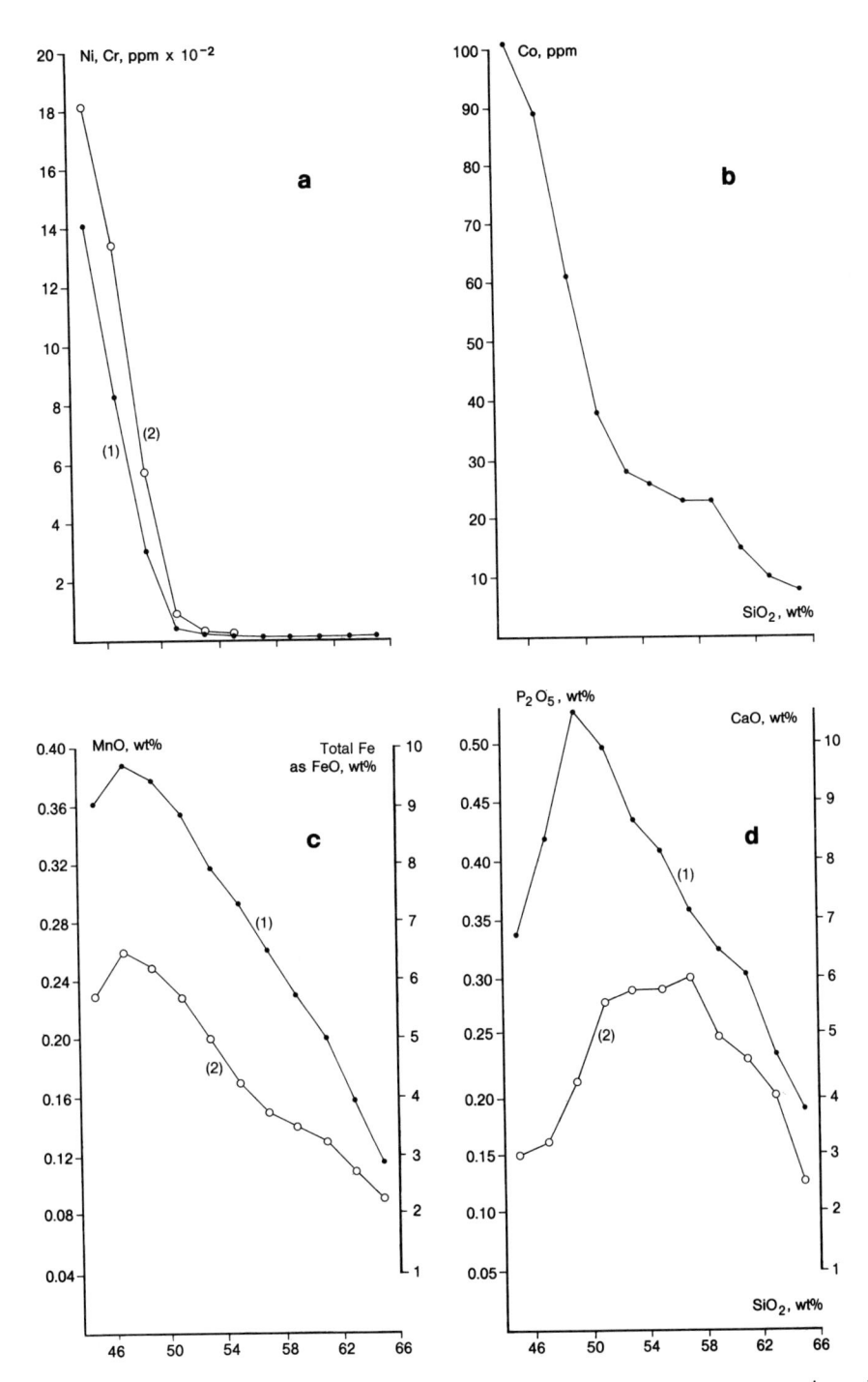

Abundance patterns in the crystallizing melt 289

Fig. 21.2. Variation diagrams showing change in mean abundances of trace elements with respect to two per cent intervals of whole-rock SiO$_2$ for (a) nickel [curve (1)] and chromium [curve (2)] (both given as ppm × 10^{-2}); (b) cobalt; (c) total iron as FeO (1) and MnO (2); (d) CaO (1) and P$_2$O$_5$ (2);

290 *Ore Elements in Arc Lavas*

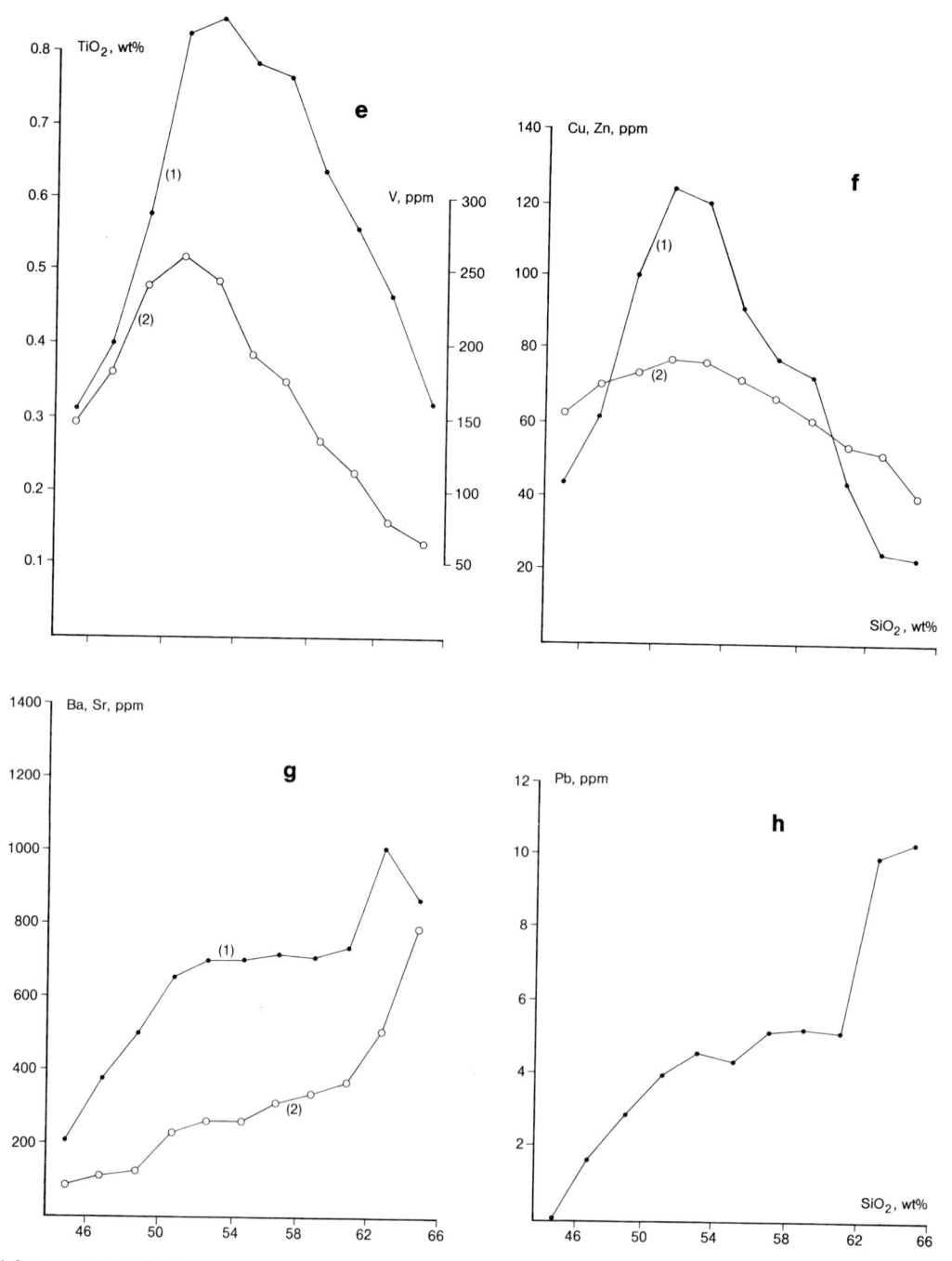

Fig. 21.2. (*cont.*) (e) TiO$_2$ (1) and vanadium (2); (f) copper (1) and zinc (2); (g) strontium (1) and barium (2); and (h) lead. These diagrams differ from the analogous variation diagrams of Chapters 5 to 20 in that they include the Rogerson *et al.* (1989) Bougainville lava analyses: each curve thus represents 241 analyses, yielding increased statistical reliability as indicated in Fig. 5.2. Quantities are given in weight per cent and parts per million by mass as indicated.

reinforced by those of plagioclase and hornblende in the later stages of evolution, and phosphorus is removed both in apatite and the silicate structures.

Titanium and vanadium. As with the preceding elements, the initial increase in titanium and vanadium (Fig. 21.2(e)) may be attributed to the copious crystallization of early olivine and clinopyroxene containing relatively low abundances of these two elements (Table 21.2). Both curves (Fig. 21.2(e)) achieve distinct maxima (TiO_2 = 0.85 per cent; V = 263 ppm), vanadium at 50–2, TiO_2 at 52–4 per cent SiO_2; both then decrease rapidly at higher SiO_2 levels owing to the crystallization/subtraction of spinel, hornblende, later pyroxenes, and biotite. Biotite, with mean TiO_2 ≈3.5 per cent, is second only to spinel as a titanium acceptor on a per unit of weight basis. Its sporadic occurrence and very low modal abundance, however, make it an insignificant subtracter of titanium in an absolute sense.

Zinc and copper. Abundances of zinc describe a very gentle curve with a somewhat muted maximum (≈76–8 ppm Zn) in the 50–4 per cent SiO_2 interval (Fig. 21.2(f), curve (2)). In a qualitative sense the early increase in zinc may be attributed to the crystallization–subtraction of high-Mg:low Zn olivine and clinopyroxene, and the later decrease to the crystallization of more Fe-rich, higher-Zn, olivines and clinopyroxenes and, more particularly, to the crystallization of magnetites.

As pointed out in Chapter 6, the behaviour of copper is not, however, so readily explained by crystallization subtraction. This element develops a pronounced and almost symmetrical maximum at Cu = 126 ppm:SiO_2 = 52–4 per cent in the Solomons lava series (Fig. 21.2(f), curve (1)). The left-hand side of the curve may readily be accounted for by the copious crystallization/subtraction of Cu-poor olivine and clinopyroxene (mean Cu ≈30–40 ppm; Table 21.2). The steep decline in copper represented by the right-hand side of the curve does not, on the other hand, appear to be attributable to crystallization processes. Crystallization of all the principal silicate species — olivine, clinopyroxene, orthopyroxene, hornblende, and, particularly, plagioclase — involves subtraction of material containing *less* copper than the coexisting melt and should lead to a continuing upward trend in the Cu:SiO_2 curve. The spinels and some biotites contain more copper than the melt, but their modal abundance is too small to account for the steep and proportionately substantial decrease in copper indicated by Fig. 21.2(f). As noted in Chapter 6, the qualitative indications are that some process additional to crystal subtraction is involved. This is an important possibility in the present context and merits a brief digression.

Digression on copper. The tendency for copper to be relatively low in the most mafic members of an igneous series, to rise to a maximum in the mafic intermediate members, and then to decrease progressively with increase in felsic character as the series evolves, has been recognized for a long time. Wager and Mitchell (1951) noted the phenomenon in the Skaergaard intrusion, attributing it to non-acceptance of copper by the earliest-formed silicates, inducing initial concentration in the melt, followed by sulphide formation leading to subtraction. Their diagram (Wager and Mitchell 1951, p. 171) is identical in principle with Cu–SiO_2 curves of Figs. 6.12 and 21.2. This sequestration of copper by sulphide in the Skaergaard layers was confirmed in detail by Wager, Vincent, and Smales (1957). Carmichael (1964) found a similar copper peak in the lavas of Thingmuli in eastern Iceland and, as in the Solomons and other arc lavas, less pronounced peaks developed by total iron, vanadium, cobalt, titanium, manganese, and zinc. He attributed the development of such maxima, and the sequence in which they manifested themselves, to the relative octahedral site-preference energies of the elements concerned — and their resultant relative propensities for incorporation in major silicate structures and consequent subtraction from the melt. Ewart *et al.* (1973) noted the development of a copper maximum in the lavas of Tonga, and attributed its later stage decrease to abstraction in sulphide. Brown *et al.* (1977) clearly recorded the effect in their tables of West Indies lava compositions, but did not remark on it (see Chapter 6). Although it has sometimes been proposed that Cu^{2+} is readily accepted into Fe^{2+} sites and thus removed from the melt in ferromagnesian minerals (i.e. that it is a 'compatible' element with respect to the silicates) the more general, tacit, assumption has been that it has been removed as sulphide, and that it is to be regarded as 'compatible' in that sense.

Apart from the early prognostications of Wager *et al.* (1957) — and their well-substantiated deduction that the copper of the later Skaergaard stages was largely sequestered by sulphides — the first serious consideration of the behaviour of copper in the latter stages of differentiation appears to be that of Eilenberg and Carr (1981). In examining the incidence of copper in the lavas of some active Central American volcanoes, Eilenberg and Carr noted that:

(a) The only other trace-element with which copper showed a correlation was vanadium.
(b) As the principal habitat of vanadium was magnetite, this presumably constituted the chief host of copper, and indeed the magnetite of a lava containing ≈140 ppm Cu was found to contain ≈325 ppm Cu.
(c) Copper behaved as an incompatible element in those lavas containing little magnetite and augite.

(d) Copper behaved as a compatible element in those lavas containing prominent magnetite and augite.
(e) Mass balance calculations indicated, however, that subtraction of copper by magnetite and augite was insufficient to account for the observed decrease in this element; separation of a copper-bearing sulphide phase was — on chemical and microscopical evidence — unlikely.
(f) Copper was a prominent component of the Central American volcanic gases and sublimates.
(g) The evidence therefore indicated that magnetite and augite fractionation *combined with* loss in the vapour phase were 'plausible mechanisms controlling the abundance of Cu in these lavas'. (Eilenberg and Carr 1981, p. 2248).

Unaware of the work of Eilenberg and Carr at the time, the present writer (Stanton 1987; see also Stanton 1990, 1991) had noted in the Solomons lavas the very low copper contents of the silicates other than biotite, the high copper content, but low modal abundance, of the magnetite, the paucity or absence of sulphide, and the contrast in the copper content of inclusion *vis-à-vis* groundmass glass alluded to in Chapter 6. It had been found that the glass inclusions within feldspars of some eight Solomons felsic basalts and basaltic andesites contained a mean of ≈1900 ppm Cu, whereas the groundmass glass of the same lavas contained a mean of ≈150 ppm Cu. It appeared that the small quantities of glass trapped in plagioclase feldspars preserved a high copper concentration developed in the melt at about the felsic basalt–basaltic andesite stage of lava evolution, whereas the melt now represented by the groundmass glass — untrapped, and free to de-gas — had lost the major part of its copper in a volatile phase. This led the author to propose that the pronounced diminution of copper in the more felsic members of the Solomons series, impossible to account for on the basis of fractional crystallization alone, resulted from a combination of fractional crystallization *and* loss in the volatile phase — essentially the same process as that put forward by Eilenberg and Carr some six years earlier.

More recently, Lowenstern *et al.* (1991) have analysed carbon dioxide- and chlorine-bearing bubbles in phenocrysts in several highly felsic (rhyolitic) lavas, and have found evidence for 'extreme partitioning' of copper into magmatic vapour phases. They noted that the observed inclusions represented 'small blebs of melt trapped in growing phenocrysts (crystals) at high temperature and pressure in the magma chamber. The phenocrysts act as pressure vessels during eruption, preventing the inclusions from outgassing and thereby preserving samples of the preeruptive melt and its volatile components'.

(Lowenstern *et al.* 1991, p. 1405). They found inclusion copper contents of up to 300 ppm, compared with ≈3 ppm Cu in leaked inclusions and matrix glass. In general the copper occurred as sulphide within the fluid of the inclusions, and vapour:melt partition coefficients calculated from the analyses ranged from ≈200 to ≈1000. Lowenstern *et al.* considered their results to induced that early vapour saturation can occur; 'that crystallization-induced volatile saturation (second boiling) may not be necessary for the production of Cu-rich fluids' (Lowenstern *et al.* 1991, p. 1405): a conclusion with which, on the basis of the Solomons data, the author concurs.

Further evidence from glass inclusions in plagioclase. Prior to the publication of the work of Lowenstern *et al.* (1991), the author had embarked on a wider investigation of the incidence of copper in inclusion *vis-à-vis* groundmass glasses. Results of analysis of the glasses of the Solomons big-feldspar basalts–basaltic andesites were so unequivocal that it seemed that a broader investigation, concerned with the possible generality of the feature, was merited.

1. Mount Erebus, Antarctica: The metallic component of the Mount Erebus exhalations has recently come to prominence with the observation of Hecht (1991) that they are producing gold crystals. Particles 0.1–20 micrometres across have been found — apparently sublimed — in the gas itself, and others, up to 60 micrometres across, in the nearby snow.
   Periodically the basaltic melt of the Erebus lava lake exhibits a scum of large feldspar crystals (≈1 cm across), which when sampled adhere to each other via congealed glass. Of the sample available to the author, all the plagioclase feldspars contained abundant glass inclusions which, with the matrix glass, were readily analysed by electron microprobe. Not surprisingly, in terms of the degassing hypothesis, both categories of glass were found to contain little copper. Of the 60 analyses (30 of each category of glass), many contained copper below the detection limit (≈30 ppm Cu). Calculated mean values were: inclusion glass, 10 ppm Cu; matrix glass, 7 ppm Cu. Apparently almost all copper had been lost to the melt prior to the formation of the feldspars, and so the trapped (inclusion) glass contained only very slightly more copper than that representing the melt at the time of collection from the surface of the lake.
2. Chichijima volcano, Bonin Islands, Japan: Professor R. J. Arculus of the University of New England kindly supplied the author in 1992 with a sample of two-pyroxene dacite from the boninitic series of the island of Chichijima (Table 21.3). This also contained sparse

phenocrystic feldspar exhibiting well-preserved glass inclusions and scattered phenocrystic magnetite, all set in a groundmass consisting substantially of glass. The material was clearly appropriate for comparing, by electron microprobe analysis, compositions of inclusion versus groundmass glass.

Fifty well-preserved glass inclusion in feldspars, and 65 sites (raster size = $12 \times 12\mu$) representing matrix glass surrounding those feldspars, yielded the results given in Table 21.3 and Fig. 21.3.

The abundance of copper is clearly highest in the glass included in the feldspars: 1.7 times that in the whole rock and 2.6 times that in the groundmass glass; Cu/Zn ratios decrease from 2.06 in included glass to 1.0 in groundmass glass.

It might be suggested that this very large decrease in copper, and relatively small decrease in zinc, in groundmass as compared with included glass reflects some preferential incorporation of copper vis-à-vis zinc in the pyroxenes and magnetite of the parent dacite. Microprobe analysis indicates, however, that the reverse is the case: it is the *zinc* that is preferentially incorporated in these minerals. Orthopyroxene and clinopyroxene contain ≈218 ppm and 91 ppm Zn respectively, and both contain <45 ppm mean Cu. The mean abundance of zinc in the magnetite is ≈403 ppm, and of copper, ≈36 ppm.

Table 21.3 Chemical composition, including copper and zinc abundances, of Chichijima boninitic dacite and its plagioclase-inclusion and groundmass glasses

	1	2	3
n	1	50	65
SiO_2	68.29	70.82	73.98
TiO_2	0.28	0.24	0.22
Al_2O_3	13.68	10.61	11.05
FeO*	6.34	6.77	4.92
MnO	0.11	0.12	0.09
MgO	1.38	1.41	0.38
CaO	4.90	3.46	3.06
Na_2O	2.77	0.33	0.14
K_2O	1.47	1.37	1.05
P_2O_5	0.05	0.03	0.02
S	0.02	0.007	0.002
H_2O§	0.61	4.83	5.09
Total	100.00	100.00	100.00
Cu	105	183	71
Zn	73	89	71

1 = Chichijima dacite, whole rock; 2 = mean of 50 glass inclusions within plagioclase phenocrysts; 3 = mean of 65 analyses of glass enclosing the relevant plagioclase crystals. n = number of analyses; FeO* = total iron as FeO; H_2O§ = sum of imputed water and allowance for Fe_2O_3; major elements given in per cent by mass; Cu and Zn in parts per million by mass.

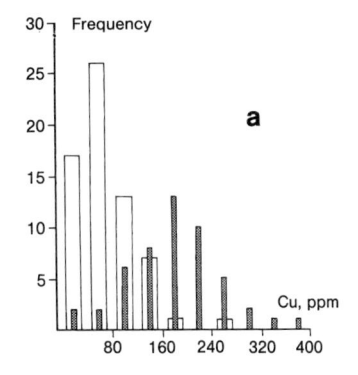

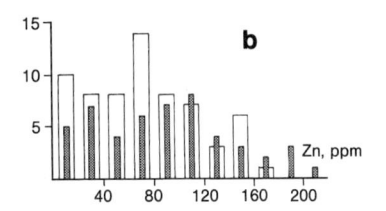

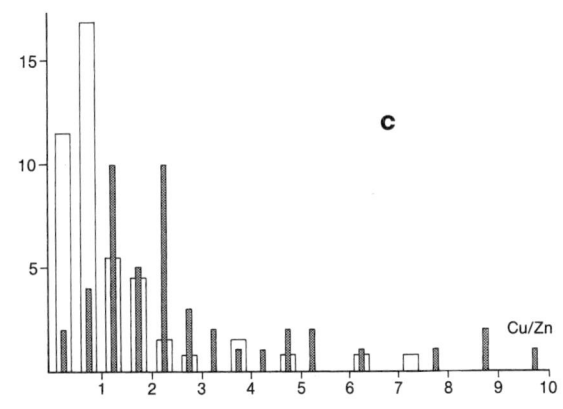

Fig. 21.3. Frequency distribution of (a) abundances of copper in 50 glass inclusions within feldspar crystals (filled bars; mean = 183 ppm Cu) and in 65 domains of adjacent enclosing groundmass glass (open bars; mean = 71 ppm Cu); (b) abundances of zinc in the same glass inclusions (filled bars; mean = 89 ppm Zn) and the same domains of groundmass glass (open bars; mean = 71 ppm Zn); and (c) corresponding copper/zinc ratios: inclusion glass, filled bars, mean = 2.06; groundmass glass, open bars, mean = 1. Material: boninitic dacite from the island of Chichijima (Bonin Islands) supplied by R. J. Arculus.

We thus find that whereas zinc is incorporated in the ferromagnesian phenocrysts to a readily measurable extent, it is little reduced in the late stages of the melt (the groundmass glass) relative to the melt existing earlier (the included glass). In contrast, copper, which is incorporated in the phenocrysts to a much lesser extent, is greatly reduced in the later melt (groundmass glass) relative to the earlier (included glass).

The inevitable conclusion is that copper has been lost to the lava system, and that this, presumably, has occurred by emission in a volatile phase.

Elements that are not accepted, or are accepted to only a very limited extent, by the principal minerals

Strontium. As in the case of the preceding group (group 2), the elements now considered represent a range, albeit a narrow one, of patterns of abundance behaviour. Strontium illustrates the point.

It was noted in Chapter 10 that Sr^{2+} (ionic radius 0.127 nm) is not only very similar in size to K^+ (0.133 nm), and hence readily accepted into K^+ sites in feldspars, but also has a conspicuous affinity for the plagioclase structure. As a result it enters abundantly into the Solomons plagioclases, particularly those crystallizing with relatively high potassium contents towards the felsic end of the evolutionary series. The series appears, however, not to have reached the stage at which plagioclase was subtracted from the melt. The result is that while strontium has been substantially captured by the late-stage plagioclase, the plagioclase has remained in the lava fraction in which it crystallized — and thus the strontium has not been 'subtracted' from the melt to any significant extent within the evolutionary spectrum concerned. (It is possible that the decrease in strontium between the 62–4 per cent SiO_2 and >64 per cent SiO_2 class intervals of Fig. 21.2(g) is, as with Pb and despite the inclusion of the Bougainville analyses, a statistical artefact resulting from the small number of samples available at the high-SiO_2 limit. It may on the other hand be genuine and reflect the beginning of loss of strontium to the melt at $SiO_2 = \approx 62$–4 per cent).

Barium. As barium is continually sequestered into the residual melt, and as strontium is increasingly accommodated in the later, more K-rich plagioclase, the pattern of barium abundance in the whole rock tends to mimic that of strontium (curve (2), Fig. 21.2(g)). It must be emphasized, however, that this muted parallelism does not reflect similar behaviour. Strontium tends to go into plagioclase, barium to remain in the melt. The processes responsible for the similarity in abundance behaviour of the two elements are the parallel crystallization of progressively more K-rich, and hence Sr-rich, plagioclase, and the development of a progressively more felsic, Ba-rich melt (see also Chapter 12, Fig. 12.2).

Lead. We have seen that, with its similar ionic radius (Pb^{2+}, 0.133 nm) and remarkable geochemical affinity for potassium sites, lead tends to partition in sympathy with potassium in K-bearing phases, chief among which is plagioclase. In fact, lead partitions more strongly and more systematically into this mineral than does potassium (Fig. 21.4). Mean $K_D^{feld:melt}$ for the 75 ultimate purity feldspar separations is 0.19 for potassium and 0.49 for lead (Figs. 21.4(b) and (d)). K_D values for both elements increase proceeding from the olivine–pyroxene-basalts to the hornblende-andesites (Table 21.4). Figures 21.4(a) and (c) indicate that feldspar lead:melt lead relations are somewhat more systematic than those developed by potassium. K_D^{Pb} feldspar:melt = 1 in four pairings of the 75, and >1 (K_D^{Pb} = 2.20) in one andesite from Savo. All 75 pairings gave K_D^K feldspar:melt <1. Apart from that for the feldspars of the olivine–pyroxene-basalts, mean K_D^{Pb} and K_D^{Ba} feldspar:melt are notably similar through the Solomons lava suite as a whole (Table 21.4).

These minor differences aside, lead, with barium and potassium, is consistently partitioned into the melt and hence steadily increases in abundance in the whole rocks (Fig. 21.2(h)). The marked similarity in abundance patterns exhibited by lead and strontium is at first sight surprising, given that mean K_D^{Sr} >1 and mean K_D^{Pb} <1 in the feldspars. This quite remarkable parallelism is, however, due to the apparent lack of subtraction (remarked on above) of the feldspars from the melts from which they have crystallized. Whereas subtraction of early olivines leads to gross reduction of nickel in the melt and hence in the later members of the lava series,

Table 21.4 Large ion mean partition coefficients developed by plagioclase feldspars of the principal lava groups of the Solomon Islands Younger Volcanic Suite

Lava group	K_D^K	K_D^{Sr}	K_D^{Ba}	K_D^{Pb}
Olivine–clinopyroxene-basalts	0.11	1.74	0.24	0.33
Big-feldspar lavas	0.15	2.12	0.45	0.47
Hornblende-andesites	0.22	2.07	0.56	0.54

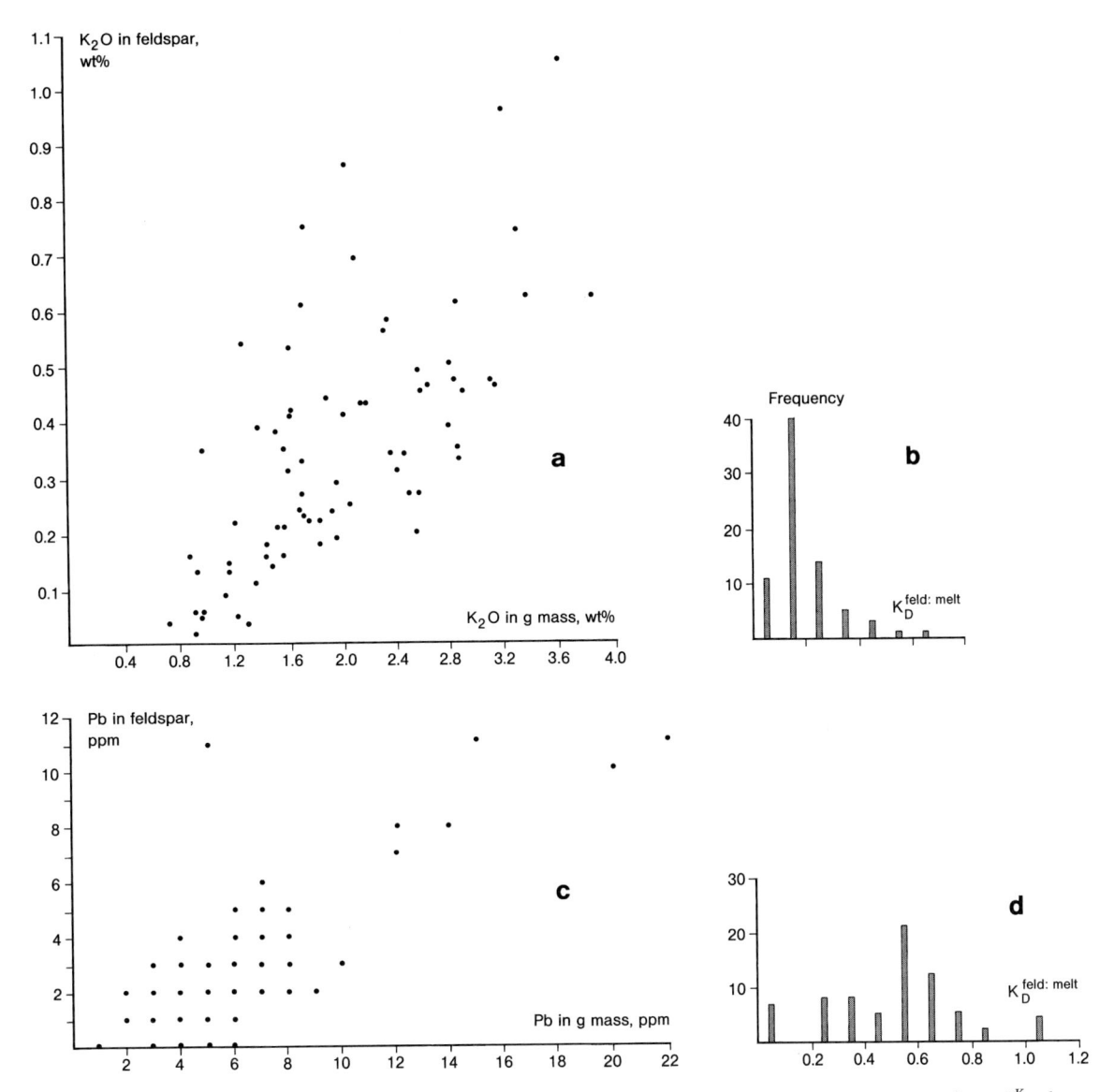

Fig. 21.4. (a) Plagioclase feldspar K$_2$O v. groundmass K$_2$O; (b) frequency distribution of individual feldspar:groundmass K_D^K values; (c) feldspar lead v. groundmass lead; and (d) frequency distribution of individual feldspar:groundmass K_D^{Pb} values for the 75 'ultimate purity' feldspar separations of the Solomons suite.

the tendency for the plagioclases to remain *in situ* leads to retention and increase in strontium together with K, Ba, and Pb in the whole rocks. One element partitions into the crystals, the other three into the melt — but crystals and melt tend to remain together. As in the case of the strontium decrease, the flattening in the lead curve at SiO$_2$ = 62–4 per cent may be a statistical artefact (which clearly requires further investigation) or it may constitute the first sign, midway through the dacite range, of subtraction of lead from the melt. The remarkable parallelism of the patterns of increase of strontium and lead up to the 62–4 per cent SiO$_2$ range suggests that their divergence at SiO$_2$ = 62–4 per cent may be real, and that the two elements are beginning at part company as the melt achieves a rhyo-dacitic composition.

Potassium and rubidium. The marked similarity of abundance behaviour exhibited by strontium, barium, and lead, and the likelihood that this is due to their common predisposition for occupying K⁺ sites, calls for a brief examination of the behaviour of the other two important large ion lithophile elements using the full 241 analyses of the Solomons cum Bougainville suite.

The relevant SiO_2 variation diagrams are given in Fig. 21.5. These confirm, with much greater statistical reliability and consequent smoothing of curves, the patterns observed in Chapter 20. The behaviour of the five large-ion elements of Figs. 21.2 and 21.5 — four lithophile, the other strongly chalcophile — is remarkably similar in the interval $44 < SiO_2 < 60$ per cent. At about this point the behaviour of the five begins to change and then diverge. The statistical reliability of these changes may be judged by the reader by reference to Fig. 5.2. Up to and including the SiO_2 interval 60–2 per cent, sample numbers are relatively high and trace-element estimations are therefore relatively reliable. At $SiO_2 > 62$ per cent sample numbers are, however, lower (an accurate reflection of the low abundance of such felsic lavas in the Solomons assemblage) and thus statistical reliability becomes lower just where important developments in lava composition look as if they may be taking place. The fact that all the five elements exhibit a change of abundance behaviour at about the same stage of lava evolution does not provide substantiation: it simply reflects similarity of behaviour among the large-ion elements.

These uncertainties accepted, it may be seen that the five elements present a spectrum of behaviour at whole-rock $SiO_2 > 58$–60 per cent. Rubidium (Fig. 21.5(a)) exhibits a flat peak at 58–64 per cent SiO_2 and then decreases sharply. Potassium (and zirconium) (Fig. 21.5(b)

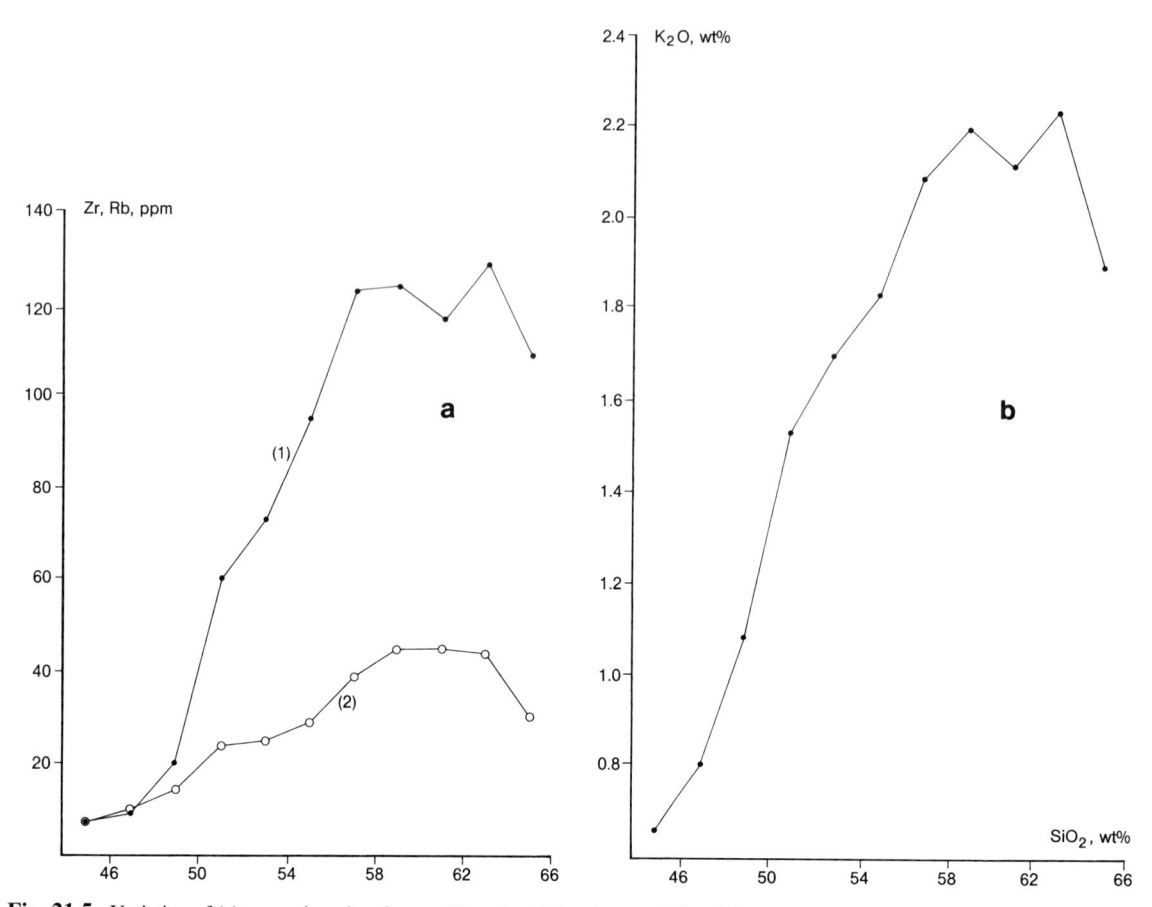

Fig. 21.5. Variation of (a) mean zirconium [curve (1)] and rubidium [curve (2)] and (b) mean potassium with respect to 2 per cent intervals of whole-rock SiO_2 in the 241 lavas of the Solomons and Bougainville suites.

and (a)) also display rather flat maxima, in the 56–64 per cent SiO$_2$ range, and then decrease in much the same way as does rubidium. Strontium and lead (Figs. 21.2(g) and (h)) show remarkable parallelism to 62–4 per cent SiO$_2$, at which stage strontium decreases but lead continues to increase at a reduced rate. (Both these elements show a pronounced flattening in their rates of increase through the SiO$_2$ range ≈50–60 per cent, both then increase sharply, their trends diverging at SiO$_2$ >64 per cent). Barium (Fig. 21.2(g)) is continuing to increase — and indeed is doing so at an increased rate — at the highest SiO$_2$ levels.

Non-ionic elements, randomly accepted as interstitials and maintaining a relatively constant abundance through a differentiation series

Two elements that Vincent (1974) suggested might fall in this class are gold and silver. Having regard to their relative inertness under magmatic conditions, and to the size of their atoms, two possible influences on their abundance patterns are readily apparent. They might, as proposed by Vincent, behave as non-ionic elements and lodge as interstitial uncharged atoms within almost any phase that crystallizes. In addition both have atomic radii almost identical with the ionic radius of Ba^{2+} (0.143 nm) (Au = 0.1442 nm; Ag = 0.1444 nm; Emsley 1989). These two influences might tend to induce concentration of the gold and silver in the later fractions of the melt with the large-ion, incompatible, elements.

The limited evidence provided by the 28 samples of the 'condensed suite' (Chapter 20) suggests, however, that perhaps neither of these possibilities holds for the Solomons lava series. As noted in Chapter 20, silver appears to rise to a peak abundance (0.96 ppm) at 51–2 per cent whole-rock SiO$_2$, and gold (8.5 ppb) at c. 54–5 per cent SiO$_2$. The relevant diagrams show considerable scatter, but do not provide a case for uniform distribution of gold and silver through the lava series. As would be expected from these metal:SiO$_2$ relations, gold and silver exhibit random relationships with barium (as a representative of the large-ion elements).

None of the elements considered appears to conform with Vincent's (1974) Group 4 in the Solomons lava series. This is not to say, however, that further investigation may not show some that do.

CRYSTAL SUBTRACTION AND ELEMENT ABUNDANCE

Visual inspection of the diagrams of Fig. 21.2 yields a qualitative indication that the abundances of all the relevant elements except copper might be accounted for by crystallization and subtraction of the principal Solomons minerals. The rise in the abundance of copper (Fig. 21.2(f)) to a peak of ≈125–30 ppm at 50–4 per cent is readily attributable to the formation of early, low-Cu, olivine and clinopyroxene, but its decrease from this peak at higher SiO$_2$ values cannot be accounted for by the subtraction of any of the principal Solomons silicates. Copper contents of all these are lower than that of the melt, and their subtraction, in a closed system, would in all cases tend only to increase the abundance of this element in the system. The only minerals containing abundances of copper sufficient to deplete the melt are the spinels, with their mean copper content of c. 100–200 ppm. Spinel abundance in the Solomons lavas is, however, too low to account for the decrease in copper in those lavas containing >54 per cent SiO$_2$. On the other hand, qualitative consideration of the abundance behaviour of all other elements in Fig. 21.2 indicates that these might be accounted for by crystal subtraction. This may now be examined quantitatively.

Problems of this kind are now commonly investigated by computation based on least-squares mixing procedures, and such an approach is appropriate to the present case.

It has been demonstrated in Chapters 5–19 that in addition to showing well-defined relations with change in whole-rock SiO$_2$ (as in Fig. 21.2) all the ore elements exhibit clear and systematic trends on ternary X–MgO–(Na$_2$O+K$_2$O) diagrams. In particular the olivine–clinopyroxene-basalts and the lavas of the hornblende-andesite group display pronounced linear fields that give the impression of a complementarity of some kind. Figures 21.6 and 21.7 bring together seven such pairings, and in each case it may be seen that the relevant fields develop a neat but limited overlap in the area of felsic basalt:mafic andesite (and the area generally occupied by the corresponding field of the big-feldspar lavas). Cursory inspection of the FeO–MgO–(Na$_2$O + K$_2$O) diagram (Fig. 21.6) indicates, to a first approximation, that:

1. The two trends have been generated from different parents by the subtraction of materials of quite different compositions.
2. The highly linear basaltic field may have been generated in a closed system by olivine and/or olivine–clinopyroxene fractionation in, say, a high-Mg basalt parent.
3. The linear but somewhat less well correlated field of the hornblende-andesites may have been generated (a) by magnetite–clinopyroxene–hornblende fractionation in a felsic basalt parent — a melt of composition now represented, perhaps, by the intersection

298 *Ore Elements in Arc Lavas*

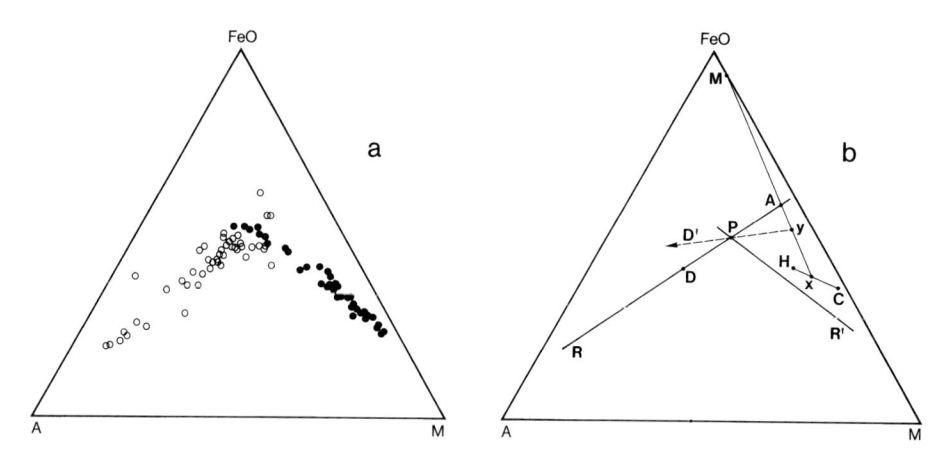

Fig. 21.6. (a) FeO–MgO–(Na$_2$O + K$_2$O) ternary relations developed by the olivine–clinopyroxene-basalts and picrites (filled circles) and the hornblende-andesite group (open circles) of the Solomons suite. The three open circles at the upper right of the andesite–dacite lineage are the three basaltic members of the hornblende-andesite group referred to in Chapter 5. These three lavas appear distinctively in a number of diagrams, and are usually the only poor 'fit' in what is generally a smooth conjunction of the two fields. (b) Simple three-component construction showing calculated regression lines for the andesite–dacite lineage (*PR*) and the basalt–picrite lineage (*PR'*). The point of intersection, *P*, is taken as a high-iron fractionation product of the basaltic lineage that then constitutes a 'secondary parent': in this case to the andesite–dacite lineage. Points *H*, *C*, and *M* represent mean hornblende, clinopyroxene, and magnetite respectively of mafic andesites, and point *x* indicates the mean modal proportion of clinopyroxene to hornblende (1.0:1.26) in these lavas. Generation of the trend *PDR* by closed system fractionation in *P* requires the generation of an accumulate falling along the line *PA*, and *A* represents the magnetite–hornblende–clinopyroxene subtractant required. The measured mean modal proportion of these three minerals in the mafic andesites is represented by the point *y*. Closed-system fractionation of *P* involving a magnetite: clinopyroxene–hornblende subtractant represented by *y* would yield accumulates falling along the line *Py* and derivatives falling along the line *PD'*. This lineage possesses a higher proportionate FeO content than does the observed lineage *PDR*: the observed 'subtractant' *y* removes insufficient FeO to generate the observed lineage *PDR* and an increment of FeO must therefore have been lost to the system.

of the basaltic and andesitic fields — in a closed system or (b) by discrete, progressive loss of iron from a basaltic parent in an open system, or (c) by some combination of (a) and (b).

Least-squares mixing computations carried out on the assumption of a closed system confirm that the basaltic lineage may indeed have been generated by olivine and/or olivine/clinopyroxene fractionation in, say, a high-Mg basalt; that the picritic members are compatible with high-Mg olivine/clinopyroxene accumulation; and that on this basis the high-Fe members constitute the felsic derivatives complementary to the picrites (see also Chapter 23).

Investigation of the possible role of fractionation in producing the andesite–dacite trend is slightly less simple. We are concerned here with determining whether closed-system fractionation involving the now-observed phenocrystic phases is — numerically — adequate to generate the various hypotholeiitic andesite–dacite trends of Figs. 21.6 and 21.7, or whether this requires at least some open-system loss. The significant ore-element-bearing minerals in the felsic basalts — the likely parents of the andesites–dacites — are hornblende, clinopyroxene, and titanomagnetite. Mixing calculations show that no combination of hornblende and clinopyroxene contains enough of the elements concerned to generate alone, without the concomitant subtraction of magnetite, the andesite–dacite trend. Magnetite, with its relatively high trace-metal content, is therefore an essential component of any subtractant, and the nature of the trends generated for each element is, assuming closed-system fractionation, dictated by magnetite:hornblende:pyroxene ratios.

On this basis, and using:

1. the mean of the six most felsic derivatives of the basaltic lineage as the hypothetical parent;
2. subtractant mineral compositions as indicated by the mean compositions of the magnetites, hornblendes, and clinopyroxenes of the more mafic andesites; and
3. total subtractant having mean weight proportions magnetite, hornblende, and pyroxene of 1:1.94: 1.54 — values indicated by density adjustment of the observed mean modal proportions of these minerals in the more mafic andesites;

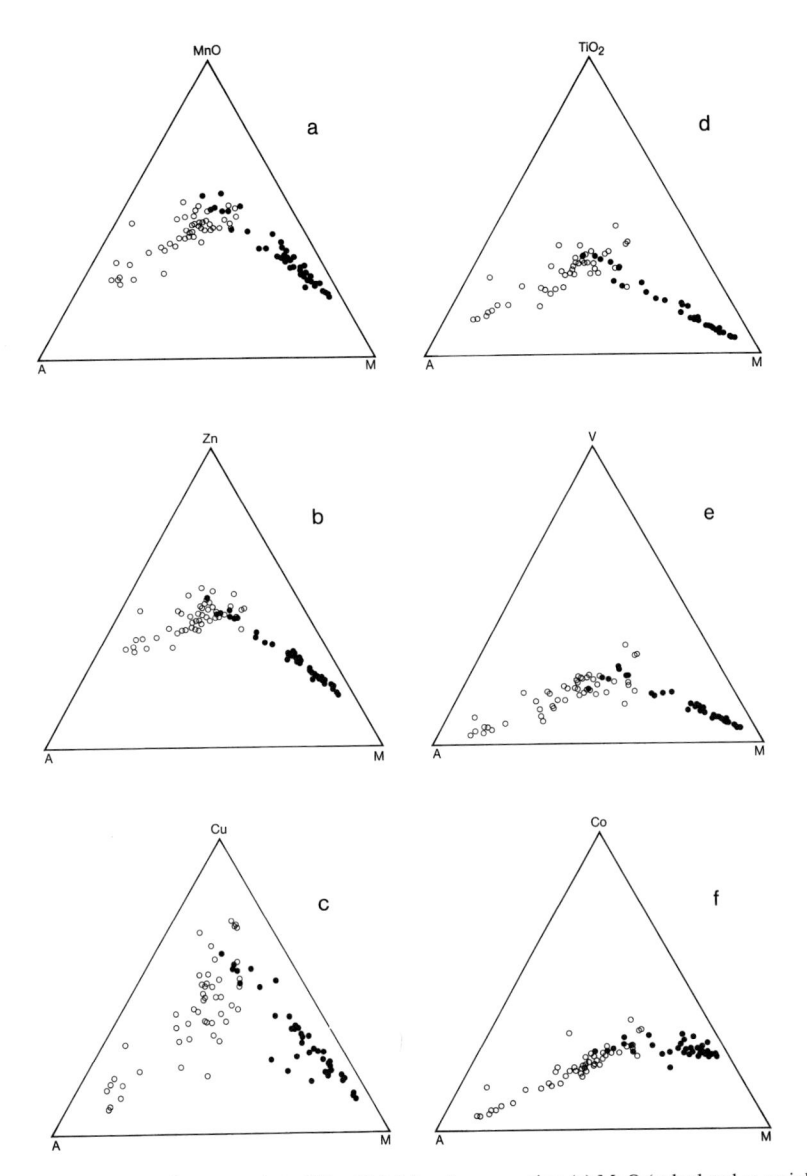

Fig. 21.7. Ternary XMA diagrams analogous to that of Fig. 21.6 (a) and representing (a) MnO (calculated as weight per cent MnO × 50); (b) zinc (ppm Zn × 10^{-1}); (c) copper (ppm Cu × 10^{-1}); (d) TiO$_2$ (weight per cent TiO$_2$ × 5); (e) vanadium (ppm V × 10^{-2}); and (f) cobalt (ppm Co × 10^{-1}). Basalt–picrite lineage shown by filled circles, hornblende-andesite group by open circles as in Fig. 21.6. Calculated magnetite–clinopyroxene–hornblende subtractants required to generate the andesite–dacite lineage by closed system fractionation of a secondary parent represented by the conjunctions of the pairs of fields are given in Table 21.5.

least-squares mixing-subtraction calculations involving the principal hypotholeiitic components — i.e., those of Figs. 21.6 and 21.7: FeO, MnO, TiO$_2$, Zn, Cu, V, and Co — gave the results shown in Table 21.5.

As pointed out above, the adequacy of a subtractant for generating the andesite–dacite trend for any given element of these seven is substantially dependent on its spinel:(hornblende + pyroxene) ratios. The observed

300 Ore Elements in Arc Lavas

Table 21.5 Observed and calculated magnetite:clinopyroxene:hornblende ratios in Solomons andesites–dacites

1 (a)	(b)	2 (a)	(b)	3	4 (a)	(b)	5
Mg:cpx:Hb	Mg:(cpx + Hb)	Mg:cpx:Hb	Mg:(cpx + Hb)		Mg:cpx:Hb	Mg:(cpx + Hb)	
1:1.54:1.94	1:3.48	1:1.62:3.44	1:5.06	FeO	1:0.84:1.05	1:1.89	3.93
				Cu	1:0.25:0.32	1:0.57	4.49
				Zn	1:1.02:1.29	1:2.31	3.85
				MnO	1:1.56:1.98	1:3.54	3.69
				TiO_2	1:1.56:1.96	1:3.52	3.69
				V	1:2.25:2.83	1:5.08	3.59
				Co	1:1.92:2.43	1:4.35	3.63

1, (a) mean magnetite:clinopyroxene:hornblende mass ratios in all Solomons mafic andesites (52 <SiO_2 <55 per cent; N.B. Some of these do not contain hornblende); (b) mean magnetite:(clinopyroxene + hornblende) mass ratios for lavas of (a);
2, (a) mean magnetite:clinopyroxene:hornblende mass ratios in all Solomons hornblende-bearing lavas containing SiO_2 <55 per cent (range, 50.68–54.88; mean = 53.54 per cent SiO_2), (b) mean magnetite:(clinopyroxene + hornblende) mass ratios for lavas of (a);
3, elements/oxides for which mineral ratios are given in cols 4(a) and (b);
4, (a) mean magnetite:clinopyroxene:hornblende mass ratios required to generate, by crystal subtraction, the observed andesite–dacite trends with respect to the components of col. 3, (b) mean magnetite:(clinopyroxene + hornblende) mass ratios as required in (a);
5, mean densities of the relevant magnetite–clinopyroxene–hornblende subtractants.

mean magnetite:clinopyroxene:hornblende mass ratio in the mafic andesites is 1.0:1.54:1.94, and the magnetite:(clinopyroxene + hornblende) ratio is thus 1.0:3.48. If the calculated required ratio for a given element is equal to or less than this, magnetite–clinopyroxene–hornblende fractionation has been adequate to generate the observed trend. If it is greater, the observed mass of magnetite relative to that of (clinopyroxene + hornblende) has been inadequate for the generation of the andesite–dacite trend, and some other process must also have contributed to the subtraction from the melt of the element concerned.

We may consider first the case in which the mean magnetite:clinopyroxene:hornblende mass ratios of the subtractant are those of 'all of the Solomons mafic andesites' as set out in columns 1(a) and 1(b) of Table 21.5, i.e. 1.0:1.54:1.94 and 1.0:3.48.

Of the seven principal elements for which the Solomons suite generates hypotholeiitic behaviour, vanadium and cobalt require magnetite:(clinopyroxene + hornblende) ratios (1.0:5.08 and 1.0:4.35 respectively) smaller than the observed (1:3.48) ratio. The andesite–dacite trend is thus more than adequately accounted for on the basis of observed phenocryst mass ratios.

Appropriate MnO (1:3.54) and TiO_2 (1:3.52) subtraction requires ratios almost precisely those observed in our 'mean mafic andesite', and trends developed with respect to these components are thus neatly accounted for by closed-system fractionation on this basis. The relevant ratios for Zn (1:2.31) and FeO (1:1.89) are distinctly less than the observed value of 1:3.48, indicating that fractional crystallization alone is inadequate to account for the diminution of these components in the melt. The ratio for Cu (1:0.62) is much lower again, indicating that closed-system fractionation alone is grossly inadequate for the generation of the hypotholeiitic trend with respect to this element on the basis of the assumed subtractant.

If, on the other hand, the mean magnetite:clinopyroxene:hornblende ratios of the subtractant be taken as those of the Solomons hornblende-bearing lavas exclusively, as set out in columns 2(a) and 2(b) of Table 21.5, i.e. 1.0:1.62:3.44 and 1.0:5.06, none of the andesite–dacite trends of Figs. 21.6 and 21.7 can be entirely accounted for by closed-system fractionation.

Our question about the adequacy of the closed-system fractionation to account for the development of hypotholeiitic patterns with respect to these seven ore elements in the Solomons suite is therefore answered as follows:

1. The highly linear basalt–picrite trends generated by the elements of Figs. 21.6 and 21.7, together with those for calcium and phosphorus, are entirely attributable to closed-system, high-magnesium, olivine–clinopyroxene fractionation (see, however, Chapter 24).
2. On the basis of the subtractant ratios of columns 1(a) and 1(b), Table 21.5, andesite–dacite trends generated by vanadium and cobalt are entirely attributable to closed-system spinel–clinopyroxene–hornblende fractionation in parental basalts such as the felsic end-members of the basalt–picrite lineage. The andesite–dacite trend with respect to MnO and TiO_2 appears to be *just* attributable to this process.

3. On the basis of the subtractant ratios of (2) above, andesite–dacite trends generated by iron and zinc (and calcium) cannot be accounted for entirely by fractional crystallization in a closed system, and such a process is quite inadequate to account for the observed decrease in copper in the andesites and dacites. Movement of these four elements to another system is indicated; i.e. their abundance patterns have developed, at least in part, in an open system. For phosphorus an accumulative phase is not readily apparent in the lavas, and so there is no petrographic basis for mixing and subtraction calculations. There is, however, no microscopical or chemical evidence indicating that a phosphate-rich accumulate might complement the conspicuous decrease in this element in the andesites and dacites, and it seems likely that the well-developed hypotholeiitic pattern of the Solomons suite with respect to phosphorus also reflects open-system loss.

4. On the basis of the subtractant ratios of columns 2(a) and 2(b) of Table 21.5, i.e. 1.0:1.62:3.44 and 1:5.06, andesite–dacite trends generated by none of the seven elements can be attributed entirely to closed-system magnetite–clinopyroxene–hornblende fractionation and at least some open system loss is indicated in all cases.

Least-squares mixing and subtraction calculations thus provide evidence that development of these seven important hypotholeiitic abundance patterns in the lava series is not attributable to a single process: olivine–clinopyroxene fractionation gave way to spinel–clinopyroxene–hornblende fractionation and, perhaps at much the same time, closed systems gave way to open ones and some elements — perhaps all — were partly lost from the regime of the melt to some phase not now represented in the lava suite.

It may reasonably be asked whether there is any ancillary evidence for a change from closed to open systems accompanying the change from basalt–picrite to andesite–dacite lineages. In fact there may: — information that follows from the above calculations, and that is included in Tables 21.5 and 21.6:

1. The mass ratios of spinel:clinopyroxene:hornblende in the subtractant required for the generation of the mean Solomons andesite–dacite (and hence for the mean Solomons basalt–andesite–dacite lineage) is, with the exception of TiO_2 and MnO, different for the different elements. Had elemental abundance patterns developed in a closed system, the ratios would have been the same for all elements.

2. It follows from this that the total mass of subtractant required for the formation of unit mass of derivative (say the average Solomons andesite–dacite above) is different for different elements. It is a simple (and obvious) numerical consequence of any closed-system segregation process that, for any given parent–derivative pairing, the mass ratio of accumulate to derivative has a single value, to which the relative concentrations of all components must conform. As the relative concentrations of the elements in the Solomons lavas do not conform with this requirement, such concentrations have not developed in a closed system.

3. Simple inspection of Figs. 21.6(a) and 21.7 indicates that, except for cobalt, the fields of the andesites–dacites are always less ordered than those of the basalt–picrites. Such degrees of order may be expressed in a quantitative way by calculating coefficients of correlation (r) with respect to perpendicular axes erected at the A apex of the ternary diagrams of Figs. 21.6 and 21.7, such that the x-axis = the side AM of the triangle. The results of such calculations for the full 14 elements of Chapter 6–19 are given in Table 21.6.

Apart from chromium and nickel, which are extremely sensitive to the slightest vagaries of fractionation, all values of r generated by the basalts–picrites are high and generally >0.9. This is compatible with closed-system fractionation involving a single parent and an essentially constant subtractant (or a numerically similar process), as already deduced for this lineage on the basis of least-squares mixing calculations.

In contrast, r values generated by the felsic basalts–andesites–dacites are (apart from cobalt) conspicuously lower and are generally ≪0.9.

In addition, as noted in Chapter 20, correlation of Rb:Pb, Zr:Pb, and $Ga:Al_2O_3$ values are generally distinctly higher (particularly in the case of $Ga:Al_2O_3$) in the more mafic olivine–pyroxene basalts than the andesitic lavas (see Figs. 20.2, 20.10, and 20.14 respectively).

Why should the two lineages behave so differently?

The mixed crystal series of olivine, clinopyroxene, and spinel in the basalts each display much greater compositional variation than do the hornblende, clinopyroxene, and spinels of the andesites–dacites (Chapter 5). This being so, it might have been expected that the fields of the basalts in the diagrams of Figs. 21.6 and 21.7 would have developed a greater degree of scatter than those of the andesites–dacites. Inspection of Figs. 21.6 and 21.7 indicates that, apart from cobalt, already referred to, exactly the opposite is the case. That is, on the basis of progressive, closed-system fractionation, those lavas — the basalts — that might have been expected to develop a

Table 21.6 Comparison of values of r, the coefficient of correlation, generated by the basaltic and andesitic lava groups in ternary diagrams involving the 14 ore elements

Pattern/element/oxide		r, olivine–clinopyroxene-basalt group	r, hornblende-andesite group
(a)	Tholeiitic		
	Sr	0.972	0.561
(b)	Hypertholeiitic		
	Ba	0.943	0.669
	Pb	0.913	0.632
(c)	Hypotholeiitic		
	Cu	0.928	0.667
	Zn	0.991	0.523
	P	0.993	0.465
	CaO	0.970	0.893
	TiO$_2$	0.981	0.780
	V	0.954	0.812
	MnO	0.972	0.748
	FeO	0.983	0.894
	Co	0.817	0.909
(d)	Nichrome		
	Cr	0.782	0.624
	Ni	0.782	0.769

relatively low level of order in their compositional variation in fact exhibit highly linear, well-correlated trends, whereas those — the andesites–dacites — that might have been expected to develop relatively more ordered patterns of variation display conspicuously *less* ordered trends.

This may imply that the segregation processes leading to the development of the andesite–dacite lineage were less simple than those producing the basalts–picrites. Such a possibility is compatible with the hypothesis that, in the generation of the andesites–dacites, spinel–pyroxene–hornblende fractionation was accompanied by some loss of components, notably copper, zinc, and iron, to another system.

Three elements exhibiting hypotholeiitic trends that might be accounted for by magnetite–pyroxene–hornblende fractionation, titanium, vanadium, and manganese, also exhibit lower coefficients of correlation among the andesites–dacites than among the basalts–picrites: perhaps our 'mean hornblende-bearing lava' subtractant is closer to the truth, and there is a component of loss here as well. In every respect the behaviour of cobalt in the andesite lineage may be accounted for by spinel–pyroxene–hornblende fractionation. The elements showing hypertholeiitic trends — lead, barium, and strontium — yield notably lower coefficients of correlation for the andesites as compared with the basalts, though the relative loss in this case must involve one or both of the alkali metals. The two elements exhibiting nichrome abundance patterns, nickel and chromium, each give essentially comparable degrees of scatter in basalts and andesites.

This returns us to the question posed early in this section: is closed-system fractionation involving the now-observed phenocrystic phases numerically adequate to generate the various hypotholeiitic andesite–dacite trends of Figs. 21.6 and 21.7, or do the latter also reflect the operation of an open system? The calculations indicate that closed-system fractionation is adequate for some elements, but not for others. They also indicate that whereas elements such as copper, zinc, and iron have been lost relative to other rock-forming components in the open system, other elements, particularly those such as lead, barium, and strontium that exhibit hypertholeiitic behaviour, have been retained and enriched in the melt through open-system loss of components such as the alkalis.

We shall consider such processes further in Chapters 23 and 24.

CONCLUDING STATEMENT

With the exception only of copper the principal trace elements of the Solomons Younger Volcanic Suite at first sight appear to conform well with the relationships between trace-element abundances and degrees of crystallization first enunciated by Wager and Mitchell (1951). Elements such as chromium and nickel, and to a lesser extent cobalt, enter early formed crystals and their abundance curves display simple, steep, negative slopes. Others, such as zinc, manganese, vanadium, and titanium (together with the 'majors', FeO and CaO) enter crystals forming somewhat later and exhibit maxima towards the middle stages of lava evolution. Yet other elements such as lead, barium, and strontium (and rubidium and zirconium) appear to 'follow' potassium and enter late-stage crystals or are progressively segregated into the residual melt, and display a pattern of increasing abundance as fractionation proceeds.

The very low concentrations of copper in the Solomons silicates preclude the development of the copper abundance curve by crystal subtraction alone. Concentrations of the other elements in the major silicates and oxides are, however, such that the development of their abundance versus crystallization patterns might be loosely attributed to the single process of fractional crystallization.

Such impressions gained from qualitative interpretation of two-component diagrams are, however, misleading. Least-squares mixing and subtraction calculations indicate that relative abundances of at least some of the elements — notably some of those displaying hypotholeiitic patterns accompanying the change from basaltic to andesite–dacite lineages — cannot have been achieved solely by fractional crystallization in a closed system. The development of andesites from basalts requires either contamination from, or differential subtraction into, other systems: the latter in a phase or phases not now present in the lavas of the Solomons suite. For the reasons given in Chapters 5 and 23, contamination appears unlikely. The alternative is subtraction in a now-absent phase. The numerical considerations, together with the observation that the andesites–dacites show a measurably higher degree of compositional disorder than do those lavas — the olivine–pyroxene-basalts and picrites — whose variation can be accounted for by closed-system processes, indicates that the now-absent phase is gas and vapour. Metals such as copper, zinc, iron, and perhaps some manganese, have been lost, relative to the alkalis and other elements, from the lava system in a gas phase. The alkalis themselves have been lost — again probably in the gas phase — relative to the hypertholeiitic elements lead, barium, and strontium. Elements such as chromium and nickel, and the hypotholeiitic elements cobalt, titanium, and vanadium, appear to have developed their abundance patterns substantially by fractional crystallization in a closed system.

Although present evidence is by no means firm (there may well be a large field here for future research), the material of this chapter seems to indicate that there may be two important points of inflection in the abundances of the ore elements as a lava series, such as that of the Solomons, evolves. The first lies in the ≈50–4 per cent SiO_2 range: the evolutionary stage at which basalts are changing to andesites. This may represent not only a change from olivine–clinopyroxene–spinel fractionation, but also a profound break from closed-system to open-system movement of lava components. The second inflection occurs at ≈58–64 per cent SiO_2: the stage at which andesites are giving way to dacites and rhyodacites, and the large-ion elements are beginning to be affected by processes other than crystal fractionation.

22

The ore elements in volcanic exhalations

If some of the ore elements are indeed lost from the melt in the volatile phase, evidence of this should appear in areas of present-day volcanism. That is, these elements should be seen to be emitted as gas, vapour, or aerosol either from lavas themselves or from orifices clearly connected with lava in a molten, active state. Such emission should also be seen to occur in association with a wide variety of lava compositions: as we saw in Chapter 3, important exhalative orebodies occur in association with a full spectrum of lava types from basalts to rhyolites. And such emission should be seen to be capable of occurring on land, with a minimum of potential for the supply of metals by processes such as sub-sea-floor convection and leaching. Much volcanism — and certainly that part of it with which this book has been concerned — occurs on and around largely isolated volcanic islands or not far from coastlines, and the possible involvement of sea water and meteoric solution must always be kept in mind (see, e.g. Symonds *et al.* 1990 on sea water in the fumarolic condensates of Augustine volcano, Alaska). Such complexities accepted, what we are concerned with in this chapter is the exhalation of potential ore-forming materials directly from the regime of molten lava — as distinct from the materials of such as the present day sea-floor 'black smokers', which may or may not have their sources in active melts. These, and the sulphide deposits they produce, are of course currently regarded as examples *par excellence* of circulating seawater debouchment and the deposition of metals derived from post-magmatic leaching. It is because of this uncertainty that deposits of this kind have deliberately not been considered in the present book, though, as noted in Chapter 4, it is hoped that our evidence may provide some contribution to the resolution of the controversy. Our present concern is with emanations demonstrably derived from molten lava at or very close to the land surface. Where sea water is involved — and it almost certainly is in many cases — it has become completely incorporated in the magmatic process: it has become part of the volcanic melt and its products.

The quantities of materials emitted in the volatile phase by subaerial volcanoes are in fact very large. This is particularly the case with andesitic–dacitic volcanoes, of which Santorin, Krakatau, Ruapehu, Katmai, Mount St Helens, and Pinatubo are recent examples. Estimates of the mass of volatiles discharged from such volcanoes are of course difficult to make because of the short-term and violent nature of their eruption. A number of attempts have, however, been made to quantify discharges, or rates of discharge, from some of the quieter, basaltic volcanoes. An early estimate of volatile discharge from one such was made by the French geologist Fouqué, who, as reported by Geikie (1903, p. 266), '... calculated that, during 100 days, one of the parasitic cones on Etna had ejected vapour enough to form, if condensed, 2 100 000 cubic metres (462 000 000 gallons) of water'. Total volatile discharge involved in eruptions such as those of Mount St Helens in 1980 and Mount Pinatubo in 1991 must have been many orders of magnitude greater than these relatively trivial emissions measured by Fouqué at Etna. Indeed, it is now recognized that the world's oceans and atmosphere — together with large quantities of less dense molecular species that have exceeded their escape velocities and departed the Earth's gravity field — have been produced by terrestrial, i.e. substantially volcanic, de-gassing (Rubey 1955; for rates of degassing see Zindler and Hart 1986, p. 527).

The principal components of the volcanic volatile phase are H_2O, CO_2, and SO_2, together with lesser quantities of H_2S, CO, HCl, HF, chlorine, fluorine, and variable 'minors' such as boron and CH_4. Proportions vary greatly from one volcano to another, with distance from the centre of activity, and with temperature and time. Contamination of juvenile products by groundwater, and, close to the volcanic orifice, by air, constitutes a perennial problem in estimating primary compositions of the volatile phase, but a first approximation to an order of relative abundance of the 'majors' is $H_2O \gg CO_2 > SO_2 > H_2S > HCl > CO > HF$. There are, however, so many variations and reversals from one volcano or volcanic province to another that, apart from the usual dominance of water, such generalizations may be difficult to sustain.

Compounds such as H_2SO_4, H_2SO_3, and NH_4Cl form by reaction at and not far below the contemporary surface. Minor and trace quantities of a wide variety of metal compounds, including compounds of all the elements considered in Chapters 6–19, are (as we have suspected) also emitted. The major part of these is lost into the air during active volcanism, but minor amounts continue to evolve via fumaroles and hot springs long after actual eruption. Such fumarolic products may be sampled as gases, sublimates, and condensates. While they may not be fully representative of the whole volatile discharge in terms of total species or proportions of these, they have provided until recently the best available indication of the incidence of the metals in the volcanic volatile phase. Early sampling was restricted to the throats and immediate vicinity of the orifice. Since the 1960s, however, aerial sampling of volcanic plumes has come to prominence, and this has yielded greatly improved insights into the discharge of volatile metal compounds during actual eruption.

Sampling of 'volcanic gases' and 'sublimates' involves several categories of material:

1. Gases and very finely particulate aerosols emitted from hot fumaroles. These are usually principally magmatic but commonly include a groundwater component and some air.
2. Sublimates, constituting direct gas → solid deposition from the gases and aerosols of (1).
3. Condensates, produced by the condensation of vapour from the fumarolic gases, the principal component of which is generally H_2O. The condensates consist largely of HCl, H_2SO_4, HF, and NH_4Cl. The acids may attack the minerals of the volcanic rock surrounding the fumarole producing a wide variety of Na, K, Al, Ca, Mg, and Fe sulphates, double sulphates, hydrated sulphates, and analogous halides, double and hydrated halides, etc. Crusts of such condensate reaction products may be mistaken for genuine sublimates, from which in fact they are quite distinct.
4. Plume aerosols: these are of course partly analogous to those of the fumaroles. They consist of fine particles of sublimate, evaporation products, fine films deposited round ultrafine silicate (commonly glass) particles, and fine products of silicate–condensate reactions occurring in the plume.

HISTORICAL: c. 1853–c. 1963

There can be little doubt that volcanic fumaroles and sublimates have attracted interest since ancient times. Sodium chloride sublimed about fumaroles of Vesuvius was used as a source of salt by the Romans, and the use of sulphur from the volcanoes of Sicily as a fumigant had been mentioned by Homer about 900 BC. Elie de Beaumont's classical paper on 'Emanations volcaniques et métallifères' (1847: see Chapters 1 and 3) indicates clearly that the metal-bearing nature of many fumarolic gases and sublimates, and their potential significance in ore formation, had been perceived before the middle of the nineteenth century.

Bunsen (1853) was one of the first to undertake quantitative investigation of fumarolic gases. His studies of fumaroles of Hekla, Krisuvik, and Reykjalidh volcanoes of Iceland indicated that N_2, O_2, H_2, CO_2, H_2S, and SO_2 were the principal gases involved but that the proportions of these might vary considerably from one centre to another. He found, for example, that the gases of Hekla contained ≈80 per cent N_2 and 2 per cent CO_2 whereas those of Krisuvik contained ≈1 per cent N_2 and 84 per cent CO_2. Among the sublimates formed about the fumaroles Bunsen observed sulphur and various metallic chlorides, particularly common salt. One sublimate, however, contained 81.68 per cent NH_4Cl, which he ascribed to the reaction of HCl from the lavas with ammonia from organic matter over which the lavas flowed.

C. Saint Claire Deville (1856) appears to have been the first to carry out systematic investigations of relationships between eruption, the development of different types of fumaroles, and the occurrence of volcanic hot springs. Based on his observations of the eruption of Vesuvius in 1855, he classified the various types of exhalative activity into five groups in order of decreasing volcanic intensity:

1. Dry fumaroles: characterized by sublimates of metallic chlorides, with variable fluorides and trace sulphate. Such fumaroles arise directly from incandescent lava, and the subliming volatiles mix with air. Some dry fumaroles emit ammonium chloride.
2. Acid fumaroles: emit water vapour mixed with HCl and lesser H_2SO_3. They commonly generate iron and copper chlorides, which are deposited round the vents.
3. Fumaroles emitting water vapour containing H_2S or elemental sulphur.
4. Mofettes: emit water vapour with CO_2.
5. Fumaroles emitting water vapour alone.

In 1858 Saint-Claire Deville and Leblanc (see Clarke 1924, p. 263) published analyses of volcanic gases of Vesuvius, Vulcano, Etna and other Mediterranean volcanoes. The principal gases found were N_2 (from air), O_2, CO_2, SO_2, and H_2S. They also noted the deposition of boric acid in some instances. They concluded that (1) the nature of the emanations at a given point varies with time elapsed since the beginning of the relevant eruption;

(2) the emanations at different points vary with their distance from the volcanic centre; and (3) in both cases the order of variation is the same.

These observations were quickly followed by those of Fouqué (1865), who studied, *inter alia*, three fumaroles at different temperatures within the crater of Vulcano. His results, given in Table 22.1, indicate very clear changes in the proportions of the principal gases with changes in temperature. In all cases HCl ≫ SO$_2$, and around the vents AsS, FeCl$_3$, and NH$_4$Cl were deposited. Another group of fumaroles at temperatures ≈100 °C were characterized by sublimation of sulphur and in some cases H$_3$BO$_3$.

Table 22.1 Analyses of gases from fumaroles, Vulcano (Fouqué 1865)

Gases	A	B	C
HCl + SO$_2$	73.8	66.00	27.19
CO$_2$	23.4	22.00	59.62
O$_2$	0.52	2.40	2.20
N$_2$	2.28	9.60	10.99
Total	100.00	100.00	100.00

Temperature at A = >350 °C; temperature at B = 250 °C; temperature at C = 150 °C. Quantities given as per cent by mass.

Of the 1865 eruption of Etna, Fouqué noted that the flowing lava itself gave rise to abundant dense fumes which, on cooling, settled on the lava crust to form saline deposits. These were found to be >90 per cent NaCl, but to contain also KCl, Na$_2$SO$_4$, and Na$_2$CO$_3$. Fouqué also found large quantities of FeCl$_3$ and NH$_4$Cl associated with the fumaroles, together with small amounts of copper compounds and traces of chlorides and oxides of manganese, cobalt, and lead. He emphasized that the more volatile substances, especially the chlorides of the heavy metals, tended to escape rapidly into the air.

Silvestri (1867) studied several eruptions of Etna, and their aftermath, and came to conclusions very similar to those of Saint-Claire Deville and Leblanc concerning relationships between fumarolic activity and the decline in intensity of volcanism:

1. The fresh, flowing lava acts as a single large 'fumarole', emitting copious white fumes from its surface. This yields a solid saline residue and a small amount of liquid containing HCl and H$_2$SO$_3$. The solid consists principally of NaCl and lesser Na$_2$CO$_3$, together with <1 per cent KCl and Na$_2$SO$_4$. In some cases the fumes contain copper chlorides which deposit as atacamite (Cu$_2$Cl(OH)$_3$) and, by oxidation, tenorite (CuO).

2. Ammonium chloride fumaroles.
   (a) Acid fumaroles, which appear mostly on the terminal walls of the lava flows, and emit HCl. They also emit FeCl$_3$, which is partly deposited as such and partly oxidized to haematite. With fall in temperature they emit H$_2$S, which on oxidation yields crystals of sulphur.
   (b) Alkaline fumaroles, which develop at lower temperature as the flow cools. These are devoid of HCl and FeCl$_3$ and deposit NH$_4$Cl.
3. Water fumaroles, which give off only water vapour mixed with air.
4. Fumaroles emitting, as the final phase of activity, water vapour and CO$_2$.

By the 1870s it had become recognized that the principal volcanic sublimates were NaCl, KCl, NH$_4$Cl, and sulphur. Describing the Vesuvian eruption of 1872, Palmieri (1873: see Fenner 1933), observed:

Not only the Vesuvian cone, but the whole adjacent country, appeared white for many days, as if covered with snow, when exposed to sunlight. This was due to the sea salt contained in the ashes with which the surface was strewn. (Fenner 1933, p. 93)

A little later, Geikie (1879, p. 241) was to remark of the exhalative component of volcanic activity in general:

With these gases and vapours are associated many substances which, sublimed by the volcanic heat, appear as deposits along crevices and surfaces, ... Besides sulphur ... there are several chlorides (particularly of sodium, and less abundantly those of iron, copper and lead), sal ammoniac, specular iron, oxide of copper (tenorite), boracic acid, and other substances. Sodium chloride sometimes appears so abundantly that wide spaces of a volcanic cone, as well as of the newly erupted lava, are crusted with salt, which can even be profitably removed by the inhabitants of the district ...

In 1887 Mallet added to the list of metals known to be deposited about volcanic fumaroles be reporting silver in volcanic ash from Cotopaxi and Tunguarua in Colorado.

By the late nineteenth century, physics had joined chemistry in the study of volcanic emanations. Libbey (1894), following Janssen's earlier spectroscopic detection of hydrogen in volcanic flames of Santorin (1867) and Kilauea (1883), detected hydrogen, CO, and probable hydrocarbons, together with sodium, copper, and chlorine in Kilauean flames. At the same time chemical studies of sublimates and lava–condensate reaction products were becoming increasingly refined. By the 1880s selenium had been found in the sublimed sulphur of the Lipari Islands, and tellurium in sulphur of both Vulcano and some Japanese volcanoes. In 1899 Bergeat reported AsS, H$_3$BO$_3$, NaCl, NH$_4$Cl, FeCl$_3$, Li$_2$SO$_4$, Na$_2$SO$_4$, K$_2$SiF$_6$, chlorides of potassium, calcium, magnesium, ferrous iron, manganese, and aluminium, together with various compounds of cobalt, tin, bismuth, lead, copper, and phosphorus, in sublimates of the Vulcano crater. At the same time he reported a wide variety of materials sublimed around the fumaroles of Stromboli. Sulphur, selenium sulphur, tellurium, realgar, NH_4Cl, and H_3BO_3 were observed abundantly. Potassium, rubidium, and caesium alums, lithium and thallium sulphates, mirabilite ($Na_2SO_4 \cdot 10H_2O$), glauberite ($CaSO_4 \cdot Na_2SO_4$), and soluble compounds of As, Fe, Zn, Sn, Bi, Pb, and Cu were present, and iodine and phosphorus were detected.

After the Vesuvius eruption of 1906 much attention was directed to the late-stage fumarolic products of this volcano, particularly by Lacroix (1907). Lacroix identified four types of fumaroles, the characteristic encrustation suites of which followed each other in order during the cooling of the source magma:

1. Early, hottest fumaroles ($T \geqslant 650$ °C) yielding principally salts of sodium and potassium (chiefly NaCl, KCl, and $K_3Na(SO_4)_2$).
2. Second-phase fumaroles ($T < 650$ °C) yielding a greater variety of sublimates including sulphides and oxides in addition to halides and sulphates.
3. A third phase characterized by $T \leqslant 300$ °C, a reduction in the variety of minerals deposited, and the dominance among these of NH_4Cl.
4. A final phase at $T < 100$ °C, in which sulphur, gypsum, and opaline silica only were deposited about the orifice.

A wide range of chlorides and double chlorides of sodium, potassium, and metals such as manganese was found, though by this time it was becoming recognized that at least some 'sublimates' were in fact lava–condensate reaction products. Among the genuine sublimates, copper chloride and sulphate, and their oxidation product CuO, were the most prominent.

Lacroix pointed out that while copper compounds are readily recognized by their colours (unless present as the white cuprous chloride Cu_2Cl_2), colourless lead chloride may easily escape detection. After the 1906 eruption of Vesuvius he found that by leaching scoriae with boiling water much lead, sodium, and potassium chloride was extracted. He also noted the presence of visible cotunnite and some crystalline galena, attributing the latter to reaction between $PbCl_2$ and H_2S. Other sulphides found at Vesuvius were sphalerite, covellite, orpiment, realgar, pyrite, and pyrrhotite. The tenorite (CuO) noted above, together with minor cuprite (Cu_2O), formed as sublimates in their own right and as oxidation products of already deposited sulphide or chloride. Various basic carbonates, chlorides, and related compounds, e.g. atacamite, azurite, and hydrozincite, were also noted among the Vesuvius sublimates and their reaction products.

Among the comprehensive list of sublimate materials he observed around the Vesuvian fumaroles Zambonini (1910) listed abundant haematite, which he attributed to the reaction

$$2\,FeCl_3 + 3H_2O \rightleftharpoons Fe_2O_3 + 6\,HCl$$

together with 'pneumtolytic' magnetite and magnesioferrite. Indeed, Vesuvius has long been recognized as notable for the large quantities of iron it has emitted in the volatile phase. Spectacular evidence of this was observed by Breithaupt, and cited by Beyschlag, Vogt, and Krusch (1914):

The intensity with which mineral formation by pneumatolysis may proceed in nature, was illustrated by the filling of a crack in lava poured out from Vesuvius in 1817. By the action of the ferric-chloride vapour upon that of water, this fissure became so rapidly filled with specularite that in 10 days a width of 3 feet was completed. (Beyschlag et al. 1914, vol. 1, p. 133)

Some six years after the great 1906 Vesuvius event, Mt Katmai, one of the volcanoes of the Alaskan Peninsular (Fig. 22.5), erupted, yielding large quantities of rhyolitic ash, lava, and pumice. One mass of rhyolitic melt was injected beneath the poorly compacted, pumiceous filling of an adjacent valley, giving rise to the development of a spectacular fumarole field named 'The Valley of Ten Thousand Smokes' by R.F. Griggs in 1916 (Griggs 1922). The valley, filled with a 100–200 ft thickness of pumice, was some 12 miles long and 4 miles wide, and the scene of some thousands of active, steaming fumaroles.

Some of the most comprehensive studies of the chemistry of fumarole deposits are the pioneering investigations of E. G. Zies, E. T. Allen, and C. N. Fenner (Allen and Zies 1923; Zies 1924, 1929; Fenner 1920, 1933) on the fumarolic encrustations and acid gases of this valley.

The early observations of Allen and Zies (1923) indicated that the steam emitted by the fumaroles contained about 0.4 per cent of a complex mixture of volatile materials, about one-half of which was made up of HCl, H_2F_2, and H_2S. Unlike much of the steam, these were largely of magmatic origin, as were the metallic compounds deposited in fissures and pumice as a result of fumarolic activity. During the early stages of fumarolic activity large quantities of magnetite (some individual masses of the order of hundreds of tons) were deposited in fissures and in the porous rhyolitic tuffs of the valley. Zies (1924, 1929) considered that magnetite resulted from the simultaneous oxidation of ferrous and ferric chlorides:

The highest steam temperature recorded in the Valley in 1919 was 645°, but the magma that has been injected under the Valley must certainly be hotter than the steam escaping from any surface fumarole in the Valley. Any substance with an appreciable vapour pressure at such temperatures can and does readily escape. On the way upwards, the volatile products are cooling and new conditions are continuously encountered ... we can readily understand from the following equations what is likely to happen when steam and the two chlorides of iron escape from the magma. As soon as the gases begin to cool the following reaction takes place from left to right:

$$Fe_2Cl_6 + 3H_2O \rightleftharpoons Fe_2O_3 + 6HCl$$
$$\text{steam}$$

and probably $FeCl_2 + H_2O \rightleftharpoons FeO + 2HCl$. It seems quite probable also that magnetite will form when steam reacts with the two chlorides, provided that the temperature is in the neighbourhood of 550°:

$$Fe_2Cl_6 + FeCl_2 + 4H_2O \rightleftharpoons Fe_3O_4 + 8HCl \text{ (Zies 1929, p. 8)}.$$

Zies presumed that the halide gas or vapour also transported very small quantities of Cu, Zn, Mn, Pb, Sn, Mo, and other trace elements, and that these were deposited as components of the growing magnetite crystal or, subsequently, as sulphide coatings on it. Copper generally occurred as crystalline Cu_2S and CuS on magnetite surfaces; Zn, Mn, Pb, Sn, and Mo within the crystal structure. A magnetite trace-element analysis by Zies (1924, 1929) is given in Table 22.2. A conspicuous feature of the analysis is the virtual absence of Ti, V, and Cr, a clear distinction between orthomagmatic spinel (see Chapters 13, 14, and 15), and this fumarolic — i.e. pneumatolytic — magnetite.

H_2S was one of the abundant gases emitted by the Valley fumaroles, and pyrite, chalcocite, covellite, sphalerite, and galena were common sulphides of the fissures and porous tuffs. Ag, Sb, As, Bi, Mo, Ti, and Ga appeared also in the sulphide encrustations. Zies (1929, p. 19) noted that

We have no idea of the order of magnitude of the amount of the sulfides, we can only infer from the obvious concentration of the minerals in our encrustations and from the extent of the magnetite deposit and the indication of its general occurrence in the Valley that the amount must have been very great. The amount of magnetite and the amount of sulfides that followed the magnetite were certainly not sufficient to form an ore body of economic importance, but it is of interest to note the great quantity of ore minerals that have been separated and concentrated in the relatively short intervals between 1912 and 1919 and between 1919 and 1923.

Varying quantities of selenium and tellurium were found in the sublimed sulphur, one analysis giving 0.13 per cent Se and 0.12 per cent Te. Some fumaroles and their sublimates were characterized by higher abundances of sulphate and fluoride relative to chloride, sulphide, and oxide. In some areas of the Valley fumarolic alteration of the pumice was accompanied by notable molybdenum deposition and some samples contained 0.1–0.5 per cent Mo as MoO_3: an enrichment factor of $\approx \times 100$ relative to the unaltered pumice. In turn, some areas of Mo-rich material contained notable quantities of bismuth and thallium. One analysis of a molybdenum-rich oxide residue gave: Cu, 5 per cent; Pb, 10 per cent; Bi, 5 per cent; Tl, 5 per cent; Sb, 2 per cent; Mo balance ≈ 73 per cent (Zies 1929, p. 30).

Zies (1929) calculated that the fumaroles and hot springs of the Valley produced $\approx 1.25 \times 10^6$ tons of HCl and $\approx 1.5 \times 10^5$ tons of HF per year, virtually all of which was contributed to the adjacent sea.

De Beaumont (1847: see Chapter 3) had seen clearly the probable analogy between fumarolic exhalation at the surface and the emanation of ore-forming fluids from granites at depth. Some 86 years later Fenner, who, in 1919 and 1923, had also worked in the Valley of Ten Thousand Smokes, was able to examine the analogy in a more quantitative way. In a very comprehensive and exhaustively argued contribution (Fenner 1933, p. 61), he traced out the manner 'in which gases may be evolved and residual solutions formed from magmas, their different behaviour and composition, and the results they may be expected to accomplish, especially in the collection, transportation, and deposition of ore ...'. Among a range of relevant aspects, he considered the importance of the halogens — particularly chlorine — in magmatic gases, the vapour pressure (volatilities) of the more common metallic chlorides, the amounts of gas that might be given off from a magma body as evidenced by the gaseous emissions of volcanoes, and the evidence for the presence of metallic halides in those gases — particularly as indicated by deposits associated with volcanic fumaroles.

Table 22.2 Trace elements in fumarolic magnetite of the Valley of Ten Thousand Smokes

Element	Quantity (per cent by mass)
Hg	nil
Pb	0.005
Bi	nil
Cu	0.23
Mo	0.04
Sn	0.004
As	nil
Sb	tr
Cr	nil
V	nil
TiO_2	0.005
Zn	0.47
Ni	0.01
Co	0.02
MnO	0.13
S	0.27

Analyst: E. G. Zies (1924). tr = trace.

Fenner's consideration of the incidence of metallic halides in volcanic gases drew heavily on the observations of others, but he also presented evidence obtained in his own investigations of the Valley of Ten Thousand Smokes. His reporting (1933, pp. 96–7) has the colour one might expect from one who has gathered his evidence under remarkable circumstances and with his own hands:

From small vents along two intersecting cracks misty steam rose. By digging with a prospecting pick, hard crusts were found a little below the surface. Specimens collected showed no magnetite, but were impregnated with a silvery gray mineral of metallic lustre. Some of this was in thin, smooth films, and some in crystals so small that only with difficulty could cubic forms be distinguished with a hand lens ... From the best looking specimen of the several collected, weighing 266 grams, the author subsequently broke a sample which appeared to be of strictly average quality, and determined lead. The result was 11.8 per cent PbO, equivalent to 12.6 per cent PbS.

In another fumarolic area, a quarter to half a mile away ... vents ... were lined with hard crusts containing sphalerite, cotunnite ($PbCl_2$), and covellite, together with much finely divided ferric oxide. The sphalerite was in relatively large crystals, the cotunnite in much smaller crystals, and the covellite in small crystals and in films of such delicacy that many were carried away by air currents...

In these fumarolic deposits the heavy metals were clearly brought up in the vapour phase, as no liquid solution had at any time been active. The work of Zies (1929) on the encrustations collected in 1919 has shown that minute amounts of a great variety of metals were widespread, and the total quantity was large, but most of them were dispersed in the porous tuff. Iron oxides, however, were more abundant, and, commonly as finely divided hematite but in places as magnetite, they had frequently been observed to line the open throats of hot fumaroles. With the waning of activity between 1919 and 1923 sulphides of lead, zinc, and copper in considerable concentration had been deposited.

The next fifteen to twenty years saw the somewhat sporadic continuation of field and chemical studies of this general kind. Zies (1938) referred again to the Valley of Ten Thousand Smokes. Naboko (1945 and later: see Naboko 1959) commenced his studies on the sublimates of the Kamchatka–Kuriles volcanoes (see below), Foshag and Henderson (1946) reported on the sublimates of Paricutin, and Shima (1957) drew attention to nickel (up to 1.64 per cent) in a sublimate of Shirane Volcano in Japan. The end of an era of some 110 years of 'on-the-ground' study of volcanic exhalation may be seen to have come to an end in 1963 with the publication of White and Waring's (1963) comprehensive review on 'Volcanic emanations'. While field studies of gases and sublimates were certainly to continue (1963 in fact saw the beginning of R. E. Stoiber's long-continued studies of the chemical products of the Central American volcanoes), these were soon to be added to by the sampling of volcanic clouds — and hence some of the earliest products of the volcanic episode — by high-flying aircraft. In this the volcanic geochemists were greatly aided by the coincident logistic interests of the cloud physicists (in their search for natural nuclei of atmospheric ice) and the environmental chemists, with their interest in distinguishing between natural and 'anthropogenic' pollution of the atmosphere and hydrosphere.

THE MODERN ERA: *c.* 1963 TO THE PRESENT

Up to this point it has been convenient to consider the investigation of volcanic exhalation on an historical basis: some 110 years of increasingly detailed and sophisticated observation and analysis of volcanic gases and sublimates as these appeared around some of the world's great volcanoes, notably those of the Mediterranean.

As will have been seen, most of these studies were concerned with individual volcanoes: with the compositions of gases and sublimates derived from fumaroles and related emissions, with the temperatures at which these were active, and with relations between stages of fumarolic activity and the proximity and progress of the associated volcanic eruption.

With the proliferation of such observations over a wider range of volcanic types and settings, and with the arrival of a capacity to carry out aerial sampling of volcanic plumes during eruption, the data bank has suddenly become much more comprehensive and the opportunities for comparative studies much greater. Investigation of exhalation products is no longer confined to emissions constituting only the dying phases of volcanic gas loss: the plume itself, at the height of eruption, can now be sampled, so that observation of volatile emission encompasses almost the full span of an eruption. Such comprehensive information is becoming available for volcanoes representing mid-ocean ridge, island-arc, and continental volcanism, and for a spectrum of types ranging from basaltic to rhyolitic.

Intra-ocean volcanoes

These include both intra-plate 'hot-spot' volcanism, such as that of Hawaii, and mid-ocean ridge eruption as exemplified by the volcanoes of Iceland and adjacent ridges.

Hawaii. Although the very first investigations of gas emission from the Hawaiian volcanoes were those of Janssen (1883) and Libbey (1894) already referred to, the history of systematic, quantitative study really begins with the pioneering work of A. L. Day and E. S. Shepherd (1913) on Kilauea. By passing an iron tube through the thin wall of a small, active cone, and connecting this through a train of glass tubes to a pump, Day and Shepherd were able to extract gases directly from the molten lava. Large quantities of water (300 ml in ≈15 min) condensed in the tubes at the beginning of the train. This water was found to contain various saline substances, probably at least partly derived from the glass, together with notable quantities of F, Cl, and SO_2. Five samples of gases yielded compositions (by volume) given in Table 22.3. Later analysis by Shepherd indicated the presence also of small quantities of argon, chlorine, and sulphur vapour.

Table 22.3 Analyses of gases from Kilauea (Day and Shepherd 1913)

Gases	1	2	3	4	5
CO_2	23.8	58.0	62.3	59.2	73.9
CO	5.6	3.9	3.5	4.6	4.0
H_2	7.2	6.7	7.5	7.0	10.2
N_2	63.3	29.8	13.8	29.2	11.8
SO_2	nil	1.5	12.8	nil	nil
Total	99.9	99.9	99.9	100.0	99.9

Day and Shepherd were concerned with the major components of the volcanic volatile fraction — water, nitrogen, carbon dioxide, etc. — not with the minor to trace quantities of metallic halides and sulphur compounds that these might bear. Interest in the composition of the associated sublimates soon arose, however, and in 1921 Washington and Merwin described material deposited during the 1920 eruption of Kilauea, which they found to contain 52.12 per cent Na_2SO_4, small quantities of calcium and potassium sulphates, and 0.94 per cent $CuSO_4$. Murata (1960) noted that four molecular bands recorded in spectrograms of volcanic flames taken during the 1960 eruption of Kilauea were the strongest emission bands of the molecule CuCl. He suggested on this basis that an appreciable part of the copper that is volatilized out of Hawaiian magmas is transported as cuprous chloride (CuCl).

In a study of a very different kind Moore and Calk (1971) noted the presence of sulphide spherules, consisting principally of Fe, S, Cu, and Ni (electron-microprobe analysis), on the walls of vesicles in 'drastically quenched' subaerial lavas of Kilauea and Kilauea Iki. Moore and Calk suggested that the vesicles formed by exsolution of a volatile phase deficient in water and composed primarily of sulphur and carbon compounds. Fe, Cu, and Ni diffused through the still-molten basalt towards already formed bubbles. Reaction of S in the gas with the three metals then produced droplets of immiscible sulphide liquid, which quenched upon cooling and formed spherules on the vesicle walls. Their work thus indicated the presence of nickel as well as copper in the Kilauean volatile phase.

Cadle et al. (1973) collected and analysed fumes generated by the lava lake Mauna Ulu, and by highly active vents of Halemaumau, two of the most active areas of Kilauea at that time. The particles were collected on polystyrene filters held by open-faced holders such that the filters were exposed directly to the fumes which were drawn through the former by a pump at about 15 litres/minute. The anions were dominated by SO_4^{2-}, with Cl^- a minor component. The principal cations were, somewhat variably, Si^{4+}, Na^+, Ca^{2+}, Mg^{2+}, and NH_4^+. Among the trace elements (Table 22.4) Cu and Pb were prominent but Ni, as far as the sampling went, was below the limit of detection. Co and Cr appeared, however, in sample 1. Perhaps surprisingly K was absent throughout, and Fe was present at very low levels.

Naughton et al. (1974) also collected from Mauna Ulu using quartz tubes packed with quartz wool that were simply lowered into the fumes above a lava fountain (Fig. 22.1). The principal components of the best sample of sublimate collected in this way were, in order of decreasing abundance: Na, Ca, Al, Fe, Mg, K, B, Si, Ti, Zn, H^+, NH^+, Cu, and Ni in the form of sulphates, chlorides, and fluorides. A complete analysis is given in Table 22.5. In contrast to the results of Cadle et al. (1973), Zn and Ni were prominent in the material collected by Naughton et al.

Table 22.4 Trace elements in particulate matter of Hawaiian volcanic fumes

	1	2	3
Mn	250	50	nd
Cr	200	nd	nd
As	24	13	nd
Sb	200	nd	nd
Fe	4000	nd	nd
Co	20	nd	nd
Ni	nil	nd	nil
Cu	900	nil	550
Cd	37	nd	180
Pb	125	nd	180
Hg	nd	100	nd
Se	nd	120	nd
B	nd	4000	nd

1 = Trace elements in particles collected from fume from Halemaumau crater on 5 April 1972 for 5 h at 15 litres/min. **2** = Ditto, on 1 December 1972 for 6.5 h at 15 litres/min. **3** = Ditto from lava lake of Mauna Ulu on 19 October 1972 for 15 min at 15 litres/min. All expressed in nanograms. nd = not determined. From Cadle et al. (1973).

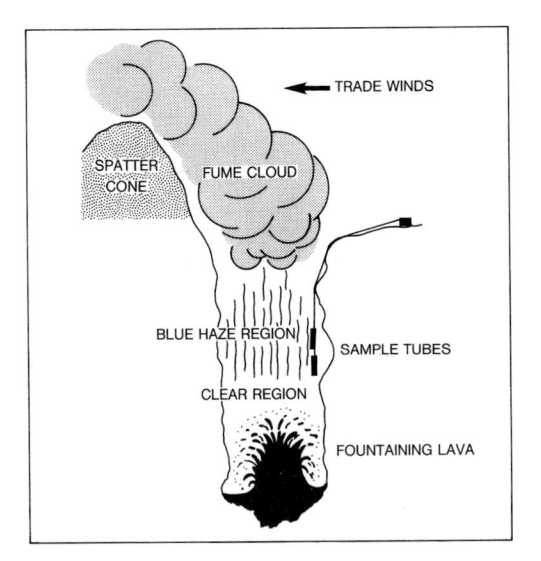

Fig. 22.1. Diagrammatic representation of lava fume development and the fume-condensate collection procedure over the lava fountain studied by Naughton *et al.*, in the summit crater of Mauna Ulu, Kilauea Volcano, Hawaii, during 1970. After Naughton *et al.* (1974).

Table 22.5 Chemical composition of Hawaiian lava fountain fumes

Major cations		Minor cations		Anions and other	
Na	4.1	NH_4^+	0.22	SO_4^{2-}	74.0
K	1.5	Ti	0.33	Cl	5.4
Ca	3.5	Zn	0.25	F	1.6
Mg	2.2	H^+	0.22	B	0.59
Al	2.8	Cu	0.12	SiO_2	0.39
Fe	2.6	Ni	0.04		

Sample no. 1 of January 1970, collected from above Mauna Ulu lava fountain (see Fig. 22.1). Quantities given as per cent by mass on an anhydrous basis. From Naughton *et al.* (1974).

In the following year Naughton *et al.* (1975) reported further work on Kilauea, noting that among the trace elements emitted during this period of sampling Cu, Ni, and ammonium ions were present sporadically in concentrations just above the limits of detection. Zn appeared to be an important trace component of the fumes.

Airborne particulate matter from the January 1983 eruption of Kilauea volcano was inadvertently collected on air filters installed at Mauna Loa Observatory to observe particles in global circulation. Zoller *et al.* (1983) analysed the fortuitously collected material and calculated enrichment factors (using Al as marker) relative to the USGS standard Hawaiian basalt BHVO-1 such that

$$EF = \frac{(X/Al)_{particles}}{(X/Al)BHVO\text{-}1}$$

where X is the concentration of the element of interest in (1) the particulate matter and (2) the standard basalt. Of the elements of present interest, Co, V, and Ti (and Ca and Fe) gave enrichment factors of 1–3, Cu and Zn factors of 10–50. The volcanic plume was thus substantially enriched in these elements in the gas phase, in particles produced by condensation in the plume, by condensate on fine dust, or in combinations of these.

Iceland. Beginning with the observations of Bunsen (1853) already referred to, Iceland and its southern offshore volcanoes of Heimaey, Surtsey, and Syrtlingur were the scene of some of the earliest volcanochemical investigations. Situated directly on, and an important part of the Mid-Atlantic Ridge, the Icelandic archipelago presents a pre-eminent view of gas-chemical emission associated with subaerial MORBs volcanism.

It is therefore perhaps appropriate that the volatile discharge of the volcano of Syrtlingur was one of the first to be investigated by aerial sampling of a volcanic plume (Fig. 22.2). McClaine *et al.* (1968) obtained 46 aerosol samples by flying through the cloud of this volcano, and a further 8 from a steaming vent on the adjacent, newly formed volcanic islet of Surtsey.

In addition to Fe and Ca among the elements of present interest, McClaine *et al.* determined Cu, Cr, P, and Ti in several of the surface vent samples. The plume aerosols also contained these elements together with commonly occurring Ni and minor Mn. Of the 46 plume aerosols, 9

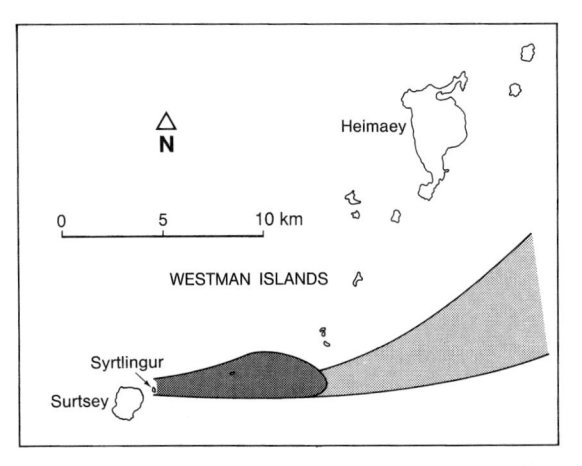

Fig. 22.2. The approximate shape and extent of the Syrtlingur plume (lighter stipple) and area of airborne sampling (heavier stipple). After McClaine *et al.* (1968).

312 *Ore Elements in Arc Lavas*

contained measurable Cu; 12, Cr; 13, P; 15, Ni; 2, Mn, and none contained measurable Ti. The authors noted the remarkable concentration of Cr and Ni relative to, e.g., Fe in the aerosols as compared with the parent lavas. Their thermodynamic analysis indicated that the chromium would be highly volatile as a hydroxy compound in spite of its low concentration in the melt. On the other hand, probe analysis indicated a common association of Cr, S, and Cl, indicating that the volatile species might be a sulpho-chloride of chromium. They observed that their thermodynamic analysis could not, however, account for the high frequency of observation of nickel relative to iron in the aerosols in view of the approximately 1:1000 ratio of Ni:Fe in the parent lavas. The best explanation they could offer for this apparent anomaly was the possible atmospheric oxidation of volatile ferrous to even more volatile ferric compounds, thus inducing a relative concentration of nickel by the selective evaporation of iron. The alternative was that, through the formation of some as yet unrecognized complex vapour species, nickel was preferentially vaporized from the melt relative to iron and other elements.

Mroz and Zoller (1975) investigated the incidence of a range of trace elements, including Zn, Mn, V, and Co, occurring in aerosols, lava-ash, and fumarole deposits collected shortly after the 1973 eruption of Heimaey. Of the elements of present interest one aerosol contained (in ng/m^3) ≈2300 ng Fe, 1800 ng Ca, 85 ng Zn, 79 ng Mn, 3.3 ng V, and 1.9 ng Co. A lava-ash sample contained ≈9.8 per cent Fe, 8.4 per cent Ca, 490 ppm Zn, 1990 ppm Mn, 270 ppm V, 34 ppm Co. Material from fumarole deposits gave ≈7 per cent Fe, 6 per cent Ca, 200 ppm Zn, 1200 ppm Mn, 210 ppm V, and 23 ppm Co. Mroz and Zoller calculated enrichment factors relative to average crust using Al as a marker* and found that Zn (and Sb, Br, and Se) were much higher in the fumarole and aerosol samples than in the lava ash. They therefore concluded that these elements had been volatilized out of the melt, possibly in the form of halides. By comparison with enrichment factors determined by others for Hawaii, and with atmospheric measurements made at the South Pole and the North Atlantic, Mroz and Zoller suggested that 'the relative amounts of enriched elements injected into the atmosphere may vary widely from volcano to volcano, depending on magma composition and eruption characteristics.' (Mroz and Zoller 1975, p. 463)

Wood *et al.* (1979) determined up to 150 ppm Cu, 130 ppm Zn, 100 ppm Ni, 1600 ppm Ti, and 1240 ppm Sr in halide encrustations associated with the Icelanditic 1970 eruption of Hekla. They also found up to 330 ppm Cu and 20 ppm Zn in encrustations produced by the tholeiitic basaltic activity of Askja of 1961, and up to 5530 ppm Cu, 1460 ppm Zn, 1300 ppm Pb, and 80 ppm Ni in sulphates formed in association with the mildly alkaline basaltic activity of Surtsey of 1963–7.

Oskarsson (1981) investigated the nature and trace-element contents of 77 halide and sulphate encrustations of five Icelandic eruptions: those of Askja 1961, Surtsey 1963–7, Hekla 1970, Heimaey 1973 (hawaiite), and Leirhnjúkur 1975–7 (tholeiitic basalt) (Fig. 22.3). One of his aims was to evaluate the mode of transport of the trace metals and, as a contribution to the understanding of magmatic ore genesis, the significance of the F/Cl ratio of the medium transporting volatile metal halides. He determined values of up to 1150 ppm Cu, 130 ppm Zn, 100 ppm Ni, 1600 ppm Ti, 160 ppm Mo, 1240 ppm Sr, and 730 ppm P in the halides and 5530 ppm Cu, 2970 Zn, 80 ppm Ni, 6200 ppm Ti, 780 ppm Mo, 130 ppm Sr, 1300 ppm Pb, and 2900 ppm P in the sulphates. The sulphate encrustations were secondary and formed by conversion of halides to sulphates by reaction with sulphur gases (SO$_2$ etc) continuously given off by the cooling lava. The chemistry of the major and trace-metal sulphates thus reflected the proportions of the halogens involved in primary transport (see below). Among the halides, fluorides were present (i.e. clearly evident) in the

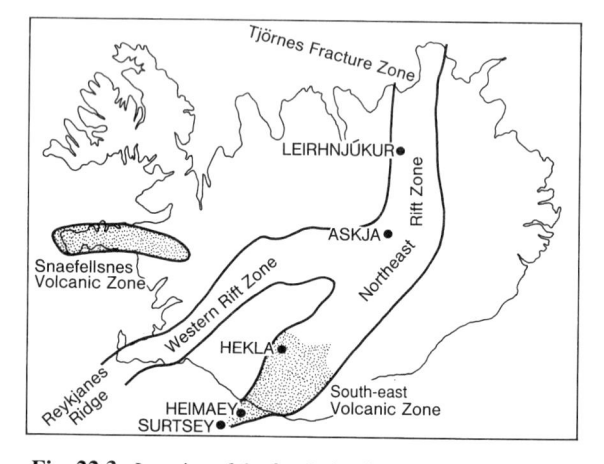

Fig. 22.3. Location of the five Icelandic volcanoes studied by Oskarsson, showing their geographical relationships with the principal rift and volcanic zones. After Oskarsson (1981).

---

* EF = $\dfrac{(X/\text{Al})_{\text{sample}}}{(X/\text{Al})_{\text{crust}}}$

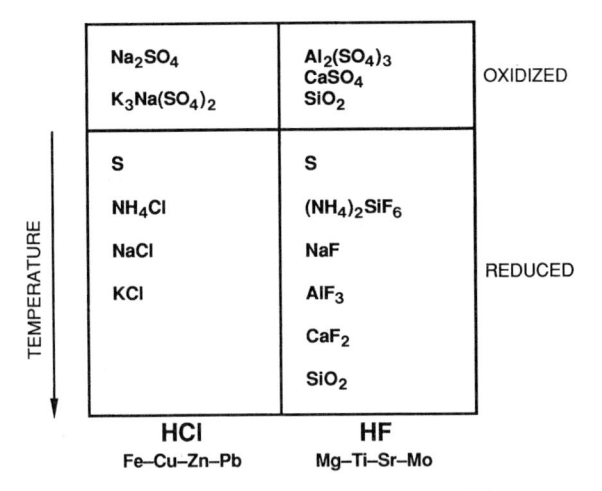

Fig. 22.4. Sublimate fractionation scheme deduced from assemblages observed around the five Icelandic volcanoes. Chlorine- and fluorine-rich and oxidized and reduced environments are as indicated. Direction of rise in temperature shown by arrow at left. Typical trace-metal assemblages are as indicated as the base of the diagram. After Oskarsson (1981).

magmatic volatiles of Surtsey, Hekla, and Heimaey, but low to absent from those of Askja and Leirhnjúkur.

Oskarsson concluded that the incidence of the trace metals in the encrustations was controlled by the F/Cl ratio of the transporting gas phase rather than by element abundances in the silicate melt (Fig. 22.4). He suggested that diffusion of hydrogen from the silicate melt during early degassing resulted in a lowering of the $p(H_2O)$. This favoured the distillation of halides rather than the halogen acids. The F/Cl ratio of the gas phase then controlled the metal chemistry of the vapour and resulting encrustation. In Cl-rich systems the alkali metals dominated and the most abundant trace metals were Fe, Cu, Zn and Pb. In F-rich systems the major cations were Ca, Al, Na, Si, and the principal trace metals were Ti, Mg, Mo, and Sr. The F/Cl ratio of the gas phase as deduced from encrustation mineral assemblages, indicated regional, i.e. tectonic and petrological, influences in the incidence of the halogens in the Icelandic magmas. The rift-zone volcanism (tholeiitic) exhibited low total halogens, very low F/Cl ratios, and higher Fe–Cu–Zn–Pb, while the off-rift volcanism (alkaline lavas) was associated with higher total halogens, high F/Cl ratios, and higher Mg–Ti–Sr–Mo among the traces.

Island-arc volcanoes

No comprehensive information is available on the compositions of exhalations associated with intra-oceanic arc volcanoes, such as those of the Solomons and Vanuatu. In 1959 the writer attempted to sample the very late stage, low-temperature exhalations of Savo, but this was unsuccessful and no further work has been done since. All arc data currently available relate to provinces, such as Japan and Indonesia, either adjacent to, or overlying, continental crust. The data necessary for direct comparison of ridge and hot-spot volcanism with that of oceanic arcs are therefore not yet available. All the examples of arc volcanoes considered below have at least some continental connection, and indeed this is substantial in some cases.

Merapi, Java (Indonesia). Although their investigations were concerned with the nature and equilibria of the abundant gases, and made no reference to the incidence of trace metals. Allard (1980, 1982) and Le Guern et al. (1982) have shown that the exhalations of Merapi are likely to have a substantial crustal component. At least some of the water is meteoric and much of the carbon and sulphur could have come from sedimentary sources. The thermodynamic calculations of Le Guern et al. indicate, however, that the gases were once at temperatures of at least 915 °C with oxygen fugacities close to that of the quartz–magnetite–fayalite buffer, and that they have not been significantly diluted by disequilibrium addition of meteoric water below these temperatures. The Merapi magma is a high-K tholeiitic andesite.

Symonds et al. (1987) collected and analysed condensates, silica-tube sublimates, and encrustations from 500–800 °C fumaroles at Merapi. Among the elements of present concern they calculated that the gases were enriched, with respect to the andesitic melt, by factors $>10^5$ in Bi and Cd; $>10^4$–10^5 in Pb, Au and W; $>10^3$–10^4 in Mo and Ag; $>10^2$–10^3 in Zn, As, and Rb, and >1–10^2 in Cu, Ni, V, Fe, Mn, Sb, and Ga. The fumaroles were transporting (principally as the chlorides) $>10^6$ g/day of S, Cl, and F; $>10^4$–10^6 g/d of Br, Al, Fe, K, and Mg; $>10^3$–10^4 g/d of Pb, As, Mo, Mn, V, W, and Sr; $<10^3$ g/d of Cu, Ni, Cr, Ga, Sb, Bi, Cd, Li, Co, and U. On the basis of a 2.65 weight per cent water loss on eruption and a gas composition deduced from the observed gas and condensate data, Symonds et al. calculated that degassing of 300 km³ of Merapi andesite would produce a total of $\approx 21\,000 \times 10^6$ tonnes H_2O; 970×10^6 t S; 74×10^6 t Cl; 3.5×10^6 t F; 0.33×10^6 t Br; 0.21×10^6 t Zn; 0.14×10^6 t K; 17×10^3 t Pb; 11×10^3 t As; 4.1×10^3 t Mo; 3.7×10^3 t Mn; 3.5×10^3 t V; 2.7×10^3 W; 1.9×10^3 t Cu; and 1.2×10^3 t Bi.

They concluded that incomplete degassing of a subvolcanic melt at 915 °C was the origin of most of the elements in the Merapi gas, though the latter was slightly contaminated with fine particulate material: either fine rock particles or products of wall-rock reactions. The metals were emitted predominantly as chloride species. As the gas cooled in the upper throat and orifice of the fumarole, it became saturated with sublimate phases that fractionated from the gas in the order of their equilibrium saturation temperatures.

Showashinzan, Hokkaido, Japan. Oana (1962) reported a semi-quantitative study of fumarole gases and sublimates, collected at different times during the period 1954–9, from the dacitic volcano of Showashinzan. Fumaroles were still active, at temperatures up to 750 °C, in 1959, some 15 years after the birth and initial eruption of the volcano.

Water condensed from high-temperature vapour was found to contain minute traces of the elements of present concern: Cu, 0.003–0.05 ppm; Zn, 0.01–0.7 ppm; Pb, 0.004–0.05 ppm; Ni, 0.0005–0.001 ppm. Abundances in the sublimates were much greater: Cu, 0.005 per cent; Zn, 0.25; Pb, 0.001–0.20. Oana commented that the major elements tended to sublime about the fumarole vents in a zonal arrangement Si–Fe–Al–Ca–Mg–Na–K corresponding to a progressive decrease in temperature. He gave no indication, however, of element–temperature relations for the heavy metals.

Kamchatka–Kuriles volcanoes. Naboko (1959) investigated fumarole–hot spring activity associated with the basaltic–andesitic volcanoes of Klyuchevskoy, Sheveluch, Koshelev, and Kikhpinych on Kamchatka Peninsular and of Golovnin and Mendeleyev on Kunashir Island. He observed that 'gas distillation' from the melt was most intense at 'the first high-temperature stage of the lava flow'. From the evidence of the sublimates he concluded that when the basalts and basaltic andesites of Klyuchevskoy were actively flowing at 800–1000 °C, and down to 500 °C, sodium and potassium chlorides were volatilizing out of the lavas in large quantities and in approximately the same proportions in which they occurred in the lava, i.e. Na/K ≈3/1. He also noted that 'Visible quantities of iron chloride was flying [i.e. volatilizing, author] out of basalt, becoming part of sublimates formed ... (Naboko 1959, p. 123). In contrast 'aluminium, magnesium and calcium were flying in insignificant quantities and only sometimes could be qualitatively determined in the first products of exhalations.'

Among the heavy metals, copper was the most conspicuous constituent of the Klyuchevskoy sublimates, in some cases constituting up to 60 per cent of the sublimate and occurring as chlorides and oxy-chlorides. Zn, Pb, Ba, Sr, Cr, V, Co, Ni, Zr, Mo, Bi, Ga, Ag, and Sn, also as chlorides, were all found at higher concentrations in the sublimates than the lavas. Si and Fe also volatilized abundantly, as fluorides.

Naboko noted that the sulphur gases were more prominent in the fumaroles of the andesitic volcano of Sheveluch than in those of the more basaltic volcano of Klyuchevskoy. Sodium and potassium were again extensively volatilized (in this case as the sulphides), but in addition up to 5 per cent CaO and 2 per cent MgO were determined in the Sheveluch sublimates. The latter were also characterized by the conspicuous occurrence of vanadium. This was emitted in sulphurous gases, the resulting sublimates exhibiting vanadium enrichment factors of up to 1200 with respect to the associated andesitic lavas. Up to 12 per cent V_2O_5 was determined in some samples. Small quantities of Zn, Pb, Ba, Sr, Zr, Mo, Sn, Ga, and Be were also present in the sublimates.

Naboko also carried out specific observations on the four satellite cones of Klyuchevskoy: Biliukai (formed in 1938), Iubileiny proryv (1946), Apakhonchich (1946), and Tuila (?):

1. Early exhalations were richer in metals than later ones. Fumarolic sublimates formed in the first year of activity of Iubileiny proryv contained more Cu, Zn, Pb, Ba, Sr, Cr, V, Ni, and other traces than did those of the second year. Sublimates of Apakhonchich fumaroles active at the time of eruption commonly contained Cu, Zn, Pb, Ba, Sr, V, Cr, Co, Ni, Bi, Te, Ga, and Li; those of Iubileiny proryv, which by that time had cooled for 1.5 years, contained these elements more rarely; those of Biliukai, cooled for 10 years, contained Cu, Ba, Sr, Cr, V, Co, Ni, Bi, and Te; and in sublimates forming on Tuila, 17 years after eruption, only Cu, Cr, and V were found.
2. The concentrations of trace elements in the given lava were dependent on its composition: in the basalts of Klyuchevskoy only Cu, Ba, V, Cr, Co, Ni, Mo, Zr, and Ga were found, whereas in the andesites of Sheveluch these elements occurred in greater abundance and were accompanied by Zn, Pb, Sn, Ag, Bi, and Be as well.
3. Concentrations of trace elements in sublimates as compared with the parent lavas is related to the composition of the latter. Enrichment in sublimate with respect to lava appeared to be greater for basaltic than for andesitic activity.

Augustine volcano, Alaska. This is an andesitic island volcano situated in Cook Inlet, Alaska (Fig. 22.5). It occurs at the geographical and tectonic merging of the island arc of the Aleutians with continental Alaska, and has, therefore, a double 'island arc-continental' affiliation. In this respect it is perhaps not fundamentally different from the three arc environments considered above, but it

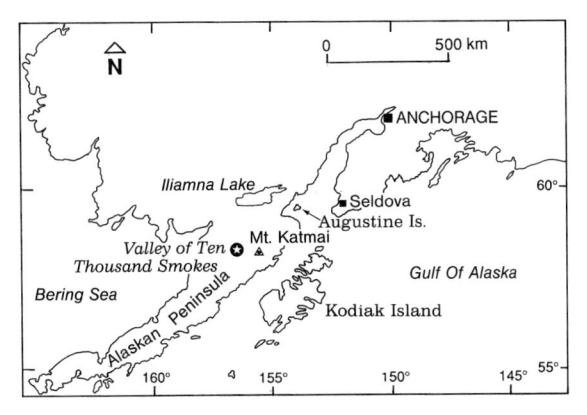

Fig. 22.5. Mount Augustine volcano (Augustine Island), Mount Katmai, and the Valley of Ten Thousand Smokes, situated at the base of the Alaskan Peninsular.

does constitute a particularly clear intermediate position between unequivocal island-arc volcanoes and those, such as Mount St Helens and the volcanoes of Central America, that occur near the edges of continents.

Augustine is one of the most active Alaskan volcanoes, having erupted six times in the past 200 years: in 1812, 1883, 1935, 1963–4, 1976, and 1986. It is of andesite–dacite type, and consists of a central dome complex surround by a volcaniclastic cone.

Samples of particulate matter of the plumes generated by the 1964 and 1976 eruptions were obtained by Lepel et al. (1978) using an aircraft specially equipped for atmospheric observation and sampling. These authors collected and analysed seven samples from the plume and, for comparison, one surface sampling of coeval andesitic ash.

Their results showed that the volatile elements Cu, Zn, Pb, Cd, Au, As, Sb, Se, Hg, S, Br, and Cl are enriched in the atmospheric samples relative to the ash by factors ranging from 10 to >1000, where

$$EF = \frac{(X/Al)_{aerosol}}{(X/Al)_{ash}}$$

and X is the abundance of the element considered. These appear to be of the same order of magnitude as enrichments observed in other volcanic areas. Samples collected early in the eruption cycle had a much greater quantity of volatile elements relative to ash than those collected later, indicating, as recognized earlier by Naboko (1959) and others, a substantial waning in the emission of volatiles with time and with progress of the eruptive cycle.

More recently Symonds et al. (1990, 1992) have made extensive samplings (1986–9) of gases emitted after the 1986 eruption. In addition to chemical analysis of the gases and fumarolic products, Symonds and his colleagues carried out extensive thermochemical calculations and have presented highly detailed accounts of the nature and conditions of derivation of the volatile products. Vent temperatures ranged from an 870 °C fumarole on the central lava dome to 650–200 °C fumaroles round the flanks of the volcano. They found the Augustine gases to be notably rich in HCl (5.3–6.0 mol per cent HCl) and sulphur (7.1 mol per cent total S), but relatively H$_2$O-poor (83.9–84.8 mol per cent H$_2$O). Condensates collected from the high-temperature lava-dome fumarole showed enrichments of 10^7 to 10^2 in Cl, Br, F, B, Cd, As, S, Bi, Pb, Sb, Mo, Zn, Cu, K, Li, Na, Si, and Ni. The lower-temperature fumaroles round the flanks of the volcano were generally less enriched in the volatile elements.

On the basis of hydrogen and oxygen isotope data, Symonds et al. (1990) concluded that the Augustine gases were mixtures of primary magmatic water, local sea water, and meteoric water. They suggested, however, that at least a component of the high chlorine content of the gases arose from a source at ≈25 km, or deeper. They also concluded (Symonds et al. 1992, p. 633) that

Because of the high volatility of metal chlorides (e.g. FeCl$_2$, NaCl, KCl, MnCl$_2$, CuCl), the extremely HCl-rich Augustine volcanic gases are favourable for transporting metals from magma. Thermochemical modelling shows that equilibrium degassing of magma near 870 °C can account for the concentrations of Fe, Na, K, Mn, Cu, Ni, and part of the Mg in the gases escaping from the dome fumaroles on the 1986 lava dome. These calculations also explain why gases escaping from the lower temperature but highly oxidized moat vents on the 1976 lava dome should transport less Fe, Na, K, Mn, and Ni, but more Cu; oxidation may also account for the larger concentrations of Zn and Mo in the moat gases.

Continental volcanoes

This category includes volcanism not far from (active) continental edges and having spatial relations with older island-arc structures (e.g. the North Island of New Zealand and the western edge of the Americas), mid-continent volcanoes such as those of the Italian Mediterranean, and those of rift valleys. There is now a substantial bank of data on the volatile products of the first two groups.

Mount St Helens. The plume resulting from the catastrophic eruption of this andesitic volcano on 18 May 1980 was sampled by Vossler et al. (1981) on the following day, 19 May, and by Phelan et al. (1982) in September of that year. The results of Vossler et al. thus represent material ejected very soon after the beginning of activity; those of Phelan et al. represent a somewhat more mature stage of the volcanic cycle.

Vossler et al. deduced that much of the finely particulate material that they analysed (obtained from altitudes of 13.1 and 17.7 km in the plume) represented the physical break-up of the mountain top rather than new lava. The higher-altitude aerosols revealed, however, significant enrichment in Zn, Cd, Sb, and S and the authors suggested:

1. moderately volatile species are vaporized in the high-temperature zone and, upon cooling, condense preferentially on small particles (volcanic glass?) that have greater surface area to mass ratios;
2. the higher-altitude plume clouds contain a relatively higher proportion of such newly generated magmatic material, and hence greater amounts of condensed magmatic volatiles.

Apart from that for cadmium, enrichment factors for 17.7 km material were, however, not high: Cd = 360; Zn = 2.8; Sb = 2.6; S = 2.35; Fe = 1.46; Ti = 1.56; V = 1.38; Co = 0.89.

The material analysed by Phelan et al. (1982) was collected on a single aircraft flight on 22 September 1980, by which time the volcano was emitting a stable plume to altitudes between 2 and 3 km. Calculated enrichment factors* for some elements of present concern were: Zn = 600 ± 720; Ba = 15 ± 9; V = 3.7 ± 2; Cd = 1700 ± 1200; Sb = 4600 ± 3700; Cr = 1300 ± 1900, this last a somewhat doubtful measurement. Phelan et al. also estimated the volcanic particle flux of some of the volatile elements of the St Helens plume (Table 22.6). The estimate of 1.0 tonne of zinc/day appears to be possibly significant from the point of view of the volcanogenic contribution of this major component of exhalative orebodies. This must, however, be seen in the light of the estimate by Phelan et al. of 0.8 tonne/day of arsenic — normally one of the very minor components of volcanic exhalative ores.

White Island, New Zealand. The White Island volcano is an active, andesitic volcano in the Bay of Plenty, a conspicuous indentation in the coastline of the North Island of New Zealand. The volcano lies within the White Island Trench, a tectonic feature situated parallel to and just to the west of the Tonga–Kermadec Ridge, and extending south into the Rotorua–Taupo graben. The latter

Table 22.6 Composition of Mount St Helens plume, 1980 eruption

Element	Mt St Helens, Sept. 1980, tonnes/day* 1	Global volcanic release tonnes/year§ 2
Cl	3.5	35 000
Zn	1.0	10 000
As	0.8	8000
Hg (vap.)	0.090	900
Hg (part.)	> 0.003	> 30
Sb	0.030	300
Se	0.010	100
Cd	0.005	50
Au	0.003	30

Estimated volcanic flux of some elements from **1**, the St Helens plume and **2**, the global volcanic emission. (*) using average elemental concentrations from co-axial plume samples and assuming an SO_2 flux from St Helens of 1000 tonnes/day and (§) assuming a global volcanic SO_2 release of 10^7 tonnes/year. From Phelan et al. (1982).

is the most highly active tectono-volcanic zone of New Zealand and is the locus of abundant modern volcanism, of which the White Island volcano is one manifestation.

The island is currently the site of quite intense high-temperature fumarolic and hot spring activity, particularly in an area of the crater known as Donald Mound (Hamilton and Baumgart 1959).

Sampling by Healy and Rothbaum in 1949 (see Hamilton and Baumgart 1959) yielded several metal-bearing encrustations. A sample of sulphur containing 26 per cent iron oxide residue also contained 0.5 per cent Pb and 0.03 per cent V, and a particularly Fe-rich portion gave 1.0 per cent H_3BO_3. A second Fe-rich sample gave 1–5 per cent H_3BO_3, 0.1 per cent Pb, and 0.02 per cent Cu. An encrustation from the lip of a vent on Donald Mound consisted principally of NaCl, up to 10 per cent KCl, iron oxides, and ≈ 1.0 per cent Cu. A light yellow portion of this sample was found to contain NaCl, KCl, $CuSO_4 \cdot 5H_2O$, 1.0 per cent Pb, 0.3 per cent H_3BO_3, 0.3 per cent As, 0.3 per cent Zn, and 0.5 per cent Sn. A sample of grey, impure sulphur contained 0.3 per cent As, and an SiO_2–$Fe(OH)_3$ deposit from an associated hot spring gave 0.4 per cent Cu.

Hamilton and Baumgart (1959) comment:

These findings indicate that in the vicinity of Donald Mound the magma is so near the surface that the volatile matter coming up

---

*Obtained from the formula $\mathrm{EF} = \dfrac{(X/Al)_{plume}}{(X/Al)_{ash}}$

from it is carrying the chlorides of sodium and potassium directly into the atmosphere ---. In addition, the steam in the fumaroles at Donald Mound is bringing up appreciable amounts of the heavy metals that are found in magmatic ore deposits.

Central American volcanoes. Knowledge of the chemical products of Central American volcanism is due largely to the efforts of R. E. Stoiber, who, with students and other close associates, notably W. I. Rose, has investigated fumarolic gases and sublimates associated with all the major volcanoes from Santiaguito (Guatemala) in the north to Irazu (Costa Rica) in the south (Fig. 22.6).

The most common sublimate minerals — and those constituting the major part of most sublimate assemblages — are simple and mixed sulphates of Na, K, Al, Fe, Ca, Mg, and NH_4^+. Chlorides and, to a lesser extent fluorides, also occur, as do oxidation products such as Fe, Cu, V, and Mo oxides. Stoiber and Rose (1974) conclude that their (very comprehensive) studies of Central American high-temperature fumarolic vents and associated encrustations indicate that in spite of the large number of minerals found, all can be reasonably explained on the basis of a fumarolic gas consisting essentially of H_2O, SO_2, HCl, HF, and possibly CO_2 gases, and very much smaller concentrations of volatile (and aerosol?) cations such as Na, K, Zn, Cu, Pb, V, and Mo.

In many instances encrustations of different minerals are deposited in a succession of concentric zones, which may be sharply defined or gradational, related to changes of temperature and $p(O_2)$ at various distances from the vent. Encrustation suites were found to be most complex and abundant in the earliest stages of a fumarole, just after the flow had come to rest. Cooling produced clearly observable inward migration of encrustation zones and a general simplification of the mineral assemblage.

Stoiber and Rose (1974) considered there to be several distinctive associations of particular elements with the fumaroles of particular volcanoes (Table 22.7), notably copper at the volcanoes of Arenal and Pacaya, zinc–lead at Santiaguito, and vanadium at Izalco (e.g. Stoiber and Rose 1974; Hughes and Stoiber 1985). Izalco was in fact

Fig. 22.6. Volcanoes of Central America, the chemical products of which have been the subject of sustained studies by R. E. Stoiber and his associates [see, e.g. Stoiber and Rose (1974); Hughes and Stoiber (1985)].

the site of the first known occurrence of crystalline, naturally occurring V_2O_5 (Stoiber and Dürr, 1964), and later yielded the four vanadates $Cu_5V_2O_{10}$, $\beta Cu_2V_2O_7$, $Cu_3V_2O_8$, and $(Na,K)_x V_x^{4+} V_{6-x}^{5+} O_{15}$ (where $0.54 < x < 0.90$). These vanadium-rich sublimates were associated with the eruption of basalts containing 266 ppm V. Hughes and Stoiber (1985) concluded that they formed by sublimation from magmatic gas rather than by fissure-wall alteration of the basalt, and that V^{3+} had been evolved from the melt as either or both of VF_3 and VOCl.

The Italian volcanoes: Vesuvius, Etna, Stromboli. The abundance and general nature of the fumarolic products of these three classical areas are too well known to

Table 22.7 Distinctive elements of fumaroles associated with several Central American volcanoes. From Stoiber and Rose (1974)

Volcano	Distinctive elements in encrustation suite
Arenal, Costa Rica	Cu
Cerro Negro, Nicaragua	Cu, Mo, Ti, K, Al
Izalco, Honduras	V, Cu, F, Ti
Pacaya, Guatemala	Cu
Santiaguito, Guatemala	Mo, Zn, Pb, As, Al

require any detailed recapitulation, (although all three occur in a broadly 'continental' setting, Etna is currently regarded as being of 'hot spot' affiliation, Vesuvius and perhaps Stromboli of arc association). As well as abundant halides and sulphates of Na, K, Fe, Ca, and Mg, a variety of compounds of Cu, Zn, Pb, As, Sb, Bi, and Sn and lesser Rb, Cs, B, I, Te, and P are also well known. Fenner (1933) noted that 'Observers of Etna and Vesuvius have considered that copper is the fumarolic metal characteristic of Etna, and lead of Vesuvius, though at each place both metals were present' (Fenner 1933, p. 97). These Mediterranean volcanoes have also been noteworthy for their extensive deposits of sublimed sodium and potassium halides, and for the large quantities of iron transported as the chlorides and deposited (by reaction with atmospheric O_2) as haematite, magnetite, and other spinels (see earlier in this chapter). As with Cu and Pb, there is some indication that other elements may tend to evolve from different centres in systematically different proportions. Again Fenner has remarked, 'Although the magma at Stromboli is poor in potash, KCl was preponderant over NaCl. At Vesuvius, on the other hand, the fumarolic deposits derived from the highly potassic magma are commonly characterized by a large excess of NaCl. Such variations are not easy to explain.' (Fenner 1933, p. 94)

Although such qualitative aspects of the Italian sublimates were well-established by the early twentieth century, it was not until 1978 that the plume of Etna was sampled by modern methods. Buat-Menard and Arnold (1978) collected samples of atmospheric particulate matter, by air filtration at ground level, from various plumes in the Etna crater. Their results are given in Table 22.8. (Buat-Menard and Arnold collected only the atmospheric particulate matter, so that elements present in the gas phase were not determined: the calculated discharge F_V for each element as given in Table 22.8 thus constitutes a lower limit for the discharge of any given element). The authors noted:

1. All concentrations are significantly higher in hot vent material than in the main plume.
2. For all samples a significant correlation ($r > 0.8$) is observed between the particulate Cl concentration and those of Mn, Fe, Co, Ni, Na, K, Cs, Br, and Sb.

Table 22.8 Mean atmospheric concentrations and enrichment factors of elements contained in particles emitted in Mount Etna plumes and vent fumes of the 1976 eruption

Element	(A) ($\mu g\ m^{-3}$)	(B) ($\mu g\ m^{-3}$)	EF (A)	EF (B)	F_V (tonnes/day)
S	1400	3100	2.2×10^4	2.4×10^5	420
Cl	990	5000	3.1×10^4	7.8×10^5	300
K	94	5900	8.9	5.8×10^3	28
Na	87	3900	15	3.4×10^3	26
Ca	69	36	6.8	17	20
Br	21	260	3.6×10^4	2.1×10^6	6.3
Al	20	3.8	1.0	1.0	6.0
Fe	10	290	0.7	110	3.0
Zn	10	78	580	2.3×10^4	3.0
Cu	3.2	243	240	9.0×10^4	1.0
Se	2.1	4.5	1.8×10^5	1.8×10^6	0.63
Mn	1.3	13	5.9	280	0.39
Pb	1.2	78	390	1.3×10^5	0.36
As	0.35	1.2	800	1.3×10^4	0.11
Ni	0.33	6.6	17	1.8×10^3	0.10
Hg	0.25	0.50	1.3×10^4	1.2×10^5	7.5×10^{-2}
Cd	0.092	30	1.9×10^3	3.0×10^6	2.8×10^{-2}
V	0.079	0.96	2.4	140	2.4×10^{-2}
Cr	0.067	0.24	2.8	49	2.0×10^{-2}
Sb	0.032	0.038	660	3.8×10^3	1.0×10^{-2}
Ag	0.030	0.69	1.6×10^3	2.0×10^5	9.0×10^{-3}
Co	0.027	0.25	4.4	200	8.0×10^{-3}
Cs	0.014	1.7	20	1.1×10^4	4.0×10^{-3}
Au	0.0080	0.050	8.6×10^3	2.6×10^5	2.4×10^{-3}
Sc	0.0018	0.0016	0.3	1.5	5.4×10^{-4}

(A) = main plume samples, (B) = hot vent samples; F_V = calculated discharge from Mount Etna to the atmosphere; EF = enrichment factor. Elements listed in order of decreasing concentration in main plume (A); F_V based on concentrations in main plume (A). From Buat-Ménard and Arnold (1978).

3. For all samples a similar correlation exists between particulate S concentration and those of V, Pb, Hg, As, and Se.
4. Cu and Zn exhibit an intermediate behaviour (confirmed by SEM–EMP observations) indicating that they are present in both halide and sulphur compounds.

Bergametti et al. (1984) investigated gaseous and particulate emissions from Mount Etna by aerial sampling of its plume in September 1983. The study was concerned with meteorological and environmental aspects as well as with geochemistry, and the distance of the sampling point from the crater varied between 9 and 275 km. The sampling altitude range was from 2900 to 3800 m, and appropriate background measurements were made. Table 22.9 gives fluxes (tonnes/day) calculated on the basis of the 1983 samplings and on an earlier set of samples collected in June 1980, and compares these results with those of Buat-Menard et al. (1978).

THEORETICAL AND EXPERIMENTAL ASPECTS

Although there had been long-standing speculation about the nature of the compounds in which metals escaped from their source magmas, and the form and fluid environment in which they were transported from source to their destination in a developing ore deposit, Fenner (1933) was one of the first to combine volcano-chemical observation with principles of modern chemistry to begin to resolve these problems. Implicit in the latter were considerations of solid–liquid–gas equilibria, the requirements and implications of the Law of Mass Action in pneumatolytic systems, and the volatilities of compounds observed in the volcanic–geothermal milieu.

Fenner considered, quite simply, that because the metallic chlorides possessed relatively high volatilities, and because direct observation of volcanic fumaroles showed their presence, it seemed most likely that chlorides, and, to a somewhat lesser degree, fluorides, were the most important compounds in which the metals escaped from the melt. He noted (Table 22.10) the high vapour pressures of the chlorides of some of the more common metals of fumarole sublimates: Zn, Pb, Cd, Sn, As, Sb, Bi, and Ni. He observed that $FeCl_3$ and $AlCl_3$ were among the compounds with the highest vapour pressures, and because of this and the mass-action effect caused by their large concentration in the magma, they might be expected to volatilize in large quantities. This was evidently the case with Fe, but not so much with Al; Fenner suggested that this might be due to the reaction of $AlCl_3$ with H_2O to form the oxide, except where the ratio of HCl to water vapour was unusually high. He drew attention to the fact that although Ca was common and

Table 22.9 Mount Etna particulate fluxes, tonnes/day (after Bergametti et al. 1984)

Component	1983 activity (normal activity)	1983 activity (ash fall episode)	1980 activity (particles $< 2\mu$)	1980 activity (total)	1976 activity (Buat-Ménard)
SO_2	2275–2745	3245	4560	4560	3500
SO_4^{2-}	29	16	nd	nd	nd
S	14	7	13	27	420
Al	1.1	13	1.3	24	6
Si	6	30	5.6	54.5	nd
P	0.21	0.55	2.3	6.4	nd
Cl	5.4	0.6	1.4	10	300
K	11	3.5	11	23.5	28
Ca	1	7	0.83	33	20
Ti	0.1	0.8	0.015	4.7	nd
Mn	nd	nd	0	0.45	0.39
Fe	1	6.2	2.1	60	3
Cu	nd	nd	0.09	0.3	1
Zn	nd	nd	0.01	0.01	3
Se	nd	nd	0.05	0.05	0.63
Br	nd	nd	7.9	12	6.3
Pb	nd	nd	0.11	0.85	0.36
Bi	nd	nd	0.12	0.12	nd

nd = no data available.

Table 22.10 Values of vapour pressures of common metallic chlorides used by Fenner (1933)

Substance	Temperature (°C)	Vapour pressure (mmHg)
LiCl	1023.4	36.7
NaCl	992.3	15.2
NaCl	1146.9	92.6
Cu_2Cl_2	988.4	132.5
Cu_2Cl_2	1114.3	248.0
AgCl	1035.6	16.7
AgCl	1155.9	70.7
$NaAgCl_2$	1019.6	61.8
$NaAgCl_2$	1195.3	134.3
$ZnCl_2$	739.5	844.5
$CdCl_2$	1012.8	1075.0
$BaCl_2$	1213.6	8.5
$CaCl_2$	1137.6	< 4.0
$MgCl_2$	1000.7	24.9
$MgCl_2$	1127.4	90.4
$PbCl_2$	993.2	1083.2
$MnCl_2$	1005.0	168.0
$MnCl_2$	1150.2	565.5
$FeCl_2$	994.9	593.0
$NiCl_2$	987.0	743.3
$CoCl_2$	983.3	418.9
$CuCl_2$	523.6	849.0 = dissoc. pressure
$SnCl_2$	641.0	972.4
$AlCl_3$	181.3	960.1
$FeCl_3$	318.4	872.6 partly decomp.
$AsCl_3$	124.0	785.8
$SbCl_3$	226.4	779.5
$BiCl_3$	468.8	1209.4

frequently abundant in volcanic sublimates, this might be seen as surprising in view of the very low volatility of the chloride. In the light of the early experiments of Fouqué (1865) he suggested, however, that $CaCl_2$ might be very effectively transported by water vapour. Certainly the alkali metal chloride volatilities are low, and yet they are clearly transported in large amounts in fumarolic gases. Perhaps the volatile transport of these is also facilitated by the presence of water vapour or perhaps, as suggested by Fenner, more highly volatile double chlorides are involved.

One feature of Table 22.10 not alluded to by Fenner is the relatively low volatility of cuprous chloride. It will be recalled that Murata (1960) found prominent lines of cuprous chloride, CuCl, in the emission spectra of Kilaueau volcanic flames.

Using data on enthalpies, entropies, and free energies, Krauskopf (1957) attempted an investigation of the capacity of a gas phase to transport significant amounts of metals at 600 °C. In doing so he emphasized the fact that there was much more to the problem than comparisons of volatilities of compounds taken in isolation: any given situation was greatly complicated by the existence and nature of coexisting phases and by equilibrium reactions involving them. Krauskopf's concern was with the intrusive (i.e. plutonic) environment and with temperatures (≈600 °C) and pressures (≈1000 atm) appropriate to that. This is somewhat different from the volcanic environment of present concern, where temperatures may be taken to be 800–1000 °C and pressures no more than a very few atmospheres. Nonetheless Krauskopf's approach and results are valid — and instructive — in principle. He considered the volatilities of the relevant elements, oxides, sulphides, and chlorides and concluded, *inter alia*:

1. The low vapour pressures of most free metals and of their sulphides preclude the significant transport of metals as metal vapour or as sulphide.
2. All the common metals for which data are available are most volatile as chlorides, except gold (the chloride of which is decomposed at 600 °C) and copper (most volatile as the sulphide). The maximum amounts of the chlorides present in magmatic vapour are in general less than the measured volatilities because of equilibrium with the solid sulphide and with H_2S and HCl in the vapour.
3. Maximum concentrations of most of the metals considered are in the range $10^{-1.3}$–$10^{-4.3}$ g/litre, adequate to transport sufficient metal to form large ore deposits. However, not all of these metals would volatilize in substantial quantities at 600 °C. The very low volatilities of Cu, Ag, and Au chlorides, together with 'anomalies' in relative volatilities such as the greater volatility of lead than manganese in this combination, indicate that no hypothesis of simple vapour transport can satisfactorily account for the formation and constitution of most ore deposits.
4. Gases must nevertheless play an important role in transporting metals wherever a magmatic gas phase forms in substantial quantity, if for no other reason than that some of the heavy metals *will necessarily vaporize* at 600 °C given the opportunity to do so. The gas phase may in fact be more important than is indicated by the closely defined thermodynamical approach, either because of the physical transport of solid particles in the gas or absorbed on gas–liquid interfaces, or because of the solvent action of highly compressed water vapour.

Krauskopf's reference to the possible importance of water vapour as a carrier recalls the suspicions of Fouqué (1865) and Fenner (1933) that water vapour might be important in facilitating the volatile transport of calcium and sodium.

It may be appropriate here to draw attention to an investigation that seems to have aroused little attention and almost no interest, but which may have considerable relevance to the volatile transport of metals. Martin and Piwinski (1969) carried out a series of experiments concerned with the well-known tendency of iron to occur abundantly in fumarole sublimates and in contact-metamorphic concentrations. Their hypersolidus and subsolidus hydrothermal experiments on a variety of calc-alkaline plutonic and volcanic rocks at between 1.25 and 10 kb water pressure demonstrated, quite spectacularly, that iron fractionates readily into the aqueous vapour phase. Experimental runs designed to investigate transport phenomena along temperature gradients showed that, at above 1.25 kb, iron was invariably abstracted from the charge, transported in the medium of the water vapour, and deposited at the cool end of the vessel as haematite. One such experiment, with 500 mg of standard diabase W-1 and 1 g of distilled deionized water held at 700 °C and 5 kbar for 14 days, yielded abundant haematite crystals in the temperature band 470–515 °C towards the cool end of the tube. Similar material held for 14 days at 10 kbar lost approximately 6 per cent of its total iron (expressed as Fe_2O_3): enough to produce 22×10^6 tons of transported Fe_2O_3 from 1 km^3 of this particular rock.

Martin and Piwinski emphasized the necessity of a free vapour phase, but observed that the presence or absence of melt, the involvement of a specific rock type, or the presence of chloride, fluoride, sulphide, or carbonate anions did not seem mandatory for the leaching and transport of iron in the vapour.

At about the same time Holland (1972) was reconsidering the long-standing problem of relations between granites, volatiles, and base-metal ore deposits. In a series of experiments investigating the composition of aqueous solutions in equilibrium with silicate melts of granitic composition he showed that zinc, manganese, and probably lead are strongly partitioned into the aqueous phase, and that the partition ratio is approximately proportional to the square of the chloride concentration of the aqueous phase. The experiments indicated that 22 per cent of the initial zinc content of the melt was removed when the initial water content of the latter was 1.8 per cent; 28 per cent when the initial water content = 3.6 per cent; 30 per cent when initial H_2O = 5.4 per cent; and that further increase in water content of the melt had little further effect in removing zinc. The amount of zinc that could be extracted from the granite by the volatile phase was, however, very sensitive to the original chloride content of the granite; at 0.7 wt per cent chloride, 78 per cent of the original zinc went into the volatile phase, and at 1.05 wt per cent chloride, 97 per cent of the zinc moved to the volatile phase. Holland pointed out that while these chloride contents were high, they were not inconsistent with those of the fluid inclusions found in many intrusive rocks associated with subsurface hydrothermal base-metal deposits. He concluded that the extraction of these metals from the melt was related to the initial chloride and water content of the latter, and on the time relationship between vapour-phase separation and crystallization of the melt. Given favourable circumstances in these respects, zinc and manganese could be transferred almost quantitatively from the melt to the vapour phase.

Recent experiments by D. A. C. Manning and his collaborators have added considerably to knowledge of melt:-vapour behaviour of many ore metals in the presence of the halogens, boron, and other potentially volatile elements and compounds. (It should be noted here that currently developing studies in this field owe an almost overwhelming debt to that giant among students of melt:volatile phenomena, C. W. Burnham.) Experimental studies of the behaviour of fluorine, boron, and lithium (Manning and Pichavant 1987) in granitic systems have shown that the presence of these volatiles leads to depolymerization of the melts, and hence to fluxing and concomitant alteration of physical and chemical properties of the latter. Their presence thus influences both phase equilibria and element partitioning behaviour; volatile-rich melts are commonly able to persist, as melts, to late stages of magmatic differentiation and at the same time to concentrate incompatible elements, including some of the ore metals, into residual magmas.

Like Holland (1972), Manning and his associates have been concerned principally with granitic systems and intrusive rather than extrusive environments. The metals of particular interest have been those widely regarded as characteristic of the volatile-rich granitic milieu: tungsten, tin, molybdenum, uranium, copper, etc. There is, however, no reason why their results should not be fully relevant, in principle, to the volcanic environment and the metals dominant in it.

Experiments of Manning (1984) and Manning and Henderson (1984) provide an example. These were carried out to determine the effect of aqueous solution composition on the partitioning behaviour of tungsten in granitic melt-vapour systems at 800 °C and 1 kbar. With chloride and phosphate solutions, tungsten partitions strongly into the aqueous phase, whereas with fluoride, carbonate, and borate solutions, and H_2O alone, it partitions in favour of the melt. In the case of chloride solutions, the fluid:melt partition coefficients for W exhibit a marked positive correlation with chloride concentration, perhaps indicating that at low chloride concentrations W–Cl complexes with low Cl:W ratios [such as associated equivalents of $(WO_3)_2Cl^-$] may be the transporting agents. In contrast, at higher chloride concentrations complexes with high Cl:W ratios, such as WOCl, WCl$_6$, and associated ionic equivalents, may predominate. Manning and Henderson (1984) note that there is no evidence to suggest that fluoride, borate, or carbonate complexes are important transporters of tungsten under the conditions of their experiments. Indeed, solutions of carbonate and fluoride complexes appear to become increasingly ineffective for tungsten transport with increasing anion concentration. On the other hand phosphate-bearing solutions appear to be as effective as NaCl brines for the transport of tungsten. Manning (1984) concludes (1) that a tungsten-bearing hydrothermal phase may be derived during the exsolution of chloride brine from crystallizing granitic melts, and (2) that phosphate-bearing solutions of igneous origin, or derived from sedimentary sequences encountered by the relevant granite, may transport tungsten — and, possibly, other metals — very effectively.

At about the same time, Candela and Holland (1984) examined the partitioning of copper and molybdenum between a highly siliceous (SiO$_2$ = 78.29 wt per cent) synthetic melt and aqueous fluids at 750 °C and 1.4 kbar, with f_{o2} controlled by nickel–nickel oxide or quartz–fayalite–magnetite buffers. They found the partition coefficient of copper $D_{Cu}^{v:l}$ to be proportional to the chloride concentration in the vapour phase, the most probable aqueous species being CuCl (vide Murata (1960) on copper in the spectra of Kilauean flames). In contrast $D_{Mo}^{v:l}$ was shown to be independent of chloride concentration up to at least 4.6 moles of chloride/kg of solution.

Candela and Holland concluded from this that (1) chlorine is important in promoting the partitioning of copper into aqueous phases from silicate melts; (2) in contrast analogous partitioning of molybdenum is independent of the chlorine content of magmas and their associated aqueous phases; (3) this profound difference in partitioning behaviour of the two metals in chloride-rich melt–vapour systems has substantial implications for the separation of the two metals at the magmatic hydrothermal stage. They then extended these results (Candela and Holland 1986) to explain the strong tendency towards separation of copper and molybdenum in porphyry systems. In the light of work by Andriambololona and Dupuy (1978) they considered copper as a compatible element,* and examined the efficiencies with which copper and molybdenum could be removed from silicate melts in the presence of a chloride-rich aqueous phase. They concluded that copper is concentrated so efficiently into a moderately to highly saline aqueous phase that vapour extraction from the melt appears to be a reasonable explanation for the concentration of Cu porphyry deposits. Again on the assumption that copper is a compatible element in crystallizing igneous systems,* Candela and Holland conclude that the most efficient extraction of copper results when the volatiles are evolved early in the crystallization of the melt, i.e. before copper is incorporated in the crystalline phases (presumably chiefly sulphides) and hence rendered unavailable to a potential volatile phase. The value of $D_{Mo}^{v:l}$ is small relative to $D_{Cu}^{v:l}$ at moderate to high chloride concentrations, the transfer of Mo from melt to volatile phase is less efficient, and so the two metals tend to segregate at the melt vapour stage.

More recently Candela (1989a and b) has considered in greater detail the general problems of mass transfer of chloride and trace metals between melt and vapour. Again, he has been concerned with melts of highly felsic (rhyolitic) composition, and the behaviour of copper based on the assumption that it behaves as a compatible element. The relevant thermodynamic and mass-transfer calculations are outside the scope of the present contribution, but Figs. 22.7 and 22.8 show relations between the proportion of initial water remaining in the melt and chlorine and copper melt/vapour concentrations respectively under the stated experimental conditions. Variables such as the amount of water required to saturate the melt, the amount of water actually in the melt, and Cl/H$_2$O ratios, clearly have a profound influence on melt:vapour concentrations (partitioning) under any given T-total pressure conditions.

Since about 1985 there has been what can only be described as an explosive development of interest in, and investigation of, magmatic gas and the degassing of volcanic systems. This recent activity has included increasingly-quantitative measurement of both elemental and isotopic abundances in fumarolic products, increasing application of thermodynamics and thermochemical calculations in quantifying melt–gas–vapour systems, more detailed examination of fluid inclusions as samples of older or deeper fluids, and the use of these observations and reconstructions as the basis for experimentation. Principal areas of concern are the derivation of various components of the gases — to what extent these might be intratelluric, of crustal or meteoric origin, or the products of reaction and 'gas leaching' of the volcanic edifice itself

---

*This is at variance with present findings so far as the silicates are concerned; it would be valid, however, if sulphide is present in significant quantities: author.
*Although, so far as the principal silicates are concerned, copper behaves as an incompatible element, present findings would tend to accentuate the efficacy of the process that Candela and Holland propose.

The ore elements in volcanic exhalations 323

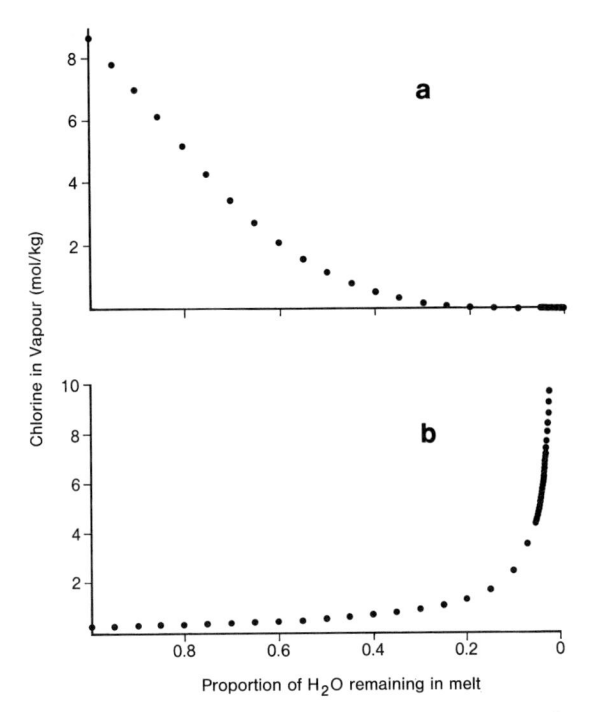

Fig. 22.7. Concentration of chlorine in successive aliquots of vapour evolved from the melt phase during second boiling as a function of the proportion of water remaining in the melt, for (a) initial [H_2O] = 1.0 weight per cent; final [H_2O] = 6.0 weight per cent; Cl/H_2O mass ratio = 0.1; and (b) initial [H_2O] = 2.0 weight per cent; saturation [H_2O] = 2.0 weight per cent; Cl/H_2O mass ratio = 0.1. The trend exhibited in (a) holds qualitatively for vapour evolution at pressures prevailing at depths >3–4 km; that in (b) is characteristic for vapour evolution at lower pressures corresponding to depths <3–4 km. After Candela (1989).

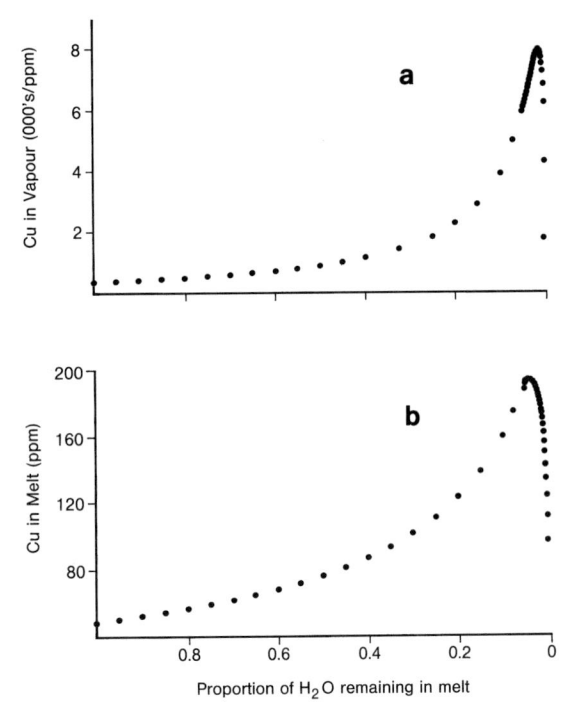

Fig. 22.8. The concentration of copper in successive aliquots of (a) vapour evolved from a melt and (b) the melt itself during second boiling of the latter for initial [Cu] in melt = 50 ppm; initial [H_2O] = 2.0 weight per cent; saturation [H_2O] = 2.0 per cent; Cl/H_2O mass ratio = 0.1. After Candela (1989).

(see, e.g. Reed 1992 in Hedenquist 1992b) — the physical and chemical nature of the emissions, and the quantities of materials evolved during the successive stages of the volcanic cycle. This current burst of activity in what might best be termed as 'chemical volcanology' is encapsulated by Hedenquist (1992b), to which the serious student of volcanic ore formation is referred.

CONCLUDING STATEMENT

Although scope for further work is immense, studies now spanning almost 140 years have shown clearly that large quantities of H_2O, CO_2, SO_2, H_2S, CO, HCl, HF, Fe, Ca, Na, and K, together with variable amounts of elements such as Cu, Zn, Pb, Sn, V, Cr, Co, Ni, As, Sb, Bi, Sr, B, and P, are emitted by volcanoes and their fumaroles in the gas/vapour phase. Few attempts have been made so far to estimate total volcanic fluxes of the minor metals, though amounts of ≈1 ton/day of zinc have been determined from plume measurements above Mount St Helens and Mount Etna.

No clear tie between tectonic setting and the amounts of various metals emitted has so far become apparent. Copper appears ubiquitous throughout the full range of volcano–tectonic settings and lava associations. There is some indication that zinc and lead are more prominent in island-arc and continental volcanic environments than in intra-oceanic situations. There is similarly an indication that nickel and chromium are more prominent in the Icelandic and Hawaiian volatile products than in those of arc and continental environments, but at least small amounts of these metals appear among volcanic volatiles of all tectonic and petrological settings. Observation of conspicuous vanadium has so far been limited to arc (Kuriles) and active continental margin (Central America) environments. Like copper, iron and manganese are virtually ubiquitous components of the volatile phase. Insufficient attention has been paid to the incidence of barium, strontium, titanium, and phosphorus for any statement on their affiliations to be made, though there is some indication that they may be present in greater quantity in volatiles of andesitic (?) environments of arcs and continents than in mid-ocean settings.

Some volcanic centres exhibit what may be referred to as 'volcanochemical individuality', i.e. although commonly occurring close to other volcanoes, they generate exhalations the compositions of which are distinctive in one or more respects. This may involve distinctive elements, or elements in common but generated in different proportions. Observations are generally qualitative, but there are now a number of them. As remarked by Fenner (1933), it has long been considered that copper is the more notable fumarolic element of Etna, and lead of Vesuvius, though at each place both metals are present in the sublimates. Oskarsson's observations of the Icelandic volcanoes indicated that those of the South-east Rift Zone (Surtsey, Heimaey, and Hekla) were characterized by a dominance of chloride over fluoride and Fe–Cu–Zn–Pb among the sublimate metals, whereas those of the North-east Rift Zone (Askja and Leirhnjúkur) were characterized by a dominance of fluoride over chloride and the sublimate metals Mg–Ti–Sr–Mo. Of the Central American volcanoes we have the observations of Stoiber and his colleagues of the remarkable incidence of vanadium in the sublimates of Izalco volcano, that copper is a distinctive element in the sublimates of Arenal and Pacaya volcanoes, and molybdenum in those of Cerro Negro and Santiaguito. Borisenko has drawn attention to the notable appearance of vanadium in the sublimates of Bezymianny volcano, and in the present volume the author has shown that of the two contemporaneous and interdigitating volcanoes of Savo and Gallego in the Solomons, the lavas of the former are distinctly the richer in lead, potassium, and rubidium, whereas of the latter are conspicuous for their higher thorium.

While there can be little doubt that some of these variations in exhalative metal abundances represent primary differences in amounts of the trace elements in the volcanic melt, theoretical considerations and experimental results indicate that others are doubtless due to, or are at least influenced by, variations in anion proportions, H_2O content, and other chemical characteristics of the lavas. The relative abundances of the halides (and F/Cl ratios), sulphur gases (H_2S, SO_2, polythionates, etc), phosphates, O_2, and H_2O are the most important of these.

To what extent the metals and various components of the carrier gases are of intratelluric, juvenile derivation, derived from deeply circulating sea water in the case of marine and seaboard volcanoes, from meteoric and other shallow groundwater circulation, and from gas leaching of the volcanic edifice, is as yet far from fully resolved. A substantial element of 'polygenesis' is likely, though doubtless varying from one volcanic setting to another.

In spite of such complications it is, however, clear that volcanic eruptions and fumarolic activity are capable of yielding, in a volatile phase derived directly from the melt, substantial quantities of a large variety of metals.

23

Petrogenesis: 1: The lavas

In this chapter we consider the derivation and mode of development of the spectrum of lava types that constitute the Solomon Islands Younger Volcanic Suite. By implication this may also cast a little light on the genesis of lavas of other island arcs, particularly those of dominantly intra-oceanic affiliation. The writer is, however, well aware of the enormous effort — field, petrographic, geochemical, and experimental — already devoted to the problem of island-arc lava generation by a long line of distinguished investigators, and is more than conscious that any contribution this chapter may make is likely to be modest — and even more likely to be judged simple-minded. Apart, however, from several minor forays by the author (Stanton 1967, 1978; Stanton and Ramsay 1980), there seems to have been no comprehensive attempt to use the geochemical behaviour of the ore elements to elucidate mechanisms of lava evolution, and this chapter is offered as an initial contribution. In doing this we consider first whether the three principal lava groups are in fact a single consanguineous series, then (very briefly) their likely derivation, and finally the ways in which the variety of lava compositions may have been induced.

RELATIONSHIPS OF THE THREE LAVA GROUPS: A SINGLE SERIES?

The relationships of the three lava groups have already been considered briefly in Chapter 5, and the provisional acceptance of close petrological ties has constituted much of the basis for the form of treatment used in the geochemical chapters 6–20. Evidence relating to connections between the three lava groups may be found in the field, in their mineralogies and chemical features, and in isotope abundances.

Field relationships

As indicated in Chapter 5, both spatial and temporal ties are close. Most of the volcanic cones of New Georgia are composed of lavas of more than one group, if in widely varying proportions. Picrite basalts, olivine–clinopyroxene-basalts, and hornblende-andesites all constitute the major elements of one or more individual, and still clearly recognizable, cones. All the principal volcanic edifices provide examples of this phenomenon. Although the main (Paraso) volcano of Vella Lavella is composed principally of hornblende-andesite, it also possesses significant amounts of olivine–pyroxene-basalt, some of this verging on picritic composition. Kolombangara is essentially an olivine–pyroxene-basalt volcano, but picrites (modal olivine >60 per cent in some cases, with olivine Mg numbers ≈Fo_{90}) also occur, and there are several prominent satellite cones composed of hornblende-andesite. The main island of New Georgia exhibits the full gamut of lava and pyroclastic compositions, from the most highly magnesian picrites of the Kolo area to the dacite of Mt Rani. The latter appears as the central spine of an edifice composed principally of olivine–clinopyroxene basaltic materials. Although the Kolo volcanic centre constitutes the principal occurrence of picrites in the Solomons, its eastern flank exhibits several small satellite cones composed of hornblende-andesite. Rendova and Vangunu, like the main island of New Georgia, exhibit virtually the full range of compositions from picrite to hornblende-andesite and occasional dacitic materials. While Gallego and Savo are dominantly hornblende-andesite volcanoes, both have produced small amounts of clinopyroxene-rich basalts similar to those of the more mafic cones of New Georgia. The single volcanic cone of the small island of Mborokua, between Gatukai and the Russell Islands, consists of olivine–clinopyroxene- and big-feldspar basalts.

The only lava type not to occur as the principal component of a well-preserved cone is, in spite of its substantial abundance overall, the big-feldspar lava group. On the evidence of field distribution patterns, this may be the oldest of the three groups, its major (though certainly not its sole) development occurring prior to that of the more recent volcanic edifices.

The broad features of field occurrence are thus:

1. Members of the three principal lava groups occur in intimate spatial association through the major part of New Georgia, from Vella Lavella to Vangunu.

2. There is a common association of the lava spectrum picrite:olivine–clinopyroxene-basalt:hornblende (pyroxene)-andesite in the young, still well-preserved, cones.

3. For the most part the big-feldspar lavas do not form an intrinsic part of these youngest cones, which appear to have spilled over them. Picrites and olivine–clinopyroxene-basalts have also been found as occasional dykes in the big-feldspar lavas. All this suggests that the big-feldspar lavas constitute a somewhat older group. The occurrence of olivine–pyroxene- and big-feldspar basalts together in the small cone of Mborokua indicates, however, that if there is a tendency to an age difference this is slight and involves an overlap.

Spatial and temporal relations of the three lava groups are thus so close as to indicate that all are likely to be part of a single 'magmatic family'.

Mineralogy

Several mineralogical features indicate that the three lava groups are consanguineous.

1. The conspicuous green diopsidic clinopyroxene that constitutes such a distinctive feature of the basalts runs as a thread through the whole spectrum of basalts, big-feldspar lavas, and hornblende-andesites as far as ≈60–2 per cent whole-rock SiO_2. Although intensity of colour is less in the more felsic rocks — the clinopyroxene of the andesites is generally a paler, more 'watery', green than that of the basalts — the general consistency of Mg numbers (Fig. 5.3) and systematic, gradational, changes in trace-element abundances with change in lava type from mafic basalt to felsic andesite (Chapters 6–20) all indicate that these pyroxenes constitute a continuum.

2. The large feldspar laths of the big-feldspar basalts and basaltic andesites are of course the characteristic and most prominent mineralogical feature of this group. Feldspars of similar appearance and composition, though of generally smaller size, appear, however, in some of the more mafic basalts, and careful observation reveals their presence among the much more numerous, smaller, 'moth-eaten' feldspars of many of the andesites. One interpretation of this is that development of feldspars of this kind began in some of the olivine–pyroxene-basalts, and reached a maximum in the big-feldspar lavas. Some of these large feldspars were then inherited by the andesites as the latter formed from the evolving melt. While such feldspars do not yield a 'thread' of quite such prominence and continuity as the clinopyroxenes, they do seem to constitute an element of continuity through the lava suite as a whole — and hence to indicate a petrological continuum.

3. In ranging from chromites to titano-magnetites the spinels, like the clinopyroxenes, exhibit conspicuously *gradational* compositional change with change in host lava from picrite basalt to hornblende-andesite. Following the very rapid complementary changes in Cr_2O_3 and TiO_2 abundances in the spinels of the more mafic basalts, these two components then follow abundance patterns that appear to be continuous from basalt to felsic andesite (Fig. 5.16). Spinel MgO and Al_2O_3 follow an even more regular pattern of decrease (Fig. 5.16), as does Cu (Fig. 6.5). Zn, Mn, Fe_2O_3, and (FeO + Fe_2O_3) in the spinels display substantially smooth, *continuous* patterns of increase as the containing lavas change from mafic basalts to felsic andesites, and FeO shows a very clear consistency in lavas covering the 50 >64 per cent SiO_2 range.

There is thus a remarkable coherency, or continuity, of major, minor, and trace-element abundances in the spinels as the full spectrum of host lava types is traversed.

4. The relatively small compositional range of the olivines, combined with the fact that they are barely represented in the andesites (i.e. that the only lithological 'boundary' spanned is that of olivine– clinopyroxene-basalt:big-feldspar basalt) limits the usefulness of this species in examining links among all three lava groups. There is, however, a general continuity of compositional change from the olivines of the picrites to those of the basaltic andesites (Fig. 5.3), and so the evidence of the olivines does not contradict that of the clinopyroxenes, feldspars, and spinels. The olivines of the more felsic (i.e. basaltic andesite) lavas do not seem to show any clear textural differences from those of the picrites and mafic basalts:larger idiomorphic to subidiomorphic, and generally smaller, rounded olivines occur in all of these lava types. Both morphologies occur throughout: olivine abundance is simply greater in the picrites. The only differences noted by the author — a qualitative observation — is that the brownish, inclusion-laden olivines (see Chapter 5) are most numerous in the picrites.

Lithochemistry

Again, there appear to be several significant features.

1. Complementarity of the three lava groups in Harker and ternary plots. The remarkable smoothness of trends generated by major, minor, and trace elements

on variation (Harker) diagrams has been noted repeatedly in the preceding text. The latter all span the SiO$_2$ interval <46 to >64 per cent, i.e. a range of ≈23 per cent, from picrites to dacites, and hence provide a substantial range for any inherent, and petrogenetically significant, discontinuities to appear. By 'discontinuity' is meant numerical, and hence graphical, inflexions or offsets that might reflect the presence of quite distinct magma series, or the contamination of a melt by extraneous materials. In the present case the olivine–pyroxene-basalts and the hornblende-andesites might represent two quite different magma series, or the andesites might be products of contamination of the basalts by sea-floor material through some hypothetical process such as subduction. The first possibility is likely to manifest itself by the appearance of two distinct and separate curves, the second by the appearance of a sharp discontinuity. In the case of Al$_2$O$_3$, MgO, Na$_2$O, K$_2$O (Fig. 5.5), and all 14 elements of Chapters 6–19, the relevant Harker diagrams exhibit no such features. It would be surprising if the remarkably smooth variation curves of, for example, Cr, Ni, Co, Sr, and Ba (see Fig. 21.2) concealed the existence of more than one magma series, or the significant contamination of a single melt. Certainly some of the diagrams (eg FeO, CaO, TiO$_2$, V) display maxima rather than simple, single trends of increase or decrease, but such maxima are generally smooth and ordered (perhaps remarkably so considering the relatively small number of samples involved) and do not show the sort of sharp break or offset that might be expected to result from the representation of more than one set of distinct materials. In addition the trends of decrease on the right-hand (higher SiO$_2$) side of the maxima in the variation diagrams are generally smooth and continuous. Those for FeO and MnO (Fig. 21.2(c)) are essentially continuous from SiO$_2$ = 48–50 per cent (basalts) to >64 per cent (dacites). It is not easy to see two separate lava series, or categories of contaminated and uncontaminated materials, yielding single curves of such remarkable consistency.

The evidence of the ternary XMA diagrams appears to be compatible with that of the Harker diagrams in indicating that the three lava groups constitute a continuum (Stanton and Ramsay 1980). Although the ternary diagrams for elements generating all of ni-chrome, hypo-, and hypertholeiitic patterns involve inflexions of fields with change from basalt to andesite, such inflexions are usually ordered, indicting that the different lava groups are complementary rather than unrelated. In general the low alkali limits of the andesitic fields abut, and fit snugly against, the high alkali–low MgO limits of the basalt–picrite fields. The big-feldspar basalt and basaltic andesite fields tend to fit over the respective junctions of the other two (Figs. 21.6 and 21.7; see also relevant figures in Chapters 6–20). Strontium, the only element generating a simple tholeiitic pattern, shows a quite neat linear progression of fields corresponding to the three lava groups (Figs. 10.14 and 10.15).

2. Zr:Hf and Rb:K$_2$O relations. It will be recalled that the 'condensed series' of Chapter 20 consists of 28 lavas spanning the spectrum picrite-dacite and the compositional range 46.20 <SiO$_2$ <65.36 per cent and 1.06 <MgO <28.07 per cent. Notwithstanding this very substantial variation, a single plot of all Zr:Hf values yields a single straight line (Fig. 20.9) with a coefficient of correlation r = 0.96. An analogous plot of Rb v. K$_2$O yields very similar fields for all three lava groups (Fig. 20.3) and coefficients of correlation of 0.96, 0.91 and 0.83 for olivine–pyroxene-basalts, big-feldspar lavas, and andesites–dacites respectively.

Isotope abundances

Minor investigations have been carried out on whole-rock ^{87}Sr/^{86}Sr (Gill and Compston 1973)* and on ^{207}Pb/^{204}Pb, ^{206}Pb/^{204}Pb, and ^{208}Pb/^{204}Pb of lead occurring in magnetites of Vangunu and Mount Gallego.

Figure 23.1 shows relationships between whole-rock SiO$_2$ and rubidium, strontium, and ^{87}Sr/^{86}Sr in the small suite of lavas analysed by Gill and Compston (1973). (Sample 412 was a crystal tuff, and is not included in Fig. 23.1.) The SiO$_2$ range of the suite (≈10 per cent) is small but represents the full span obtained in the early collection (Stanton and Bell 1969, reporting field work carried out in 1959 and 1963–4). Although sampling for the present investigation yielded an SiO$_2$ range more than double this (>20 per cent) the earlier suite, though limited, does span all three lava groups from picrite (Stanton and Bell 1969; No. 364, SiO$_2$ = 46.29, MgO = 23.01 per cent) to andesite (No. 349, SiO$_2$ = 55.60, MgO = 5.44 per cent).

---

*Although accidentally unacknowledged by these authors, the Solomons Islands samples were in fact collected by Stanton in 1959 and given to Compston for strontium isotope analysis in 1965. The 'sample numbers' given by Gill and Compston are in fact the author's 1959 field collection numbers: see Stanton and Bell (1963, 1969).

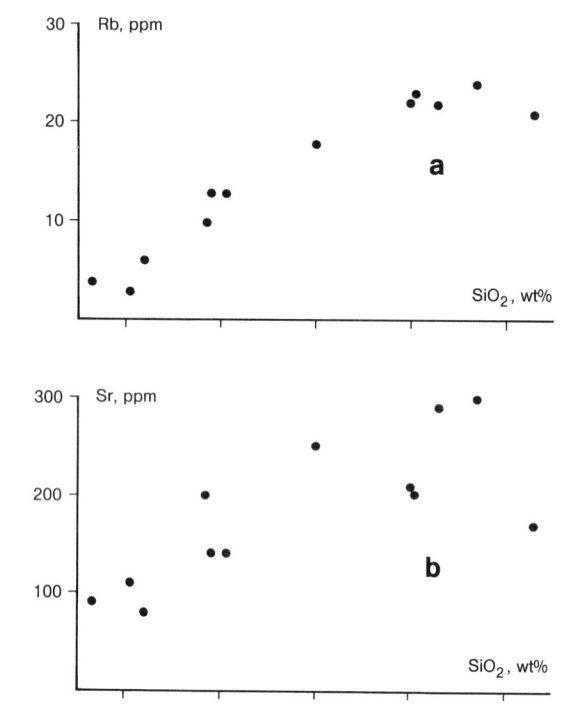

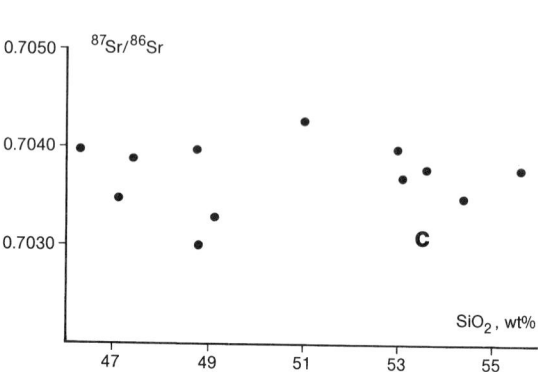

Fig. 23.1. Rb, Sr, and $^{87}Sr/^{86}Sr$ systematics of 12 New Georgia basalts and andesites collected by the author in 1959 (Stanton and Bell 1969) and analysed by Gill and Compston (1973). SiO$_2$ data from Stanton and Bell. (a) Rb v. SiO$_2$ for the 12 samples, (b) Sr v. SiO$_2$, and (c) $^{87}Sr/^{86}Sr$ v. SiO$_2$. (Sample no. 412 analysed by Gill and Compston has since been found to have been affected by the intense late-stage hydrothermal alteration of the Kele River mineralized zone, and is omitted from this figure).

As far as a small group of 12 samples may be taken to be reliable, Gill and Compston's results indicate that $^{87}Sr/^{86}Sr$ ratios do not vary significantly from picrite to andesite. The three most mafic lavas of Fig. 23.1 have mean SiO$_2$ = 46.92 per cent, mean $^{87}Sr/^{86}Sr$ = 0.7037; the three most felsic have mean SiO$_2$ = 54.52 per cent and mean $^{87}Sr/^{86}Sr$ = 0.7038. On the basis of the limited data the difference is not significant.

The lead isotope data (Table 23.1) are even more sparse. Sample 59/81 is a magnetite concentrate from a big-feldspar basalt of Vangunu; sample 9/80 is a magnetite from a Mount Gallego andesite. The small differences in corresponding lead isotope ratios are 0.2, 0.2, and 0.06 per cent for $^{206}Pb/^{204}Pb$, $^{207}Pb/^{204}Pb$, and $^{208}Pb/^{204}Pb$ respectively, and hence of the order of experimental error. On the basis of this admittedly very limited evidence there is therefore no significant isotopic difference between the magnetite-included leads of the big-feldspar basalts and the hornblende-andesites.

PROVENANCE OF THE SUITE

The question of the derivation, or provenance, of the lava and pyroclastic rocks of the Suite seems fairly clear from general features of its regional geological setting and tectonics.

As noted in the Introduction, an important reason for choosing the Solomons lavas for an investigation of the present kind was that the arc is *intra-oceanic*: that is, it appears to be built entirely upon *oceanic*, not continental, crust. This carried with it the advantage that at no stage of their history would the lavas have been exposed to continental crustal contamination, so keeping to a minimum the number of influences affecting their compositions and hence the number of variables involved in geochemical interpretation.

Current tectonic interpretations (Fig. 1.3) place the Solomons arc between an extinct northern, SW-dipping subduction zone coincident with the Kilinaelau–North Solomons Trench and a now-active southern, NE-dipping subduction zone coincident with the New Britain–San Cristobal Trench. Earthquake distribution maps indicate that the New Georgia Group constitutes an area of relatively shallow and relatively infrequent activity occurring between areas of deep and intense activity in the vicinities of Guadalcanal and Bougainville. All three domains are characterized by Recent and Modern development of the Solomon Islands Younger Volcanic Suite, and currently active or quiescent volcanism. The islands of New Georgia are, however, the only ones on which picritic lavas have been found, and here they are prominent. They are also the only member of the Solomons group to lie close to a major tectonic intersection: in this case the abutting of the Woodlark Spreading Ridge against the medial, arched portion of the New Britain–San Cristobal Trench System of Fig. 1.3. This spreading

Table 23.1 Isotope ratios of lead occurring as a trace element in magnetite separated from two lavas of the Solomon Islands Younger Volcanic Suite

Sample	Pb (ppm)	$^{206}Pb/^{204}Pb$	$^{207}Pb/^{204}Pb$	$^{208}Pb/^{204}Pb$
Sol. 9/80 (Vangunu: basalt of the BFL group)	44.2	17.926	15.548	37.733
Sol. 59/81 (Mt Gallego: hornblende-andesite)	1.54	17.969	15.579	37.754

Analyses performed by C. M. Fanning of the Research School of Earth Sciences, Australian National University, on magnetic:heavy liquid separations produced by R. L. Stanton. CMF has provided analytical information as follows: The sample powders were repeatedly washed with distilled ethanol and acetone in an ultrasonic until the supernatant liquid remained clear. Approximately 50 mg were dissolved in PFA screw-capped vials using concentrated HCl. There was some residue remaining in sample 9/80 which was completely dissolved with HF and HNO_3. The samples were dried then completely redissolved in c. 3N HCl, split into unspiked and spiked aliquots with a ^{208}Pb tracer added to the latter. The Pb was extracted on standard anion-exchange columns using HBr. The Pb was loaded by the standard silica gel–phosphoric acid technique on single Re filaments and the isotopic compositions were measured with a Finnigan MAT 261 mass spectrometer in the static multicollection mode. The data have been corrected for the laboratory-induced Pb blank, measured at 140 pg, and for mass fractionation. Errors in the Pb concentration are 0.5 per cent (95 per cent confidence limits) and 0.1 per cent or better for the $^{206}Pb/^{204}Pb$ and $^{207}Pb/^{204}Pb$ ratios, and 0.15 per cent for the $^{208}Pb/^{204}Pb$ ratios.

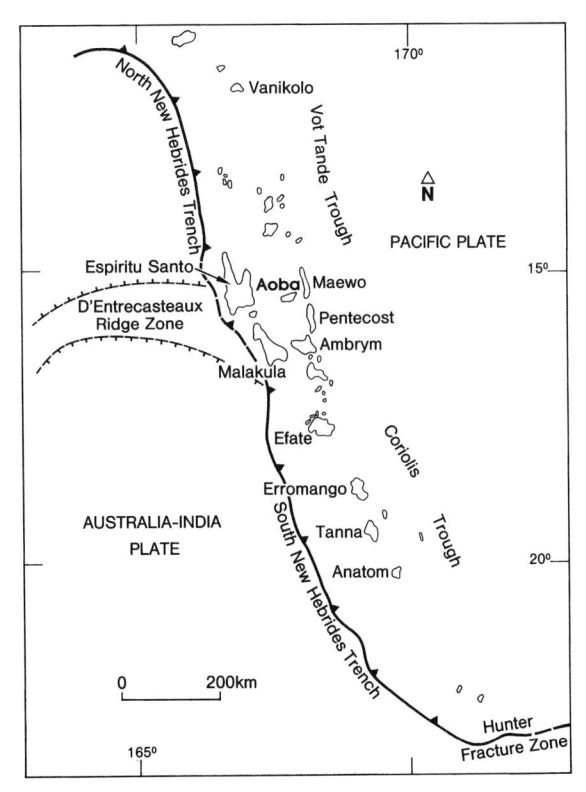

Fig. 23.2. Current tectonic synthesis of the Vanuatu volcanic arc, showing the position of Aoba (site of abundant development of picrites similar to those of New Georgia) opposite the intersection of the New Hebrides Trench and the d'Entrecasteaux Ridge. (Adapted from Greene, Macfarlane, and Wong 1988). Compare with Fig. 1.3.

ridge–trench intersection is, as has been widely recognized for the past 15 years, remarkably reminiscent of that between the d'Entrecasteaux Ridge and the Vanuatu Trench occurring in the northern part of the Vanuatu chain. This latter intersection is more or less opposite the island of Aoba (Fig. 23.2), which as it happens is also the site of prominent picritic lava development (Gorton 1974, 1977; Eggins 1993). While any suggestion that there may be some fundamental tie between active ridge–trench intersection and prominent picritic basalt eruption can be no more than tentative at this stage, the possibility that there may be some relationship — involving seismic activity and appropriate partial melting at ≈100–150 km — may merit consideration as petrotectonic studies progress. Be this as it may, we may note these two rather similar tectonic–petrological associations of the southwest Pacific, and their possible connotations concerning the New Georgia picrites and their associates of the Solomons Younger Volcanic Suite.

DEVELOPMENT OF THE PRINCIPAL LAVA GROUPS

The many relevant diagrams of Chapters 5–21 indicate that the big-feldspar basalts are in some way a bridging group between the olivine–clinopyroxene-basalts and the hornblende-andesites and dacites: a state of affairs first noted by Stanton and Ramsay (1980). The big-feldspar basalts are, however, unlikely to be parental to the olivine–pyroxene basalts: in terms of Figs. 5.20 and 21.6, crystallization-subtraction of olivine and clinopyroxene to produce picrites and olivine–pyroxene-basalts as accumulates would inevitably have led to the development of a complementary series of iron-rich tholeiites, but these do not appear. On the other hand there does seem to be a complete gradation, by diminution of olivine and pyroxene and increasing prominence of tabular feldspars, from the olivine–pyroxene-basalts to the more mafic members of the big-feldspar lava group (Stanton and Bell 1969; this volume, Chapter 5). The author therefore inclines to the view that the latter — variably modified by minor clinopyroxene–magnetite fractionation — may constitute the most felsic limit of the olivine–pyroxene-basalt series.

In an analogous but converse way the more mafic big-feldspar lavas are unlikely to be parental to the hornblende-andesites by processes of fractionation alone, as demonstrated in Chapter 21. Such processes would necessarily lead to the development of a complementary series of iron-rich–alkali-poor accumulates, which, like the iron-rich tholeiites, do not appear. In many diagrams, e.g. most of the ternary XMA diagrams of Chapters 6–20, the field of points representing the big-feldspar lavas appears as an almost non-directional 'cloud' between the highly directed fields, or trends, of olivine–pyroxene-basalts and hornblende-andesites. The prominent development of hornblende constitutes a marked difference between the big-feldspar lavas and the most mafic hornblende-andesites but, as pointed out earlier in this chapter, the common possession of the prominent green clinopyroxenes, some feldspars of similar composition and crystal habit, and some features of whole-rock chemistry seem to indicate that the two rock groups are not unrelated.

The olivine–clinopyroxene-basalts

The ternary XMA fields generated by these lavas are remarkably linear. As we have seen (Table 21.6), those for Zn and P, for example, possess linearities so closely defined as to yield coefficients of correlation, r, = 0.991 and 0.993 respectively. This points clearly to the development of the series by closed-system fractionation of classical Bowen type, or by a process of partial melting yielding a linear series of compositions.

In the former case, crystallization of olivine and clinopyroxene of composition $\approx Fo_{88}$ and $\approx En_{88}$ respectively would have led to the development of high-MgO picritic and ankaramitic accumulates and complementary, lower-MgO, olivine–pyroxene-basalt — and perhaps big-feldspar basalt — derivatives. In terms of the principles set out in Chapters 5 and 21, the composition of the relevant parent (or parents: see below) must now be embedded somewhere along the linear fields of the ternary diagrams of Chapters 5–20 respectively:

separation of crystals from 'parental' material, and the concomitant development of high-MgO accumulates and lower MgO derivatives, would have taken place in a closed system, with no gains or losses with respect to any other system.

In the latter case the lava series results from a number of episodes of partial melting, each capable of yielding melt batches of different, but linearly related, compositions.

The actual process might, of course be compound: different degrees of partial melting may yield batches of different compositions, such differences subsequently being accentuated or diminished by crystal fractionation.

The matter is clearly part of the very much larger problem of basalt petrogenesis, in turn one of the major concerns of the whole of igneous petrology. The writer is under no illusion as to the profundity or difficulty of the problem, particularly as this pertains to the picrites and other high-MgO olivine-basalts. This latter debate has, in the words of Clarke and O'Hara (1979) 'rumbled on for at least fifty years, but … is scarcely a trivial one'. Its crux lies in the question 'If a lava is high in MgO (say >14 per cent) and is rich in olivine phenocrysts, does this reflect a primary high-MgO melt, or simply the fractionation and accumulation of olivine?'

Previous work. Serious consideration of the problem probably began with Bowen (1928), who noted that all the highly magnesian lavas he observed were olivine-phyric and hence likely to be accumulative. He proposed that no basaltic melts with normative olivine $\geq \approx 10$ per cent could exist in nature and that, on the basis of liquid-line-of-descent requirements, low-MgO basalts are primary and picrites accumulative. This line of argument was widely accepted until 1969, when Viljoen and Viljoen (1969) described high-MgO, quench-(spinifex-) textured komatiite from the Barbeton Mountain land, South Africa, and set the cat among the petrological pigeons. The quenching of very high MgO contents in these lavas indicated that *in at least some cases* high-MgO lavas were the primary materials, the low-MgO types derivatives produced by removal of olivine.

Just prior to the work of Viljoen and Viljoen, Murata and Richter (1966a and b) had demonstrated that the high-MgO members of the 1959 Kilauea Iki lavas series had been generated by olivine fractionation and closed-system accumulation. In related investigations Moore and Evans (1967) and Evans and Moore (1968) showed with similar clarity that lithological and MgO variation in the prehistoric Makaopuki tholeiitic lava lake of Hawaii was generated substantially by olivine settling. Gunn (1971) confirmed the findings of Murata and Richter (1966a) on the Kilauea Iki series using a range of trace elements to document olivine control (and, incidentally, anticipated some of the results reported in this volume). Sato (1977), using the nickel content of basaltic lavas to identify primary magmas and to measure degrees of olivine fractionation, deduced that in the case of the Kilauea 1959 series, lavas with >0.05 per cent NiO or >12.5 per cent MgO were probably accumulative, whereas those with ≈0.03–0.05 per cent NiO and 10.0–12.5 per cent MgO might represent the primary composition.

More recently Wilkinson and Hensel (1988) have investigated picrite flows from Mauna Loa and Kilauea and have concluded that these lavas are probably cumulates: calculated olivine megacryst–melt Mg–Fe partition coefficients indicated that low-pressure equilibria (K_D = 0.30–0.34) are defined only by melts of approximately 12–14 per cent MgO and this, taken in conjunction with Ni–MgO relations, indicates that the more magnesian picrites (MgO > 14–15 per cent) result from olivine accumulation and do not represent melt compositions. Wilkinson and Hensel proposed that the observed olivine enrichment resulted from post-eruptive, mechanical (flow) differentiation of extruded mushes of intratelluric cognate olivine phenocrysts (Mg ≈88) and tholeiitic melt. The authors also drew attention to the likelihood that closed-system differentiation of Hawaiian tholeiitic magmas (10–15 per cent MgO) yield picritic derivatives which, however, differ from the extrusive picrites through their distinctly higher FeO^t contents and correspondingly more Fe-rich olivines and spinels. Wilkinson and Hensel proposed that the parental magmas of both picrite suites were generated by 35–40 per cent melting of Fe-rich spinel lherzolites.

However, at about the time Murata and Richter and Evans, Moore, and Gunn were concluding that the high-MgO lavas of Hawaii were substantially accumulative, D. B. Clarke (1970) was proposing that Tertiary picritic basalts from Baffin Bay represented primary melts. MgO contents of these picrites may exceed 26 per cent, and are accompanied by high levels of Ni and Cr and low concentrations of the incompatible elements Ti, K, P, Z, Rb, Sr, Ba, and Y. Clarke considered that the bulk compositions of the Baffin Bay basalts were compatible with advanced partial melting of garnet peridotite at ≈30 kbar, i.e. at a depth of approximately 100 km, and he proposed that they consist, first, of primary picritic melts produced in this way. Derived from these in turn are some compositions produced at high pressures, and others (olivine-poor and feldsparphyric basalts) that evolved in low-pressure, high-level magma chambers.

Hart and Davis (1978), on the other hand, concluded that the high-MgO Baffin Bay basalts (together with similar lavas of Kilauea, Crozet, and Cape Verde) were accumulates. Their experimental data on olivine:liquid Ni partitioning indicated that $K_{Ni}^{ol:liquid}$ was independent of Ni concentration, and almost independent of temperature. It was, however, strongly dependent on melt composition, varying linearly with the reciprocal of melt MgO at constant temperature. They concluded that a combination of fractional crystallization and olivine accumulation processes was sufficient to model the high-MgO series of the four provinces: that the linear Ni-MgO trends, high absolute Ni concentrations, and large spread of Ni contents of the high-MgO basalts all constituted evidence that the high-MgO basalts did not represent primary liquids. They considered that their data identified parental liquids for each series, and that the MgO contents of such liquids ranged from 6 to 13 per cent. Basalts with higher MgO concentrations were accumulates.

Both Clarke and O'Hara (1979) and Elthon and Ridley (1979) disagreed with Hart and Davis, considering, *inter alia*, that modelling of compatible trace-element behaviour during fractionation was not, at that stage, rigorous enough for reliable diagnosis of processes. Clarke and O'Hara considered that the weight of geochemical and experimental evidence favoured the existence of highly magnesian (20–30 per cent MgO) primary magmas and that these were parental to low-MgO tholeiites, largely through fractional crystallization of olivine. Elthon and Ridley considered that given the current wide room for error implicit in experimental-geochemical modelling, the most reliable evidence was still that of petrological and field observation of ophiolite and komatiite terrains — and that this favoured the existence of picritic liquids. At much the same time, Elthon (1979) drew attention to the existence of four aphyric high-MgO dykes intruding the oceanic cumulate rocks of the ophiolite complex of the island of Tortuga, Chile. These contained up to 17.61 per cent MgO, and according to Elthon the textural and petrological evidence indicated that they had been intruded as liquids. He noted similarities to the high-MgO Baffin Island basalts (Clarke 1970) and concluded that the Tortuga rocks represented a primary mantle melt of MgO ≈18 per cent that constituted a parental magma for ocean-floor basalts.

Contemporaneously with the work of Hart and Davis (1978), Malpas (1978) was investigating the existence of picritic tholeiitic melts using evidence obtained from the Bay of Islands (Newfoundland) ophiolites, and Maaløe (1979) on evidence provided by Hawaiian tholeiites.

Malpas (1978) noted that the Bay of Islands lower series of ultramafic tectonites, consisting of spinel lherzolites overlain by harzburgites and representing mantle material, was cut by numerous olivine–pyroxene veins. He concluded that these represented early crystallization products of a picritic tholeiite magma derived at 18–22 kbar by approximately 23 per cent partial melting of spinel lherzolite.

Maaløe (1979) drew attention to the earlier observations and deductions of:

1. MacDonald and Katsura (1961), who suggested that the MgO content of the primary Kilauean tholeiitic magma ≈14.6 per cent.
2. Murata and Richter (1966b), who found tholeiitic glass of the Kilauea Iki 1959 eruption containing 10.2 per cent MgO.
3. MacDonald (1968), who suggested that oceanites containing ≈20 per cent MgO were primary and who demonstrated that the three basaltic magma series of Hawaii converge towards oceanite.
4. D. B. Clarke (1970: see above), who showed that lavas of Baffin Bay containing up to 29 per cent MgO were primary, and concluded that high-MgO tholeiitic compositions commonly represented primary melts rather than olivine fractionation/accumulation.

Maaløe concluded that the range in MgO content of primary Hawaii tholeiites is from at least 13 per cent to over 20 per cent, and that the total possible range in MgO content for primary tholeiitic magmas is from 8–9 per cent to about 20 per cent.

More recently Wilkinson (1985), in considering the possible composition of the mantle beneath Hawaii, took, as part of the basis of his calculations, an 1852 Mauna Loan picrite containing 15.51 per cent MgO as the assumed derivative of partial melting. This is just a little higher than the estimate of 12–14 per cent MgO made subsequently (Wilkinson and Hensel 1988: see above).

The problem of the true nature of picrites and related high-MgO lavas — primary melts or crystal accumulates? — thus appears to be not fully solved. The field evidence, including that of chilled margins, minor undifferentiated intrusions, the occurrence of quench, e.g. spinifex, textures, and of high-MgO aphyric lavas, indicates that melts containing up to almost 30 per cent MgO may exist in nature. Combinations of field, microscopical, and geochemical evidence indicates on the other hand that some picritic series are likely to have developed principally by olivine fractionation. The experiments of Green and Ringwood (1967) and others show that about 40 per cent partial melting of mantle material at ≈30 kbar (≈100 km depth) yields picritic liquids of ≈30 per cent normative olivine. If, as pointed out by Clarke (1970), early fractionation of such melt material is eliminated or minimized by rapid ascent, high MgO basaltic compositions may survive to the surface, or to temporary impounding in shallow, sub-volcanic magma chambers, where low-pressure fractionation may occur.

On this basis, variable degrees of partial melting combined with the surging of new batches of melt into already fractionating sub-volcanic magma chambers would inevitably lead to a wide range of picritic and related basaltic compositions — perhaps to spectra of compositions such as that now seen in the New Georgia picrites, ankaramites, and related olivine–clinopyroxene-basalts.

The Solomons lavas. As indicated in numerous diagrams in the preceding text (see, in particular, Chapters 5, 15, and 19) the spectrum of olivine–clinopyroxene-rich basalts of the Solomon Islands Younger Volcanic Suite covers the full range from ≈6–29 per cent MgO. Repeated efforts by the present writer to find lavas with MgO >29 per cent have, however, failed, and it is probable that material containing > ≈30–1 per cent MgO has not been emitted at the surface. This is not out of keeping with the findings of Clarke (1970) on the Baffin Bay and similar picrites, and may represent some kind of limiting partial melting composition.

Simple olivine-addition computations using the hypothetical parent-melt and olivine compositions set out in columns 1 and 2 of Table 23.2 indicate that, in terms of the elements listed, the compositions of all of the olivine–clinopyroxene-basalts containing MgO > 14.42 per cent may be generated by the accumulation of olivine of composition Fo_{88} in parent melt containing SiO_2 = 50.03 per cent and MgO = 14.42 per cent. (This hypothetical parent represents the mean of two of the Solomons olivine–clinopyroxene-basalts having very similar compositions and constituting (1) the lower MgO limit of a

Table 23.2 Compositions of 'parent basalt' and 'accumulative olivine' used in least-squares mixing calculation, and mean composition of olivines of Solomons olivine–clinopyroxene-basalts containing 14–16 per cent MgO

	1	2	3
SiO_2	50.03	40.54	39.63
TiO_2	0.60		
Al_2O_3	11.84		
FeO*	9.17	11.60	17.01
MnO	0.17	0.16	0.33
MgO	14.42	47.70	43.02
CaO	10.50		
Na_2O	2.07		
K_2O	1.01		
P_2O_5	0.22		
Total	100.03	100.00	99.99

1 = 'parent basalt' (this Marovo basalt has been used as an hypothetical 'parent' for calculation purposes only; there is no suggestion that this might be the actual parent to the spectrum of New Georgia picrites and olivine-basalts: see text); **2** = high-magnesium olivine of a Marovo picrite; **3** = mean composition of olivines of Solomons olivine–clinopyroxene-basalts containing 14–16 per cent MgO (this is very close to the composition of the olivine of the basalt of col. 1); FeO*, total Fe as FeO; all analyses recalculated to 100 on an anhydrous basis.

Table 23.3 Composition of groundmass of Kohinggo Island picrite

	1	2
SiO_2	51.04	51.56
TiO_2	0.42	0.42
Al_2O_3	12.27	12.39
Fe_2O_3	2.49	2.51
FeO	3.40	3.43
MnO	0.12	0.12
MgO	14.72	14.88
CaO	10.94	11.05
Na_2O	1.95	1.97
K_2O	1.21	1.22
P_2O_5	0.26	0.26
NiO	0.09	0.09
Cr_2O_3	0.10	0.10
Total	99.01	100.00

1 = separated groundmass of picrite, no. 63/80, as analysed;
2 = ditto, recalculated to 100. This is the groundmass of the Kohinggo Island picrite of which the whole-rock analysis is given in column 3, Table 5.1.

series of 30 lavas exhibiting a relatively uniformly progressive increase in MgO to 28.55 per cent; (2) material immediately above a 'break' of about 1.5 per cent MgO in this component of basalt composition, below which MgO varies less regularly and Al_2O_3 attains notably higher values. Although it happens to contain an amount of MgO almost identical with that of the most magnesian of the Solomons groundmasses (Table 23.3) and with Wilkinson and Hensel's (1988) deduced Hawaiian parent, it has not been chosen on these grounds.)

While an arithmetical procedure of this kind, representing as it does a very simple model of igneous olivine accumulation, is capable of generating the major element constitutions of all of the Solomons high-MgO basalts, it carries with it two corollaries:

1. The process of olivine subtraction required to provide for the postulated olivine addition must yield, in a closed system, an Mg-impoverished, Fe-enriched derivative tholeiitic lava series (Figs. 5.20(c), 21.6(b)) in accordance with the principles set out in Chapters 5 and 21.
2. The amount of olivine of composition Fo88 required to be added to that already present in the postulated parent in order to produce the most magnesian of the Solomons basalts (containing 28.55 per cent MgO) is 73.79 per cent of the original olivine, yielding a product lava containing 42.5 per cent olivine by mass.

Concerning (1): as already noted, derivative lavas of this kind have not been found as members of the Solomon Islands Younger Volcanic Suite. Concerning (2): on the basis of point counting (using 2531 points), this particular picrite contains 36.3 per cent modal, or ≈39.5 per cent by mass, olivine of mean composition Fo_{88}. The amount of olivine actually present in the lava is thus 3.0 per cent by mass less than that indicated by computation to be required to generate the observed MgO content: given the difference in sizes of the samples used for chemical and modal analysis, it is likely that such a small discrepancy is no more than a sampling artefact.

While olivine-addition calculations of this kind are unquestionably arithmetically precise, they constitute a somewhat crude approach. A conceptually more refined, if numerically slightly less precise, method of attack lies in the consideration of individual lava–olivine compositional relationships using methods only slightly different from those employed by preceeding investigators of the picrite problem.

What appear to be significant features of olivine, whole rock-groundmass, and olivine–whole-rock chemistry in the Solomons high MgO basalts are:

1. Olivine core compositions range up to Fo_{94}: core compositions of Fo_{90}–Fo_{92} are, though not abundant, quite common. In some cases rim compositions of Fo_{90} have been measured.
2. Picrite groundmasses have been found to contain as much as 14.72 per cent MgO (14.88 recalculated: Table 23.3): a figure very close to the estimate of 12–14 per cent for a Mauna Loa–Kilauea parent by Wilkinson and Hensel (1988), and indicating a Solomons parental melt containing a *minimum* of approximately 14–15 per cent MgO.
3. Olivine Mg v. whole rock Mg (Fig. 23.3(a)) indicates a positive, linear ($y = 0.573x + 38.0$), reasonably well correlated ($r = 0.80$), relationship throughout. Thus increase in whole-rock Mg is accompanied by *continuous* increase in olivine Mg: there is no indication of a sharp break in slope, with olivine Mg remaining relatively constant above, say, Mg ≈75 (whole rock MgO ≈15 per cent), as might be expected to result from olivine accumulation in a parent melt of ≈15 per cent MgO.
4. In an analogous way the most Ni-rich lavas contain the most Ni-rich olivine (Fig. 23.3(b): the relationship here may be slightly curved rather than linear as in the case of Mg). That is, the high-Ni lavas have not developed this feature simply by the accumulation of unusually large quantities of olivine of relatively constant Ni content: the olivines themselves contain high Ni where they occur in high-Ni lavas, again indicating, 'on average' for each lava, a general approach to olivine–melt equilibrium (Figs. 23.3(c) and (d)).

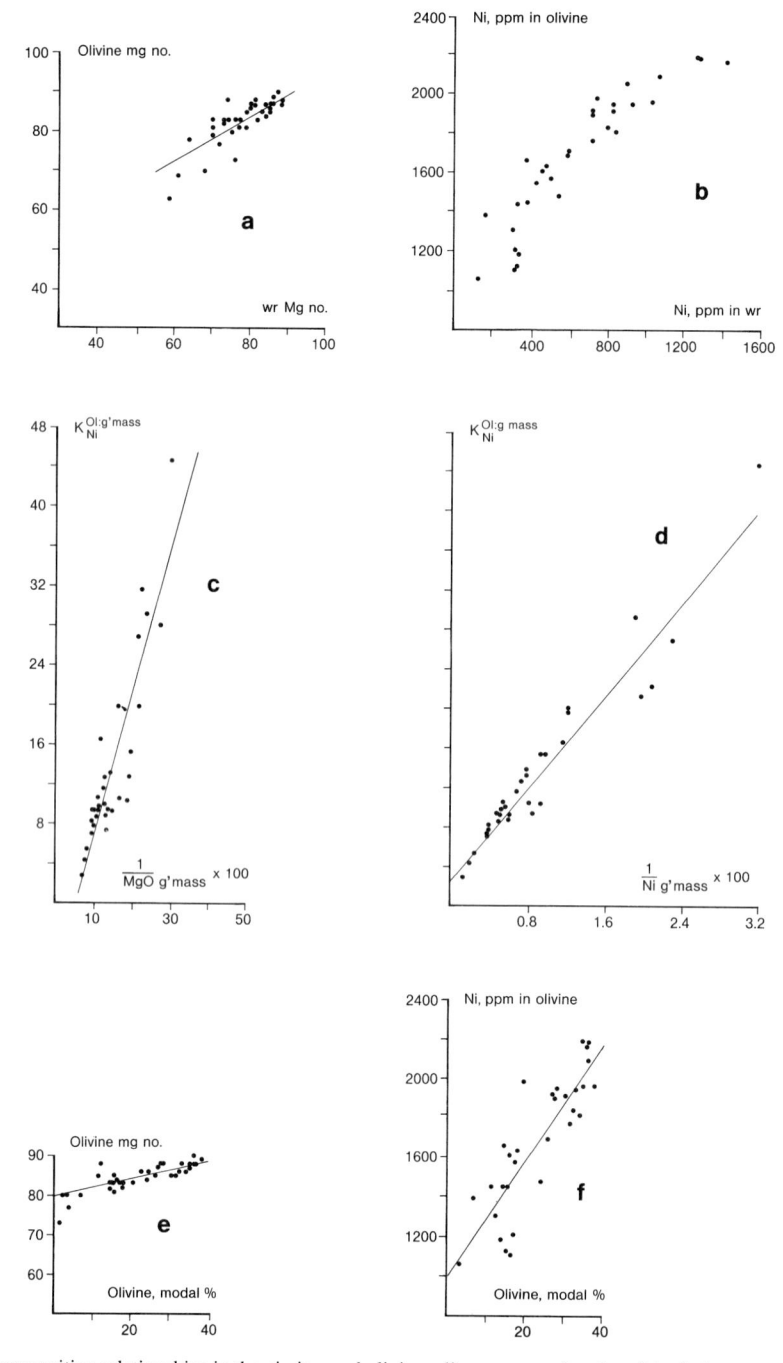

Fig. 23.3. Olivine composition relationships in the picrites and olivine–clinopyroxene-basalts of the Solomons suite: (a) olivine v. whole-rock Mg numbers; (b) olivine and whole-rock nickel abundance relations; (c) relationships between olivine:groundmass partition coefficients with respect to nickel and the reciprocal of groundmass MgO × 100; (d) relationships between olivine:groundmass partition coefficients with respect to nickel and the reciprocal of groundmass nickel × 100; (e) olivine mg number v. modal per cent olivine; and (f) concentration of nickel in olivine v. modal per cent olivine.

5. Olivine Mg is linearly related to modal abundance (Fig. 23.3(e): $y = 0.217x + 80$; $r = 0.80$). Clearly this is not the pattern that would be developed if high whole-rock values of MgO were generated simply by the accumulation of increasing amounts of olivine, which however was of relatively uniform Mg.
6. Similarly, olivine Ni is positively, and essentially linearly, related to olivine modal abundance (Fig. 23.3(f): $y = 29.21x + 986.3$, $r = 0.87$) — not the relationship that would appear if increase in whole-rock Ni resulted simply from increasing concentration of olivine of essentially constant Ni content i.e. a process of olivine accumulation.
7. Figure 23.4, which shows the Ni–MgO fields of picrites:olivine–clinopyroxene-basalts, and of their contained olivines, indicates that the very clear and linear trend line of the former (see also Fig. 19.16) is unlikely to have been generated by subtraction/addition of their olivines from and to lavas of ≈ 15 per cent MgO.

All this seems to indicate that those Solomons picritic basalts containing, say, >15 per cent MgO are not simply the products of accumulation of olivine crystallized in, and in equilibrium with, a 'parent melt' of MgO ≈ 14–15 per cent. Those lavas containing the highest abundances of whole-rock MgO and Ni exhibit these features not only because they contain more olivine than do lower MgO–Ni lava types, but also because their olivines contain significantly higher MgO and Ni. Whole-rock:olivine compositions are clearly correlated throughout, and 'on average' (see Chapter 5) the olivines appear to have crystallized in a fairly close chemical relationship (though probably not in full equilibrium) with the lavas in which they now occur. It follows that, at least in large part, the high-MgO:high-Ni lavas of the Solomons Suite represent melts, and that these probably ranged up to ≈ 30 per cent MgO and ≈ 1250 ppm Ni.

The writer therefore suspects that, at least in many cases of picritic volcanism, a single, identifiable, primary parent is a Will o' the Wisp. Some suites, such as those of Hawaii, do appear to result from a substantially constant degree of partial melting, followed by fractionation and olivine accumulation — essentially as deduced by Wilkinson and Hensel (1988) and earlier investigators. Others, such as those of the Solomons, Baffin Bay, Vanuatu, and Grenada, probably result principally from the crystallization of the products of differing degrees of partial melting. As a world-wide phenomenon such primary melts probably contain up to about 30 per cent MgO, and were generated by up to ≈ 40 per cent partial melting of lherzolitic mantle at up to ≈ 30 kbar pressure and ≈ 100 km depth. Each batch of partial melt, i.e.

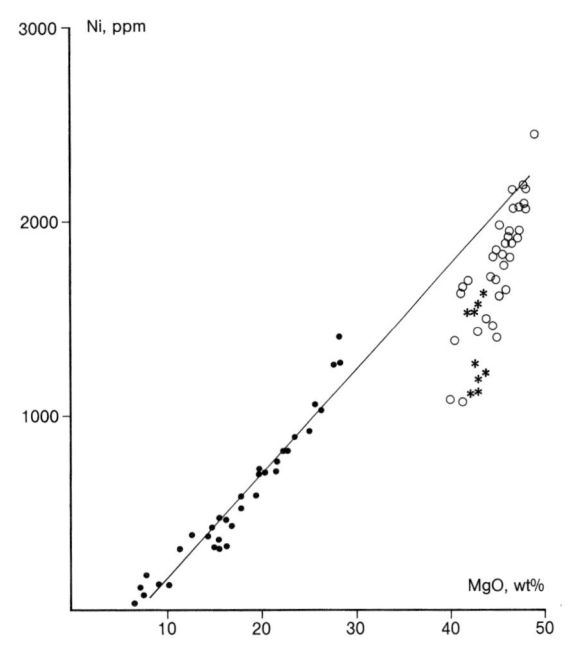

Fig. 23.4. Nickel–MgO relationships in 34 picrites and olivine–clinopyroxene-basalts (filled circles), and the olivines of those basalts (open circles; X-ray fluorescence analyses of mineral separations supplemented with electron-microprobe analyses). The nine asterisks represent the olivines of all of those Solomons olivine–clinopyroxene-basalts containing 14–16 per cent MgO, i.e. those Solomons basalts corresponding to basalts postulated as parents to picritic lavas elsewhere (ref. analysis 3, Table 23.2). The pronouncedly linear field of the lavas can be accounted for by olivine accumulation only if the olivine accumulate is composed exclusively of the most Ni–MgO rich olivines, and hence cannot be accounted for simply by the accumulation of the olivines of the 14–16 per cent MgO basalts. These relationships indicate that the range of basalt compositions cannot have developed primarily by olivine–(high-Mg clinopyroxene) fractionation.

primary magma, ascended rapidly to a sub-volcanic magma chamber or to direct eruption. In the former case the new melt would have been added to, and turbulently mixed with, partly fractionated material already residing in the chamber. The resulting magma surge would have led to the more-or-less simultaneous eruption of partly differentiated material from the upper parts of the sub-volcanic reservoir. The final assemblage of erupted lavas would thus include the directly erupted, primary products of variable partial melting, secondary products of minor fractionation (almost certainly involving polybaric fractionation of the general kind proposed by Cox and Bell 1972), and tertiary products of mixing of primary and partially fractionated melts. The author suspects that it is a complex of processes of this general kind that has led to the relationships of Fig. 23.3 and to the patterns of ore-element distribution now found in the picrites, olivine–clinopyroxene-basalts, and more mafic big-feldspar basalts of the Solomon Islands Younger Volcanic Suite.

The big-feldspar lava group

It was noted repeatedly in Chapters 5 to 20 that ternary XMA fields representing the big-feldspar lava group tend to appear as 'clouds of points' rather than the well-defined trends or lineages so characteristically developed by the olivine–clinopyroxene-basalts and the hornblende-andesites. Where some indication of a lineage can be perceived this is usually abbreviated and diffuse. The possibility that they may have acquired their present compositions as a result of fractional crystallization is therefore not so amenable to investigation by least squares mixing methods.

Simple addition–subtraction calculations indicate, however, that the group is likely to have developed by clinopyroxene:magnetite-dominated fractionation. Whereas whatever fractionation may have occurred in the basalt–picrite series was dominated by high-Mg olivine (with the lesser influence of high-Mg clinopyroxene), that inducing the development of the big-feldspar basalts and basaltic andesites would have been dominated by less magnesian clinopyroxenes, with lesser contributions by more fayalitic olivine and magnetite — the latter achieving its greatest development in this lava group.

It seems most probable that the parents to this lava group have been members of the more felsic, Fe-rich spectrum of the basalt-picrite lineage: basalts containing 50–2 per cent SiO_2. Minor clinopyroxene (–magnetite)-dominated fractionation then induced the development of (i) accumulates rich in phenocrystic clinopyroxene and magnetite, relatively small amounts of phenocrystic olivine and plagioclase, and containing $SiO_2 \approx 48$–50 per cent and (ii) derivatives containing typical prominent plagioclase, lesser clinopyroxene and magnetite, essentially nil olivine, and $SiO_2 \approx 53$–6 per cent. Stemming from this, mean SiO_2 for the group as a whole is ≈ 53 per cent, and overall composition is as set out in Tables 5.1 and 5.2.

The hornblende-andesite group

In contrast to the volcanic-arc picrites, which seem to have been first observed in 1959, the 'orogenic' andesites have been known and recognized as an important petrogenetic group for almost 160 years.

Hornblende–, plagioclase-bearing porphyritic lavas collected in the Chilean and Bolivian Andes by members of several early German scientific expeditions were examined by Leopold von Buch in Berlin, and were named 'andesites' by him in 1835. It was subsequently realized that many andesites also contained pyroxene, and that this might be ortho- or clinopyroxene or both. By early in the twentieth century it had become clear that some andesites had distinct tectonic affiliations, and in 1912 Marshall drew attention to what he saw as the 'andesite line', separating the andesitic province of the circum-Pacific orogens from the dominating basalts of the Pacific basin itself (Marshall 1912).

Increasing suspicion that at least some varieties of andesite might have quite distinct and profound tectonic affiliations and implications led to increasing interest in the lava group. Given that they were generally porphyritic, included varying proportions of the above minerals in their assemblages, and contained ≈ 52–62 per cent SiO_2, progressively greater attention was paid to their chemical constitutions and normative mineral assemblages, and to mechanisms by which the commonly occurring lineage basalt–andesite–dacite–(rhyolite) might be generated in the orogenic, i.e. volcanic arc, environment.

Previous work. The relevant literature is now very large and its detailed consideration far beyond the scope of the present contribution. (The interested reader is referred to Gill (1981) and Baker (1982) for comparatively up-to-date introductions to, and reviews of, the history of ideas.) A few remarkable contributions stand out, however, and these may be briefly considered using as a general background the first three diagrams of Fig. 5.20.

After the work of Bowen (1928) on the chemical evolution of crystallizing melts, much attention was paid to the possibility that andesites (–dacites–rhyolites) might be products of different degrees of fractional crystallization of basalt. This however, immediately encountered the 'volume problem'; that observed volumes of andesites appeared far greater than could be accounted for by fractionation of the apparently smaller volumes of available basalt (see, however, Baker 1968). Petrogenetic controversy then revolved about two possibilities: (1) fractional crystallization, with its apparent volume problem, and (2) sialic crustal contamination of basalt — which seemed incapable of accounting for the andesites of intra-oceanic arcs, such as those of the Solomons, Vanuatu, the Marianas, and Bonin festoons. Tilley (1950) drew attention to the fact that the 'basalt' of orogenic zones was not a single entity but included a number of distinct varieties, and he proposed that sialic contamination of such basalts led to the formation of andesites. In the same year Kuno (1950), investigating the lavas of Hakone volcano in Japan, recognized two principal lava series which he categorized on the basis of their pyroxenes; a pigeonitic series, characterized by a single clinopyroxene in the groundmass, and a hypersthenic series, characterized by ortho- and clinopyroxene in the groundmass. Kuno considered the former to have been essentially tholeiitic and generated by fractional crystallization (see Fig. 5.20(b), trend B–Tb), the latter to be essentially calc-alkaline and generated by the assimilation of alkali-rich material, i.e. granite (see Fig. 5.20(a), trend B–R).

In spite of what appeared to be inherent weaknesses, these two theories of andesite genesis — sialic contamination of basalt, fractional crystallization of basalt — held centre stage for most of the period 1950–65. The former was, however, soon eliminated on the grounds that many andesite-bearing island arcs were not associated with sialic crust, and on the corroborating evidence of strontium isotope ratios. The latter has clung to life rather more tenaciously; magnetite (e.g. Osborne 1959, 1962, 1969), amphibole (e.g. Yoder and Tilley 1962; T. H. Green and Ringwood 1968a; Holloway and Burnham 1972), and garnet–pyroxene (e.g. Green and Ringwood 1968a, b 1972; Green 1972; Ringwood 1974) fractionation have all been proposed as crystal-subtraction mechanisms by which the calc-alkaline trend might be generated. Put very briefly, such proposals encounter the following difficulties:

1. Magnetite subtraction produces a trend of quite different direction from that of the calc-alkaline basalt–andesite–dacite–rhyolite series (Stanton 1967, 1978; Stanton and Ramsay 1980).
2. Hornblende contains about the same amount of ($Na_2O + K_2O$) as any proposed basaltic parent, and its subtraction is therefore not capable of driving the calc-alkaline trend towards higher alkali contents found in andesites, dacites, and rhyolites. In addition the common FeO/MgO ratios of volcanic hornblendes are generally too low for the generation of calc-alkaline trends.
3. The garnets that would necessarily constitute a large part of any garnet–pyroxene accumulate are in fact rare in the island arc lava–cumulate regime.
4. While appropriate accumulative mixtures of magnetite–hornblende–pyroxene (–garnet) certainly could generate calc-alkaline trends, there is no clear evidence — field, geophysical, or petrological — indicating the requisite segregation of large masses of such material.

While fractional crystallization is still proposed as a mechanism contributing to the generation of volcanic-arc andesites, its limitations have led to the more modern proposal of partial melting: that hydrous partial melting of quartz-eclogite in the depth range 100–150 km may lead to the formation of dacitic melt, which may then mix with overlying basaltic material. Such mixed magmas may subsequently fractionate and segregate to yield the various members of the calc-alkaline series.

While such a derivation is of course possible, the hypothesis has the disadvantage that it is not amenable to testing. The mechanism requires the formation of a residuum at such depth that it cannot be observed. The only possible means of exposure is as inclusions in some of the derived lavas. Some coarse-grained, generally gabbroic inclusions do appear in some island-arc calc-alkaline lavas (and certainly in those of the Solomons), but to the writer's knowledge they are of compositions inappropriate for the development of calc-alkaline trends (see Stanton and Ramsay 1980).

A fourth possible mechanism, hinted at by many investigators and examined in some detail by the present writer (Stanton 1967, 1978; Stanton and Ramsay 1980), is that of volatile transfer and/or volatile loss. Such a process involves, in general terms, the differential movement of some of the major elements in a gas or vapour phase. In spite of the fact that elements such as iron and the alkalis have been found as prominent components of many volcanic sublimates and plumes, volcanic petrologists have given little credence to suggestions that volatile loss of iron, calcium, and other elements might contribute to the conversion of basalt to andesite and dacite.

Further possibilities lie in combinations of these processes. It has been suggested in the previous section that the Solomons picrites and the spectrum of associated olivine–clinopyroxene-basalts derive from variable partial melting, fractionation of the products, and mixing of fractionated material with new partial melt. A substantially analogous history of andesite formation might be proposed: hydrous partial melting, variable fractional crystallization, and, in some instances, mixing of the products with new batches of partial melt, followed by differential loss in the volatile phase.

The Solomons lavas. In this latter connection there is, however, one piece of evidence that (it might be said) demands to be taken into account. This is the whole array of constitutional similarities, continuities, and connections between olivine-basalt and hornblende-andesite lineages, as these are developed in the Solomon Islands, demonstrated in Chapters 5 to 21. These ties (some of which have already been emphasized in Chapter 21) are far too close and too numerous to be dismissed as coincidences. The two lineages as these appear in the Solomons (and, it would appear, in other arcs) seem to have some quite fundamental genetic connection. Possibilities are:

1. The two have a common basaltic parent, of composition indicated by the points and areas of constitutional commonality indicated in the various geochemical diagrams, each lineage being developed by its own distinctive trend of variable partial melting. The olivine-basalts would presumably have been generated by anhydrous partial melting, the hornblende-andesites by hydrous partial melting. A derivation of this kind is, however, very difficult to sustain. The possibility that picrites might be generated by partial melting of a somewhat felsic basaltic parent has no support in theory or experiment. If the andesites–dacites have been derived by partial melting of the felsic basalts there is little or no evidence of this in inclusions of the required ferro-gabbro residuum (see (2) and (3) below).

2. The two have a common basaltic parent as above, each lineage being developed by its own distinctive trend of fractionation. In this case the high-Mg basalts and picrites would have developed as accumulates by olivine (–clinopyroxene) segregation; the hornblende-andesites–dacites would have developed by hornblende–pyroxene–magnetite subtraction.

In this case, and in terms of the principles demonstrated in Figs. 5.20 and 21.6, the high-Mg basaltic lineage would have developed without the required complementary derivatives (tholeiitic basalts and andesites), the andesitic lineage without the required complementary accumulates (mafic ferro-gabbros and their lava equivalents). It appears that this hypothesis also cannot be sustained.

3. The basalt–picrite lineage has developed by the process of variable partial melting–minor fractionation as indicated earlier in this chapter, the most felsic, Fe-rich, basaltic products of this process then constituting a secondary parent from which, by hornblende–pyroxene–magnetite subtraction, the andesites and dacites developed as a derivative lineage.

We have already concluded that this explanation of the basalt–picrite lineage is compatible with available evidence. However, in the case of the andesite–dacite lineage:

(a) The appropriate hornblende–pyroxene–magnetite accumulates do not seem to occur (see (2) above); extensive field-work carried out over more than forty years has revealed no occurrence of hornblende–pyroxene–magnetite-rich lavas analogous with the olivine-rich picrites as potential accumulates (see also (c) below).

(b) Least squares mixing calculations, using:
  (i) the mean of the six most felsic basalts of the basalt–picrite lineage as hypothetical parent (Table 23.4, col. 1),
  (ii) the mean Solomons andesite–dacite as the felsic daughter (derivative) product (Table 23.4, col. 2), and
  (iii) appropriate mean compositions of magnetite, pyroxene, hornblende, and plagioclase (Table 23.4, cols 3–6 respectively) as mixing components

Table 23.4 Accumulate–parent–derivative relations and the generation of the basalt–andesite–dacite lineage of the Solomons suite

	1	2	3	4	5	6	7	8	9	10	11	12	13
SiO_2	50.59	56.79		50.96	43.94	53.35	57.95	57.17	0.78	1.35	45.97	45.66	47.76
TiO_2	0.79	0.63	5.78	0.53	1.63	0.04	0.64	0.49	0.15	23.44	1.10	1.01	1.00
Al_2O_3	17.72	17.66	4.04	3.17	11.66	29.03	18.02	18.14	−0.13	0.72	17.89	18.01	14.57
FeO^*	9.37	6.21	82.65	8.63	12.20	0.75	6.34	6.26	0.08	1.26	12.63	12.29	11.00
MnO	0.19	0.13	0.60	0.32	0.26		0.13	0.21	−0.07	53.85	0.18	0.25	0.19
MgO	5.12	3.41	2.53	15.01	13.65	0.29	3.48	3.61	−0.13	3.74	6.72	6.72	9.31
CaO	10.07	7.61		20.88	11.80	12.13	7.76	7.89	−0.13	1.68	12.47	12.45	12.91
Na_2O	2.91	4.11		0.47	2.39	4.36	4.19	3.11	1.08	25.78	2.82	1.88	2.63
K_2O	1.34	1.29			0.45	0.21	1.32	2.56	−1.24	93.94	0.23	1.38	0.52
P_2O_5	0.27	0.17					0.17	0.56	−0.39	229.41		0.36	1.10
Total	98.37	98.01	95.60	99.97	97.98	100.16	100.00	100.00	3.55§		100.01	100.01	99.99

1 = hypothetical parent basalt (mean of the six most felsic basalts of the olivine–clinopyroxene-basalt group); **2** = hypothetical derivative (mean of all Solomons andesites and dacites); **3, 4, 5, 6** = magnetite, clinopyroxene, hornblende and plagioclase used in least-squares mixing and subtraction calculations (means of each as they occur in the mafic andesites and big-feldspar lavas of the Solomons suite); **7** = derivative (col. **2** recalculated to 100 per cent (i.e. anhydrous); **8** = composition of derivative as obtained by least-squares mixing calculation; **9** = deviation of calculated values (col. 8 from analysed values (col. 7); **10** = deviation as percentage of analysed values (col. 7); **11** = bulk composition of accumulate complementary to derivative (col. **8**) as determined from least-squares mixing calculation; **12** = same, as determined using the method of Stanton (1967); **13** = mean composition of 15 mafic inclusions occurring in Solomons andesites and dacites (recalculated to anhydrous basis); FeO^* = total iron as FeO; §= sum of squares of deviations. Least-squares calculations by R. J. Arculus.

yield the felsic daughter product of Table 23.4, col. 8, and the mafic 'accumulate' of Table 23.4, col. 11. While these results are, by definition, derived from a line of best fit (obtained by minimizing the sum of the squares of the distances of the points from the required regression line), $\Sigma R^2 = 3.545$ (col. 9), and the 'best fit' is in fact a poor one. Col. 10 gives such residual as a percentage of the analysed value of the relevant component in the average Solomons andesite–dacite. Those for the alkalis, phosphorus, manganese, and (perhaps surprisingly in view of the data and conclusions of Chapter 21) titanium are large. The calculated bulk composition of the complementary accumulate (col. 11) may be compared with that derived by the method of Stanton (1967), set out in column 12. The latter constitutes a precise determination of the composition of the accumulate that must be generated as a complement to the andesite–dacite if the latter has formed from the parent of column 1, given an initial derivative:accumulate mass ratio of 0.45. Clearly the bulk composition as indicated by the least squares estimation departs grossly from this with respect to the alkalis and manganese (phosphorus was not an imputed component of the minerals involved in the least squares calculation: cols 3–6) and, to a lesser extent, titanium.

Numerical modelling of major element variation based on a fractional crystallization mechanism thus indicates that such a process cannot alone account for the basalt–andesite–dacite lineage as represented by the Solomons suite.

(c) The proportions of magnetite:clinopyroxene:-hornblende:plagioclase that provide the best fit on the basis of the least squares mixing calculation of (b) are 1:1.89:3.32:5.53. The observed proportions of the four minerals in the more mafic andesites (those that would have generated the crystals that, by subtraction, would have induced the development of the felsic andesites and dacites: we use the mean of all hornblende-bearing andesite (-basalts) containing SiO_2 <55 per cent (range 50.86–54.88; mean, 53.54 per cent SiO_2) are 1:2.52:5.6:11.38. Calculated magnetite:(clinopyroxene + hornblende) ratios are thus 1:5.21 compared with 1:8.12 observed.

Two points arise from this: (i) there are large discrepancies between the two sets of ratios, indicating that those existing in the felsic basalts–mafic andesites are not appropriate for the generation of the observed basalt–andesite–dacite trend, and (ii) as noted in Chapter 21, the observed magnetite: (pyroxene + hornblende) ratio is *too low* to generate, by fractional crystallization alone, that lineage.

(d) It might reasonably be suggested that the hornblende-rich inclusions well known to occur in the Solomons andesites and dacites (Stanton and Ramsay 1980) could represent the naturally occurring mafic complement to the latter. This is not, however, supported by chemical analysis of the inclusions. Table 23.3, column 13, represents the mean of 15 inclusions from the andesites and dacites of Savo; clearly Al_2O_3, MgO, and the alkalis differ substantially from the required values indicated in column 12.

(e) It was shown in Chapter 21 (see Table 21.5) that the proportions of magnetite, pyroxene, and hornblende that would have had to have been subtracted from a basaltic parent to generate the basalt–andesite–dacite lineage with respect to the ore elements varied greatly from one element to another in most cases. Fractional crystallization operating as the sole process of differentiation would necessarily have generated trace-element trends that, at least broadly, conformed in all cases with the mean magnetite–pyroxene–hornblende subtraction that generated the accumulate–derivative lava assemblages overall.

Given that — with the many constitutional ties presented among the data of Chapters 5–20 — the basalt–picrite and andesite–dacite lineages almost certainly have a fundamental genetic relationship, and given that the most felsic, iron-rich members of the former are most likely to have been parental to the latter, the andesites–dacites do not seem attributable to either partial melting or (as an exclusive process), fractional crystallization of basaltic material of this kind.

This brings us to the possibility of a hybrid mechanism, and the evidence of Chapters 21 and 22.

In Chapter 21 it was shown that modal magnetite in the most mafic andesites was, in proportion to hornblende and pyroxene, too low to account for the relatively low concentrations of elements such as iron, copper, and zinc in the andesite–dacite lineage as a whole and it was suggested that discrete loss in the gas/vapour phase might be a process of depletion additional to that of magnetite–pyroxene–hornblende subtraction. This has been reiterated, in more general terms, in 3(c)(ii) above. In Chapter 22 it was shown that many elements, among them major rock-forming components such as sodium, potassium, calcium, and magnesium as well as iron and the less abundant elements such as copper, zinc, and lead, could be observed in abundance in exhalative sublimates and plumes of modern volcanoes.

It may therefore be proposed that the felsic basalt–andesite–dacite (–rhyolite) lineage of the Solomons suite (and analogous suites, such as those of Vanuatu, the

Lesser Antilles and the Aleutians, from which hornblende has crystallized) is generated by three principal processes:

1. hydrous partial melting at >100 km depth, yielding felsic, relatively high-Fe, basaltic melt essentially similar in its chemical composition to the more felsic members of the basalt–picrite lineage, but containing amphibole in addition to clinopyroxene;
2. rapid ascent to the sub-volcanic-volcanic environment: the system undergoes rapid decompression, amphibole is destabilized and reacts with the melt, fractional crystallization proceeds; and
3. movement of some materials to the volatile phase, these escaping from the melt under open system volcanic conditions.

There are, of course, other island arc andesite:dacite-bearing suites, such as those of the Izu-Honshu, Mariana, and New Britain arcs, in which orthopyroxene rather than hornblende appears. This reflects, at least in large part, lower proportionate sodium and water, and higher temperatures, of the melts from which orthopyroxene crystallized — melts now represented in the Solomon Islands Younger Volcanic suite by the Simbo lavas. (It should also be remembered that many of the Solomons hornblende andesites contain small quantities of orthopyroxene.) This is another, if closely related, problem.

CONCLUDING STATEMENT

All field, mineralogical and chemical evidence indicates that the basalt–picrite (olivine–clinopyroxene-basalt) and andesite–dacite ('hornblende-andesite') lineages of the Solomons suite have some fundamental genetic tie. It is proposed that both have their origins in partial melting of the mantle at ≥30 kbar, i.e. at ≥100 km depth. The Solomons basaltic lineage is generated as a series of high-magnesium basalt–picritic melts under anhydrous conditions, and these undergo varying degrees of fractional crystallization (including polybaric fractionation) and re-mixing during ascent and residence in the sub-volcanic magma chamber. The various members of the big-feldspar lava group appear likely to constitute the most felsic products of these processes, modified to a limited extent by clinopyroxene (-magnetite) dominated fractionation. The parent to the Solomons andesite–dacite lineage is, on the other hand, generated as a hornblende-bearing basaltic melt under hydrous conditions: a hornblende-bearing, more hydrous, analogue of the relatively felsic, iron-rich basalts from which the big-felspar lavas were generated. With ascent and decompression, the hornblende begins to lose its volatiles and react with the melt, the latter fractionates by magnetite–pyroxene–hornblende subtraction and, at the same time, a component of major elements (e.g. Fe, Ca, Mg, Na, K) and minor elements (e.g. Cu, Zn, Pb, As, Sb, Bi) is lost to the system in the volatile phase.

On this basis the propensity of a basalt to transform to an andesite is determined primarily in the mantle by the hydrous nature of the partial melting that produces the parent hornblende-bearing basaltic melt. The actual basalt → andesite transformation, however, takes place beneath, and at, the volcanic orifice, where volatile loss combines with fractional crystallization to deflect compositional change from closed-system basalt → tholeiitic basalt to open-system basalt → andesite → dacite.

Much more work is required to determine whether greater quantities of the major and minor metals are emitted in the volatile phase by andesitic–dacitic volcanoes than by evolved tholeiitic volcanoes. If they are, is this simply because the more volatile-rich andesitic volcanoes have a correspondingly greater capacity for discharging their metals as volatile compounds, or is it because *hydrous* partial melting — in contrast to partial melting proceeding under essentially *anhydrous* conditions — has induced a greater degree of partitioning of metals from solid mantle to developing melt?

Perhaps both factors contribute. Certainly, as we saw in the early chapters, it is the andesites and dacites rather than the basalts with which the very largest metallic orebodies occur. One suspects that this is a problem in which volcanic ore genesis and the most fundamental considerations of volcanic petrogenesis are inextricably linked.

24

Petrogenesis: 2: Exhalative ores

Having considered the derivation of the various components of the Solomon Islands Younger Volcanic Suite and, in particular, having deduced that the generation of the andesitic and dacitic rocks has been influenced by exhalative loss of some of the major elements of the magmas in question, it is now appropriate to examine the possible place of the exhalative ores in the overall petrogenetic process.

The very large copper–gold orebody of Panguna is contained in sub-volcanic andesitic rocks of the Younger Volcanic Suite on Bougainville. Indications of similar mineralization, containing also conspicuous molybdenum, have been found on Vangunu (Royle, pers. comm. 1987), and minor occurrences of copper and gold associated with the development of andesitic to dacitic rocks of the Younger Volcanics are widespread from Guadalcanal to Bougainville. Although in this chapter we are concerned with general principles of exhalative ore generation as distinct from local features of ore formation in the Solomons, it is clear that, as in many older mineralized volcanic districts, the development of the andesites and dacites of this very young province has been accompanied by the evolution of significant quantities of some of the more important 'ore elements'.

BROAD CONSIDERATIONS

In order to consider the possible place of exhalative ore formation in the context of volcanic silicate petrogenesis it is helpful to review briefly the principal features of such ores and their occurrence.

1. The ores characteristically contain some elements and compounds and equally characteristically do *not* contain others. Of the 14 elements of Chapters 6–19 those that are commonly present in noteworthy quantities are Fe, Cu, Zn, Pb, Ba, Ca, Mn, and P; those that are conspicuous by their absence or very low amount are Ni, Cr, V, Ti, and Sr. Cobalt occasionally appears at low, but distinctly higher than trace, levels.

Iron occurs variously as sulphide, silicate, carbonate, and oxide; copper as sulphide; zinc as sulphide, oxide, and silicate; lead as sulphide and silicate; barium as sulphate and silicate; calcium as carbonate, silicate, fluoride, and phosphate; manganese as sulphide (principally as [ZnFeMnCd]S), silicate, carbonate, and mixed oxide; phosphorus as the phosphate, apatite, and probably as a trace in silicates. Any cobalt tends to occur as a trace in iron sulphides, and as Co and Co–Fe sulpharsenide and arsenide. An intriguing feature is that although lead is well known to form a highly insoluble sulphate, and although barium and calcium sulphates are commonly intimately associated with lead in exhalative ores, lead never occurs as the sulphate. Even in instances where small patches of lead-rich ore occur completely isolated in massive barite, the lead always occurs entirely as galena. While iron and zinc commonly — and sometimes abundantly — occur in non-sulphide combinations, their principal incidence within most exhalative orebodies is as sulphides.

2. Exhalative ore type is related to associated lava type in that the ores tend to become more chemically complex as the volcanic rocks become more felsic. Those associated with basaltic rocks are relatively simple and consist of iron or iron and copper sulphides. As a broad tendency, zinc begins to appear in the ores as the lavas evolve into basaltic andesites, and commonly dominates over copper when the association is andesitic. Lead and barium join iron, copper, and zinc in andesitic–dacitic–rhyodacitic–rhyolitic environments. It is, however, important to recognize that (1) the elements appear to be *added*, not substituted, in sequence: i.e. Fe, Fe + Cu, Fe + Cu + Zn etc. — not Fe, then Cu, then Zn ..., each element occurring alone in its particular ore:lava associations, and (2) the development of the most felsic rocks is not necessarily accompanied by the incidence of the full spectrum of metals; e.g. it is far from uncommon for rhyodacitic rocks to be accompanied by Fe + Cu + Zn sulphides, without significant lead. For example, mafic basalts to

basaltic andesites of Cyprus are accompanied by Fe, Fe + Cu, and Fe + Cu + Zn sulphide assemblages to constitute a typical 'Cyprus-type' association. Almost precisely similar sulphide assemblages, though on a much larger scale, are, however, associated with rhyodacites and rhyolites in the Iberian Pyrite Belt. The evidence of some hundreds of volcanic-associated exhalative orebodies (we are not considering modern sea-floor sulphide accumulations at this juncture) appears to indicate that whereas mafic basaltic environments have the potential for little more than Fe and Fe + Cu sulphide accumulations, more felsic environments have the *potential*, which may or may not be realized, for the development of much higher proportions of zinc and large quantities of lead sulphides.

Although one orebody (that of Outokumpu in Finland) of ultramafic–mafic volcanic association and undoubted exhalative origin contains significant Ni, this element normally occurs as no more than a trace in volcanic exhalative deposits. Similarly, chromium and vanadium are essentially absent. Cobalt is usually present only at trace level, though occasionally it appears as an exploitable 'minor' element. Arsenic, antimony, and bismuth sulphide and sulphosalt compounds (and the native elements) occur as trace minerals, appearing in greater amount and with increasing mineralogical complexity as the ore:lava association becomes more felsic.

3. The maximum size of exhalative deposits appears to increase as the associated volcanic rocks become more felsic and the pyroclastic component becomes more abundant (see Chapters 2 and 4; also Stanton 1967, 1978; Stanton and Ramsay 1980). At the basaltic end of the spectrum the pyroclastic component and hence, presumably, the proportionate loss of gas from the magma is minor, and the associated pyritic Cu and Cu + Zn lenses are most commonly quite small. Many contain only a few tens to a few thousands of tons of ore. Worked orebodies generally range in the interval $0.5 \times 10^6 - 5.0 \times 10^6$ tons. A few, as in Cyprus, Japan, and Norway fall in the approximate range $10 \times 10^6 - 20 \times 10^6$ tons, but these are uncommon and represent about the maximum known size for deposits of this ore association. The Løkken orebody of the Trondheim district of Norway is exceptional in containing about 25×10^6 tons of pyritic copper ore.

In the andesite–dacite range the volcanic rocks become, conspicuously, largely pyroclastic. The accompanying exhalative ores are most commonly Cu–Zn with zinc dominant, and the largest orebodies attain sizes of the order 10×10^6 to 30×10^6 tons and may be larger. Where the volcanic rocks are dacitic to rhyolitic they appear to be almost entirely pyroclastic, and to have incurred massive, highly explosive, loss of volatiles. The ores of this association commonly contain much lead as well as zinc and copper (zinc is usually the dominant metal and copper may be quite subordinate), and the deposits may be very large: many of them are in the $20 \times 10^6 - 50 \times 10^6$ tons range and the largest $>100 \times 10^6$ tons.

As in the case of lava type:ore type relationships, the proposal of a tie between volcanic rock composition and orebody size calls for caution rather than a sweeping statement. We are referring to the sizes of the *largest* deposits that the different environments seem capable of producing: basaltic, say, $\approx 20 \times 10^6$ tons; andesitic, $\approx 50 \times 10^6$ tons; dacitic–rhyolitic, $\approx 100-200 \times 10^6$ tons. Just as the dacite–rhyolite associated ores commonly contain lead but may contain none, the maximum size of the orebodies may be very large — but there are many deposits of this association that are very small. We have been concerned with *maximum* compositional complexity of the ores, and *maximum* size of the deposits concerned.

4. As a broad class the exhalative ores exhibit a number of time-space-compositional relationships.

On the grandest scale there appear to be three great episodes of formation:

(a) The Archaean, $\approx 2500-700$ Ma: a period dominated by (though not entirely confined to) andesite-associated, pyritic Cu–Zn ores as represented by the Archaean deposits of Central and Eastern Canada and Western Australia.

(b) The Lower Proterozoic, $\approx 1600-1850$ Ma: characterized by a number of giant deposits whose composition is dominated by zinc and lead; e.g. Broken Hill, NSW; Aggenys and Gamsberg, Namaqualand; Balmat, NY; Sullivan, BC; Mount Isa and Century, Queensland; Macarthur River, Northern Territory of Australia.

(c) The Lower to Middle Palaeozoic, $\approx 550-300$ Ma: polymetallic deposits ranging from pyritic Cu ores to complex Cu–Zn–Pb deposits; e.g. those of the Caledonides of Britain and Norway, the Iberian Pyrite Belt, the Appalachians of eastern North America, the Eastern Highlands Belt of Australia, and the Urals in Russia.

Ores substantially similar to those of (c) have been developing again since early Tertiary times (e.g. the ores of Cyprus and the Kuroko deposits of Japan) and it is possible that the Earth is once more passing through a period of notable exhalative ore formation. Indeed the Kuroko ores provide, on a smaller space–time scale, a striking example of the sharply episodic nature of much volcanic exhalative mineralization: although the deposits are distributed through a province extending over a distance of more than 300 km, all were formed in a single interval of ≈250,000 years during the later Nishikurosawan period, some 13 Ma.

A feature of many volcanic ore provinces is that individual deposits occur in clusters related to major andesitic to rhyolitic marine-volcanic rises, and to individual volcanic edifices within these. The 'basement' — and commonly by far the most abundant — volcanic material is tholeiitic basalt, over which lies a volumetrically relatively minor carapace of andesitic–dacitic–rhyolitic volcanic rocks, pyroclastic materials, and derived sediments. In such cases of large-scale bimodal volcanism the basaltic component of the pile usually contains little in the way of concentrated sulphide other than isolated small pods of exhalative pyrite/pyrrhotite/chalcopyrite, although much veinlet and disseminated sulphide may occur. The transition to the overlying calc-alkaline rocks may not be sharp (some alternation of basalt with andesite–dacite lava and pyroclastic units is common) but it is after this break in the nature and style of volcanism that major and numerous *concentrations* of sulphides appear. It is a striking feature of such provinces (e.g. the Archaean of Canada, Australia, Finland, India) that the voluminous basalts (principally lavas) contain little concentrated sulphide whereas the volumetrically relatively minor calc-alkaline rocks (largely pyroclastic) commonly contain numerous, and in some cases very large, polymetallic exhalative orebodies. This inverse relationship between volcanic rock type/volume and ore occurrence in many mineralized volcanic provinces is so marked as to be of virtually certain significance.

Not all exhalative ore provinces exhibit the overall dominance of volcanic rocks displayed in many of these Archaean domains: in other Archaean provinces and in many of the Lower Proterozoic and Palaeozoic ore regions the volcanic members are no more than important components of eugeosynclinal sequences. In all cases, however, there is a very strong tendency for the exhalative ores of a district or region to be restricted to just one or two 'horizons' of the relevant volcanic or volcanic–sedimentary stratigraphic sequence (Stanton 1955*b*). Where the ores are associated with andesites and more felsic volcanism they commonly exhibit the grouping about volcanic centres noted above. Deposits associated with basaltic terrains, on the other hand, show the almost invariable strong stratigraphic affiliations, but associations with particular centres are usually not so obvious. Where deposits are distributed round a volcanic centre, their constitutional features may show a concentric arrangement; e.g. at Noranda, Quebec, pyritic base-metal deposits tend to occur around the central area of the volcanic pile, whereas pyritic gold deposits are prominent about the periphery. Where orebodies occur on two preferred horizons rather than one, it is common for the lower group to be associated with more mafic rock types, e.g. basalts and basaltic andesites, and to be relatively Cu-rich, and for the upper group to be associated with more felsic rocks, e.g. dacites and rhyodacites, and to be more Zn:Pb-rich, a feature first noted by Stanton (1955*b*).

Finally, on the smallest scale, many exhalative deposits exhibit their own within-deposit chemical stratigraphy. Perhaps significantly this tends to parallel the spectrum of ore types from Fe to Fe + Cu + Zn + Pb + Ba. Many Cyprus-type deposits are pyrite-rich towards their base, proportionate copper increasing with stratigraphic height. Similarly deposits with a major zinc component tend to a stratigraphic order Fe-rich at base, becoming more Cu-rich with height, and finally more Zn-rich in the upper zones. The more complex ores exhibit the tendency Fe-rich at base, grading to more Cu-rich, and then Zn → Pb → Ba at the stratigraphic top — a feature first observed by Kraume *et al.* (1955) in their studies of the Rammelsberg deposits.

DISCUSSION

There is clearly a great deal of field and geochemical evidence that may bear on the origin of exhalative ores of volcanic affiliation. It would be surprising if this did not yield at least some indication as to whether the ores were derived dominantly by sub-sea-floor leaching or by active magmatism. One uses the word 'dominantly' advisedly, for it must be regarded as likely that both processes, here and there, contribute in different ways and to different degrees in the ultimate derivation of the metals and sulphur of the ores. There however, are, so many consistencies in the nature and occurrence of exhalative deposits that some basic 'principal process' is indicated.

We may now consider the evidence and the indications as to what this might be:

1. *Episodical nature of ore formation.* Apart from the fact that they are all associated with volcanic rocks in one way or another, perhaps the principal feature of the occurrence of these ores is that their incidence is episodic. They do not occur randomly through Earth history: from the grand scale of geological eras, to the smaller scale of periods (to which the Kuroko ores, for example, are confined), to the smallest scale of local volcanic stratigraphy, their incidence is, as we have just seen, staccato. The occurrence of the pyritic Cu–Zn

ores of the Archaen, the stratigraphic ties of the Iberian Pyrite Belt, Cyprus, and the Japanese Green Tuff, and the remarkable association of ore occurrence with the top of the Millenbach Andesite horizon at Noranda, all illustrate the principles in descending order of scale.

This very clear and virtually ubiquitous tendency for ore occurrence to be tied to relatively brief intervals of time seems to indicate an association with a phenomenon of episodic nature: with somewhat sudden and relatively short-lived 'events'. Were the deposits the result of essentially continuous sub-sea-floor convection and leaching, it might be expected that there would be a relatively even distribution through time — from the scale of geological eras to that of the individual volcanic pile. This is clearly not the case (the deposits are always tightly clustered in time), suggesting that exhalational and ore formation result, in each instance, from some comparatively brief, clearly defined event, such as a volcanic episode, rather than from a long-continued process such as sea-floor leaching.

2. *Ore type, lava type, and relative lava volumes.* As pointed out in Chapter 4, the relationship between ore type, lava type, maximum orebody size, and the relative volumes of the lava types with which the deposits are associated are the inverse of those that might have been anticipated. This state of affairs appears quite fundamental and merits brief recapitulation:

The total volume of basalts in the marine environment is many orders of magnitude greater than that of marine andesites, dacites, and rhyolites. Basalts contain approximately twice as much trace Cu + Zn + Pb (say 150 ppm) as dacites (say 70–80 ppm). By far the major part of the marine basalts is erupted on to the deep sea floor and remains on and beneath the ocean floor for geologically long periods of time. The marine andesites, dacites, and rhyolites, on the other hand, generally do not appear until the volcanic edifice approaches the surface of the sea, and for the most part their eruption is initiated in relatively shallow water just prior to volcanism becoming subaerial. Thus the marine basalts bulk enormously larger, contain significantly more base metal, and are exposed to sea-floor leaching to a far greater degree than the materials of the andesite–rhyolite series. This would suggest, on the basis of the leaching hypothesis, that the exhalative orebodies associated with basalts should be far larger and more numerous than those associated with andesites, dacites, and rhyolites.

Exactly the opposite is the case.

Apart from the very few exceptions in Japan, Cyprus, and Norway referred to in Chapter 4 and above, the basalt-associated deposits are relatively small in size and few in number. The world's great volcanogenic orebodies are almost invariably associated with the development of rocks of the andesite–rhyolite series and their pyroclastic and volcaniclastic derivatives.

3. *Ore occurrence, pyroclastic rocks, and volcanic degassing.* This brings us to the pyroclastic component of the volcanic pile, and the evolution of volatiles during eruption.

The enormous volumes of sea-floor basalt carry little vesiculation and appear to have undergone minimal degassing. (There have been many studies of submarine basalt vesiculation and degassing: both certainly occur but are relatively minor.) The much smaller volumes of marine andesite–dacite–rhyolite, on the other hand, are very largely pyroclastic, and at the dacite–rhyolite stage almost entirely so. Thus the andesites–rhyolites degas to a much greater degree than the basalts on a per unit volume basis.

Combining (2) and (3):

(a) The largest exhalative orebodies are not associated with the relatively huge volumes of deep-sea basalt that are exposed for very long periods to sea-water leaching: they are associated with the relatively small volumes of andesite–rhyolite erupted into shallower water and little exposed to sea-water leaching.

(b) The basalts have lost little gas and hence, presumably, those volatile compounds of metals that we know to be emitted in volcanic gases. The andesites–dacites–rhyolites, on the other hand, have lost much gas and, presumably, much of their original content of volatile metal compounds.

Taken together these two lines of evidence seem to indicate that the metals of the ores are related, not to potential for leaching, but to the volcanic, i.e. active magmatic, emission of gas.

4. *Relative abundances of copper, zinc, and lead in lavas* vis-à-vis *exhalative ores.* As we have noted on a number of occasions in the preceding pages, the characteristic non-ferrous metal of the basaltic exhalative milieu is copper: the ores are of Cyprus type. While zinc is present at least as a trace in virtually all such deposits, and in some cases sphalerite may be a prominent constituent, copper is normally the dominant metal after iron. Zinc is most commonly minor, and lead essentially absent. On the other hand, the exhalative ores of andesite–dacite–rhyolite association are conspicuously rich in zinc, commonly possess abundant lead, and contain relatively minor copper.

However, as we observed in Chapter 4, basalts contain as much or more zinc than andesites, dacites, and rhyolites, and, what is more, contain it in at least partly more leachable form, i.e. in olivine. It is therefore not easy to see, on the basis of a leaching hypothesis, why major accumulations of basaltic lavas should not commonly yield zinc deposits just as large as — indeed much larger than — those associated with some andesite-to-rhyolite sequences of relatively small volumes.

While the essential absence of lead from basalt-associated Cyprus-type ores is generally attributed — on the basis of derivation by leaching — to the near absence of lead from, and high Cu/Pb ratios in, basalts, this somewhat glib line of argument may not bear careful inspection. Using lavas from the Solomons suite as an example, all basalts (SiO_2 <52 per cent) contain mean Pb = 2.9 ppm, an all dacites–rhyolites (SiO_2 >62 per cent) contain mean Pb = 10.9 ppm, yielding the ratio Pb_{d-r}/Pb_b = 3.76. While Pb_b is thus only ≈0.27 Pb_{d-r}, it is not easy to see, on the basis of leaching, why many exhalative orebodies associated with dacites–rhyolites may contain very large quantities of lead, whereas those associated with large basalt piles contain essentially none.

5. *The 'nickel problem'*. The greater abundance of nickel than of copper and zinc in basalts, coupled with the near-absence of nickel from most exhalative ores, looks as if it may be one of the more decisive discriminants of source and process.

While there are indeed just a very few exhalative nickel deposits in some of the Archaean 'greenstone belts', their number is minute compared with those of Cu–Zn–Pb in the same terranes, and they occur only with very high-Mg basaltic lavas. The other two great epochs of exhalative ore formation, the Lower Proterozoic and the Lower to Middle Palaeozoic, appear to be devoid of significant exhalative nickel.

There is no reason to believe that, in a sub-sea-floor leaching environment, nickel would not be released with zinc (and copper) from olivines, spinels, and pyroxenes and that it would not travel with them in the relevant hydrothermal solutions. Further, nickel is a highly chalcophile element — more so than iron and much more so than zinc — and could be expected to form highly stable and insoluble sulphides wherever contributed to the sea-floor hydrothermal sulphide-forming locale.

All of this it conspicuously does not do: not only is nickel almost absent from ancient exhalative deposits, it is also virtually absent from all currently known sea-floor hydrothermal deposits. Unless an explanation for this is found, the absence of nickel from most exhalative ores, and particularly those of Cyprus type, must be seen as powerful evidence against any major role for leaching in the formation of deposits of this kind.

Looked at from the point of view of magmatic fractionation, on the other hand, the non-appearance of nickel in most exhalative ores is as might have been expected. As a result of the early subtraction of nickel in olivine and its concomitant dramatic impoverishment in the remaining melt, little nickel remains in the latter by the time differentiation has produced andesitic rocks, and particularly their groundmasses. This is accentuated in the later differentiates (dacites and rhyolites), in which nickel becomes almost nil. On this basis nickel is virtually absent from the late-stage, volatile-rich liquid and there is almost none available to enter a late volatile phase. The development of exhalative nickel concentrations is therefore not favoured by the volcanic magmatic process.

The incidence of nickel thus appears to conform well with what might be expected from magmatic derivation, and very poorly with what might be expected to result from leaching. Indeed, one suspects that it may deliver the *coup de grâce* to the hypothesis of leaching as a major contributor of the metals to exhalative deposits.

6. *The 'cobalt problem'*. Although the properties, behaviour, and incidence of cobalt are very similar to those of nickel there are some critical, if slight, differences, and these may provide some subtle evidence concerning exhalative sulphide derivation.

Cobalt's remarkably consistent abundance of ≈40–4 ppm in both MORBs and arc basalts is ≈0.4 × Ni in MORBs and ≈0.3 × Ni in arc basalts: and, to keep 'scale of abundance' in mind, ≈0.5–0.6 × both Cu and Zn in these materials. Like nickel, cobalt occurs principally in olivine and spinel, although its incidence in olivine diverges from that of nickel in being virtually independent of MgO/FeO ratios.

As a result of their similarities of chemical properties and mineralogical affiliations, it might be expected that the two elements would exhibit closely similar behaviour in almost any sub-sea-floor leaching– hydrothermal transport regime. The marked instability of olivine under low-temperature hydrothermal conditions would ensure the release of substantial quantities of nickel and cobalt together. The very high solubilities of nickel and cobaltous chlorides and sulphates would facilitate their ready, and essentially simultaneous, solution and transport. Their highly chalcophile nature and the insolubility of their sulphides and sulphosalts would induce rapid co-precipitation in any likely site of sea-floor hydrothermal sulphide deposition. Furthermore, if the leaching sources were basaltic, it might be expected that the abundance of cobalt in the resulting exhalative sulphide accumulation would be, if rather less than, at least of the same order as, that of, for example, zinc.

The fact that, in spite of these factors apparently favouring leaching, cobalt (like nickel) occurs in only minute quantities in exhalative ores and modern sea-floor sulphide accumulations, seems to indicate that leaching processes have not made major contributions to exhalative ore formation.

Considered in a magmatic context, the behaviour of cobalt during fractionation is at first sight very similar to that of nickel: it is abstracted rapidly from the melt in olivine and early spinel. However, the two metals diverge proportionately owing to (1) the extreme propensity for nickel to enter the earliest formed, high-Mg olivines and (2) the almost total lack of any relationship between MgO/FeO ratios in olivine and the tendency for cobalt to enter it. As a result, a larger proportion of original cobalt than of nickel remains in the melt at, say the 52–4 per cent whole-rock SiO_2 interval, at which stage basalts are changing to andesites (Fig. 21.2(a) and (b)). That is, while negligible nickel remains in the melt by the time exhalation begins, some cobalt persists. Perhaps this is why many exhalative ores contain sufficient cobalt for the development of small but discrete quantities of cobalt minerals, whereas nickel occurs as little more than a trace in the iron sulphides.

7. *Chromium and vanadium in lavas and exhalative ores.* Although the depression of titanium is a striking feature of some exhalative deposits (see (8) below), the possible relevance of this to understanding processes of ore formation is far from clear. On the other hand, the general absence of chromium and vanadium from these ores looks as if it may be highly significant and amenable to analysis.

As we have seen, the principal repositories of chromium in basaltic and andesitic rocks are spinel and clinopyroxene, both of which develop very high values of K_D with respect to this element in the residual melts. A large proportion of marine basalts contain 200–300 ppm chromium, and there is thus much more chromium available for leaching than, for example, copper, zinc, lead, cobalt, or nickel. According to the nature of the leaching solutions it might be expected that chromium would be transported, at low Eh–low pH, as chromic chloride or chromic sulphate. That chromium, where available, might be transported in significant quantities in hydrothermal fluids has been proposed by Treloar (1987). Of the Outokumpu Cr-bearing silicate veins Treloar remarks:

> These veins are obviously of an unusual composition, but suggest that under certain conditions a sulfide-bearing hydrothermal fluid is capable of transporting substantial amounts of Cr ...in addition to the base metals, both Cr and Ni are substantially enriched within the cordierite–orthoamphibole rocks of the stockwork ... If these rocks represent the fluid pathways of the ore-forming fluids it is probable that Cr and Ni were being scavenged at the same time as the base metals and transported with them within the hydrothermal ore-forming fluid. (1987, p. 883).

On discharge at the sea floor the chromium might enter the structures of neoforming clays or phyllosilicates, or precipitate as a component of exhalative sedimentary oxides. That it may be leached from basalts (chromian clinopyroxenes) and redeposited in phyllosilicates (fuchite) has been demonstrated by Jack (1989): see Chapter 15).

To the best of the writer's knowledge there is, however, no known example of chromium as a major component of an exhalative ore, and its incidence in modern hydrothermal sediments is at no more than trace level. The reason for this is not obvious on the basis of leaching.

On the other hand, chromium is, of the 14 elements discussed in Chapters 6–19, the second most rapidly captured by the earliest-formed crystals of a solidifying basaltic melt. It is in fact subtracted from the melt almost as rapidly as nickel (Fig. 21.2(a)) and is removed almost quantitatively by the time a significant volatile phase could develop. This appears to be consistent with the fact that chromium is generally inconspicuous in, or absent from volcanic exhalations (see, however, the work on the plume of Syrtlingur by McClaine *et al.* (1968), referred to in Chapter 22). Under such circumstances it is unlikely to be available for incorporation in any late-stage, metal-bearing gas phase, and hence to be contributed to the sea floor for deposition as exhalative oxide or as a component of layered silicates.

The principal hosts of vanadium in the Solomons lavas are titanomagnetite and hornblende. Mean vanadium in the former is ≈3650 ppm and in the latter is ≈434 ppm. The mean vanadium content of MORBs ≈260–270 ppm and of the Solomons basalts is ≈235 ppm.

Like chromium, vanadium is rarely if ever reported as a component of exhalative ores. The minor vanadium, occurring as a spectacular range of vanadiferous minerals, associated with the Hemlo gold deposit of Ontario (Chapter 14) is the only (possible) exhalative deposit known to the author in which vanadium appears in notable quantity. There are few data on vanadium in the sediments associated with modern seafloor hydrothermal discharges. Embley *et al.* (1988) have reported quantities up to 740 ppm in altered sea-floor basalts, where the unaltered parents contain ≈300 ppm vanadium. Judging from its mobility in low-temperature groundwater and aquifer environments, as evidenced by its common occurrence in sandstone-type U–V–Cu deposits, any noteworthy quantities of vanadium in exhalative ores should be revealed in their gossans. The spectacular colours of many vanadates should render its presence obvious even where it occurs in only small quantities. However, in spite of these potentially very sensitive indicators vanadium rarely if ever appears and it must be concluded that, in spite of its relatively high abundance in basaltic and andesitic lavas, it is not present in any significant amount in volcanic exhalative ore deposits.

It might be expected that solution and transport of vanadium should occur at least as readily under hydrothermal leaching conditions as under those prevailing in the aquifers involved in the movement of components of U–V–Cu deposits — and in the latter environments vanadium is certainly highly mobile. Under leaching conditions it is most likely that vanadium would be transported as one or more of the chlorides VCl_4, VCl_3 or VCl_2, or as the vanadate ion $(VO_4)^{3-}$ (as in vanadic acid, H_3VO_4, cf. H_3PO_3). With discharge of the hydrothermal solutions into the higher Eh–pH environment of the sea floor, the vanadium might precipitate as $VOCl$, $VO(OH)$, metal vanadates, and/or phosphovanadates, or as a component of clays and phyllosilicates such as roscoelite. If T–Eh–pH at the immediate site of exhalative sulphide deposition were not conducive to the formation of such compounds, it might be expected that the appropriate conditions would be encountered in the vicinity of the margins of the sulphide accumulation, thus leading to the development of vanadium-bearing chemical–sedimentary haloes about the sulphide lenses — as manganese oxy and other compounds are well known to do.

The absence of such materials from exhalative ores and their immediate host rocks indicates once again a likely lack of significant sub-sea-floor leaching.

Considered in the context of its behaviour in the Solomons lava series, however, such non-appearance is not surprising. As we have seen, although vanadium develops a hypotholeiitic trend with progressive fractionation, this can (just) be accounted for by the change from olivine–clinopyroxene–spinel to hornblende–clinopyroxene–spinel subtraction accompanying the onset of andesite–dacite formation (Chapter 21). This being the case, the steady decrease in vanadium relative to $(Na_2O + K_2O)$ reflects an essentially closed-system process, and any loss to a volatile phase — at least in the evolution of the Solomons lavas — is likely to have been very minor to insignificant. On this basis vanadium does not appear among the products of exhalation for the simple reason that it did not enter the magmatic gas phase *ab initio*.

Thus, whereas the absence of vanadium may be difficult to account for on the basis of sea-water leaching, it appears, at first sight, to be exactly what might be expected were the exhalations of active magmatic derivation.

There is, however, one complicating area of evidence: that of modern volcanic exhalations (Chapter 22). Although most volcanic plume aerosols and many sublimate samples — representing the entire range of on-land volcanic environments — contain little if any vanadium, it is an important component of the sublimates of Izalco volcano in Honduras (Hughes and Stoiber 1985) and of Bezymianny volcano in Kamchatka (Borishenko *et al.* 1970; Borishenko 1972), and it has been noted in sporadic trace quantities elsewhere. In view of this it might have been expected that *some* sulphide ores derived from exhalations of active magmatic source would, here and there, display elevated vanadium. That it does not is, on the basis of a magmatic gas hypothesis, perhaps just a little surprising.

8. *The 'titanium problem'*. It is becoming apparent (Chapter 13 and section 7 above) that a marked depression of titanium, in spectacularly close combination with aluminium, is a notable feature of a variety of exhalative ores and modern sea-floor hydrothermal sediments, e.g. the northern Juan de Fuca Ridge (see Goodfellow and Blaise 1988, Fig. 11, p. 688). It is quite possible that this reflects a 'swamping' of normal, very slow, detrital sedimentation by the rapid local deposition of sulphides and associated hydrothermal precipitates contributed to the sea floor in the ore-forming locale. Whether this is the case or not, it does indicate that the titanium (and aluminium) contribution from the hydrothermal source was distinctly lower than the normal sedimentary flux as reflected in the compositions of the enclosing, non-ore, strata.

The principal Ti-bearing minerals of the Solomons lavas are spinel (titanomagnetite) and hornblende. Like vanadium, titanium exhibits a somewhat muted hypotholeiitic pattern that can be approximately accounted for by a change from olivine–clinopyroxene–spinel to hornblende–clinopyroxene–spinel fractionation accompanying the onset of andesite–dacite formation. Its abundance behaviour in the Solomons lavas thus indicates that titanium in the cooling melt is captured entirely by crystals and would be unavailable for incorporation —and loss — in a volatile phase.

As titanium is a relatively immobile element it is unlikely to be incorporated and transported in a hydrothermal leaching phase. Thus both leaching and magmatic volatile loss mechanisms appear capable of accounting for the commonly low Ti content of exhalative ores and modern hydrothermal sediments.

There is, however, as with vanadium, one complication: where active magmatism — volcanism — can be observed generating a gas-aerosol phase quite independently of any possible sea-water leaching contribution, titanium is almost invariably a component (Chapter 22). The behaviour and incidence of titanium in the magmatic gas phase, and the possible significance of this in the development of low-Ti exhalative ores, clearly requires much further consideration.

9. *The problem of within-deposit chemical stratigraphy.* That the metallic components of exhalative ores are dominated by iron, copper, zinc, and lead — not by nickel, cobalt, and other metals that occur in volcanic rocks and are capable of forming insoluble sulphides — is generally taken for granted and left unremarked. It is a noteworthy feature, none the less. The tendency for the four elements to occur in Fe, Fe + Cu, Fe + Cu + Zn, and Fe + Cu + Zn + Pb ± Ba is also striking, as are the affiliations of these groupings with stratigraphy (both between and within individual deposits) and the evolution of the associated volcanic rock types. As we have seen, the associated tendency for maximum chemical (and mineralogical) complexity and maximum orebody size to be inversely related to the volume of the volcanic rock type with which they occur is also a remarkable feature — as is the apparent direct relationship between size, complexity, and the proportionate amount of gas evolved by the associated lavas.

Copper and zinc occur in about equal concentration in the Solomons basalts (100 and 75 ppm respectively) and in MORBs (71 and 84 ppm). Mean lead in the Solomons basalts is ≈2.9 ppm, and in MORBs ≈0.6–0.8 ppm. In the lavas of the Solomons suite copper tends to be slightly higher in the groundmass than the whole rocks of the more mafic members, attains essentially similar maxima in whole rocks and groundmasses, and is a little higher in the whole rocks than in the groundmasses of the more felsic members (Fig. 6.12). Zinc is higher in the whole rocks than the groundmasses throughout (Fig. 7.23) and lead is slightly higher in the groundmasses than whole rocks almost throughout (Fig. 8.10; the single exception is probably a statistical artefact).

Maximum copper in the Solomons whole rocks and groundmasses occurs in the 52–4 per cent whole-rock SiO_2 interval, and this is followed by a sharp and sustained decrease. The apparent similarity of the two maxima must, however, by viewed in the light of the copper contents of the glass inclusions in some of the basaltic–andesitic feldspars. It will be recalled (Chapter 6) that such inclusions within the feldspars of some of the big-feldspar lavas contain much higher levels of copper (mean ≈1900 ppm or 0.19 per cent Cu) than do the corresponding glasses *enclosing* those feldspars, indicating that in at least some instances copper concentrated in the melt to a much greater extent than is now indicated by its abundance in the groundmasses. This seems to have occurred, perhaps fleetingly, more or less just as basalt was about to give way to andesite.

Maximum zinc in the whole rocks (79 ppm) is, like copper, attained at 52–4 per cent whole-rock SiO_2, but that in the groundmasses (61 ppm) takes the form of a plateau between 48–56 per cent SiO_2 (Fig. 7.23). Zinc, like copper, tends to notably higher abundance in the glass inclusions within feldspars than in the glasses enclosing the latter. However, this elevation of zinc in the slightly earlier glasses is, at its mean of ≈330 ppm, not nearly so pronounced as for copper, but it is a quite unambiguous feature none the less. The mean ratio (Zn in inclusion glass/Zn in groundmass glass) for the basalt–andesite lavas on which this detailed electron-microprobe analysis for the basalt–andesite lavas on which this detailed electron-microprobe analysis was carried out is 330.6/97.2 = 3.40. This compares with an analogous ratio of 12.74 for copper in the same inclusion and groundmass glasses. Assuming that the inclusion glasses represented earlier melt compositions, there has clearly been a dramatic decline in both copper and zinc in the melt over a very short interval of the full span of Solomons lava evolution.

Lead appears to undergo sustained increase with increase in whole-rock SiO_2 overall, though there may be a beginning of flattening of the curve at ≈62–4 per cent whole-rock SiO_2: the onset of the formation of dacites.

Although Large (1992) has recently suggested an ingenious alternative (based on an interplay between temperature–solubility relationships of the major ore minerals, and the evolutionary pattern of temperature rise and fall in the prevailing hydrothermal system), the writer still inclines to the view that the relative stratigraphic position of these three non-ferrous metals in the relevant exhalative sulphide deposits probably indicates some kind of paragenesis, or time sequence of discharge from the exhalative source. This was first suggested by Kraume et al. (1955) as a result of their observations of the chemical layering of the Rammelsberg orebodies: 'The chemical compositon of the solutions changed so that at first iron, then copper, zinc, lead, and finally barium were present in larger concentration. The supplementary and trace elements also display a certain sequence' (Kraume et al. 1955, p. 352). The generality of their observations has been confirmed by both within- and between-deposit stratigraphy in many exhalative ores and ore districts.

From the point of view of 'ease of release' it is not obvious why a leaching process should lead to this observed stratigraphic:paragenetic sequence of Fe–Cu–Zn–Pb–Ba and to the development of the ore types involved. In the crystalline phase, copper is low in the silicates ($\approx$30 ppm) and much higher in the magnetite (mean $\approx$185 ppm; maximum $\approx$1000 ppm Cu). Most of the copper thus occurs in the groundmass, much of the remainder in magnetite. Zinc is relatively concentrated in the crystalline phase — in olivine (mean Zn $\approx$200 ppm) and magnetite (mean Zn $\approx$490 ppm) — and is less abundant in the groundmass. Lead occurs in magnetite (mean $\approx$15 ppm), K-bearing plagioclase (mean $\approx$2.8 ppm), and groundmass (mean $\approx$5.4 ppm Pb). It is likely that the groundmasses, i.e. glass and microcrystalline matter, would be the component most susceptible to leaching, hence perhaps favouring the release of copper and lead over that of zinc. As, however, much of the latter occurs in olivine, a highly unstable component in hydrothermal environments, overall susceptibility to release from the silicates is probably much the same for the three metals.

The principal habitat of the three elements on a weight-for-weight basis is clearly the spinel (magnetite), though this constitutes only $\approx$1–3 modal per cent of the spectrum of Solomons lavas. The principal potential leaching sources for copper and lead in basaltic lavas are therefore groundmass and magnetite; and those for zinc are olivine and magnetite. Where more felsic lavas are involved we may add the opaque rims of hornblende as a source of zinc. Although it is a very effective host for all three metals, biotite is such a minor component of oceanic (including arc) lavas that it could not be a significant source of metals for leaching.

Any leaching of a lava pile constituting a 'passive source' of these three metals is thus likely to yield solutions carrying all of them. From the point of view of *source* it is not easy to see why copper should be leached and transported in preference to zinc to form the almost exclusively Fe–Cu assemblage of the more mafic basalt-associated ores of Cyprus type. Certainly, as noted above, Cyprus-type ores overall contain zinc and in some instances this may be a substantial component, but there remain examples, particularly in association with more mafic basalts, in which zinc is present as no more than a trace to minor component and copper is the overwhelmingly dominant base metal. As leaching and transport of copper may require higher temperatures ($\approx$400 °C: e.g. Seyfried and Janeck 1985; Seyfried et al. 1988) than those for zinc it may be, of course, that both have been leached and transported in solutions at $T \geqslant 400$ °C, but that the two metals have been separated by a sudden *fall* in temperature: copper has been precipitated, zinc carried on in solution, perhaps to be deposited elsewhere. If, however, we visualize high-temperature hydrothermal solutions discharging into marine or lacustrine bottom waters it is difficult to see how such a sharp change in T, Eh, and pH could fail to precipitate both metals, and in the same order of abundance.

We may now consider the possibility that progressive magmatic degassing is responsible for the time–stratigraphic relationships of the three metals.

There is no doubt that copper, zinc, and lead all escape from sub-volcanic chambers in the high-temperature gas phase, quite independently of any such process as sub-sea-floor hydrothermal leaching (Chapter 22). Copper is virtually ubiquitous in present-day volcanic flames, plume gases, and aerosols, and in fumarole sublimates — in association with the whole range of volcanic environments from mid-ocean basaltic through to arc and continental rhyolitic activity. Exhalation of zinc and lead has also been observed through the full span of volcanic environments though — perhaps a somewhat subjective judgement — they are not so prominent in mid-ocean basaltic settings and achieve their greatest incidence in more felsic arc and continental volcanoes.

As noted above, the Solomons lava series indicates a rapid rise in copper content from a 'parental' basalt $\approx$80 ppm at SiO_2 $\approx$48 per cent to Cu $\approx$120–130 ppm at SiO_2 $\approx$50–4 per cent and (as indicated by the high levels of copper in the glass inclusions within feldspars) at least a brief interval, at about the stage where basalt gives way to andesite, during which the copper concentration in the melt was very much higher than 120–130 ppm: on average, perhaps $\approx$2000 ppm. This is followed by a rapid decline to Cu $\approx$75 ppm at 56–8 per cent SiO_2 and Cu $\approx$30 ppm at 60–2 per cent SiO_2 (Figs 6.12 and 21.2). Because this decrease cannot be accounted for by crystal subtraction it seems almost certain that it has been lost, as we have already concluded (Chapter 21), in a volatile phase: a likelihood elegantly supported by the contrasted copper contents of the glass inclusions and groundmass glasses.

The plateau developed by the zinc maximum (61 ppm) in the Solomons groundmasses (as representing the progressively more felsic liquids) extends to 56 per cent whole-rock SiO_2, at which point zinc in the melt declines, at least partly in the volatile phase (Chapter 21). As for copper, there is probably a fleeting rise of concentration of zinc in the melt to much more than 61 ppm, probably ⩾330 ppm in the 52–6 per cent SiO_2 interval.

Lead increases in the Solomons liquids at least to 16 ppm at 60–2 per cent SiO_2, though this may not represent a final maximum.

We thus see that for the Solomons lava series, copper has begun to decrease in the melt (probably in the volatile phase) at $SiO_2 \approx 52$–4 per cent (basaltic andesite), zinc at $SiO_2 \approx 54$–6 per cent (andesite), lead at $SiO_2 \approx 62$ per cent (dacite). This is a remarkably good fit (the reader may think it almost too good to be true) for the ore type:lava type associations, metal groupings, and between- and within-deposit chemical–stratigraphic sequencing that have been set out in the preceding pages. It must, however, be kept in mind that all we have shown here is the point in the evolution of the lava series at which each metal achieves its maximum abundance in the melt, and from which it then declines. As the work of Burnham, Holland, Manning, Candela, and others have shown (Chapter 22), the timing and scale of metal evolution from the melt are influenced by many factors other than this.

A further factor that may influence the behaviour of zinc (and manganese) in particular is the notable capacity of magnetite and hornblende opaque rims to capture and 'store' these elements (Chapter 7). While the opaque rims derive from decompression-oxidation, hornblende–melt reaction, and volume-for-volume replacement of the hornblende, zinc and manganese are much more abundant in rim than in parent hornblende (see, e.g. Fig. 7.9), and their major part must come from the surrounding melt. At the dacitic stage of Solomons lava evolution magnetite is the principal host for zinc (>1200 ppm) and manganese (>9000 ppm), and it may incorporate substantial lead (to ≈45 ppm). A large proportion of the zinc and manganese in the hornblende rims is almost certainly located in the magnetite component of the latter.

Increase in the abundance of water in the melt as this evolves, followed by release of pressure, may then lead to the phenomenon described by Martin and Piwinski (1969: see Chapters 21 and 23): a rapid breakdown of Fe-bearing compounds in the melt and the movement of Fe to the aqueous volatile phase. This would lead to the liberation of zinc, manganese, and lead from the magnetite and zinc and manganese from the hornblende rims, perhaps also to go to the volatile phase and to be released from the melt as exhalative components.

If such a process occurred at the andesitic to dacitic stage of melt evolution its onset would post-date the principal period of copper evolution from the melt, so favouring the observed stratigraphic relations between the metals. Such a proposal is of course no more than an hypothesis that might be used to design further observation and experiment. However, while the incidence of the three metals in exhalative ores is complex it is also remarkably systematic, and the writer inclines to the view that a solution to the problem is more likely to lie in magmatic processes than in those of leaching.

10. *Barium–strontium abundances in lavas and exhalative ores.* In view of their many chemical similarities, the remarkably contrasted incidence of the alkaline earth elements barium and strontium in exhalative ores looks as if it may well constitute a vital clue in determining whether the materials of the ores have been derived through active magmatic or passive leaching processes.

As we have noted already (Chapters 2 and 9) barium is a common — and commonly abundant — element of exhalative ores, particularly of those of high Zn–Pb content. It is now well recognized that Ba-rich layers tend to form the uppermost parts of many chemically stratified exhalative orebodies either as essentially separate layers at the top of the ore-stratigraphic sequence, or as a barium mineral concentration in the upper more lead-rich layers of the sulphide mass itself. The barium occurs principally as barite ($BaSO_4$) though it may occur as a component of K-feldspar and celsian, Ba-rich biotite, and so on. It is also a common, and frequently abundant, component of modern sea-floor exhalative deposits, again chiefly as barite.

Strontium, on the other hand, is not a notable constituent of exhalative ores and, indeed, the writer is unaware of any description of an exhalative orebody in which strontium has been reported at higher than trace level. Nor has it been noted as other than a trace component of modern sea-floor deposits.

Both barium and strontium constitute 'major traces' in basaltic and andesitic lavas, and the chemical properties and behaviour of the two are very similar. Both form highly soluble chlorides and both form sulphides. Of their sulphates that of strontium is sparingly soluble, that of barium essentially insoluble.

Strontium is clearly the more abundant of the two elements in all basaltic and andesitic rocks. Mean Sr/Ba ratios in MORBs and arc basalts are essentially identical at 2.20/1 and in arc andesites = 1.30. In the case of the Solomons lavas, Sr/Ba in the olivine–clinopyroxene-basalts = 4.04; in the big-feldspar lavas = 2.50; and in the hornblende-andesites = 2.28. The incidence of both elements is low in the Solomons ferromagnesian silicates and spinels, but strontium tends to partition strongly into the plagioclase feldspars, with mean $K_D^{plag:melt}$ = 2.01. Barium, on the other hand, tends to partition in favour of the melt throughout, with mean $K_D^{plag:melt}$ = 0.43. Thus the major part of the barium in the lava occurs in the groundmass, with much of the remainder in the more K-rich plagioclases. The disposition of strontium is the reverse: the major part in the more K-rich plagioclases, most of the remainder in the groundmass (ref. Fig. 12.2).

In the context of a leaching hypothesis two facts stand out: (1) strontium is 2–4 times more abundant than barium in all the principal Solomons lava groups; and (2) both elements form soluble chlorides. Certainly, through its greater tendency to segregate in the groundmass, barium is relatively concentrated in that component of the rock most susceptible to leaching. However, though proportionately reduced in the groundmass owing to its preferential incorporation in plagioclase, the absolute amount of strontium in the groundmass remains greater than that of barium. Because of the relatively low abundance of plagioclase in the olivine–clinopyroxene-basalts, mean Sr/Ba in their groundmasses is, at 4.20, essentially the same as in the whole-rocks. With increased crystallization of plagioclase, mean ratios in the big-feldspar lava (Sr/Ba = 2.07) and hornblende-andesite (1.63) groundmasses are reduced with respect to the corresponding whole-rocks, but abundant strontium remains, and mean Sr/Ba ratios persist at values >1 until the point at which dacites develop.

If the two elements were leached from basalts and basaltic andesites by hydrothermal convection, it might be expected that the two would be delivered to the point of sea-floor discharge in about the same order of abundance, and as their chlorides. Both would then react with any available SO_4 to form sulphates, or adsorb on to neoforming clays, phyllosilicates, or zeolitic materials, later to be incorporated into diagenetic and metamorphic silicates.

As $BaSO_4$ is essentially insoluble, $SrSO_4$ 'sparingly soluble' ($\approx$1 g/l at 15 °C) in water, the precipitation of Ba would be favoured in the presence of SO_4^{2-}. In spite of this somewhat greater solubility of $SrSO_4$, the common abundant presence of the other alkaline earth sulphates, $CaSO_4$ and $CaSO_4.2H_2O$, in exhalative ores would, however, suggest (1) that at least minor $SrSO_4$ would be fixed in the sediment, or (2) that Sr would be incorporated as a substantial minor element in $BaSO_4$, or (3) that it would be incorporated in developing carbonate or silicate.

Such does not appear to be the case. Although very large quantities of $BaSO_4$ (and $CaSO_4$) and Ba- and Ca-rich silicates are common associates of exhalative sulphides (and indeed in some cases exceed the sulphides in amount) strontium is rarely if ever reported. Where it does occur as a significant trace, this is always in abundances more than an order of magnitude less than Ba. Whitford et al. (1992) determined strontium abundances in some 20 barite samples from several volcanic exhalative ore environments of the Cambrian Mount Read Volcanics of western Tasmania. Samples represented both stratiform and veinlet material, and exhibited a range of strontium abundances from 0.16 to 1.16 per cent Sr, with a mean of 0.76 per cent Sr (the stratiform samples gave a range of 0.45–1.0, and mean of 0.73, per cent Sr). Mean Sr/Ba in these 'ore' barites was therefore $\approx$0.013:1, which compares with a mean of $\approx$2.5:1 in the Solomons lava suite: a ratio in the exhalative barites just 0.0052 that in a suite of lavas that are probably close to being modern analogues of the Mount Read Volcanics.

If the barium of exhalative ores is incorporated in ore solutions by the hydrothermal leaching of accumulated lavas, it is not easy to see show it has been segregated from, and deposited almost entirely independently of, the originally more abundant and closely associated strontium.

11. *The incidence of phosphorus in lavas and ores.* Phosphorus is a common minor component of exhalative orebodies, particularly in some of the major Lower Proterozoic deposits and their associated iron-rich exhalites (Chapter 11). Its incidence is, however, erratic: it occurs in abundances up to $\approx$20 per cent (as P_2O_5) in parts of the Broken Hill deposit of New South Wales, but is present at no more than a very minor level in most Archaean and Palaeozoic exhalative ores. It is generally not conspicuous in modern hydrothermal aprons, though slightly 'anomalous' abundances appear in some cases: for example, reporting on hydrothermal sedimentation near the Rodriguez triple junction of the Indian Ocean seafloor, Herzig and Plüger (1988) noted an average of 0.62 per cent P_2O_5, in 'hydrothermally influenced sediments' associated with basalts containing mean P_2O_5 = 0.06 per cent. Where phosphate is a conspicuous component of exhalative ores (say 1–5 per cent mean P_2O_5: e.g. Broken Hill, NSW; Pegmont,

Queensland; Aggenys and Gamsberg, Namaqualand) and their associated exhalites, it is normally sharply confined within the relevant sulphide- and/or oxide-rich domains or parts of these. It does not extend away from the orebody along or across bedding, indicating that supply and deposition of the phosphate have been intrinsic parts of the hydrothermal event —and that the other components of the ores have not simply been 'dumped' into a more extensive phosphatic sediment accumulating independently at the time of hydrothermal discharge. As it is likely that collophane and related phosphates would be stable in most exhalative sedimentary environments, its presence or absence in any given exhalative ore is probably substantially a reflection of supply.

Given that the phosphorus of volcanic rocks is accommodated in all of silicates, spinel and apatite (the apatite occurring almost entirely as a groundmass component) it is likely that it would be leached and transported in low pH, convecting hydrothermal solutions. On this basis, and in view of the relative constancy of circulation/exhalation conditions along deep-sea-floor mid-ocean ridges, it might have been expected that *supply* of phosphorus (as PO_4^{3-} and as complex phosphate ions) to MOR sites would be relatively consistent. The data are sparse and inconsistent. Embley et al. (1988), in their investigation of an extinct hydrothermal system on the Galapagos Ridge, found an average of 0.44 per cent P_2O_5 in 19 unaltered basalts, 0.61 per cent in 12 altered materials, whereas Hannington and Scott (1988) found an average of only 63 ppm P in the material from a hydrothermal spire on Juan de Fuca Ridge, and Goodfellow and Blaise (1988) a tendency for a decrease in phosphorus in hydrothermal sediments relative to their surroundings in hemipelagic sediment of the northern Juan de Fuca Ridge. What little data there are thus indicate that the incidence of phosphate in modern hydrothermal environments is erratic and low. Whether this reflects leaching or magmatic discharge is beyond speculation at present.

In the Solomons magmatic system phosphorus behaves as a hypotholeiitic element: a beautifully linear (closed-system fractionation or variable partial melting) P–MgO–($Na_2O + K_2O$) pattern in the olivine–clinopyroxene-basalts, and a diffuse, erratic, pattern of changed fractionation or loss in the hornblende-andesites (Figs. 11.9, 11.10). On the P–SiO_2 variation diagram it displays a marked preference for melt over solid phases at low SiO_2, shows a sharp peak in the whole rock at 50 <SiO_2 <52 per cent and in the groundmass at 52 <SiO_2 <54 per cent (Figs. 11.8 and 21.2). This evidence, together with least-squares mixing and related calculations, indicates the likelihood of loss of phosphorus accompanying the development of the hornblende-andesites: a process initiated in the felsic basalt– basaltic andesite stage.

Coupled with this is the fact that phosphorus is well known to be a volatile component of magmatic (volcanic) systems and that it is an almost ubiquitous component of modern volcanic aerosols and plumes.

Thus, while the evidence is slight and inconclusive, and while it would be surprising if at least some phosphorus were not leached and transported from volcanic rocks by convecting hydrothermal solutions, the indications are that leaching is probably not the process of supply of large quantities of phosphorus to exhalative depositional sites. As noted above, it might have been expected that leaching would have given rise to moderately consistent levels of phosphorus in the ores, but in fact its abundance is highly erratic. On the other hand, this erratic occurrence, the pattern of loss developed in the Solomons suite, together with the known propensity of phosphorus to escape from volcanic systems in the volatile phase, seems to indicate an active magmatic source for at least the major part of this element in exhalative ores.

12. *The incidence of iron, calcium, and manganese in lavas and ores*. Of the three, iron is a ubiquitous major component of exhalative ores, calcium a common and commonly abundant component, and manganese a common minor-to-major component. In some cases, as at Broken Hill and Aggenys, all three elements are major components of the ores.

All are prominent elements of the modern sea-floor exhalative environment, and it is likely that all are susceptible to sea-floor leaching of basaltic materials. In volcanic systems, all display hypotholeiitic abundance patterns and all behave essentially similarly on X–SiO_2 variation diagrams: patterns compatible with loss in a magmatic gas phase. All are prominent components of modern volcanic aerosols and sublimates, confirming the potential of each to be involved in active volatile loss.

The three elements are thus likely to be of little use in discriminating between leaching and active magmatic processes in exhalative ore formation. Exhalative concentrations of all of them may well be produced by either process — and it may well be that in some cases both mechanisms make a contribution.

SUMMARY OF DISCUSSION: ACTIVE MAGMATISM VERSUS PASSIVE LEACHING

The world of exhalative ore genesis theory has become captivated by the modern sulphide accumulations now known to be associated with sea-floor spreading ridge and related tectonic features. These deposits are generally seen as analogues of the Cyprus-type deposits of older terranes, and are thought to demonstrate — in the modern setting — the principal processes of exhalative ore formation. Almost by definition the association is essentially basaltic: apart from a very few occurrences such as those of the Guaymas Basin and Escanaba Trough (where there are partial sedimentary settings) the sulphide accumulations are built on mid-ocean ridge basaltic substrates, and are associated with effusive rather than pyroclastic activity.

Is this the geological locale of most of the world's major volcanic exhalative ore deposits? The answer is, quite clearly, no. Even if orebodies of Cyprus-type in older terrains have formed on the deep sea floor (which is far from certain), they constitute only a very small proportion of the world's total bulk of volcanic-associated exhalative ores. Further, an early island arc rather than a deep-sea-floor origin has been suggested for the Cyprus deposits themselves and others like them (see Miyashiro 1973; Stanton 1978; Stanton and Ramsay 1980). In particular Stanton (1978) has suggested that rather than being of deep-sea-floor affiliation, many deposits of Cyprus type may have formed in association with tholeiitic basaltic volcanism constituting the early stages of island-arc development.

We find in fact that most of the world's major exhalative orebodies are associated with felsic — or bimodal — volcanism. The smallest orebodies, constituting a very small part of the total bulk of exhalative ores, occur in association with huge volumes of basalt, whereas the largest orebodies and those accounting for a very large part of the total bulk of exhalative ores are associated with relatively small volumes of andesite, dacite, and rhyolite. The basalts with which the Cyprus-type deposits occur have been exposed on the sea floor for long periods of time; the felsic rocks with which the large Cu–Zn–Pb deposits are associated are usually erupted into shallow water and are submerged for only a relatively short time. The basalts are chiefly lavas and have undergone little degassing; the felsic rocks are largely pyroclastic and many have degassed copiously and explosively. In all cases — from basaltic to the most felsic exhalative ore provinces and districts — groups of related orebodies tend strongly to occur on or very close to a 'preferred' horizon in the relevant volcanic or volcanic–sedimentary stratigraphy; that is, on both large and small time-scales, exhalative ore formation is episodic: it appears to be a distinct, relatively brief 'event', rather than a long-continued process.

All of this seems to suggest that the exhalation that has led to the formation of major orebodies has been some kind of special degassing event — one that occurred only once or perhaps twice in the accumulation of a given volcanic pile — associated with felsic volcanism, or a change from basaltic to felsic activity.

Given that, on average, basalts contained about 1.5 times as much nickel as copper or zinc, and about 0.5 times as much cobalt as copper or zinc, perhaps one of the most remarkable features of Cyprus-type deposits is their very low nickel and cobalt content. The same may be observed of chromium. As we have seen, these are subtracted from the melt (in the order Ni, Cr, Co) at the very earliest stages of fractionation and thus would not be available for incorporation in a volatile phase when this developed in the later stages of magmatic differentiation. That is, if the exhalative copper ores were derived through magmatic degassing the absence of these three metals from their assemblages would be as expected. On the other hand, as all three would be freely available for leaching over long periods of time, their near-absence from the ores may constitute decisive evidence against the significant operation of this process in basaltic environments.

In an analogous way the marked paucity of strontium, as compared with the conspicuous abundance of barium (and calcium), in many exhalative ores of more felsic association may constitute an important discriminator of process. The tendency for plagioclase to act as a geochemical sieve — strongly incorporating strontium and just as strongly excluding barium — leads to the concentration of the latter in the residual melt, there to be available for transfer into the late-stage volatile phase. Judging from the evidence of Sr/Ba abundances to the dacite stage as evidenced by the Solomons lavas, Sr/Ba fractionation probably accelerates greatly as the final (rhyolitic) stages of magmatic differentiation are approached. If this is the case (and if we neglect depositional influences), the common occurrence of barium in, and the conspicuous near-absence of strontium from, exhalative environments and the general occurrence of barite layers as the final stage of the paragenesis of the relevant ores may constitute a very good 'fit' between magmatic processes and exhalative ore geochemistry. On the other hand, having in mind the incidence of strontium as a 'major' trace element in most volcanic rocks, it is not easy to see how the leaching process could yield layers of abundant barium and calcium sulphate in many exhalative ores, and no — or quite inconspicuous — associated strontium compounds.

The behaviour of vanadium may also provide a discriminant. Vanadium is a 'medium-level' trace element in basaltic rocks and is likely to be soluble in most hydrothermal solutions, but it is conspicuous by its near-absence from exhalative ores. We have seen that although it is a hypotholeiitic element in the Solomons lava series, this can be accounted for almost entirely by crystal subtraction from the melt, and thus without loss to a gas phase. This, together with the likelihood that vanadium would be readily leached and transported by circulating hydrothermal solutions, may indicate that its non-appearance in exhalative ores is due to lack of magmatic release (in the gas phase) and hence reflects an active magmatic, rather than a late-stage leaching, process.

This is supported in large part by observation of modern volcanic emissions. Few volcanic aerosols and sublimates contain significant vanadium — with two spectacular exceptions: the volcanoes of Izalco and Bezymianny. While for the most part there appears to be a good correspondence between magmatic processes and the near-absence of vanadium from exhalative ores, perhaps we should keep in mind the possibility, as a rare but not unexpected phenomenon, of a vanadium-rich exhalative deposit.

Although several attempts (e.g. Seyfried and Janeck 1985) have been made to explain ore type–lava type relations, and relative abundances of copper, zinc, and lead, on the basis of leaching processes, this aspect clearly requires greater consideration than it has had so far. Why, on the basis of a leaching mechanism, should copper be the first of the non-ferrous metals to be exhaled? Why not zinc, which might be expected to be released earlier and to be more soluble at lower temperatures? Why is it that the world's great exhalative zinc deposits occur with felsic rocks that contain relatively little zinc and occur in relatively small volume, whereas those rocks that contain a greater abundance of zinc and occur in huge volume — the sea-floor basalts — apparently generate relatively small concentrations of zinc? Why, on the basis of leaching, do many exhalative deposits show chemical stratification, and this of such remarkably constant kind? We have seen that copper and zinc are lost from the melt at a relatively early stage of fractionation and that the (upwardly progressing) exhalative ore stratigraphy of (Fe)–Cu–Zn–Pb–Ba corresponds with the stages of fractionation at which each of these elements begins to be lost from the melt.

Such relations between (1) rock type and size and constitution of the associated orebodies and (2) the principal elements involved in exhalative ore stratigraphy thus probably conform better with magmatic than with leaching processes — but this avenue requires much further investigation.

PROGNOSTICATIONS ON THE POSSIBLE INFLUENCE OF MAGMATIC PROCESSES ON PATTERNS OF EXHALATIVE ORE FORMATION

It will now be clear that it is the writer's present view that the essential absence of chromium, nickel, and vanadium and only very minor incidence of cobalt in exhalative ores, the apparent fractionation of barium *vis-à-vis* strontium in the ore-forming process, and the various relations between volcanic rock type and Cu–Zn–Pb abundances and ore stratigraphy, all point to the generation of the relevant elements by magmatic rather than leaching processes. It must of course be emphasized that we have looked only at the abundances of the elements in the volcanic rocks, their minerals, and their groundmasses. From this we have tried to deduce whether these patterns of abundance could be accounted for by fractionation alone, or whether they also required a quantum of loss in a gas phase. Apart form the brief considerations of Chapter 22, we have not considered the nature of such a gas phase, nor the influence of different anions on the solid–liquid–gas partitioning of the various individual elements. That is, we have tried to estimate whether and when, in the fractionation path, elements were lost from the melt to a hypothetical volatile phase; we have not attempted to deduce *how* (i.e. in what anionic combination) they were lost. In other words, we have attempted to solve just half the problem. The other half — chiefly the considerations of melt-vapour partitioning and the influence of different anions and anionic complexes on this — remains to be elucidated by researches such as those of Burnham, Holland, Candela, Manning, and others in this field.

This accepted, let us look at the ways in which sub-volcanic solid–liquid, i.e. crystal–melt, systems might deliver the most appropriate quantities of the 'exhalative ore elements' to the anions and anionic complexes beginning to concentrate as the magma system evolves.

1. *Clinopyroxene–plagioclase fractionation a prerequisite for exhalative ore generation?* The most inhospitable trace-element acceptors, and hence the principal concentrators of these elements in the melt, are high-Mg clinopyroxene and low-K plagioclase.

While such pyroxene accepts significant chromium, and the plagioclase significant strontium, copious precipitation of these two mixed-crystal series tends to partition continuously all the other trace elements, particularly the chalcophile metals, into the melt. Clinopyroxene and plagioclase are indeed the great 'concentrators of the chalcophiles': as the principal influence in the initial concentration of copper, zinc and lead, their prolonged and abundant crystallization must have a profound influence on, and is probably a prerequisite for, the generation of exhalative base metal orebodies of the kinds now observed.

2. *A high-Mg parent the most effective separator of nickel and chromium from copper and zinc?* As nickel and chromium tend to be strongly partitioned into the high-Mg members of the various ferromagnesian mixed crystal series, and copper and zinc into the more Fe-rich members, the higher the initial Mg/Fe ratios of the basaltic parent the greater its propensity to 'bury' nickel and chromium in early accumulates and to preserve the availability of copper and zinc in the melt for incorporation in a later vapour phase — and hence to generate exhalative ores with the metal assemblages most commonly observed.

3. *Relatively high early SiO_2 activity a concentrator of zinc in the melt?* Prolonged crystallization of olivine down the evolutionary path of a lava series may lead to the formation of highly fayalitic members with their high propensity for incorporating zinc ($\approx$300–400 ppm). A relatively high SiO_2 activity in the melt leading to early substitution of clinopyroxene ($\approx$20–50 ppm Zn) for olivine would, however, lead to a much lower rate of abstraction of zinc from the melt, and indeed to its concentration in the melt. Early substitution of clinopyroxene for olivine in this way would barely effect Ni abstraction, for this is performed almost entirely by the very earliest formed, high-Mg olivines.

4. *Jahn–Teller distortion the key to the behaviour of copper?* Copper in its divalent state (Cu^{2+}; ionic radius = 0.072 nm) and zinc (Zn^{2+}; 0.083 nm) have similar charges and essentially similar ionic radii. It might therefore be suspected that the two would have rather similar propensities for substituting in the octahedral sites of the principal ferromagnesian silicates. The data of Chapters 6 and 7 show, however, that this is clearly not so; although copper is slightly more abundant than zinc in the lavas, its incidence in all the ferromagnesian silicates and in magnetite is generally very much lower than that of zinc (Table 6.5). It appears to be slightly higher in the plagioclases, though this is at a very low level and may be questionable (*vide* the fine glass inclusions considered in Chapters 6 and 21).

On the basis of crystal-field theory the electronic structure of Cu^{2+} is found to be a degenerate configuration when in an octahedral environment. As the structure is non-linear, it must distort to eliminate this degeneracy, and does so by taking either an elongated or compressed (flattened) form: the Jahn–Teller effect referred to in Chapter 6. This may lead to a poor fit in the octahedral sites of ferromagnesian and spinel structures, and hence to a strong tendency for the exclusion of Cu^{2+} from these minerals. Such an effect may be accentuated by the accelerated oxidation of the melt in the evolutionary interval $\approx$48–52 per cent SiO_2 observed in the Solomons suite (see Chapters 5, 17, and 21): concomitant oxidation of Cu^+ to Cu^{2+} would render an increased proportion of total copper ions susceptible to Jahn–Teller distortion, the two phenomena compounding to reduce further the compatibility of copper with respect to the silicates.

5. *The rapid development of a high-Ca plagioclase melt structure a mechanism for the sudden expulsion of copper at the basalt → andesite transition?* We have seen that the reluctance of the olivines and clinopyroxenes to accept copper leads to the development of mean melt copper concentrations of $\approx$120–30 ppm at $\approx$52–4 per cent whole-rock SiO_2. These mean values are, however, derived from averaging large numbers of whole-rock and groundmass separations over 2 per cent SiO_2 intervals, and must be seen as likely to be somewhat insensitive to any changes in the melt of a fleeting nature. This is indicated by the copper contents of the glass inclusions in the plagioclases, which for the Solomons lavas contain mean Cu $\approx$1900 ppm: more than order of magnitude greater than the maxima of Figs. 6.12 and 21.2. This evidence, together with that of Lowenstern *et al.* (1991) and others derived from fluid inclusions, may indicate that certain copper-bearing melts achieve very high copper concentrations (perhaps $> \approx$2000 ppm Cu) over relatively brief periods of time and essentially at the stage where basalts are transforming to andesites.

In their metallurgical contribution on the effect of Al_2O_3 and CaO on the solubility of copper in FeO–Fe_2O_3–SiO_2–Al_2O_3 slags at 1300 °C, See and Rankin (1983) showed that equilibrium copper contents may be reduced by lime addition. At 1300 °C the addition of about 9 per cent by mass of CaO to a slag containing about 15 per cent by mass of Al_2O_3 reduces copper solubility by an amount corresponding to a $CuO_{0.5}$ activity increase from 3.32 to 4.43. Such addition of

CaO presumably led to the development of a component of plagioclase structure in the melt, this in turn leading to an overall reduction in copper solubility.

It may therefore be suggested that a large increase in copper concentration, closely followed by a decrease in copper solubility, may be brought about by early copious olivine–clinopyroxene crystallization closely followed, at about the basalt → andesite stage, by the rapid, widespread, development of plagioclase structure in the melt. This then leads to the fleeting development of very high copper concentrations in any remaining unstructured melt, the exceeding of copper solubility, and a greatly heightened but short-lived availability of copper for fractionation into any developing vapour phase.

6. *Plagioclase a critical factor in the concentration of lead?* The ferromagnesian minerals are, of course, well known to be unreceptive hosts to lead and the present investigation bears this out. They are therefore highly efficient concentrators of lead in the melt. Plagioclase (or 'the feldspars'), on the other hand have been seen as more receptive hosts and hence as possible abstractors of lead from the melt. We have, however, shown (Chapter 8) that, in the Solomons lavas at least, $K_D^{plag:melt}$ is always <1 and hence that plagioclase crystallization tends to enrich the melt in lead. Large-scale, late-stage crystallization of low-K plagioclase thus accentuates the melt lead concentration initiated by earlier crystallization of olivine, pyroxene, and hornblende. This may well be the reason why lead-rich exhalative orebodies are associated with calc-alkaline, rather than K-rich alkaline, volcanism. In the latter, lead is readily incorporated in abundant K-feldspar, is removed from the melt, and is thus unavailable for incorporation in a volatile phase. In the former, lead is rejected by low-K plagioclase, is enriched in the melt, and hence is available in elevated quantities for incorporation in the gas/vapour phase. Much the same argument clearly applies to barium.

7. *An important role for a high melt oxidation potential and magnetite in the storage and concentrated release of iron, zinc, manganese, lead and copper?* A high oxidation potential in the melt will, clearly, favour the formation of magnetite. As this is a receptive host for the four trace metals it will be a particularly effective agent for their impoverishment in the melt. A low oxidation potential will, on the other hand, favour the crystallization of the (Fe^{2+}) ferromagnesian silicates; these are generally rather poorer hosts for the metals and their crystallization will tend to maintain or only slightly decrease the concentrations of the latter in the melt.

At first sight this implies that a high oxidation potential and abundant magnetite development would inhibit the concentration of the four elements in the evolving melt and hence the generation of a metal-rich volatile phase and resultant exhalative orebodies. This may well be the case in many instances.

There are, however, two circumstances that may induce the late-stage release of the trace metals from such magnetite: (1) the onset of conditions sufficiently oxidizing to cause the conversion of magnetite to haematite, and (2) the formation of an aqueous vapour phase.

Sidhu, Gilkes, and Posner (1980, 1981) have shown that solid-state oxidation to maghemite and haematite of microcrystalline synthetic magnetite doped with small substitutions of Cu, Zn, Cr, Mn, Co, and Ni is accompanied by a redistribution of these elements when haematite forms as the higher oxide. Maghemite retains most of the substituents, but the Cu, Zn, Co, and Ni are ejected when haematite develops as the oxidation product because these divalent ions are incompatible with the haematite structure. Although the shapes of the ejection curves were somewhat different, Sidhu *et al.* obtained essentially analogous results using natural magnetites.

Many of the Solomons andesites are the usual grey of this rock type, but a significant proportion are evenly and distinctively pink: the result of what appears to have been an episode of relatively intense magmatic oxidation and the conversion of magnetite to haematite. Unfortunately the data are insufficient to yield an unequivocal indication of metal loss in these oxidized lavas: closer investigation is called for.

In the above case the metals would be released into the melt via rejection from haematite, then partitioning to the volatile phase. In other cases, however, breakdown of magnetite in an aqueous environment may yield the divalent metals directly to the volatile phase. If Martin and Piwinski's (1969) experimental results can be taken as an indication of the behaviour of iron in the presence of an aqueous phase in large subvolcanic systems, accumulation of large concentrations of the metals in magnetite may become highly susceptible to late-stage, and possibly rapid, exhalative loss. Sudden addition of water to, or its exsolution from, such a system might well induce a rapid dissolution of accumulated magnetite (and of hornblende and its rims) with concomitant release of 'stored' Fe, Zn, Mn, Pb, and Cu into the aqueous vapour phase. This just might constitute the kind of metal-releasing 'event' — associated with large scale degassing and pyroclastic rock formation — referred to earlier in this chapter.

8. *A long closed-system fractionation path favourable to concentration of the metals in high-iron tholeiites?* The very substantial iron enrichment in some MORBs and related lavas (see Figs. 17.1(e) and (f) and in some of the tholeiitic basalts erupted in the early stages of volcanic island-arc development (Stanton and Ramsay 1980) is accompanied by similarly extended tholeiitic behaviour by zinc (Figs. 7.22(e) and (f)) and manganese (though not by copper on present evidence).

This phenomenon is illustrated with quite remarkable clarity by the high-Fe, high-Zn (unaltered) abyssal basalts of the Galapagos Ridge (Embley *et al.* 1988; Perfit, pers. comm. 1990). Figure 24.1 shows XMA relations with respect to FeO, Zn, MnO, and Cu (using the same factors as those employed in the analogous figures in Chapters 17, 7, 16, and 5 respectively) generated by 29 samplings of these lavas. Total iron as FeO (Fig. 24.1(a)) reveals the extreme iron enrichment that may be attained by some of the Galapagos tholeiites. Zinc (24.1(b)) displays similar enrichment, but an even more highly correlated, linear, tholeiitic distribution of points. MnO (Fig. 24.1(c)) behaves similarly in principle, but the tholeiitic distribution is not nearly so well defined. Copper (Fig. 24.1(d)) displays a poorly correlated linear field of hypotholeiitic type; i.e. copper *decreases* with increase

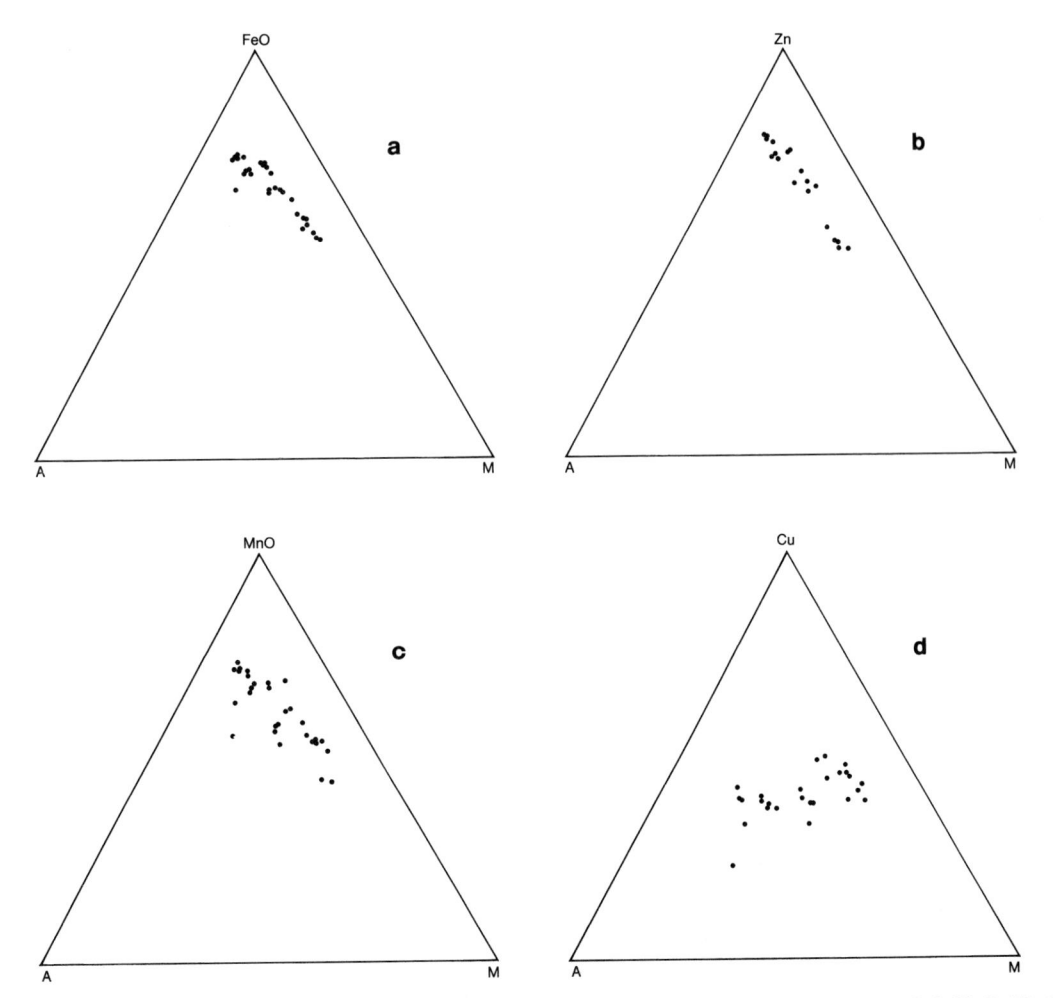

Fig. 24.1. Ternary relationships generated by 29 Eastern Galapagos Rift basalts with respect to (a) total iron as FeO–MgO–(Na$_2$O + K$_2$O); (b) Zn–MgO–(Na$_2$O + K$_2$O) (Zn calculated as ppm Zn $\times$ 10^{-1}); (c) MnO–MgO–(Na$_2$O + K$_2$O) (MnO calculated as per cent MnO $\times$ 50); and (d) Cu–MgO–(Na$_2$O + K$_2$O) (Cu calculated as ppm Cu $\times$ 10^{-1}). The parallelism between iron and zinc, and hence the highly tholeiitic nature of these basalts with respect to zinc as well as iron, is clear.

in $(Na_2O + K_2O)/MgO$, in contrast to the three elements with respect to which the Galapagos suite is tholeiitic, which increase with increase in this ratio. On the basis of the principles discussed in Chapters 21 and 23, it appears that whereas all of iron, zinc, and manganese are incorporated in crystals (almost certainly largely magnetite) and develop closed-system patterns of abundance, copper is lost from the Galapagos basalts, at least from the earliest stages of solidification).

As indicated in Chapters 7 and 16, Zn–Fe and Mn–Fe correlations in the Solomons lavas are so good that it is clear that much of their zinc and manganese occurs in magnetite — particularly in groundmasses and the more felsic members. The implication is that the maintenance of a closed-system and at the same time an oxidation potential favourable to the continued crystallization of spinel should lead, as in section (7) above, to the progressive incorporation and storage of zinc and manganese in magnetite (Fig. 7.13), perhaps to the point where very significant quantities of these two metals are involved. Sudden availability of a substantial aqueous volatile phase would then lead to the rapid dissolution of the magnetite and release of iron, zinc, lead (magnetite being the principal acceptor of lead), and manganese into the gas phase. Lead (and barium) would also have followed their own, but simultaneous, course, accompanying potassium into the increasingly (K + Pb + Ba)-rich residual melt. Perhaps associated mechanisms of this general kind, involving prolonged fractionation paths terminated by the sudden destabilization of metal-rich phases, may be the beginning of an explanation of the common ore association Pb–Zn–Mn–Ba.

9. *Phosphorus an agent in the exhalation of lead?* The not uncommon association of apatite with lead-rich exhalative ores indicates the possibility of some functional relationship between the phosphorus and lead in source processes. Although phosphorus behaves as an unequivocally hypotholeiitic element in the Solomons lava series (Figs. 11.9, 11.10) its abundance behaviour is rather inconsistent from one volcanic province to another (Figs. 11.8, 11.10). Lead, on the other hand, is fairly consistently tholeiitic to hypertholeiitic. The degree to which their maximum abundances in the melt might coincide is therefore difficult to perceive. It has, however, been proposed (Stanton *et al.* 1978) that at least some of the phosphorus of the late magmatic stage may occur as pentasulphide (see Chapter 22 and Manning's 1984 suggestion that some late magmatic tungsten may be in the form of phosphorus heteropolytungstate,

$H_2PW_{12}O_{40}$). Phosphorus pentasulphide may react with lead fluoride:

$$P_2S_5 + 3PbF_2 \rightarrow 3PbS + 2PSF_3,$$

excess P_2S_5 may hydrolyse:

$$P_2S_5 + 8H_2O \rightarrow 2H_3PO_4 + 5H_2S,$$

and if calcium bicarbonate is present in the overlying body of water at the point of exhalative discharge, apatite is formed:

$$5H_3PO_4 + PSF_3 + 10Ca(HCO_3)_2 + 4H_2O \rightarrow$$
$$9CaO.CaF_2.3P_2O_5 + HF + H_2S + 20H_2CO_3 + 6H_2O$$

and excess calcium bicarbonate combines with HF yielding fluorite:

$$Ca(HCO_3)_2 + 2HF \rightarrow CaF_2 + 2H_2CO_3$$

10. *Timing of formation of the aqueous volatile phase: a critical factor in the magmatic concentration of exhalative metals?* The spectacular loss of copper from the melt at an early stage of the latter's evolution, and the common occurrence of copper-bearing sulphide in basaltic vesicles, indicate that a large part of the sub-volcanic quantum of copper — and particularly that of Cyprus-type deposits — is probably lost during first boiling. Iron and some zinc are lost at this stage also. Loss of zinc, lead, and associated elements at a later stage may result from second boiling or, as indicated above, from catastrophic ingress of water — perhaps seawater — into the shallow subvolcanic magma chamber, with concomitant breakdown of magnetite.

Be this as it may, the two common hydrous minerals of volcanic rocks — hornblende and, particularly, biotite — are receptive hosts for the trace metals and their formation leads to substantial scavenging of the latter from the melt. As we have seen from the very limited incidence of biotite in the Solomons dacites, this mineral is an extraordinarily effective scavenger of the trace metals and its development to, say, 5 modal per cent in a 'dacitic andesite' would lead to the virtually quantitative removal of the residual trace metals, including barium, from such a melt. Hornblende would disgorge its trace-element content by the formation and dissolution of its opaque derivative, but trace elements once incorporated in the more stable biotite (and the relevant partition coefficients are >10 in many instances) would be permanently lost to the liquid–gas system. The prevention of any substantial degree of biotite crystallization is therefore vital to the development of a high concentration of the ore elements in the residual melt and derived gas phases. This presumably requires that H_2O activity in the melt does not reach the point where biotite nucleates: i.e.

that the primary H$_2$O content of the magma is not sufficiently high or, as is probably the more important case, the aqueous volatile phase begins to escape from the magma chamber before significant nucleation and growth of biotite.

CONCLUDING STATEMENT

Some twenty years ago it was emphasized (Stanton 1972) that the 'origin' of many ores could, and perhaps should, be resolved into three fairly clearly defined segments: derivation, transport and deposition. No group of ore deposits illustrates this more clearly than do the exhalative–sedimentary ores of volcanic affiliation: the 'volcanic massive sulphides'.

In this chapter we have considered just one aspect of the first of the three segments: derivation. That the principal constituents of the ores are derived in one way or another from volcanic rocks is not in question. What has been a matter of conjecture, however, is the timing, mechanism, and circumstance of extraction of the ore elements from the volcanic source. Have these components been derived from active sub-volcanic melts via magmatic degassing, or have they come from relatively much older, long-erupted, principally basaltic lavas by sub-sea-floor leaching? We have not been concerned with processes of collection or transport by degassing or leaching, and we have not considered depositional processes and effects in the development of the ores and their associations. Our interest has lain in the patterns of abundance of the ore elements in a single series of lavas covering a large compositional span, the influence of magmatic processes in the development of these patterns, and the ways in which the latter might help in discriminating between active magmatism and leaching as processes responsible for the derivation of the ores.

The near-absence of nickel from exhalative ores (except for a few Archaean deposits associated, perhaps significantly, with picritic and komatiitic lavas) looks as if it may be a critical piece of evidence against leaching, as does the very low level of cobalt. Both metals occur in basalts in the same order of amount as copper and zinc, both should be readily released into and transported by hydrothermal solutions, and both are chalcophile and form insoluble sulphides — and yet, of the thousands of exhalative orebodies now known, almost none contains significant nickel or cobalt. Chromium and vanadium similarly occur as substantial traces in basaltic rocks and should be capable of ready leaching and transport. Both are lithophile but, if present in the hydrothermal solutions, should be readily localized in clays, phyllosilicates, or silicate–oxide iron formations of the hydrothermal apron.

Neither is, however, a common constituent of exhalative ores. Strontium is probably the most abundant of all the trace elements of basaltic and andesitic rocks; it should be readily leached and transported in the hydrothermal medium, and might be expected to precipitate as sulphate, carbonate, or as a component of K-bearing sedimentary–diagenetic phyllosilicates and feldspars. It is, however, conspicuously almost absent from exhalative orebodies — in quite dramatic contrast to its closely related alkaline-earth element barium which, although of lesser abundance than strontium in volcanic lavas is, with Fe, Cu, Zn, Pb, and S, one of the six principal elements of exhalative ores.

Such features are, however, readily related to the progressive crystallization of a basalt–andesite–dacite–rhyolite lava system as represented by the Solomons Suite. The essential absence of Ni, Co, Cr, V, and Sr from the orebodies can be plausibly attributed to their capture by crystals as these grew in the melt, as can the separation of strontium from barium. The early appearance of copper in the ores appears to correspond with evidence of its early loss from the melt. Conversely, the late appearance of lead and barium in the ores appears to correspond with their tendency to late concentration in the melt. Features such as the close tie of zinc and manganese with the fayalitic component in olivine, of copper in clinopyroxene and magnetite with change in whole-rock SiO$_2$, and of lead and barium with potassium and rubidium in both whole rocks and minerals, all seem to tie the incidence of the ore elements to the systematics of volcanic crystallization and residual melt formation.

Such geochemical features, together with relations between ore type and lava type, between ore occurrence and volcanic stratigraphy and facies, and between orebody size, felsic index of the associated lithologies, and proportionate pyroclastic development (volatile loss), all seem to point to an active magmatic, rather than passive leaching, connection between exhalative ores and their associated volcanic rock.

This being the case, and given the presence of significant quantities of the ore elements in the primary melt, the compositional features of magmas most conducive to concentration of those elements most characteristic of exhalative orebodies are:

1. high initial Mg/Fe ratio, to favour the formation of Mg- rather than Fe^{2+}-rich ferromagnesian minerals;
2. high SiO$_2$ activity, to favour the crystallization of clinopyroxene rather than olivine;
3. high Ca activity, to favour the formation of clino- rather than orthopyroxene, and of Ca-rich plagioclase;
4. low K, to inhibit the formation of K-feldspars and/or K-rich plagioclase;

5. high oxygen fugacity, to favour crystallization of spinel and the development of hornblende rims, both as 'stores' of metals for late-stage, sudden release on the development of an aqueous volatile phase, and

6. release of the aqueous volatile phase prior to the nucleation and growth of abundant biotite — a highly efficient, late-stage, trace-element scavenger.

Epilogue

We set out at the beginning of this volume to consider the derivation of exhalative ores: those deposits, particularly of the base metal sulphides, that have formed on the sea floor, or on the bottoms of lakes and lagoons, as products of hot spring and related exhalative activity. We noted that until about 1970 such activity was assumed to be volcanic in the sense that it derived directly from the active sub-volcanic melt, but that since about that time this idea had largely given way to the view that the emanations were the products of sub-sea and sub-lake-floor hydrothermal circulation (of sea and meteoric water), and concomitant leaching of much older, long-congealed volcanic rocks. The earlier hypothesis thus held that the volcanic melt played a highly active role; the second, that much older lavas played an entirely passive one. While one suspects that both mechanisms may have contributed to different degrees from one deposit to another, the merits and demerits of the two hypotheses have been strongly contested during the ensuing twenty years. Perhaps strangely, although many major exhalative deposits occur in volcanic-arc and related orogenic environments and are associated with felsic volcanism, much recent investigation has been concerned with ocean-ridge environments and their basaltic volcanism. Additionally, much of the relevant research has been concerned with the isotopic constitutions of the major anions (those of oxygen and sulphur), together with the cations hydrogen and carbon; relatively little has been done on the incidence of the metals themselves in volcanic rocks, or their geochemical behaviour in the volcanic and leaching milieux. And finally, much investigation has been oblique rather than direct, concerning itself more with the nature of the ore solutions than with the geochemistry of the metals themselves.

The geochemistry of the metals has been the province of the present volume: to examine the systematics of the principal ore-forming cations — metallic, non-metallic; chalcophile, lithophile — in the lavas of a modern volcanic arc, and to consider the active melt:passive leaching hypotheses in the light of these.

While the investigation must be seen as little more than an initial foray into a large and complex field, it does show that there are remarkably systematic relationships between the various abundance patterns developed by the ore elements, and the evolution of the lava series in which they occur. Some of these ties, such as the direct relationships between the abundance of nickel and high Mg-highly mafic lavas, and of lead with lavas rich in K-rich feldspars and glass, have been well known in at least a qualitative way for a long time. Others, such as the various patterns of maxima developed by copper, zinc, and phosphorus, and the hypertholeiitic behaviour of lead and barium, were not known, or had been perceived but not well substantiated.

The leaching mechanism as the dominant provider of the metals of exhalative deposits of the volcanic environment probably receives its *coup de grâce* through the paucity of nickel and cobalt in most deposits of this kind. Nickel is slightly more abundant than copper in most marine basalts; its principal habitat (olivine) is the basaltic component probably most susceptible to hydrothermal breakdown and release of its constituents; and nickel and copper possess similar tendencies to form sulphides in reducing environments. Cobalt abundances in such basalts are usually about one-half to two-thirds those of copper — and hence of very much the same order of magnitude — and its source and affinity for sulphide sulphur are essentially the same as those of nickel. And yet well over 99 per cent of all exhalative orebodies are essentially devoid of nickel, and few contain cobalt at above trace levels. On the basis of a leaching hypothesis it is also difficult to see why chromium and vanadium — and perhaps titanium — are not more prominent among the silicates and oxides associated with many exhalative ores, and why barium (and calcium) should be such prominent components while strontium is invariably, and conspicuously, almost absent. These are quite simple and elementary considerations, but they are tests that any leaching hypothesis must negotiate. Present evidence seems to indicate that they are unlikely to do this.

On the other hand, the possibility that a fractionating, degassing, subvolcanic melt is the principal source of the metals appears to 'fit' the patterns of exhalative ore compositions quite well. Nickel, chromium, and to a slightly lesser extent cobalt, are removed (in olivine and spinel) from the melt so rapidly that they are essentially unavailable at the time of formation of a volatile phase, and are thus not delivered to the site of ore deposition. Copper, zinc, and lead reach their maximum concentrations in the melt at the basalt–basaltic andesite, andesitic, and dacitic–rhyodacitic stages of magmatic evolution respectively: a state of affairs that seems to correspond rather prettily with the lava-type associations and stratigraphic affiliations of the exhalative copper, copper–zinc, and lead–zinc ores respectively. The capacity of plagioclase to operate as a geochemical sieve, selectively capturing strontium and excluding barium —thus progressively impoverishing the melt in the former and enriching it in the latter — may begin to account for the absence of strontium and abundance of barium in many exhalative orebodies. The phosphorus pentasulphide–lead fluoride reaction, or others related to it, may provide an explanation of the not uncommon association of phosphorus (as abundant apatite) with lead in some major ore deposits and their associated exhalites. As for chromium, the virtual absence of titanium and vanadium from (and the greatly variable abundance of manganese in) exhalative ores may reflect the almost quantitative subtraction of these metals from the melt in spinels.

The fact — based on a very large bank of data on mineral compositions — that fractional crystallization cannot account for *the whole* of the reduction of copper, zinc, manganese, iron, calcium, and other elements observed in the Solomons basalt–andesite–dacite lineage, points to the likelihood of the loss of an additional increment in the volatile phase. In a qualitative way this accords well with observations of metal loss via modern fumaroles and volcanic plumes, and the common occurrence of these elements in exhalative deposits.

However, while such a combination of fractionation and volatile loss may be the beginning of a plausible explanation of many volcanic ore associations, it, too, has its problems. We have made much of the point that lead is progressively enriched in the melt and increases from ≈ 0 in the mafic basalts to ≈ 5 ppm in felsic andesites and ≈ 10 ppm in the Solomons dacites, but it must be remembered that the mean nickel abundances in the same felsic andesites and dacites are 8.5 and 6.6 ppm (and cobalt 23 and 7 ppm) respectively. While fractionation — in particular plagioclase crystallization — certainly leads to a progressive decrease of strontium and concomitant increase in barium in the residual melt, mean strontium (≈ 870 ppm) remains greater than mean barium (≈ 790 ppm) even in the most evolved of the Solomons lavas: those dacites containing >64 per cent SiO_2. Why, if significant amounts of zinc, lead, and manganese are stored in magnetite, and then released as a result of the decomposition of magnetite in a late-stage aqueous volatile phase, are not titanium and vanadium similarly released, to find their way into the silicates and oxides associated with exhalative ores? Why are the ratios of copper, zinc, and lead to iron in most exhalative sulphide ores so very much higher than they are in any lava from which they are likely to be derived? Many exhalative copper ores possess a Cu/Fe ratio of $\approx 1:10$, but in most basaltic lavas it is <1:1000; the Zn/Fe ratio in some exhalative ores is >1, but in most basaltic lavas it is, like Cu/Fe, <1000. The 'anomaly' is even greater with respect to lead in many lead-bearing orebodies. (And why does the lead of these orebodies occur virtually entirely as the sulphide, never as the sulphate? Sulphate has clearly been abundant in many exhalative environments — *vide* the presence of large quantities of barite and anhydrite in many deposits — and lead sulphate is highly insoluble.)

These and related questions cannot be lightly dismissed. If the principal source of the metals is in fact the active melt, ore type–lava relations stem from factors additional to, and much more complex than, the simple raising or lowering of concentrations of those metals by fractionation and more-or-less constant volatile loss. It seems very likely that a role equal to or more important than that of concentrations may be played by the anions: what anions are present and at what concentrations; what ionic complexes may be generated under conditions prevailing in and close to the fractionating melt; and at what stages of the degassing process it is that the critical anions achieve their maximum activities. The substantial extension of the work of investigators such as Stoiber, Rose, Symonds, Reed, and Anderson on volcanic fluid equilibria, and of Burnham, Holland, Candela, and Manning on experimental melt–gas–vapour relationships is clearly essential for the resolution of questions such as these.

What might be some of the major questions stemming from our brief look at the behaviour of the ore elements in island arc lavas?

Does the fact that it is *copper* among the base metals that dominates porphyry deposits and appears earliest in the exhalative ores ultimately stem from the Jahn–Teller distortion of the Cu^{2+} ion: is it for this reason that, in contrast to Zn^{2+}, it is not accepted above a very low level in the octahedral sites of the major ferromagnesian silicates, that it builds up to very high concentrations in the melt by the time basaltic andesites begin to develop, and is then suddenly rejected from the residual melt as this acquires a high-Ca plagioclase structure at the basalt → andesite phase of lava evolution?

Do lead-rich exhalative ores tend to occur in association with calc-alkaline rocks because of the latter's paucity of potassium, a resultant scarcity of crystal sites receptive to lead, and the consequent high availability of this metal at the late, volatile-rich stages of subvolcanic melt evolution?

Is it volatile loss of the alkali metals — particularly potassium — at the dacitic stage of lava evolution that lead to the hypertholeiitic patterns of barium and lead?

Is the occurrence of iron, calcium, the alkalis, and other major rock-forming elements in volcanic sublimates and plumes simply a minor manifestation of what are in fact very significant losses of these substances in the volatile phase during explosive episodes? The hypotholeiitic behaviour of iron and calcium in island-arc lavas is remarkably similar to that of zinc and manganese (and phosphorus). The latter three are known to be lost from the melt in the volatile phase, with corresponding modification in melt composition. Does it follow that significant quantities of iron and calcium are also lost in the volatile phase and that this, too, induces significant changes in melt composition. In other words, if volatile loss is an important factor in the development of the patterns of incidence of many of the trace metals of the melt, might it be important also in modifying major element abundances, and hence in inducing some of the constitutional changes that we have long used to define the evolution of island-arc and other lava suites? If so, does the change from a basaltic to an andesitic–dacitic lineage in a hypotholeiitic system reflect a change from olivine–clinopyroxene–(spinel) fractionation in a *closed system* to hornblende–clinopyroxene–spinel fractionation in an *open system*? It has been one of the author's principal themes that most of the world's largest exhalative orebodies, and their sometimes very substantial associated 'exhalite' formations, occur in andesitic–rhyolitic terrains. Together with the greatly increased explosiveness and pyroclastic nature of andesitic eruptions, all this indicates that the change from basaltic to andesitic activity is accompanied by greatly increased exhalative material transfer from the melt to the Earth's surface environment. To emphasize the question asked above, could such exhalative transfer involve, in combination with fractional crystallization, quantities of materials large enough to lead to the development of hypotholeiitic basalt–andesite–dacite rhyolite lineages? Many of those searching for 'volcanic massive sulphide' deposits regard the occurrence of shallow-water marine andesitic, dacitic, and rhyolitic rocks as a promising sign for ore occurrence. Could it be that the features that make these particular rocks such good indicators of past exhalative activity result, in fact, from the exhalative processes that led to the formation of the ores? Most petrologists have been reluctant to consider any significant role for volatile loss in the generation of felsic lavas, but the time may have come for that possibility to be given more serious thought — and to be investigated by experiment.

If there is a tendency towards a relationship between stages of lava evolution on one hand and the amount and composition of emitted volatiles on the other, perhaps we should be paying more attention to observing and comparing the volatile products of what might be referred to as *petrotectonic categories of volcanoes*. Perhaps we should be comparing and contrasting the volatile emissions of mid-ocean volcanoes (e.g. Hawaii) with those of arc volcanoes, and among the latter, emissions associated with each of basaltic, andesitic, and dacitic eruptions.

There are of course other questions. We have noted remarkably large short-range variations in the compositions of some of the minerals of the Solomons suite, particularly the spinels. Does this mean that many observed volcanic mineral assemblages depart significantly from true equilibrium: that they result from turbulent mixing of crystals formed in different magmatic environments, and that in many instances quenching has preserved what are metastable assemblages on a large scale? Should we, indeed, be re-examining current assumptions that most volcanic mineral assemblages represent a close approach to chemical equilibrium in the melt? Perhaps the mineral assemblages of deep-sea-floor basalts — materials evolved under relatively uniform, quiet conditions, and with a minimum of opportunity for rapid volatile loss — do reflect a close approach to such equilibrium. Basalts of arc volcanoes such as those of the Solomons, on the other hand, with their opportunities for subvolcanic turbulence and mixing, may yield a 'freezing-in' of assemblages that have not grown in equilibrium and have not had time to develop it. Andesites, dacites, and rhyolites, as products of highly turbulent, substantially explosive, open-system, crystal–melt–gas regimes, may develop and preserve assemblages that depart even further from a true equilibrium state. This general possibility calls for much further investigation: initially by comprehensive microprobe analysis of natural materials followed, if appropriate, by experiment.

How does subvolcanic, particularly continental, contamination effect the patterns of abundance of the trace and minor elements we have been considering? We have, as noted in Chapter 1, used the Solomons lavas for an initial examination of abundance patterns particularly because, as products of *intra-oceanic* arc volcanism, these lavas have not been exposed to contamination by continental crustal materials, and therefore represent a minimum of geochemical complexity. Much of the ultimate purpose of the Solomons analysis has, however, been to aid the interpretation of geochemical patterns in older, perhaps mineralized, volcanic-arc provinces, many of which have formed at least partly on continental substrates. These bring in the complexities we avoided by investigating the Solomons suite, but such complications must be resolved if we are to understand the geochemical patterns of many older arc terrains. Perhaps we should, for the time being, continue in the Recent to Modern milieu, proceeding by the progressive addition of the continental component. Beginning with intra-oceanic arcs such as those of the Solomons, Vanuatu, and the Marianas, we might move to examine the lava geochemistry of Fiji, where continental crust is in its earliest stages of development — or the southern extremities of the Tonga–Kermadec and Lesser Antilles arcs, and the eastern section of the Aleutian arc, where volcanism is known to traverse continental crustal material. From here our examination might progress to the (Mesozoic)–Tertiary–Recent volcanic provinces of Western Alaska, the North Island of New Zealand, eastern New Guinea, Japan, and of the various islands of Indonesia, in all of which provinces continental crust —some of it at least as old as Palaeozoic — is well established. By investigating in a systematic way such as this what are probably the progressively increasing geochemical complexities of arcs of progressively increasing continental context, we may be able to establish a reliable basis for elucidating the geochemical patterns of older mineralized arc structures such of those of the Appalachians, the Urals and Eastern Australia of Palaeozoic age, and the Superior and Yilgarn provinces of the Archaean of Canada and Australia.

The author is all too conscious that, throughout the present volume, he has 'left the reader dangling' at 64 per cent SiO_2: almost all the variation diagrams have as their final class interval '>64 per cent SiO_2': just the point at which the abundance patterns of the large-ion incompatible elements, apart from barium, look as if they may be about to undergo significant change. The M:SiO_2 curves for lead, potassium, rubidium, (zirconium), and strontium, all of which generate positive slopes as far as ≈64–6 per cent SiO_2, appear to show signs of flattening or changing to negative slopes at about this stage. Termination of the curves, and the relatively low statistical reliability of the relevant points, at this dacitic phase of lava evolution is of course due to the fact that lavas containing more than ≈64–6 per cent SiO_2 are uncommon in the Solomons. Sampling frequency is a fairly accurate reflection of lava abundance. From the point of view of large-ion element abundance patterns, those lavas containing SiO_2 >64 per cent are clearly critical, and special efforts should perhaps be made in the future to collect and examine them. Judging from the work of others, such lavas are also not abundant in the Vanuatu and Lesser Antilles arcs, but they may be in those of Tonga–Kermadec

and the Aleutians — and perhaps others. Should we be looking much more carefully at these felsic lavas — the dacites, rhyodacites, and rhyolites — to perceive the ultimate fate of lead, potassium, rubidium, zirconium, and strontium — and, perhaps at a slightly later stage, of barium?

In 1949, as a very young man, the writer was a fascinated listener at a lecture given by W. R. Browne on 'Metallogenetic epochs and ore regions in the Commonwealth of Australia'. Browne showed how many ore deposits could be grouped into provinces, how the ores of a given province were often substantially contemporaneous, and how many of these space–time groupings were characterized by the conspicuous presence of particular elements such as gold or lead, or tungsten or tin. In the following year (1950), while working in the Solomon Islands, it occurred to the author that perhaps many of the great metallogenetic provinces of which Browne had spoken were in fact old island-arc structures that had become parts of continents: that ore formation in such cases was a result of early volcanic rather than much later plutonic processes, and that the ores themselves displayed their many space–time–composition consanguinities because they were all an intrinsic part of the volcano–sedimentary evolution and palaeogeographical development of a particular parent arc.

More than forty years later much of this is now regarded as self-evident. We are, however, little wiser concerning the processes underlying the association of arcs, andesites, and exhalative ores. Perhaps we can now see that the suggestion that exhalative ores *of this particular association* are products of sea-water leaching is likely to be a red herring. It seems much more likely that such ores are — primarily — products of partial melting and degassing of the mantle. If so, we can begin to see individual arcs, their andesites, and their ores as interconnected manifestations of great but finite *episodes* and *domains* of partial melting of the mantle. The common geochemical distinctiveness of many arc lava suites, and the constitutional consanguinities, the size, and the abundance of the ore deposits, may reflect not only heterogeneity of the mantle but variations — from one epoch and from one mantle domain to another — in conditions under which partial melting took place. The degree of hydration, and the abundance of other volatiles, may well affect mantle:melt partitioning, and hence the geochemical distinctiveness of the lavas and the metallogenetic individuality of the ores.

That some provinces are notable for the incidence of tin, or of lead, or of gold or other metals, and that this distinctiveness may have survived 're-working' by geological processes over several epochs, is well known. That some provinces such as the Iberian Pyrite Belt and the

Northern Appalachians contain vastly more ore than otherwise almost identical provinces such as the Lachlan Belt of south-eastern Australia and the Brito-Norwegian Caledonides has also been recognized for a long time. Is this sheer chance? Is it because some provinces have provided environments more conducive to the precipitation of particular metals and to sulphide accumulation generally? Or does it reflect some fundamental differences at source: some critical variation in conditions of formation, and the ultimate nature, of the parent partial melt?

If the 'individuality' of many volcanic–metallogenetic provinces does in fact reflect fundamental variation in the nature and conditions of partial melting from one mantle domain to another, it is likely that this induces the basic pattern of constitution and abundance. Upon this are superimposed the effects of sub-volcanic fractionation, of volatile loss at the volcanic orifice, and finally the conditions under which the metallic compounds accumulate on the sea-floor.

It might be said that 'the jury is still out' on the question of leaching versus magmatism and related problems. In this author's view the jury has not yet been given more than a small fraction of the quantitative evidence that the volcanic materials are capable of providing. If the present volume does little more than scratch the surface, it may encourage others to look more closely at the incidence of the metals and their companion anions in lavas and volcanic exhalations.

If, as the author now strongly suspects, the principal source of the metals is indeed the active melt, what are the implications of this in ore-genesis theory and mineral exploration? Do the highly productive exhalative ore provinces, for example those of the Northern Appalachians and the Iberian Pyrite belt, which contain large numbers of large orebodies, owe their existence in the first place to particularly propitious partial melting processes? Do, for example, more highly hydrous partial melting events lead not only to the incorporation of greater amounts of potential volatiles in the melt, but also to the partitioning of greater quantities of the metals, the halogens, and sulphur into the melt? Do the conspicuously less productive provinces, such as those of south-eastern Australia and the Brito-Norwegian Caledonides, which contain smaller numbers of much smaller orebodies, reflect less hydrous partial melting with lesser partitioning of the ore elements, and their carrier anions, from the solid mantle into the developing melt?

Much recent currency has been given to the proposition that those volcanic melts that have fractionated most extensively are likely to be the most 'fertile'. This is little more than a repetition of the statements of Lindgren and his contemporaries made almost a century ago, and now seems likely to be 'only half the story'. The evidence of the Solomons materials indicates that concentration of the metals during the fractionation phase of lava evolution depends not only on the *extent* of this process, but also on the *route* that it follows. A low (SiO_2–CaO–O_2):high (FeO–K_2O) basaltic melt is likely to yield olivine–orthopyroxene–high-K plagioclase assemblages and hence to favour the capture of most of the ore elements in the silicates. A somewhat higher (SiO_2–CaO–O_2):lower (FeO–K_2O) basaltic parent is, on the other hand, likely to generate clinopyroxene–low-K plagioclase assemblages, and hence to favour the concentration of the metals in the melt. Perhaps it was such a fractionation pattern that led to the development of such high concentrations of zinc and iron in the tholeiitic basalts and andesites of the Galapagos sea-floor. The process of fractionation may, however, go too far: the copious development of biotite, with its extraordinary capacity for scavenging the metals from the melt from which it forms, in a potentially ore-bearing volcanic pile is something that no mineral explorer would wish to see.

Perhaps the most fundamental corollary of our examination of relationships between exhalative ore deposits and volcanic rocks in the arc environment is that the origin of the ores, and the generation of the 'calc-alkaline' (hypotholeiitic) andesite–dacite lineages, are inextricably linked. The elucidation of the origins of the orogenic andesites is not only an important aspect of igneous petrology; it is fundamental to an understanding of orogenic exhalative ores. One suspects that the solving of either one of these problems will automatically provide the solution to the other.

Bibliography

This bibliography is intended to give a comprehensive but not exhaustive cover of relevant literature. Details of less important and more obscure works cited are obtainable from the appropriate listed references. Information on many older works is given particularly by F. W. Clarke (1924), C. N. Fenner (1933), and V. M. Goldschmidt (1954).

Allard, P. (1980). Proportions des isotopes ^{13}C et ^{12}C du carbone emis à haute temperature par un dôme andesitique en cours de croissance, Le Merapi (Indonesia). *Comptes Rendus de l'Academie des Sciences, Paris, Series D*, **291**, 613–16.

Allard, P. (1982). Stable isotope composition of hydrogen, carbon and sulphur in magmatic gases from rift and island arc volcanoes. *Bulletin Volcanologique*, **45**, 269–71.

Allen, E. T., and Zies, E. G. (1923). A chemical study of the fumaroles of the Katmai region. Geophysical Laboratory of the Carnegie Institution of Washington, Paper No. 485. *National Geographic Society: contributed technical papers. Katmai Series*, Number 2.

Anderson, A. T. (1992). Subvolcanic degassing of magma. In *Magmatic contributions to hydrothermal systems*. Extended abstracts of the Japan–US seminar held at Kagoshima and Ebino, November 1991, Geological Survey Report No. 279 (ed. J. W. Hedenquist), pp. 12–15.

Anderson, A. T., and Greenland, L. P. (1969). Phosphorus fractionation diagram as a quantitative indicator of crystallization differentiation of basaltic liquids. *Geochimica et Cosmochimica Acta*, **33**, 493–505.

Andriambololona, R., and Dupuy, C. (1978). Répartition et comportement des éléments de transition dans les roches volcaniques: cuivre et zinc. *Bulletin du BRGM (deuxième série)*, Section II, No. 2, 121–38.

Ansted, D. T. (1857). On some remarkable mineral veins. *Quarterly Journal of the Geological Society of London*, **13**, 240–54.

Arculus, R. J. (1973). The alkali basalt, andesite of Grenada, Lesser Antilles. Unpublished Ph.D. thesis. University of Durham.

Arculus, R. J. (1976). Geology and geochemistry of the alkali basalt–andesite association of Grenada, Lesser Antilles island arc. *Bulletin of the Geological Society of America*, **87**, 612–24.

Arculus, R. J. (1991). The evolution of Fuji and Hakone volcanoes, Honshu, Japan. *National Geographic Research and Exploration*, **7**, 276–309.

Arculus, R. J., and Curran, E. B. (1972). The genesis of the calc-alkaline rock suite. *Earth and Planetary Science Letters*, **15**, 255–62.

Arculus, R. J., Delong, S. E., Kay, R. W., Brooks, C., and Sun, S-S. (1977). The alkalic rock suite of Bogoslof Island, eastern Aleutian Arc, Alaska. *Journal of Geology*, **85**, 177–86.

Arndt, N. T. (1977). Partitioning of nickel between olivine and ultrabasic and basic komatiite liquids. *Carnegie Institution of Washington Year Book*, **76**, 553–7.

Baker, P. E. (1968a). Petrology of Mt Misery volcano, St Kitts, West Indies. *Lithos*, **1**, 124–50.

Baker, P. E. (1968b). Comparative volcanology and petrology of the Atlantic island arcs. *Bulletin Volcanologique*, **32**, 189–206.

Baker, P. E. (1982b). Evolution and classification of orogenic volcanic rocks. In *Andesites* (ed. R. S. Thorpe), Wiley, Chichester.

Baker, P. E. (1984). Geochemical evolution of St Kitts and Montserrat, Lesser Antilles. *Journal of the Geological Society of London*, **141**, 401–11.

Baker, P. E., Buckley, F., and Padfield, T. (1980). Petrology of the volcanic rocks of Saba, West Indies. *Bulletin Volcanologique*, **43**, 337–46.

Bauerman, H., and Foster, C. le N. (1869). On the occurrence of celestine in the nummulitic limestone of Egypt. *Quarterly Journal of the Geological Society of London*, **25**, 40–7.

Bence, A. E., and Taylor, S. R. (1977). Petrogenesis of Mid-Atlantic Ridge basalts at DSDP leg 37 holes 332A and 332B from major and trace element geochemistry. In *Initial Reports of the Deep Sea Drilling Project*, **37**, pp. 705–10. United States Government Printing Office, Washington.

Bender, J. F., Hodges, F. N., and Bence, A. E. (1978). Petrogenesis of basalts from the project FAMOUS area: experimental study from 0 to 15 kilobars. *Earth and Planetary Science Letters*, **41**, 277–302.

Bender, J. F., Langmuir, C. H., and Hanson, G. N. (1984). Petrogenesis of basalt glasses from the Tamayo region, East Pacific Rise. *Journal of Petrology*, **25**, 213–54.

Bergametti, G., Martin, D., Carbonnelle, J., Faivre-Pierret, R., and vie le Sage, R. (1984). A mesoscale study of the elemental composition of aerosols emitted from Mt Etna volcano. *Bulletin Volcanologique*, **47**, 1107–14.

Bergeat, A. (1899). Cited by Fenner, C. N. (1933), p. 95.

Berlin, R., and Henderson, C. M. B. (1969). The distribution of strontium and barium between the alkali feldspar, plagioclase and groundmass phases of porphyritic trachytes and phonolites. *Geochimica et Cosmochimica Acta*, **33**, 247–55.

Bernier, L. R. (1990). Vanadiferous zincian–chromian hercynite in a metamorphosed basalt hosted alteration zone, Atik Lake, Manitoba. *Canadian Mineralogist*, **28**, 37–50.

Berry, L. G., and Mason, B. (1959). *Mineralogy*. Freeman, San Francisco.

Beyschlag, F., Vogt, J. H. L., and Krusch, P. (1916). *The deposits of the useful minerals and rocks* (trans. S. J. Truscott). Macmillan, London.

Bischoff, J. L., and Dickson, F. W. (1975). Seawater–basalt interaction at 200 °C and 500 bars: implications for origin of sea-floor heavy-metal deposits and regulation of seawater chemistry. *Earth and Planetary Science Letters*, **25**, 385–97.

Borishenko, L. F. (1972). The new mineral shcherbinaite. Cited by Hughes, J. M. and Stoiber, R. E. (1985), p. 285.

Borishenko, L. F., Serafimova, E. K., Kazakova, M. E., and Shumyatskaya, N. O. (1970). First find of crystalline V_2O_5 in the products of volcanic eruption on Kamchatka. *Doklady Akademii Nauk SSSR*, **193**, 683–6.

Bostrom, K., and Peterson, H. T. (1966). Precipitates from hydrothermal exhalations on the East Pacific Rise. *Economic Geology*, **61**, 1258–65.

Bougault, H., and Hekinian, R. (1974). Rift valley in the Atlantic Ocean near 36° 50′ N; petrology and geochemistry of basaltic rocks. *Earth and Planetary Science Letters*, **24**, 249–61.

Bougault, H., Dmitriev, L., Schilling, J. G., Sobolev, A., Joron, J. L., and Needham, H. D. (1988). Mantle heterogeneity from trace elements: MAR triple junction near 14° N. *Earth and Planetary Science Letters*, **88**, 27–36.

Bouglise, G de la, and Cumenge, E. (1885). Cited by Wilson, I. F. (1955), p. 83.

Bowen, N. L. (1928). *The evolution of the igneous rocks*. Princeton University Press.

Bowen, N. L. (1933). The broader story of magmatic differentiation, briefly told. In *Ore deposits of the Western States (Lindgren Volume)*, pp. 106–128, American Institute of Mining and Metallurgical Engineers, New York.

Brophy, J. G. (1987). The Cold Bay volcanic centre, Aleutian Volcanic Arc. *Contributions to Mineralogy and Petrology*, **97**, 378–88.

Brothers, R. N., and Searle, E. J. (1970). The geology of Raoul Island, Kermadec Group, south west Pacific. *Bulletin Volcanologique*, **34**, 7–37.

Brown, G. M. (1956). The layered ultrabasic rocks of Rhum, Inner Hebrides. *Philosophical Transactions of the Royal Society of London, Series B*, **24**, 1–53.

Brown, G. M., Holland, J. G., Sigurdsson, H., Tomblin, J. F., and Arculus, R. J. (1977). Geochemistry of the Lesser Antilles volcanic island arc. *Geochimica et Cosmochimica Acta*, **41**, 785–801.

Brown, W. R. (1949). Metallogenetic epochs and ore regions in the Commonwealth of Australia. *Journal and Proceedings of the Royal Society of New South Wales*, **83**, 96–113.

Bryan, W. B. (1979). Regional variation and petrogenesis of basalt glasses from the FAMOUS area, Mid-Atlantic ridge. *Journal of Petrology*, **20**, 293–325.

Bryan, W. B., and Moore, J. G. (1977). Compositional variations of young basalts in the Mid-Atlantic Ridge valley near lat. 36° 49′ N. *Bulletin of the Geological Society of America*, **88**, 556–70.

Bryan, W. B., Thompson, G., and Michael, P. J. (1979). Compositional variation in a steady-state zoned magma chamber: Mid-Atlantic ridge at 36° 50′ N. *Tectonophysics*, **55**, 63–85.

Bryan, W. B., Thompson, G., Frey, F. A., Dickey, J. S., and Roy, S. (1977). Petrology and geochemistry of basement rocks recovered on leg 27, DSDP. In *Initial Reports of the Deep Sea Drilling Project*, 37. United States Government Printing Office, Washington, pp. 695–703.

Bryan, W. B., Thompson, G., and Ludden, J. N. (1981). Compositional variation in normal MORB from 22°–25° N: Mid-Atlantic Ridge and Kane fracture zone. *Journal of Geophysical Research*, **86**, 11 815–36.

Buat-Ménard, P., and Arnold, M. (1978). The heavy metal chemistry of atmospheric particulate matter emitted by Mount Etna volcano. *Geophysical Research Letters*, **5**, 245–48.

Bunsen, R. W. (1851). Cited by Fenner, C. N. (1933), p. 73.

Bunsen, R. W. (1853). Cited by Clarke, F. W. (1924), p. 262.

Burnham, C. W. (1967). Hydrothermal fluids at the magmatic stage. In *Geochemistry of hydrothermal ore deposits* (ed. H. L. Barnes), pp. 34–76. Holt, Rinehart and Winston, New York.

Burnham, C. W. (1979a). The importance of volatile constituents. In *The evolution of igneous rocks* (ed. H. S. Yoder Jr.), pp. 439–82. Princeton University Press.

Burnham, C. W. (1979b). Magmas and hydrothermal fluids. In *Geochemistry and hydrothermal ore deposits* (ed. H. L. Barnes). John Wiley, New York.

Burns, R. G. (1970). *Mineralogical applications of crystal field theory*. Cambridge University Press.

Burns, R. G., and Burns, V. M. (1974). Nickel. In *Handbook of geochemistry* (ed. K. H. Wedepohl). Springer-Verlag, Berlin.

Byers, F. M. (1961). Petrology of three volcanic suites, Umnak and Bogoslof Islands, Aleutian Islands, Alaska. *Bulletin of the Geological Society of America*, **72**, 93–128.

Cadle, R. D., Wartburg, A. F., Pollock, W. H., Gandrud, B. W., and Shedlovsky, J. P. (1973). Trace constituents emitted to the atmosphere by Hawaiian volcanoes. *Chemosphere*, **6**, 231–4.

Campbell, I. H., Franklin, J. M., Gorton, M. P., Hart, T. R., and Scott, S. D. (1981). The role of subvolcanic sills in the generation of massive sulfide deposits. *Economic Geology*, **76**, 2248–53.

Candela, P. A. (1989a). Magmatic ore-forming fluids: thermodynamic and mass-transfer calculations of metal concentrations. In *Ore deposition associated with magmas*, Reviews in Economic Geology, vol. 4 (ed. J. A. Whitney and A. J. Naldrett), pp. 203–21.

Candela, P. A. (1989b). Felsic magmas, volatiles, and metallogenesis. In *Ore deposition associated with magmas*. Reviews in Economic Geology, Vol. 4 (ed. J. A. Whitney and A. J. Naldrett), pp. 223–33.

Candela, P. A., and Holland, H. D. (1984). The partitioning of copper and molybdenum between silicate melts and aqueous fluids. *Geochimica et Cosmochimica Acta*, **48**, 373–80.

Candela, P. A., and Holland, H. D. (1986). A mass transfer model for copper and molybdenum in magmatic hydrothermal systems. *Economic Geology*, **81**, 1–19.

Carmichael, I. S. E. (1964). The Petrology of Thingmuli, a Tertiary volcano in Eastern Iceland. *Journal of Petrology*, **5**, 435–60.

Carmichael, I. S. E. (1967). The mineralogy and petrology of the volcanic rocks from Leucite Hills, Wyoming. *Contributions to Mineralogy and Petrology*, **15**, 24–66.

Carmichael, I. S. E., Turner, F. J., and Verhoogen, J. (1974). *Igneous petrology*. McGraw-Hill, New York.

Carmichael, J. and McDonald, A. (1961). The geochemistry of some natural acid glasses from the North Atlantic Tertiary volcanic province. *Geochimica et Cosmochimica Acta*, **25**, 189–222.

Carr, M. H., and Turekian, K. K. (1961). The geochemistry of cobalt. *Geochimica et Cosmochimica Acta*, **23**, 9–60.

Carstens, C. W. (1920). Cited by Stanton, R. L. (1984), p. 1433.

Carstens, C. W. (1923). Der unterordovizische Vulkanhorizont in dem Trondheimsgebiet (mit besonderer Berüchsichtigung der in ihm auftretenden Kiesvorkommen). *Norsk Geologisk Tidsskrift*, **7**, 185–269.

Clark, L. A. (1971). Volcanogenic ores: comparison of cupriferous pyrite deposits of Cyprus and Japanese kuroko deposits. In *Society of Mining Geologists of Japan, Special Issue* **3**, 206–15.

Clarke, D. B. (1970). Tertiary basalts of Baffin Bay: possible primary magma from the mantle. *Contributions to Mineralogy and Petrology*, **25**, 203–24.

Clarke, D. B., and O'Hara, M. J. (1979). Nickel, and the existence of high-MgO liquids in nature. *Earth and Planetary Science Letters*, **44**, 153–8.

Clarke, F. W. (1910). Analyses of rocks and minerals from the laboratory of the United States Geological Survey, 1880–1908. Analyses of igneous and crystalline rocks. *Bulletin of the United States Geological Survey*, **419**, 13–181.

Clarke, F. W. (1924). *The data of geochemistry* (5th edn). Bulletin. United States Geological Survey, 770.

Clarke, F. W., and Steiger, G. (1914). The relative abundance of several metallic elements. *Journal of the Washington Academy of Science*, **4**, 58.

Clarke, F. W., and Washington, H. S. (1924). The composition of the earth's crust. *U.S. Geological Survey Professional Paper* 127.

Coats, R. R. (1952). Magmatic differentiation in Tertiary and Quaternary volcanic rocks from Adak and Kanaga Islands, Aleutian Islands, Alaska. *Geological Society of American Bulletin*, **63**, 485–514.

Cole, J. W. (1982). Tonga–Kermadec–New Zealand. In *Orogenic andesites and related rocks* (ed. R. S. Thorpe), pp. 245–58. Wiley, Chichester.

Collins, J. J. (1950). Summary of Kinoshita's Kuroko deposits of Japan. *Economic Geology*, **45**, 363–76.

Collot, J.-Y., Green, H. G., and Stokking, H. B. (1992). Summary and conclusions, DSDP Leg 134 (Vanuatu). In *Proceedings of the Ocean Drilling Program, Initial Reports*, **134**, 561.

Conway, E. J. (1945). Mean losses of Ca, Na, etc. in one weathering cycle and potassium removal from the ocean. *American Journal of Science*, **243**, 583–605.

Cooper, J. A., and Richards, J. R. (1969). Lead isotope measurements in sediments from Atlantis II and Discovery Deep areas. In *Hot brines and recent heavy metal deposits in the Red Sea* (ed. E. T. Degens and D. A. Ross). Springer-Verlag, New York.

Corliss, J. B. (1971). The origin of metal-bearing submarine hydrothermal solutions. *Journal of Geophysical Research*, **76**, 8128–38.

Cornwall, H. R., and Rose, H. J., (1957). Minor elements in Keweenawan lavas, Michigan. *Geochimica et Cosmochimica Acta*, **12**, 209–24.

Correns, C. W. (1978). Titanium. In *Handbook of geochemistry* (ed. K. H. Wedepohl). Springer-Verlag, Berlin.

Cotton, F. A., and Wilkinson, G. (1972). *Advanced inorganic chemistry* (3rd edn). John Wiley and Sons, New York.

Coulson, F. I. (1985). Solomon Islands. In *The ocean basins and margins*. Vol. 7A. *The Pacific Ocean* (ed.

A. E. M. Nairn, F. G. Stehli, and S. Uyeda). Plenum Press, New York.

Cox, K. G. and Bell, J. D. (1972). A crystal fractionation model for the basaltic rocks of the New Georgia Group, British Solomon Islands. *Contributions to Mineralogy and Petrology*, **37**, 1–13.

Craig, H. (1969). Geochemistry and origin of the Red Sea brines. In *Hot brines and recent heavy metal deposits in the Red Sea* (ed. E. T. Degens and D. A. Ross), pp. 209–42. Springer-Verlag, New York.

Crook, T. (1914). The genetic classification of ore deposits. *Mineralogical Magazine*, **17**, 55–85.

Crook, T. (1933). *History of the theory of ore deposits*. Thomas Murby, London.

Curtis, C. D. (1964). Applications of crystal field theory to the inclusion of trace elements in minerals during magmatic crystallization. *Geochimica et Cosmochimica Acta*, **28**, 389–402.

Cuturic, N., Kafol, N., and Karamata, S. (1966). Lead contents in K-feldspars of young igneous rocks of the Dinarides and neighbouring areas. In *Origin and distribution of the elements* (ed. L. H. Ahrens). Pergamon Press, Oxford.

Daly, R. A. (1933). *Igneous rocks and the depths of the earth*. McGraw-Hill, New York.

Day, A. L., and Shepherd, E. S. (1913). Water and volcanic activity. *Bulletin of the Geological Society of America*, **24**, 573–606.

de Beaumont, E. (1847). Note sur les émanations volcaniques et métallifères. *Bulletin de la Société Géologique de France*, 4, 1429–533.

de la Beche, H. (1851). *The geological observer* (1st edn). Thomas Murby, London.

Des Cloiseaux, N. (1880). Cited by Clarke, F. W. (1924), p. 709.

Deville, C. Sainte-Claire. (1856). Cited by Clarke, F. W. (1924), p. 263.

Deville, C. Sainte-Claire, and Leblanc, F. (1858). Cited by Clarke, F. W. (1924), p. 263.

Dickey, J. S., Frey, F. A., Hart, S. R., and Watson, E. B. (1977). Geochemistry and petrology of dredged basalts from the Bouvet triple junction, south Atlantic. *Geochimica et Cosmochimica Acta*, **41**, 1105–18.

Dissanayake, C. B., and Vincent, E. A. (1972). Zinc in rocks and minerals from the Skaergaard intrusion, East Greenland. *Chemical Geology*, **9**, 285–97.

Doe, B. R., and Tilling, R. I. (1967). The distribution of lead between coexisting K-feldspar and plagioclase. *American Mineralogist*, **52**, 805–16.

Dosso, L., Bougault, H., Beuzart, P., Calvez, J-Y., and Joron, J-L. (1988). The geochemical structure of the south-east Indian ridge. *Earth and Planetary Science Letters*, **88**, 47–59.

Dostal, J., Capedri, S., and de Albuquerque, C. A. R. (1976–7). Calc-alkaline volcanic rocks from N. W. Sardinia: evaluation of a fractional crystallization model. *Bulletin Volcanologique*, **40**, 1–7.

Dostal, J., Elson, C., and Walker, J. A. (1979). Distribution of lead, silver and cadmium in some igneous rocks and their constituent minerals. *Canadian Mineralogist*, **17**, 561–7.

Dostal, J., Dupuy, C., Carron, J. P., le Guen de Kerneizon, M., and Maury, R. C. (1983). Partition coefficients of trace elements: application to volcanic rocks of St Vincent, West Indies. *Geochimica et Cosmochimica Acta*, **47**, 525–33.

Drake, M. J., and Weill, D. F. (1975). Partitioning of Sr, Ba, Ca, Y, Eu^{2+}, Eu^{3+}, and other REE between plagioclase feldspar and magmatic liquid: an experimental study. *Geochimica et Cosmochimica Acta*, **39**, 689–712.

Dudas, M. J., Harward, M. E., and Schmitt, R. A. (1973). Identification of dacitic tephra by activation analysis of their primary mineral phenocrysts. *Quaternary Research*, **3**, 307–15.

Duke, J. M. (1976). Distribution of the period four transition elements among olivine, calcic clinopyroxene and mafic silicate liquid: experimental results. *Journal of Petrology*, **17**, 499–521.

Duncan, A. M. (1978). The trachybasaltic volcanics of the Adrano area, Mount Etna, Sicily. *Geological Magazine*, **115**, 273–85.

Duncan, A. R., and Taylor, S. R. (1969). Trace element analyses of magnetites from andesitic and dacitic lavas from Bay of Plenty, New Zealand. *Contributions to Mineralogy and Petrology*, **20**, 30–3.

Dunkley, P. N. (1983). Volcanism and the evolution of the ensimatic Solomon Islands arc. In *Arc volcanism: physics and tectonics* (ed. D. Shimozuru and I. Yokoyama). Terra Scientific Publishing Company, Tokyo.

Dunn, T. (1987). Partitioning of Hf, Lu, Ti, and Mn between olivine, clinopyroxene and basaltic liquid. *Contributions to Mineralogy and Petrology*, **96**, 476–84.

Eggins, S. M. (1989). The origin of primitive ocean island and island arc basalts. Unpublished Ph.D. thesis. University of Tasmania.

Eggins, S. M. (1993). Origin and differentiation of picritic arc magmas, Ambae (Aoba), Vanuatu. *Contributions to Mineralogy and Petrology*, **114**, 79–100.

Ehrenberg, H., Pilger, A., and Schroder, F. (1954). *Das Schwefelkies-Zinkblende-Schwerspatlager von Meggen (Westfalen)*. Niedersächsisches Landesamt Bodenforschung, Monograph 7, Hanover.

Eilenberg, S., and Carr, M. J. (1981). Copper contents of lavas from active volcanoes in El Salvador and adjacent regions in Central America. *Economic Geology*, **76**, 2246–8.

Elthon, D. (1979). High magnesia liquids as the parental magma for ocean floor basalts. *Nature*, **278**, 514–18.

Elthon, D., and Ridley, W. I., (1979). Comments on 'The partitioning of nickel between olivine and silicate melt' by S. R. Hart and K. E. Davis. *Earth and Planetary Science Letters*, **44**, 162–4.

Embley, R. W., Jonasson, I. R, Perfit, M. R., Franklin, J. M., Tivey, M. A., Malahoff, A., Smith, M. F., and Francis, T. J. G. (1988). Submersible investigation of an extinct hydrothermal system on the Galapagos Ridge: sulfide mounds, stockwork zone, and differentiated lavas. In *Seafloor hydrothermal mineralization* (ed. T. J. Barrett and J. L. Jambor). *Canadian Mineralogist*, **26**, 517–39.

Emsley, J. (1989). *The elements*. Clarendon Press, Oxford.

Engel, A. E. J., Engel, C. G., and Havens, R. G. (1965). Chemical characteristics of oceanic basalts and the upper mantle. *Bulletin of the Geological Society of America*, **76**, 719–34.

Erlank, A. J., and Kable, E. J. D. (1976). The significance of incompatible elements in Mid-Atlantic Ridge basalts from 45° N with particular reference to Zr/Nb. *Contributions to Mineralogy and Petrology*, **54**, 281–91.

Evans, B. W., and Moore, J. G. (1968). Mineralogy as a function of depth in the prehistoric Makaopuhi tholeiitic lava lake, Hawaii. *Contributions to Mineralogy and Petrology*, **17**, 85–115.

Ewart, A. (1976). A petrological study of the younger Tongan andesites and dacites, and the olivine tholeiities of Niua Fo'ou Island, SW Pacific. *Contributions to Mineralogy and Petrology*, **58**, 1–21.

Ewart, A. (1982). The mineralogy and petrology of Tertiary–Recent orogenic volcanic rocks: with special reference to the andesitic–basaltic compositional range. In *Andesites* (ed. R. S. Thorpe). Wiley, Chichester.

Ewart, A., and Bryan, W. B., (1972). The petrology and geochemistry of the igneous rocks from Eua, Tongan Islands. *Bulletin of the Geological Society of America*, **83**, 3281–98.

Ewart, A., and Bryan, W. B. (1973). The petrology and geochemistry of the Tongan Islands. In *The Western Pacific: island arcs, marginal seas, geochemistry* (ed. P. J. Coleman). University of Western Australia Press, Perth.

Ewart, A., and Stipp, J. J. (1968). Petrogenesis of the volcanic rocks of the Central North Island, New Zealand, as indicated by a study of Sr^{87}/Sr^{86} ratios, and Sr, Rb, K, U and Th abundances. *Geochimica et Cosmochimica Acta*, **32**, 699–735.

Ewart, A., and Taylor, S. R. (1969). Trace element geochemistry of the rhyolitic volcanic rocks, central North Island, New Zealand. Phenocryst data. *Contributions to Mineralogy and Petrology*, **22**, 127–46.

Ewart, A., Taylor, S. R., and Capp, A. C. (1968). Trace and minor element geochemistry of the rhyolitic volcanic rocks, central North Island, New Zealand. *Contributions to Mineralogy and Petrology*, **18**, 76–104.

Ewart, A., Bryan, W., and Gill, J. (1973). Mineralogy and geochemistry of the younger volcanic islands of Tonga, SW Pacific. *Journal of Petrology*, **14**, 429–65.

Ewart, A., Brothers, R. N., and Mateen, A. (1977). An outline of the geology and geochemistry, and the possible petrogenetic evolution of the volcanic rocks of the Tonga–Kermadec–New Zealand Island Arc. *Journal of Volcanology and Geothermal Research*, **2**, 205–50.

Faure, G. (1978). Strontium. In *Handbook of geochemistry* (ed. K. H. Wedepohl). Springer-Verlag, Berlin.

Fenner, C. N. (1920). The Katmai Region, Alaska, and the Great Eruption of 1912. *Journal of Geology*, **28**, 560–606.

Fenner, C. N. (1926). The Katmai magmatic province. *Journal of Geology*, **34**, 673–772.

Fenner, C. N. (1933). Pneumatolytic processes in the formation of minerals and ores. In *Ore deposits of the Western States (Lindgren Volume)*, pp. 58–105, American Institute of Mining and Metallurgical Engineers, New York.

Finlow-Bates, T., and Large, D. E. (1978). Water depth as major control on the formation of submarine exhalative ore deposits. *Geologisches Jahrbuch*, D**30**, 27–39.

Fischer, K. (1972). Strontium. In *Handbook of geochemistry* (ed. K. H. Wedepohl). Springer-Verlag, Berlin.

Flower, M. F. J., Robinson, P. T., Schmineke, H.-U., and Ohnmacht, W. (1977). Petrology and geochemistry of igneous rocks, DSDP leg 37. In *Initial Reports of the Deep Sea Drilling Project*, **37**, pp. 653–79. United States Government Printing Office, Washington.

Ford, C. E., Russell, D. G., Craven, J. A., and Fisk, M. R. (1983). Olivine–liquid equilibria: temperature, pressure and composition dependence of the crystal/liquid cation partition coefficients for Mg, Fe^{2+}, Ca and Mn. *Journal of Petrology*, **24**, 256–65.

Foshag, W. F., and Gonzalez, J. (1956). Birth and development of Paracutin Volcano. *Bulletin. United States Geological Survey*, 965–D. United States Government Printing Office, Washington.

Foshag, W. F., and Henderson, (1946). Primary sublimates at Parícutin Volcano (Mexico). *American Geophysical Union Transactions*, **27**, 685–6.

Fouqué, F. (1865). Cited by Clarke, F. W. (1924), pp. 264–5.

Frazer, G. D., and Barnett, H. F. (1959). Geology of the Delarof and westernmost Andreanof Islands, Aleutian Islands, Alaska. *Bulletin of the United States Geological Survey*, **1028**–I, 211–48.

Frey, F. A., Bryan, W. B., and Thompson, G. (1974). Atlantic Ocean floor: geochemistry and petrology of basalts from Legs 2 and 3 of the Deep Sea Drilling Project. *Journal of Geophysical Research*, **79**, 5507–27.

Frey, F. A, Dickey, J. S., Thompson, G., Bryan, W. B., and Davies, H. L. (1980). Evidence for heterogeneous primary MORB and mantle sources, NW Indian Ocean. *Contributions to Mineralogy and Petrology*, **74**, 387–402.

Fuchs, E. (1886). Cited by Wilson, I. F. (1955), p. 83.

Fukuchi, N., and Tsujimoto, K. (1902). Ore beds in the Misaka Series. *Journal of the Geological Society of Tokyo*. See Collins, J. J. (1950), p. 367.

Fyfe, W. S. (1964). *Geochemistry of solids: an introduction*. McGraw-Hill, New York.

Galkin, X. (1910). Chemische Untersuchung einiger Hornblenden und Augite aus Basalten der Röhn. *Neues Jahrbuch für Mineralogie, Geologie, und Paläontologie*, **29**, 681–718.

Garlick, G. D., and Dymond, J. R. (1970). Oxygen isotope exchange between volcanic materials and ocean water. *Bulletin of the Geological Society of America*, **81**, 2137–42.

Gast, P. W. (1968). Trace element fractionation and the origin of tholeiitic and alkaline magma types. *Geochimica et Cosmochimica Acta*, **32**, 1057–86.

Geikie, A. (1879). Geology. In *Encyclopaedia Britannica* (9th edn.), vol. 10.

Geikie, A. (1903). *Textbook of geology*, Vol. 1 (4th edn). Macmillan, London.

Gill, J. (1981). *Orogenic andesites and plate tectonics*. Springer-Verlag, Berlin.

Gill, J. B., and Compston, W. (1973). Strontium isotopes in island arc volcanic rocks. In *The Western Pacific-island arcs, marginal seas, geochemistry* (ed. P. J. Coleman). University of Western Australia Press, Perth.

Gmelin, L. (1965). *Handbuch der Anorganischen Chemie*. Phosphor, Teil A. Verlag Chemie, Weinheim.

Gmelin (1968). See Vanadium: abundance in common igneous rock types, by S. Landergren. In *Handbook of geochemistry*, ed. K. H. Wedepohl, Springer-Verlag, Berlin.

Goldberg, E. D. (1976). Rock volatility and aerosol composition. *Nature*, **260**, 128–9.

Goldschmidt, V. M. (1937). The principles of distribution of chemical elements in minerals and rocks. *Journal of the Chemical Society*, **33**, 655–73.

Goldschmidt, V. M. (1938). Cited by Goldschmidt, V. M. (1954), p. 485.

Goldschmidt, V. M. (1954). *Geochemistry* (ed. A. Muir). Clarendon Press, Oxford.

Goodfellow, W. D., and Blaise, B. (1988). Sulfide formation and hydrothermal alteration of hemipelagic sediment in Middle Valley, northern Juan de Fuca Ridge. *In* Seafloor hydrothermal mineralization (ed. T. J. Barrett and J. L. Jambour). *Canadian Mineralogist*, **26**, 675–96.

Gorton, M. P. (1974). The geochemistry and geochronology of the New Hebrides. Unpublished Ph.D. thesis. Australian National University.

Gorton, M. P. (1977). The geochemistry and origin of Quaternary volcanism in the New Hebrides. *Geochimica et Cosmochimica Acta*, **41**, 1257–70.

Green, D. H., and Ringwood, A. E. (1967). The genesis of basaltic magmas. *Contributions to Mineralogy and Petrology*, **15**, 103–90.

Green, T. H. (1972). Crystallization of calcalkaline andesite under controlled high pressure hydrous conditions. *Contributions to Mineralogy and Petrology*, **34**, 150–66.

Green, T. H., and Ringwood, A. E. (1968a). Genesis of the calc-alkaline igneous rock suite. *Contributions to Mineralogy and Petrology*, **18**, 105–62.

Green, T. H., and Ringwood, A. E. (1968b). Origin of garnet phenocrysts in calc-alkaline rocks. *Contributions to Mineralogy and Petrology*, **18**, 163–74.

Green, T. H., and Ringwood, A. E. (1972). Crystallization of garnet-bearing rhyodacite under high pressure hydrous conditions. *Journal of the Geological Society of Australia*, **19**, 203–12.

Greene, H. G., Macfarlane, A., and Wong, F. L. (1988). Geology and offshore resources of Vanuatu — introduction and summary. In *Geology and offshore resources of Pacific island arcs — Vanuatu region* (ed. H. G. Green and F. L. Wong), Circum-Pacific Council for Energy and Mineral Resources Earth Science Series, Vol. 8, pp. 1–25.

Griggs, R. F. (1922). The Valley of Ten Thousand Smokes. National Geographic Society, Washington.

Grout, F. F. (1910). Keweenawan copper deposits. *Economic Geology*, **5**, 471–6.

Grove, T. L., and Bryan, W. B. (1983). Fractionation of pyroxene-phyric MORB at low pressure: an experimental study. *Contributions to Mineralogy and Petrology*, **84**, 293–309.

Grove, T. L., and Kinzler, R. J. (1986). Petrogenesis of andesites. In *Annual Review of Earth and Planetary Sciences* (ed. G. W. Wetherill), **14**, 417–54.

Groves, D. I., Barrett, F. M., Binns, R. A., Martson, R. J., and McQueen, K. G. (1976). A possible volcanic–

exhalative origin for lenticular nickel sulfide deposits of volcanic association, with special reference to those in Western Australia: Discussion of paper by J. Lusk. *Canadian Journal of Earth Sciences*, **13**, 1646–50.

Gunn, B. M. (1971). Trace element partitioning during olivine fractionation of Hawaiian basalts. *Chemical Geology*, **8**, 1–13.

Gunn, B. M., Coy-Yll, R., Watkins, N. D., Abranson, C. E., and Nougier, J. (1970). Geochemistry of an oceanite-ankaramite-basalt suite from East Island, Crozet Archipelago. *Contributions to Mineralogy and Petrology*, **28**, 319–39.

Gunn, B. M., Abranson, C. E., Nougier, J., Watkins, N. D., and Hajash, A. (1971). Amsterdam Island, an isolated volcano in the southern Indian Ocean. *Contributions to Mineralogy and Petrology*, **32**, 79–92.

Gunn, B. M., Watkins, N. D., Trzcienski, W. E., and Nougier, J. (1975). The Amsterdam–St Paul volcanic province, and the formation of low Al tholeiitic andesites. *Lithos*, **8**, 137–49.

Häkli, T. A., and Wright, T. L. (1967). The fractionation of nickel between olivine and augite as a geothermometer. *Geochimica et Cosmochimica Acta*, **31**, 877–84.

Hamilton, W. M., and Baumgart, I. L. (1959). White Island. *New Zealand Department of Scientific and Industrial Research Bulletin* 127.

Hannington, M. D., and Scott, S. D. (1988). Mineralogy and geochemistry of a hydrothermal silica-sulfide-sulfate spire in the caldera of Axial Seamount, Juan de Fuca Ridge. In *Seafloor hydrothermal mineralization* (ed. T. J. Barrett and J. L. Jambour). *Canadian Mineralogist*, **26**, 603–25.

Harris, D. C. (1989). Mineralogy and geochemistry of the Hemlo gold deposit, Hemlo, Ontario. *Canadian Geological Survey Economic Geology Report* No. 38.

Hart, S. R, and Brooks, C. (1974). Clinopyroxene–matrix partitioning of K, Rb, Cs, Sr and Ba. *Geochimica et Cosmochimica Acta*, **38**, 1799–806.

Hart, S. R., and Davis, K. E. (1978). Nickel partitioning between olivine and silicate melt. *Earth and Planetary Science Letters*, **40**, 203–19.

Hart, S. R., and Davis, K. E. (1979). Reply to D. B. Clarke and M. J. O'Hara: Nickel, and the existence of high MgO liquids in nature. *Earth and Planetary Science Letters*, **44**, 159–61.

Hartmann, P. (1969). Can Ti^{4+} replace Si^{4+} in silicates? *Mineralogical Magazine*, **37**, 366–9.

Haslam, H. W. (1968). The crystallization of intermediate and acid magmas at Ben Nevis, Scotland. *Journal of Petrology*, **9**, 84–104.

Heaton, T. H. E., and Sheppard, S. M. F. (1977). Hydrogen and oxygen isotope evidence for sea-water–hydrothermal alteration and ore deposition, Troodos complex, Cyprus. In *Volcanic processes in ore genesis: proceedings of a joint meeting of the Volcanic Studies Group of the Geological Society of London and the Institution of Mining and Metallurgy held in London on 21 and 22 January, 1976*. Institution of Mining and Metallurgy and Geological Society, London.

Hecht, J. (1991). Antarctic gold dust. *New Scientist*, **131** (7), 14.

Hedenquist, J. W. (1992a). Recognition of magmatic contributions of active and extinct hydrothermal systems. In *Magmatic contributions to hydrothermal systems*. Extended abstracts of the Japan–US seminar held at Kagoshima and Ebino, November, 1991, *Geological Survey of Japan Report* No. 279 (ed. J. W. Hedenquist), pp. 68–79.

Hedenquist, J. W. (1992b). Magmatic contributions to hydrothermal systems *and* The behaviour of volatiles in magma. Extended abstracts of the Japan–US seminar on 'Magmatic contributions to hydrothermal systems' held at Kagoshima and Ebino, November 1991, and abstracts of the 4th symposium on deep crustal fluids 'The behaviour of volatiles in magma' held at Tsukuba, November 1991 (ed. J. W. Hedenquist). *Geological Survey of Japan Report* No. 279.

Hegeman, F. (1948). Uber sedimentäre Lagerstätten mit submarines vulcanischen Stoffzufuhr. *Fortschritte der Mineralogie, Kristallographie, und Petrographie*, **27**, 54–5.

Heier, K. S. (1960). Petrology and geochemistry of high-grade metamorphic rocks on Langöy, Northern Norway. *Norges geologiske undersøkelse*, **207**.

Heier, K. S. (1962). Trace elements in feldspars — a review. *Norsk Geologisk Tidsskrift*, **42**, 415–54.

Heier, K. S., Palmer, P. D., and Taylor, S. R. (1967). Comment on the Pb distribution in southern Norwegian Precambrian alkali feldspars. *Norsk Geologisk Tidsskrift*, **47**, 185–9.

Hekinian, R. (1968). Rocks from the mid-oceanic ridge in the Indian Ocean. *Deep-Sea Research*, **15**, 195–213.

Hekinian, R. (1974). Petrology of igneous rocks from leg 22 in the northeastern Indian Ocean. In *Initial Reports of the Deep Sea Drilling Project*, 22, pp. 413–36. United States Government Printing Office, Washington.

Hekinian, R., and Walker, D. (1987). Diversity and spatial zonation of volcanic rocks from the East Pacific Rise near 21° N. *Contributions to Mineralogy and Petrology*, **96**, 265–80.

Hekinian, R., Moore, J. G., and Bryan, W. B. (1976). Volcanic rocks and processes of the Mid-Atlantic Ridge rift valley near 36 ° 49′ N. *Contributions to Mineralogy and Petrology*, **58**, 83–110.

Helland, A. (1873). *Forekomster av Kise i visse skifere i Norge: Christiana*. University of Munchen, Universitets-program.

Henderson, P. (1968). The distribution of phosphorus in the early and middle stages of fractionation of some basic layered intrusions. *Geochimica et Cosmochimica Acta*, **32**, 897–911.

Henderson, P., and Dale, I. M. (1969). The partitioning of selected transition element ions between olivine and groundmass of oceanic basalts. *Chemical Geology*, **5**, 267–74.

Hendry, D. A. F., Chivas, A. R., Reed, F. J. B., and Long, J. V. P. (1981). Geochemical evidence for magmatic fluids in porphyry Cu mineralization. Part II. Ion-probe analysis of Cu contents of mafic minerals, Koloula Igneous Complex. *Contributions to Mineralogy and Petrology*, **78**, 404–12.

Hendry, D. A. F., Chivas, A. R., Long, J. V. P., and Reed, F. J. B. (1985). Chemical differences between minerals from mineralizing and barren intrusions from some North American porphyry Cu deposits. *Contributions to Mineralogy and Petrology*, **89**, 317–29.

Herzig, P. M., and Plüger, W. (1988). Exploration for hydrothermal activity near the Rodriguez Triple Junction, Indian Ocean. In *Seafloor hydrothermal mineralization* (ed. T. J. Barrett and J. L. Jambour). *Canadian Mineralogist*, **26**, 721–36.

Higuchi, H., and Nagasawa, H. (1969). Partitioning of trace elements between rock-forming minerals and the host volcanic rocks. *Earth and Planetary Science Letters*, **2**, 281–7.

Hillebrand, W. F. (1900). Distribution and quantitative occurrence of vanadium and molybdenum in rocks of the United States. *Bulletin of the United States Geological Survey*, **167**, 49–55.

Hitchcock, C. H. (1878). *The geology of the Ammonoosuc mining district* (an extract from the geology of New Hampshire). New Hampshire Geological Survey, Concord, New Hampshire.

Holland, H. D. (1972). Granites, solutions, and base metal deposits. *Economic Geology*, **67**, 281–301.

Holloway, J. R., and Burnham, C. W. (1972). Melting relations of basalt with equilibrium water pressures less than total pressure. *Journal of Petrology*, **13**, 1–29.

Holmes, A. (1969). *Principles of physical geology* (2nd edn). Nelson, London.

Holyk, W. (1956). Mineralization and structural relations in northern New Brunswick. *Northern Miner*, **41**, 27.

Howie, R. A. (1955). The geochemistry of the charnockite series of Madras, India. *Transactions of the Royal Society of Edinburgh*, **62**, 725–68.

Hughes, J. M., and Stoiber, R. E. (1985). Vanadium sublimates from the fumaroles of Izalco Volcano, El Salvador. *Journal of Volcanology and Geothermal Research*, **24**, 282–91.

Humler, E., and Whitechurch, H. (1988). Petrology of basalts from the Central Indian Ridge: estimates of frequencies and fractional volumes of magma injections in a two-layered reservoir. *Earth and Planetary Science Letters*, **88**, 169–81.

Humphris, S. E., and Thompson, G. (1978). Trace element mobility during hydrothermal alteration of oceanic basalts. *Geochimica et Cosmochimica Acta*, **42**, 127–36.

Huntingdon, A. T. (1977). Mount Etna sublimates. In *United Kingdom Research on Mount Etna* (ed. A. T. Huntingdon, G. P. L. Walker, and C. R. Argent), pp. 51–2. The Royal Society, London.

Hutchinson, R. W. (1973). Volcanogenic sulphide deposits and their metallogenic significance. *Economic Geology*, **68**, 1223–46.

Iida, C., Kuno, H., and Yamasaki, K. (1961). Trace elements in minerals and rocks of the Izu–Hakone region, Japan. Part I. Olivine. *Journal of Earth Sciences, Nagoya University*, **9**, 1–13.

Irvine, T. N., and Kushiro, I. (1976). Partitioning of Ni and Mg between olivine and silicate liquids. Carnegie Institution of Washington Year Book, 75, 668–75.

Irving, A. J. (1978). A review of experimental studies of crystal/liquid trace element partitioning. *Geochimica et Cosmochimica Acta*, **42**, 743–70.

Jack, D. J. (1989). Hellyer host rock alteration. Unpublished M.Sc. thesis. University of Tasmania.

Jakes, P., and White, A. J. R. (1972). Major and trace element abundances in volcanic rocks of orogenic areas. *Bulletin of the Geological Society of America*, **83**, 29–40.

James, H. L. (1954). Sedimentary facies of iron formation. *Economic Geology*, **49**, 235–93.

Janssen, J. (1867). Cited by Clarke, F. W. (1924), p. 269.

Janssen, J. (1883). Cited by Clarke, F. W. (1924), p. 269.

Jasmund, K., and Seck, H. A. (1964). Geochemische Untersuchungen an Auswürflingen (Gleesiten) des Laacher-See-Gebietes. *Beiträge zur Mineralogie und Petrographie*, **10**, 275–314.

Kaplan, I. R., Sweeney, R. E., and Nissenbaum, A. (1969). Sulfur isotope studies on Red Sea geothermal brines and sediments. In *Hot brines and recent heavy metal deposits in the Red Sea* (ed. E. T. Degens and D. A. Ross), Springer-Verlag, New York.

Karamata, S. (1969). Lead in sanidines from quartz-latites from Zvecan area, Yugoslavia. *Zavod za Geoloska i Geofizicka Istrazivanja Bulletin* Series A, **27**, 267.

Kashintsev, G. L., and Rudnik, G. B. (1976). New data on basalts of the east Indian Ocean Ridge. *International Geology Review*, 18, 1165–72.

Kay, R. W. (1977). Geochemical constraints on the origin of Aleutian magmas. In *Island arcs, deep sea trenches and back-arc basins*, Maurice Ewing Series 1 (ed. M. Talwani and W. C. Pitman), pp. 229–42. American Geophysical Union, Washington.

Kay, R. W. (1978). Aleutian magnesian andesites:melts from subducted Pacific Ocean crust. *Journal of Volcanology and Geothermal Research*, **4**, 117–32.

Kay, R. W. (1980). Volcanic arc magmas: implications of a melting-mixing model for element recycling in the crust–upper mantle system. *Journal of Geology*, **88**, 497–522.

Kay, R. W., Hubbard, N. J., and Gast, P. W. (1970). Chemical characteristics and origin of oceanic ridge volcanic rocks. *Journal of Geophysical Research*, **75**, 1585–615.

Kay, S. M., and Kay, R. W. (1982). Tectonic controls on tholeiitic and calc-alkaline magmatism in the Aleutian Arc. *Journal of Geophysical Research*, **87**, 4051–72.

Kay, S. M., and Kay, R. W. (1985). Aleutian tholeiitic and calc-alkaline magma series. I: the mafic phenocrysts. *Contributions to Mineralogy and Petrology*, **90**, 276–90.

Kay, S. M., Kay, R. W., Brueckner, H. K., and Rubenstone, J. L. (1983). Tholeiitic Aleutian Arc plutonism: the Finger Bay pluton, Adak, Alaska. *Contributions to Mineralogy and Petrology*, **82**, 99–116.

Kempe, D. R. C. (1973). Basalts from the southern Indian Ocean. *Transactions of the American Geophysical Union*, **54**, 1008–11.

Kimura, K. (1925). The chemical investigation of Japanese minerals containing rarer elements. V. Analyses of fergusonite of Hagata, hegatalite of Hagata, and oyamalite of Oyama, Iyo province. *Japanese Journal of Chemistry*, **2**, 81–5.

Koch, A. (1888). Cited by Clarke, F. W. (1924). p. 589.

Koistinen, T. J. (1981). Structural evolution of an early Proterozoic strata-bound Cu–Co–Zn deposit, Outokumpu, Finland. *Transactions of the Royal Society of Edinburgh: Earth Sciences*, **72**, 115–58.

Korringa, M. K., and Noble, D. C. (1971). Distribution of Sr and Ba between natural felspar and igneous melt. *Earth and Planetary Science Letters*, **11**, 147–51.

Koritnig, S. (1965). Geochemistry of phosphorus–I. The replacement of Si^{4+} by P^{5+} in rock forming silicates. *Geochimica et Cosmochimica Acta*, **29**, 361–71.

Koritnig, S. (1978). Phosphorus. In *Handbook of geochemistry* (ed. K. H. Wedepohl). Springer-Verlag, Berlin.

Kovalenko, V. I., Ryabchivkov, I. D., and Antipin, V. S. (1986). Empirical formulas for the temperature and mineral-composition dependence of strontium and barium distribution coefficients in magmatic rocks. *Geochemistry International*, **23**, 153–70.

Kraume, E., Dahlgrun, F., Ramdohr, P., and Wilke, A. (1955). *Die Erzlager des Rammelsberges bei Goslar*. Niedersächsisches Landesamt Bodenforschung, Monograph 8, Hanover.

Kraus, E. H. (1904). The occurrence of celestite near Syracuse, N. Y., and its relation to the Vermicular Limestones of the Salina epoch. *American Journal of Science*, **18**, 30–39.

Kraus, E. H., and Hunt, W. F. (1906). The origin of sulphur and celestite at Maybee, Michigan. *American Journal of Science*, **21**, 237–44.

Krauskopf, K. B. (1957). The heavy metal content of magmatic vapor at 600 °C. *Economic Geology*, **52**, 786–807.

Kuno, H. (1950). Petrology of Hakone volcano and the adjacent areas, Japan. *Geological Society of America Bulletin*, **61**, 957–1020.

Kushiro, I. (1960). Si–Al relation in clinopyroxenes from igneous rocks. *American Journal of Science*, **258**, 548–54.

Lacroix, A. (1907). Les minéreaux de fumarolles de l'éruption du Vésuve en avril 1906. *Bulletin de la Société Minéralogique de France*, **30**, 219–66.

Landergren, S. (1954). Phosphorus. In *Geochemistry*, Goldschmidt, V. M. (1954), p. 454–67.

Large, R. R. (1977). Chemical evolution and zonation of massive sulfide deposits in volcanic terrains. *Economic Geology*, **72**, 549–72.

Large, R. R. (1990). The gold-rich seafloor massive sulphide deposits of Tasmania. *Geologische Rundschau*, **79**, 265–78.

Large, R. R. (1992). Australian volcanic-hosted massive sulfide deposits: features, styles, and genetic models. *Economic Geology*, **87**, 471–510.

Leeman, W. P. (1974). Experimental determination of partitioning of divalent cations between olivine and basaltic liquid. Unpublished Ph.D. thesis. University of Oregon.

Leeman, W. P. (1979). Partitioning of Pb between volcanic glass and coexisting sanidine and plagioclase feldspar. *Geochimica et Cosmochimica Acta*, **43**, 171–5.

Leeman, W. P., and Lindstrom, D. J. (1978). Partitioning of Ni^{2+} between basaltic and synthetic melts and olivines — an experimental study. *Geochimica et Cosmochimica Acta*, **42**, 801–16.

Leeman, W. P., Ma, M. S., Murali, A. V., and Schmitt, R. A. (1978a). Empirical estimation of magnetite-liquid distribution coefficients for some transition elements. *Contributions to Mineralogy and Petrology*, **65**, 269–72.

Leeman, W. P., Ma, M. S., Murali, A. V., and Schmitt, R. A. (1978b). Empirical estimation of magnetite–liquid distribution coefficients for some transition elements. *Contributions to Mineralogy and Petrology*, **66**, 429.

Le Guern, F., Gerlach, T. M., and Nohl, A. (1982). Field gas chromatograph analyses of gases from a glowing dome at Merapi Volcano, Java, Indonesia, 1977, 1978, 1979. *Journal of Volcanology and Geothermal Research*, **14**, 223–45.

Lemarchand, F., Villemant, B., and Calas, G. (1987). Trace element distribution coefficients in alkaline series. *Geochimica et Cosmochimica Acta*, **51**, 1071–81.

Lepel, E. A., Stefansson, K. M., and Zoller, W. H. (1978). The enrichment of volatile elements in the atmosphere by volcanic activity: Augustine Volcano 1976. *Journal of Geophysical Research*, **83**, 6213–20.

le Roex, A. P., and Dick, H. J. B. (1981). Petrography and geochemistry of basaltic rocks from the Conrad fracture zone on the America–Antarctica Ridge. *Earth and Planetary Science Letters*, **54**, 117–38.

le Roex, A. P., Dick, H. J. B., Erlank, A. J., Reid, A. M., Frey, F. A., and Hart, S. R. (1983). Geochemistry, mineralogy and petrogenesis of lavas erupted along the southwest Indian ridge between the Bouvet triple junction and 11° east. *Journal of Petrology*, **24**, 267–318.

le Roex, A. P., Dick, H. J. B., Reid, A. M., Frey, F. A., Erlank, A. J., and Hart, S. R. (1985). Petrology and geochemistry of basalts from the American–Antarctic Ridge, Southern Ocean: implications for the westward influence of the Bouvet mantle plume. *Contributions to Mineralogy and Petrology*, **90**, 367–80.

le Roex, A. P., Dick, H. J. B., Gulen, L., Reid, A. M., and Erlank, A. J. (1987). Local and regional heterogeneity in MORB from the Mid-Atlantic Ridge between 54.5° S and 51° S: evidence for geochemical enrichment. *Geochimica et Cosmochimica Acta*, **51**, 541–55.

Lesher, C. M., and Campbell, I. H. (1987). Trace element geochemistry of ore-associated and barren, felsic metavolcanic rocks in the Superior province, Canada: Reply to discussion by Whitford and Cameron (1987). *Canadian Journal of Earth Sciences*, **24**, 1500–1.

Lesher, C. M., and Campbell, I. H. (1990). Trace element geochemistry of felsic metavolcanic rocks associated with massive polymetallic base-metal sulphide deposits. *Australian National University, Research School of Earth Sciences Annual Report*, 1990, p. 90.

Lesher, C. M., Goodwin, A. M., Campbell, I. H., and Gorton, M. P. (1986). Trace-element geochemistry of ore-associated and barren, felsic metavolcanic rocks in the Superior Province, Canada. *Canadian Journal of Earth Sciences*, **23**, 222–37.

Libbey, W. (1894). Gases in Kilauea. *American Journal of Science*, **47**, 371–2.

Lindstrom, D. J. (1976). Experimental study of the partitioning of the transition metals between clinopyroxene and coexisting silicate liquids. Unpublished Ph.D. thesis. University of Oregon.

Lindstrom, D. J., and Weill, D. F. (1978). Partitioning of transition metals between diopside and coexisting silicate liquids. *Geochimica et Cosmochimica Acta*, **42**, 817–31.

Longhi, J., Walker, D., and Hayes, J. F. (1978). The distribution of Fe and Mg between olivine and lunar basaltic liquids. *Geochimica et Cosmochimica Acta*, **42**, 1545–58.

Lowenstern, J. B., Mahood, G. A., Rivers, M. L., and Sutton, S. R. (1991). Evidence for extreme partitioning of Cu into a magmatic vapor phase. *Science*, **252**, 1402–5.

Lusk, J. (1976a). A possible volcanic exhalative origin for lenticular nickel sulfide deposits of volcanic association, with special reference to those in Western Australia. *Canadian Journal of Earth Sciences*, **13**, 451–8.

Lusk, J. (1976b). A possible volcanic–exhalative origin for lenticular nickel sulfide deposits of volcanic association, with special reference to those of Western Australia: Reply. *Canadian Journal of Earth Sciences*, **13**, 1651–3.

Maaløe, S. (1979). Compositional range of primary tholeiitic magmas evaluated from major element trends. *Lithos*, **12**, 59–72.

McClaine, L. A., Allen, R. V., McConnell, R. K., and Surprenant, N. F. (1968). Volcanic smoke clouds. *Journal of Geophysical Research*, **73**, 5235–46.

McKay, G. A., and Weill, D. F. (1976). Petrogenesis of KREEP. In *Proceedings of the Seventh Lunar Science Conference*, 2427–47.

McKay, G. A., and Weill, D. F. (1977). KREEP petrogenesis revisted. In *Proceedings of the Eighth Lunar Science Conference*, 2339–55.

Macdonald, G. A. (1968). Composition and origin of Hawaiian lavas. *Geological Society of America Memoir* **115**, 477–522.

Macdonald, G. A., and Katsura, T. (1961). Variations in the lava of the 1959 eruption in Kilauea Iki. *Pacific Science*, **15**, 358–69.

Macdonald, G. A., and Katsura, T. (1964). Chemical composition of Hawaiian lavas. *Journal of Petrology*, **5**, 82–133.

Machatschki, F. (1931). In Koritnig, S. (1965), p. 361.

Maclaurin, J. S. (1911). Occurrence of pentathionic acid in natural water. *Proceedings of the Chemical Society*, **27**, 10–12.

Mallet, J. W. (1887). Cited by Clarke, F. W. (1924), p. 274.

Malpas, J. (1978). Magma generation in the upper mantle, field evidence from ophiolite suites, and application to the generation of oceanic lithosphere. *Philosophical Transactions of the Royal Society, London, A.*, **288**, 527–46.

Manheim, F. T. (1978). Molybdenum. In *Handbook of geochemistry* (ed. K. H. Wedepohl). Springer-Verlag, Berlin.

Manning, D. A. C. (1984). Volatile control of tungsten partitioning in granitic melt–vapour systems. *Transactions of the Institution of Mining and Metallurgy (Section B: Applied Earth Science)*, **93**, B185–9.

Manning, D. A. C., and Henderson, P. (1984). The behaviour of tungsten in granitic melt-vapour systems. *Contributions to Mineralogy and Petrology*, **86**, 286–93.

Manning, D. A. C., and Pichavant, M. (1988). Volatiles and their bearing on the behaviour of metals in granitic systems. In *Recent advances in the geology of granite-related mineral deposits* (ed. R. P. Taylor and D. F. Strong), pp. 13–24. Canadian Institute of Mining and Metallurgy Special Volume 39.

Marshall, P. (1912). Presidential Address to the Australasian Association for the Advancement of Science. *Reports of the Australasian Association for the Advancement of Science*, **13**, 90–9.

Martin, R. F., and Piwinski, A. J. (1969). Experimental data bearing on the movement of iron in an aqueous vapour. *Economic Geology*, **64**, 798–803.

Martin, R. L. (1976). Coordination, topology and structure in transition metal oxides. *Journal and Proceedings, Royal Society of New South Wales*, **109**, 137–50.

Mason, B. (1966). *Principles of geochemistry* (3rd edn). Wiley, New York.

Mason, B., and Berggren, Th. (1941). A phosphate bearing spessatite garnet from Wodgina, Western Australia. *Geologiska Föreningens i Stockholm Förhandlingar*, **63**, 413–18.

Mercy, E., and O'Hara, M. J. (1967). Distribution of Mn, Cr, Ti, and Ni in co-existing minerals of ultramafic rocks. *Geochimica et Cosmochimica Acta*, **31**, 2331–42.

Miller, A. R. (1964). Highest salinity in the world ocean? *Nature*, **203**, 590.

Miller, A. R. (1969). Atlantis II account. In *Hot brines and recent heavy metal deposits in the Red Sea* (ed. E. T. Degens and D. A. Ross), Springer-Verlag, New York.

Miller, A. R., Densmore, C. D., Degen, E. T., Hathaway, J. C., Manheim, F. T., McFarlin, P. F., Pocklington, R., and Jokela, A. (1966). Hot brines and recent iron deposits in deeps of the Red Sea. *Geochimica et Cosmochimica Acta*, **30**, 341–59.

Miller, W. H. (1842). On the specific gravity of sulphuret of nickel. *London, Edinburgh and Dublin Philosophical Magazine and Journal of Science*, **20**, 378–9.

Miyashiro, A. (1973). The Troodos ophiolitic complex was probably formed in an island arc. *Earth and Planetary Science Letters*, **19**, 218–24.

Moore, J. G., and Calk, L. (1971). Sulfide spherules in vesicles of dredged pillow basalt. *American Mineralogist*, **56**, 476–488.

Moore, J. G., and Evans, B. W. (1967). The role of olivine in the crystallization of the prehistoric Makaopuhi tholeiitic lava lake Hawaii. *Contributions to Mineralogy and Petrology*, **15**, 202–3.

Moore, R. K., and White, W. B. (1971). Intervalence electron transfer effects in the spectra of the melanite garnets. *American Mineralogist*, **56**, 826–49.

Mottle, M. J., and Holland, H. D. (1978). Chemical exchange during hydrothermal alteration of basalt by sea-water–I. Experimental results for major and minor components of seawater. *Geochimica et Cosmochimica Acta*, **42**, 1103–15.

Moxham, R. L. (1960). Minor element distribution in some metamorphic pyroxenes. *Canadian Mineralogist*, **6**, 522–45.

Moxham, R. L. (1965). Distribution of minor elements in coexisting hornblendes and biotites. *Canadian Mineralogist*, **8**, 204–40.

Mroz, E. J., and Zoller, W. H. (1975). Composition of atmospheric particulate matter from the eruption of Heimaey, Iceland. *Science*, **190**, 461–3.

Muir, I. D., and Tilley, C. E. (1964). Basalts from the northern part of the Rift zone of the Mid-Atlantic Ridge. *Journal of Petrology*, **5**, 409–34.

Murata, K. J. (1960). Occurrence of CuCl emission in volcanic flames. *American Journal of Science*, **258**, 769–772.

Murata, K. J., and Richter, D. H. (1961). Magmatic differentiation in the Uwekehuna laccolith, Kilauea caldera, Hawaii. *Journal of Petrology*, **2**, 424–37.

Murata, K. J., and Richter, D. H. (1966a). The settling of olivine in Kilauean magmas as shown by lavas of the 1959 eruption. *American Journal of Science*, **264**, 194–203.

Murata, K. J., and Richter, D. H. (1966b). *Chemistry of the lavas of the 1959–60 eruption of Kilauea volcano, Hawaii*. U.S. Geological Survey Professional Paper No. 537A.

Mysen, B. O. (1978). Experimental determination of nickel partition coefficients between liquid, pargasite, and garnet peridotite minerals and concentration limits of behaviour according to Henry's Law at high pressure and temperature. *American Journal of Science*, **278**, 217–43.

Mysen, B. O., and Kushiro, I. (1976). Partitioning of iron, nickel and magnesium between metal, oxide, and silicates in the Allende meteorite as a function of f_{O_2}. *Carnegie Institution of Washington Year Book*, **75**, 678–84.

Naboko, S. I. (1959). Volcanic exhalations and products of their reactions as exemplified by Kamchatka-Kuriles volcanoes. *Bulletin Vulcanologique*, **20**, 121–36.

Nagasawa, H., and Schnetzler, C. C. (1971). Partitioning of rare earth, alkali and alkaline earth elements between phenocrysts and acidic igneous magma. *Geochimica et Cosmochimica Acta*, **35**, 953–68.

Naughton, J. J., Lewis, V. A., Hammond, D., and Nishimoto, D. (1974). The chemistry of sublimates collected directly from lava fountains at Kilauea Volcano, Hawaii. *Geochimica et Cosmochimica Acta*, **38**, 1679–90.

Naughton, J. J., Lewis, V. Thomas, D., and Finlayson, J. B. (1975). Fume compositions found at various stages of activity at Kilauea volcano, Hawaii. *Journal of Geophysical Research*, **80**, 2963–6.

Neumann, E.-R. (1974). The distribution of Mn^{2+} and Fe^{2+} between ilmenites and magnetites in igneous rocks. *American Journal of Science*, **274**, 1074–88.

Nockolds, S. R., and Mitchell, R. L. (1948). The geochemistry of some Caledonian plutonic rocks: a study in the relationship between the major and trace elements of igneous rocks and their minerals. *Transactions of the Royal Society of Edinburgh*, **61**, 533–75.

Noddack, I., and Noddack, W. (1931). Cited by Goldschmidt, V. M. (1954), p. 491.

Noddack, I., and Noddack, W. (1934). Cited by Goldschmidt, V. M. (1954). p. 260.

Noll, A. (1934). Cited by Goldschmidt, V. M. (1954), p. 243.

Oana, S. (1962). Volcanic gases and sublimates from Showashinzan. *Bulletin Vulcanologique*, **24**, 49–51.

Oftedahl, C. (1958). A theory of exhalative–sedimentary ores. *Geologiska Föreningens Stockholm Förhandlingar*, **80**, 1–19.

O'Hara, M. J. (1977). Geochemical evolution during fractional crystallization of a periodically refilled magma chamber. *Nature*, **266**, 503–07.

O'Hara, M. J., and Mathews, R. E. (1981). Geochemical evolution in an advancing, periodically replenished, periodically tapped, continuously fractionated magma chamber. *Journal of the Geological Society of London*, **138**, 237–77.

Ohashi, R. (1920). On the origin of the kuroko of the Kosaka copper mine, northern Japan. *Akita Mining College Journal*, **1**, 11–27.

Ohmoto, H., and Rye, R. O. (1974). Hydrogen and oxygen isotopic compositions of fluid inclusions in the Kuroko deposits, Japan. *Economic Geology*, **69**, 947–53.

Onuma, N., Higuchi, H., Wakita, H., and Nagasawa, H. (1968). Trace element partition between two pyroxenes and the host lava. *Earth and Planetary Science Letters*, **5**, 47–51.

Osborne, E. F. (1959). Role of oxygen pressure in the crystallization and differentiation of basaltic magma. *American Journal of Science*, **257**, 609–47.

Osborne, E. F. (1962). Reaction series for subalkaline igneous rocks based on different oxygen pressure conditions. *American Mineralogist*, **47**, 211–26.

Osborne, E. F. (1969). The complimentariness of orogenic andesite and alpine periodotite. *Geochemica et Cosmochimica Acta*, **33**, 307–24.

Óskarsson, N. (1981). The chemistry of Icelandic lava incrustations and the latest stages of degassing. *Journal of volcanology and geothermal research*, **10**, 93–111.

Pan, Y., and Fleet, M. E. (1989). Cr-rich calc-silicates from the Hemlo area, Ontario. *Canadian Mineralogist*, **27**, 565–77.

Pan, Y., and Fleet, M. E. (1991). Mineral chemistry and geochemistry of vanadian silicates in the Hemlo gold deposit, Ontario, Canada. *Contributions to Mineralogy and Petrology*, **109**, 511–25.

Pan, Y., and Fleet, M. E. (1992). Calc-silicate alteration in the Hemlo gold deposit, Ontario: mineral assemblages, P–T–X constraints, and significance. *Economic Geology*, **87**, 1104–20.

Partington, J. R. (1943). *A text-book of inorganic chemistry* (5th edn). Macmillan, London.

Paster, T. P., Schauwecker, D. S., and Haskin, L. A. (1974). The behaviour of some trace elements during solidification of the Skaergaard layered series. *Geochimica et Cosmochimica Acta*, **38**, 1549–77.

Patterson, E. M. (1952). A petrochemical study of Tertiary lavas of north-east Ireland. *Geochimica et Cosmochimica Acta*, **2**, 283–99.

Patterson, E. M., and Swaine, D. J. (1955). A petrochemical study of Tertiary tholeiitic basalts: the Middle Lavas of the Antrim Plateau. *Geochimica et Cosmochimica Acta*, **8**, 173–81.

Pearce, J. A., and Norry, M. J. (1979). Petrogenetic implications of Ti, Zr, Y, and Nb variations in volcanic rocks. *Contributions to Mineralogy and Petrology*, **69**, 33–47.

Peltola, E. (1978). Origin of Precambrian copper sulfides of the Outokumpu district, Finland. *Economic Geology*, **73**, 461–77.

Phelan, J. M., Finnegan, D. L., Ballantine, D. S., and Zoller, W. H. (1982). Airbourne aerosol measurements

in the quiescent plume of Mount St Helens: September, 1980. *Geophysical Research Letters*, **9**, 1093–6.
Philpotts, J. A., and Schnetzler, C. C. (1970). Phenocryst–matrix partition coefficients for K, Rb, Sr and Ba, with applications to anorthosite and basalt genesis. *Geochimica et Cosmochimica Acta*, **24**, 307–22.
Plimer, I. R. (1981). Water depth — a critical factor for exhalative ore deposits. *Bureau of Mineral Resources Journal of Australian Geology and Geophysics*, **6**, 293–300.
Plimer, I. R., and Finlow-Bates, T. (1978). Relationship between primary iron sulphide species, sulphur source, depth of formation and age of submarine exhalative sulphide deposits. *Mineralium Deposita*, **13**, 339–410.
Powers, H. A. (1955). Composition and origin of basaltic magma of the Hawaiian Islands. *Geochimica et Cosmochimica Acta*, **7**, 77–107.
Printz, M. (1967). Geochemistry of basaltic rocks: trace elements. In *Basalts, the Poldervaart treatise on rocks of basaltic composition* (ed. H. H. Hess and A. Poldervaart). Interscience Publishers, New York.
Puchelt, H. (1972). Barium. In *Handbook of geochemistry* (ed. K. H. Wedepohl). Springer-Verlag, Berlin.
Ramberg, H., and de Vore, G. (1951). The distribution of Fe^{++} and Mg^{++} in coexisting olivine and pyroxenes. *Journal of Geology*, **59**, 193–210.
Ringwood, A. E. (1955). The principles governing trace element distribution during magmatic crystallization. Part I. The influence of electronegativity. *Geochimica et Cosmochimica Acta*, **7**, 189–202.
Ringwood, A. E. (1956). Melting relationships of Ni–Mg olivines and some geochemical implications. *Geochimica et Cosmochimica Acta*, **10**, 297–303.
Ringwood, A. E. (1970). Petrogenesis of Apollo II basalts and implications for lunar origin. *Journal of Geophysical Research*, **75**, 6453–79.
Ringwood, A. E. (1974). The petrological evolution of island arc systems. *Journal of the Geological Society of London*, **130**, 183–204.
Robertson, J. D. (1894). *Missouri Geological Survey Reports*, Volume VII, p. 480. (See Sandell and Goldich 1943).
Roeder, P. L. (1974). Activity of iron and olivine solubility in basaltic liquids. *Earth and Planetary Science Letters*, **23**, 397–410.
Rogerson, R. J., Hilyard, D. B., Finlayson, E. J., Johnson, R. W., and McKee, C. O. (1989). *The geology and mineral resources of Bougainville and Buka Islands, Papua New Guinea*. Geological Survey of Papua New Guinea, Memoir 16.
Rose, W. I. (1967). Scavenging of volcanic aerosols by volcanic ash: atmospheric and volcanic implications. *Geology*, **5**, 621–4.

Rubey, W. W. (1955). Development of the hydrosphere and atmosphere, with special reference to probable composition of the early atmosphere. In *Crust of the Earth* (ed. A. Poldervarrt), pp. 631–50. Geological Society of America Special Paper 62.
Rutherford, M. J. (1993). Experimental petrology applied to volcanic processes. *Eos*, **74**, 50, 55.
Rye, R. O., and Ohmoto, H. (1974). Sulfur and carbon isotopes and ore genesis: a review. *Economic Geology*, **69**, 826–42.
Sakuyama, M., and Kushiro, I. (1979). Vesiculation of hydrous andesitic melt and transport of alkalis by separated vapor phase. *Contributions to Mineralogy and Petrology*, **71**, 61–6.
Sandell, E. B., and Goldich, S. S. (1943). The rarer metallic constituents of some American igneous rocks. *Journal of Geology*, **51**, 99–115, 167–89.
Sangster, D. F. (1968). Relative sulphur isotope abundances of ancient seas and stratabound sulphide deposits. *Proceedings of the Geological Association of Canada*, **19**, 79–91.
Sasaki, A. (1970). Seawater sulfate as a possible determinant of some stratabound sulfide ores. *Geochemical Journal*, **4**, 41–51.
Sato, H. (1977). Nickel content of basaltic magmas: identification of primary magmas and a measure of the degree of olivine fractionation. *Lithos*, **10**, 113–20.
Sawkins, F. J. (1986). Some thoughts on the genesis of kuroko-type deposits. In *Geology in the real world — the Kingsley Dunham volume* (ed. R. W. Nesbitt and I. Nichol). Institution of Mining and Metallurgy, London.
Sawkins, F. J., and Kowalik, J. (1981). The source of ore metals at Buchans: magmatic versus leaching models. *Geological Association of Canada Special Paper* No. 22, pp. 255–68.
Schröpfer, L. (1968). Titanium. In *Handbook of geochemistry*, ed. K. H. Wedepohl, Springer-Verlag, Berlin.
Schuchert, C., and Dunbar, C.O. (1934). *Stratigraphy of Western Newfoundland. Memoirs*. Geological Society of America, No. 1.
See, J. B., and Rankin, W. J. (1983). Effect of Al_2O_3 and CaO on solubility of copper in silica-unsaturated iron silicate slags at 1300 °C. *Transactions of the Institution of Mining and Metallurgy*, **92**, C9–C13.
Seifert, S., O'Neill, H. St. C., and Brey, G. (1988). The partitioning of Fe, Ni and Co between olivine, metal and basaltic liquid: an experimental and thermodynamic investigation, with application to the composition of the lunar core. *Geochimica et Cosmochimica Acta*, **52**, 603–16.

Seward, T. M. (1971). The distribution of transition elements in the system $CaMgSi_2O_6$–$Na_2Si_2O_5$–H_2O at 1,000 bars pressure. *Chemical Geology*, **7**, 73–95.

Seyfried, W. E., and Janecky, D. R. (1985). Heavy metal and sulfur transport during subcritical and supercritical hydrothermal alteration of basalt: Influence of fluid pressure and basalt composition and crystallinity. *Geochimica et Cosmochimica Acta*, **49**, 2545–60.

Seyfried, W. E., Berndt, M. E., and Seewald, J. S. (1988). Hydrothermal alteration processes at mid-ocean ridges: constraints from diabase alteration experiments, hot spring fluids and composition of the earth's crust. In *Seafloor hydrothermal mineralization* (ed. T.J . Barrett and J. L. Jambour). Canadian Mineralogist, **26**, 787–804.

Shaw, D. M. (1953). The camouflage principle of trace element distribution in magmatic minerals. *Journal of Geology*, **61**, 142–51.

Shaw, D. M. (1970). Trace element fractionation during anatexis. *Geochimica et Cosmochimica Acta*, **34**, 237–43.

Sheppard, S. M. F. (1977). Identification of the origin of ore-forming solutions by the use of stable isotopes. In *Volcanic processes in ore genesis: proceedings of a joint meeting of the Volcanic Studies Group of the Geological Society of London and the Institution of Mining and Metallurgy held in London on 21 and 22 January, 1976*. Institution of Mining and Metallurgy and Geological Society, London.

Shibata, T., and Fox, P. J. (1975). Fractionation of abyssal tholeiites. Samples from the Oceanographer fracture zone (35° N, 35° W), *Earth and Planetary Science Letters*, **27**, 62–72.

Shibata, T., Thompson, G., and Frey, F. A. (1979). Tholeiitic and alkali basalts from the mid-Atlantic ridge at 43° N. *Contributions to Mineralogy and Petrology*, **70**, 127–41.

Shima, M. (1957). A new sublimate containing nickel, found in a fumarole of an active volcano. *Science Research Institute (Tokyo) Journal*, **51**, 11–14.

Shimizu, N. (1974). An experimental study of the partitioning of K, Rb, Cs, Sr and Ba between clinopyroxene and liquid at high pressures. *Geochimica et Cosmochimica Acta*, **38**, 1789–98.

Shiraki, K. (1978). Chromium. In *Handbook of geochemistry* (ed. K. H. Wedepohl). Springer-Verlag, Berlin.

Sidhu, P. S., Gilkes, R. J., and Posner, A. M. (1980). The behaviour of Co, Ni, Zn, Cu, Mn and Cr in magnetite during alteration to maghemite and hematite. *Soil Science Society of America Journal*, **44**, 135–8.

Sidhu, P. S., Gilkes, R. J., and Posner, A. M. (1981). Oxidation and ejection of nickel and zinc from natural and synthetic magnetite. *Soil Science Society of America Journal*, **45**, 641–4.

Sillitoe, R. H. (1972). Formation of certain massive sulphide deposits at sites of sea-floor spreading. *Transactions of the Institution of Mining and Metallurgy*, Section B, **81**, B141–8.

Sillitoe, R. H. (1973). Environments of formation of volcanogenic massive sulfide deposits. *Economic Geology*, **68**, 1321–5.

Silvestri, O. (1967). Cited by Clarke, F. W. (1924), p. 266.

Simkin, T., and Smith, J. V. (1970). Minor element distribution in olivine. *Journal of Geology*, **78**, 304–25.

Simonetti, A., and Bell, K. (1993). Isotopic disequilibrium in clinopyroxenes from nephelinitic lavas, Napak volcano, eastern Uganda. *Geology*, **21**, 243–6.

Snyder, J. K. (1959). Distribution of certain elements in the Duluth complex. *Geochimica et Cosmochimica Acta*, **16**, 243–77.

Spence, C. D., and de Rosen-Spence, A. F. (1975). The place of sulfide mineralization in the volcanic sequence at Noranda, Quebec. *Economic Geology*, **70**, 90–101.

Spooner, E. C. T. (1977). Hydrodynamic model for the origin of the ophiolitic cupriferous pyrite ore deposits of Cyprus. In *Volcanic processes in ore genesis: proceedings of a joint meeting of the Volcanic Studies Group of the Geological Society of London and the Institution of Mining and Metallurgy held in London on 21 and 22 January, 1976*. Institution of Mining and Metallurgy and Geological Society, London.

Spooner, E. C. T., and Fyfe, W. S. (1973). Sub-sea-floor metamorphism, heat and mass transfer. *Contributions to Mineralogy and Petrology*, **42**, 287–304.

Stanton, R. L. (1954). Lower Palaeozoic mineralisation and features of its environment near Bathurst, central western New South Wales. Unpublished Ph.D. thesis. University of Sydney.

Stanton, R. L. (1955*a*). The genetic relation between limestone, volcanic rocks, and certain ore deposits. *Australian Journal of Science*, **17**, 173–5.

Stanton, R. L. (1955*b*). Lower Paleozoic mineralisation near Bathurst, New South Wales. *Economic Geology*, **50**, 681–714.

Stanton, R. L. (1958). Abundances of copper zinc and lead in some sulfide deposits. *Journal of Geology*, **66**, 484–502.

Stanton, R. L. (1960). General features of the conformable 'pyritic' orebodies. Part I — field association. Part II — mineralogy. *Transactions of the Canadian Institute of Mining and Metallurgy*, **63**, 22–36.

Stanton, R. L. (1961). Geological theory and the search for ore. *Mining and Chemical Engineering Review*, **53**, 48–55.

Stanton, R. L. (1967). A numerical approach to the andesite problem. *Koninkle Nederlandse Akademie van Wetenschappen, Proceedings* Series B, 70, 176–216.

Stanton, R. L. (1972). *Ore petrology.* McGraw-Hill, New York.

Stanton, R. L. (1976). Petrochemical studies of ore environment, Broken Hill, New South Wales, Australia: — 1 constitution of 'banded iron formations'. *Transactions of the Institution of Mining and Metallurgy*, Section B, **85**, B 33–46.

Stanton, R. L. (1978). Mineralization in island arcs, with particular reference to the south-west Pacific region. Introductory lecture, William Smith Meeting, Geological Society of London, February 1977. *Proceedings of the Australasian Institute of Mining and Metallurgy*, **268**, 9–19.

Stanton, R. L. (1983). Stratiform ores and metamorphic processes — some thoughts arising from Broken Hill. *Proceedings of conference held at Broken Hill, July 1983*, pp. 11–28. Australasian Institute of Mining and Metallurgy.

Stanton, R. L. (1985). Stratiform ores and geological processes. *Royal Society of New South Wales, Journal and Proceedings*, **118**, 77–100.

Stanton, R. L. (1987). Magmatic evolution and exhalative ores: evidence from the SW Pacific. In *Proceedings of the Pacific Rim Congress 87.* Australasian Institute of Mining and Metallurgy, pp. 591–5.

Stanton, R. L. (1990). Magmatic evolution and the ore type-lava type affiliations of volcanic exhalative ores. In *Geology of the mineral deposits of Australia and Papua New Guinea — the Haddon King Volume* (ed. F. E. Hughes), pp. 101–7. Australasian Institute of Mining and Metalldurgy, Melbourne.

Stanton, R. L. (1991). Understanding volcanic massive sulfides — past, present and future. In *Historical perspectives of genetic concepts and case histories of famous discoveries* (ed. R. W. Hutchinson and R. I. Grauch). Economic Geology Monograph 8, pp. 82–95.

Stanton, R. L., and Bell, J. D. (1963). New Georgia Group — a preliminary geological statement. In *The British Solomon Islands Geological Record*, **2**, 35–6.

Stanton, R. L., and Bell, J. D. (1969). Volcanic and associated rocks of the New Georgia Group, British Solomon Island Protectorate. *Overseas Geology and Mineral Resources*, **10**, 113–45.

Stanton, R. L., and Ramsay, W. R. H. (1980). Exhalative ores, volcanic loss, and the problem of the island arc calc-alkaline series. *Norges Geologiske Undersøkelse*, **360**, 9–57.

Stanton, R. L., and Russell, R. D. (1959). Anomalous leads and the emplacement of lead sulfide ores. *Economic Geology*, **54**, 588–607.

Stanton, R. L., Roberts, W. P. H., and Chant, R. A. (1978). Petrochemical studies of the ore environment at Broken Hill, NSW. 5 — Major element constitution of the lode and its interpretation. *Proceedings of the Australasian Institute of Mining and Metallurgy*, **266**, 51–78.

Steiger, G. (1924). Cited by Clarke, F. W. (1924), p. 641.

Stoiber, R. E., and Dürr, F. (1964). Vanadium in the sublimates, Izalco Volcano, El Salvador. *Geological Society of America Special Paper*, **76**, p. 159.

Stoiber, R. E., and Rose, W. I. (1970). The geochemistry of Central American volcanic gas condensates. *Bulletin of the Geological Society of America*, **81**, 2891–912.

Stoiber, R. E., and Rose, W. I. (1974). Fumarole incrustations at active Central American volcanoes. *Geochimica et Cosmochimica Acta*, **38**, 495–516.

Stueber, A. M. (1969). Abundances of K, Rb, Sr and Sr isotopes in ultramafic rocks and minerals from western North Carolina. *Geochimica et Cosmochimica Acta*, **33**, 543–53.

Stueber, A. M. (1978). Strontium. In *Handbook of geochemistry* (ed. K. H. Wedepohl). Springer-Verlag, Berlin.

Stueber, A. M., and Goles, G. G. (1967). Abundances of Na, Mn, Cr, Sc, and Co in ultramafic rocks. *Geochimica et Cosmochimica Acta*, **31**, 75–93.

Stueber, A. M., and Ikramuddin, M. (1974). Rubidium, strontium and the isotopic constitution of strontium in ultramafic nodule minerals and host basalts. *Geochimica et Cosmochimica Acta*, **38**, 207–216.

Sun, S-S., Nesbitt, R. W., and Sharaskin, A. Y. (1979). Geochemical characteristics of mid-ocean ridge basalts. *Earth and Planetary Science Letters*, **44**, 119–38.

Symonds, R. B., Rose, W. I., Reed, M. H., Lichte, F. E, and Finnegan, D. L. (1987). Volatilization, transport and sublimation of metallic and non-metallic elements in high temperature gases at Merapi Volcano, Indonesia. *Geochimica et Cosmochimica Acta*, **51**, 2083–101.

Symonds, R. B., Rose, W. I., Gerlach, T. M., Briggs, P. H., and Harmon, R. S. (1990). Evaluation of gases, condensates, and SO_2 emissions from Augustine volcano, Alaska: the degassing of a Cl-rich volcanic system. *Bulletin Volcanologique*, **52**, 355–74.

Symonds, R. B., Reed, M. H., and Rose, W. I. (1992). Origin, speciation, and fluxes of trace element gases at Augustine Volcano, Alaska: Insights into magma degassing and fumarolic processes. *Geochimica et Cosmochimica Acta*, **56**, 633–57.

Taborszky, F. K. (1962). Geochemie des Apatits in Tiefengesteinen am Beispiel des Odenwaldes. *Beiträge zur Mineralogie und Petrographie*, **8**, 354–92.

Tauson, L. V. (1965). Factors in the distribution of trace elements during the crystallization of magmas. *Physics and chemistry of the Earth*, **6**, 219.

Taylor, G. R. (1977). The ophiolite terrain and volcanogenic mineralisation of the Florida Islands, Solomon Islands. Unpublished Ph.D. thesis. University of New England.

Taylor, S. R. (1964). Abundance of chemical elements in the continental crust: a new table. *Geochimica et Cosmochimica Acta*, **28**, 1273–85.

Taylor, S. R. (1965). The application of trace element data to problems in petrology. In *Physics and chemistry of the Earth*, (ed. L. H. Ahrens, K. Rankama, and S. K. Runcorn) vol. 6, p. 133.

Taylor, S. R., Capp, A. C., and Graham, A. L. (1969a). Trace element abundances in andesites: Saipan, Bougainville and Fiji. *Contributions to Mineralogy and Petrology*, **23**, 1–26.

Taylor, S. R., Kaye, M., White, A. J. R., Duncan, A. R., and Ewart, A. (1969b). Genetic significance of Co, Cr, Ni, Sc, and V content of andesites. *Geochimica et Cosmochimica Acta*, **33**, 275–86.

Thode, H. G., and Monster, J. (1965). Sulphur isotope geochemistry of petroleum, evaporites, and ancient seas. *American Association of Petroleum Geologists Memoir* **4**, pp. 367–77.

Thomas, H. H., and MacAlister, D. A. (1909). *The geology of ore deposits*. Edward Arnold, London.

Tiller, K. G. (1959). The distribution of trace elements during differentiation of the Mt Wellington dolerite sill. *Papers and Proceedings of the Royal Society of Tasmania*, **93**, 153.

Tilley, C. E. (1950). Some aspects of magmatic evolution. *Quarterly Journal of the Geological Society of London*, **106**, 37–61.

Tillmanns, E. (1972). Barium hexatitanate, $BaTi_6O_{13}$. *Crystal Structure Communications*, **1**, 1.

Tinoco, M. (1885). Cited by Wilson, I. F. (1955), p. 82.

Treloar, P. J. (1987a). The Cr-minerals of Outokumpu — their chemistry and significance. *Journal of Petrology*, **28**, 867–86.

Treloar, P. J. (1987b). Chromian muscovites and epidotes from Outokumpu. *Mineralogical Magazine*, **51**, 593–9.

Treloar, P. J., Koistinen, T. J., and Bowes, D. R. (1981). Metamorphic development of cordierite–amphibole rocks and mica schists in the vicinity of the Outokumpu ore deposit, Finland. *Transactions of the Royal Society of Edinburgh: Earth Sciences*, **72**, 201–15.

Urabe, T. (1987). Kuroko deposit modelling based on magmatic hydrothermal theory. *Mining Geology*, **37**, 159–76.

Urabe, T., and Sato, T. (1978). Kuroko deposits of the Kosaka mine, northeast Honshu, Japan — products of submarine hot springs on Miocene sea floor. *Economic Geology*, **73**, 161–79.

Van Hise, C. R., and Leith, C. K. (1911). *The geology of the Lake Superior region*. United States Geological Survey Monograph No. 52.

Vedder, J. G., and Colwell, J. B. (1989). Introduction to the geology and offshore resources of the central and western Solomon Islands and eastern Papua New Guinea. In *Circum-Pacific Council for Energy and Mineral Resources Earth Science Series*, Vol. 12, pp. 1–6.

Verhoogen, J. (1962). Distribution of titanium between silicates and oxides in igneous rocks. *American Journal of Science*, **260**, 211–20.

Viljoen, M. J., and Viljoen, R. P. (1969). Evidence of the existence of a mobile extrusive magma. In Upper Mantle Project. *Geological Society of South Africa Special Publication* No. 2, pp. 87–112.

Villemant, B., Jaffrezic, H., Joron, J-L., and Treuil, M. (1981). Distribution coefficients of major and trace elements; fractional crystallization in the alkali basalt series of Chaîne de Puys (Massif Central, France). *Geochimica et Cosmochimica Acta*, **45**, 1997–2016.

Vincent, E. A. (1974). Trace elements in minerals from the Skaergaard gabbroic intrusion, East Greenland: a general summary. *Revue de la Haute Auvergne*, **44**, 441–70.

Vogt, J. H. L. (1889). Cited by Stanton, R. L. (1984), p. 1432.

Vogt, J. H. L. (1890). Cited by Stanton, R. L. (1984), p. 1432.

Vogt, J. H. L. (1923). Nickel in igneous rocks. *Economic Geology*, **18**, 307–53.

Vogt, J. H. L. (1931a). Cited by Goldschmidt, V. M. (1954), p. 457.

Vogt, J. H. L. (1931b). Cited by Goldschmidt, V. M. (1954), p. 647.

Von Engelhardt, W. (1936). Die Geochemie des Barium. *Chemie der Erde*, **10**, 187–246.

Von Hevesy, G., Alexander, E., and Würstlin, K. (1930). Cited by Goldschmidt, V. M. (1954), p. 485.

Von Hevesy, G., Merkel, A., and Würstlin, K. (1934). Cited by Goldschmidt, V.M. (1954), p. 621.

Vossler, T., Anderson, D. L., Aras, N. K., Phelan, J. M., and Zoller, W. H. (1981). Trace element composition of the Mount St Helens plume: stratospheric samples from the 18 May eruption. *Science*, **211**, 827–30.

Wager, L. R. (1960). The major element variation of the layered series of the Skaergaard intrusion and a re-estimation of the average composition of the hidden layered series and of successive residual magmas. *Journal of Petrology*, **1**, 364–98.

Wager, L. R. (1963). The mechanism of adcumulus growth in the layered series of the Skaergaard intrusion. *Mineralogical Society of America Special Paper 1, Symposium on Layered Intrusions*, pp. 1–9.

Wager, L. R., and Brown, G. M. (1968). *Layered igneous rocks*. Oliver and Boyd, Edinburgh.

Wager, L. R., and Mitchell, R. L. (1951). The distribution of trace elements during strong fractionation of basic magma — a further study of the Skaergaard intrusion, east Greenland. *Geochimica et Cosmochimica Acta*, **1**, 129–208.

Wager, L. R., and Mitchell, R. L. (1953). Trace elements in a suite of Hawaiian lavas. *Geochimica et Cosmochimica Acta*, **3**, 217–23.

Wager, L. R., Vincent, E.A., and Smales, A. A. (1957). Sulphides in the Skaergaard Intrusion. *Economic Geology*, **52**, 855–903.

Warden, A. J. (1970). Evolution of Aoba caldera volcano, New Hebrides. *Bulletin Volcanogique*, **34**, 107–40.

Washington, H. S., and Merwin, H. E. (1921). Aphthitalite from Kilauea. *American Mineralogist*, **6**, 121–5.

Watson, E. B. (1977). Partitioning of manganese between forsterite and liquid. *Geochimica et Cosmochimica Acta*, **41**, 1363–74.

Wedepohl, K. H. (1953). Untersuchungen zur Geochemie des Zincs. *Geochimica et Cosmochimica Acta*, **3**, 93–142.

Wedepohl, K. H. (1956). Untersuchungen zur Geochemie des Bleis. *Geochimica et Cosmochimica Acta*, **10**, 69–148.

Wedepohl, K. H. (1974a). Copper. In *Handbook of geochemistry* (ed. K. H. Wedepohl). Springer-Verlag, Berlin.

Wedepohl, K. H. (1974b). Lead. In *Handbook of geochemistry* (ed. K. H. Wedepohl). Springer-Verlag, Berlin.

Wedepohl, K. H. (1975). The contribution of chemical data to assumptions about the origin of magmas from the mantle. *Fortschritte der Mineralogie*, **52**, 141.

Wedepohl, K. H. (1978a). Zinc. In *Handbook of geochemistry* (ed. K. H. Wedepohl). Springer-Verlag, Berlin.

Wedepohl, K. H. (1978b). Manganese. In *Handbook of geochemistry* (ed. K. H. Wedepohl). Springer-Verlag, Berlin.

Weed, W. H. (1902). Copper deposits of the Appalachian states. *Bulletin of the United States Geological Survey*, **213**, 181–5.

Weed, W. H. (1911). Copper deposits of the Appalachian states. *Bulletin of the United States Geological Survey*, 455.

Wells, R. C. (1924). Cited by Clarke, F. W. (1924), p. 641.

Westrich, H. R., and Gerlach, T. M. (1992). Magmatic gas source for the stratospheric SO_2 cloud from the June 15, 1991, eruption of Mount Pinatubo. *Geology*, **20**, 867–70.

White, D. E., and Waring, G. A. (1963). Data of geochemistry: Volcanic emanations. *United States Geological Survey Professional Paper 440-K*, 1–29.

Whitford, D. J., and Cameron, M. A. (1987). Trace element geochemistry of ore-associated and barren, felsic metavolcanic rocks in the Superior province, Canada: Discussion of paper by Lesher *et al.* (1986). *Canadian Journal of Earth Sciences*, **24**, 1498–1500.

Whitford, D. J., Korsch, K. J., and Solomon, M. (1992). Strontium isotope studies of barites: implications for the origin of base metal mineralization in Tasmania. *Economic Geology*, **87**, 953–9.

Whitney, J. A. (1989). Origin and evolution of silicic magmas. In *Ore deposition associated with magmas*, Reviews in Economic Geology, Vol. 4 (ed. J. A. Whitney and A. J. Naldrett), pp. 183–201.

Whitney, J. D. (1854). *Metallic wealth of the United States, described and compared with that of other countries*. Lippincott, Grambo & Co., Philadelphia.

Wilkinson, J. F. G. (1985). Undepleted mantle composition beneath Hawaii. *Earth and Planetary Science Letters*, **75**, 129–38.

Wilkinson, J. F. G., and Hensel, H. D. (1988). The petrology of some picrites from Mauna Loa and Kilauea volcanoes, Hawaii. *Contributions to Mineralogy and Petrology*, **98**, 326–45.

Wilson, I. F. (1955). Geology and mineral deposits of the Boleo copper district, Baja California. *United States Geological Survey Professional Paper* No. 273.

Winchell, N. H., and Winchell, H. V. (1889). On a possible chemical origin of the iron ores of the Keewatin in Minnesota. *Proceedings of the American Association for the Advancement of Science, Toronto Meeting, 2nd September 1889.*

Winchell, N. H., and Winchell, H. V. (1891). Iron ores of Minnesota. *Minnesota Geological Survey Bulletin* No. 6.

Yoder, H. S., and Tilley, C. E. (1962). Origin of basalt magmas: an experimental study of natural and synthetic rock systems. *Journal of Petrology*, **3**, 342–532.

Zambonini, F. (1910). Cited by Fenner, C. N. (1933), p. 71.

Zies, E. G. (1924). Hot springs of the Valley of Ten Thousand Smokes. *Journal of Geology*, **32**, 303–310.

Zies, E. G. (1929). *The Valley of Ten Thousand Smokes. 1. The fumarolic incrustations and their bearing on ore deposition. 2. The acid gases contributed to the sea during volcanic activity.* Geophysical Laboratory of the

Carnegic Institution of Washington, Paper No. 693. National Geographic Society: contributed technical papers. Katmai Series, Number 4.

Zies, E. G. (1938). The concentration of the less familiar elements through igneous and related activity. *American Journal of Science*, **35**, 385–404.

Zindler, A., and Hart, S. (1986). Chemical geodynamics. In *Annual review of earth and planetary sciences* (ed. G.W. Wetherill), **14**, 493–571.

Zoller, W. H., Parrington, J. R., and Kotra, J. M. P. (1983). Iridium enrichment in airbourne particles from Kilauea Volcano: January, 1983. *Science*, **222**, 1118–21.

Author index

Allard, P. 313
Allen, E.T. 307
Anderson, A. T. 142
Andriambololona, R. 322
Ansted, D.T. 11
Arculus, R.J. 282
Arndt, N.T. 256
Arnold, M. 318

Baas-Becking, L.G.M. 17
Baker, P.E. 336
Bauerman, H. 120
Baumgart, I.L. 316
Bell, J.D. 5, 30, 164, 327, 335
Bender, J.F. 163, 213–14
Bergametti, G. 319
Bergeat, A. 306
Berggren, Th. 138
Berlin, R. 114
Bernier, L.R. 170
Berry, L.G. 65
Beyschlag, F. 307
Bischoff, J.L. 20
Blaise, B. 347, 352
Borishenko, L.F. 347
Bostrom, K. 13, 106
Bougault, H. 66–7, 87, 89, 163, 193–4, 213–14, 237–8, 256
Bouglise, G. de la 11
Bowen, N.L. 2, 5, 18, 49, 330, 336
Brooks, C. 127
Brown, G.M. 55, 73, 90, 224, 280–2, 287, 291
Browne, W.R. 364
Bryan, W.B. 59, 62, 163
Buat-Ménard, P. 318–19
Bunsen, R.W. 305
Burnham, C.W. 321, 337, 362
Burns, R.G. 249, 280
Burns, V.M. 249, 280

Cadle, R.D. 310
Calk, L. 310
Campbell, I.H. 23
Candela, P.A. 54, 322–3, 362
Carmichael, I.S.E. 49, 114, 291

Carmichael, J. 59, 97
Carr, M.H. 229, 231
Carr, M.J. 292
Carstens, C.W. 12
Clark, L.A. 13–15
Clarke, D.B. 331–2
Clarke, F.W. 54, 155, 169, 183, 204, 228, 244, 249, 305
Colwell, J.B. 5
Compston, W. 327
Conway, E.J. 135
Cooper, J.A. 19
Corliss, J.B. 19–20, 24
Cornwall, H.R. 59, 250
Correns, C.W. 158
Cotton, F.A. 58
Cox, K.G. 335
Craig, H. 19
Cumenge, E. 11
Curtis, C.D. 280
Cuturic, N. 97

Dale, I.M. 237, 255
Daly, R.A. 49
Davis, K.E. 331
Day, A.L. 309
de Beaumont, E. 1–2, 10, 18, 305, 308
de la Beche, H. 1, 10–11, 18
de Rosen-Spence, A.F. 24
Des Cloiseaux, N. 244
Deville, C. Sainte-Claire 305
de Vore, G. 249
Dickson, F.W. 20
Dissanayake, C.B. 79
Doe, B.R. 97, 101
Dostal, J. 66–7, 87, 89, 194, 256
Drake, M.J. 114
Dudas, M.J. 215, 238
Duke, J.M. 193, 213, 214, 237–8, 256
Dunn, T. 213, 214
Dupuy, C. 322
Dürr, F. 317
Dymond, J.R. 20

Eggins, S.M. 329
Ehrenberg, H. 12–13

Eilenberg, S. 292
Elthon, D. 331
Embley, R.W. 347, 352, 357
Emsley, J. 97, 155, 169, 245, 264, 276
Engelhardt, W. von 106
Evans, B.W. 330
Ewart, A. 59, 62, 114, 126–7, 187, 250, 256, 291

Faure, G. 120, 132
Fenner, C.N. 2, 5, 18, 307, 309, 318, 324
Finlow-Bates, T. 16
Fischer, K. 122
Fleet, M.E. 170
Foshag, W.F. 309
Foster, C. le N. 120
Fouqué, F. 306
Frey, F.A. 184
Fuchs, E. 11
Fukuchi, N. 11
Fyfe, W.S. 20, 59

Galkin, X. 138
Garlick, G.D. 20
Gast, P.W. 120, 132
Geikie, A. 304, 306
Gill, J. 336
Gill, J.B. 327
Gmelin, L. 143, 180
Goldich, S.S. 54–5, 69, 73, 75–6, 78, 87, 97, 228, 244–5, 247
Goldschmidt, V.M. 5, 56, 73, 75–6, 78, 97, 106, 111, 120, 135, 137–8, 155, 169, 175, 183–4, 204, 206, 217, 228, 244–5, 272, 280
Goles, G.G. 228
Goodfellow, W.D. 347, 352
Gorton, M.P. 77, 111, 329
Green, D.H. 337
Green, T.H. 332, 337
Greene, H.G. 329
Greenland, L.P. 142
Griggs, R.F. 307
Grout, F.F. 54
Grove, T.L. 163
Gunn, B.M. 87, 255, 330

Häkli, T.A. 255–6
Hamilton, W.M. 316
Hannington, M.D. 352
Harris, D.C. 170
Hart, S. 304
Hart, S.R. 127, 331
Hartmann, P. 157
Haslam, H.W. 114
Heaton, T.H.E. 20–1
Hecht, J. 292
Hedenquist, J.W. 323
Hegeman, F. 12–13
Heier, K.S. 101
Hekinian, R. 66–7, 87, 89, 163, 193–4, 213–14, 237–8, 256
Helland, A. 11
Henderson, C.M.B. 114
Henderson, P. 137, 140, 142, 237, 255, 321
Hendry, D.A.F. 66–7
Hensel, H.D. 331–3
Herzig, P.M. 351
Hevesy, G. von 155, 169, 204
Higuchi, H. 195, 215
Hillebrand, W.F. 169
Hise, C.R. van 12
Hitchcock, C.H. 11
Holland, H.D. 59, 321–2, 362
Holloway, J.R. 337
Holyk, W. 13, 24
Howie, R.A. 101
Hughes, J.M. 317, 347
Hunt, W.F. 120
Hutchinson, R.W. 8, 14–15

Iida, C. 59
Irvine, T.N. 256
Irving, A.J. 66

Jack, D.J. 346
Jakes, P. 184
James, H.L. 15
Janecky, D.R. 349
Janssen, J. 309
Jasmund, K. 79

Kaplan, I.R. 17, 19
Karamata, S. 97
Katsura, T. 332
Kimura, K. 138
Koch, A. 120
Koistinen, T.J. 244
Koritnig, S. 137–9, 142
Korringa, M.K. 127
Kovalenko, V.I. 114, 126–7
Kowalik, J. 22
Kraume, E. 8, 12–13, 16, 349
Kraus, E.H. 120
Krauskopf, K.B. 320
Kuno, H. 336–7
Kushiro, I. 158, 256

Lacroix, A. 307
Landergren, S. 143
Large, D.E. 16
Large, R.R. 15–16, 348
Leblanc, F. 305
Leeman, W.P. 97, 101, 163, 256
Le Guern, F. 313
Leith, C.K. 12
Lemarchand, F. 237–8
Lepel, E.A. 315
le Roex, A.P. 184
Lesher, C.M. 23
Libbey, W. 306, 309
Lindgren, W. 7
Lindstrom, D.J. 163, 215, 238, 256
Longhi, J. 193, 213
Lowenstern, J.B. 292
Lusk, J. 244

Maaløe, S. 331–2
MacAlister, D.A. 11
McClaine, L.A. 311, 346
Macdonald, G.A. 59, 97, 332
Macfarlane, A. 329
Machatschki, F. 138
McKay, G.A. 127
Mallet, J.W. 306
Malpas, J. 331
Manheim, F.T. 276
Manning, D.A.C. 321–2, 362
Marshall, P. 336
Martin, R.F. 321, 350, 356
Martin, R.L. 125
Mason, B. 65, 138, 169
Mathews, R.E. 282
Mercy, E. 249–50
Merwin, H.E. 310
Miller, A.R. 13
Miller, W.H. 244
Mitchell, R.L. 54, 66–7, 87, 126–7, 172, 215, 228–9, 237–8, 249, 255–6, 280, 291, 303
Miyashiro, A. 353
Monster, J. 19
Moore, J.G. 310, 330
Moore, R.K. 157
Moxham, R.L. 123
Mroz, E.J. 312
Murata, K.J. 310, 330, 332
Mysen, B.O. 256

Naboko, S.I. 309, 314
Nagasawa, H. 114, 123, 126, 195, 215
Naughton, J.J. 310–11
Noble, D.C. 127
Nockolds, S.R. 229
Noddack, I. 75, 169
Noddack, W. 75, 169
Noll, A. 120
Norry, M.J. 163

Oana, S. 314
Oftedahl, C. 13
O'Hara, M.J. 249–50, 282, 331
Ohashi, R. 11, 12
Ohmoto, H. 20, 21
Onuma, N. 127, 213, 237
Osborne, E.F. 337
Óskarsson, N. 312

Pan, Y. 170
Partington, J.R. 155
Paster, T.P. 66–7, 87
Patterson, E.M. 54, 267
Pearce, J.A. 163
Peltola, E. 244
Peterson, H.T. 13, 106
Phelan, J.M. 315–16
Philpotts, J.A. 111, 114, 126–7
Pichavant, M. 321
Piwinski, A.J. 321, 350, 356
Plimer, I.R. 16
Plüger, W. 351
Prinz, M. 169, 184
Puchelt, H. 106, 111, 123

Ramberg, H. 249
Ramsay, W.R.H. 4, 26, 325, 329, 337, 353
Rankin, W.J. 355
Reed, M.H. 323, 362
Richards, J.R. 19
Richter, D.H. 330, 332
Ridley, W.I. 331
Ringwood, A.E. 193, 249, 280, 332, 337
Robertson, J.D. 75
Roeder, P.L. 213
Rogerson, R.J. 3, 27
Rose, H.J. 59, 250
Rose, W.I. 317, 362
Royle, D.Z. 341
Rubey, W.W. 304
Russell, R.D. 13, 19
Rye, R.O. 20–1

Sandell, E.B. 54–5, 69, 73, 75–6, 78, 87, 97, 228, 244–5, 247
Sangster, D.F. 19, 21
Sasaki, A. 21
Sato, H. 331
Sato, T. 21–2
Sawkins, F.J. 22
Schnetzler, C.C. 111, 114, 126–7, 215
Schröpfer, L. 157
Scott, S.D. 352
Seck, H.A. 79
See, J.B. 355
Seifert, S. 237, 256
Seward, T.M. 214, 256
Seyfried, W.E. 349
Shaw, D.M. 120, 132, 280

Author index

Shepherd, E.S. 309
Sheppard, S.M.F. 20–1
Shibata, T. 137
Shima, M. 309
Shimizu, N. 114
Shiraki, K. 184, 187, 191
Sidhu, P.S. 356
Sillitoe, R.H. 14
Silvestri, O. 306
Simkin, T. 206
Smith, J.V. 206
Snyder, J.K. 59
Spence, C.D. 24
Spooner, E.C.T. 20
Stanton, R.L. 4, 13–15, 19, 24, 26, 30, 65, 148, 164, 203, 292, 325, 327, 329, 337, 339, 353
Steiger, G. 54
Stoiber, R.E. 317, 347, 362
Stueber, A.M. 126, 228
Swaine, D.J. 53, 267
Symonds, R.B. 315, 362

Taborszky, F.K. 138
Tauson, L.V. 123
Taylor, S.R. 101, 114, 123, 126–7, 169, 187, 228–9, 256
Thode, H.G. 19
Thomas, H.H. 11

Tilley, C.E. 336–7
Tilling, R.I. 97, 101
Tillmanns, E. 157
Tinoco, M. 11
Treloar, P.J. 170, 346
Tsujimoto, K. 11
Turekian, K.K. 229, 231

Urabe, T. 21–2, 24

Vargas Bedemar, E.R. 10, 12
Vedder, J.G. 5
Verhoogen, J. 158
Viljoen, M.J. 330
Viljoen, R.P. 330
Villemant, B. 127, 237–8, 256
Vincent, E.A. 59, 79, 280–1, 287, 291, 297
Vogt, J.H.L. 11, 135, 244–5, 247, 249, 307
von Engelhardt, W., *see* Engelhardt, W. von
von Hevesy, G., *see* Hevesy, G. von
Vossler, T. 315–16

Wager, L.R. 54, 66–7, 87, 126–7, 172, 215, 228, 237–8, 249, 255–6, 280–1, 287, 291, 303, 343

Waring, G.A. 309
Washington, H.S. 54, 155, 169, 183, 204, 228, 244, 310
Watson, E.B. 213
Wedepohl, K.H. 59, 76, 78–9, 97–8, 204, 217, 228
Weed, W.H. 8
Weill, D.F. 114, 127, 238, 256
Wells, R.C. 54
White, A.J.R. 184
White, D.E. 309
White, W.B. 157
Whitford, D.J. 351
Whitney, J.D. 11
Wilkinson, G. 58
Wilkinson, J.F.G. 331–3
Wilson, I.F. 11
Winchell, H.V. 18
Winchell, N.H. 18
Wright, T.L. 255–6

Yoder, H.S. 337

Zambonini, F. 307
Zies, E.G. 307, 309
Zindler, A. 304
Zoller, W.H. 311–12

Subject index

andesites, *see* hornblende-andesite group; big-feldspar lavas
andesites, two pyroxene 37–8
antimony, in Solomons lavas 265, 274–5
arsenic, in Solomons lavas 265, 274–5

barium
 in arc lavas 57, 109–10, 115–18
 chemical properties of 58, 110–11
 in crustal materials 108
 in crystal inclusions 114
 in crystal:melt partitioning 114–15
 in exhalative ores 8, 108, 350–1
 in MORBs 54, 108–9, 115, 117
 in silicates 61, 111–13, 294
 in Solomons whole rocks 57, 115–18
 in spinels 61, 113–14
basaltic andesites, *see* big-feldspar lavas; hornblende-andesite group
basalts, *see* big-feldspar lavas; olivine–clinopyroxene-basalts; picrites
big-feldspar lavas
 distribution of 26, 28–9
 general nature of 30–1
 lead isotope abundances 328–9
 major element chemistry 31–6
 petrogenesis of 336
 principal minerals in 34, 36–49
 strontium isotope abundances 327–8
 trace element patterns in 289–90
biotite
 principal trace elements in 61, 286–7
 in Solomons lavas 42
bismuth, in Solomons lavas 265, 274–5

cadmium, in Solomons lavas 265, 277
calcium
 in arc lavas 57, 151–3
 chemical properties of 58, 149–51
 in exhalative ores 148–9
 in MORBs 54, 151, 153
 in silicates 61, 290–1
 in Solomons whole rocks 57, 150–3
 in spinels 61

chlorine, in Solomons lavas 265–6
chromium
 in arc lavas 57, 185–6, 195–200
 chemical properties of 58, 187
 in crustal materials 183–4
 in crystal:melt partitioning 193–5
 in exhalative ores 183, 346–7
 in MORBs 54, 184–5, 199
 in the plume of Syrtlingur volcano, Iceland 311–12
 in silicates 61, 187–92, 288
 in Solomons whole rocks 57, 195–200
 in spinels 61, 192–3, 288
clinopyroxene
 principal trace elements in 61, 285–6
 in Solomons lavas 28–9, 34, 37–9
cobalt
 in arc lavas 57, 230–1
 chemical properties of 58, 231
 in crustal materials 228–9
 in crystal:melt partitioning 236–9
 in exhalative ores 228, 345–6
 in MORBs 54, 229–30, 242
 in silicates 61, 231–6, 288
 in Solomons whole rocks 57, 239–43
 in spinels 61, 231, 236
copper
 in arc lavas 55–7, 69–74
 chemical properties of 56, 58–9
 in crustal materials 53, 55, 58
 in crystal inclusions 61–2, 291–4
 in crystal:melt partitioning 66–9
 in exhalative ores 8–9, 53, 341–2, 348–50
 in Galapagos tholeiites 357–8
 in Hawaiian lavas 53–4
 Jahn–Teller distortion of 58–9, 355, 362
 in lavas of north-eastern Ireland 53
 in lavas of Skye 53–4
 in MORBs 54–5, 71–2
 in silicates 59–62, 291–4
 in Skaergaard intrusion 53–4
 in Solomons whole rocks 57, 69–74
 in spinels 59, 61, 62–6

crystal subtraction and element abundance 297–302, 330–6, 337–40

d'Entrecasteaux Ridge 329

exhalative ores
 of Baja California 11
 of Broken Hill 148–9, 154, 203–4
 of Buchans 22
 of Cyprus type 8–9, 13–15
 definition of 7
 evolution of ideas on 10–17
 of the Iberian Pyrite Belt 9, 24
 of Kuroko type 8–9, 12–14, 24
 of the Lachlan Fold Belt 13, 24
 of Meggen 12–13
 of Mount Morgan 11–12
 of New Brunswick 13, 24
 of Rammelsberg 12–13
 types and volcanic associations 8–9

fluorine, in Solomons lavas 265–6

Galapagos Rift basalts 357–8
gallium, in Solomons lavas 265, 277
gold
 in Mount Erebus gases 292
 in Solomons lavas 265, 276

hafnium, in Solomons lavas 265, 272–3
hornblende
 opaque rims to 40–3
 principal trace elements in 61, 286
 in Solomons lavas 28, 34, 39–43
hornblende-andesite group
 distribution of 26, 28–9
 general nature of 31–2
 lead isotope abundances in 328–9
 major element chemistry 31–6
 petrogenesis of 336–40
 principal minerals of 37–49

Subject index 389

Kilinaelau–North Solomons Trench 5, 328

lava series nomenclature
  calc-alkaline 49–51
  hypertholeiitic 49–51
  hypotholeiitic 49–51
  nichrome 49–51
  tholeiitic 49–51
lead
  in arc lavas 57, 96–7, 102–7
  chemical properties of 58, 97
  in crustal materials 96
  in crystal:melt partitioning 101–2
  in exhalative ores 8–9, 96, 344–5, 348–50
  isotopes 328–9
  in MORBs 54, 96
  in silicates 61, 97–100, 294–5
  in Solomons whole rocks 57, 102–7
  in spinels 61, 98, 100–1

manganese
  in arc lavas 57, 205–6, 217–20
  chemical properties of 58, 205–6
  in crustal materials 204
  in crystal:melt partitioning 213–15
  in exhalative ores 203–4, 352
  in Galapagos tholeiites 357–8
  in MORBs 54, 204–5, 218–19
  in silicates 61, 206–11, 215–17, 288
  in Solomons whole rocks 57, 217–22
  in spinels 61, 207, 211–13, 216, 288
molybdenum, in Solomons lavas 265, 275–6

New Britain–San Cristobal Trench 5, 328
nickel
  in arc lavas 57, 247–8, 257–62
  chemical properties of 58, 247, 249
  in crustal materials 244–5
  in crystal:melt partitioning 255–6
  in exhalative ores 244, 345
  in MORBs 54, 245–7, 257, 261–2
  in olivine 61, 250–1
  in silicates 61, 249–54, 287–8
  in Solomons whole rocks 57, 257–63
  in spinels 61, 254–5, 287–8

olivine
  cobalt in 61, 231–3, 283
  manganese in 61, 206–8, 283
  nickel in 61, 249–51, 283
  principal trace elements in 61, 283–4
  in Solomons lavas 28, 34, 36–7, 61, 333–4

olivine–clinopyroxene-basalts
  distribution of 26, 28–9
  general nature of 29–30
  major element chemistry 30–6
  petrogenesis of 330–6
  principal minerals in 34, 36–49
  strontium isotope abundances 327–8
  trace element patterns in 289–90
orthopyroxene
  principal trace elements in 61, 286–7
  in Solomons lavas 37–9
oxidation index 34–5, 226–7

partial melting 335, 337, 340
phosphorus
  in arc lavas 57, 137–8, 143–6
  chemical properties of 58, 138–9
  in crustal materials 135, 137
  in crystal:melt partitioning 142
  in exhalative ores 135, 148–9, 351–2, 358
  in MORBs 54, 136–8, 145–6
  in silicates 61, 139–42
  in Solomons whole rocks 57, 143–6
  in spinels 61
picrites
  of Baffin bay 331
  of the Bay of Islands 331
  distribution of 26, 28–9
  general nature of 29–30
  of Hawaii 330–1, 332
  of Kohinggo Island 32
  of the Kolo (Marovo) area 30, 32, 325
  major element chemistry of 30–6
  petrogenesis of 330–6
  principal minerals of 34, 36–49
  of Tortuga 331
  trace element patterns in 289–90
plagioclase
  principal trace elements in 61, 286
  in Solomons lavas 34, 43–5

rubidium, in Solomons lavas 265–71, 296–7

silver, in Solomons lavas 265, 276
Solomon Islands Older Volcanic Suite 4
Solomon Islands Younger Volcanic Suite 4, 26–7
  distribution 26–7
  major element chemistry 32–6
  principal minerals 36–49
  principal rock types 27–32
spinel
  principal trace elements in 61, 284–5
  in Solomons lavas 34, 45–9

strontium
  in arc lavas 57, 121–2, 129–33
  chemical properties of 58, 122–3
  in crustal materials 120
  in crystal inclusions 128
  in crystal:melt partitioning 126–8
  in exhalative ores 120, 344–5
  isotope abundances 327–8
  in MORBs 54, 120–1, 129, 132
  in silicates 61, 123–5, 294
  in Solomons whole rocks 57, 129–33
  in spinels 61, 126
sulphur, in Solomons lavas 264–6

thorium, in Solomons lavas 278
titanium
  in arc lavas 57, 156–7, 164–8
  chemical properties of 58, 157
  in crustal materials 155
  in crystal:melt partitioning 162–4
  in exhalative ores 154, 347–8
  in MORBs 54, 155–6, 164, 166–7
  in silicates 61, 158–61, 291
  in Solomons whole rocks 57, 164–8
  in spinels 61, 158, 161–2, 291
tungsten, in Solomons lavas 265, 277

uranium, in Solomons lavas 265, 273, 278

vanadium
  in arc lavas 57, 170–1, 177–81
  chemical properties of 58, 171
  in crustal materials 169
  in crystal:melt partitioning 176–7
  in exhalative ores 169–70, 346–7
  in fumaroles of Bezymianny volcano, Kamchatka 347
  in fumaroles of Izalco volcano, Honduras 317, 347
  in MORBs 54, 170, 180–1
  in silicates 61, 172–5, 291
  in Solomons whole rocks 57, 177–81
  in spinels 61, 173, 175–6, 291
Vanuatu Trench 329
volcanic aerosols 305
volcanic condensates 305
volcanic gases 305
volcanic plumes 305
volcanic sublimates 305
volcanoes
  Askja, Iceland 312–13
  Augustine, Alaska 314–15
  Bezymianny, Kamchatka 347
  Central American volcanoes 317, 324
  Etna, Sicily 304–6, 317–19, 323–4
  Heimaey, Iceland 311–13

Subject index

hornblende-andesite group *(cont.)*
  strontium isotope abundances in 327–8
  trace element patterns in 289–90

iron
  in arc lavas 57, 224–6
  chemical properties of 58, 223–4
  in exhalative ores 8–9, 223, 352
  in MORBs 54, 224–5
  in Solomons whole rocks 57, 224–6

Kilinaelau–North Solomons Trench 5, 328

lava series nomenclature
  calc-alkaline 49–51
  hypertholeiitic 49–51
  hypotholeiitic 49–51
  nichrome 49–51
  tholeiitic 49–51
lead
  in arc lavas 57, 96–7, 102–7
  chemical properties of 58, 97
  in crustal materials 96
  in crystal:melt partitioning 101–2
  in exhalative ores 8–9, 96, 344–5, 348–50
  isotopes 328–9
  in MORBs 54, 96
  in silicates 61, 97–100, 294–5
  in Solomons whole rocks 57, 102–7
  in spinels 61, 98, 100–1

manganese
  in arc lavas 57, 205–6, 217–20
  chemical properties of 58, 205–6
  in crustal materials 204
  in crystal:melt partitioning 213–15
  in exhalative ores 203–4, 352
  in Galapagos tholeiites 357–8
  in MORBs 54, 204–5, 218–19
  in silicates 61, 206–11, 215–17, 288
  in Solomons whole rocks 57, 217–22
  in spinels 61, 207, 211–13, 216, 288
molybdenum, in Solomons lavas 265, 275–6

New Britain–San Cristobal Trench 5, 328
nickel
  in arc lavas 57, 247–8, 257–62
  chemical properties of 58, 247, 249
  in crustal materials 244–5
  in crystal:melt partitioning 255–6
  in exhalative ores 244, 345

  in MORBs 54, 245–7, 257, 261–2
  in olivine 61, 250–1
  in silicates 61, 249–54, 287–8
  in Solomons whole rocks 57, 257–63
  in spinels 61, 254–5, 287–8

olivine
  cobalt in 61, 231–3, 283
  manganese in 61, 206–8, 283
  nickel in 61, 249–51, 283
  principal trace elements in 61, 283–4
  in Solomons lavas 28, 34, 36–7, 61, 333–4
olivine–clinopyroxene-basalts
  distribution of 26, 28–9
  general nature of 29–30
  major element chemistry 30–6
  petrogenesis of 330–6
  principal minerals in 34, 36–49
  strontium isotope abundances 327–8
  trace element patterns in 289–90
orthopyroxene
  principal trace elements in 61, 286–7
  in Solomons lavas 37–9
oxidation index 34–5, 226–7

partial melting 335, 337, 340
phosphorus
  in arc lavas 57, 137–8, 143–6
  chemical properties of 58, 138–9
  in crustal materials 135, 137
  in crystal:melt partitioning 142
  in exhalative ores 135, 148–9, 351–2, 358
  in MORBs 54, 136–8, 145–6
  in silicates 61, 139–42
  in Solomons whole rocks 57, 143–6
  in spinels 61
picrites
  of Baffin bay 331
  of the Bay of Islands 331
  distribution of 26, 28–9
  general nature of 29–30
  of Hawaii 330–1, 332
  of Kohinggo Island 32
  of the Kolo (Marovo) area 30, 32, 325
  major element chemistry of 30–6
  petrogenesis of 330–6
  principal minerals of 34, 36–49
  of Tortuga 331
  trace element patterns in 289–90
plagioclase
  principal trace elements in 61, 286
  in Solomons lavas 34, 43–5

rubidium, in Solomons lavas 265–71, 296–7

silver, in Solomons lavas 265, 276
Solomon Islands Older Volcanic Suite 4
Solomon Islands Younger Volcanic Suite 4, 26–7
  distribution 26–7
  major element chemistry 32–6
  principal minerals 36–49
  principal rock types 27–32
spinel
  principal trace elements in 61, 284–5
  in Solomons lavas 34, 45–9
strontium
  in arc lavas 57, 121–2, 129–33
  chemical properties of 58, 122–3
  in crustal materials 120
  in crystal inclusions 128
  in crystal:melt partitioning 126–8
  in exhalative ores 120, 344–5
  isotope abundances 327–8
  in MORBs 54, 120–1, 129, 132
  in silicates 61, 123–5, 294
  in Solomons whole rocks 57, 129–33
  in spinels 61, 126
sulphur, in Solomons lavas 264–6

thorium, in Solomons lavas 278
titanium
  in arc lavas 57, 156–7, 164–8
  chemical properties of 58, 157
  in crustal materials 155
  in crystal:melt partitioning 162–4
  in exhalative ores 154, 347–8
  in MORBs 54, 155–6, 164, 166–7
  in silicates 61, 158–61, 291
  in Solomons whole rocks 57, 164–8
  in spinels 61, 158, 161–2, 291
tungsten, in Solomons lavas 265, 277

uranium, in Solomons lavas 265, 273, 278

vanadium
  in arc lavas 57, 170–1, 177–81
  chemical properties of 58, 171
  in crustal materials 169
  in crystal:melt partitioning 176–7
  in exhalative ores 169–70, 346–7
  in fumaroles of Bezymianny volcano, Kamchatka 347
  in fumaroles of Izalco volcano, Honduras 317, 347
  in MORBs 54, 170, 180–1
  in silicates 61, 172–5, 291
  in Solomons whole rocks 57, 177–81
  in spinels 61, 173, 175–6, 291
Vanuatu Trench 329
volcanic aerosols 305

volcanic condensates 305
volcanic gases 305
volcanic plumes 305
volcanic sublimates 305
volcanoes
   Askja, Iceland 312–13
   Augustine, Alaska 314–15
   Bezymianny, Kamchatka 347
   Central American volcanoes 317, 324
   Etna, Sicily 304–6, 317–19, 323–4
   Heimaey, Iceland 311–13
   Hekla, Iceland 305, 312–13
   Izalco, Honduras 317, 347
   Kamchatka–Kuriles 314, 324
   Kilauea, Hawaii 306, 309–11
   Krakatau, Indonesia 304
   Krisuvik, Iceland 305
   Leirhnjúkur, Iceland 312–13
   Merapi, Indonesia 313–14

Mount Gallego, Solomon Is 35–6, 99–100, 103, 137, 266, 268
Mount Katmai, Alaska 304, 307–9, 315
Mount St. Helens, USA 304, 315–6, 323
Pinatubo, Philippines 304
Reykjalidh, Iceland 305
Ruapehu, New Zealand 304
Santorin 304
Savo, Solomon Is 35–6, 99–100, 103, 137, 266, 268
Showashinzan, Japan 314
Stromboli 317–18
Surtsey, Iceland 311–13
Syrtlingur, Iceland 311–13
Vesuvius, Italy 305–7, 317–18, 324
Vulcano, Italy 306
White Island, New Zealand 316–17

Woodlark spreading ridge 5, 328–9

zinc
   in arc lavas 57, 77, 90–4
   chemical properties of 58, 77–8
   in crustal materials 58, 75–6
   in crystal inclusions 61–2, 293–4
   in crystal:melt partitioning 87–90
   in exhalative ores 8–9, 75, 344–5, 348–50
   in Galapagos tholeiites 357–8
   in MORBs 54, 76–7, 90, 92–3
   in silicates 61, 78–84, 291
   in Solomons whole rocks 57, 90–4
   in spinels 61, 79, 84–6, 291
zirconium, in Solomons lavas 265, 271–5, 296–7